Thermodynamics and Exergy Analysis for Engineers

Thermodynamics and Exergy Analysis for Engineers

RAJINDER PAL

Professor of Chemical Engineering
University of Waterloo
Ontario, Canada

CWP
Central West Publishing

This edition has been published by Central West Publishing, Australia

For more information about the books published by Central West Publishing, please visit https://centralwestpublishing.com

Disclaimer

Every effort has been made by the publisher and author while preparing this book; however, no warranties are made regarding the accuracy and completeness of the content. The publisher and author disclaim without any limitation all warranties as well as any implied warranties about sales, along with fitness of the content for a particular purpose. Citation of any website and other information sources does not mean any endorsement from the publisher and author. For ascertaining the suitability of the contents contained herein for a particular lab or commercial use, consultation with the subject expert is needed. In addition, while using the information and methods contained herein, the practitioners and researchers need to be mindful for their own safety, along with the safety of others, including the professional parties and premises for which they have professional responsibility. To the fullest extent of law, the publisher and author are not liable in all circumstances (special, incidental, and consequential) for any injury and/or damage to persons and property, along with any potential loss of profit and other commercial damages due to the use of any methods, products, guidelines, procedures contained in the material herein.

NATIONAL LIBRARY OF AUSTRALIA

A catalogue record for this book is available from the National Library of Australia

ISBN (print): 978-1-922617-25-5

To the memory of
Shri Khushal Chand and Muni Harmilap Ji

Contents

Preface

Thermodynamics is a difficult subject to understand as many concepts are abstract in nature. I have taught this subject to our third year chemical engineering students at the University of Waterloo many times. Although a number of excellent textbooks are available on thermodynamics, I have felt the need for a new textbook where the material is presented in order and style such that it is easily comprehensible to second and third year engineering students.

This book consists of eighteen chapters. Chapter one presents introductory concepts and definitions along with a brief discussion of historical development of thermodynamics. Chapters two and three cover the first law of thermodynamics. Chapter two is devoted to the first law for control mass or closed systems and Chapter three is devoted to the first law for control volume or open (flow) systems. The second law of thermodynamics for closed systems is presented in Chapter four. Chapter five is devoted to the second law for open systems with applications. Thermodynamics of compressible and incompressible flows in ducts and pipes is covered in depth in Chapter six. Chapter seven is devoted to estimation of volumetric and thermodynamic properties of fluids. Chapters eight to ten provide in-depth coverage of power cycles, internal combustion engines, and refrigeration cycles. Chapters eleven and twelve are devoted to vapor-liquid phase equilibrium of ideal and non-ideal systems. Chapter thirteen provides in-depth coverage of chemical reaction equilibrium. Work and entropy analysis of closed and open systems is presented along with the Gouy-Stodola theorem in Chapter fourteen. Due to the importance of exergy and exergy analysis in many practical applications, the last four chapters (Chapters fifteen to eighteen) are fully devoted to this topic. The available textbooks in thermodynamics rarely provide satisfactory coverage of exergy and exergy analysis of processes.

Each chapter begins with the student learning outcomes. Near the end of each chapter, a tutorial session is included where a number of problems are solved to illustrate the applications of concepts and theory covered in the chapter. Notation is included separately in each chapter. For the benefit of students, practice problems are included at the end of each chapter.

It is hoped that the students will be able to understand the concepts and material presented in this textbook without difficulty. As thermodynamics is at the core of engineering practice, it is important that students develop a liking for this subject. Understanding and liking of a subject go hand in hand. Students are inspired to pursue a particular field at a more advanced level only if they are able to develop a good understanding and liking for the subject.

Finally, I would like to thank my wife Archana and my children Anuva and Arnav for their love and constant support.

Waterloo, October 2021 Rajinder Pal

Author

Rajinder Pal is a Professor in the Department of Chemical Engineering at the University of Waterloo since 1991, and a Fellow of the Chemical Institute of Canada. He received degrees in Chemical Engineering from I.I.T. Kanpur (B.Tech. 1981) and the University of Waterloo (Ph.D. 1987), followed by a post-doctoral fellowship at the University of Alberta. His research interests include complex fluid rheology, interfacial and transport phenomena, physical properties of composite materials, and non-equilibrium thermodynamics. He has published more than 150 refereed journal articles, 4 books, and 5 book chapters on these and related subjects, including 8 papers in engineering education journals. He has taught over ten different undergraduate and graduate courses related to equilibrium staged separation processes, fluid mechanics, heat and mass transfer, physical chemistry and thermodynamics.

Professor Pal has been the recipient of several awards including the University of Waterloo prestigious Distinguished Teacher Award (2020), Sandford Fleming Foundation Teaching Excellence Award (twice 2001 and 2017), an Outstanding Performance Award from the University of Waterloo (2009), and the Syncrude Canada Innovation Award from the Canadian Society for Chemical Engineering (1998). He is a Fellow of the Chemical Institute of Canada since 1999. He serves on several journal editorial boards and was an Associate Editor of the Canadian Journal of Chemical Engineering for twelve years. He is a registered Professional Engineer in the province of Ontario, Canada.

Chapter 1

Introductory Concepts and Definitions

1.1 LEARNING OUTCOMES

Upon completion of this chapter, students will be able to:

- Define thermodynamics and explain the historical background of its development
- Describe the practical importance of thermodynamics
- Define and explain introductory thermodynamic concepts such as: system, surroundings, state property, extensive and intensive properties, cyclic process, equilibrium, Gibb's phase rule, work, heat, exact and inexact differentials, internal energy, enthalpy, reversible and irreversible processes
- State the four laws of thermodynamics
- Calculate expansion and compression work

1.2 INTRODUCTION

Thermodynamics is a branch of engineering and physical sciences that deals with the conversion of heat into mechanical energy or work. A more general definition of thermodynamics can be given as follows: It is a branch of engineering and physical sciences that is concerned with the interconversion and interrelationship of different forms of energy and the behavior of macroscopic systems in terms of certain basic quantities such as pressure, temperature, etc.

Steam engine (see Figure 1.1) is considered to be one of the first applications or products of thermodynamics. In a steam engine, coal is burnt to release the heat energy. The heat is used to boil water in the boiler and generate high pressure steam. The high-pressure steam moves the reciprocating pistons connected to the wheels of the locomotive and thus generate motion (mechanical energy). Steam engines are highly inefficient heat engines in that nearly 80 to 90 percent of the energy produced from coal is wasted away. Only about 10 to 20 percent of the coal energy is converted into mechanical energy (kinetic energy) of the train.

Steam engine driven train – one of the first applications of Thermodynamics

Figure 1.1 Steam engine driven train.

Much of the thermodynamics was developed during the nineteenth century. Table 1.1 summarizes the historical background of the development of Thermodynamics and Figure 1.2 shows the pictures of some of the founders of thermodynamics.

Table 1.1 Historical background of the development of Thermodynamics [1]

Date	Event
1798	Count Rumford (Benjamin Thompson) began the quantitative study of the conversion of work into heat by means of his famous cannon-boring experiments.
1799	Sir Humphry Davy studied the conversion of work into heat by means of his ice-rubbing experiments
1824	Sadi Carnot published his famous thesis "Reflections on the Motive Power of Fire," which includes the new concept of cycle and the principle that the efficiency of a reversible cycle engine operating between two heat reservoirs depends only on the temperatures of the reservoirs and not on the working substance.
1842	Mayer proposed the principle of conservation of energy.
1847	Helmholtz formulated the principle of conservation of energy, independent of Mayer.
1843-1848	James Prescott Joule laid the experimental foundation of the first law of thermodynamics by performing experiments to establish the equivalence of work and heat. In honor of this great scientist, we use Joule to denote the mechanical equivalent of heat.
1848	Lord Kelvin (William Thomson) established an absolute temperature scale based on the Carnot cycle.
1850	Rudolf J. Clausius was probably the first to see that there were two basic principles: the first and second laws of thermodynamics. He also introduced the concept of internal energy.
1865	Clausius stated the first and second laws of thermodynamics in two lines: 1. The energy of the universe is constant. 2. The entropy of the universe tends toward a maximum.
1873-1878	Josiah Willard Gibbs published his monumental work " On the Equilibrium of Heterogeneous Substances," which extends thermodynamics in a general form to heterogeneous systems and chemical reactions. His work includes the important concept of chemical potential. The concept of "availability" was first introduced by Gibbs.
1889	Gouy, apparently quite independently, developed the concept of usable energy
1897	Max Planck stated the second law of thermodynamics in the following form: "It is impossible to construct an engine which, working in a complete cycle, will produce no effect other than the raising of a weight and the cooling of a heat reservoir."
1898	Stodola independently developed the concept of "free technical energy" applicable to steady flow processes
1931	Onsager established his celebrated "reciprocal relations"
1947	Prigogine developed a consistent phenomenological theory of irreversible processes
1956	Rant coined the term "exergy" for the first time, thus replacing terms such as availability, available energy, usable energy etc.

Figure 1.2 Some of the founders of thermodynamics.

Today thermodynamics is an extremely vast and complex area of engineering and physical sciences. Engineers use the principles of thermodynamics in the analysis and design of a wide variety of energy systems including power plants, automobiles, air conditioning systems, refrigeration systems, rockets and jet engines [2]. Thermodynamics also plays an important role in the analysis and design of fuel cells, chemical reactors, compressors, pumps, absorption columns, distillation columns, and other process equipment.

It would be appropriate to quote Peter Atkins here [3]:

"Do not think that thermodynamics is only about steam engines: it is about everything. The concepts did indeed emerge during the nineteenth century when steam was the hot topic of the day, but as the laws of thermodynamics became formulated and their ramifications explored it became clear that the subject could touch an enormously wide range of phenomena, from the efficiency of heat engines, heat pumps, and refrigerators, taking in chemistry on the way, and reaching as far as the processes of life".

Thus, a thorough understanding of thermodynamics is indispensable for the graduating engineers. It is no wonder that thermodynamics is an essential part of the global engineering curricula [2,4,5].

1.3 THERMODYNAMICS IS A DIFFICULT SUBJECT

Thermodynamics is considered to be one of the most difficult and abstract disciplines of the engineering and physical sciences [6]. It is appropriate to quote Arnold Sommerfeld, well-known German theoretical physicist, here [2]:

"Thermodynamics is a funny subject. The first time you go through it, you don't understand it at all. The second time you go through it, you think you understand it, except for one or two small points. The third time you go through it, you know you don't understand it, but by that time you are so used to it, so it doesn't bother you anymore".

Engineering students all over the world face difficulties in learning thermodynamics. This is a well-documented fact in the literature [2,4-6, 7-24]. The national passing rates in thermodynamics are generally low as indicated by a survey of the National Council of Examiners for Engineering and Surveying [7]. Table 1.2 presents the national average scores in the morning thermodynamics section of the fundamental engineering exam. The scores range anywhere from 29 to 71%. The mechanical engineering students happen to score the highest and the civil engineering students score the lowest in this survey.

Some researchers have described thermodynamics as the gateway course [25] in engineering in the sense that a good correlation exists between the student performance in thermodynamics and other courses in the curriculum. Extensive research has been carried out on the effectiveness of different pedagogical approaches to enhance the teaching and learning of thermodynamics [2]. Among the various approaches used are: traditional lecture style approach, active learning techniques, computer-based instruction, virtual lab – a web-based student learning tool for thermodynamic concepts, interactive multimedia e-book, and blended learning approach. Each approach has its own pros and cons.

Table 1.2 National Average Scores in *thermodynamics* section of the engineering exam [7]

Year	Civil	Mechanical	Electrical
2002	43%	63%	44%
2003	44%	N/A	N/A
2004	29%	57%	N/A
2005	44%	57%	N/A
2006	48%	62%	47%
2007	51%	63%	N/A
2008	49%	61%	50%
2009	53%	71%	N/A
2010	32%	45%	N/A
2011	42%	56%	43%
2012	58%	70%	61%

1.4 CONCEPTS AND DEFINITIONS

1.4.1 Thermodynamic System and Surroundings

Thermodynamic system is simply a collection of matter that is selected for analysis. The bounding surface of the system is called the *system boundary*. Everything external to the system boundary is the system's *surroundings*. A thermodynamic system, by definition, is a *closed system* in that it always contains the same matter. The thermodynamics analysis of such closed systems is often referred to as a system analysis or control mass analysis. The system can interact with its surroundings via exchange of heat and work through its boundary. A special case of a closed system is an *isolated system* that does not interact with its surroundings in any way.

1.4.2 Control Volume

Control volume is a region in space that is selected for analysis. The thermodynamic analysis of flow processes is done using the control volume approach. The bounding surface (boundary) of the control volume is permeable to mass transfer, that is, mass can enter and leave the control volume through openings. The terms control volume, control region, and open system all mean the same and are used interchangeably.

1.4.3 Thermodynamic Property

Thermodynamic property of a system is any quantifiable macroscopic characteristic of a system whose value depends on the current condition or state of the system. Examples of thermodynamic properties are pressure, temperature, volume, energy, and mass. Note that the properties of a system depend only on the current condition or state of the system and are independent of how the system is arrived at the current condition. The thermodynamic properties are also called *state properties* or *state functions* as they specify the state or condition of the system.

1.4.4 Extensive and Intensive Properties

Thermodynamic properties are divided into two groups: extensive and intensive. An *extensive* property is one which depends on the size or extent of the system. Extensive property is additive. If a system is divided into a number of sub-systems, the value of the extensive property for the whole system is the sum of the values of that property for all the sub-systems. Mass and volume are examples of extensive properties. An *intensive* property is one which does not depend on the size or extent of the system. It is not additive. Examples of intensive properties are temperature and pressure. A special type of intensive property is a *specific property* which is defined as the ratio of extensive property to the corresponding mass.

1.4.5 Thermodynamic State

The thermodynamic state of a system is the condition of the system as described by the values of its properties. As the properties of a system are interrelated, only a small number of properties can be specified independently. Thus, the state of a system is completely specified by specifying only the independent properties. Once the independent properties are specified, all other properties of the system are automatically fixed.

1.4.6 Thermodynamic Process

When a system undergoes a change in thermodynamic state from some initial state to some final state, it is said to have undergone a thermodynamic process. The succession of states through which the system passes is called the *path* of the process.

1.4.7 Cyclic Process

When a system undergoes some processes and finally returns to its initial state, it is said to have undergone a cyclic process. An important point to note is that in any cyclic process, the change in any state property is zero as the system returns to its initial state.

1.4.8 Thermodynamic Equilibrium

A system is said to be in thermodynamic equilibrium when its properties are not observed to change with time. Not only the system properties are time-invariant, the system also has no tendency towards a change in state, that is, the driving forces for changes are zero at equilibrium. A system in thermodynamic equilibrium satisfies the conditions of mechanical, thermal, chemical, and phase equilibrium.

Mechanical equilibrium: When the pressure is uniform inside the system (no internal pressure gradients) and the pressure of the system is balanced by the pressure exerted by the surroundings, the system is said to be in mechanical equilibrium.

Thermal equilibrium: When the temperature is uniform inside the system (no internal temperature gradients) and the temperature of the system is equal to the temperature of the surroundings, the system is said to be in thermal equilibrium.

Chemical equilibrium: When the chemical composition of the system is time-invariant and has no tendency towards change, the system is said to be in chemical equilibrium.

Phase equilibrium: If the system consists of two or more phases, it is said to be in phase equilibrium if there is no tendency for phase transformation, that is, the amounts and compositions of the various phases are time-invariant.

1.4.9 Gibbs Phase Rule

The thermodynamic intensive state of a pure single-phase substance is fully specified when two intensive properties are fixed. For example, the intensive state of a pure gas is fully specified when temperature and pressure are fixed. The *phase rule* tells us the number of independent variables that must be specified in order to specify the intensive state of a system at equilibrium. For non-reacting systems, it is given as follows:

$$F = C - \pi + 2 \tag{1.1}$$

where F is the degrees of freedom of the system, C is the number of chemical species, and π is the number of phases present in the system.

As an example, consider a system consisting of a mixture of steam and liquid water at equilibrium. How many degrees of freedom does the system have?

$$F = 1 - 2 + 2 = 1 \tag{1.2}$$

In this case, there is only one degree of freedom. Thus, we need to specify either temperature or pressure (not both) in order to completely specify the intensive state of the system. If we specify $T = 100^oC$, the pressure is automatically fixed at 101.33 kPa.

1.4.10 Work

When a force acts on a body and the body undergoes displacement, work is said to be done on the body by the force (see Figure 1.3). Mathematically, the work can be expressed as:

$$\delta W = \vec{F} \cdot \overrightarrow{dl} \tag{1.3}$$

where δW is differential amount of work, $\vec{F}$ is the force applied on the body, $\overrightarrow{dl}$ is the differential displacement of the body, and the sign $\cdot$ represents the dot product between two vectors. Upon integration over the total displacement of L, Equation (1.3) gives:

$$W = \int_0^L \vec{F} \cdot \overrightarrow{dl} \tag{1.4}$$

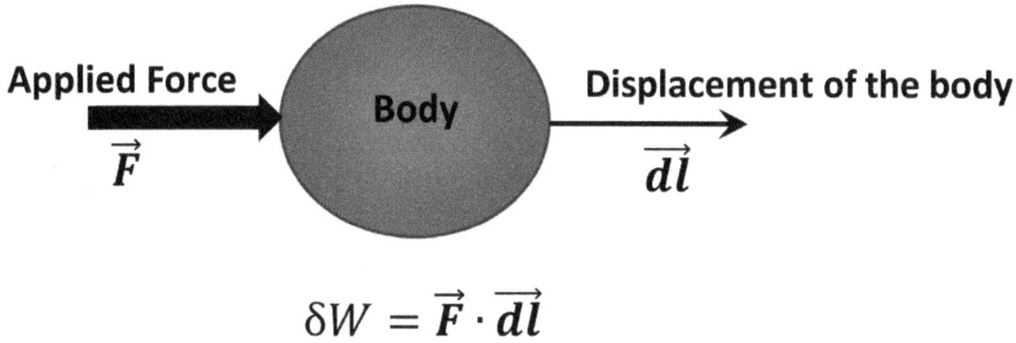

Figure 1.3 Work done by force acting on a body.

In thermodynamics, we are often interested in expansion and compression types of work. Consider expansion of a gas from some initial volume V_i to final volume V_f , as shown in Figure 1.4. What is the work done by the gas on the surroundings?

Let gas inside the piston/cylinder assembly be our system (closed system). The work done by the gas on the surroundings can be calculated using the force-displacement expression, Equation (1.3). We assume that there is no friction between the piston and cylinder walls and that the expansion of the gas is carried out *quasi-statically* so that pressure inside the system is uniform. The upward force on the piston from the gas side is:

$$F = PA_{piston} \tag{1.5}$$

where P is the gas pressure and A_{piston} is the area of the piston. As force and displacement are in the same direction, Equation (1.3) gives:

$$\delta W = \vec{F} \cdot \vec{dl} = PA_{piston}dl \tag{1.6}$$

Replacing $A_{piston}dl$ by differential increase in volume dV, Equation (1.6) reduces to:

$$\delta W = PdV \tag{1.7}$$

Thus, the total work done by the system (gas) is:

$$W = \int_{V_i}^{V_f} P\,dV \tag{1.8}$$

In order to calculate W from Equation (1.8), the relationship between gas pressure and the gas volume must be known.

The same expressions, Equations (1.7) and (1.8), also hold for a compression process. In the case of compression of gas, W is negative. Thus,

$W > 0$, when work is done by the system on the surroundings (1.9a)
$W < 0$, when work is done on the system by the surroundings (1.9b)

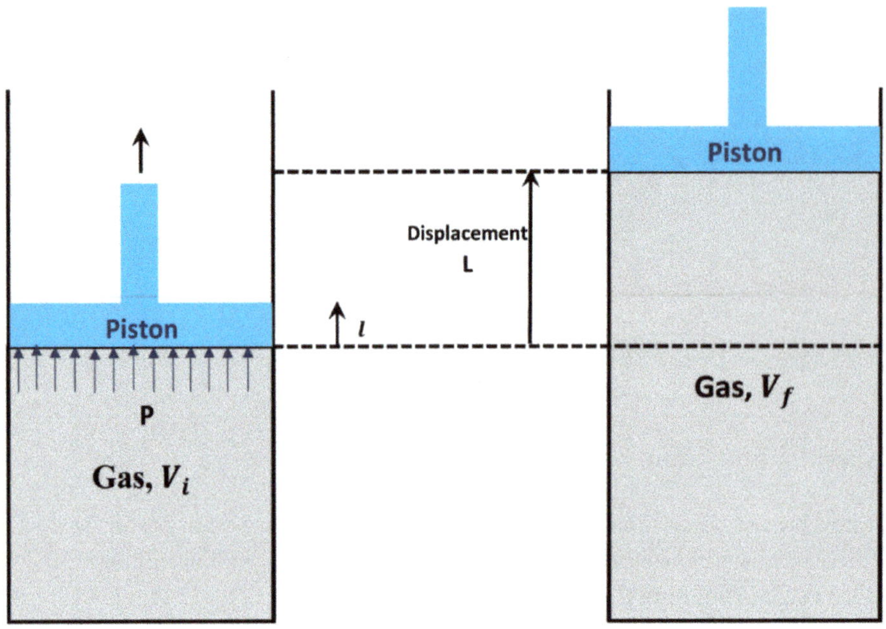

Figure 1.4 Expansion of gas from initial volume V_i to final volume V_f.

It should be noted that work is energy-in-transit, that is, it has meaning only during the transfer process. Systems do not possess work, they possess energy. Furthermore, work produced by a system is a *path-dependent* quantity. Due to its path-dependent nature, the differential of work is an *inexact* differential, denoted as δW. The differentials of thermodynamic properties are *exact* differentials, denoted as *d*(property); for example, dT, dP, and dV, etc. The integration of inexact differential cannot be evaluated until the path is specified whereas the integral of exact differential is independent of the path. Thus,

$$\int_1^2 \delta W = W \text{ (not } W_2 - W_1) \tag{1.10}$$

$$\int_1^2 dP = P_2 - P_1 \tag{1.11}$$

1.4.11 Heat

When system and surroundings are at different temperatures and they are in thermal contact with each other, energy flows between them by virtue of temperature difference. This energy which flows by virtue of temperature difference is called heat and it flows in the direction of decreasing temperature. If the system's temperature is higher than that of the surroundings, energy flows from system to the surroundings and vice versa. Note that heat, like work, is energy-in-transit and has meaning only during the transfer process. Systems do not possess heat. Like work, heat is also a path-dependent quantity, and its differential is inexact, denoted as δQ. The integral of heat is expressed as:

$$\int_1^2 \delta Q = Q \text{ (not } Q_2 - Q_1) \tag{1.12}$$

1.4.12 Internal Energy

Internal energy of a thermodynamic system is the total microscopic energy associated with molecules and atoms contained in the system. It consists of microscopic kinetic energy and microscopic potential energy. The microscopic kinetic energy is energy associated with translation, rotation, and vibration of molecules.

If the system consists of monoatomic gas, only translational kinetic energy of atoms is relevant. The microscopic potential energy is energy stored within molecules and atoms.

Internal energy of a system is an extensive thermodynamic property. Like any other thermodynamic property, it depends only on the current state of the system and is independent of how the system is arrived at the current state. Figure 1.5 gives the detailed picture of different components of internal energy.

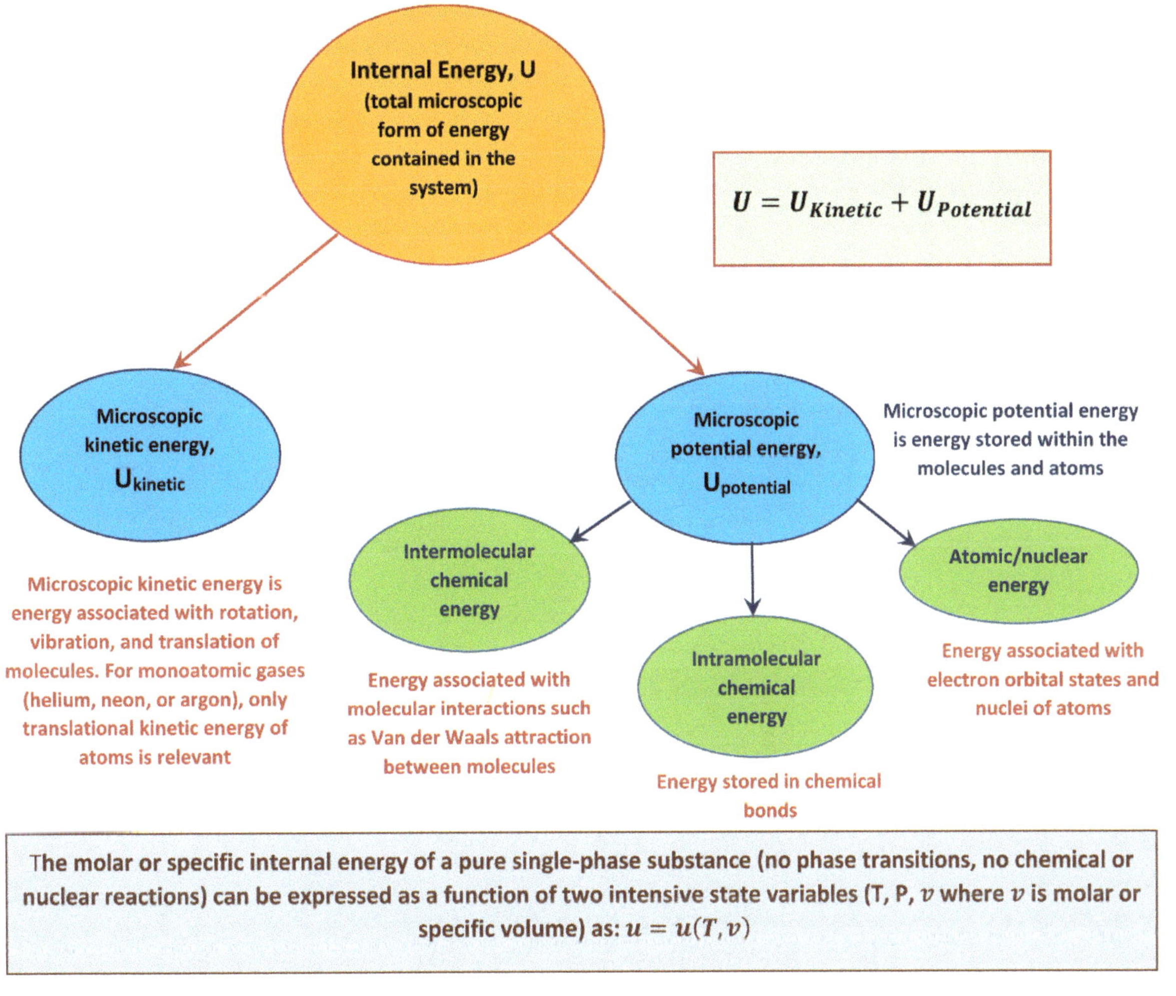

Figure 1.5 Internal energy of a thermodynamic system.

1.4.13 Enthalpy

The enthalpy (H) of a system is defined as:

$$H = U + PV \tag{1.13}$$

where U is the internal energy, P and V are pressure and volume of the system, respectively. Enthalpy is an extensive thermodynamic property. It appears in many thermodynamic analyses as we will see in later chapters.

1.4.14 Reversible and Irreversible Processes

A *reversible process* is one in which a finite change in the state of the system is carried out such that the system always remains in thermodynamic equilibrium during the process. In other words, the system progresses through an infinite sequence of intermediate equilibrium states to reach the final state. The driving forces for changes are zero in a reversible process. The most important characteristic of a reversible process is that the system and its surroundings both can be restored to their original states by reversing the process, that is, by following the same path in the reverse direction. Such a cyclic reversible process, where the same path is followed in the forward and backward directions reversibly, does not leave any trace behind. No one would know if such a cyclic process ever took place. Another important characteristic of reversible processes (cyclic or non-cyclic) is that there is no generation of entropy or destruction of exergy in such processes.

Obviously, a reversible process is an idealized or hypothetical process, impossible to achieve in practice. How can one realize a finite change in the state of the system without disturbing the system from equilibrium? The only way we can approach a reversible process is to carry out the process *quasi-statically* meaning that the process is carried out such that at any instant of the process, the system departs from thermodynamic equilibrium only infinitesimally. Thus, the driving forces for changes are differential (very small) in a quasi-static process. A quasi-static or quasi-equilibrium process can approach the truly reversible process in the limit of driving forces becoming zero.

An *irreversible process* is one in which the driving forces for changes are finite. That is, during the process the system is disturbed from equilibrium by finite changes. An important characteristic of an irreversible process is that it cannot be reversed. In other words, the system and its surroundings cannot be restored to their original states. There occurs a significant generation of entropy and destruction of exergy in irreversible processes. All real processes are irreversible processes.

1.4.15 Reversible Isothermal Expansion of a Gas

As an example of a reversible process, let us consider *reversible* isothermal expansion of a gas. Our system is a gas contained in a piston/cylinder assembly, as shown in Figure 1.6. The piston/cylinder assembly is placed in a constant temperature bath maintained at temperature of T_0.

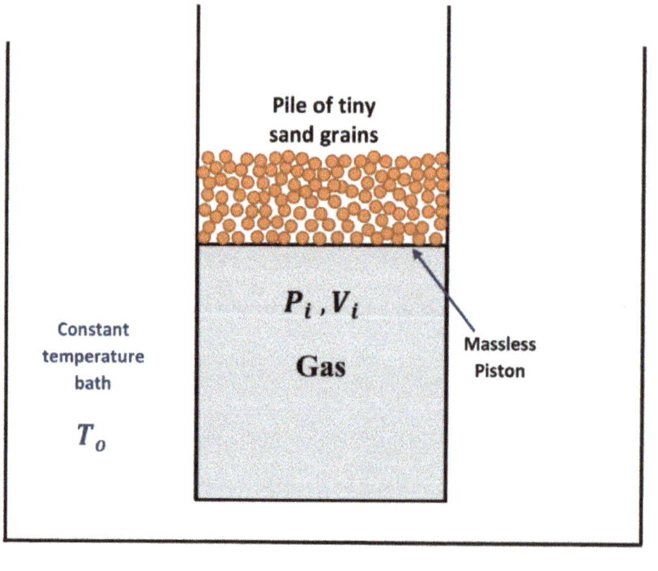

Figure 1.6 Gas contained in a piston/cylinder assembly at constant temperature.

The initial equilibrium state of the system (gas) is: T_o , V_i, P_i where V_i is the initial volume and P_i is the initial pressure of the gas. The pressure of the gas is balanced by the weight of the sand grains as:

$$P_i = Mg/A_{piston} \tag{1.14}$$

where M is the total mass of sand grains, g is acceleration due to gravity, and A_{piston} is the area of the piston.

How do we carry out reversible isothermal expansion of this gas? We begin the process by removing one tiny sand grain. When a tiny sand grain is removed, it creates a slight imbalance of forces on the piston. The piston moves upward allowing the gas to expand slightly. The system reaches a new equilibrium state in a short duration with new uniform pressure and volume. If we continue to remove one sand grain at a time and allow the system to reach equilibrium after each sand grain removal, we can reach the final equilibrium state as shown in Figure 1.7. Thus, the system passes through a large sequence of intermediate equilibrium states before reaching the final equilibrium state. At no stage of the process, the system deviates from equilibrium more than an infinitesimal amount. The thermodynamic properties of the system such as temperature and pressure are all uniform at the intermediate equilibrium states. This is how the gas can be expanded in a quasi-static or reversible fashion. The smaller the mass of the sand grains, the closer we are to achieving reversibility.

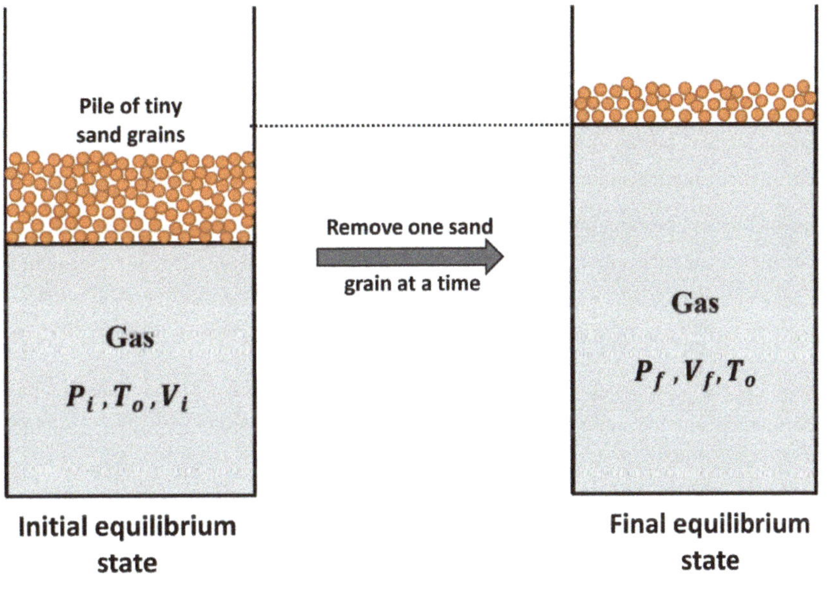

Figure 1.7 Quasi-equilibrium isothermal expansion of gas.

Let us now calculate the amount of work that is produced by the system in this reversible isothermal expansion of gas. Equation (1.8) can be used to calculate the work produced. It is simply the area under the P versus V curve, as shown in Figure 1.8.

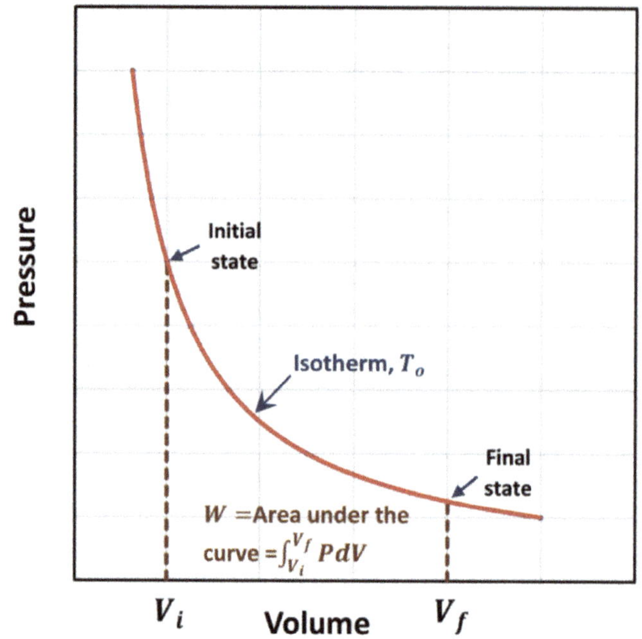

Figure 1.8 Work produced in isothermal expansion of gas.

Assuming gas to be an ideal gas, the pressure is given as:

$$P = \frac{nRT_o}{V} \tag{1.15}$$

where n is the number of moles and R is the universal gas constant. Upon substitution of Equation (1.15) in Equation (1.8), we obtain:

$$W = nRT_o \int_{V_i}^{V_f} \frac{1}{V} dV = nRT_o ln\left(\frac{V_f}{V_i}\right) \tag{1.16}$$

This is the work produced in reversible isothermal expansion of an ideal gas. Note that this is the *maximum possible work* that can be produced for the given change in state of the system.

We would now like to determine if any entropy is generated in the reversible isothermal expansion of ideal gas. The concept of entropy and its calculation are covered in detail in Chapters 4 and 5. For now, we simply accept the discussion about entropy. Entropy, like energy, is a state property. It is a measure of the degree of disorderliness of a system. It is not a conserved quantity. It may be generated during the process.

The amount of entropy generated (S_G) is given as:

$$S_G = \Delta S_{system} + \Delta S_{surroundings} \tag{1.17}$$

The entropy change of the system (ideal gas) can be calculated as:

$$\Delta S_{system} = nRln\left(\frac{V_f}{V_i}\right) \tag{1.18}$$

The entropy change of the surroundings (constant temperature bath) can be calculated as:

$$\Delta S_{surroundings} = -Q/T_o \tag{1.19}$$

where Q is the heat transferred to the system from the surroundings. The amount of heat transferred to the system (ideal gas) is equal to the amount of work produced at constant temperature. Thus,

$$\Delta S_{surroundings} = -Q/T_o = -nRT_o ln\left(\frac{V_f}{V_i}\right)/T_o = -nRln\left(\frac{V_f}{V_i}\right) \tag{1.20}$$

From Equations (1.17), (1.18) and (1.20), it follows that:

$$S_G = 0 \tag{1.21}$$

Thus, no entropy is generated in a reversible process. This is a very important feature of all reversible processes. In fact, Pal [26] defines a reversible process in terms of entropy generation as follows:

"A reversible process is one which does not generate any entropy anywhere in the system or in the surroundings. In other words, there occurs no net increase in the disorder (entropy) of the thermodynamic universe due to a reversible process".

1.4.16 Irreversible Isothermal Expansion of a Gas

We now consider *irreversible* isothermal expansion of a gas, see Figure 1.9. The gas is enclosed in a piston/cylinder assembly which is placed in a constant temperature bath maintained at temperature of T_o. The system (gas) is the same as in reversible isothermal expansion of a gas (Figure 1.6). However, the pressure of the gas is now balanced by two chunks of masses (M_1 and M_2), instead of a large pile of sand grains used in reversible expansion (Figure 1.6). The piston is assumed to be of negligible mass and there is no friction between the piston and the cylinder walls. The initial state of the system is the same as in reversible expansion of gas (Figure 1.6). The total mass M_1+M_2 is equal to the mass of the pile of sand grains shown in Figure 1.6.

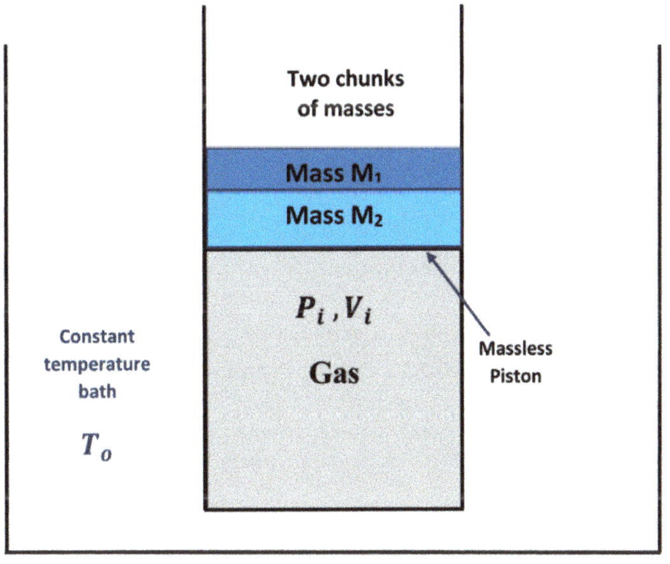

Figure 1.9 Gas contained in a piston/cylinder assembly at constant temperature.

The pressure of gas in the initial equilibrium state can be expressed in terms of the masses as:

$$P_i = (M_1 + M_2)g/A_{piston} \tag{1.22}$$

To carry out irreversible isothermal expansion of the system (gas), we simply remove mass M_1 from top of the piston, see Figure 1.10. As soon as mass M_1 is removed, it creates a large imbalance of forces on the piston and the piston shoots up. After some oscillations, the piston eventually settles in a new position. In the final equilibrium state, the pressure inside the gas is as follows:

$$P_f = M_2 g/A_{piston} \tag{1.23}$$

This final pressure P_f is the same as in reversible expansion of gas, shown in Figure 1.7.

Although the initial and the final equilibrium states in reversible and irreversible expansion processes are the same, there are major differences in the manner in which the two processes are carried out. In reversible expansion, the system was always very close to equilibrium during the process. The properties of the system were well defined and uniform throughout the system. In the present irreversible process (Figure 1.10), the system is disturbed from equilibrium in a major way. When mass M_1 is removed, the gas undergoes transient phenomena involving turbulence. The properties of the system such as temperature and pressure are no longer uniform during the transient expansion process. The properties are defined only in the initial and final equilibrium states.

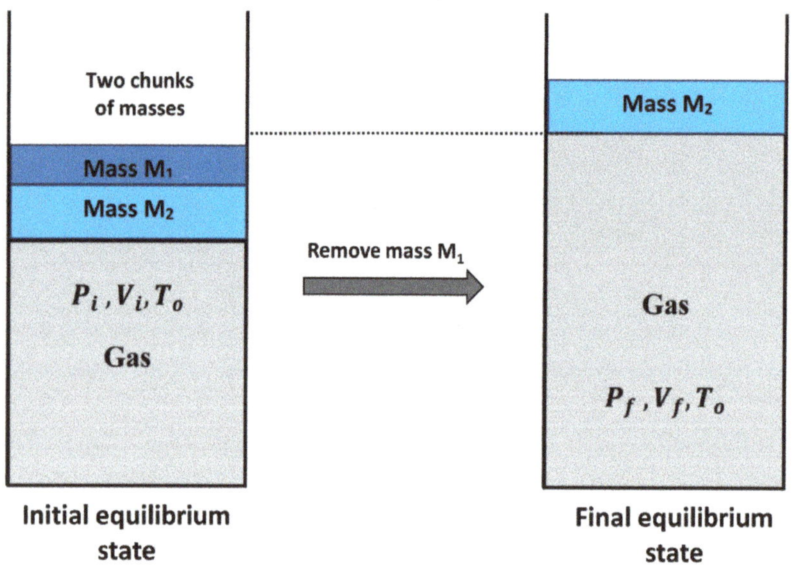

Figure 1.10 Irreversible isothermal expansion of gas.

What is the work produced in this irreversible expansion of a gas? Equation (1.8) is no longer applicable as properties such as pressure are not uniform during the process. However, work is transferred to the surroundings as the mass M_2 is raised in a gravitational field. Thus, the amount of work delivered to the surroundings is simply the increase in potential energy of the surroundings, given as:

$$W = M_2 g(V_f - V_i)/A_{piston} = P_f(V_f - V_i) \tag{1.24}$$

Note that in the final equilibrium state, the upward force on the piston due to gas pressure P_f is balanced by the downward force on the piston due to weight $M_2 g$. Also, during the expansion, the weight is raised in a gravitation field through a distance of $(V_f - V_i)/A_{piston}$.

The work produced by the system in irreversible isothermal expansion is shown in Figure 1.11. It is shown as colored rectangle. The work produced in irreversible expansion is much smaller as compared with reversible expansion between the same initial and final states. Obviously, work is lost in irreversible expansion. The work lost is as follows:

$$W_{lost} = W_{rev} - W_{irr} = nRT_o ln\left(\frac{V_f}{V_i}\right) - P_f(V_f - V_i) \tag{1.25}$$

Using the ideal gas law, Equation (1.25) can be re-written as:

$$W_{lost} = nRT_o\left[ln\left(\frac{V_f}{V_i}\right) - \left(1 - \frac{V_i}{V_f}\right)\right] \tag{1.26}$$

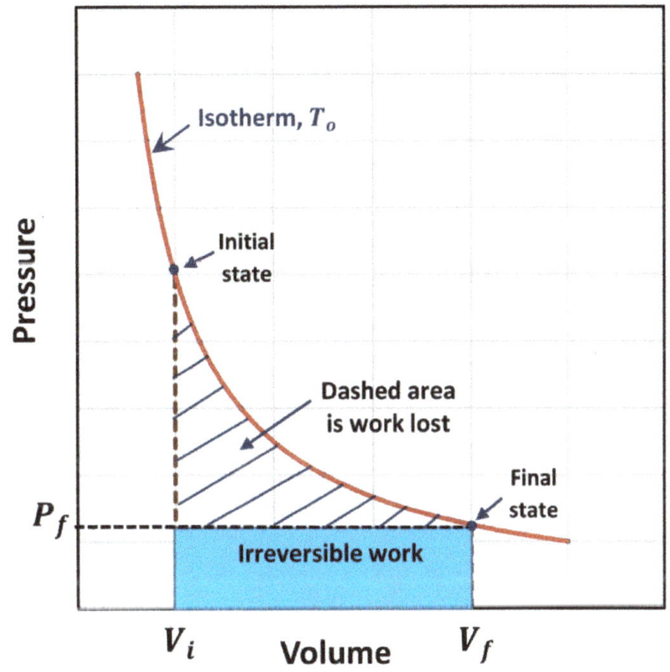

Figure 1.11 Work produced and work lost in irreversible expansion.

Thus, irreversible processes are characterized by destruction of work potential or exergy [26]. The work produced is lower than a reversible process for the same change in states.

We would now like to determine if any entropy is generated in the irreversible isothermal expansion of ideal gas. The amount of entropy generated (S_G) is given by Equation (1.17). The entropy change of the system (ideal gas) is the same as before, given by Equation (1.18). However, the amount of heat transferred to the system is much lower now as work produced is less:

$$Q = W = P_f(V_f - V_i) \tag{1.27}$$

Thus, the decrease in entropy of the surroundings is much lower:

$$\Delta S_{surroundings} = -Q/T_o = -P_f(V_f - V_i)/T_o = -nR\left(1 - \frac{V_i}{V_f}\right) \tag{1.28}$$

Combining Equations (1.18) and (1.28), amount of entropy generated (S_G) is given as:

$$S_G = \Delta S_{system} + \Delta S_{surroundings} = nR\left[ln\left(\frac{V_f}{V_i}\right) - \left(1 - \frac{V_i}{V_f}\right)\right] \quad (1.29)$$

Figure 1.12 shows the plot of S_G/nR versus V_f/V_i. The entropy generation occurs in both irreversible expansion and compression. The entropy generation increases with the increase in deviation of volume ratio V_f/V_i from unity.

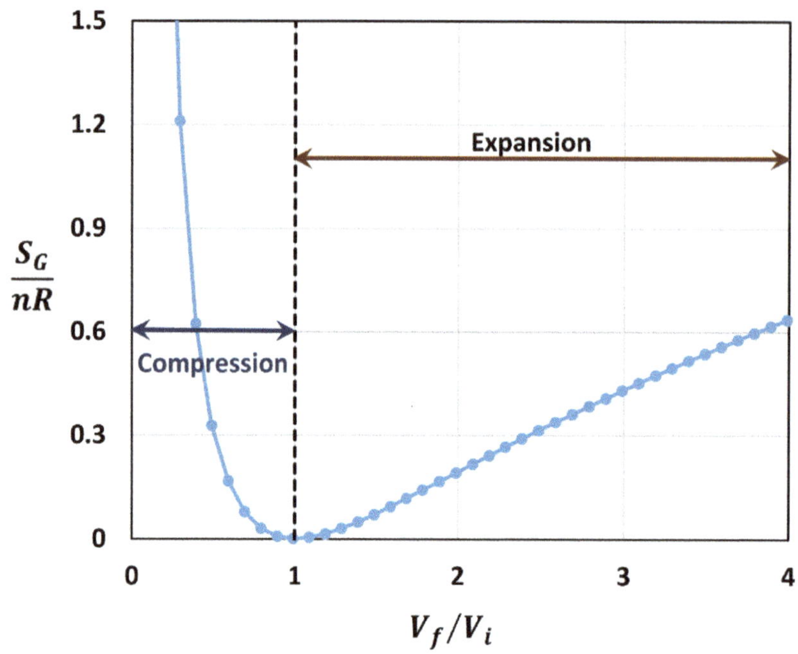

Figure 1.12 Entropy generation in irreversible expansion and compression of ideal gas.

Thus, irreversible processes are accompanied by destruction of work potential and generation of entropy [26]. We can also say that irreversible processes result in an increase in the disorderliness of the thermodynamic universe.

1.5 FOUR LAWS OF THERMODYNAMICS

There are four laws of thermodynamics. They are described here briefly. The first and the second laws are covered in detail in the chapters to follow.

1.5.1 Zeroth Law of Thermodynamics

Consider three thermodynamic systems A, B, and C. The zeroth law states that if A and B are in thermal equilibrium, and A and C are in thermal equilibrium, then B and C are also in thermal equilibrium with each other (see Figure 1.13).

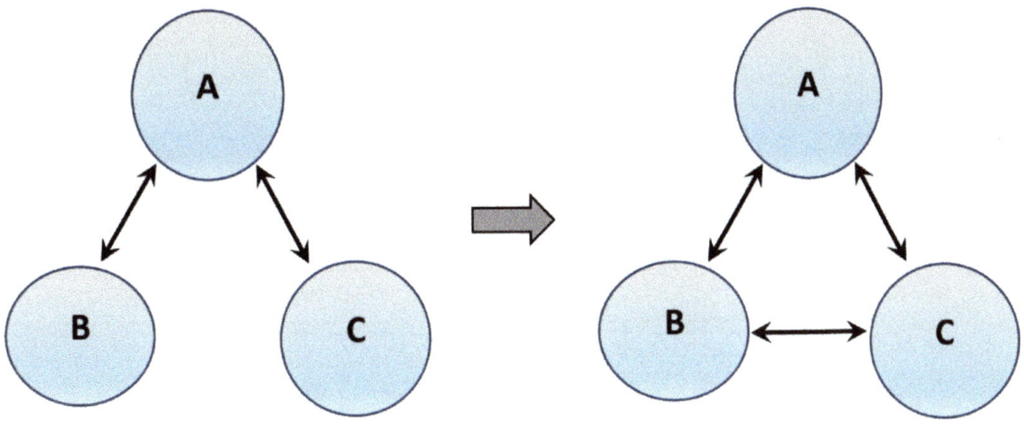

Figure 1.13 Illustration of zeroth Law of thermodynamics. Double arrow indicates thermal equilibrium. Note that two systems are in thermal equilibrium if there is no flow of heat between them when they are in thermal contact.

It is the zeroth law of thermodynamics that implies the existence of a thermodynamic intensive property of a system called *temperature*. Thus, the zeroth law could also be stated as: There exists an intensive thermodynamic property of a system called temperature. For systems to be in thermal equilibrium with each other, the temperatures of the systems must be the same.

1.5.2 First Law of Thermodynamics

One way to express the first law of thermodynamics is as follows: There exists an extensive thermodynamic property of a system called *Energy*. The energy of an isolated system is constant. In other words, energy can neither be created nor destroyed.

1.5.3 Second Law of Thermodynamics

There exits an extensive thermodynamic property of a system called *Entropy*. The entropy of an isolated system never decreases.

1.5.4 Third Law of Thermodynamics

The entropy of a system approaches a constant value as the absolute temperature of the system approaches zero. For a pure, perfectly crystalline substance, the entropy is zero at absolute temperature of zero (0 K).

NOTATION

A_{piston}	piston cross-sectional area
C	number of components (chemical species) in phase rule
d	differential of exact quantity
F	degrees of freedom in phase rule
$\vec{F}$	force vector
g	acceleration due to gravity
H	enthalpy
dl	differential displacement
L	total displacement
M	mass

n	number of moles
P	pressure
Q	heat
R	universal gas constant
S	entropy
S_G	amount of entropy generated
S_{system}	entropy of system
$S_{surroundings}$	entropy of surroundings
T	temperature
u	molar or specific internal energy
U	internal energy
$U_{kinetic}$	microscopic internal energy
$U_{potential}$	microscopic potential energy
v	molar or specific volume
V	volume
V_i	initial volume
V_f	final volume
W	work
W_{irr}	irreversible work
W_{lost}	work lost in a process due to irreversibilities
W_{rev}	reversible work

Greek Letters

δ	differential of inexact quantity

REFERENCES

1. Huang, F.F., *Engineering Thermodynamics: Fundamentals and Applications,* Macmillan, New York, 2nd edition, 1988.
2. Mulop, N., Yosof, K.M., Tasir, Z., "A review on enhancing the teaching and learning of thermodynamics", *Procedia-Social and Behavioral Sciences* 2012; 56:703-712.
3. Atkins, P., *Four laws that drive the universe*, Oxford University Press, Oxford, 2007.
4. Abu-Mualweh, H.I., "Portable experimental apparatus for demonstrating thermodynamics principles", *Int. J. Mech. Eng. Edu.* 2004; 32:223-231.
5. Hassan, O., Mat, R., "A comparative study of two different approaches in teaching thermodynamics", In proceedings of the 2005 regional conference on engineering education, Dec. 12-13, 2005, Johor, Malaysia.
6. Fuchs, H.U., "Thermodynamics: A misconceived theory", In proceedings of the second seminar on misconceptions and educational strategies in science and mathematics, July 26-29, 1987, Cornell University, Ithaca, New York.
7. Dukhan, N., Schumack, M., "Understanding the continued poor performance in thermodynamics as a first step toward an instructions strategy", In proceedings of 120th ASEE annual conference & exposition, June 23-26, 2013, Atlanta.
8. Patron, F., Conceptual understanding of thermodynamics: a study of undergraduate and graduate students, Ph.D. thesis, Purdue University, 1997.
9. Junglas, P., "Simulation programs for teaching thermodynamics", *Global J. Eng. Edu.* 2006; 10: 175-180.
10. Meltzer, D.E., "Investigation of student learning in thermodynamics and implications for instruction in chemistry and engineering", In proceedings of physics education research conference, July 26-27, 2006, Syracuse, New York.
11. Meltzer, D.E., "Investigation of students' reasoning regarding heat, work, and the first law of thermodynamics in an introductory calculus-based general physics course", *Am. J. Phys.* 2004; 72: 1432-1446.
12. Anderson, E.E., Taraban, R., Sharman, M.P., "Implementing and assessing computer-based active learning materials in introductory thermodynamics", *Int. J. Eng. Edu.* 2005; 21: 1168-1176.

13. Cotignola, M.I., Bordogna, C., Punte, G., Cappannini, O.M., "Difficulties in learning thermodynamic concepts: are they linked to the historical development of this field?", *Science & Education* 2002; 11: 279-291.
14. Liu, Y., "Development of instructional courseware in thermodynamics education", *Comput. Appl. Eng. Edu.* 2011; 10(1): 115-124.
15. Bullen, P., Russel, M., "A blended learning approach to teaching first year engineering degree students", In proceedings of international conference on engineering education-ICEE 2007, Sept. 3-7, 2007, Coimbra, Portugal.
16. Baher, J., "How articulate virtual labs can help in thermodynamics education: a multiple case study", In proceedings of 28th Annual Frontiers in Education conference, 1998, Vol 2, 663-668.
17. Kelly, G., "A powerful virtual learning environment", In proceedings of ASCILITE conference, Dec. 8-11, 2002, Auckland, New Zealand.
18. Huang, M., Gramoll, K., "Online interactive multimedia for engineering thermodynamics", In proceedings of American society for engineering education annual conference & exposition, June 20-23, 2004, Salt Lake City.
19. Cox, A.J., Belloni, M., Dancy, M., Christian, W., "Teaching thermodynamics with physlets in introductory physics", *Physics Education* 2003; 38: 433-440.
20. Forbus, K.D., Whalley, P.B., Everett, J.O., Ureel, L., Brokowski, M., Baher, J., Kuehne, S.E., "Cyclepad: An articulate virtual laboratory for engineering thermodynamics", *Artificial Intelligence* 1999; 114: 297-347.
21. Chaturvedi, S., Abdel-Salam, T., Kasinadhuni, O., "Virtual Assembly – a web-based student learning tool for thermodynamics concepts related to multistaging in compressors and turbines", In proceedings of international conference on engineering education (ICEE 2007), Sept. 3-7, 2007, Coimbra, Portugal.
22. Colbourn, W.G., Lindauer, G.C.A., "A flexible multimedia instructional material module for introductory thermodynamics", *J. Eng. Edu.* 1994; 83: 271-277.
23. Grigull, U., "Students views on learning thermodynamics" In proceedings of the workshop on the second law of thermodynamics", August 27-30, 1990, Erciyes University, Kayseri, Turkey.
24. Costa, V.A.F., "Some considerations and reflections on engineering thermodynamics teaching and learning", In proceedings of the 3rd international conference of the Portuguese society for engineering education (CISPEE), June 27-29, 2018, Aveiro, Portugal.
25. Manteufel, R.D., "A spiral approach for mechanical engineering thermodynamics", In proceedings of the ASME international mechanical engineering conference, November 14-19, 1999, Nashville, Tennessee.
26. Pal, R., "Conceptual issues related to reversibility and reversible work produced in closed and open flow systems", *Education for Chemical Engineers* 2017; 19: 29-37.

TUTORIAL PROBLEMS WITH SOLUTION

Problem 1

A gas is contained in a piston/cylinder assembly. The diameter of the piston is 0.5 m. To balance the pressure of the gas, a mass of 160 kg is placed on the piston. The piston is of negligible mass. The atmospheric pressure is 101.325 kPa.

a) Determine the pressure of the gas.
b) What is the work done by the gas if the gas is expanded quasi-statically by heating such that the piston and weight are raised by a distance of 1 m against gravity. The acceleration due to gravity is 9.81 m/s^2.

Solution

a) Pressure of the gas

Force acting on the piston (gas):

$$F = Mg + P_{atm}\frac{\pi}{4}D^2 = 160 \times 9.81 + 101.325 \times 10^3 \times \frac{\pi}{4}(0.5)^2 = 1569.6 + 19895.12$$
$$= 21464.72\ N$$

Thus, the pressure of the gas is:

$$P = \frac{F}{\frac{\pi}{4}D^2} = \frac{21464.72}{\frac{\pi}{4}(0.5)^2} = 109318.91\ Pa = 109.32\ kPa$$

b) What is the work done by the gas?

For quasi-static expansion,

$$W = \int PdV = P\Delta V$$

Volume change is

$$\Delta V = 1 \times \frac{\pi}{4}(0.5)^2 = 0.19635\ m^3$$

$$W = P\Delta V = 109.32 \times 0.19635 = 21.46\ kJ$$

Thus, the work done by gas in quasi-static expansion is $21.46\ kJ$.

Note that work can also be calculated as:

$$W = F \times distance = 21464.72 \times 1 = 21464.72J = 21.46\ kJ$$

Problem 2

Gas contained in a piston-cylinder assembly is heated and the gas expands quasi-statically such that $PV = C$ where P is pressure, V is total volume, and C is a constant. The initial pressure and volume of the gas are 210 kPa and 1.75 m^3, respectively. The final pressure of the gas is 120 kPa.

a) Determine the final volume of the gas.
b) What is the work done by the gas?

Solution

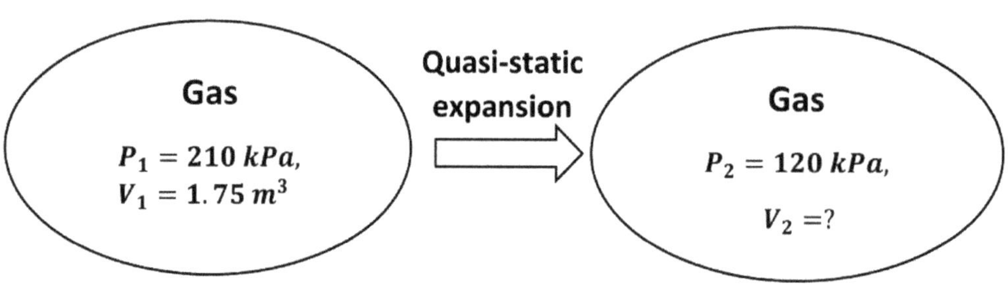

a) Final volume

$$C = P_1V_1 = 210 \times 1.75 = 367.5\ kPa.m^3$$

$$V_2 = C/P_2 = 367.5/120 = 3.0625\ m^3$$

b) Work done by gas

For quasi-static expansion,

$$W = \int PdV = \int_{V_1}^{V_2} \frac{C}{V} dV = Cln\left(\frac{V_2}{V_1}\right) = 367.5 \times ln\left(\frac{3.0625}{1.75}\right) = 205.66\ kJ$$

Problem 3

One kg of air undergoes a cyclic process as shown below. All the steps are carried out quasi-statically. What is the net work of the process? You can assume that the air behaves as an ideal gas. The molecular weight of air is 28.97 g/mol.

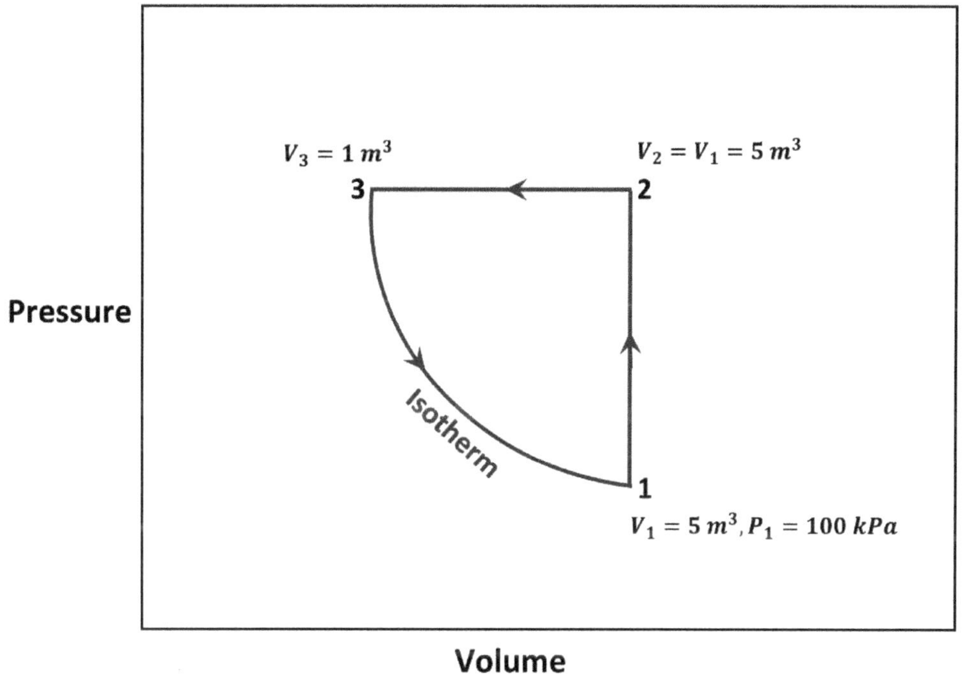

Solution

Number of moles of air $= n = \frac{1000}{28.97} = 34.518\ moles$

$$T_1 = \frac{P_1V_1}{nR} = \frac{100 \times 1000 \times 5}{34.518 \times 8.314} = 1742.27\ K$$

$$P_3 = P_1V_1/V_3 = 100 \times 5/1 = 500\ kPa$$

Process $1 \rightarrow 2$:

$W_{1\rightarrow2} = 0$ (no volume change)

Process $2 \rightarrow 3$:

$W_{2\rightarrow3} = P_3(V_3 - V_2) = 500(1 - 5) = -2000\ kJ$

Process 3 → 1:

$$W_{3\to1} = nRT_1 ln\left(\frac{V_1}{V_3}\right) = 34.518 \times 8.314 \times 1742.27 \times ln\left(\frac{5}{1}\right) = 804720.99J = 804.721\ kJ$$

$$Net\ Work = W_{1\to2} + W_{2\to3} + W_{3\to1} = 0 - 2000 + 804.721 = -1195.28\ kJ$$

Note that the net work is negative indicating that work is absorbed by the process.

PRACTICE PROBLEMS

Problem 1

An ideal gas is compressed reversibly in a piston/cylinder assembly such that the volume decreases from 800 cm^3 to 360 cm^3. The temperature is held constant at 50 °C. The initial pressure of the gas is 200 kPa. Calculate the work done by the gas.

Problem 2

Consider reversible expansion of an ideal gas such that $PV^{1.4} = Constant$. Develop the expression for work done by the gas in going from state 1 (P_1, V_1) to state 2 (P_2, V_2).

Problem 3

One kg of air undergoes a cyclic process as shown below. All the steps are carried out quasi-statically. What is the net work of the process? Show calculations of all steps. You can assume that the air behaves as an ideal gas. The molecular weight of air is 28.97 g/mol.

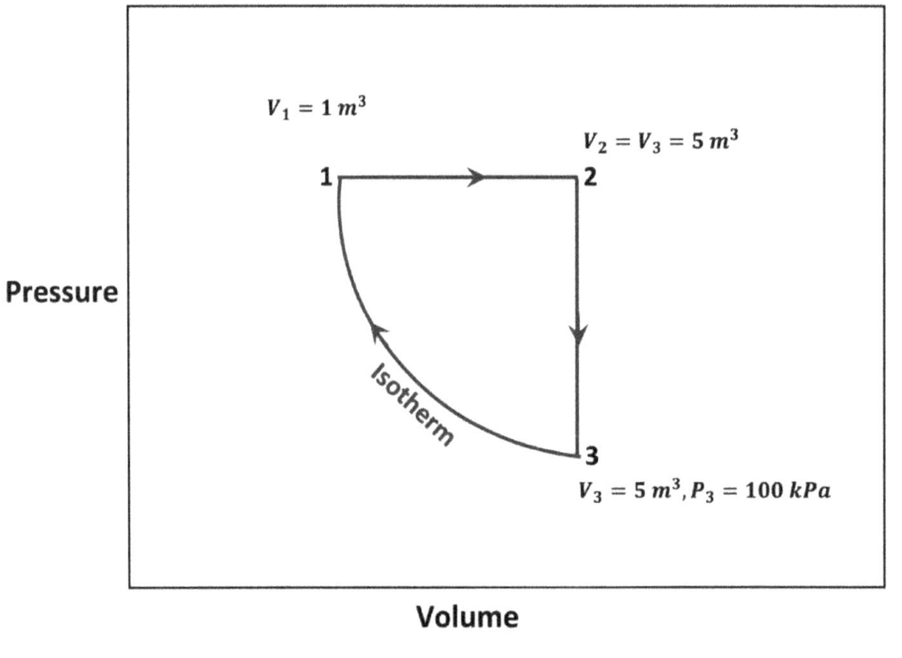

Problem 4

Ten moles of an ideal gas expand irreversibly from an initial volume of 0.1 m^3 to 1 m^3. A constant external pressure of 500 kPa acts on the gas during the expansion process. Calculate the work done by the gas.

Problem 5

One kg of air undergoes a cyclic process as shown below. All the steps are carried out quasi-statically. What is the net work of the process? Show calculations of all steps. You can assume that the air behaves as an ideal gas. The molecular weight of air is 28.97 g/mol. Note that: $V_1 = V_3 = 1\ m^3$ and $P_3 = P_2 = 120\ kPa$.

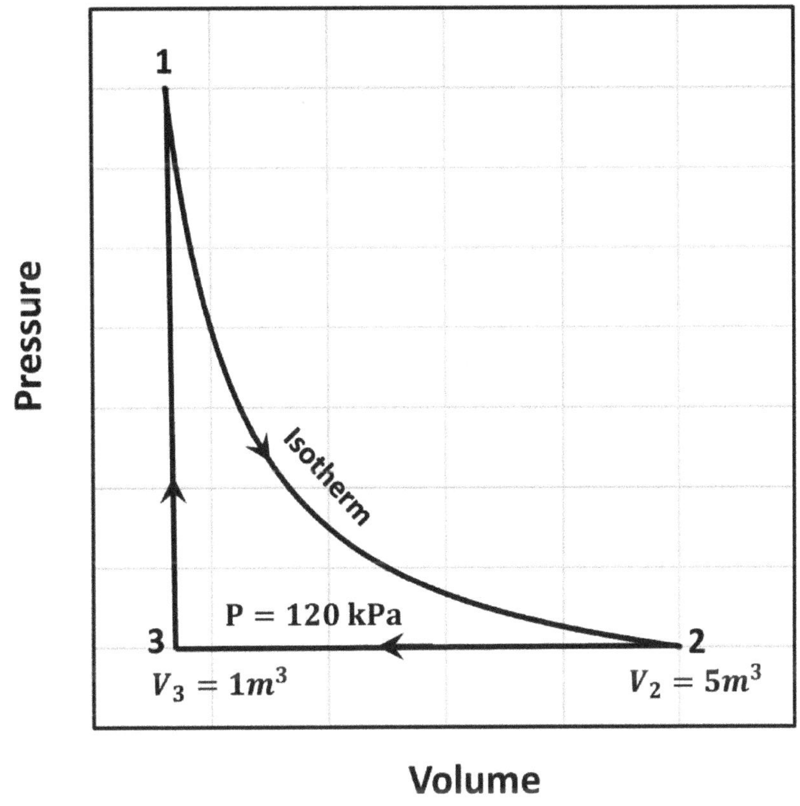

Chapter 2

First Law of Thermodynamics for Closed Systems (Control Mass)

2.1 LEARNING OUTCOMES

Upon completion of this chapter, students will be able to:

- State the first law for closed systems undergoing non-cyclic and cyclic processes
- Apply the first law for closed systems to different non-cyclic processes such as isochoric, isobaric, isothermal, and adiabatic processes
- Calculate internal energy and enthalpy changes for homogeneous substances of constant composition
- Apply the first law for closed systems to cyclic processes

2.2 INTRODUCTION

The first law of thermodynamics is simply the *principle of conservation of energy*. Energy cannot be created or destroyed. In other words, energy just cannot appear from nowhere. Neither can it just disappear from anywhere. The total amount of energy is constant. However, it may change from one form to another. The first law of thermodynamics places no restriction on conversion of one form of energy into another. All forms of energy are equal in the eyes of the first law.

When applied to a thermodynamic universe (system + surroundings), the first law of thermodynamics can be expressed as:

$$E_{system} + E_{surroundings} = Constant \tag{2.1}$$

where E is the total energy. During any process, the change in the total energy of the universe is zero. Thus,

$$\Delta E_{system} + \Delta E_{surroundings} = 0 \tag{2.2}$$

$$\Delta E_{system} = -\Delta E_{surroundings} \tag{2.3}$$

If some energy disappears from the surroundings, it must simultaneously appear in the system in exactly the same amount although its form could be different. The first law of thermodynamics allows us to calculate the changes in total energy of the system and the amounts of energy transferred between the system and the surroundings.

2.3 DIFFERENT FORMS OF ENERGY

Different forms of energy that may be involved in a thermodynamic process are:

a) *Heat (Q)*: Heat may be exchanged between the system and the surroundings
b) *Work (W):* Work may be exchanged between the system and the surroundings
c) *Macroscopic Kinetic Energy:* When the system is in motion, it has bulk or macroscopic kinetic energy given as: $E_k = mV^2/2$, where m is the mass of the system and V is its velocity

d) *Macroscopic Potential Energy:* The macroscopic potential energy of a system is the energy that a system possesses due to its location in a gravitational field. Moving the system from one location to another in a gravitational field could change its potential energy. It is given as: $E_p = mgh$, where g is acceleration due to gravity (9.8 m/s^2 in the vicinity of the earth's surface), and h is the elevation of the system in a gravitation field.
e) *Internal Energy (U):* As noted in Chapter one, internal energy of a system is the total microscopic energy contained in the system. It includes microscopic kinetic energy associated with molecules and atoms and microscopic potential energy stored within atoms and molecules.

2.3.1 Sign Convention for Heat and Work Exchange

In this book, the following sign convention is used for work and heat interactions between the system and the surroundings:

Work (W): $W > 0$ when it done *by* the system on the surroundings; $W < 0$ when it is done *on* the system by the surroundings.

Heat (Q): $Q > 0$ when it transferred *to* the system from the surroundings; $Q < 0$ when it is transferred *from* the system to the surroundings.

2.3.2 Different Units of Energy

In the International System of Units (SI), the unit of energy in all forms is Joule (J). Other units of energy are calories (cal) and BTU. 1 cal = 4.184 J and 1 BTU=252 cal = 1.055 kJ

2.4 DIFFERENT FORMS OF THE FIRST LAW

2.4.1 First Law for an Isolated System

An *isolated system* is a special case of closed systems that does not interact with its surroundings in any way. It is cut off from the surroundings. No mass and energy can enter or leave the system. Thus, the total energy of an isolated system is constant according to the first law:

$$\Delta E_{system} = -\Delta E_{surroundings} = 0 \tag{2.4}$$

As the total energy of the system consists of kinetic, potential and internal energies, Equation (2.4) can be re-written as:

$$\Delta E_{system} = \Delta\left(U + E_k + E_p\right) = 0 \tag{2.5}$$

Thus, the first law for an *isolated* system is:

$$\Delta U + \Delta E_k + \Delta E_p = 0 \tag{2.6}$$

If the system is fixed in space with no bulk motion, Equation (2.6) simplifies to:

$$(\Delta U)_{isolated\ system} = 0 \tag{2.7}$$

2.4.2 First Law for a Closed System

No mass enters or leaves a closed system. However, heat and work interactions with the surroundings can take place (see Figure 2.1). The first law for a closed system is:

$$\Delta E_{system} = Q - W \tag{2.8}$$

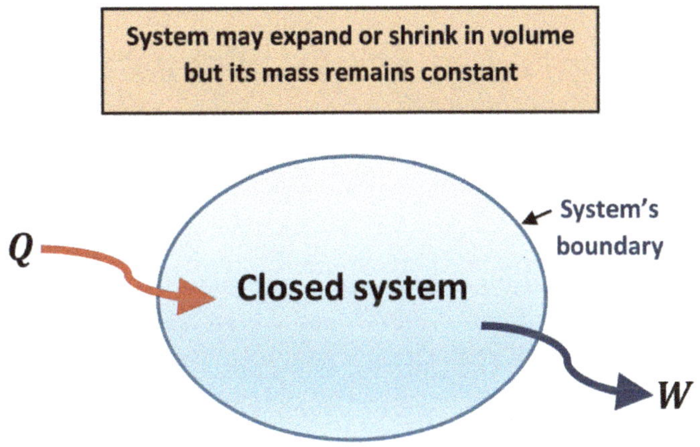

Figure 2.1 Closed system exchanging heat and work with the surroundings.

Assuming that the system is fixed in space with no bulk motion, Equation (2.8) simplifies to:

$$\Delta U = Q - W \tag{2.9}$$

The differential form of the first law for a closed system is:

$$dU = \delta Q - \delta W \tag{2.10}$$

Note that U is a state property and hence the differential of U is exact, denoted as dU. Heat and work are path-dependent quantities, and their differentials are inexact differentials, denoted as δQ and δW, respectively.

2.4.3 First Law for a Closed System Undergoing a Cyclic Process

Starting from the differential form of the first law, Equation (2.10), we can express the first law for a cyclic process as:

$$\oint dU = \oint \delta Q - \oint \delta W \tag{2.11}$$

where the integral symbol implies an integration over a complete cycle. As internal energy is a state property, its cyclic integral is zero. Thus,

$$\oint \delta Q = \oint \delta W \tag{2.12}$$

$$Q_{cycle} = W_{cycle} \tag{2.13}$$

Equations (2.12) and (2.13) are the mathematical statements of the first law for a cyclic process. The net work done by the system is equal to the net transfer of heat to the system when a system undergoes a cyclic process. Note that Equations (2.12) and (2.13) are valid even when the system possesses macroscopic kinetic and potential energies. Like internal energy, the total energy of a system is also a state property and therefore, its cyclic integral is zero.

2.5 APPLICATION OF FIRST LAW TO ISOCHORIC AND ISOBARIC PROCESSES

2.5.1 Isochoric Process

In an isochoric process, the volume of the system is constant throughout the process (see Figure 2.2). As the volume of the system is constant during the process, there is no expansion or compression work. Thus,

$$W = 0 \tag{2.14}$$

Note that it is only the expansion/compression work that is zero. We can always transfer work to the system in a constant volume process through other mechanisms such as stirring.

The first law in the form of Equation (2.9) gives:

$$Q = \Delta U \tag{2.15}$$

Thus, the amount of heat exchanged between the system and the surroundings is equal to the internal energy change of the system. It is interesting to note this is true whether the heat is exchanged reversibly or irreversibly.

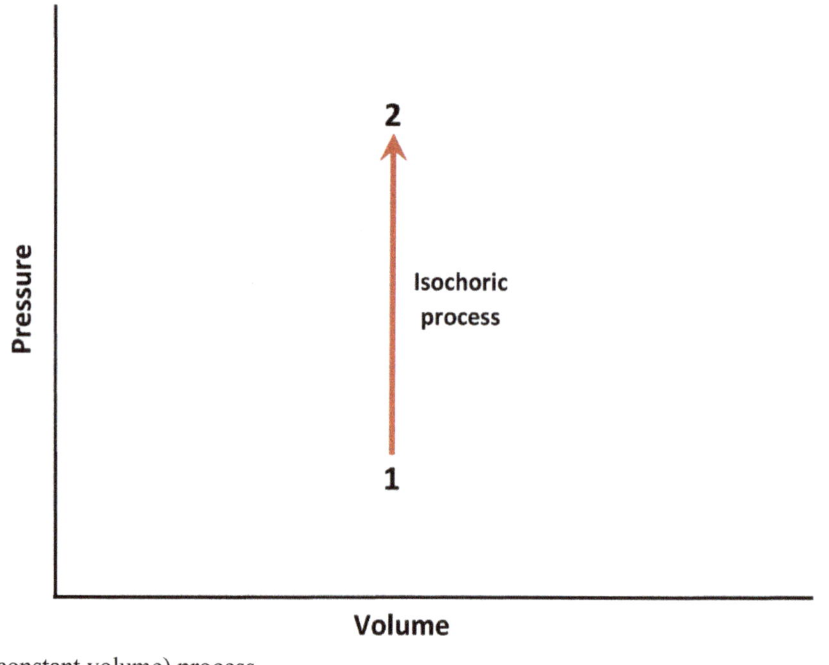

Figure 2.2 Isochoric (constant volume) process.

Let us now consider a variation of the constant volume process (see Figure 2.3). The initial and final volumes are the same. But volume is maintained constant throughout the process only along path I. Along path II (assumed reversible), the volume is not constant during the process. Only the initial and final states have the same volume. What is the difference between these two processes?

- ΔU is the same along paths I and II as U is a state property.
- Q is a path-dependent quantity. It is different along different paths I and II.
- Only along path I, $Q = \Delta U$.
- Along path II, $Q \neq \Delta U$
- Work is zero along path I

- Work is non-zero along path II. $W < 0$ along path II. It is simply negative of the area of rectangle bounded by red lines.
- Along path II, $Q = W + \Delta U$

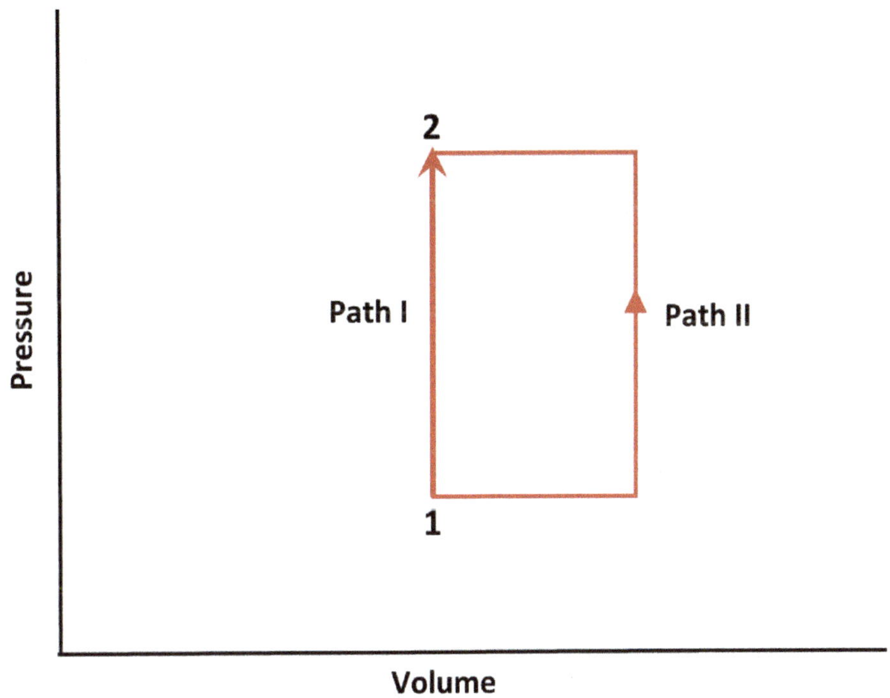

Figure 2.3 Processes with same volumes in the initial and final states.

2.5.2 Isobaric Process

In an isobaric process, the pressure of the system is constant throughout the process (see Figure 2.4). Assuming that the process is quasi-static, the work interaction is given as:

$$\delta W = PdV \tag{2.16}$$

From the first law,

$$dU = \delta Q - \delta W = \delta Q - PdV \tag{2.17}$$

Upon re-arrangement, Equation (2.17) gives:

$$\delta Q = dU + PdV \tag{2.18}$$

For a constant process, the change in enthalpy is:

$$dH = dU + PdV \tag{2.19}$$

From Equations (2.18) and (2.19), we get:

$$\delta Q = dH \tag{2.20}$$

Upon integration,

$$Q = \Delta H \tag{2.21}$$

Thus, for a quasi-static, constant-pressure, closed-system process, the heat exchanged is equal to the enthalpy change of the system.

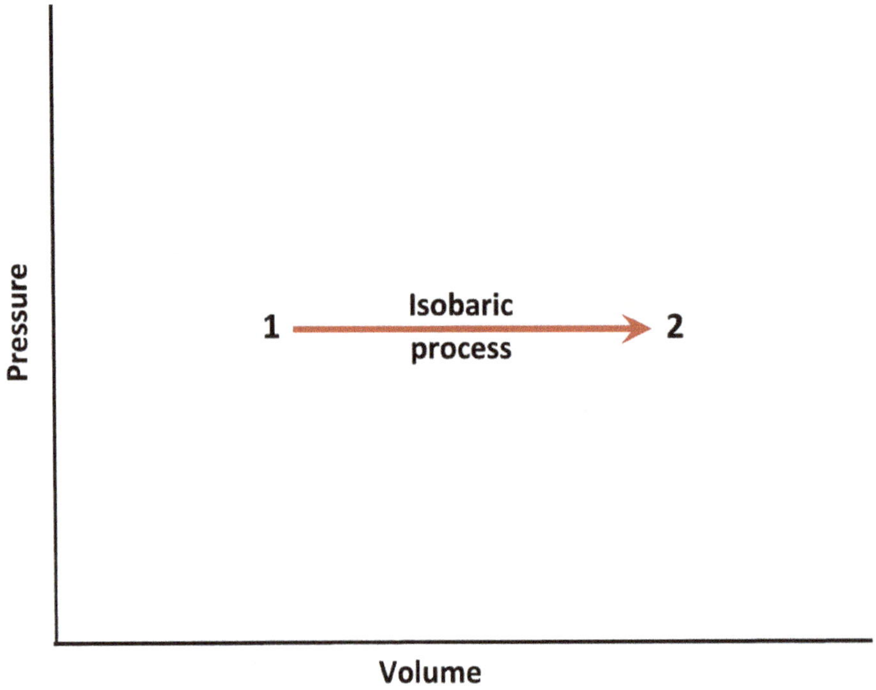

Figure 2.4 Isobaric (constant pressure) process.

Let us now consider a variation of the constant pressure process (see Figure 2.5). The initial and final pressures are the same. But pressure is maintained constant throughout the process only along path I. Along path II (assumed quasi-static), the pressure is not constant during the process. Only the initial and final states have the same pressure. What is the difference between these two processes?

- ΔH is the same along paths I and II as H is a state property.
- Q is a path-dependent quantity. It is different along different paths I and II.
- Only along path I, $Q = \Delta H$.
- Along path II, $Q \neq \Delta H$
- Work is $P_I(V_2 - V_1)$ along path I (where P_I is the constant pressure along path I)
- Work is not $P_I(V_2 - V_1)$ along path II. W along path II is $P_{II}(V_2 - V_1)$, where P_{II} is the constant pressure along path II

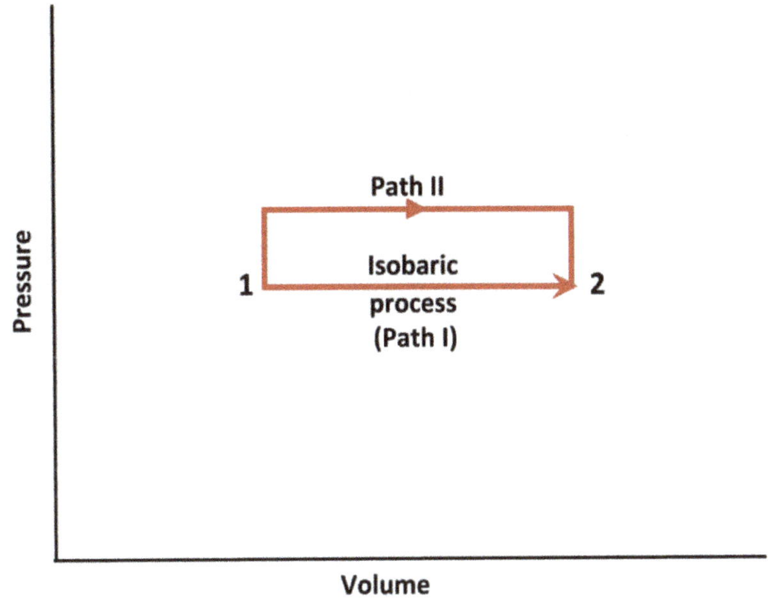

Figure 2.5 Quasi-static processes with same pressures in the initial and final states.

2.6 CALCULATION OF INTERNAL ENERGY AND ENTHALPY CHANGES

The calculation of internal energy and enthalpy changes in thermodynamics processes are important. For example, in an isochoric process, the amount of heat transferred is equal to internal energy change of the system. In a quasi-static isobaric process, the amount heat transferred is equal to the enthalpy change of the system.

2.6.1 Internal Energy Changes

For a homogeneous substance of constant composition (pure gas, pure liquid or pure solid), the Gibbs phase rule gives two degrees of freedom to completely specify the intensive state of the system. Thus, the molar or specific internal energy of a substance can be expressed as a function of two state variables such as temperature and molar or specific volume:

$$u = u(T, v) \tag{2.22}$$

where u is molar or specific internal energy and v is molar or specific volume. From Equation (2.22):

$$du = \left(\frac{\partial u}{\partial T}\right)_v dT + \left(\frac{\partial u}{\partial v}\right)_T dv \tag{2.23}$$

By definition, the constant-volume molar heat capacity is given as:

$$c_v = \left(\frac{\partial u}{\partial T}\right)_v \tag{2.24}$$

From Equations (2.23) and (2.24):

$$du = c_v dT + \left(\frac{\partial u}{\partial v}\right)_T dv \tag{2.25}$$

Upon integration,

$$\Delta u = \int_{T_1}^{T_2} c_v dT + \int_{v_1}^{v_2} \left(\frac{\partial u}{\partial v}\right)_T dv \tag{2.26}$$

When $v_1 = v_2$, Equation (2.26) simplifies to:

$$\Delta u = \int_{T_1}^{T_2} c_v dT \tag{2.27}$$

This result is valid for all homogeneous substances of constant composition provided that $v_1 = v_2$.

For ideal gas,

$$\left(\frac{\partial u}{\partial v}\right)_T = 0 \tag{2.28}$$

Consequently, the internal energy of an ideal gas is a function of temperature only and Equation (2.27) is valid for an ideal gas regardless of the process. It is not necessary that $v_1 = v_2$.

2.6.2 Enthalpy Changes

For a homogeneous substance of constant composition (pure gas, pure liquid or pure solid), the molar or specific enthalpy of a substance can be expressed as a function of two state variables such as temperature and pressure:

$$h = h(T, P) \tag{2.29}$$

where h is molar or specific enthalpy. From Equation (2.29):

$$dh = \left(\frac{\partial h}{\partial T}\right)_P dT + \left(\frac{\partial h}{\partial P}\right)_T dP \tag{2.30}$$

By definition, the constant-pressure molar heat capacity is given as:

$$c_p = \left(\frac{\partial h}{\partial T}\right)_P \tag{2.31}$$

From Equations (2.30) and (2.31):

$$dh = c_p dT + \left(\frac{\partial h}{\partial P}\right)_T dP \tag{2.32}$$

Upon integration,

$$\Delta h = \int_{T_1}^{T_2} c_p dT + \int_{P_1}^{P_2} \left(\frac{\partial h}{\partial P}\right)_T dP \tag{2.33}$$

When $P_1 = P_2$, Equation (2.33) simplifies to:

$$\Delta h = \int_{T_1}^{T_2} c_p dT \tag{2.34}$$

This result is valid for all homogeneous substances of constant composition provided that $P_1 = P_2$.

For ideal gas,

$$\left(\frac{\partial h}{\partial P}\right)_T = 0 \tag{2.35}$$

Consequently, the enthalpy of an ideal gas is a function of temperature only and Equation (2.34) is valid for an ideal gas regardless of the process. It is not necessary that $P_1 = P_2$.

2.7 APPLICATION OF FIRST LAW TO ISOTHERMAL AND ADIABATIC PROCESSES

2.7.1 Isothermal Process

In an isothermal process, the temperature of the system is held constant throughout the process. Assuming that the process is quasi-static (reversible), the work produced in an isothermal expansion is shown in Figure 2.6. It is simply the area under the isotherm, given as:

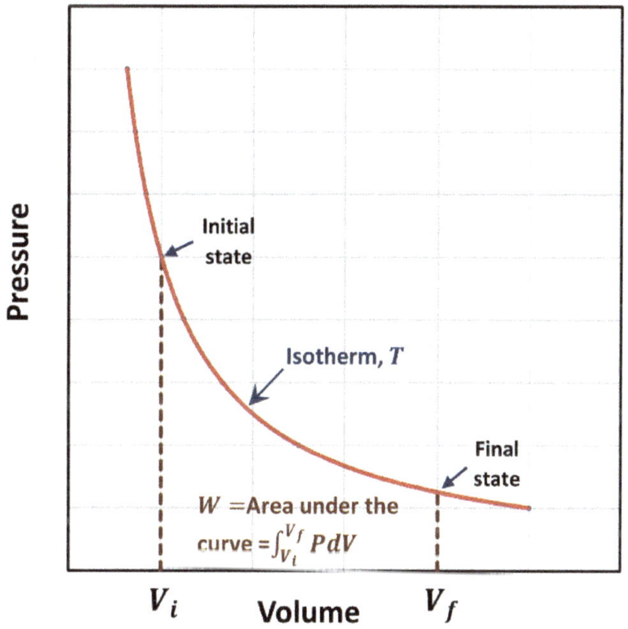

Figure 2.6 Work produced in a quasi-static isothermal expansion of a material.

$$W = \int_{V_1}^{V_2} PdV \ (T = \text{constant}) \tag{2.36}$$

The first law of thermodynamics, that is, Equations (2.9) and (2.10), are applicable. We are assuming that there are no macroscopic kinetic and potential energy changes involved. The internal energy change is given by Equations (2.25) and (2.26) for any material (homogeneous of constant composition). However, if the material is an *ideal gas*, the internal energy change is zero as internal energy of an ideal gas depends only on temperature:

$$\Delta U = 0 \ (\text{ideal gas}, T = \text{constant}) \tag{2.37}$$

The first law in finite-change form, Equation (2.9), simplifies to:

$$Q = W \tag{2.38}$$

Assuming that the process is quasi-static (reversible),

$$W = \int_{V_i}^{V_f} PdV = nRT \int_{V_i}^{V_f} \frac{dV}{V} = nRTln\left(\frac{V_f}{V_i}\right) \quad (2.39)$$

From Equations (2.38) and (2.39), it follows that for a reversible isothermal expansion/compression of an ideal gas:

$$Q = W = nRTln\left(\frac{V_f}{V_i}\right) = nRTln\left(\frac{P_i}{P_f}\right) \quad (2.40)$$

Let us now consider a variation of the constant temperature process (see Figure 2.7). The initial and final temperatures are the same. But temperature is maintained constant throughout the process only along path I. Along path II (assumed quasi-static), the temperature is not constant during the process. Only the initial and final states have the same temperature. What is the difference between these two processes?

- ΔU and ΔH are the same along paths I and II (regardless of the working material) as U and H are state properties. For ideal gas, $\Delta U = 0$ and $\Delta H = 0$ as temperatures are the same in the initial and final states. Recall that internal energy and enthalpy are functions of temperature only for ideal gas.
- Q and W are path-dependent quantities. They are different along different paths I and II. Work produced along path II is much larger than that produced along path I.
- If the material is an ideal gas, $Q = W$ for both processes. Along path I,

$$Q = W = nRTln\left(\frac{V_f}{V_i}\right) \quad (2.41)$$

Along path II,

$$Q = W = P_i(V_f - V_i) \quad (2.42)$$

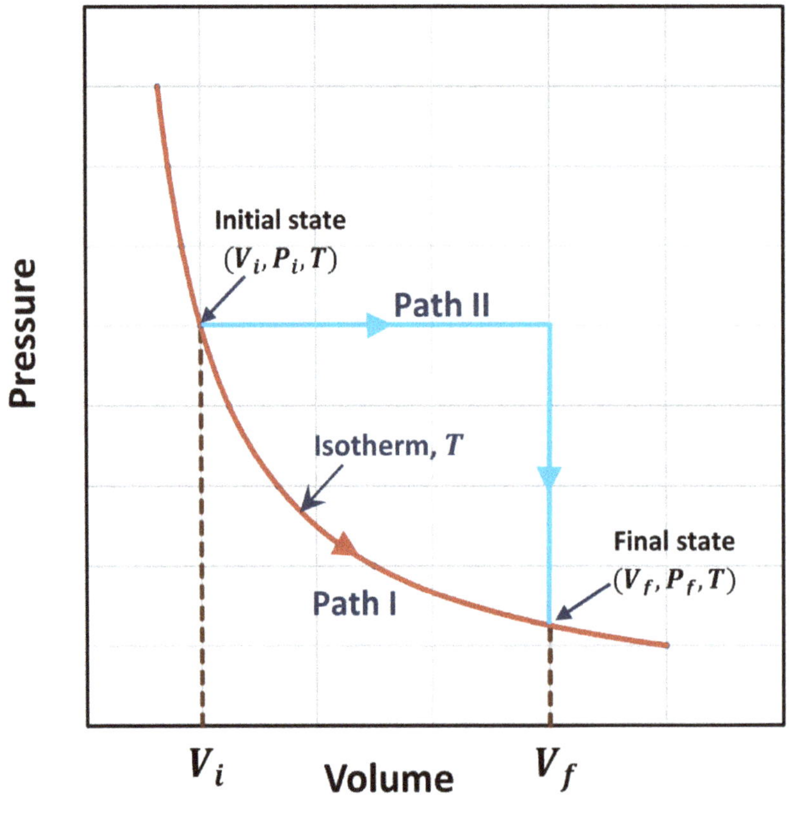

Figure 2.7 Quasi-static processes with same temperatures in the initial and final states.

2.7.2 Adiabatic Process

A process in which there is no heat exchange permitted between the system and its surroundings is called an adiabatic process. As $\delta Q = 0$, the first law gives:

$$dU = -\delta W \tag{2.43}$$

$$\Delta U = -W \tag{2.44}$$

Thus, for all adiabatic processes between a given pair of end states, the amount of the work exchanged is the same as ΔU is the same. In other words, the amount of work done on or by the system undergoing an adiabatic process between two fixed states does not depend on the details of the adiabatic process.

For quasi-static adiabatic process, the change in molar internal energy can be expressed as:

$$du = -Pdv \tag{2.45}$$

If the material is ideal gas, we can re-write Equation (2.45) as:

$$c_v dT = -\frac{RT}{v} dv \tag{2.46}$$

Upon re-arrangement and integration, we get:

$$\int_{T_1}^{T_2} c_v \frac{dT}{T} = -\int_{v_1}^{v_2} \frac{R}{v} dv \tag{2.47}$$

Assuming c_v to be independent of temperature,

$$c_v ln\left(\frac{T_2}{T_1}\right) = -Rln\left(\frac{v_2}{v_1}\right) \tag{2.48}$$

For an ideal gas,

$$c_p - c_v = R \tag{2.49}$$

Therefore, the heat capacity c_v of an ideal gas can be expressed in terms of the heat capacity ratio γ (defined as c_p/c_v) as:

$$c_v = R/(\gamma - 1) \tag{2.50}$$

From Equations (2.48) and (2.50):

$$ln\left(\frac{T_2}{T_1}\right) = (1-\gamma)ln\left(\frac{v_2}{v_1}\right) \tag{2.51}$$

Upon simplification, Equation (2.51) gives:

$$\frac{T_2}{T_1} = \left(\frac{v_1}{v_2}\right)^{\gamma-1} \tag{2.52}$$

Equation (2.52) implies that:

$$Tv^{\gamma-1} = \text{constant} \tag{2.53}$$

Using the ideal gas law, Equation (2.53) gives another useful relation:

$$Pv^{\gamma} = \text{constant} \tag{2.54}$$

Equations (2.53) and (2.54) are important relationships valid for reversible adiabatic expansion/compression of an ideal gas of constant heat capacity ratio γ.

The work produced in a reversible adiabatic expansion of an ideal gas

$$W = -\Delta U = -nc_v(T_2 - T_1) \tag{2.55}$$

From Equations (2.50) and (2.55), we get:

$$W = \frac{nR}{\gamma-1}(T_1 - T_2) = \frac{P_1V_1 - P_2V_2}{\gamma-1} \tag{2.56}$$

From Equations (2.54) and (2.56), it can be readily shown that:

$$W = \frac{P_1V_1}{\gamma-1}\left[1 - \left(\frac{P_2}{P_1}\right)^{\frac{\gamma-1}{\gamma}}\right] = \frac{nRT_1}{\gamma-1}\left[1 - \left(\frac{P_2}{P_1}\right)^{\frac{\gamma-1}{\gamma}}\right] \tag{2.57}$$

2.8 CALCULATION OF WORK IN REAL PROCESSES

The calculation of work in real processes is important from a practical point of view. We can readily calculate work exchanged between the system and the surroundings in reversible processes using the principles of thermodynamics. However, real processes are irreversible in nature. How do we calculate work in a real processes? We can calculate work in real processes provided that we know the efficiency of the real process.

If the process produces work (as in expansion), the reversible work is the maximum work that can be produced for a given change in state. In this case, the actual work produced for the same change in state is reversible work multiplied by the process efficiency.

If the process requires work (as in compression), the reversible work is the minimum work required to accomplish the given change in state. In this case, the actual work required for the same change in state is reversible work divided by the process efficiency.

NOTATION

c_p	constant pressure molar heat capacity
c_v	constant volume molar heat capacity
d	differential of exact quantity
E	total energy
E_k	macroscopic kinetic energy
E_p	macroscopic potential energy
E_{system}	total energy of system
$E_{surroundings}$	total energy of surroundings
g	acceleration due to gravity
h	elevation of the system in a gravitational field
H	enthalpy
m	mass
n	number of moles

P	pressure
Q	heat
Q_{cycle}	net heat transfer in a cyclic process
R	universal gas constant
T	temperature
u	molar or specific internal energy
U	internal energy
v	molar or specific volume
V	velocity or volume
W	work
W_{cycle}	net work in a cyclic process

Greek Letters

γ	heat capacity ratio, c_p/c_v
Δ	difference operator
δ	differential of inexact quantity

TUTORIAL PROBLEMS WITH SOLUTION

Problem 1

One mole of an ideal gas at 80 °C and 550 kPa undergoes reversible adiabatic expansion. The heat capacity ratio γ is 1.4. The pressure of the gas in the final state is 100 kPa.

a) Determine the initial volume, final volume, and final temperature of the gas.
b) What is the work done by the gas?

Solution

a)

$$V_1 = \frac{RT_1}{P_1} = \frac{8.314 \times (80 + 273.15)}{550 \times 10^3} = 5.338 \times 10^{-3}\ m^3$$

$$P_1 V_1^\gamma = P_2 V_2^\gamma$$

$$V_2 = \left(\frac{P_1}{P_2}\right)^{1/\gamma} V_1 = \left(\frac{550}{100}\right)^{1/1.4} \times 5.338 \times 10^{-3} = 18.04 \times 10^{-3}\ m^3$$

$$T_2 = \frac{P_2 V_2}{R} = \frac{100 \times 10^3 \times 18.04 \times 10^{-3}}{8.314} = 216.98\ \text{K}$$

b)

$$W = \frac{R}{\gamma - 1}(T_1 - T_2) = \frac{8.314}{1.4 - 1}(353.15 - 216.98) = 2830.29\ \text{J}$$

$$Q = 0$$

Problem 2

Ten moles of an ideal gas undergo a cyclic process involving three steps as indicated in the table below. Compute the missing values.

Process	$Q(kJ)$	$W(kJ)$	$\Delta U(kJ)$
$1 \rightarrow 2$	100	50	?
$2 \rightarrow 3$	?	-20	?
$3 \rightarrow 1$	?	100	-90

Solution

Apply first law to process $1 \rightarrow 2$:

$$\Delta U_{12} = Q_{12} - W_{12} = 100 - 50 = 50 \text{ kJ}$$

As internal energy is a state property,

$$\oint dU = 0\,;\, \Delta U_{12} + \Delta U_{23} + \Delta U_{31} = 0 \;;\; \Delta U_{23} = -\Delta U_{31} - \Delta U_{12} = -(-90) - 50 = 40 \text{ kJ}$$

Apply first law to process $2 \rightarrow 3$:

$$\Delta U_{23} = Q_{23} - W_{23}\,; \qquad Q_{23} = W_{23} + \Delta U_{23} = -20 + 40 = 20 \text{ kJ}$$

Apply first law to process $3 \rightarrow 1$:

$$Q_{31} = W_{31} + \Delta U_{31} = 100 - 90 = 10 \text{ kJ}$$

Check: $Q_{cycle} = W_{cycle}$

$$Q_{cycle} = 100 + 20 + 10 = 130 \text{ kJ}\,;\; W_{cycle} = 50 - 20 + 100 = 130 \text{ kJ}$$

Problem 3

Five moles of an ideal gas expand reversibly from $P_1 = 10\ bar$ and $T_1 = 500\ K$ to $P_2 = 1\ bar$ by each of the following processes: a) Isothermal; b) isochoric; and c) adiabatic.
Calculate $\Delta U, \Delta H, Q,$ and W for each process.
The heat capacities of the gas are: c_p=7R/2 and c_v =5R/2.

Solution

a) Isothermal Process

$$P_1 = 10\ bar\,,\ T_1 = T_2 = 500\ K,\ P_2 = 1\ bar$$

$\Delta U = 0$, $\Delta H = 0$ (note that for ideal gas, internal energy and enthalpy changes are zero when temperature is constant)

From first law,

$$Q = W = \int_{V_1}^{V_2} \frac{nRT}{V} dV = nRTln\left(\frac{V_2}{V_1}\right) = 5 \times 8.314 \times 500 \times ln\left(\frac{P_1}{P_2}\right) = 20785 \times ln\left(\frac{10}{1}\right) = 47859.23 \text{ J}$$

b) Isochoric Process

$$P_1 = 10\ bar\ ,\ T_1 = 500\ K,\ P_2 = 1\ bar$$

$$T_2 = \left(\frac{P_2}{P_1}\right)T_1 = \left(\frac{1}{10}\right) \times 500 = 50\ \text{K}$$

$$\Delta U = nc_v(T_2 - T_1) = 5 \times 2.5 \times 8.314(50 - 500) = -46766.25\ \text{J}$$

For a constant volume process, $Q = \Delta U = -46766.25\ \text{J}$

$$\Delta H = nc_p(T_2 - T_1) = 5 \times 3.5 \times 8.314(50 - 500) = -65472.75\ \text{J}$$

For a constant volume process, $W = 0$

c) Adiabatic Process

$$P_1 = 10\ bar\ ,\ T_1 = 500\ K,\ P_2 = 1\ bar$$

$$\frac{T_2}{T_1} = \left(\frac{v_1}{v_2}\right)^{\gamma-1};\quad \frac{P_2}{P_1} = \left(\frac{v_1}{v_2}\right)^{\gamma}$$

$$\frac{T_2}{T_1} = \left(\frac{P_2}{P_1}\right)^{(\gamma-1)/\gamma} = \left(\frac{1}{10}\right)^{(1.4-1)/1.4} = 0.51795$$

$$T_2 = 0.51795 \times 500 = 258.98\ K$$

$$\Delta U = nc_v(T_2 - T_1) = 5 \times 2.5 \times 8.314(258.98 - 500) = -25048\ \text{J}$$

$$\Delta H = nc_p(T_2 - T_1) = 5 \times 3.5 \times 8.314(258.98 - 500) = -35067.2\ \text{J}$$

For adiabatic process, $Q = 0$

From first law: $W = -\Delta U = 25948$ J

Problem 4

One mole of an ideal gas with c_p =7R/2 and c_v =5R/2 undergoes the following cyclic process reversibly:

a) It is compressed adiabatically from an initial state of 47 °C and 165 kPa to 550 kPa.
b) It is then cooled to 47 °C at a constant pressure of 550 kPa.
c) It is finally expanded isothermally to the original state.

Sketch the cyclic process on a PV diagram.
For each of the three processes and for the whole cycle, calculate $Q, W, \Delta U,$ and ΔH.
Repeat the calculations when the same changes of state are accomplished irreversibly with an efficiency of 75 percent.

Solution

The cyclic process is shown in the figure below on a PV diagram.

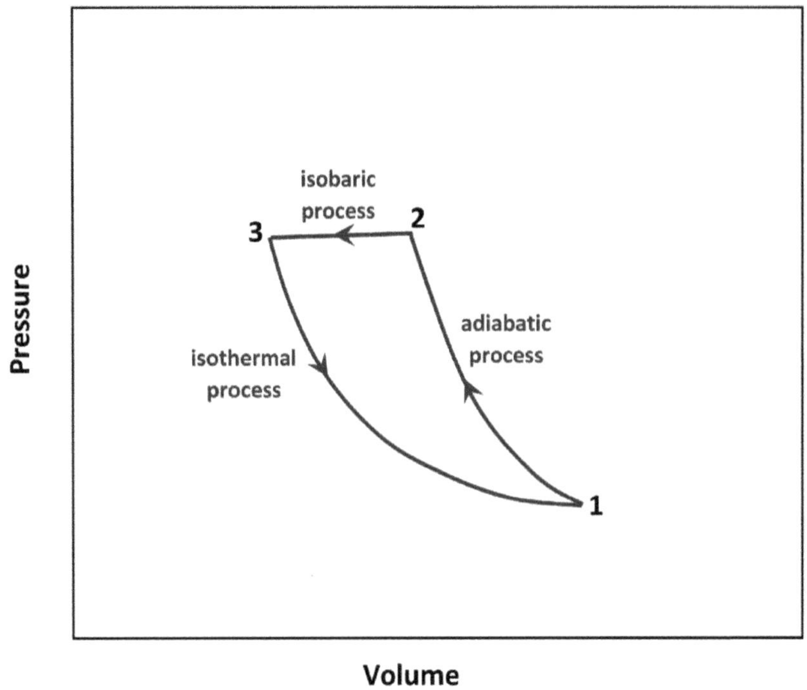

Let us first determine the P, V, T information in each state.

State 1:

$$T_1 = 47 + 273.15 = 320.15 \text{ K}; \ P_1 = 165 \text{ kPa}$$

$$V_1 = \frac{nRT_1}{P_1} = \frac{1 \times 8.314 \times 320.15}{165 \times 10^3} = 0.016132 \, m^3$$

State 2:

$$P_2 = 550 \text{ kPa} \, ; \ P_1 V_1^{\gamma} = P_2 V_2^{\gamma} \, ; \gamma = c_p/c_v = 7/5 = 1.4$$

$$V_2 = \left(\frac{P_1}{P_2}\right)^{1/\gamma} V_1 = \left(\frac{165}{550}\right)^{1/1.4} \times 0.016132 = 0.0068266 \, m^3$$

$$T_2 = \frac{P_2 V_2}{nR} = \frac{550 \times 10^3 \times 0.0068266}{1 \times 8.314} = 451.60 \text{ K}$$

State 3:

$$P_3 = P_2 = 550 \text{ kPa}; \ T_3 = T_1 = 47 + 273.15 = 320.15 \text{ K}$$

$$V_3 = \frac{nRT_3}{P_3} = \frac{1 \times 8.314 \times 320.15}{550 \times 10^3} = 0.0048395 \, m^3$$

Now we can calculate the desired quantities $Q, W, \Delta U$, and ΔH for each process and for the whole cycle.

Process $1 \rightarrow 2$ *: Reversible adiabatic process*

$$Q = 0\ ;\ \Delta U = -W = nc_v\,(T_2 - T_1) = 1 \times 2.5 \times 8.314 \times (451.60 - 320.15) = 2732.19\text{ J}$$

$$\Delta H = nc_p\,(T_2 - T_1) = 1 \times 3.5 \times 8.314 \times (451.60 - 320.15) = 3825.1\text{ J}$$

Process 2 → 3 : Reversible isobaric process

$$Q = \Delta H = nc_p\,(T_3 - T_2) = 1 \times 3.5 \times 8.314 \times (320.15 - 451.60) = -3825.1\text{ J}$$

$$W = P_3 \Delta V = 550 \times 10^3 \times (V_3 - V_2) = 550 \times 10^3 \times (0.0048395 - 0.0068266) = -1092.91\text{ J}$$

$$\Delta U = nC_v\,(T_3 - T_2) = 1 \times 2.5 \times 8.314 \times (320.15 - 451.60) = -2732.19\text{ J}$$

Process 3 → 1 : Reversible isothermal process

$$\Delta U = 0\ ;\ \Delta H = 0$$

$$Q = W = \int_{V_3}^{V_1} \frac{nRT_1}{V}\,dV = nRT_1 ln\left(\frac{V_1}{V_3}\right) = 1 \times 8.314 \times 320.15 \times ln\left(\frac{0.016132}{0.0048395}\right) = 3204.70\text{ J}$$

Summary of Calculations

Process	ΔU**(J)**	ΔH**(J)**	Q**(J)**	W**(J)**
1 → 2	2732.19	3825.1	0	-2732.19
2 → 3	-2732.19	-3825.1	-3825.1	-1092.91
3 → 1	0	0	3204.70	3204.70
Cycle	0	0	-620.4	-620.4

As expected, $\oint dU = 0$, $\oint dH = 0$, $Q_{cycle} = W_{cycle} = -620.4$ J

There occurs a net absorption of work in this cyclic process.

Repeat the calculations when the same changes of state are accomplished irreversibly with an efficiency of 75 percent.

Since the changes of state are the same, ΔU and ΔH will remain the same. Only Q and W will change.

Process	ΔU**(J)**	ΔH**(J)**	Q_{rev}**(J)**	W_{rev}**(J)**	W_{irrev}**(J)**	Q_{irrev}**(J)**
1 → 2	2732.19	3825.1	0	-2732.19	-3642.92	-910.73
2 → 3	-2732.19	-3825.1	-3825.1	-1092.91	-1457.21	-4189.4
3 → 1	0	0	3204.70	3204.70	2403.52	2403.52
Cycle	0	0	-620.4	-620.4	-2696.61	-2696.61

Note that:
$W_{irrev} = W_{rev}/0.75$ for work-absorbing process and $W_{irrev} = 0.75 \times W_{rev}$ for work-producing process. The first law is applicable to the irreversible process. Thus,

$$Q_{irrev} = W_{irrev} + \Delta U$$

The total work required to carry out the cyclic process irreversibly is -2696.61 J. This is much higher than the total work required to carry out the cyclic process reversibly (-620.4 J).

PRACTICE PROBLEMS

Problem 1

An ideal gas at 515 K and 1900 kPa expands to seven times its initial volume of 0.5 m^3 as follows:

- By a reversible isothermal process.
- By a reversible adiabatic process.
- By an adiabatic irreversible process in which the expansion occurs against the restraining pressure of 95 kPa.

Calculate for each case, the final temperature, pressure and the work done by the gas. $c_p = 23$ J/mol.K.

Problem 2

One mole of an ideal gas undergoes two steps as described below:

- The gas is cooled at constant volume from $T_1 = 900$ K and $P_1 = 3$ bar to $T_2 = 300$ K.
- The gas is then heated at constant pressure until its temperature reaches 900 K.

a) Sketch the process on a PV diagram.

b) If we replace this two-step process by a single step involving isothermal expansion of the gas from $T_1 = 900$ K and $P_1 = 3$ bar to some final pressure P_2, What is the value of P_2 that makes the work of the proposed single-step process equal to the existing two-step process. You can assume that all processes are reversible and the heat capacities of the gas are: c_p =7R/2 and c_v =5R/2.

Problem 3

For each of the following reversible processes, determine the missing information. The working fluid is 15 moles of an ideal gas.

Process	Q (kJ)	W (kJ)	ΔU (kJ)	ΔH (kJ)	T_2 (°C)	T_1 (°C)	P_2 (kPa)	P_1 (kPa)	V_2 (m^3)	V_1 (m^3)
Isothermal	55				95		60			
Isochoric				90	250		180			
Isobaric	90					180		600		
Adiabatic						40			0.15	0.50

Problem 4

An ideal gas undergoes the same change in state from $V_1 = 10\ m^3$ and $P_1 = 0.5$ bar to $V_2 = 0.9\ m^3$ and $P_2 = 9$ bar through different reversible processes described below:

- Isothermal compression
- Adiabatic compression followed by isobaric cooling
- Adiabatic compression followed by isochoric cooling
- Isochoric heating followed by isobaric cooling
- Isobaric cooling followed by isochoric heating

a) On a single PV diagram, sketch the paths of all the processes

b) Calculate $\Delta U, \Delta H, Q,$ and W for each of the processes.

The heat capacities of the gas are: c_p =7R/2 and c_v =5R/2.

Problem 5

One kg of some unknown material undergoes a cyclic process involving four steps as indicated in the table below. Compute the missing values.

Process	$Q(kJ)$	$W(kJ)$	$\Delta U(kJ)$
$1 \rightarrow 2$	-150	?	10
$2 \rightarrow 3$	750	?	?
$3 \rightarrow 4$	?	500	300
$4 \rightarrow 1$	10	?	-1000

Problem 6

Five moles of an ideal gas are compressed adiabatically from 1 bar and 35°C to 4.7 bar. The process is irreversible. The work required is 37 percent more than that required for a reversible adiabatic compression from the same initial state (1 bar and 35°C) to a final pressure of 4.7 bar. Determine the final temperature of the gas. The heat capacities of the gas are: $c_p = 7R/2$ and $c_v = 5R/2$.

Chapter 3

First Law of Thermodynamics for Control Volume (Open Systems)

3.1 LEARNING OUTCOMES

Upon completion of this chapter, students will be able to:

- Develop the first law for control volume (open systems)
- Develop mechanical energy and Bernoulli equations
- Apply the first law for control volume to nozzles and diffusers, turbines, compressors and pumps, throttling devices, heat exchangers, and pipe flow
- Apply the first law for control volume to cyclic processes
- Explain perpetual motion machine of first kind
- Solve problems related to the application of first law for control volume to non-cyclic and cyclic processes

3.2 INTRODUCTION

Engineering processes usually involve flow of material from one plant unit to another. In flow processes, it is more convenient and beneficial to adopt the *control volume* approach (instead of the system or control mass approach used in Chapter 2). The control volume approach is also referred to as *open system* approach. Control volume is simply a region in space selected for analysis. The bounding surface of the control volume is called the *control surface*. Mass and energy can enter or leave the control volume through the control surface. Energy can enter or leave the control volume through the control surface by means of heat and work. It can also enter or leave the control volume through mass flow; mass entering or leaving the control volume carries with it energy. It should be noted that, in general, the control volume could be moving and deforming. However, we will restrict our analysis to fixed and non-deforming control volumes, referred to as simply stationary control volumes.

Consider flow of a fluid through a stationary control volume. The macroscopic balance of any entity (mass or energy) over the control volume can be expressed as:

$$Input - Output + Generation - Destruction = Accumulation \tag{3.1}$$

3.3 MASS BALANCE

According to the *principle of conservation of mass*, mass can neither be generated nor destroyed. Thus, mass balance for a control volume is given as:

$$Input - Output = Accumulation \tag{3.2}$$

In terms of rates, the mass balance for a control volume can be written as:

$$\begin{pmatrix} Rate\ at \\ which\ mass \\ enters\ CV \end{pmatrix} - \begin{pmatrix} Rate\ at \\ which\ mass \\ leaves\ CV \end{pmatrix} = \begin{pmatrix} Rate\ of \\ accumulation \\ of\ mass \end{pmatrix} \tag{3.3}$$

where "*CV*" refers to control volume.

Referring to control volume shown in Figure 3.1, the mass balance can be expressed as:

$$\dot{m}_1 - \dot{m}_2 = \frac{d(m)_{CV}}{dt} \tag{3.4}$$

where $\dot{m}$ is the total mass flow rate and m is mass of fluid. At steady state, the storage or accumulation of mass inside the control volume is zero. Thus,

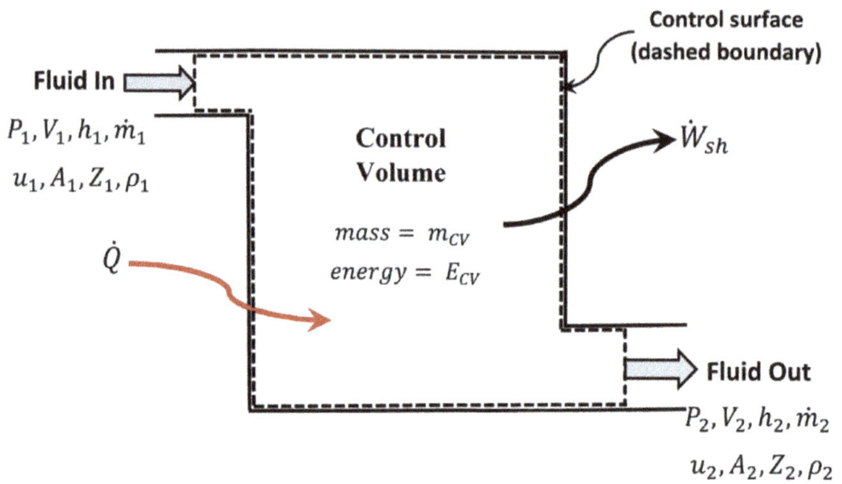

Figure 3.1 Control volume with one inlet and one outlet (P is pressure, V is velocity, h is specific enthalpy, $\dot{m}$ is mass flow rate, u is specific internal energy, A is cross-sectional area, Z is elevation, ρ is density, $\dot{Q}$ is rate of heat transfer, $\dot{W}_{sh}$ is rate of shaft work, m is mass, and E is total energy).

$$\dot{m}_1 = \dot{m}_2 = \dot{m} = constant \tag{3.5}$$

This steady state mass balance can be generalized to control volume with multiple inlets and multiple outlets as follows:

$$\Sigma_{INLETS}\ (\dot{m}_i) = \Sigma_{OUTLETS}\ (\dot{m}_i) \tag{3.6}$$

3.4 ENERGY BALANCE

Consider the flow process shown in Figure 3.1. Fluid enters the control volume at some condition and leaves the control volume at a different condition. The variables relevant to energy balance are indicated at the inlet and exit conditions of the control volume shown in Figure 3.1. The rate of heat input to the control volume is $\dot{Q}$. The control volume is producing shaft work at a rate of $\dot{W}_{sh}$.

According to the *principle of conservation of energy*, energy cannot be generated or destroyed. Thus, energy balance for a control volume is:

$$\begin{pmatrix} Rate\ at \\ which\ energy \\ enters\ CV \end{pmatrix} - \begin{pmatrix} Rate\ at \\ which\ energy \\ leaves\ CV \end{pmatrix} = \begin{pmatrix} Rate\ of \\ accumulation \\ of\ energy \end{pmatrix} \tag{3.7}$$

Referring to Figure 3.1, the various terms in energy balance equation, Equation (3.7), can be written as:

$$\begin{pmatrix} Rate\ at \\ which\ energy \\ enters\ CV \end{pmatrix} = \dot{m}_1 \left[u_1 + \frac{1}{2}V_1^2 + gZ_1\right] + \dot{Q} + (P_1A_1V_1) \tag{3.8}$$

$$\begin{pmatrix} Rate\ at \\ which\ energy \\ leaves\ CV \end{pmatrix} = \dot{m}_2 \left[u_2 + \frac{1}{2}V_2^2 + gZ_2\right] + \dot{W}_{sh} + (P_2A_2V_2) \tag{3.9}$$

$$\begin{pmatrix} Rate\ of \\ accumulation\ of \\ energy\ inside\ CV \end{pmatrix} = \frac{dE_{CV}}{dt} = \frac{d(mu)_{CV}}{dt} \tag{3.10}$$

where variables u, V, Z, A, P refer to specific internal energy, fluid velocity, elevation, cross-sectional area of flow, and pressure, respectively. Note that the last term on the right hand side of Equation (3.8) or Equation (3.9), that is, the PAV term, is the flow work (work needed to push the fluid into and out of the control volume).

Thus, the energy balance, Equation (3.7), becomes:

$$\left\{\dot{m}_1 \left[u_1 + \frac{1}{2}V_1^2 + gZ_1\right] + \dot{Q} + (P_1A_1V_1)\right\} - \left\{\dot{m}_2 \left[u_2 + \frac{1}{2}V_2^2 + gZ_2\right] + \dot{W}_{sh} + (P_2A_2V_2)\right\} = \frac{d(mu)_{CV}}{dt} \tag{3.11}$$

In this energy balance equation, $\dot{Q}$ is positive if it enters the control volume and negative if it leaves the control volume. $\dot{W}_{sh}$ is positive if it flows out of the control volume and negative if it flows into the control volume.

Assuming steady state, Equation (3.11) can be written as:

$$\dot{m}\left[(u_2 - u_1) + \frac{1}{2}(V_2^2 - V_1^2) + g(Z_2 - Z_1) + \left(\frac{P_2}{\rho_2} - \frac{P_1}{\rho_1}\right)\right] = \dot{Q} - \dot{W}_{sh} \tag{3.12}$$

Making use of the definition of specific enthalpy: $h = u + P/\rho$, Equation (3.12) can be expressed as:

$$\dot{m}\left[(h_2 - h_1) + \frac{1}{2}(V_2^2 - V_1^2) + g(Z_2 - Z_1)\right] = \dot{Q} - \dot{W}_{sh} \tag{3.13}$$

This is the *first law of thermodynamics for control volume* assuming steady state and one inlet / one outlet. Equation (3.13) can be re-written as:

$$\dot{m}\left[\Delta h + \frac{1}{2}\Delta V^2 + g\Delta Z\right] = \dot{Q} - \dot{W}_{sh} \tag{3.14}$$

For multiple inlets and outlets, Equation (3.13) can be generalized as follows:

$$\Sigma_{OUTLETS}\ \dot{m}_i(h_i + \frac{1}{2}V_i^2 + gZ_i) - \Sigma_{INLETS}\ \dot{m}_i(h_i + \frac{1}{2}V_i^2 + gZ_i) = \Sigma\dot{Q} - \dot{W}_{sh} \tag{3.15}$$

where $\sum \dot{Q}$ is the total rate of heat input to the control volume (heat may be added at different locations of the control surface).

3.4.1 Mechanical Energy Equation

In many flow processes, engineers are concerned with *mechanical energy* balance instead of the total energy balance. The mechanical energy consists of kinetic energy, potential energy, shaft work and flow work. It is important to note that mechanical energy is not conserved. It is the total energy that is conserved. During a real flow process, part of the mechanical energy is converted into internal energy due to compressibility of fluid and part of it is dissipated into frictional heating of fluid due to non-zero viscosity of fluid. Thus, the mechanical energy balance can be expressed as:

$$\begin{pmatrix} Rate\ at \\ which\ ME \\ enters\ CV \end{pmatrix} - \begin{pmatrix} Rate\ at \\ which\ ME \\ leaves\ CV \end{pmatrix} - \begin{pmatrix} Rate\ at\ which\ ME\ is\ converted \\ into\ internal\ energy\ within\ CV \\ due\ to\ fluid\ compresibility \end{pmatrix} - \begin{pmatrix} Rate\ at\ which\ ME \\ is\ dissipated\ due\ to \\ friction\ in\ fluid \end{pmatrix} = \begin{pmatrix} Rate\ of \\ accumulation \\ of\ ME \end{pmatrix} \tag{3.16}$$

where "*ME*" refers to mechanical energy. Consider the control volume of Figure 3.1. The rate at which *ME* enters the *CV* is:

$$\begin{pmatrix} Rate\ at \\ which\ ME \\ enters\ CV \end{pmatrix} = \dot{m}_1\left[\frac{1}{2}V_1^2 + gZ_1\right] + (P_1A_1V_1) \tag{3.17}$$

The rate at which *ME* leaves the *CV* is:

$$\begin{pmatrix} Rate\ at \\ which\ ME \\ leaves\ CV \end{pmatrix} = \dot{m}_2\left[\frac{1}{2}V_2^2 + gZ_2\right] + \dot{W}_{sh} + (P_2A_2V_2) \tag{3.18}$$

The rate at which *ME* is converted into internal energy within *CV* due to fluid compression is:

$$\begin{pmatrix} Rate\ at\ which\ ME\ is\ converted \\ into\ internal\ energy\ within\ CV \\ due\ to\ fluid\ compressibility \end{pmatrix} = -\int P\left(\nabla\cdot\vec{V}\right)dV \tag{3.19}$$

The rate at which *ME* is dissipated due to friction in fluid is:

$$\begin{pmatrix} Rate\ at\ which\ ME \\ is\ dissipated\ due\ to \\ friction\ in\ fluid \end{pmatrix} = \dot{F}_{loss} \tag{3.20}$$

Assuming steady state,

$$\begin{pmatrix} Rate\ of \\ accumulation\ of \\ energy\ inside\ CV \end{pmatrix} = 0 \tag{3.21}$$

$$\dot{m}_1 = \dot{m}_2 = \dot{m} \tag{3.22}$$

The mechanical energy balance, Equation (3.16) gives:

$$\dot{m}\left[\frac{1}{2}(V_2^2 - V_1^2) + g(Z_2 - Z_1) + \left(\frac{P_2}{\rho_2} - \frac{P_1}{\rho_1}\right)\right] + \dot{W}_{sh} - \int P(\nabla \cdot \vec{V})\, dV + \dot{F}_{loss} = 0 \tag{3.23}$$

It has been shown [1] that:

$$\dot{m}\,\Delta(P/\rho) - \int P(\nabla \cdot \vec{V})\, dV = \dot{m}\int_1^2 \frac{dP}{\rho} \tag{3.24}$$

Combining Equations (3.23) and (3.24), we get the following *mechanical energy* equation:

$$\dot{W}_{sh} + \dot{F}_{loss} + \dot{m}[\Delta(V^2/2) + g\Delta Z] + \dot{m}\int_1^2 \frac{dP}{\rho} = 0 \tag{3.25}$$

On a unit mass basis, Equation (3.25) gives:

$$\Delta(V^2/2) + g\Delta Z + \int_1^2 \frac{dP}{\rho} + W_{sh} + F_{loss} = 0 \tag{3.26}$$

Note that the mechanical energy equation could also be obtained directly from the total energy equation, Equation (3.14), by realizing that:

$$\dot{Q} = \dot{m}\left[\Delta h - \int_1^2 \frac{dP}{\rho} - F_{loss}\right] \tag{3.27}$$

From the total energy balance, Equation (3.14), and Equation (3.27), the mechanical energy equation, Equation (3.25), is readily obtained.

For *incompressible fluids*, the mechanical energy equations, Equation (3.25) and (3.26), become:

$$\dot{W}_{sh} + \dot{F}_{loss} + \dot{m}[\Delta(V^2/2) + g\Delta Z + \Delta P/\rho] = 0 \tag{3.28}$$

$$\Delta(V^2/2) + g\Delta Z + \Delta P/\rho + W_{sh} + F_{loss} = 0 \tag{3.29}$$

3.4.2 Bernoulli Equation

In the special case of fluid flow where there is no friction and no shaft work and the fluid is incompressible, the mechanical energy equation, Equation (3.29), becomes the celebrated *Bernoulli equation*:

$$\left[\frac{\Delta P}{\rho} + \Delta(V^2/2) + g\Delta Z\right] = 0 \tag{3.30}$$

$$\frac{P_2}{\rho} + \frac{V_2^2}{2} + gZ_2 = \frac{P_1}{\rho} + \frac{V_1^2}{2} + gZ_1 \tag{3.31}$$

3.5 APPLICATIONS OF FIRST LAW TO NON-CYCLIC FLOW PROCESSES

3.5.1 Nozzles and Diffusers

Nozzle is a device of varying cross-sectional area that is used to increase the velocity of a flowing fluid. Figure 3.2 shows a schematic diagram of a subsonic nozzle. The velocity increases and pressure decreases as cross-sectional area in the direction of flow decreases.

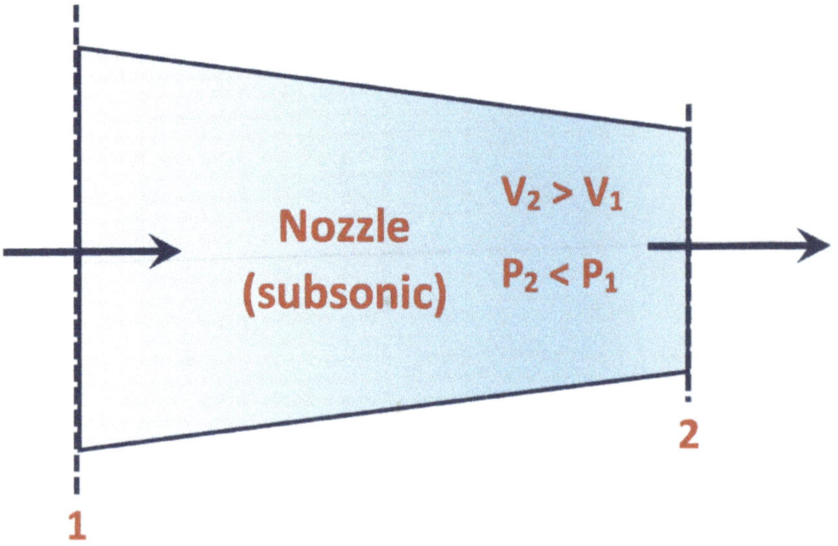

Figure 3.2 Subsonic nozzle.

There is no shaft work involved and heat transfer is negligible. From the first law of thermodynamics, Equation (3.13), we have:

$$\left[(h_2 - h_1) + \frac{1}{2}(V_2^2 - V_1^2) + g(Z_2 - Z_1)\right] = 0 \tag{3.32}$$

Neglecting potential energy change, Equation (3.32) gives:

$$V_2^2 = V_1^2 - 2(h_2 - h_1) \tag{3.33}$$

Thus, velocity of the fluid is increased at the expense of enthalpy.

Given enthalpy change, one can calculate the exit velocity using Equation (3.33). Once exit velocity is known, one can determine the nozzle exit diameter using mass balance as:

$$\dot{m} = \rho_1 A_1 V_1 = \rho_2 A_2 V_2 \tag{3.34}$$

where A is the cross-sectional area of the nozzle. From Equation (3.34):

$$\frac{D_2}{D_1} = \sqrt{\left(\frac{\rho_1}{\rho_2}\right)\left(\frac{V_1}{V_2}\right)} \tag{3.35}$$

where D is the nozzle diameter.

Diffuser is a device of varying cross-sectional area that is used to decrease the velocity of a flowing fluid. Figure 3.3 shows a schematic diagram of a subsonic diffuser. The velocity decreases and pressure increases as cross-sectional area in the direction of flow increases. All the equations applicable to nozzle are also applicable to diffuser.

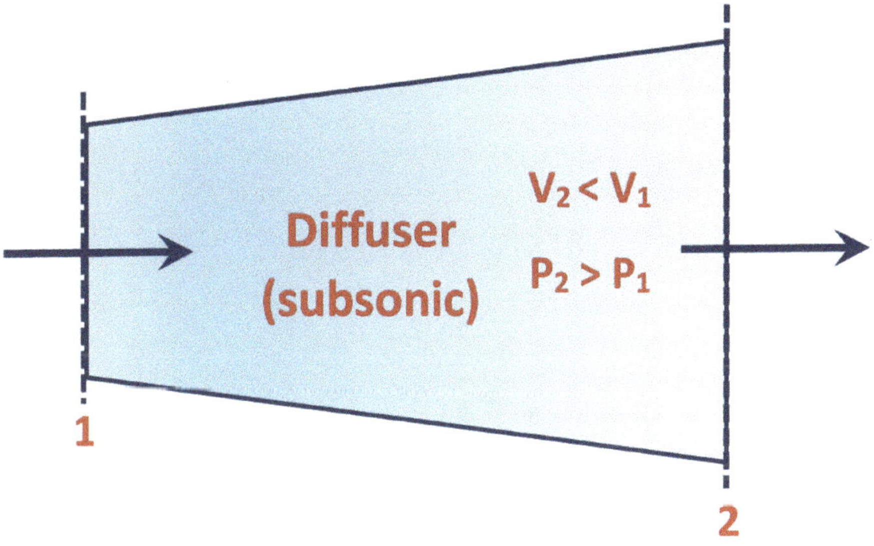

Figure 3.3 Subsonic diffuser.

3.5.2 Turbines

A turbine is a device which produces shaft work using the internal energy of the fluid. Turbines are used extensively in steam power plants. A high pressure superheated steam enters the turbine and expands to a low exit pressure. Inside the turbine, there are a number of nozzles in which the fluid is expanded to generate high-velocity fluid streams that impinge on the blades attached to a rotating shaft (see Figures 3.4 and 3.5). Thus, the internal energy of the fluid entering the turbine is first converted into kinetic energy which in turn is converted into shaft work.

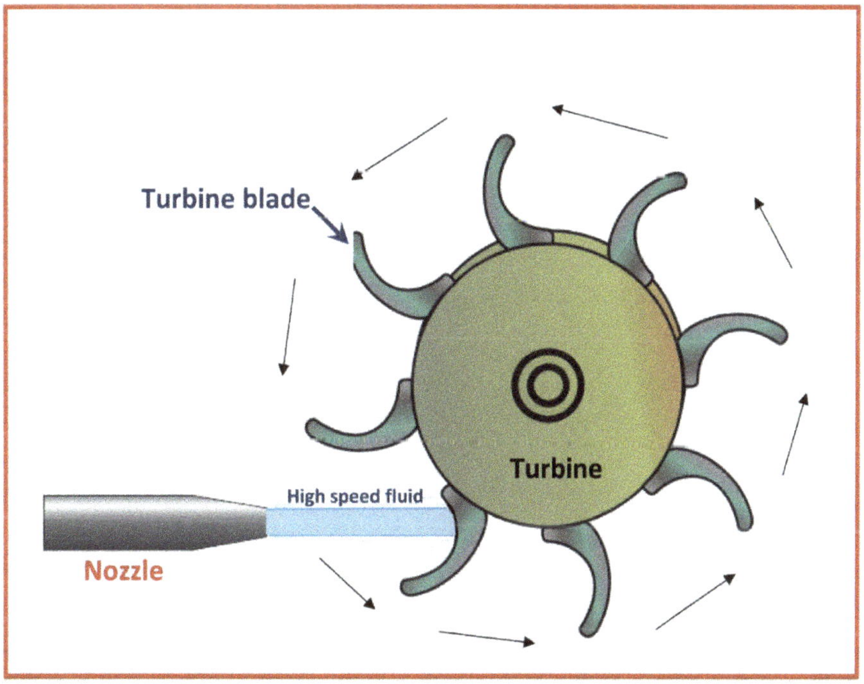

Figure 3.4 Shaft work is produced when a high velocity fluid stream impinges on the blades of a turbine.

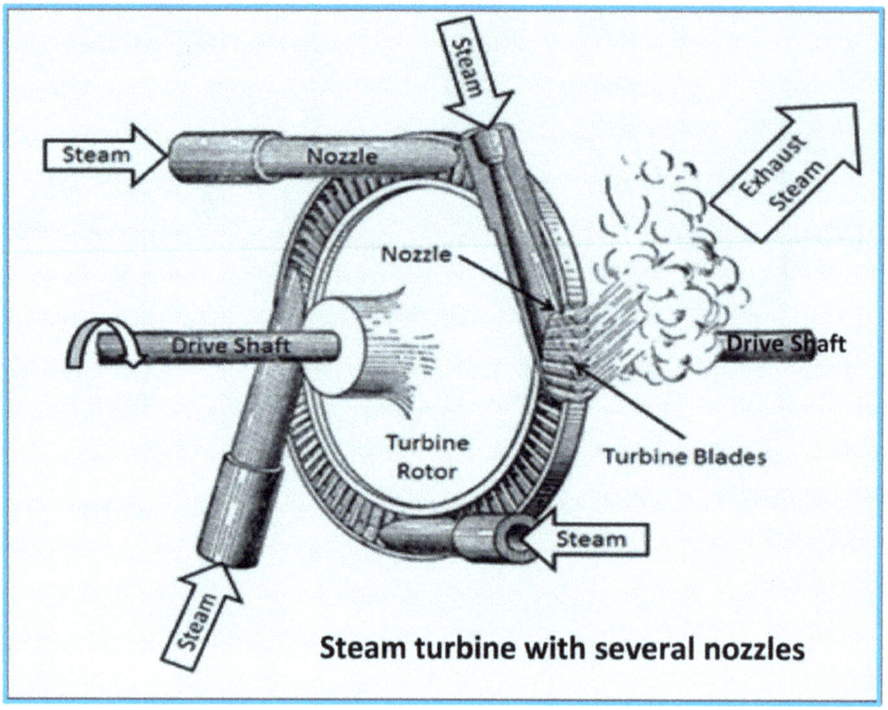

Figure 3.5 Turbine with multiple nozzles.

A schematic diagram of a turbine is shown in Figure 3.6. The first law of thermodynamics for control volume, Equation (3.13), can be applied to calculate the amount of shaft work produced by the turbine. Assuming steady state and neglecting kinetic and potential energy changes and heat transfer, Equation (3.13) simplifies to:

$$-\dot{W}_{sh} = \dot{m}(h_2 - h_1) \tag{3.36}$$

On a unit mass basis,

$$-W_{sh} = (h_2 - h_1) \tag{3.37}$$

Thus, work produced by a turbine can be calculated if the change in enthalpy of the fluid is known.

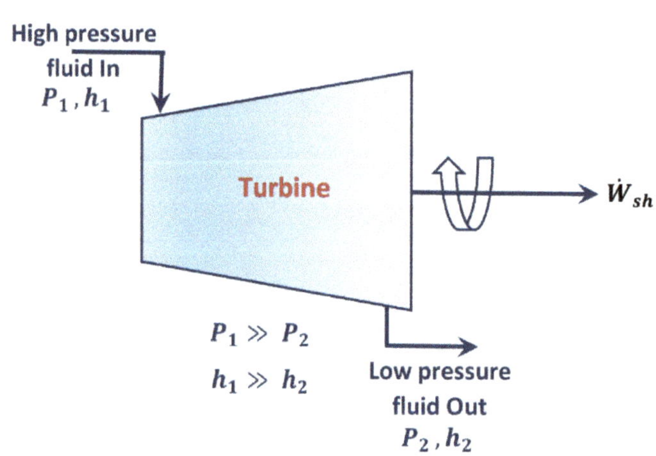

Figure 3.6 Schematic diagram of a turbine.

3.5.3 Compressors and Pumps

Compressors are devices in which mechanical energy (work) is added to raise the pressure of a *gas* passing through them (see Figure 3.7). *Pumps* are similar devices in which mechanical energy is added to raise the pressure of a *liquid* passing through them. The power required by the compressor can be calculated from Equation (3.36) neglecting kinetic and potential energy changes and heat transfer. Note that $\dot{W}_{sh}$ is negative for compressors. For pumps through which liquid (incompressible fluid) flows, Equation (3.28) gives the following expression for the power required (neglecting kinetic and potential energy changes and heat transfer):

$$-\dot{W}_{sh,pump} = \dot{m}\left[\frac{P_2-P_1}{\rho}\right] \tag{3.38}$$

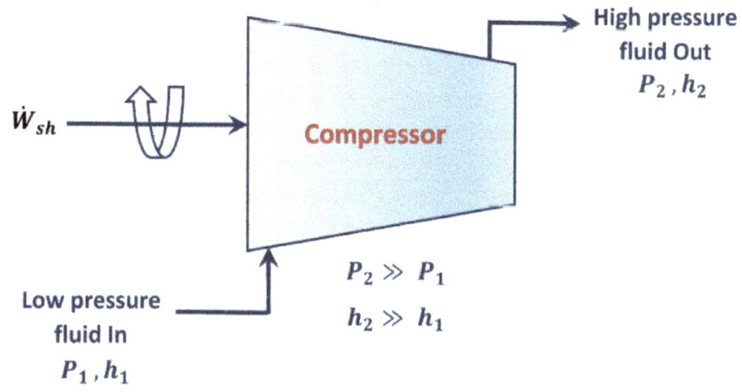

Figure 3.7 Schematic diagram of a compressor.

3.5.4 Throttling Devices

Throttling devices are devices in which a sharp reduction in pressure of the fluid occurs adiabatically. The kinetic and potential energy changes of the fluid are negligible and there is no shaft work. From the first law, Equation (3.13), it follows that:

$$h_2 = h_1 \tag{3.39}$$

Thus, throttling is an isenthalpic process. Examples of throttling devices are: a tube consisting of porous plug (Figure 3.8) and a pipe consisting of partially opened valve (Figure 3.9). In a throttling device, a constriction is placed in the path of fluid flow to cause a sharp reduction in pressure.

Throttling phenomenon is widely used for cooling of fluids in refrigeration processes. A large drop in pressure at constant enthalpy generally results in a considerable drop in temperature of the fluid for most real gases.

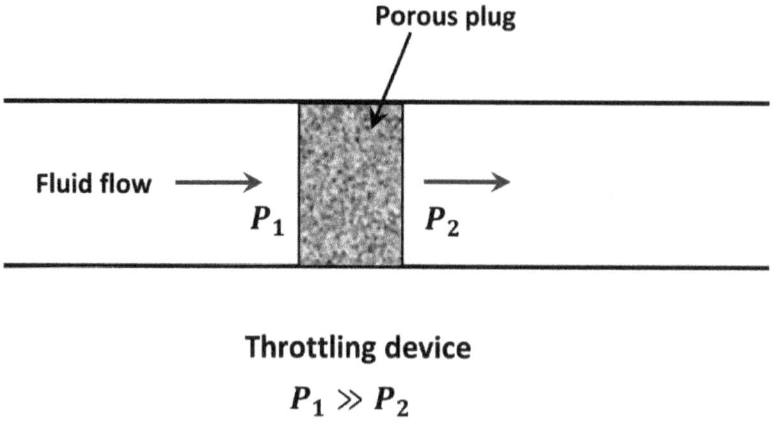

Figure 3.8 Porous plug type throttling device.

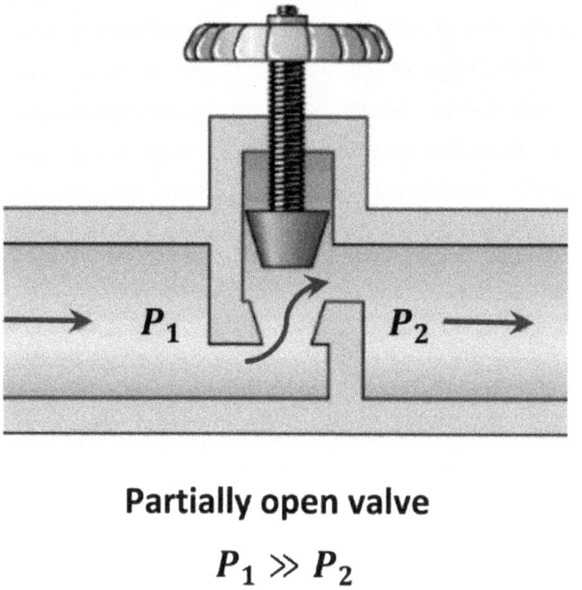

Figure 3.9 Partially open valve type throttling device.

3.5.5 Heat Exchangers

Heat exchangers are used extensively in industrial applications to transfer energy between fluids at different temperatures. Energy is transferred from a hot fluid stream to a cold fluid stream as they pass through the heat exchanger. A commonly used heat exchanger is a double pipe heat exchanger shown schematically in Figure 3.10.

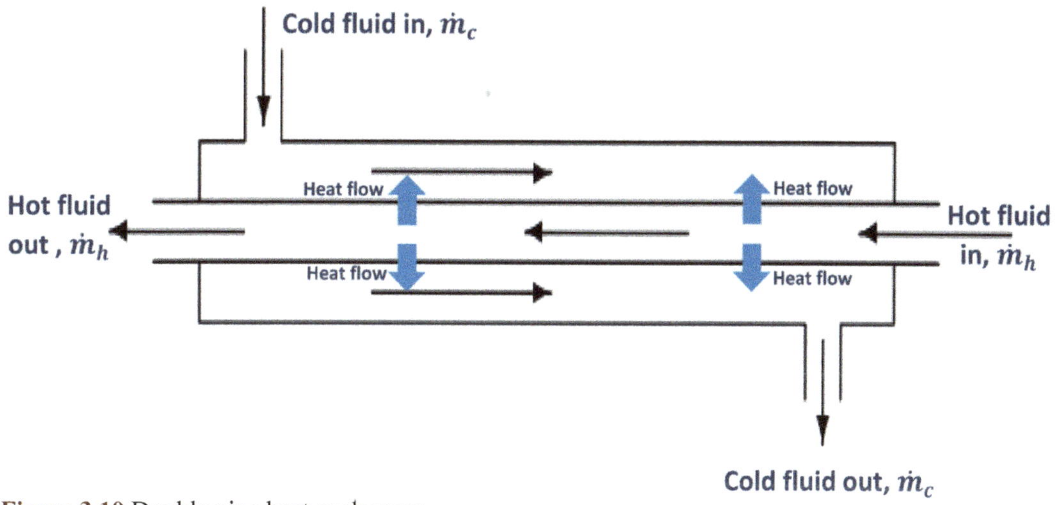

Figure 3.10 Double pipe heat exchanger.

The amount of heat exchanged between the fluids can be calculated from the first law for control volume, Equation (3.13) assuming that kinetic and potential energy changes are negligible and $\dot{W}_{sh} = 0$:

$$\dot{Q} = \dot{m}_h(h_2 - h_1)_{hot\ fluid} \tag{3.40}$$

$$\dot{Q} = \dot{m}_c(h_2 - h_1)_{cold\ fluid} \tag{3.41}$$

Assuming that there is no heat loss to the surroundings, the heat lost by the hot fluid is equal to the heat gained by the cold fluid. In Equation (3.40), $\dot{Q}$ will have a negative value as heat is lost by the hot fluid. $\dot{Q}$ will have a positive value in Equation (3.41) as heat is gained by the cold fluid. The sum of two $\dot{Q}$'s is zero:

$$\dot{m}_h(h_2 - h_1)_{hot\ fluid} + \dot{m}_c(h_2 - h_1)_{cold\ fluid} = 0 \tag{3.42}$$

3.5.6 Pipe Flow

Flow through pipes is an everyday occurrence (see Figure 3.11). Pipelines are used to transport water, oil, gas and many more fluidic substances from one place to another. Pipelines that transport oil and gas are big business.

When fluid flows through a pipe, mechanical energy is dissipated into frictional heating of fluid due to non-zero viscosity of fluid. The dissipation of mechanical energy is reflected in loss of pressure of the fluid as it flows through the pipe. Assuming fluid to be incompressible and no potential and kinetic energy changes, and $\dot{W}_{sh} = 0$, the mechanical energy equation, Equation (3.28), gives:

$$\dot{F}_{loss} = -\dot{m}[\Delta P/\rho] \tag{3.43}$$

Figure 3.11 Flow of water in a pipe.

Thus, pressure loss occurs in pipeline flow of fluids due to friction. The mechanical energy loss due to friction, and hence pressure loss, is usually expressed as [2]:

$$F_{loss} = \dot{F}_{loss}/\dot{m} = 4f(L/D)(\bar{V}^2/2) \tag{3.44}$$

where F_{loss} is mechanical energy loss in friction per unit mass of fluid, f is Fanning friction factor, L is length of pipe, D is pipe diameter, and $\bar{V}$ is average velocity of fluid in the pipe. In order to calculate the mechanical energy loss in pipeline flows, the value of friction factor f is required.

In *laminar* flow of Newtonian fluids ($Re \leq 2100$), the Fanning friction factor is related to Reynolds number through the following theoretical relationship:

$$f = 16/Re \tag{3.45}$$

where Re is defined as:

$$Re = \rho D\bar{V}/\mu \tag{3.46}$$

where μ is fluid viscosity.

In *turbulent* flow of Newtonian fluids, friction factor is function of Reynolds number and relative roughness of pipe (ϵ/D). For hydraulically smooth pipes ($\epsilon/D \to 0$), the friction factor depends only on Re in turbulent regime. The following semi-empirical equation, often referred to as von Karman – Nikuradse equation, describes the f $vs.Re$ turbulent behavior of Newtonian fluids in smooth pipes very well [3]:

$$1/\sqrt{f} = 4log_{10}\left(Re\sqrt{f}\right) - 0.40 \tag{3.47}$$

where f is the Fanning friction factor. The von Karman – Nikuradse equation is not explicit in friction factor. A number of explicit f $vs.Re$ relations are available in the literature. One of the popular ones is the Blasius friction factor equation for turbulent flow of Newtonian fluids in smooth pipes:

$$f = 0.079/Re^{0.25} \tag{3.48}$$

Equation (3.48) is accurate over a Reynolds number range of $3000 \leq Re \leq 100{,}000$. For turbulent flow in rough pipes, the following Colebrook equation [4,5] is widely accepted:

$$1/\sqrt{f} = -4log_{10}\left(\frac{\epsilon/D}{3.7} + \frac{1.26}{Re\sqrt{f}}\right) \tag{3.49}$$

This Colebrook equation is implicit in Fanning friction factor. A number of explicit f $vs.$ Re relations are available in the literature for turbulent flow of Newtonian fluids in rough pipes [6-8]. An explicit equation which is very accurate for turbulent flow of Newtonian fluids in rough pipes is as follows:

$$1/\sqrt{f} = -4log_{10}\left[\frac{\epsilon/D}{3.7} - \frac{5.02}{Re}log_{10}\left(\frac{\epsilon/D}{3.7} - \frac{5.02A}{Re}\right)\right] \quad (3.50)$$

$$A = log_{10}\left(\frac{\epsilon/D}{3.7} + \frac{13}{Re}\right) \quad (3.51)$$

This equation was originally proposed by Zigrang and Sylvester [9].

3.6 APPLICATIONS OF FIRST LAW TO CYCLIC FLOW PROCESSES

Figure 3.12 shows a schematic of a simple steam power plant. The working fluid (H_2O) circulates through a closed loop under steady-state condition. The liquid water at high pressure (state 1, Figure 3.12) enters the boiler. The water is heated in the boiler and the discharge from the boiler is high pressure superheated steam (state 2, Figure 3.12). The superheated steam is expanded in a turbine to produce the shaft work. The discharge from the turbine is wet steam at low pressure (state 3, Figure 3.12). In the condenser, the steam is condensed to liquid water by removing heat using cooling water. The discharge from the condenser is liquid water at low pressure (state 4, Figure 3.12). The liquid water is finally pressurized using a pump and sent back to the boiler (state 1, Figure 3.12). Thus, the working fluid undergoes a cyclic process on a continuous basis. In this cyclic process, there is a net shaft work production and a net heat absorption.

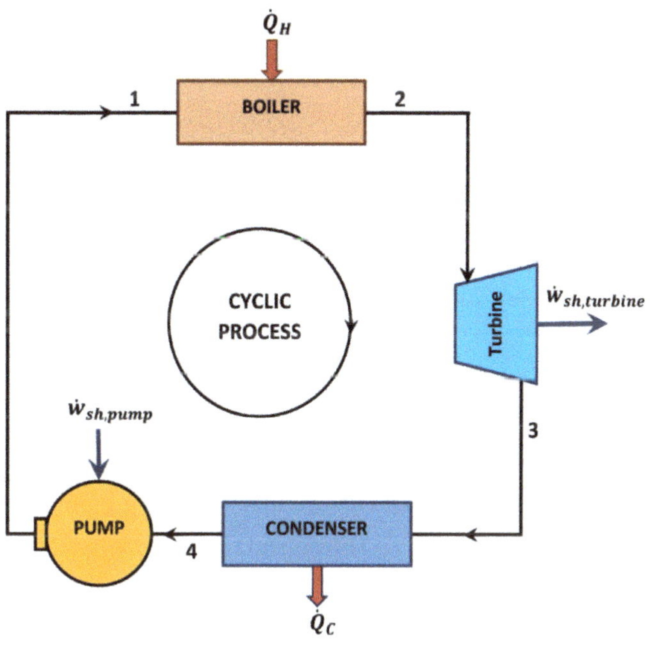

Figure 3.12 Simple steam power plant.

We will now apply the first law, Equation (3.13), to different components of this cyclic process and to the whole cycle.

3.6.1 Boiler

Referring to Figure 3.12 and neglecting kinetic/potential energy changes and shaft work ($\dot{W}_{sh} = 0$), the first law gives:

$$\dot{Q}_H = \dot{m}(h_2 - h_1) \tag{3.52}$$

Thus, the heat supplied to the boiler goes into enthalpy increase of the fluid.

3.6.2 Turbine

The work produced in a turbine (Figure 3.12), neglecting kinetic/potential energy changes and heat transfer, is as follows according to the first law (Equation (3.13)):

$$-\dot{W}_{sh,turbine} = \dot{m}(h_3 - h_2) \tag{3.53}$$

There occurs a large reduction in enthalpy of the fluid in the turbine and shaft work is produced. Note that $\dot{W}_{sh,turbine} > 0$.

3.6.3 Condenser

In the condenser (Figure 3.12), heat is rejected to the surroundings. According to the first law, neglecting kinetic/potential energy changes and shaft work, we have:

$$\dot{Q}_C = \dot{m}(h_4 - h_3) \tag{3.54}$$

3.6.4 Pump

The work required in a pump (Figure 3.12), neglecting kinetic/potential energy changes and heat transfer, is as follows according to the first law:

$$-\dot{W}_{sh,pump} = \dot{m}(h_1 - h_4) \tag{3.55}$$

There occurs an increase in enthalpy of the fluid in the pump and shaft work is required. Note that $\dot{W}_{sh,pump} < 0$.

3.6.5 Whole Cycle

We add Equations (3.52) to (3.55) to find heat and work effects for the whole cycle. Thus,

$$\dot{Q}_H - \dot{W}_{sh,turbine} + \dot{Q}_C - \dot{W}_{sh,pump} = \dot{m}[(h_2 - h_1) + (h_3 - h_2) + (h_4 - h_3) + (h_1 - h_4)] \tag{3.56}$$

Upon simplification, Equation (3.56) gives:

$$\dot{Q}_H + \dot{Q}_C = \dot{W}_{sh,turbine} + \dot{W}_{sh,pump} \tag{3.57}$$

Alternatively,

$$\dot{Q}_{cycle} = \dot{W}_{sh,cycle} \tag{3.58}$$

3.7 PERPETUAL MOTION MACHINE OF FIRST KIND

Any device, engine, machine or process that violates the first law of thermodynamics is called a *Perpetual Motion Machine of First Kind.* According to the first law, energy cannot be created or destroyed. As an example, Figure 3.13 shows a *Perpetual Motion Machine of First Kind.* In this cyclic process, there is a perpetual creation of energy (electric power plus heat $\dot{Q}_C$) without any energy input from the outside. This violates the first law.

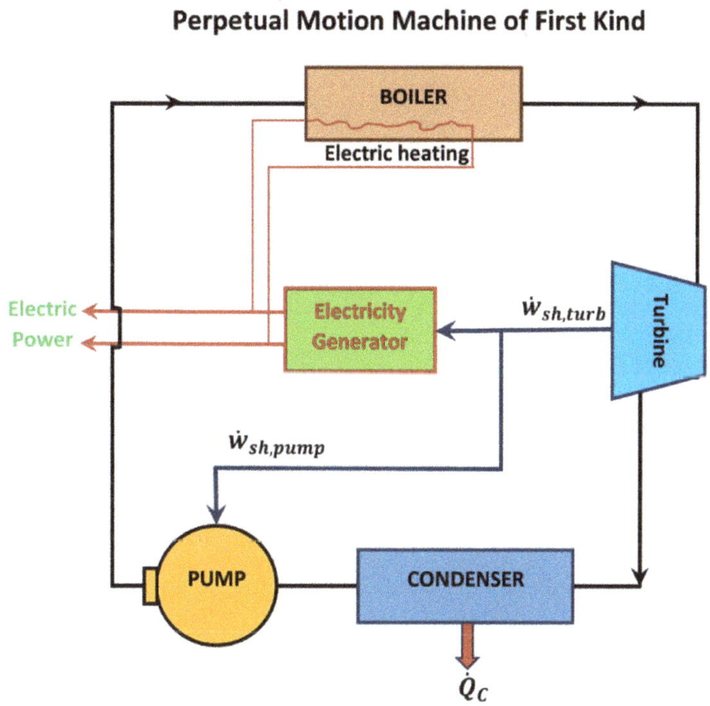

Figure 3.13 Steam power plant as a Perpetual Motion Machine of First Kind.

NOTATION

A cross-sectional area or parameter in Equation (3.51)
D pipe inner diameter
E total energy
f friction factor
F_{loss} mechanical energy loss per unit mass of fluid
$\dot{F}_{loss}$ rate of mechanical energy dissipation
g acceleration due to gravity
h specific enthalpy
L length of pipe
m mass
$\dot{m}$ mass flow rate
P pressure
$\dot{Q}$ heat transfer rate
$\dot{Q}_C$ heat transfer rate in condenser
$\dot{Q}_H$ heat transfer rate in boiler
$\dot{Q}_{cycle}$ net heat transfer rate in a cyclic process
Re Reynolds number
t time
u specific internal energy
V velocity or volume
$\vec{V}$ velocity vector

W_{sh}	shaft work per unit mass of fluid
$\dot{W}_{sh}$	rate of shaft work
$\dot{W}_{sh,cycle}$	net rate of shaft work in a cyclic process
Z	elevation

Greek Letters

Δ	difference operator
ϵ	pipe roughness
μ	fluid viscosity
ρ	fluid density

REFERENCES

1. Bird, R.B., Stewart, W.E., Lightfoot, E.N., *Transport Phenomena*, John Wiley & Sons, New York, 2nd edition, 2007, Chapter 7.
2. Pal, R., "Teaching fluid mechanics and thermodynamics simultaneously through pipeline flow experiments", *Fluids* 2019; 4: Article 103, 1-18.
3. Welty, J.R., Wicks, C.E., Wilson, R.E., Rorrer, G., *Fundamentals of Momentum, Heat, and Mass Transfer*, John Wiley & Sons, New York, 4th edition, 2001.
4. Colebrook, C.F., "Turbulent flow in pipes with particular reference to the transition region between smooth and rough pipe laws", *J. of Inst. Civil Eng.* 1939; 11: 133-156.
5. Colebrook, C.F., White, C.M., "Experiments with fluid friction in roughened pipes", *Proc. Roy. Soc. Lond. Ser. A. Maths. Phys. Sci.* 1937; 161: 367-381.
6. Asker, M., Turgut, O.E., Coban, M.T., "A review of non-iterative friction factor correlations for the calculation of pressure drop in pipes", *Bitlis Eren Univ J Sci Technol.* 2014; 4: 1-8.
7. Brkic, D., "Review of explicit approximations to the Colebrook relation for flow friction", *J. Pet. Sci. Tech.* 2011; 77:34-48.
8. Fang, X., Xu, Y., Zhou, Z., "New correlations of single-phase friction factor for turbulent pipe flow and evaluation of existing single-phase friction factor correlations", *Nuclear Eng. Design* 2011; 241: 897-902.
9. Zigrang, D.J., Sylvester, N.D., "Explicit approximations to the solution of Colebrook's friction factor equation", *AIChE J.* 1982, 28: 514-515.

TUTORIAL PROBLEMS WITH SOLUTION

Problem 1

A liquid flows through a horizontal pipeline expansion system shown in the figure below. The inlet velocity of liquid is 4.5 m/s and the inlet diameter is 1.8 cm. Assuming no heat transfer or shaft work, Calculate:

a) Enthalpy change of liquid if the exit diameter is 3.6 cm.
b) Enthalpy change of liquid if the exit diameter is 7.2 cm.
c) Maximum change in enthalpy for the pipeline expansion.

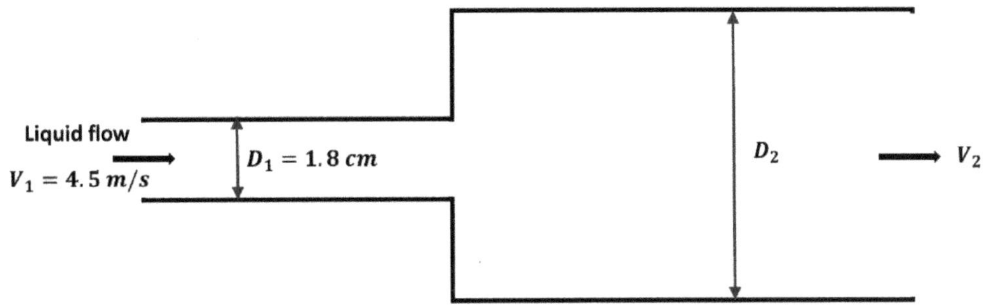

Solution

In this problem, the kinetic energy of fluid is being converted into enthalpy of the fluid.

Apply first law for open systems (control volume):

$$\dot{m}\left[\Delta h + \frac{1}{2}\Delta V^2 + g\Delta Z\right] = \dot{Q} - \dot{W}_{sh}$$

On a unit mass basis,

$$\left[\Delta h + \frac{1}{2}\Delta V^2 + g\Delta Z\right] = Q - W_{sh}$$

As $Q = 0$ and $W_{sh} = 0$, and the system is horizontal $\Delta Z = 0$, we get:

$$\Delta h = -\frac{1}{2}\Delta V^2 = -\frac{1}{2}(V_2^2 - V_1^2)$$

From mass balance,

$$\dot{m} = \rho_1 A_1 V_1 = \rho_2 A_2 V_2$$

Assuming liquid to be incompressible, $\rho_1 = \rho_2$, the mass balance gives:

$$V_2 = \left(\frac{D_1}{D_2}\right)^2 V_1$$

Consequently,

$$\Delta h = -\frac{1}{2}(V_2^2 - V_1^2) = \frac{V_1^2}{2}\left[1 - \left(\frac{D_1}{D_2}\right)^4\right]$$

a)

$$D_1 = 1.8\ cm, \qquad D_2 = 3.6\ cm, V_1 = 4.5\ m/s$$

$$\Delta h = \frac{V_1^2}{2}\left[1 - \left(\frac{D_1}{D_2}\right)^4\right] = \frac{(4.5)^2}{2}\left[1 - \left(\frac{1.8}{3.6}\right)^4\right] = 9.492 \text{ J/kg}$$

b)

$$D_1 = 1.8\ cm, \qquad D_2 = 7.2\ cm, V_1 = 4.5\ m/s$$

$$\Delta h = \frac{V_1^2}{2}\left[1 - \left(\frac{D_1}{D_2}\right)^4\right] = \frac{(4.5)^2}{2}\left[1 - \left(\frac{1.8}{7.2}\right)^4\right] = 10.085 \text{ J/kg}$$

c)

$$D_1 = 1.8\ cm, \qquad D_2 = \infty, V_1 = 4.5\ m/s$$

$$\Delta h = \frac{V_1^2}{2}\left[1 - \left(\frac{D_1}{D_2}\right)^4\right] = \frac{(4.5)^2}{2}[1 - 0] = 10.125 \text{ J/kg}$$

$$\text{Note that } (\Delta h)_{max} = \frac{V_1^2}{2}$$

Problem 2

A liquid flows through a horizontal heating coil which is heated from the outside by superheated steam. The inlet velocity of liquid is 2.6 m/s. As the liquid passes through the coil, it undergoes phase change from liquid to vapor. The fluid leaving the coil is superheated vapor and its velocity is 165 m/s. What is the heat transferred to the fluid per unit mass of fluid? The enthalpies of the fluid at the inlet and outlet are given as: $h_1 = 245$ kJ/kg and $h_2 = 2545$ kJ/kg.

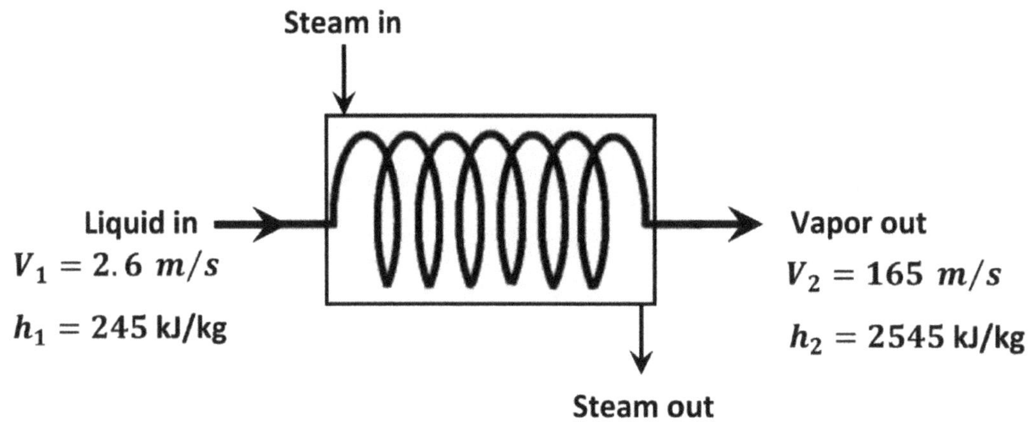

Solution

From first law for control volume:

$$\left[\Delta h + \frac{1}{2}\Delta V^2 + g\Delta Z\right] = Q - W_{sh}$$

As $W_{sh} = 0$, and the system is horizontal $\Delta Z = 0$, we get:

$$Q = \Delta h + \frac{1}{2}(V_2^2 - V_1^2) = (2545 - 245) \times 10^3 + \frac{1}{2}(165^2 - 2.6^2) = 2300000 + 13609.12$$
$$= 2313609.12\,\text{J/kg} = 2313.61\,\text{kJ/kg}$$

Problem 3

An ideal gas flows through a converging nozzle with an inlet diameter of 4.5 cm. At the inlet, the temperature is 300 °C and the pressure is 650 kPa. The gas velocity at the inlet is 25 m/s. At the exit of the nozzle, the temperature is 250 °C and the pressure is 300 kPa. What is the velocity of the gas at the nozzle exit? What is the diameter of the nozzle at the exit?
The molecular weight of the gas is 29 g/mol and the heat capacity is $c_p = 7R/2$.

Solution

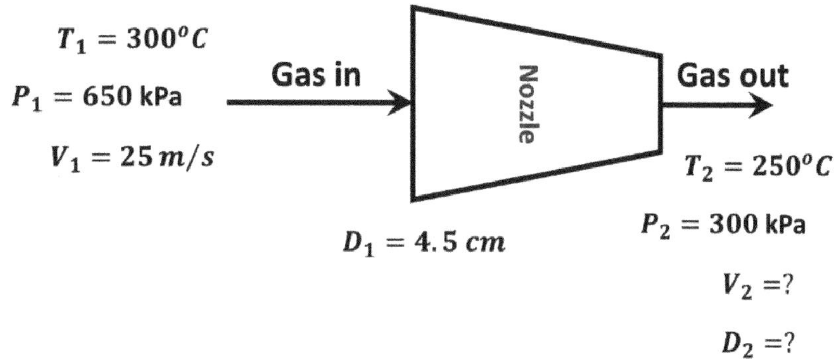

For ideal gas,

$$\Delta h = c_p(T_2 - T_1) = 3.5 \times 8.314 \times (250 - 300) = -1454.95 \text{ J/mol} = -1454.95/29$$
$$= -50.17 \text{ kJ/kg}$$

First law for control volume (open systems):

$$\left[\Delta h + \frac{1}{2}\Delta V^2 + g\Delta Z\right] = Q - W_{sh}$$

As $Q = 0, W_{sh} = 0$, and the system is horizontal $\Delta Z = 0$, we get:

$$V_2^2 = V_1^2 - 2\Delta h = (25)^2 - 2(-50.17 \times 10^3) = 100965$$

$$V_2 = \sqrt{100965} = 317.75 \text{ m/s}$$

Exit diameter of the nozzle:

Using mass balance,

$$\dot{m} = \rho_1 A_1 V_1 = \rho_2 A_2 V_2$$

Since $A = \pi/4D^2$,

$$\left(\frac{D_2}{D_1}\right)^2 = \frac{\rho_1}{\rho_2}\frac{V_1}{V_2}$$

The density of ideal gas is given as:

$$\rho = \frac{PM}{RT}$$

where M is molecular weight. Thus,

$$\left(\frac{D_2}{D_1}\right)^2 = \frac{P_1}{P_2}\frac{V_1}{V_2}\frac{T_2}{T_1} = \left(\frac{650}{300}\right)\left(\frac{25}{317.75}\right)\left(\frac{250 + 273.15}{300 + 273.15}\right) = 0.155598$$

$$\frac{D_2}{D_1} = 0.39446$$

$$D_2 = 0.39446 \times 4.5 = 1.775 \text{ cm}$$

Problem 4

A gas turbine expands 3.5 kg/s of air from 9 bar and 850 °C to 1 bar adiabatically. Calculate the exhaust temperature and the power output of the turbine. You can assume that the process is reversible. The heat capacity ratio γ is 1.4 and c_p =1005 J/kg.K.

Solution

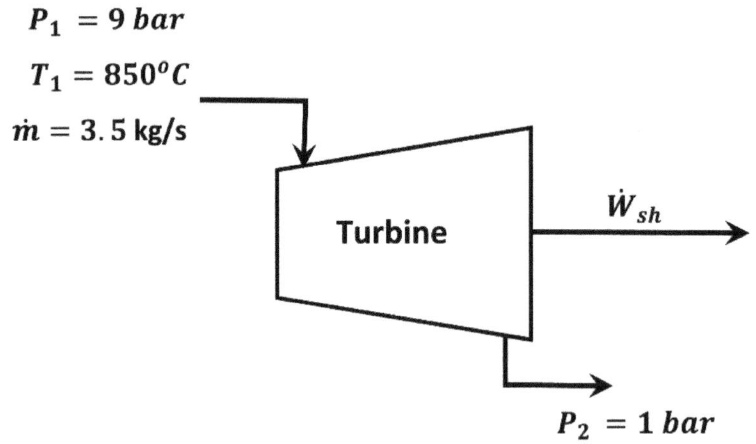

We will treat air as an ideal gas. For reversible adiabatic expansion of an ideal gas,

$$\frac{T_2}{T_1} = \left(\frac{P_2}{P_1}\right)^{(\gamma-1)/\gamma} = \left(\frac{1}{9}\right)^{(1.4-1)/1.4} = 0.5338$$

$$T_2 = 0.5338 \times (850 + 273.15) = 599.54 \text{ K}$$

First law for open systems,

$$\dot{m}\left[\Delta h + \frac{1}{2}\Delta V^2 + g\Delta Z\right] = \dot{Q} - \dot{W}_{sh}$$

Neglecting kinetic and potential energy changes and $\dot{Q} = 0$, the first law gives:

$$\dot{W}_{sh} = -\dot{m}[\Delta h] = -\dot{m}c_p\Delta T = -\dot{m}c_p(T_2 - T_1) =$$
$$-3.5 \times 1005 \times (599.54 - 1123.15) = 1841798.18\ W = 1.842\ MW$$

Problem 5

Air at 550 K and 1 atm is mixed continuously in a mixing device at a flow rate of 3 mol/s with another air stream at 395 K and 1 atm at a flow rate of 1.5 mol/s. The product air stream leaving the mixing device is at 515 K and 1 atm. Determine the rate of heat transfer to the mixing device.
You can assume air to be an ideal gas with a heat capacity of c_p=7R/2.

Solution

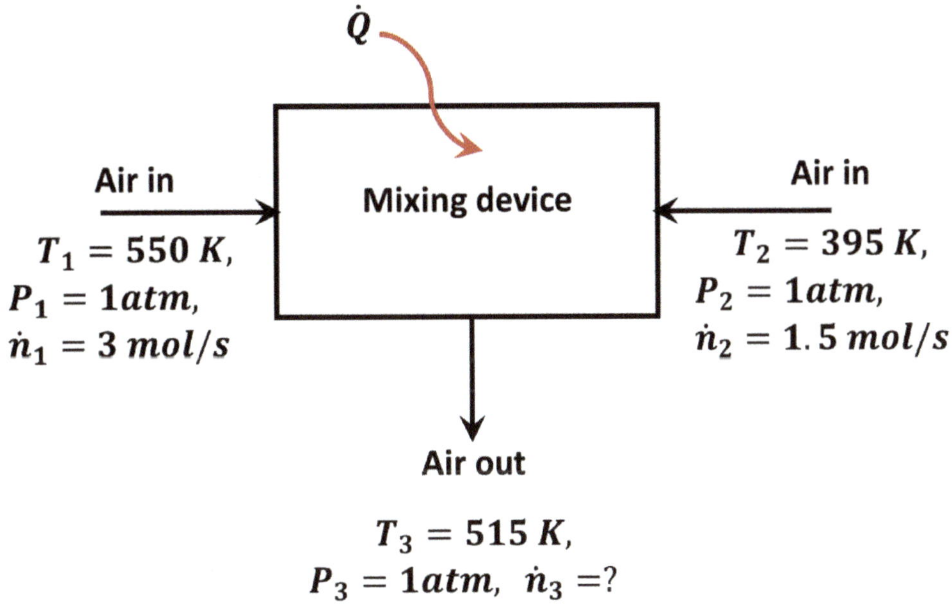

Mass (mole) balance:

$$\textstyle\sum_{INLETS}\ (\dot{n}_i) = \sum_{OUTLETS}\ (\dot{n}_i)$$

$$\dot{n}_3 = \dot{n}_1 + \dot{n}_2 = 3 + 1.5 = 4.5 \text{ mol/s}$$

First law for control volume (neglecting kinetic and potential energy changes):

$$\sum_{OUTLETS} (\dot{n}_i h_i) - \sum_{INLETS} (\dot{n}_i\, h_i) = \dot{Q} - \dot{W}_{sh}$$

$$\dot{n}_3 h_3 - (\dot{n}_1 h_1 + \dot{n}_2 h_2) = \dot{Q}$$

$$(\dot{n}_1 + \dot{n}_2)h_3 - (\dot{n}_1 h_1 + \dot{n}_2 h_2) = \dot{Q}$$

$$\dot{n}_1(h_3 - h_1) + \dot{n}_2(h_3 - h_2) = \dot{Q}$$

For ideal gas,

$$\Delta h = c_p \Delta T$$

Consequently,

$$\dot{Q} = \dot{n}_1 c_p (T_3 - T_1) + \dot{n}_2 c_p (T_3 - T_2) = 3 \times 3.5 \times 8.314 \times (515 - 550) + 1.5 \times 3.5 \times 8.314 \times (515 - 395) - 2182.425 \text{ J/s}$$

Problem 6

Water at 20 °C flows through a 2 inch (0.0508 m) ID cast-iron pipe with an average velocity of 1 m/s. The pipe is 35 m long and there is an increase in elevation of 1 m from the inlet to the exit of the pipe. Calculate the power needed to generate this flow. The density and viscosity of water are: 998.21 kg/m^3 and 1.002 mPa.s, respectively.

Solution

Let us first calculate the Reynolds number:

$$Re = \rho D\bar{V}/\mu = 998.21 \times 0.0508 \times 1/1.002 \times 10^{-3} = 50{,}607.85$$

Thus, the flow is highly turbulent as $Re \gg 2100$.

Now we apply the mechanical energy equation (Equation (3.28)):

$$\dot{W}_{sh} + \dot{F}_{loss} + \dot{m}[\Delta(V^2/2) + g\Delta Z + \Delta P/\rho] = 0$$

As the velocity is same at the inlet and outlet of the pipe, $\Delta(V^2/2) = 0$. Also, the pressure is atmospheric pressure at the inlet and outlet of the pipe, $\Delta P/\rho$. The mechanical energy simplifies to:

$$\dot{W}_{sh} = -\dot{F}_{loss} - \dot{m}[g\Delta Z]$$

where

$$\dot{F}_{loss} = 4\dot{m}f(L/D)(\bar{V}^2/2)$$

We can calculate the friction factor f using Equation (3.50):

$$1/\sqrt{f} = -4log_{10}\left[\frac{\epsilon/D}{3.7} - \frac{5.02}{Re}log_{10}\left(\frac{\epsilon/D}{3.7} - \frac{5.02A}{Re}\right)\right]$$

$$A = log_{10}\left(\frac{\epsilon/D}{3.7} + \frac{13}{Re}\right)$$

$Re = 50{,}607.85$, $\epsilon/D = 0.0055$. Upon solving for f, we get $f = 8.237 \times 10^{-3}$. The mass flow rate is:

$$\dot{m} = \frac{\pi}{4}D^2\bar{V}\rho = \frac{\pi}{4}(0.0508)^2 \times 1 \times 998.21 = 2.0232 \text{ kg/s}$$

Thus,

$$\dot{F}_{loss} = 4\dot{m}f(L/D)(\bar{V}^2/2) = 4 \times 2.0232 \times 8.237 \times 10^{-3}(35/0.0508)(1^2/2) = 22.964 \text{ W}$$

$$\dot{W}_{sh} = -\dot{F}_{loss} - \dot{m}[g\Delta Z] = -22.964 - 2.0232[9.81 \times 1] = -42.81\ W$$

PRACTICE PROBLEMS

Problem 1

Consider the steam power plant shown in Figure 3.12. The mass flow rate of fluid circulation in the loop is 75 kg/s. The specific enthalpies of circulating fluid at different locations of the loop are as follows:
$h_1 = 195$ kJ/kg ; $h_2 = 3500$ kJ/kg; $h_3 = 2300$ kJ/kg; $h_4 = 188$ kJ/kg

Calculate the following:

a) Heat transfer rate in the boiler.
b) Power output of the turbine.
c) Heat transfer rate in the condenser.
d) Power requirement of the pump.

Problem 2

Consider the following open system:

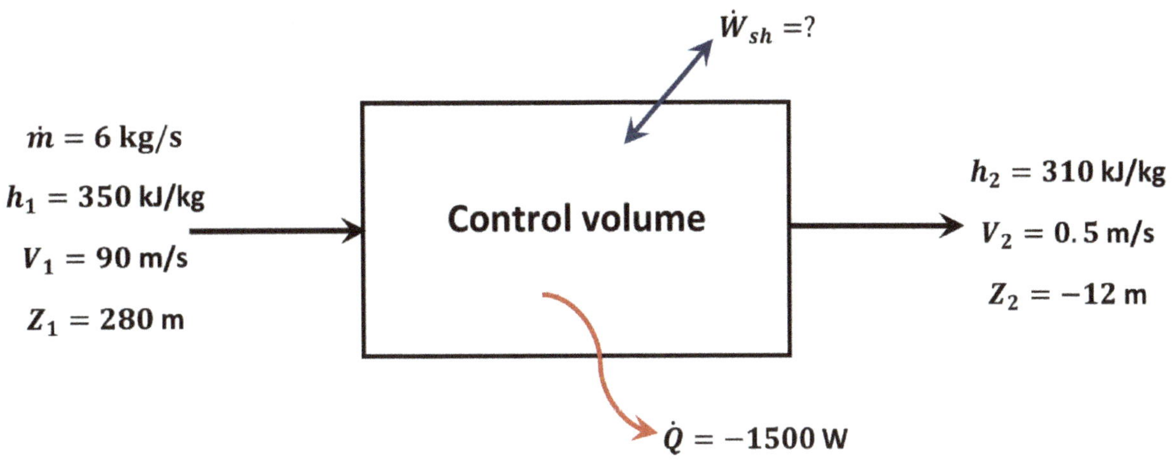

Based on the given information, calculate the shaft work. Assume steady state.

Problem 3
An ideal gas flows through a heat exchanger. The molar flow rate of the gas is 25 mol/s. The temperature of the gas is increased from 100 to 500 °C. The heat capacity of the gas is given as:

$$\frac{c_p}{R} = A + BT + CT^2 + DT^{-2}$$

where T is in K, $A = 6.094, B = 1.056 \times 10^{-3}, C = 0$, and $D = 2.028 \times 10^5$.
What is the rate of heat transfer?

Problem 4
An ideal gas is compressed adiabatically from 25 °C and 110 kPa to 518 kPa. Assuming the process to be reversible, calculate the discharge temperature of the gas and estimate the work requirement of the compressor. The heat capacities of the gas are: $c_p = 7R/2$ and $c_v = 5R/2$.

Problem 5
Three streams of air are mixed continuously in a mixing device. The flow rates and conditions of streams entering the device are specified as follows:
Stream 1: $\dot{n}_1 = 5$ mol/s , $T_1 = 640$ K, $P_1 = 1$ atm
Stream 2: $\dot{n}_2 = 3.5$ mol/s , $T_2 = 440$ K, $P_2 = 1$ atm
Stream 3: $\dot{n}_3 = 4.5$ mol/s , $T_3 = 500$ K, $P_1 = 1$ atm
The product air stream leaving the mixing device is at 550 K and 1 atm. You can assume air to be an ideal gas with a heat capacity $c_p = 7R/2$.
Determine the rate of heat transfer in the mixing device.

Problem 6
Water is flowing through a pipeline system shown in the figure below. The inlet pressure of the fluid is 50 kPa. The discharge pressure is 150 kPa. The exit section of the pipe is 5 m above the inlet section. The pump supplies a mechanical energy of 165 J/kg to the fluid. The pipe diameter is uniform throughout, and the flow is highly turbulent. The fluid density is 998.21 kg/m^3. Calculate the frictional loss in J/kg for the pipeline system. Take g=9.81 m/s^2.

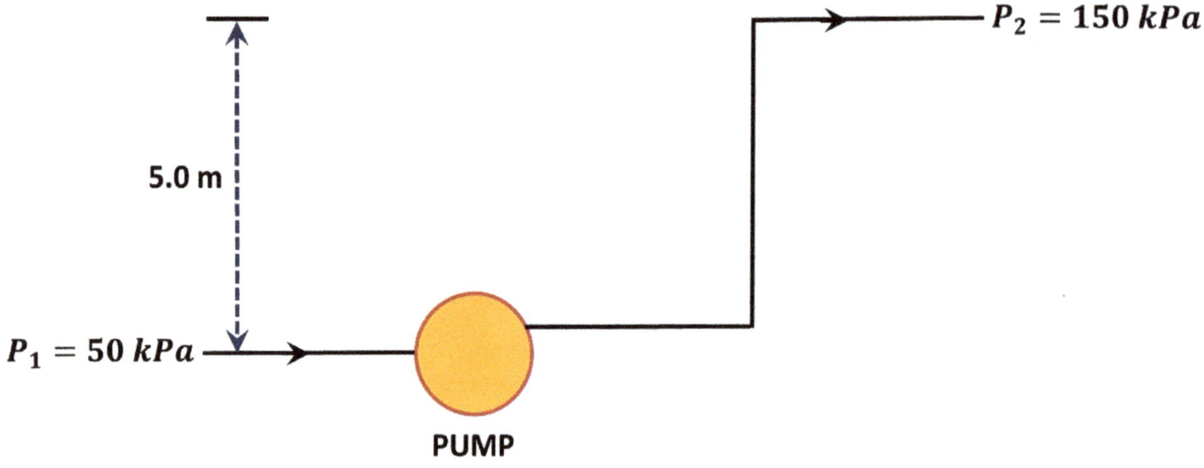
5.0 m
$P_2 = 150\ kPa$
$P_1 = 50\ kPa$
PUMP

Chapter 4

Second Law of Thermodynamics for Closed Systems (Control Mass)

4.1 LEARNING OUTCOMES

Upon completion of this chapter, students will be able to:

- Explain the second law of thermodynamics using the Kelvin-Planck and Clausius statements
- Show the equivalence of Kelvin-Planck and Clausius statements of second law
- Describe the Carnot cycle and derive the expression for its thermal efficiency
- State and prove the Carnot theorems
- Explain the Clausius inequality
- Define entropy and explain generation of entropy due to irreversibilities
- Describe the principle of entropy increase
- Explain the physical significance of entropy using the Boltzmann relation for entropy
- State the second law of thermodynamics in terms of entropy change of the universe
- Calculate entropy changes of gases, liquids and solids
- Solve problems related to the application of second law to heat engines and non-flow processes

4.2 INTRODUCTION

"The law that entropy always increases, holds, I think, the supreme position among the laws of Nature. If someone points out to you that your pet theory of the universe is in disagreement with Maxwell's equations — then so much the worse for Maxwell's equations. If it is found to be contradicted by observation — well, these experimentalists do bungle things sometimes. But if your theory is found to be against the second law of thermodynamics I can give you no hope; there is nothing for it but to collapse in deepest humiliation." – Sir Arthur Stanley Eddington

The first law of thermodynamics does not distinguish between different forms of energy. All forms of energy are *equal* in the eyes of the first law. One form of energy (for example, heat) can be fully converted into another form (for example, mechanical energy). It does not place any restrictions on energy conversion other than that the total quantity of energy in the universe remains constant. However, the second law of thermodynamics indicates that all forms of energy are *not equal.* According to the second law, energy possesses both *quality* and quantity where quality of energy is defined in terms of its ability to produce work. A low quality energy is not capable of producing as much work as a high quality energy. Furthermore, a low quality energy cannot be converted completely into a high quality energy. As an example, consider conversion of heat into mechanical energy. According to the first law, heat can be completely converted into mechanical energy. The second law says "No". Heat is an inferior form of energy and has low quality. Mechanical energy, on the other hand, is a superior form of energy and has high quality. While mechanical energy can be fully converted into heat, the reverse is not true, that is, heat cannot be fully converted into mechanical energy.

Another important implication of the second law is that "*processes occur in a certain direction*". Thus, the second law is directional in nature. It allows us to predict the direction of spontaneous or natural changes. As an example, consider heat interaction between two material bodies at different temperatures (see Figure

4.1). When two bodies at different temperatures are brought into thermal contact with each other, heat will flow between them. In which direction will the heat flow? From high temperature body to low temperature body, or from low temperature body to high temperature body? The second law of thermodynamics provides the answer. It predicts that heat will flow in the direction of decreasing temperature, that is, from body A to body B in Figure 4.1. This is consistent with our everyday experience that heat indeed flows from a region of high temperature to a region of low temperature. This was a simple example where our day to day experience was sufficient to determine the direction of heat flow. However, in many complex processes, where our experience is lacking, we have to rely on the second law of thermodynamics to determine the direction of spontaneous changes.

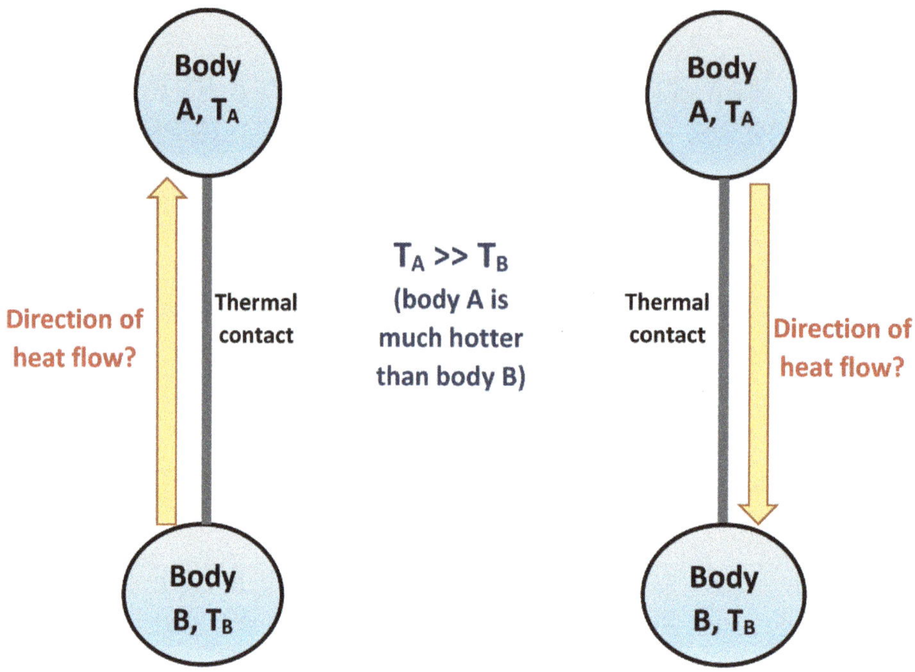

Figure 4.1 Which is the correct direction of heat flow? From B to A or from A to B? The second law tells us that the correct direction of heat flow is from A to B (figure on the right).

4.3 CLASSICAL STATEMENTS OF THE SECOND LAW

The second law of thermodynamics could be stated in several different but equivalent ways. The classical statements of the second law are the *Kelvin–Planck statement* and the *Clausius statement*. Before we describe the Kelvin–Planck and Clausius statements of the second law, it is necessary to define the terms such as heat reservoir, heat engine and thermal efficiency of a heat engine.

4.3.1 Heat Reservoir

Heat reservoir is simply a large body of uniform temperature. The removal or addition of unlimited quantities of heat to the reservoir has no appreciable effect on its temperature as its heat capacity is large. Thus, it is a constant temperature heat source or sink.

4.3.2 Heat Engine

Heat engine is a cyclic device that absorbs heat from a heat reservoir and converts part of the heat absorbed into work (mechanical energy).

4.3.3 Thermal Efficiency of a Heat Engine

The thermal efficiency of a heat engine is defined as:

$$\eta = W_{cycle}/Heat\ Input \tag{4.1}$$

4.3.4 Kelvin-Planck Statement of Second Law

The Kelvin–Planck statement of the second law says that "*It is impossible for a system operating in a cycle and interacting with a single heat reservoir to produce a positive amount of work in the surroundings*". According to the Kelvin–Planck statement, it is impossible to build a heat engine shown schematically in Figure 4.2 where heat absorbed from a heat reservoir is completely converted into work without altering the properties of the system. Note that the thermal efficiency of such an engine is one. Thus, the Kelvin-Planck statement implies that it is impossible to build a heat engine with a thermal efficiency of 100%.

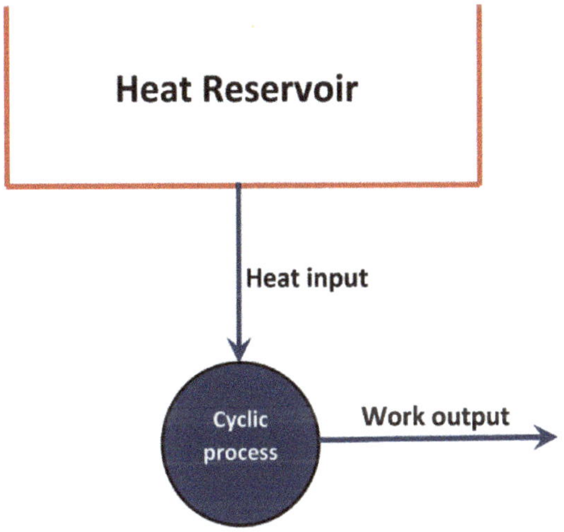

Figure 4.2 Impossible heat engine.

Mathematically, the Kelvin–Planck statement can be expressed as:

$$W_{cycle} \leq 0 \qquad \text{(single heat reservoir)} \tag{4.2}$$

where the system communicates thermally only with a single heat reservoir. The sign convention for work (W) used in this book is W = positive, if it is produced by the system (flows out of the system to surroundings) and W = negative, if it is absorbed by the system (flows into the system from surroundings). Thus, no cyclic process is possible where $W_{cycle} > 0$ using a single heat reservoir. The equality in Equation (4.2) is valid for a reversible process and the inequality is valid for an irreversible process. It should also be noted that if the cyclic process is irreversible, work flows from surroundings to the system ($W_{cycle} < 0$). As $Q_{cycle} = W_{cycle}$, the work that flows into the system is rejected as heat into the heat reservoir.

The Kelvin-Planck statement of second law can be expressed in an alternative way as: "*No process is possible whose sole effect is the conversion of heat absorbed by a system completely into work done by the system*". Note that heat absorbed by a system can be completely converted into work if that is not the sole effect of the process. For example, in a reversible isothermal expansion of an ideal gas, the heat absorbed by the system (ideal gas) is completely converted into work. However, the system properties are altered at the same time.

The heat engine shown in Figure 4.3 does not violate the second law. Now there are two heat reservoirs - hot and cold heat reservoirs. The system absorbs heat from a hot reservoir, converts part of it into work, and rejects part of the heat absorbed to a cold heat reservoir. The thermal efficiency of this heat engine is less than unity.

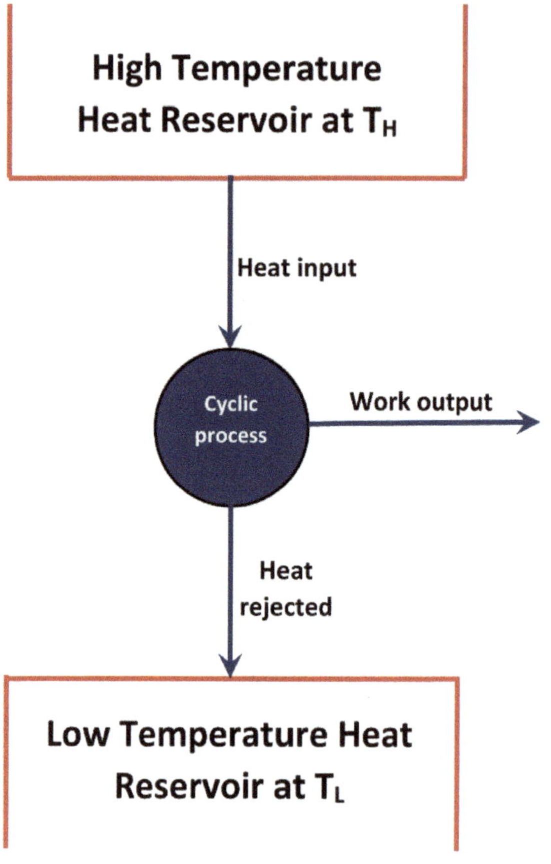

Figure 4.3 Heat engine that does not violate second law.

4.3.5 Clausius Statement of Second Law

The Clausius statement of the second law says that "*It is impossible for a system operating in a cycle to have its sole effect the transfer of heat from a low temperature heat reservoir to a high temperature heat reservoir*". According to the Clausius statement, it is impossible to construct a device based on the scheme shown schematically in Figure 4.4 where the sole result of the process is the transfer of heat from a cold body at low temperature (T_L) to a hot body at high temperature (T_H).

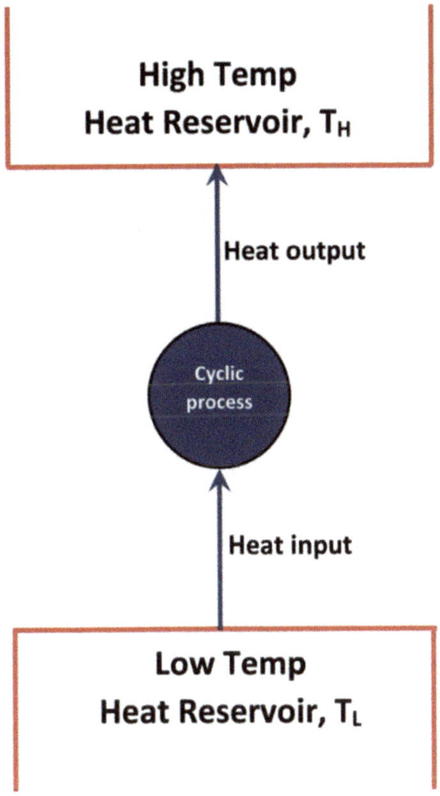

Figure 4.4 Impossible scheme.

The Clausius statement of second law can be expressed in an alternative way as: "*Heat always flows spontaneously in the direction of decreasing temperature*". When hot and cold bodies are brought into thermal contact with each other, heat flows spontaneously from a hot body to a cold body.

The scheme shown in Figure 4.5 does not violate the second law. Now the heat is transferred from a cold body at low temperature (T_L) to a hot body at high temperature (T_H) but that is not the sole effect of the process. The surroundings is losing work in order to carry out the process.

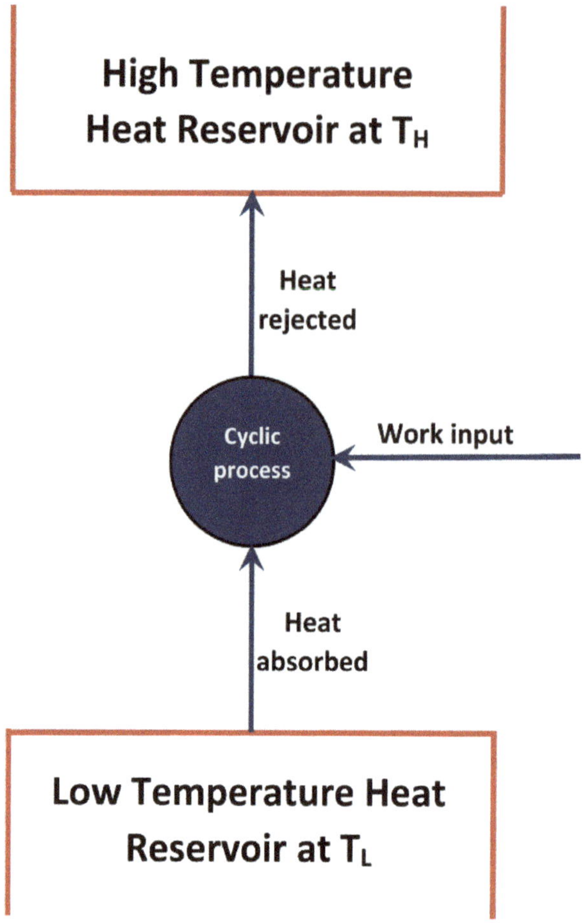

Figure 4.5 Scheme that does not violate the second law.

4.3.6 Equivalence of Kelvin-Planck and Clausius Statements

The Kelvin-Planck and Clausius statements of second law are equivalent. It can be readily shown that a violation of Kelvin-Planck statement implies a violation of the Clausius statement and vice versa.

Consider the cyclic processes shown in Figure 4.6 interacting with the same heat reservoirs. The cyclic process I violates the Kelvin-Planck statement, since it is producing work while receiving heat from a single heat reservoir at T_H. Let the cyclic process II remove heat Q_L from a low temperature heat reservoir at T_L and transfer heat in the amount of $Q_H + Q_L$ to a high temperature reservoir at T_H. Note that cyclic process II does violate the second law. The combination of the two cyclic processes I and II can be viewed as a cyclic process III that transfers heat in the amount of Q_L from a low temperature heat reservoir at T_L to a high temperature heat reservoir at T_H without requiring any work input from outside. The cyclic process III is in clear violation of the Clausius statement. Therefore, a violation of the Kelvin-Planck statement implies a violation of the Clausius statement.

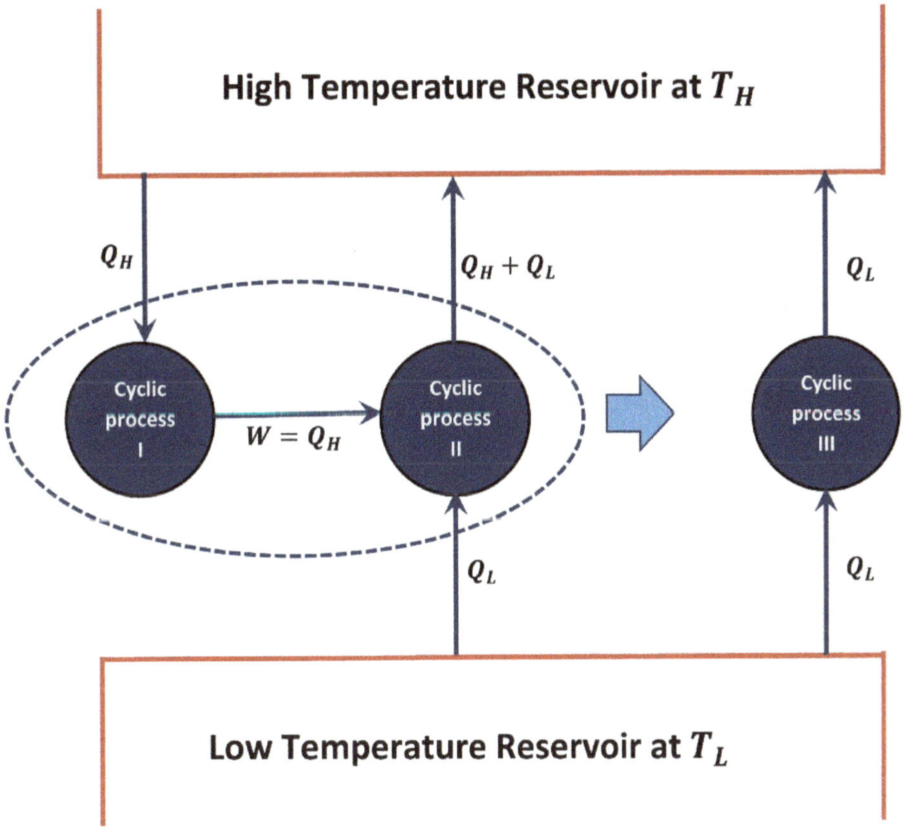

Figure 4.6 Violation of Kelvin-Planck statement implies violation of Clausius statement.

In a similar manner, one can show that a violation of Clausius statement results in the violation of the Kelvin-Planck statement. Consider the cyclic processes shown in Figure 4.7 interacting with the same heat reservoirs. The cyclic process I violates the Clausius statement, since the sole effect of this cyclic process is the transfer of heat Q_1 from a cold heat reservoir at T_L to a hot heat reservoir at T_H. Let the cyclic process II absorb heat Q_2 ($Q_2 > Q_1$ in magnitude) from a high temperature heat reservoir at T_H and convert part (Q_2-Q_1) of that heat to work and reject the remaining part Q_1 to a low temperature reservoir at T_L. Note that cyclic process II does violate the second law. The combination of the two cyclic processes I and II can be viewed as a cyclic process III that produces work while receiving heat from a single heat reservoir at T_H. The cyclic process III is in clear violation of the Kelvin-Planck statement. Therefore, a violation of the Clausius statement implies a violation of the Kelvin-Planck statement.

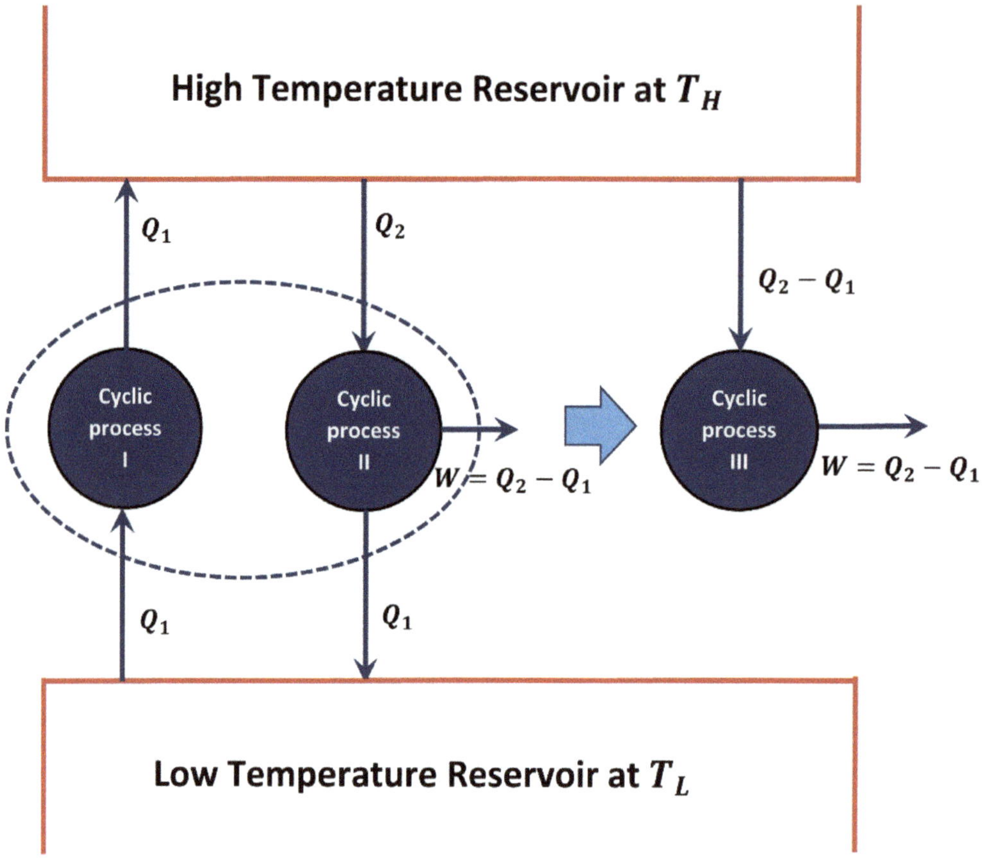

Figure 4.7 Violation of Clausius statement implies violation of Kelvin-Planck statement. Note that $Q_2 > Q_1$ in magnitude.

4.4 DISCOVERY OF A STATE PROPERTY CALLED ENTROPY

According to the Kelvin-Planck statement of second law, no heat engine can have a thermal efficiency of 100 percent. Some of the heat extracted by the heat engine from a hot reservoir must be rejected to a cold reservoir. Only a part of the heat extracted from a hot reservoir can be converted into work. Although it is impossible to have a thermal efficiency of a heat engine to be 100 percent as it violates the second law, it is of engineering interest to know if there is any upper limit on the thermal efficiency of feasible or possible heat engines which do not violate the second law. Do feasible or possible heat engines which communicate thermally with hot and cold heat reservoirs, and hence do not violate the second law, have a maximum in thermal efficiency? Interestingly, a heat engine that operates *reversibly* between two temperature levels (hot and cold heat reservoirs) happens to have the maximum efficiency. The thermal efficiency of an *irreversible* heat engine is always less than that of a *reversible* heat engine operating between the same temperature levels. We will explore these points and many other interesting aspects related to heat engines and the second law in the discussion to follow.

4.4.1 Carnot Heat Engine

Of all the possible or feasible heat engines operating between hot and cold heat reservoirs, the Carnot heat engine (see Figure 4.8) is the *most efficient*. It is a unique heat engine that produces work from heat using a *reversible cyclic process* that involves heat interactions with two heat reservoirs - hot and cold heat reservoirs. It is named after Sadi Carnot (1796-1832), a French military engineer, who devised it.

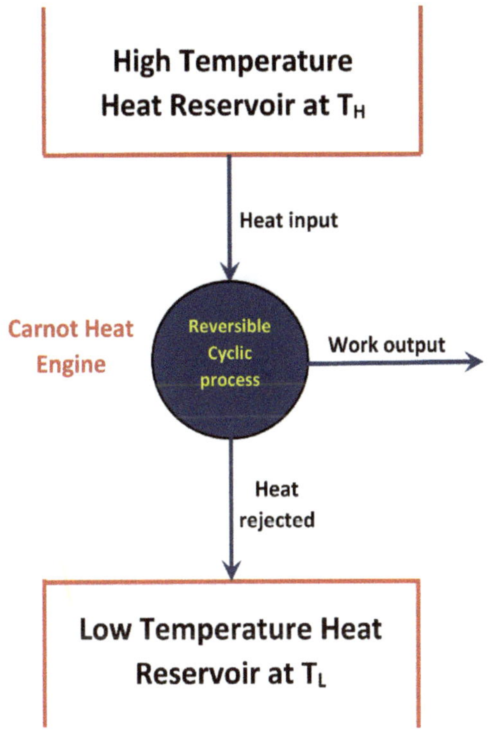

Figure 4.8 Carnot heat engine.

Figure 4.9 shows the Carnot cycle on a P-V diagram. The working material (system) of the Carnot heat engine undergoes this cyclic process to produce work. The Carnot cycle consists of four *reversible* steps:

1 → 2 : Reversible adiabatic compression of working material
2 → 3 : Reversible isothermal expansion of working material
3 → 4 : Reversible adiabatic expansion of working material
4 → 1 : Reversible isothermal compression of working material

The system absorbs heat Q_H along the path **2 → 3** from a high temperature heat reservoir at T_H and rejects some heat Q_L along the path **2 → 3** to a low temperature heat reservoir at T_L. No heat is exchanged along adiabatic paths **1 → 2** and **3 → 4.** The net work produced by the system in executing the cycle is the area enclosed by cycle **1 → 2 → 3 → 4 → 1.**

It is important to realize that there is one and only one reversible cyclic process that can operate between two heat reservoirs (hot and cold), and that is the Carnot cycle. There cannot be any other reversible cyclic process operating between two heat reservoirs. All other reversible cyclic processes (other than the Carnot cycle) must involve more than two heat reservoirs. Thus, Carnot cycle is a unique reversible cycle that absorbs heat from a hot reservoir and rejects part of the heat absorbed to a cold reservoir while producing work at the same time.

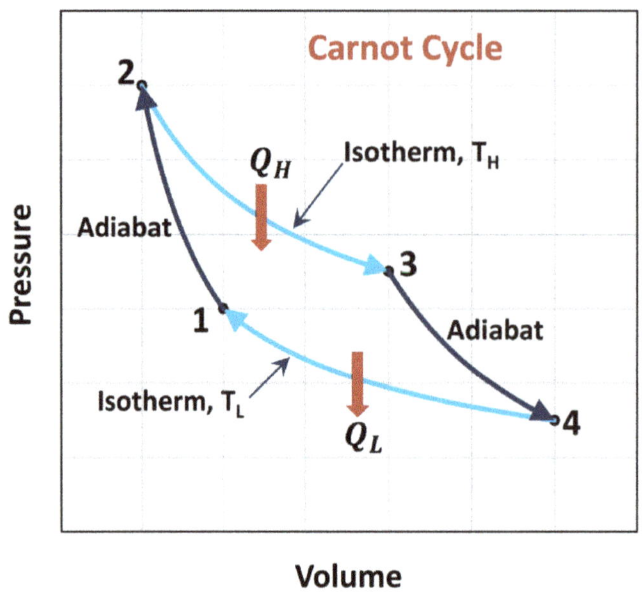

Figure 4.9 Carnot cycle on P-V diagram.

Figure 4.10 demonstrates the execution of a Carnot cycle for gas as a working material. The gas contained in a piston/cylinder assembly undergoes four reversible processes corresponding to four steps of a Carnot cycle. The piston and cylinder walls are assumed to be thermally non-conducting. The heat enters or leaves the system (gas) through the base of the cylinder. The gas is compressed reversibly and adiabatically in process 1-2. In this process, work is done on the gaseous system to compress it to high pressure and there is no heat exchange involved. The gas is then expanded isothermally and reversibly in process 2-3. In this process, heat Q_H is absorbed by the system from high temperature heat reservoir at T_H and work is produced by the system. In the next step, process 3-4, the gas is expanded reversibly and adiabatically and work is produced without any heat exchange. Finally, in process 4-1, the gas is compressed isothermally and reversibly. In this process, work is required and heat Q_L is rejected by the system to a cold heat reservoir at T_L. This completes the execution of a Carnot cycle. There is a net production of work by the system and some of the heat that is absorbed from a hot reservoir is rejected to a cold heat reservoir.

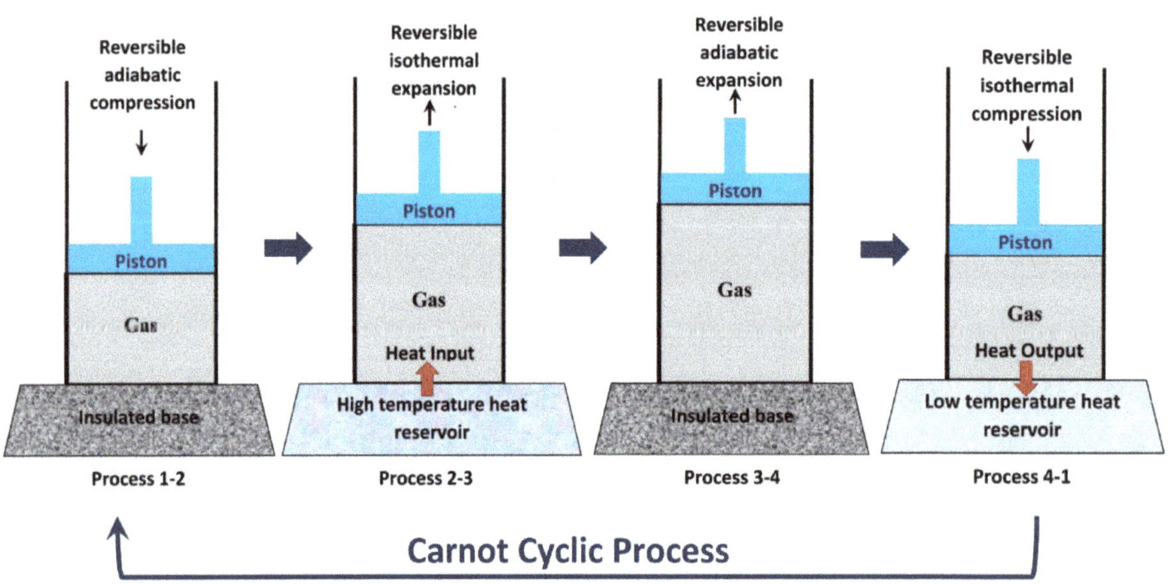

Figure 4.10 Gas contained in a piston/cylinder assembly executing a Carnot cycle.

4.4.2 Thermal Efficiency of a Carnot Cycle

The Carnot heat engine does not violate the Kelvin-Planck statement of second law. As required by the second law, it communicates thermally with hot and cold heat reservoirs. Only part of the heat absorbed from a hot reservoir is converted to work. Thus, the thermal efficiency of a Carnot heat engine is less than 100 percent.

Let us now determine the thermal efficiency of a Carnot cycle using the working material as an ideal gas (system). The thermal efficiency of a heat engine is defined in Equation (4.1). It is the ratio of net work produced by the system in executing the cycle divided by the heat input to the cyclic process.

Referring to Figure 4.9, the amounts of heat and work interactions between the system (ideal gas) and the surroundings, and the internal energy changes of the system, upon execution of the Carnot cycle are as follows:

1 → 2 : Reversible adiabatic compression of working material

$$Q_{12} = 0\ ;\ \Delta U_{12} = -W_{12} = nc_v(T_H - T_L) \tag{4.3}$$

2 → 3 : Reversible isothermal expansion of working material

$$\Delta U_{23} = 0\ ; W_{23} = Q_{23} = nRT_H ln\left(\frac{V_3}{V_2}\right) \tag{4.4}$$

3 → 4 : Reversible adiabatic expansion of working material

$$Q_{34} = 0\ ;\ \Delta U_{34} = -W_{34} = nc_v(T_L - T_H) \tag{4.5}$$

4 → 1 : Reversible isothermal compression of working material

$$\Delta U_{41} = 0\ ; W_{41} = Q_{41} = nRT_L ln\left(\frac{V_1}{V_4}\right) \tag{4.6}$$

Note that $Q_{23} = Q_H$ and $Q_{41} = -Q_L$ where Q_H and Q_L are the absolute (non-negative) values.

The changes for the whole cycle are the sum of the changes in the individual steps. Thus,

$$\oint dU = \Delta U_{12} + \Delta U_{23} + \Delta U_{34} + \Delta U_{41} = nc_v(T_H - T_L) + nc_v(T_L - T_H) = 0 \tag{4.7}$$

$$\oint \delta Q = Q_{12} + Q_{23} + Q_{34} + Q_{41} = nRT_H ln\left(\frac{V_3}{V_2}\right) + nRT_L ln\left(\frac{V_1}{V_4}\right) \tag{4.8}$$

$$\oint \delta W = W_{12} + W_{23} + W_{34} + W_{41} = -nc_v(T_H - T_L) + nRT_H ln\left(\frac{V_3}{V_2}\right) + nc_v(T_L - T_H) + nRT_L ln\left(\frac{V_1}{V_4}\right) = nRT_H ln\left(\frac{V_3}{V_2}\right) + nRT_L ln\left(\frac{V_1}{V_4}\right) \tag{4.9}$$

As expected, the cyclic change in the state property U is zero and $Q_{cycle} = W_{cycle}$ in agreement with the first law.

The thermal efficiency of the Carnot cycle is:

$$\eta = W_{cycle}/Q_{23} = \frac{\left[nRT_H ln\left(\frac{V_3}{V_2}\right) + nRT_L ln\left(\frac{V_1}{V_4}\right)\right]}{nRT_H ln\left(\frac{V_3}{V_2}\right)} = 1 + \frac{T_L ln\left(\frac{V_1}{V_4}\right)}{T_H ln\left(\frac{V_3}{V_2}\right)} \quad (4.10)$$

We can now make use of the following relations for a reversible adiabatic process involving idea gas of constant heat capacity ratio γ :

$$T_H V_2^{\gamma-1} = T_L V_1^{\gamma-1} \quad (4.11)$$

$$T_H V_3^{\gamma-1} = T_L V_4^{\gamma-1} \quad (4.12)$$

From Equations (4.11) and (4.12), we can get:

$$\frac{V_3}{V_2} = \frac{V_4}{V_1} \quad (4.13)$$

Upon substitution of Equation (4.13) into Equation (4.10), we have:

$$\eta = 1 + \frac{T_L ln\left(\frac{V_1}{V_4}\right)}{T_H ln\left(\frac{V_4}{V_1}\right)} = 1 - \frac{T_L}{T_H} \quad (4.14)$$

Thus the thermal efficiency of a Carnot cycle is dependent only on the temperatures of the hot and cold heat reservoirs. With the increase in hot reservoir temperature T_H, the efficiency of a Carnot cycle (engine) increases as shown in Figure 4.11.

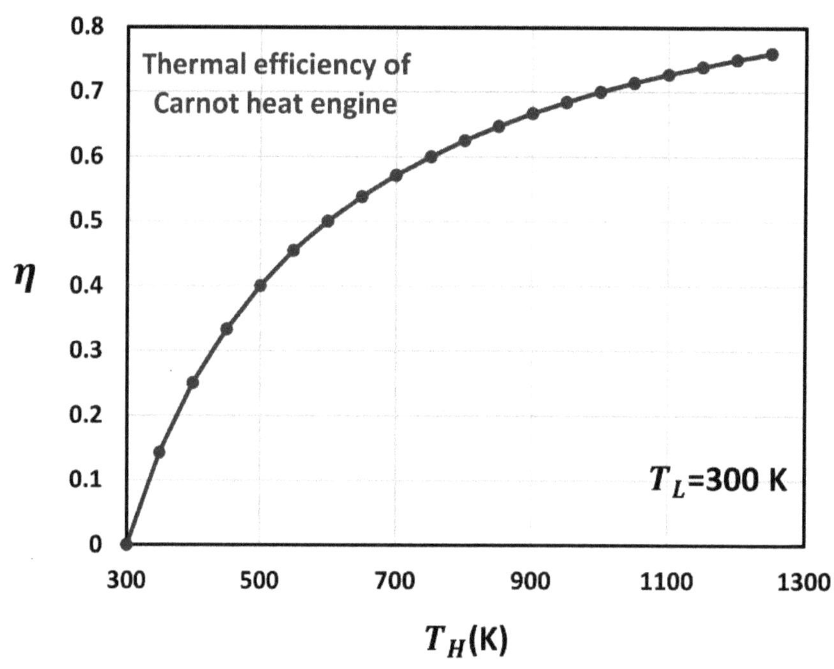

Figure 4.11 Variation of thermal efficiency of a Carnot heat engine with hot reservoir temperature keeping cold reservoir temperature constant at 300 K.

According to the Carnot efficiency formula, Equation (4.14), the work produced in a cycle is $Q_H\left(1 - \frac{T_L}{T_H}\right)$ where Q_H is the heat supplied by the hot reservoir. This result allows us to define the *quality* of heat. Not all the heat supplied by the hot reservoir can be converted into work. Only a fraction, that is, $\left(1-\frac{T_L}{T_H}\right)$, of the heat supplied can be converted into work. Thus $\left(1-\frac{T_L}{T_H}\right)$ is the quality factor of heat. Higher the temperature of the hot reservoir, higher is the quality of heat available from the hot reservoir.

4.4.3 The Cyclic Integral $\oint \frac{\delta Q}{T}$ for Carnot Cycle

The thermal efficiency of a Carnot heat engine or cycle (Figure 4.9) can be expressed as:

$$\eta = W_{cycle}/Heat\ Input = (Q_{23}+Q_{41})/Q_{23} = 1+(Q_{41}/Q_{23}) \tag{4.15}$$

Note that $W_{cycle} = Q_{cycle}$. Comparing Equation (4.14) and Equation (4.15), we have:

$$\eta = 1+(Q_{41}/Q_{23}) = 1-\frac{T_L}{T_H} \tag{4.16}$$

Upon re-arrangement, Equation (4.16) gives:

$$\frac{Q_{23}}{T_H}+\frac{Q_{41}}{T_L} = 0 \tag{4.17}$$

Equation (4.17) can also be expressed in terms of the absolute values of heat, that is, $Q_{23} = Q_H$ and $Q_{41} = -Q_L$, as:

$$\frac{Q_H}{T_H} = \frac{Q_L}{T_L} \tag{4.18}$$

Let us now determine the following cyclic integral for the Carnot cycle:

$$\oint \frac{\delta Q}{T_{Surr}} = \frac{Q_{23}}{T_H}+\frac{Q_{41}}{T_L} \tag{4.19}$$

where T_{Surr} is the surroundings (heat reservoirs) temperature. Upon comparison of Equations (4.17) and (4.19), we get:

$$\oint \frac{\delta Q}{T_{Surr}} = 0 \tag{4.20}$$

As Carnot cycle is a reversible cycle, there are no temperature gradients within the system and the system temperature (T) is the same as reservoir temperature during heat transfer, that is, $T = T_{Surr}$. Consequently, Equation (4.20) can be re-written in terms of the system temperature as:

$$\oint \frac{\delta Q}{T} = 0 \tag{4.21}$$

Thus, the cyclic integral $\oint \frac{\delta Q}{T}$ is zero for a Carnot cycle using ideal gas as a working substance.

4.4.4 What Happens if Carnot Heat Engine is Operated in Reverse Direction?

In a Carnot heat engine, the working substance executes the Carnot cycle in the direction $\mathbf{1 \rightarrow 2 \rightarrow 3 \rightarrow 4 \rightarrow 1}$, as shown in Figure 4.9. The Carnot heat engine can be operated in the opposite or reverse direction, that is, the working substance now executes the Carnot cycle in the direction $\mathbf{1 \rightarrow 4 \rightarrow 3 \rightarrow 2 \rightarrow 1}$. Carnot heat engine, operated in reverse, is called a Carnot *refrigerator* as it sucks up heat from a cold region and rejects it to a high temperature region. This is what happens in our home refrigerators. The heat leaked into the cold refrigerator box is continuously removed and rejected to a high temperature environment (room).

When the same Carnot heat engine is operated in reverse as a Carnot refrigerator, the quantities Q_H , Q_L and W remain exactly the same except that their direction is reversed, as shown in Figure 4.12.

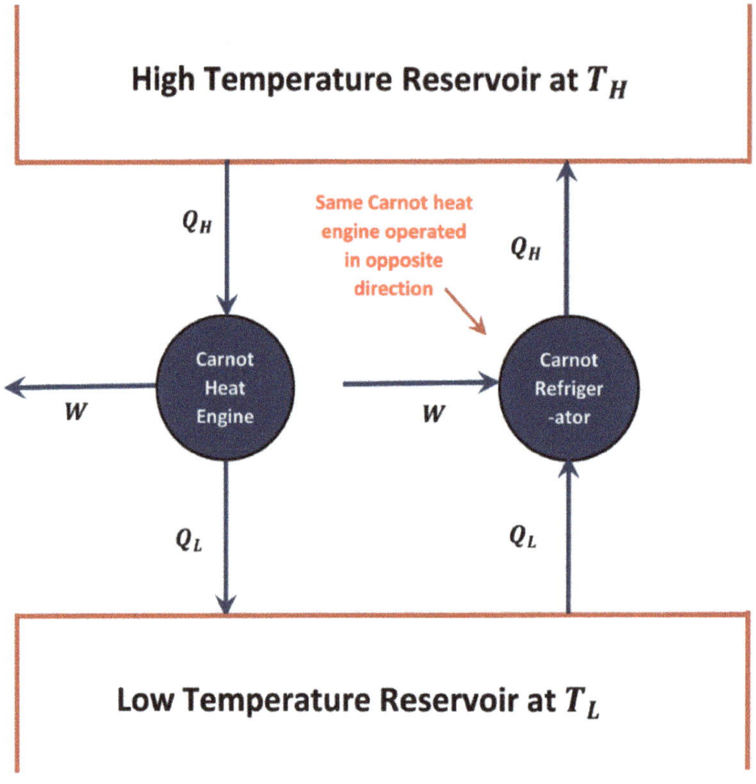

Figure 4.12 Same Carnot heat engine operated in reverse as a Carnot refrigerator.

4.4.5 Carnot Theorems

There are two Carnot theorems. They are described below.

First Carnot Theorem

The thermal efficiencies of all Carnot heat engines (reversible heat engines) operating between the same heat reservoirs (hot and cold reservoirs) are the same, independent of the nature of the working substance of the engine.

Proof of First Carnot Theorem

To prove the first Carnot theorem, we consider two Carnot heat engines operating between the same heat reservoirs, as shown in Figure 4.13. We assume that the first Carnot heat engine (CHE-1) is more efficient than the second Carnot heat engine (CHE-2). Thus, for the same heat input Q_H:

$$W_1 > W_2 \tag{4.22}$$

$$Q_{L,1} < Q_{L,2} \tag{4.23}$$

Now let the CHE-2 be reversed and operate as a refrigerator, as shown in Figure 4.14. The combination of CHE-1 and refrigerator gives a heat engine that produces work while receiving heat from a single heat reservoir in violation of the Kelvin-Planck statement of second law. Therefore, our assumption of Carnot heat engines having different efficiencies is incorrect and we can conclude that all Carnot heat engines operating between the same temperature levels must have the same efficiencies, regardless of the nature of their working substances.

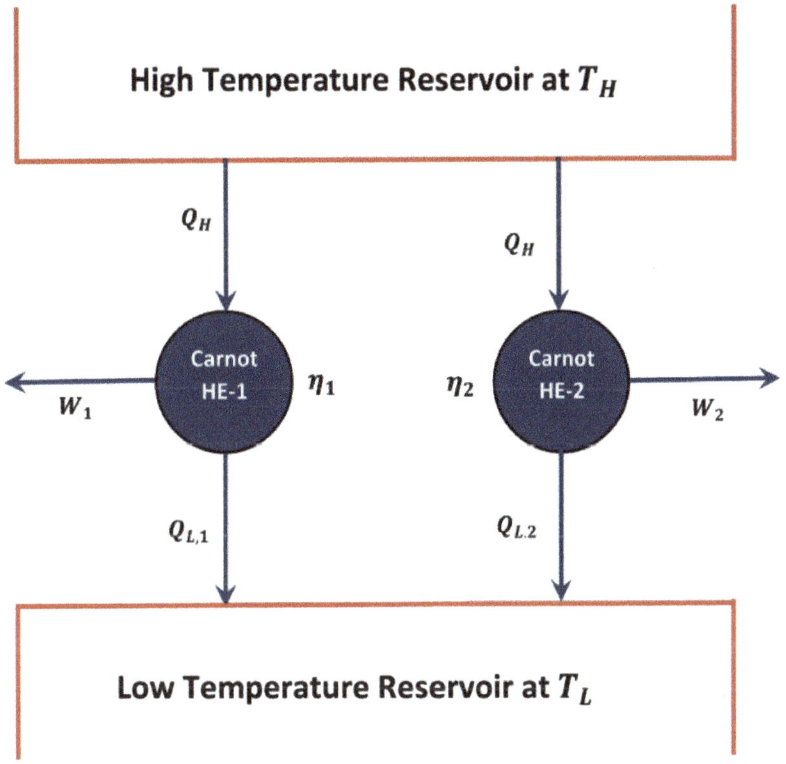

Figure 4.13 Two Carnot heat engines assumed to have different thermal efficiencies ($\eta_1 > \eta_2$).

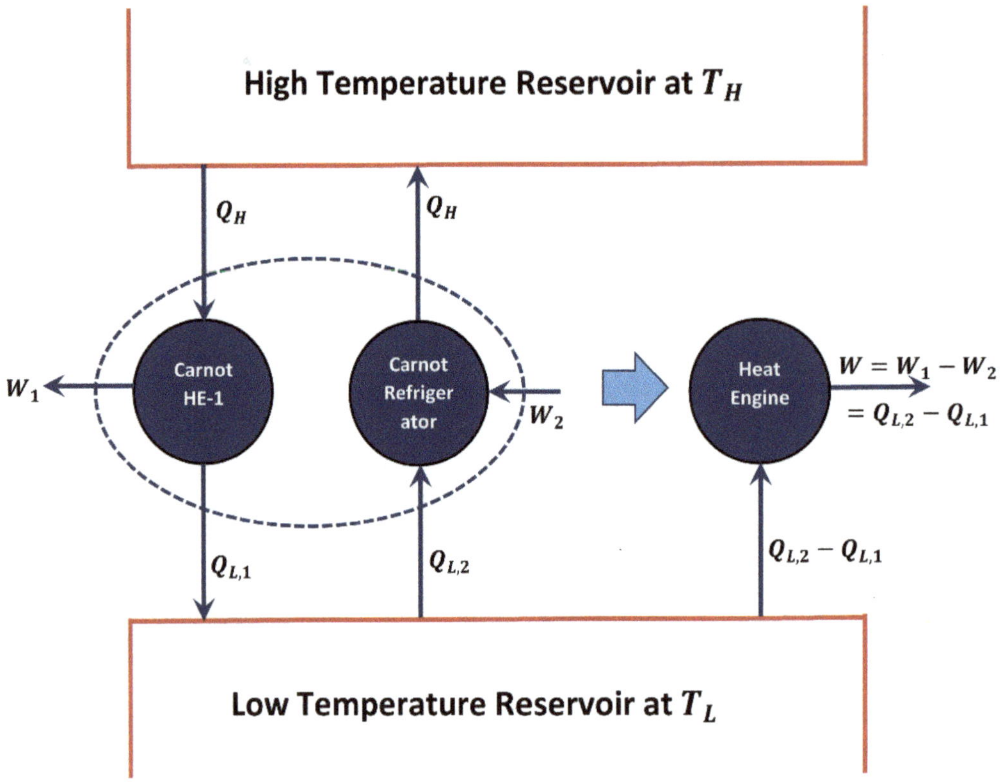

Figure 4.14 Assumption of different thermal efficiencies of Carnot heat engines leads to violation of the Kelvin-Planck statement of second law.

Second Carnot Theorem

The thermal efficiency of an irreversible heat engine is always less than that of a reversible heat engine operating between the same hot and cold heat reservoirs.

Proof of Second Carnot Theorem

To prove the second Carnot theorem, we consider two heat engines operating between the same heat reservoirs – one irreversible heat engine and one reversible (Carnot) heat engine. Each engine is supplied the same heat input Q_H. We assume that the irreversible heat engine is more efficient than the Carnot heat engine. Thus, for the same heat input Q_H:

$$W_{irr} > W_{rev} \tag{4.24}$$

$$Q_{L,irr} < Q_{L,rev} \tag{4.25}$$

Now let the reversible heat engine be reversed and operate as a Carnot refrigerator, as shown in Figure 4.15. The combination of irreversible heat engine (assumed more efficient) and Carnot refrigerator gives a heat engine that produces work while receiving heat from a single heat reservoir in violation of the Kelvin-Planck statement of second law. Therefore, our assumption of irreversible heat engine being more efficient than reversible heat engine operating between the same temperature levels is incorrect as it leads to the violation of Kelvin-Planck statement of second law.

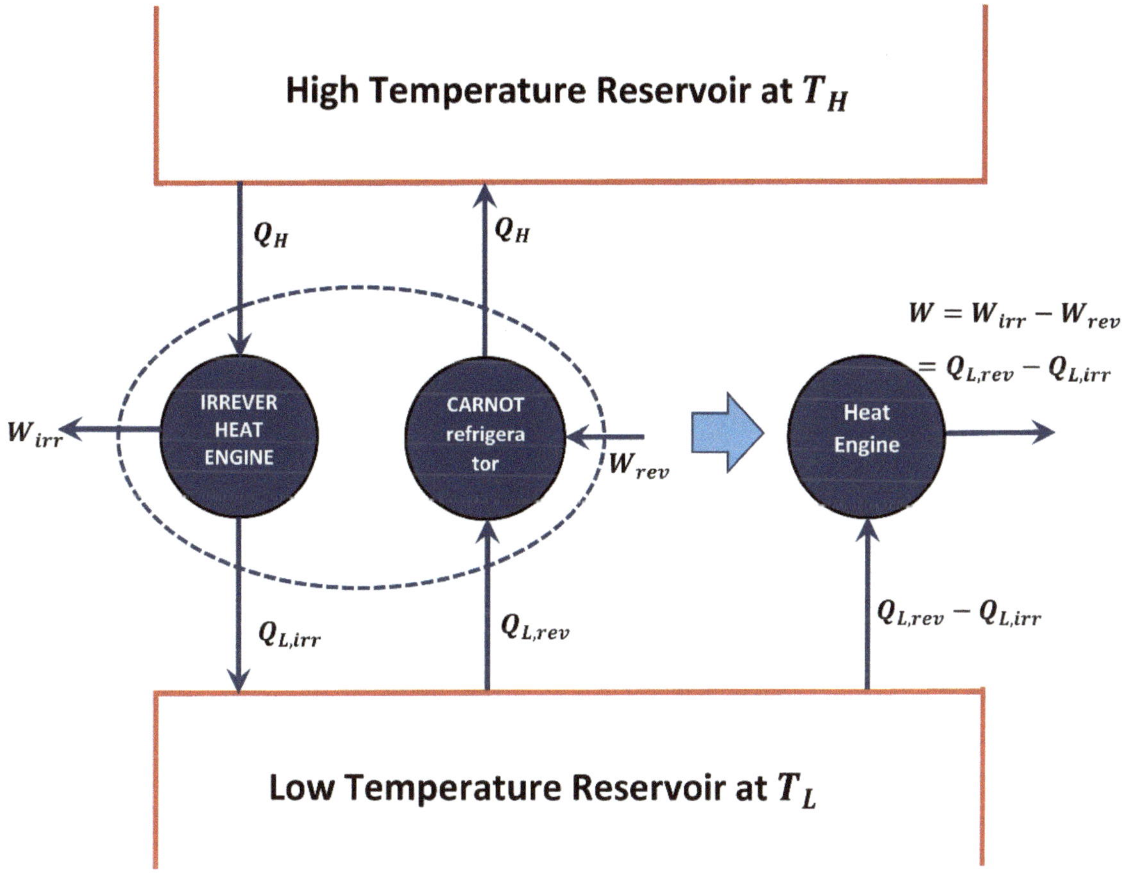

Figure 4.15 Assumption of an irreversible heat engine having thermal efficiency higher than that of the Carnot heat engine leads to violation of the Kelvin-Planck statement of second law.

4.4.6 Corollaries of Carnot Theorems

From the *first* Carnot theorem, it follows that:

$$\oint \frac{\delta Q}{T} = 0 (\text{Carnot cycle, any working substance}) \quad (4.26)$$

This result is valid for all Carnot cycles, independent of the nature of the working substance. It is a generalization of Equation (4.21).

From the *second* Carnot theorem, it follows that:

$$\oint \frac{\delta Q}{T_{Surr}} < 0 \quad (4.27)$$

This result is valid for all irreversible heat engines operating between hot and cold heat reservoirs, independent of the nature of the working substance.

Combining Equations (4.26) and (4.27), one can write:

$$\oint \frac{\delta Q}{T_{Surr}} \leq 0 \quad (4.28)$$

The equality sign applies to a reversible heat engine where $T_{Surr} = T$ and the inequality sign applies to an irreversible heat engine.

4.4.7 Clausius Inequality

Thus far our focus of attention has been heat engines – cyclic devices that produce work while communicating thermally with hot and cold heat reservoirs. Here we will discuss the *Clausius inequality* and give its proof. The *Clausius inequality* focuses on the *system* undergoing a cyclic process without giving any attention to the heat reservoirs. This does not mean that heat interactions between the system and surroundings (heat reservoirs) are ignored. The heat transfer is now taken into consideration at the *boundary* of the system itself.

The Clausius inequality can be stated mathematically as:

$$\oint \left(\frac{\delta Q}{T}\right)_b \leq 0 \tag{4.29}$$

where the subscript "b" indicates that the integrand is determined at the boundary of the system undergoing the cyclic process. Thus, δQ is the amount of heat transfer at a differential portion of the system boundary where T is the absolute temperature. The integral in the Clausius inequality is carried out over the entire boundary of the system and over the full cycle executed by the system.

The Clausius inequality is valid for any cyclic process, independent of the nature of the working material. For an *internally reversible* (no irreversibilities within a system) cyclic process, the equality sign in Equation (4.29) applies. The inequality sign applies when the cyclic process is *internally irreversible*. Note that the Clausius inequality expressed in the form of Equation (4.29) is concerned with irreversibilities within the system alone. The external irreversibilities are not considered. This does not imply that the external irreversibilities are absent. For example, there may be present a finite temperature gradient between the heat reservoir and the system boundary. In such a case, the heat transfer from heat reservoir to system boundary will be irreversible.

Proof of Clausius Inequality

Consider a closed system in combination with a Carnot (reversible) heat engine as shown in Figure 4.16. The heat engine extracts δQ_R amount of heat from the heat reservoir and produces δW_{HE} amount of work and rejects heat δQ to the system boundary where the absolute temperature is T_b.

The system produces δW amount of work. The total work produced by the combined system of Carnot heat engine and the closed system is:

$$\delta W_C = \delta W_{HE} + \delta W \tag{4.30}$$

The first law, when applied to the combined system, gives:

$$dE_C = \delta Q_R - \delta W_C \tag{4.31}$$

where E_C is the total energy of the combined system. For the Carnot heat engine, the following relationship is applicable:

$$\frac{\delta Q_R}{T_R} = \frac{\delta Q}{T_b} \tag{4.32}$$

From Equations (4.31) and (4.32), it follows that:

$$\delta W_C = \delta Q_R - dE_C = \left(\frac{\delta Q}{T_b}\right) T_R - dE_C \tag{4.33}$$

Now allow the closed system to undergo a complete cycle. As the Carnot heat engine is cyclic and the closed system is undergoing a cycle, this implies that the combined system of Carnot heat engine and closed system (see Figure 4.16) also undergoes a cyclic change. Consequently,

$$W_C = \oint \delta W_C = \oint \left(\frac{\delta Q}{T_b}\right) T_R - \oint dE_C \tag{4.34}$$

As the cyclic integral of a state property E_C is zero, Equation (4.34) simplifies to

$$W_C = T_R \oint \left(\frac{\delta Q}{T_b}\right) \tag{4.35}$$

Now we make use of the Kelvin-Planck statement of the second law. According to the Kelvin-Planck statement, a system undergoing a cyclic process and communicating thermally with a single heat reservoir, cannot produce work in the surroundings. Thus, for the combined system of Figure 4.16, we have:

$$W_C = T_R \oint \left(\frac{\delta Q}{T_b}\right) \leq 0 \tag{4.36}$$

As T_R has some positive value, Equation (4.36) gives:

$$\oint \left(\frac{\delta Q}{T}\right)_b \leq 0 \tag{4.29}$$

The equality sign is valid when the cyclic change is reversible and the "less than" sign is valid when the cyclic change is irreversible. As Carnot heat engine is reversible in nature, the only possible location of irreversibilities in the combined system of Figure 4.16 is the closed system. Thus, the Clausius inequality is applicable to closed systems undergoing cyclic processes with "equal to" sign for internally reversible cyclic processes and "less than" sign for internally irreversible cyclic processes.

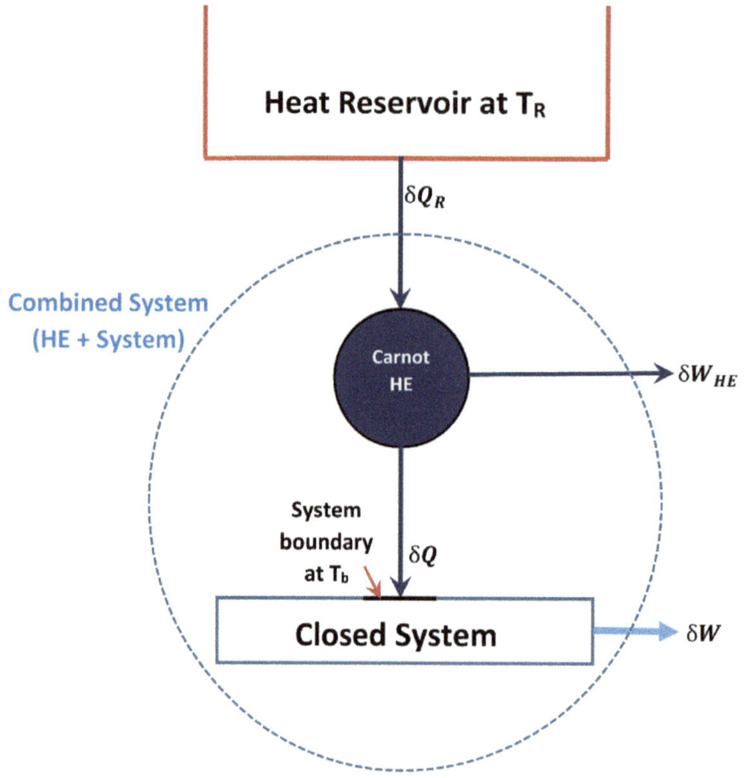

Figure 4.16 Combination of Carnot heat engine and closed system considered to prove the Clausius Inequality.

4.4.8 A New State Property called Entropy

Consider any thermodynamic property (state property) of a system, say volume (V). The cyclic integral of a state property is always zero as the initial and final states are the same. Thus,

$$\oint dV = 0 \tag{4.37}$$

In general,

$$\oint dM = 0 \tag{4.38}$$

where M is any state property.

The differential whose cyclic integral is zero is referred to as an *exact differential* and is denoted as "d". The differentials of all thermodynamic (state) properties are exact differentials. The most important characteristic of exact differentials is that their integration is independent of the path of integration and depends only on the initial and final states. Thus,

$$\int_1^2 dM = M_2 - M_1 \tag{4.39}$$

Let us now consider a closed system undergoing an internally reversible cyclic process. In this case, the Clausius inequality, Equation (4.29), gives:

$$\oint \left(\frac{\delta Q}{T}\right)_{\substack{int \\ rev}} = 0 \tag{4.40}$$

Note that for an internally reversible process (abbreviated as int rev), the temperature is uniform throughout the system and the system boundary temperature is the same as system temperature. Hence there is no need to use "b" referring to the boundary of the system.

As the cyclic integral of $\delta Q/T$ for a closed system undergoing *any* internally reversible process is zero, it implies that $\delta Q/T$ is an exact differential of some state property of the system. This new state property is called *entropy*. Thus, the defining relationship for entropy is:

$$dS = \left(\frac{\delta Q}{T}\right)_{\substack{int\\rev}} \tag{4.41}$$

As entropy is a state property, the change in entropy of a system is dependent only on the initial and final states of the system, regardless of the path taken by the system in going from the initial state to the final state. As an example, consider two processes resulting in the same change in state as shown in Figure 4.17. In one case, the change in state is caused by some internally reversible process and in the other case, the same change in state is caused by some internally irreversible process. The entropy change of the system $\Delta S = \int_1^2 dS$ is the same in both processes. However, there are major differences in the two processes as indicated below:

$$\Delta S = \int_1^2 \frac{\delta Q}{T} \text{ (internally reversible process)} \tag{4.42}$$

$$\Delta S \neq \int_1^2 \frac{\delta Q}{T} \text{ (internally irreversible process)} \tag{4.43}$$

It should be noted that entropy is an extensive property with the units of J/K.

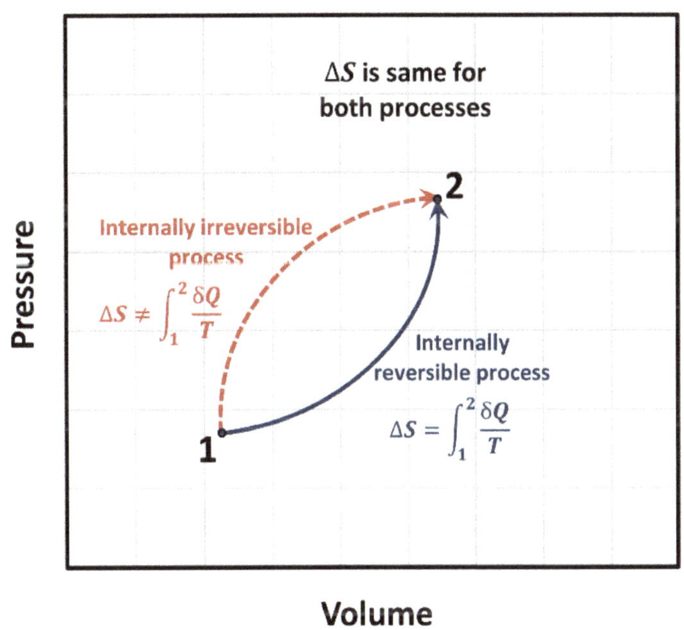

Figure 4.17 Entropy change is the same when initial and the final states are fixed, regardless of the process.

Now consider a cyclic process shown in Figure 4.18. The system first undergoes a change in state from $1 \rightarrow 2$ by some internally irreversible process. The system then returns to the original state from $2 \rightarrow 1$ by some internally reversible process. Thus, the system executes a cyclic process which is to be considered irreversible as part of it executed irreversibly. Consequently, the following Clausius inequality applies:

$$\oint \left(\frac{\delta Q}{T}\right)_b < 0 \tag{4.44}$$

When applied to the cyclic process of Figure 4.18, we have:

$$\oint \left(\frac{\delta Q}{T}\right)_b = \int_1^2 \left(\frac{\delta Q}{T}\right)_b + \int_2^1 \left(\frac{\delta Q}{T}\right)_{\substack{int\\rev}} < 0 \tag{4.45}$$

$$\int_1^2 \left(\frac{\delta Q}{T}\right)_b + \int_2^1 dS < 0 \tag{4.46}$$

$$\Delta S > \int_1^2 \left(\frac{\delta Q}{T}\right)_b \tag{4.47}$$

Thus, the change in entropy in going from state 1 → 2 is larger than $\int_1^2 \left(\frac{\delta Q}{T}\right)_b$ where the integral is evaluated at the system boundary for the internally irreversible process 1 → 2.

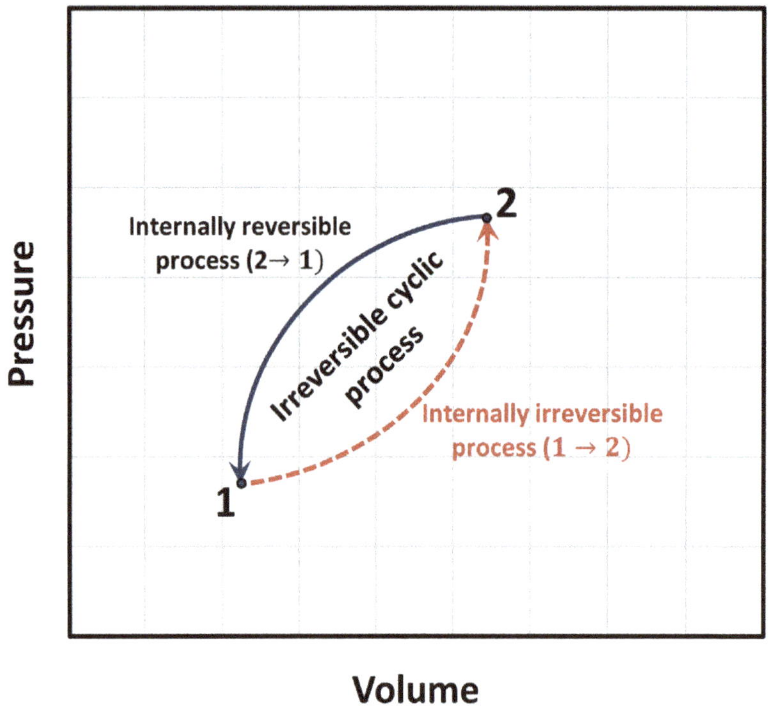

Figure 4.18 Irreversible cyclic process.

We can now interpret the inequality given by Equation (4.47) in the following manner. The integral $\int_1^2 \left(\frac{\delta Q}{T}\right)_b$ represents the net inward flow of entropy to the system through heat transfer at the boundary. The change in entropy of the system ΔS is larger than the net inward flow of entropy through the boundary due to *generation of entropy*. Thus, the inequality represented by Equation (4.47) can be reinterpreted as:

$$\Delta S = S_G + \int_1^2 \left(\frac{\delta Q}{T}\right)_b \tag{4.48}$$

where S_G is the entropy generated within the system due to irreversibilities. $S_G > 0$ when internal irreversibilities are present. $S_G = 0$ when the process is internally reversible (see Equation (4.42)). S_G can never be negative. From Equations (4.47) and (4.48), it follows that:

$$S_G = \Delta S - \int_1^2 \left(\frac{\delta Q}{T}\right)_b \geq 0 \tag{4.49}$$

The "equal to" sign is valid for internally reversible process and "greater than" sign is valid for internally irreversible process.

4.5 PRINCIPLE OF ENTROPY INCREASE

The principle of entropy increase can be stated as follows:
"*For any spontaneous process occurring within an isolated system, the entropy of the system always increases*".

Note that all spontaneous (natural or real) processes are irreversible processes. In the special case of a *reversible* process, the entropy of an isolated system remains constant.

The principle of entropy increase follows from Equation (4.49). For an isolated system, no heat and work transfers occur, that is, $\int_1^2 \left(\frac{\delta Q}{T}\right)_b = 0$. Consequently,

$$S_{G,isol} = (\Delta S)_{isol} \geq 0 \tag{4.50}$$

The entropy generated within the system due to irreversibilities goes into entropy increase of the system. There is no entropy generated in a reversible process and hence $(\Delta S)_{isol} = 0$ in this case.

4.6 ENTROPY AS A MEASURE OF DISORDERLINESS OF A SYSTEM

Entropy is a measure of randomness or disorderliness of a system. For a given macroscopic equilibrium state, there are a large number of possible microstates. While the macrostate of a system, characterized by system properties such as P, T, V, H etc., remains the same, the microstate is constantly changing as the molecules of a system are constantly redistributing in location and energy. The disorderliness of a system is characterized in terms of the number of microstates Ω available to the system for a given macroscopic equilibrium state. Higher the Ω, more disorderly is the system.

Boltzmann proposed the following relation to relate entropy to the number of microstates Ω:

$$S = k \ln \Omega \tag{4.51}$$

where k is Boltzmann's constant (=1.3806 x 10^{-23} J/K).

In a spontaneous process occurring in an isolated system, the number of available microstates to the system increases resulting in an increase in entropy.

4.7 SECOND LAW OF THERMODYNAMICS IN TERMS OF ENTROPY

The principle of entropy increase stated for an isolated system can be generalized to a closed system interacting with its surroundings via heat and work transfers. The closed system and its surroundings together form an isolated system or thermodynamic universe, as shown in Figure 4.19 below:

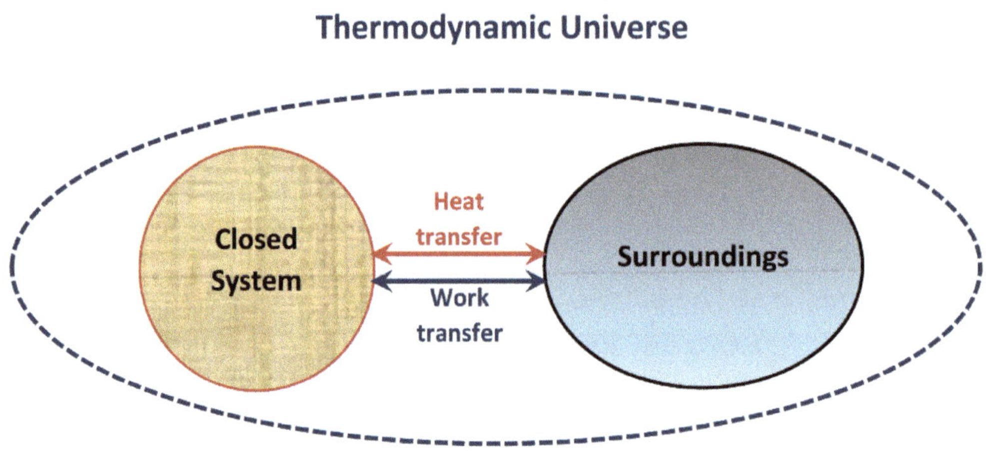

Figure 4.19 Thermodynamic universe (system plus surroundings) is an isolated system.

Thus, Equation (4.50) can be generalized as:

$$S_{G,total} = (\Delta S)_{universe} = (\Delta S)_{sys} + (\Delta S)_{surr} \geq 0 \tag{4.52}$$

Equation (4.52) is a mathematical statement of the second law in terms of entropy. Thus, the second law states that "All real (natural, spontaneous) processes are accompanied by an increase in the entropy of the universe". Note that $(\Delta S)_{sys}$ or $(\Delta S)_{surr}$ may be positive or negative but their sum will always be positive for spontaneous processes. One can also say that only those processes are feasible processes which lead to an increase in the entropy of the thermodynamic universe (system + surroundings).

4.8 ENTROPY AND TIME

As time progresses, spontaneous processes occur in the universe resulting in an increase in the entropy of the universe. In other words, entropy of the universe always increases with time. As entropy and time increase in the same direction (see Figure 4.20), it is often stated that "*entropy is an arrow of time*". Entropy measurements can be used to distinguish the past from the present. The present is more disorderly (high entropy) as compared with the past (low entropy). The future will be even more disorderly than the present.

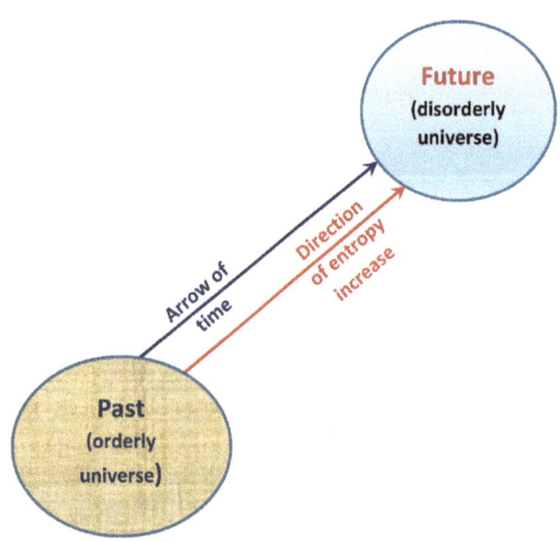

Figure 4.20 The directions of increase in entropy and progress in time are the same.

4.9 THERMODYNAMIC INTERPRETATION OF THE DEATH OF THE UNIVERSE

The entropy of the universe will continue to increase as time progresses due to spontaneous processes. Eventually, all the driving forces or gradients for spontaneous processes will disappear resulting in the final equilibrium state of the universe where the entropy is maximum. Thus, the universe will eventually end up in a so-called "*heat death*" in which everywhere the temperature will be the same. There will be no hot bodies such as our sun and no cold bodies such as our planet earth as they will eventually reach thermal equilibrium. This will be a state of maximum disorder. All the energy of the universe will end up as random motion energy of molecules and atoms. However, it may take more than 10^{100} years before the universe can reach the final equilibrium state.

4.10 CALCULATION OF ENTROPY CHANGES

The entropy change of a system can be calculated from the defining relation, Equation (4.41), provided that the process is internally reversible. Another more powerful way of calculating the entropy changes of a system is to utilize the so-called "TdS" relations described in the following section.

4.10.1 TdS Relations

Consider a closed system consisting of a pure compressible material. The system undergoes an internally reversible process. From the first law in differential form,

$$dU = \delta Q_{\substack{int \\ rev}} - \delta W_{\substack{int \\ rev}} \tag{4.53}$$

The work produced by the system in an internally reversible process is:

$$\delta W_{\substack{int \\ rev}} = PdV \tag{4.54}$$

The amount of heat absorbed by the system is:

$$\delta Q_{\substack{int \\ rev}} = TdS \tag{4.55}$$

Combining Equations (4.53) to (4.55) yields

$$TdS = dU + PdV \tag{4.56}$$

This is the *first* TdS relation. Note that the TdS relation is derived by considering an internally reversible process but the relation itself is independent of the process as it involves only state properties. Thus, the equation can be used to calculate entropy changes for any process, reversible or irreversible.

Another TdS relation can be derived using the definition of enthalpy. The differential change in enthalpy is:

$$dH = dU + PdV + VdP \tag{4.57}$$

Combining Equations (4.56) and (4.57) gives:

$$TdS = dH - VdP \tag{4.58}$$

This is the *second* TdS relation.

The first TdS relation can be used to gain some insight into entropy. For pure substances, we can express entropy as a function of two other state properties such as internal energy and volume:

$$S = S(U, V) \tag{4.59}$$

From Equation (4.59), it follows that:

$$dS = \left(\frac{\partial S}{\partial U}\right)_V dU + \left(\frac{\partial S}{\partial V}\right)_U dV \tag{4.60}$$

Upon multiplication of Equation (4.60) with T, we obtain

$$TdS = T\left(\frac{\partial S}{\partial U}\right)_V dU + T\left(\frac{\partial S}{\partial V}\right)_U dV \tag{4.61}$$

Comparison of Equation (4.61) with the first TdS relation, Equation (4.56), gives:

$$\left(\frac{\partial S}{\partial U}\right)_V = \frac{1}{T} > 0 \tag{4.62}$$

$$\left(\frac{\partial S}{\partial V}\right)_U = \frac{P}{T} > 0 \tag{4.63}$$

Since $\left(\frac{\partial S}{\partial U}\right)_V$ is always positive, it implies that the entropy of a system always increases with the increase in internal energy at constant volume. Likewise, $\left(\frac{\partial S}{\partial V}\right)_U$ is always positive. Therefore, the entropy of a system always increases upon expansion of the system at constant internal energy.

4.10.2 Entropy Change of an Ideal Gas

For ideal gas, the following relations are applicable:

$$Pv = RT \tag{4.64}$$

$$du = c_v\, dT \tag{4.65}$$

$$dh = c_p\, dT \tag{4.66}$$

$$c_p - c_v = R \tag{4.67}$$

where v, u, h, c_v, c_p are molar quantities and R is universal gas constant.

The entropy change relations for ideal gas can be derived using the TdS relations. From the first TdS relation, it follows that

$$ds = \frac{du}{T} + \frac{P}{T} dv \tag{4.68}$$

where s is the molar entropy. Making use of the ideal gas relations, Equations (4.64) and (4.65), Equation (4.68) can be rewritten as:

$$ds = c_v \frac{dT}{T} + \frac{R}{v} dv \tag{4.69}$$

Upon integration, Equation (4.69) gives:

$$\Delta s = \int_{T_1}^{T_2} c_v \frac{dT}{T} + Rln\left(\frac{v_2}{v_1}\right) \tag{4.70}$$

Assuming c_v to be independent of temperature, we get:

$$\Delta s = c_v\, ln\left(\frac{T_2}{T_1}\right) + Rln\left(\frac{v_2}{v_1}\right) \tag{4.71}$$

Equation (4.71) gives molar entropy change. For total entropy change,

$$\Delta S = nc_v\, ln\left(\frac{T_2}{T_1}\right) + nRln\left(\frac{v_2}{v_1}\right) \tag{4.72}$$

Equation (4.72) is developed using the first TdS relation. An equivalent equation can be developed using the second TdS relation as follows:

$$ds = \frac{dh}{T} - \frac{v}{T}dP \tag{4.73}$$

Making use of the ideal gas relations, Equations (4.64) and (4.66), Equation (4.73) can be rewritten as:

$$ds = c_p \frac{dT}{T} - \frac{R}{P}dP \tag{4.74}$$

Upon integration, Equation (4.74) gives:

$$\Delta s = \int_{T_1}^{T_2} c_p \frac{dT}{T} - Rln\left(\frac{P_2}{P_1}\right) \tag{4.75}$$

Assuming c_p to be independent of temperature, we get:

$$\Delta s = c_p\, ln\left(\frac{T_2}{T_1}\right) - Rln\left(\frac{P_2}{P_1}\right) \tag{4.76}$$

Equation (4.76) gives molar entropy change. For total entropy change,

$$\Delta S = nc_p\, ln\left(\frac{T_2}{T_1}\right) - nRln\left(\frac{P_2}{P_1}\right) \tag{4.77}$$

Note that Equations (4.71) and (4.76) are equivalent. One follows from the other using the heat capacity relation for ideal gas, Equation (4.67).

4.10.3 Entropy Change of Liquids and Solids

Liquids and solids can be treated as incompressible substances. For such materials, $dv = 0$ and $c_p \cong c_v = c$. Thus, the first TdS relation gives

$$ds = c\frac{dT}{T} \tag{4.78}$$

Upon integration

$$\Delta s = \int_{T_1}^{T_2} c\frac{dT}{T} \tag{4.79}$$

Assuming specific heat capacity is independent of temperature,

$$\Delta s = c\ ln\left(\frac{T_2}{T_1}\right) \tag{4.80}$$

NOTATION

c	molar or specific heat capacity of incompressible materials
c_p	molar constant pressure heat capacity
c_v	molar constant volume heat capacity
d	differential of exact quantity
E	total energy
h	molar or specific enthalpy
H	enthalpy
k	Boltzmann's constant
M	any state property
n	number of moles
P	pressure
Q	heat
Q_H	heat transfer between high temperature heat reservoir and system
Q_L	heat transfer between low temperature heat reservoir and system
Q_{irr}	heat transfer in an irreversible process
Q_{rev}	heat transfer in a reversible process
Q_{cycle}	net heat transfer in a cyclic process
δQ	differential amount of heat transfer
R	universal gas constant
s	molar or specific entropy
S	entropy
S_G	entropy generated
T	absolute temperature
T_b	absolute temperature at the boundary of the system
T_H	absolute temperature of a high temperature reservoir
T_L	absolute temperature of a low temperature reservoir
T_R	absolute temperature of a heat reservoir
T_{surr}	absolute temperature of the surroundings
u	molar or specific internal energy
U	internal energy
v	molar or specific volume
V	volume
W	work
W_{irr}	work in an irreversible process
W_{rev}	work in a reversible process
W_{cycle}	net work in a cyclic process
δW	differential amount of work

Greek Letters

Δ	difference operator
δ	differential of an inexact quantity
η	thermal efficiency of a cyclic process
Ω	number of microstates

TUTORIAL PROBLEMS WITH SOLUTION

Problem 1

Consider the following heat engine operating between the high temperature reservoir at 2015 K and a low temperature reservoir at 290 K.

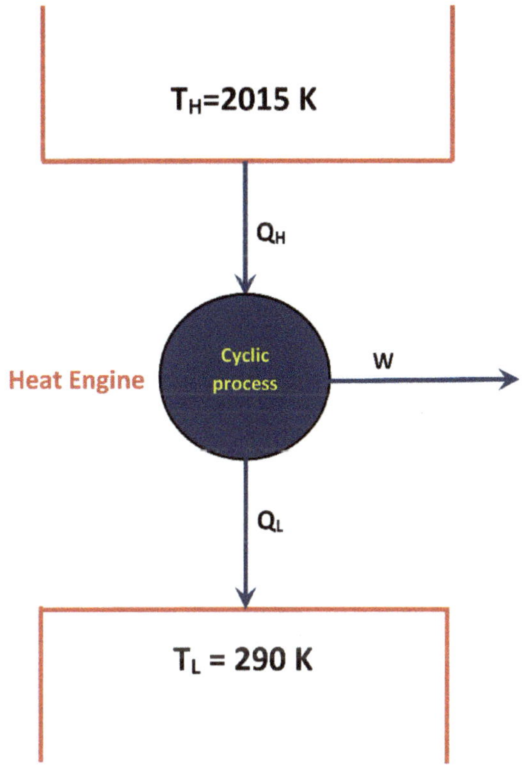

For the following cases, determine if the heat engine is reversible, irreversible, or impossible?

a) $Q_H = 1010\,\text{J}\,,\ W = 850\,\text{J}$
b) $Q_H = 1510\,\text{J}\,,\ W = 1360\,\text{J}$
c) $Q_H = 2100\,\text{J}\,,\ Q_L = -250\,\text{J}$
d) $Q_L = -290\,\text{J}\,,\ W = 1725\,\text{J}$

Solution

For a reversible (Carnot) heat engine, the thermal efficiency is given as:

$$\eta_{Carnot} = 1 - \frac{T_L}{T_H} = 1 - \frac{290}{2015} = 0.856$$

a) $Q_H = 1010\,\text{J},\ W = 850\,\text{J}$

$$\eta = \frac{W}{Q_H} = \frac{850}{1010} = 0.8416$$

This heat engine is *irreversible* as $\eta < \eta_{Carnot}$

b) $Q_H = 1510\,\text{J}\,,\ W = 1360\,\text{J}$

$$\eta = \frac{W}{Q_H} = \frac{1360}{1510} = 0.90$$

This heat engine is *impossible* as $\eta > \eta_{Carnot}$

c) $Q_H = 2100\,\text{J}\,,\ Q_L = -250\,\text{J}$

$$W = Q_H + Q_L = 2100 - 250 = 1850\,\text{J}$$

$$\eta = \frac{W}{Q_H} = \frac{1850}{2100} = 0.881$$

This heat engine is *impossible* as $\eta > \eta_{Carnot}$

d) $Q_L = -290\,\text{J}\,,\ W = 1725\,\text{J}$

$$W = Q_H + Q_L\,;\, Q_H = W - Q_L = 1725 - (-290) = 2015\,\text{J}$$

$$\eta = \frac{W}{Q_H} = \frac{1725}{2015} = 0.856$$

This is a *reversible* heat engine as $\eta = \eta_{Carnot}$

Problem 2

A power plant is generating 645 MW. The temperature of the heat source reservoir is 298 °C. The heat sink (river water) is available at 18 °C.

a) What is the maximum possible thermal efficiency of the plant?
b) What is the minimum rate at which heat is discarded to the heat sink?
c) What is the rate at which heat is discarded to the heat sink if the actual efficiency of the plant is only 57% of the maximum?

Solution

The schematic of the power plant is shown in the figure below.

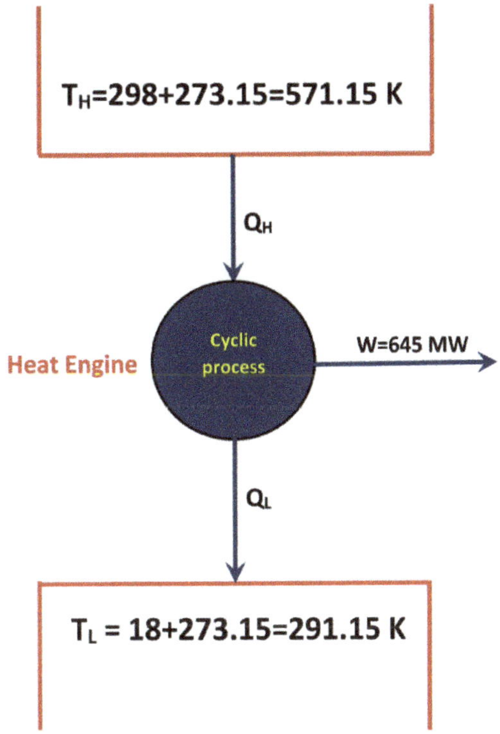

a) The maximum possible thermal efficiency of the plant is the Carnot efficiency:

$$\eta = 1 - \frac{T_L}{T_H} = 1 - \frac{291.15}{571.15} = 0.4902$$

b) The minimum rate at which heat is discarded to the heat sink?

$$W - Q_H + Q_L$$

$$\eta = \frac{W}{Q_H} = 0.4902$$

$$Q_H = \frac{W}{0.4902} = \frac{645}{0.4902} = 1315.8 \text{ MW}$$

$$Q_L = W - Q_H = 645 - 1315.8 = -670.8 \text{ MW}$$

Thus, the minimum rate at which heat is discarded to the heat sink is 670.8 MW.

c) The rate at which heat is discarded to the heat sink if the actual efficiency of the plant is only 57% of the maximum:

$$\eta = \frac{W}{Q_H} = 0.57 \times 0.4902 = 0.2794$$

$$Q_H = \frac{645}{0.2794} = 2308.52 \text{ MW}$$

$$Q_L = W - Q_H = 645 - 2308.52 = -1663.52 \text{ MW}$$

Thus, the rate at which heat is discarded to the heat sink is 1663.52 MW.

Problem 3

A Carnot heat engine uses a solar energy collector as a heat source. The temperature of the collector is 87 °C. The solar energy flux is 3.9 J/cm^2.min. The temperature of the heat sink is 20 °C. What should be the area of the solar energy collector if the work output of the heat engine is 1.5 horsepower.

Solution

The heat engine is shown schematically in the figure below.

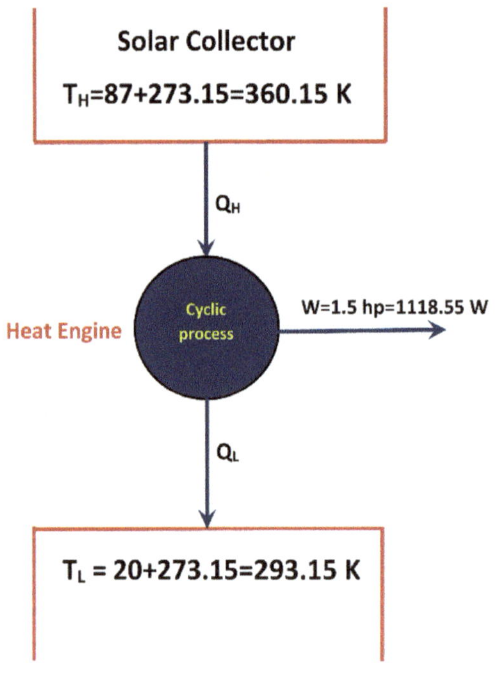

$$\eta = 1 - \frac{T_L}{T_H} = 1 - \frac{293.15}{360.15} = 0.186$$

$$\eta = \frac{W}{Q_H} = 0.186$$

$$W = 1.5 \times 745.7 = 1118.55 \text{ W}$$

$$Q_H = \frac{W}{0.186} = \frac{1118.55}{0.186} = 6013.71 \text{ W}$$

$$Q_H = Solarflux \times A$$

$$Solarflux = 3.9 \frac{\text{J}}{\text{cm}^2.\text{min}} = 3.9 \times \frac{100^2}{60} = 650 \frac{W}{m^2}$$

$$A = Q_H/Solarflux = \frac{6013.71}{650} = 9.252\ m^2$$

Problem 4

10 moles of an ideal gas undergo a two-step cyclic process. The initial temperature of the gas is 28 °C and the initial pressure is 1 atm. The steps are as follows:
Step 1: The gas expands irreversibly against zero pressure to double the volume. The initial and final temperatures are the same (that is 28 °C).
Step 2: The gas is compressed isothermally and reversibly from ½ atm and 28 °C to 1atm and 28 °C.

a) What is the value of $\oint \frac{\delta Q}{T}$?
b) Calculate ΔS for step 2. Compare ΔS to $\frac{Q}{T}$.
c) Calculate ΔS for step 1. Compare ΔS to $\frac{Q}{T}$.

Solution

The cyclic process is shown in the figure below. As a portion of the cyclic process is irreversible, the whole cyclic process is irreversible and $\oint \frac{\delta Q}{T} < 0$

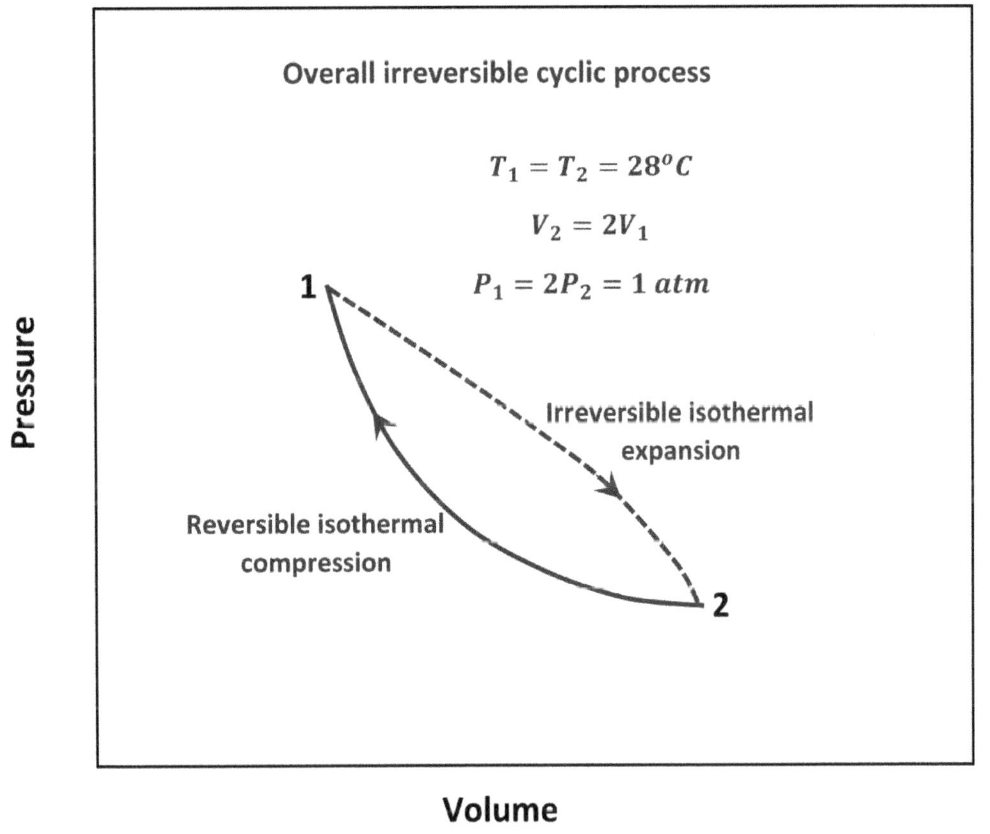

a) the value of $\oint \frac{\delta Q}{T}$

$$\oint \frac{\delta Q}{T} = \int_1^2 \frac{\delta Q}{T} + \int_2^1 \frac{\delta Q}{T}$$

For step 1 (process 1 → 2),

$$W = 0, \Delta U = 0, \text{and } Q = \Delta U + W = 0$$

Thus,

$$\int_1^2 \frac{\delta Q}{T} = 0$$

For step 2 (process 2 → 1),

$$dU = 0, \quad \delta Q = \delta W = PdV$$

$$\int_2^1 \frac{\delta Q}{T} = \int_2^1 \frac{PdV}{T} = nRln\left(\frac{V_1}{V_2}\right) = 10 \times 8.314 \times ln\left(\frac{1}{2}\right) = -57.628 \text{ J/K}$$

Thus,

$$\oint \frac{\delta Q}{T} = \int_1^2 \frac{\delta Q}{T} + \int_2^1 \frac{\delta Q}{T} = 0 - 57.628 = -57.628 \text{ J/K}$$

As expected,

$$\oint \frac{\delta Q}{T} < 0$$

b) ΔS for step 2 (process 2 → 1)? Compare ΔS to $\frac{Q}{T}$.

$$\Delta S = nc_p ln\left(\frac{T_1}{T_2}\right) - nRln\left(\frac{P_1}{P_2}\right)$$

Since $T_1 = T_2$ and $P_1 = 2P_2$,

$$\Delta S = -10 \times 8.314 \times ln(2) = -57.628 \text{ J/K}$$

$$Q = W = nRTln\left(\frac{V_1}{V_2}\right)$$

$$\frac{Q}{T} = nRln\left(\frac{V_1}{V_2}\right) = 10 \times 8.314 \times ln\left(\frac{1}{2}\right) = -57.628 \text{ J/K}$$

Thus, for a reversible isothermal process

$$\Delta S = \frac{Q}{T} = -57.628 \text{ J/K}$$

c) ΔS for step 1 (process 1 → 2)? Compare ΔS to $\frac{Q}{T}$.

$$\Delta S = nc_p ln\left(\frac{T_2}{T_1}\right) - nRln\left(\frac{P_2}{P_1}\right)$$

Since $T_1 = T_2$ and $P_1 = 2P_2$,

$$\Delta S = -10 \times 8.314 \times ln\left(\frac{1}{2}\right) = 57.628 \text{ J/K}$$

$$\frac{Q}{T} = 0$$

Thus, for an irreversible isothermal process,

$$\Delta S \neq \frac{Q}{T}$$

Problem 5

Five moles of an ideal gas are compressed adiabatically from 1 bar and 35°C to 4.7 bar. The process is irreversible. The work required is 37 percent more than that required for a reversible adiabatic compression from the same initial state (1 bar and 35°C) to a final pressure of 4.7 bar. What is the entropy change of the gas? The heat capacities of the gas are: $c_p = 7R/2$ and $c_v = 5R/2$.

Solution

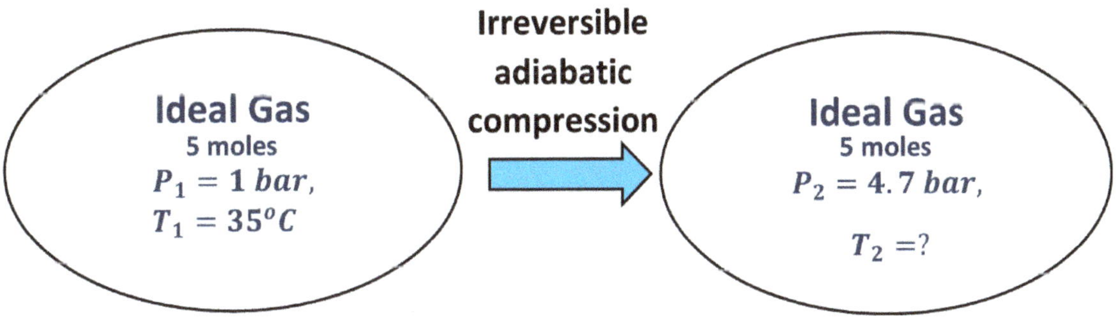

$$W_{irr} = 1.37W_{rev}$$

$$W_{rev} = \frac{nRT_1}{\gamma - 1}\left[1 - \left(\frac{P_2}{P_1}\right)^{\frac{\gamma-1}{\gamma}}\right] = \frac{5\times 8.314\times 308.15}{1.4-1}\left[1 - \left(\frac{4.7}{1}\right)^{0.4/1.4}\right] = -17807.724\ \text{J}$$

$$W_{irr} = 1.37 \times (-17807.724\ ) = -24396.58\ \text{J}$$

As the process is adiabatic, $Q = 0$. From first law for closed systems,

$$\Delta U = Q - W = 0 - (-24396.58) = 24396.58\ \text{J}$$

$$\Delta U = nc_v\Delta T = 24396.58\ \text{J}$$

$$\Delta T = T_2 - T_1 = \frac{24396.58}{5 \times 2.5 \times 8.314} = 234.75$$

$$T_2 = 308.15 + 234.75 = 542.90\ \text{K}$$

$$\Delta S = nc_p ln\left(\frac{T_2}{T_1}\right) - nRln\left(\frac{P_2}{P_1}\right) = 5 \times 3.5 \times 8.314 \times ln\left(\frac{542.90}{308.15}\right) - 5 \times 8.314 \times ln\left(\frac{4.7}{1}\right) = 18.067\ \text{J/K}$$

Problem 6

Three kg of water is heated from 5°C to 85°C at constant pressure. Calculate the entropy change. Given that: $c_p = 4181$ J/kg.K.

Solution

$$\Delta S = mc_p \, ln\left(\frac{T_2}{T_1}\right) = 3 \times 4181 \times ln\left(\frac{85+273.15}{5+273.15}\right) = 3170.76 \text{ J/K}$$

Problem 7

A company has filed for a patent related to the development of a new non-flow process in which 100 moles of air is the working material. The company claims:

a) Air undergoes a change in state from 450 K and 1.5 bar to 315 K and 1 bar.
b) The process produces 150 kJ of work.
c) An undisclosed amount of heat is transferred to a heat reservoir at 298 K.

Determine if the claimed performance is feasible. You can treat air as an ideal gas with $c_p = 7R/2$ and $c_v = 5R/2$.

Solution

The changes taking place in the process are shown below. For the process to be feasible, it must be in agreement with the laws of thermodynamics.

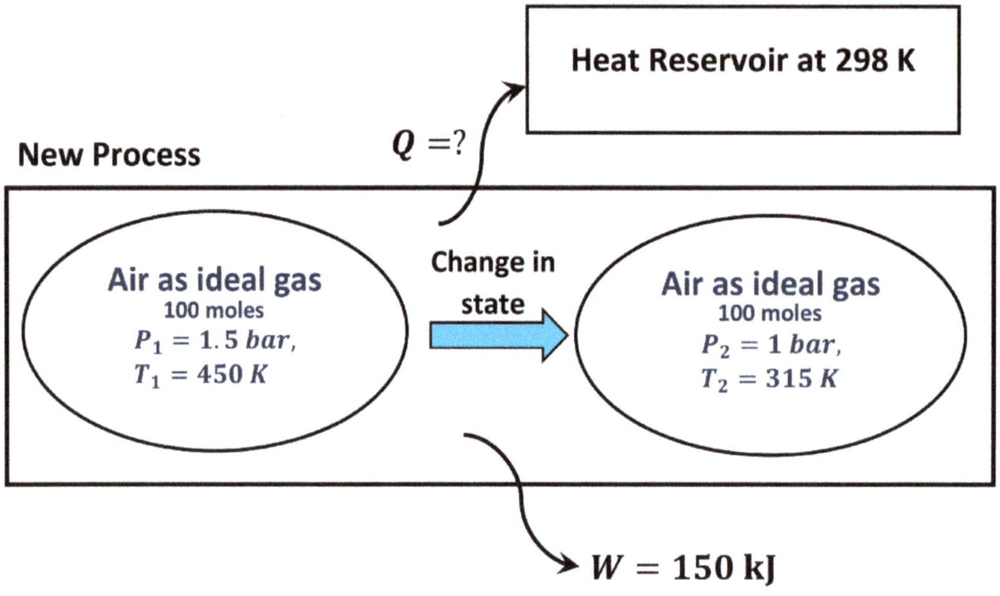

First Law:

$$\Delta U = Q - W$$

$$Q = \Delta U + W$$

$$\Delta U = nc_v\Delta T = 100 \times 2.5 \times 8.314 \times (315 - 450) = -280597.5 \text{ J} = -280.6 \text{ kJ}$$

$$Q = \Delta U + W = -280.6 + 150 = -130.6 \text{ kJ}$$

Thus, the undisclosed amount of heat transferred to the reservoir at 298 K is 130.6 kJ.

Second Law:

For all real processes,

$$(\Delta S)_{universe} = (\Delta S)_{sys} + (\Delta S)_{surr} > 0$$

$$(\Delta S)_{sys} = nc_p ln\left(\frac{T_2}{T_1}\right) - nRln\left(\frac{P_2}{P_1}\right)$$

$$= 100 \times 3.5 \times 8.314 \times ln\left(\frac{315}{450}\right) - 100 \times 8.314 \times ln\left(\frac{1}{1.5}\right) = -700.79 \text{ J/K} = -0.7008 \text{ kJ/K}$$

$$(\Delta S)_{surr} = (\Delta S)_{reservoir} = \frac{Q}{T_{reservoir}} = \frac{130.6}{298} = 0.4383 \text{ kJ/K}$$

$$(\Delta S)_{universe} = (\Delta S)_{sys} + (\Delta S)_{surr} = -0.7008 + 0.4383 = -0.2625 < 0$$

This is *impossible.* All real processes are accompanied by an increase in the entropy of the universe. Hence the proposed process is not feasible.

PRACTICE PROBLEMS

Problem 1

Which of the following two heat engines operating between the same temperature levels is a Carnot heat engine and why?

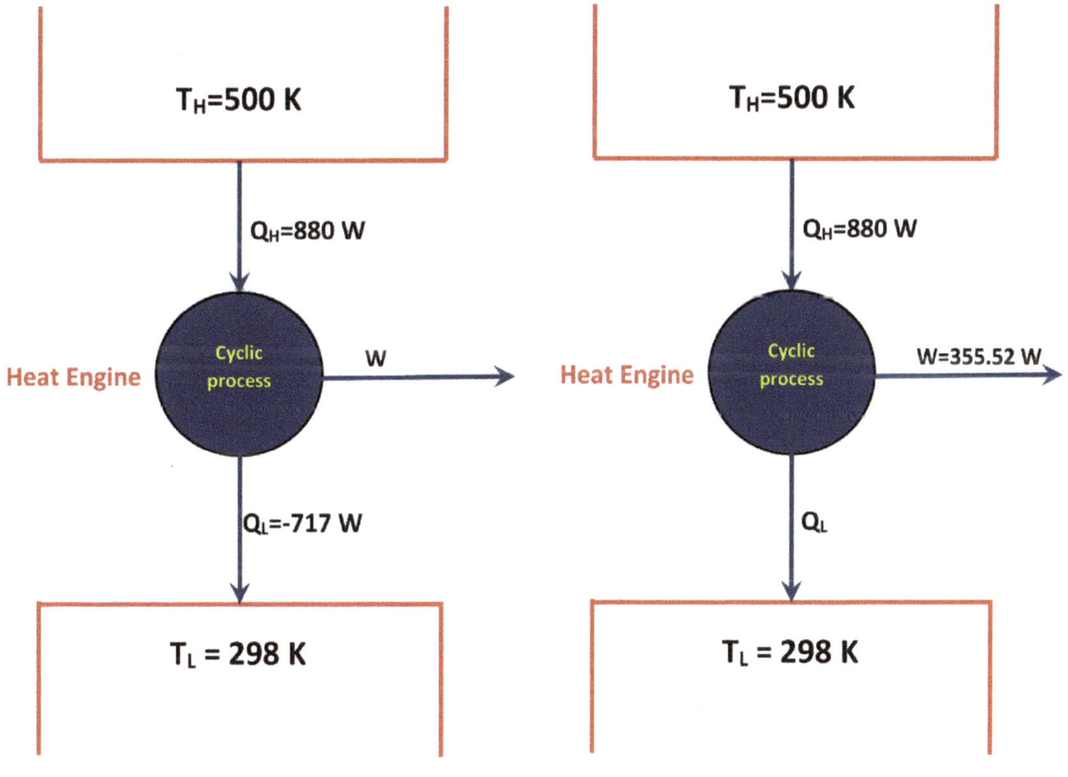

Problem 2

A heat engine is producing power at a rate of 51,500 kW. The temperatures of the heat reservoirs are 590 and 298 K. The thermal efficiency of the engine is 63% of the Carnot efficiency. Calculate the rates at which heat is absorbed from the high temperature reservoir and discarded to the low temperature heat sink.

Problem 3

One mole of an ideal gas is expanded to twice its initial volume using two different processes:

- Process (a) where the gas is expanded reversibly at a constant temperature.
- Process (b) where the gas is expanded irreversibly. However, the temperature remains the same in the initial and final states.

Calculate ΔS in each case. Discuss your results.

Problem 4

A block of solid copper with a mass of 2 kg and heat capacity of 380 J/kg.K is at a temperature of 350 °C. The block is quenched in water initially at 20 °C. What should be the mass of water so that the final temperature of block and water is 85 °C? Calculate (a) entropy change of block, (b) entropy change of water, and (c) entropy change of universe. Any comments?

Problem 5

Ten moles of air (ideal gas) are compressed isothermally from 1 bar to 4.7 bar at 35°C. The process is irreversible. The work required is 29 percent more than that required for a reversible isothermal compression from the same initial state (1 bar, 35°C) to a final pressure of 4.7 bar. The heat produced is transferred to a heat reservoir at 298 K. Calculate:

a) the heat produced,
b) the entropy change of the gas,
c) the entropy change of the reservoir,
d) the entropy change of the universe.

Problem 6

A new non-flow process is developed by a company. The company has filed for a patent in which it is claimed that:

- the working material consisting of 100 moles of air (ideal gas) undergoes a change in state from 400 K and 0.1 m^3 to 298 K and 1 m^3,
- the process produces 115 kJ of work,
- the process produces an undisclosed amount of heat that is transferred to a heat reservoir at 293 K.

Determine if the claimed performance is feasible. The heat capacities of air are: $c_p = 7R/2$ and $c_v = 5R/2$.

Problem 7

The heat source of a Carnot heat engine is 185 kg of water at 90 °C. The heat sink is a very large body of water at 5 °C. As the heat engine operates, the water at 90 °C begins to cool and finally ends up at 5 °C. What is the maximum amount of work obtainable? The heat capacity of water is
$c_p = 4181$ J/kg.K.

Chapter 5

Second Law of Thermodynamics for Control Volume (Open Systems)

5.1 LEARNING OUTCOMES

Upon completion of this chapter, students will be able to:

- Express the second law of thermodynamics for control volume (open systems)
- Carry out entropy balance for a control volume
- Apply the second law for control volume to nozzles, turbines, compressors and pumps, throttling devices, and heat exchangers
- Explain the perpetual motion machine of second kind
- Solve problems related to the application of second law for control volume to flow processes

5.2 INTRODUCTION

As noted in Chapter 3, engineering processes usually involve flow of fluids from one plant unit to another. In flow processes, it is more convenient and beneficial to adopt the *control volume* approach. In this chapter, the second law of thermodynamics is applied to a control volume and its applications in different flow processes are discussed.

The second law of thermodynamics simply states that all irreversible (i.e., real) processes, flow or non-flow, are accompanied by entropy generation in the universe.

5.3 SECOND LAW OF THERMODYNAMICS APPLIED TO A CONTROL VOLUME

Consider the flow process shown in Figure 5.1. Fluid enters and leaves the control volume through the openings. When fluid enters or leaves the control volume, it carries with it mass, energy (internal, kinetic, and potential), and entropy. The control volume is receiving heat from a heat reservoir and producing shaft work. The variables relevant to the application of second law are indicated at the inlet and exit conditions of the control volume shown in Figure 5.1.

The second law for the control volume can be expressed as:

$$\begin{pmatrix} Rate\ of \\ entropy\ generation \\ in\ the\ universe \end{pmatrix} = \begin{pmatrix} Rate\ at \\ which\ entropy \\ leaves\ CV \end{pmatrix} - \begin{pmatrix} Rate\ at \\ which\ entropy \\ enters\ CV \end{pmatrix} + \begin{pmatrix} Rate\ of \\ change\ of \\ entropy \\ inside\ \ CV \end{pmatrix} + \begin{pmatrix} Rate\ of \\ change\ of \\ entropy\ in \\ surroundings \end{pmatrix} \geq 0 \quad (5.1)$$

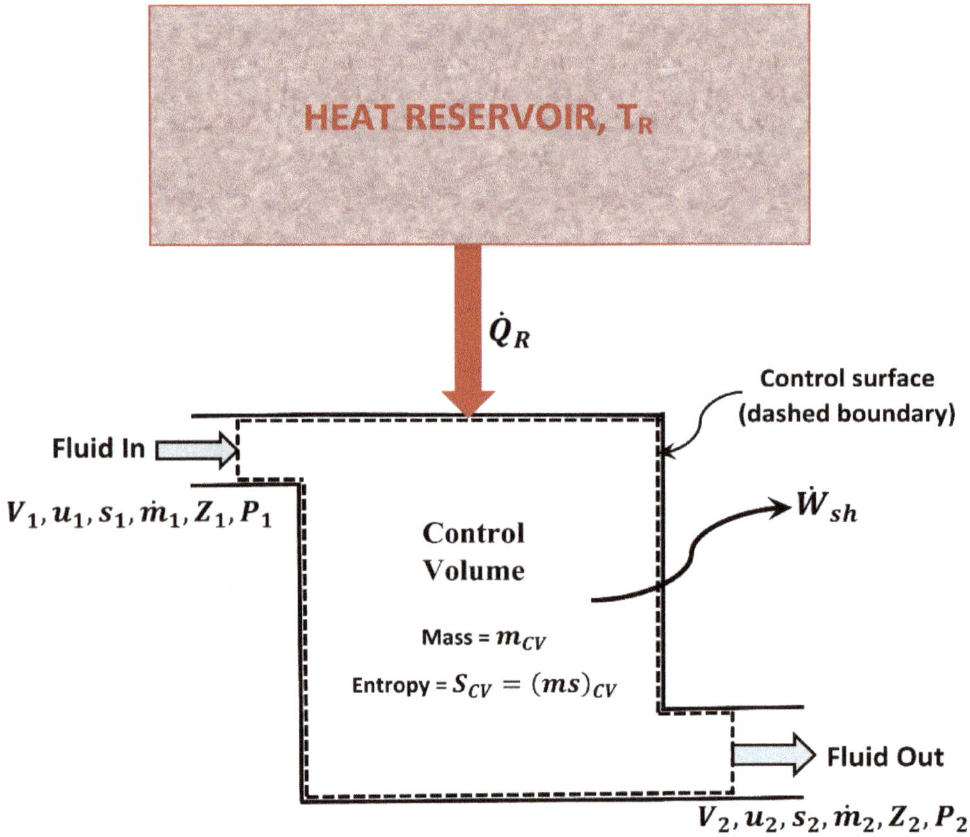

Figure 5.1 Control volume with one inlet and one outlet (P is pressure, V is velocity, u is specific internal energy, s is specific entropy, $\dot{m}$ is mass flow rate, Z is elevation, $\dot{Q}_R$ is rate of heat transfer, $\dot{W}_{sh}$ is rate of shaft work, m is mass, and S is total entropy).

Referring to Figure 5.1, the various terms of Equation (5.1) can be written as:

$$\begin{pmatrix} Rate\ of \\ entropy\ generation \\ in\ the\ universe \end{pmatrix} = \dot{S}_{G,total} \tag{5.2}$$

$$\begin{pmatrix} Rate\ at \\ which\ entropy \\ leaves\ CV \end{pmatrix} = \dot{m}_2 s_2 \tag{5.3}$$

$$\begin{pmatrix} Rate\ at \\ which\ energy \\ enters\ CV \end{pmatrix} = \dot{m}_1 s_1 \tag{5.4}$$

$$\begin{pmatrix} Rate\ of \\ change\ of \\ entropy\ inside\ CV \end{pmatrix} = \frac{dS_{CV}}{dt} = \frac{d(ms)_{CV}}{dt} \tag{5.5}$$

$$\begin{pmatrix} Rate\ of \\ change\ of\ entropy \\ in\ surroundings \end{pmatrix} = \frac{dS_{surr}}{dt} \tag{5.6}$$

where s is the specific entropy and S is the total entropy.

Thus, the second law, Equation (5.1), becomes:

$$\dot{S}_{G,total} = \dot{m}_2 s_2 - \dot{m}_1 s_1 + \frac{d(ms)_{CV}}{dt} + \frac{dS_{surr}}{dt} \tag{5.7}$$

At steady state,

$$\dot{m}_2 = \dot{m}_1 = \dot{m} \tag{5.8}$$

$$\frac{d(ms)_{CV}}{dt} = 0 \tag{5.9}$$

The rate of change of entropy of the surroundings (heat reservoir) is:

$$\frac{dS_{surr}}{dt} = -\frac{\dot{Q}_R}{T_R} \tag{5.10}$$

Thus, under steady state condition, the second law becomes:

$$\dot{S}_{G,total} = \dot{m}(s_2 - s_1) - \frac{\dot{Q}_R}{T_R} \geq 0 \tag{5.11}$$

For a control volume consisting of multiple inlets and outlets, and communicating thermally with multiple heat reservoirs, Equation (5.11) can be generalized as follows:

$$\dot{S}_{G,total} = \Sigma_{OUTLETS}\ (\dot{m}_i s_i) - \Sigma_{INLETS}\ (\dot{m}_i\, s_i) - \Sigma \frac{\dot{Q}_i}{T_i} \geq 0 \tag{5.12}$$

where T_i is the absolute temperature of ith heat reservoir and $\dot{Q}_i$ is the rate heat transfer between the control volume and ith heat reservoir. Note that $\dot{Q}_i$ is positive if it flows into the control volume.

Equation (5.12) is the most general form of the second law applicable to control volume under steady state conditions.

5.4 ENTROPY BALANCE FOR CONTROL VOLUME

In this section, we will do entropy balance for a control volume alone without any regard for the heat reservoirs with which the control volume may be communicating thermally. Consider the control volume shown in Figure 5.2. The heat transfer rate is $\dot{Q}_b$ at the boundary where the absolute temperature is T_b. The entropy transfer to the control volume accompanied by heat transfer at the boundary is $\dot{Q}_b/T_b$.

The macroscopic balance of any entity (mass, energy or entropy) over the control volume can be expressed as:

$$Input - Output + Generation - Destruction = Accumulation \tag{5.13}$$

As entropy cannot be destroyed (it can only be generated), Equation (5.13) reduces to:

$$Input - Output + Generation = Accumulation \tag{5.14}$$

In terms of rates, the entropy balance for a control volume can be written as:

$$\begin{pmatrix} Rate\ at \\ which\ entropy \\ enters\ CV \end{pmatrix} - \begin{pmatrix} Rate\ at \\ which\ entropy \\ leaves\ CV \end{pmatrix} + \begin{pmatrix} Rate\ of \\ entropy \\ generation \end{pmatrix} = \begin{pmatrix} Rate\ of \\ accumulation \\ of\ entropy \end{pmatrix} \tag{5.15}$$

The rate at which entropy enters the *CV* is:

$$\begin{pmatrix} Rate\ at \\ which\ entropy \\ enters\ CV \end{pmatrix} = \dot{m}_1 s_1 + \frac{\dot{Q}_b}{T_b} \tag{5.16}$$

The rate at which entropy leaves the *CV* is:

$$\begin{pmatrix} Rate\ at \\ which\ entropy \\ leaves\ CV \end{pmatrix} = \dot{m}_2 s_2 \tag{5.17}$$

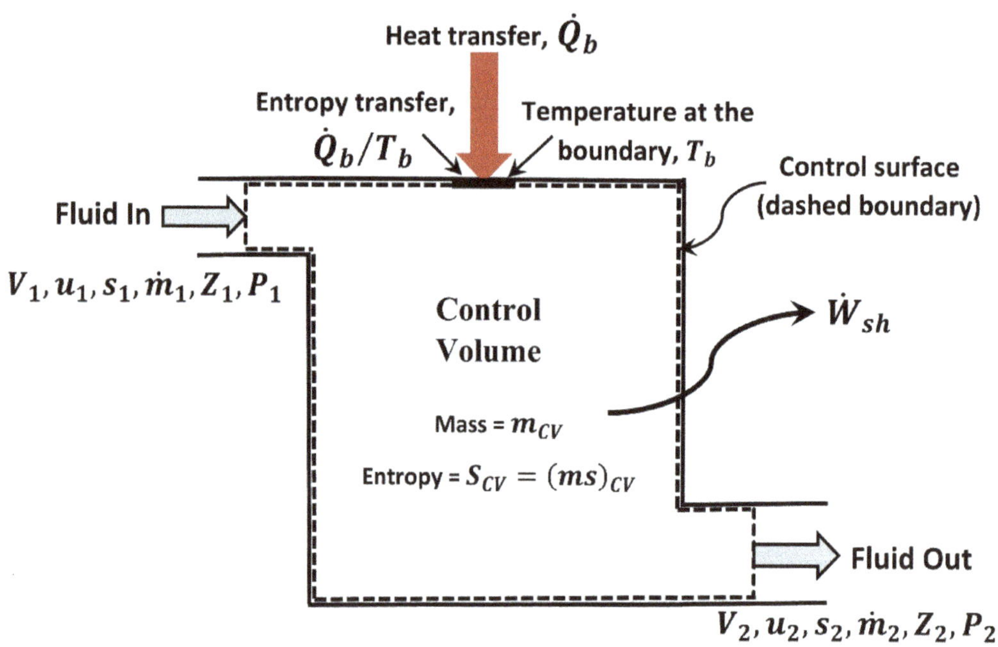

Figure 5.2 Control volume with one inlet and one outlet. Entropy transfer to the control volume accompanied by heat transfer at the boundary is $\dot{Q}_b/T_b$.

The rate at which entropy is generated inside the *CV* due to internal irreversibilities is:

$$\begin{pmatrix} Rate\ of \\ entropy \\ generation \end{pmatrix} = \dot{S}_{G,CV} \tag{5.18}$$

The rate at which entropy accumulates inside the *CV* is:

$$\begin{pmatrix} Rate\ of \\ accumulation\ of \\ entropy\ inside\ \ CV \end{pmatrix} = \frac{d(ms)_{CV}}{dt} \tag{5.19}$$

Thus, entropy balance for the control volume becomes:

$$\left(\dot{m}_1 s_1 + \frac{\dot{Q}_b}{T_b}\right) - (\dot{m}_2 s_2) + \left(\dot{S}_{G,CV}\right) = \frac{d(ms)_{CV}}{dt} \tag{5.20}$$

Upon rearrangement, Equation (5.20) gives:

$$\left(\dot{S}_{G,CV}\right) = (\dot{m}_2 s_2) - \left(\dot{m}_1 s_1 + \frac{\dot{Q}_b}{T_b}\right) + \frac{d(ms)_{CV}}{dt} \tag{5.21}$$

As entropy is always generated within the control volume due to internal irreversibilities, Equation (5.21) can be expressed in the form of an inequality as:

$$\dot{S}_{G,CV} = \dot{m}_2 s_2 - \dot{m}_1 s_1 - \frac{\dot{Q}_b}{T_b} + \frac{d(ms)_{CV}}{dt} \geq 0 \tag{5.22}$$

For a control volume with multiple inlets and outlets, Equation (5.22) can be generalized as:

$$\dot{S}_{G,CV} = \sum_{OUTLETS} (\dot{m}_i s_i) - \sum_{INLETS} (\dot{m}_i\, s_i) - \frac{\dot{Q}_b}{T_b} + \frac{d(ms)_{CV}}{dt} \geq 0 \tag{5.23}$$

Under steady state condition, there is no accumulation of entropy within the control volume. Consequently, Equation (5.22) for control volume with one inlet and one outlet, simplifies as:

$$\dot{S}_{G,CV} = \dot{m}(s_2 - s_1) - \frac{\dot{Q}_b}{T_b} \geq 0 \tag{5.24}$$

If the flow process is adiabatic, Equation (5.24) gives:

$$\dot{S}_{G,CV} = \dot{m}(s_2 - s_1) \geq 0 \tag{5.25}$$

For an internally reversible adiabatic process, it follows that:

$$\dot{S}_{G,CV} = \dot{m}(s_2 - s_1) = 0 \tag{5.26}$$

Thus, no change in entropy occurs in an internally reversible adiabatic process. Such a process is also called *isentropic* process.

5.5 APPLICATIONS OF SECOND LAW TO FLOW PROCESSES

5.5.1 Nozzles

Nozzle is a device of varying cross-sectional area that is used to increase the velocity of a flowing fluid. The fluid undergoes expansion in a nozzle in that its pressure decreases sharply. There is no shaft work involved and heat transfer is negligible. From the first law of thermodynamics for control volume, we have:

$$\left[(h_2 - h_1) + \frac{1}{2}(V_2^2 - V_1^2) + g(Z_2 - Z_1)\right] = 0 \tag{5.27}$$

Neglecting potential energy change, Equation (5.27) gives:

$$h_1 - h_2 = \frac{1}{2}(V_2^2 - V_1^2) \tag{5.28}$$

Since $V_2 \gg V_1$, Equation (5.28) simplifies to:

$$\frac{1}{2}(V_2^2) = h_1 - h_2 \tag{5.29}$$

Thus, the specific kinetic energy of the fluid leaving the nozzle is simply the decrease in enthalpy of the fluid.

It is of practical interest to know the conditions under which the specific kinetic energy of the fluid leaving the nozzle is maximum, when the fluid is expanded from a given inlet state to the outlet pressure. It is here that the second law becomes important.

The application of the second law (entropy balance over the adiabatic nozzle) on a unit mass basis gives:

$$s_{G,CV} = (s_2 - s_1) \geq 0 \tag{5.30}$$

If the adiabatic expansion of the fluid is reversible, no entropy is generated, and the process is isentropic. As the actual expansion is irreversible, entropy is generated and $s_2 > s_1$.

The process of expansion of a fluid in a nozzle is shown in Figure 5.3. As can be seen, the decrease in enthalpy, and hence kinetic energy of the fluid leaving the nozzle, is maximum when the process is carried out isentropically. In the actual expansion, entropy is generated and the decrease in enthalpy is less. The isentropic nozzle efficiency is defined as:

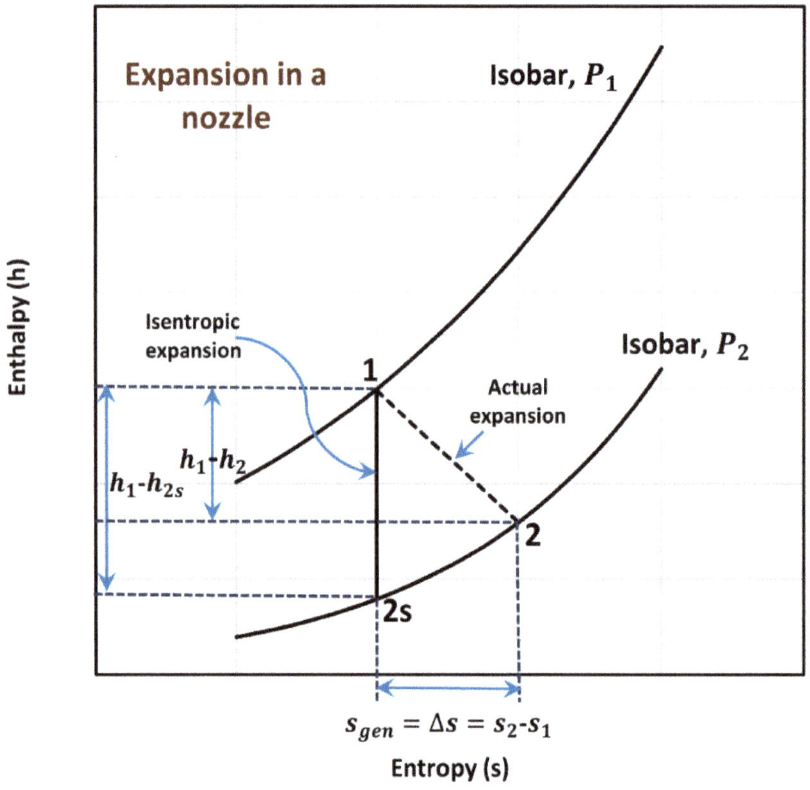

Figure 5.3 Expansion in a nozzle.

$$\varepsilon_{nozzle} = \frac{V_2^2/2}{(V_2^2/2)_s} \tag{5.31}$$

From Equations (5.29) and (5.31), it follows that:

$$\varepsilon_{nozzle} = \frac{h_1 - h_2}{h_1 - h_{2s}} \tag{5.32}$$

Thus, the nozzle efficiency is the ratio of actual enthalpy decrease to isentropic enthalpy decrease (see Figure 5.3) of the fluid. Properly designed nozzles can have efficiencies of 95% or more, indicating that internal reversibilities can be reduced to a negligible level in well-designed nozzles.

5.5.2 Turbines

A turbine is a device which produces shaft work using the internal energy of the fluid. The fluid undergoes expansion in a turbine in that its pressure decreases sharply. A schematic diagram of a turbine is shown in Figure 5.4. The first law of thermodynamics for control volume can be applied to calculate the amount of shaft work produced by the turbine. Assuming steady state and neglecting kinetic and potential energy changes and heat transfer, Equation (3.13) simplifies to:

$$-\dot{W}_{sh} = \dot{m}(h_2 - h_1) \tag{5.33}$$

On a unit mass basis,

$$W_{sh} = (h_1 - h_2) \tag{5.34}$$

Thus, work produced by a turbine is simply the decrease in enthalpy of the fluid.

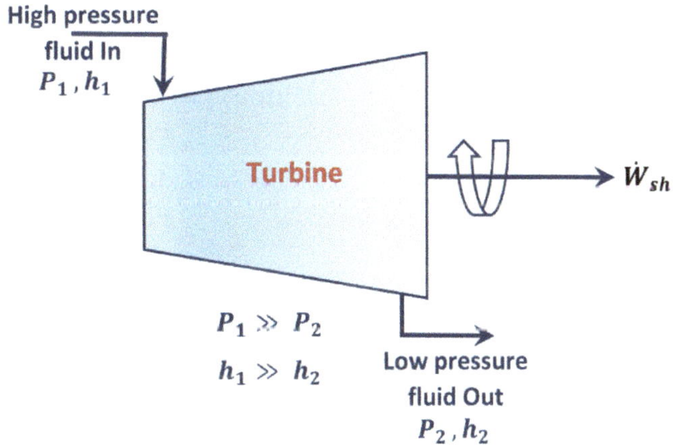

Figure 5.4 Schematic diagram of a turbine.

It is of practical interest to know the conditions under which the work produced by the turbine is maximum, when the fluid is expanded from a given inlet state to the outlet pressure. Once again, it is the second law that determines the conditions under which the work produced by the turbine is maximum.

The application of the second law (entropy balance over the adiabatic turbine) on a unit mass basis gives:

$$s_{G,CV} = (s_2 - s_1) \geq 0 \tag{5.30}$$

If the adiabatic expansion of the fluid is reversible, no entropy is generated, and the process is isentropic. As the actual expansion is irreversible, entropy is generated and $s_2 > s_1$.

The process of expansion of a fluid in a turbine is shown in Figure 5.5. As can be seen, the decrease in enthalpy, and hence the shaft work produced in a turbine, is maximum when the process is carried out isentropically. In the actual expansion, entropy is generated due to internal irreversibilities in the turbine and the decrease in enthalpy is less.

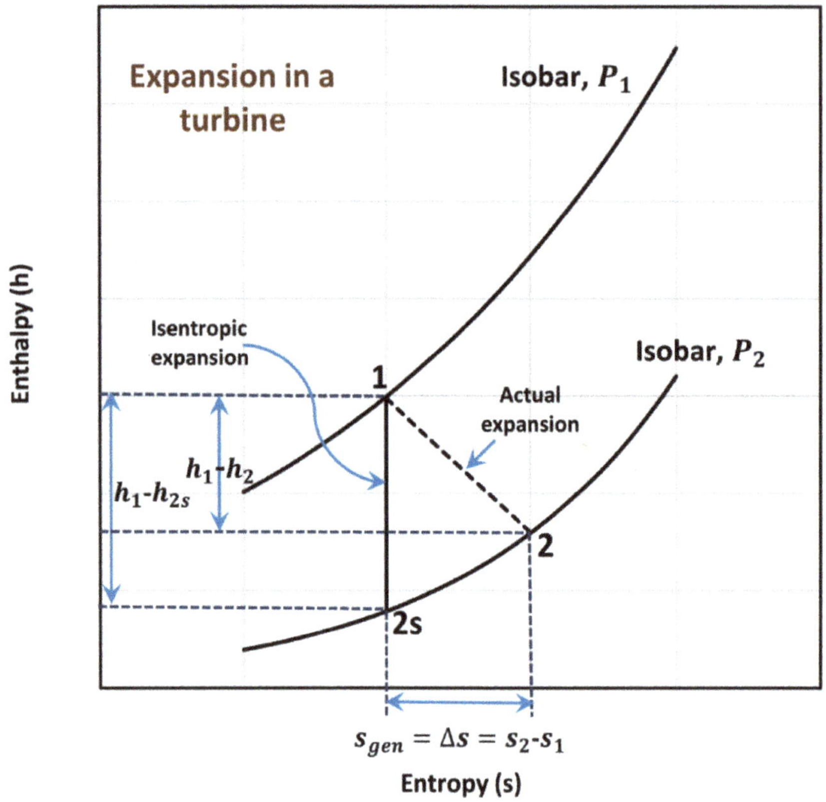

Figure 5.5 Expansion in a turbine.

The isentropic turbine efficiency is defined as:

$$\varepsilon_{turbine} = \frac{W_{sh}}{(W_{sh})_{max}} \tag{5.35}$$

From Equations (5.34) and (5.35), it follows that:

$$\varepsilon_{turbine} = \frac{h_1 - h_2}{h_1 - h_{2s}} = \frac{h_2 - h_1}{h_{2s} - h_1} \tag{5.36}$$

Thus, the turbine efficiency is the ratio of actual enthalpy decrease to isentropic enthalpy decrease (see Figure 5.5) of the fluid, when the fluid undergoes expansion from the same inlet state to the same exit pressure. The efficiency of a turbine is typically in the range of 70 to 90%.

5.5.3 Compressors and Pumps

Compressors are devices in which mechanical energy (work) is added to raise the pressure of a *gas* passing through them (see Figure 5.6). *Pumps* are similar devices in which mechanical energy is added to raise the pressure of a *liquid* passing through them. The power required by the compressor or pump can be calculated from the first law for control volume, neglecting kinetic and potential energy changes and heat transfer. Thus,

$$\dot{W}_{sh} = -\dot{m}(h_2 - h_1) \tag{5.37}$$

Note that $\dot{W}_{sh}$ is negative for compressors and pumps as it flows into the control volume. The shaft work on a unit mass basis is:

$$W_{sh} = -(h_2 - h_1) \tag{5.38}$$

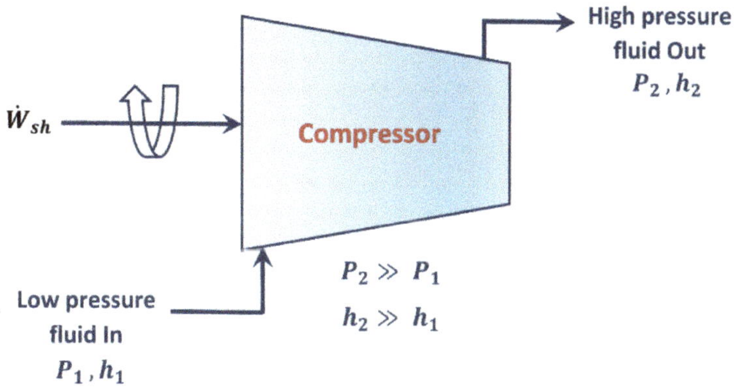

Figure 5.6 Schematic diagram of a compressor (also valid for pump).

It is of practical interest to know the conditions under which the work required by the compressor or pump is minimum, when the fluid is compressed from a given inlet state to the outlet pressure. Once again, it is the second law that determines the conditions under which the work required by the compressor or pump is minimum.

The application of the second law (entropy balance over the adiabatic compressor or pump) on a unit mass basis gives:

$$s_{G,CV} = (s_2 - s_1) \geq 0 \tag{5.30}$$

If the adiabatic compression of the fluid is reversible, no entropy is generated, and the process is isentropic. As the actual compression is irreversible, entropy is generated and $s_2 > s_1$.

The process of compression of a fluid in a compressor or pump is shown in Figure 5.7. As can be seen, the increase in enthalpy, and hence the shaft work required in a compressor or pump, is minimum when the process is carried out isentropically. In the actual compression, entropy is generated due to internal irreversibilities in the compressor or pump and the increase in enthalpy is more.

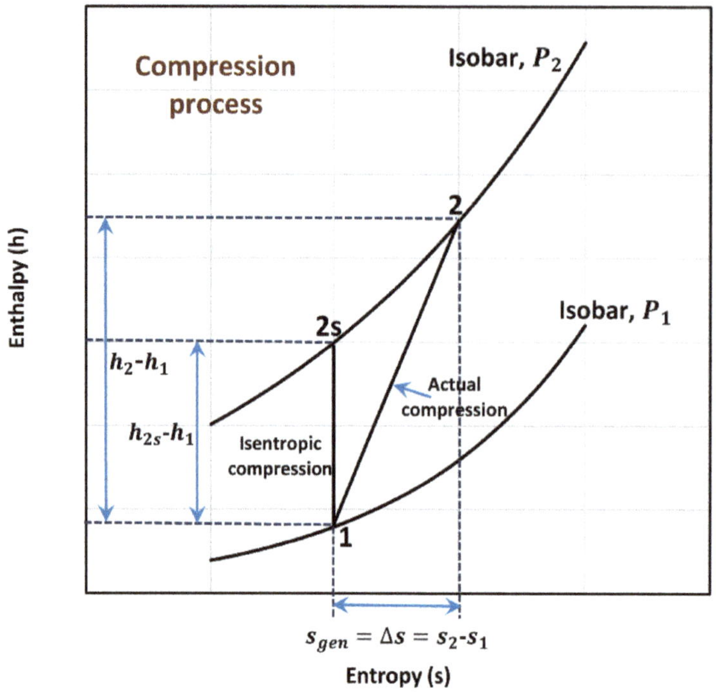

Figure 5.7 Compression process (compressors and pumps).

The isentropic efficiency of a compressor or pump is defined as:

$$\varepsilon_{turbine} = \frac{(W_{sh})_{min}}{W_{sh}} \tag{5.39}$$

From Equations (5.38) and (5.39), it follows that:

$$\varepsilon_{turbine} = \frac{h_{2s}-h_1}{h_2-h_1} \tag{5.40}$$

Thus, the compressor or pump efficiency is the ratio of isentropic enthalpy increase to actual enthalpy increase (see Figure 5.7) of the fluid, when the fluid undergoes compression from the same inlet state to the same exit pressure. The efficiency of a compressor or a pump is typically in the range of 70 to 90%.

5.5.4 Throttling Devices

Throttling devices are expansion devices in which a sharp reduction in pressure of the fluid occurs adiabatically. In a throttling device, a constriction is placed in the path of fluid flow to cause a sharp reduction in pressure. The kinetic and potential energy changes of the fluid are negligible and there is no shaft work. As noted in Chapter 3, the throttling process is an isenthalpic process $(h_2 = h_1)$. This follows from the first law of thermodynamics for control volume.

The throttling process is shown in Figure 5.8 using enthalpy-entropy diagram. As can be seen, the throttling process cannot be isentropic. As the process is inherently irreversible, there always occur a production of entropy in the process, $s_{G,CV} = (s_2 - s_1) > 0$.

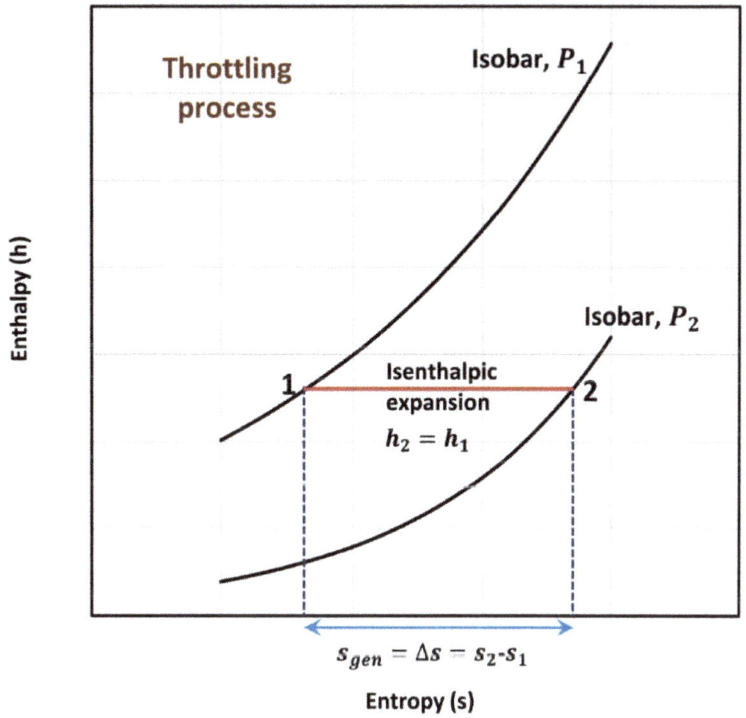

Figure 5.8 Throttling process (isenthalpic expansion).

5.5.5 Heat Exchangers

As noted in Chapter 3, heat exchangers are used extensively in industrial applications to transfer energy between fluids at different temperatures. Energy is transferred from a hot fluid stream to a cold fluid stream as they pass through the heat exchanger. Figure 5.9 shows a counter-current double pipe heat exchanger with temperature profiles. The hot fluid flows from right to left and the cold fluid flows from left to right of the heat exchanger.

As the temperature difference between hot and cold fluid is finite at any given axial position x, the heat transfer between the hot and cold fluid is irreversible in nature. Thus, it is of interest to determine the rate of entropy generation and quantify internal irreversibilities in the heat exchanger using the second law of thermodynamics.

The first law for control volume (see Figure 5.10) with multiple inlets and outlets is:

$$\Sigma_{OUTLETS}\ \dot{m}_i(h_i + \frac{1}{2}V_i^2 + gZ_i) - \Sigma_{INLETS}\ \dot{m}_i\ (h_i + \frac{1}{2}V_i^2 + gZ_i) = \Sigma\dot{Q} - \dot{W}_{sh} \tag{5.41}$$

Neglecting kinetic and potential energy changes and assuming no heat loss, we get:

$$\Sigma_{OUTLETS}\ (\dot{m}_i h_i) - \Sigma_{INLETS}\ (\dot{m}_i\ h_i) = 0 \tag{5.42}$$

Note that there is no shaft work involved in heat exchangers. Equation (5.42) when applied to the control volume of Figure 5.10 gives:

$$\dot{m}_h(h_2 - h_1)_{hot\ fluid} + \dot{m}_c(h_2 - h_1)_{cold\ fluid} = 0 \tag{5.43}$$

Assuming negligible pressure changes in the flowing streams, Equation (5.43) simplifies to:

$$\dot{m}_h c_{p,h}(T_{h,2} - T_{h,1}) + \dot{m}_c c_{p,c}(T_{c,2} - T_{c,1}) = 0 \quad (5.44)$$

There are eight variables in Equation (5.44). Given seven variables, the unknown eighth variable can be calculated.

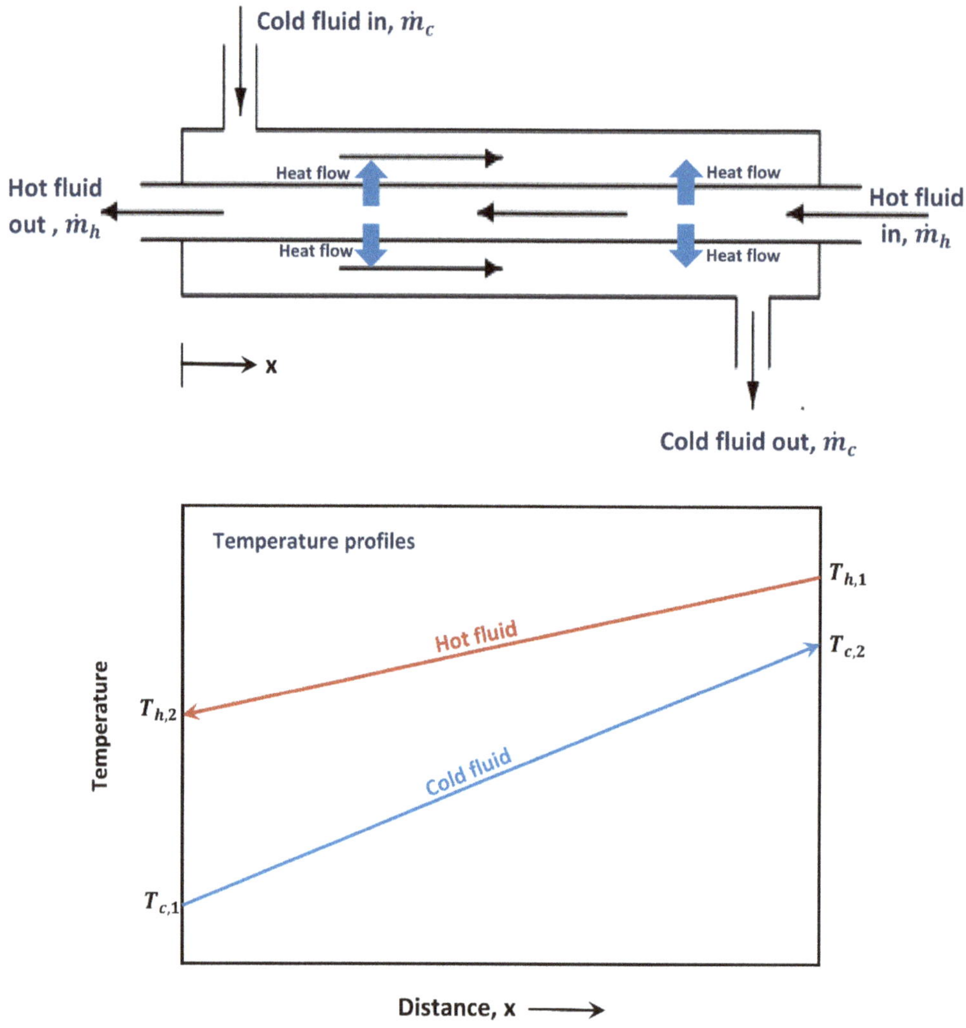

Figure 5.9 Double pipe counter-current heat exchanger with temperature profiles.

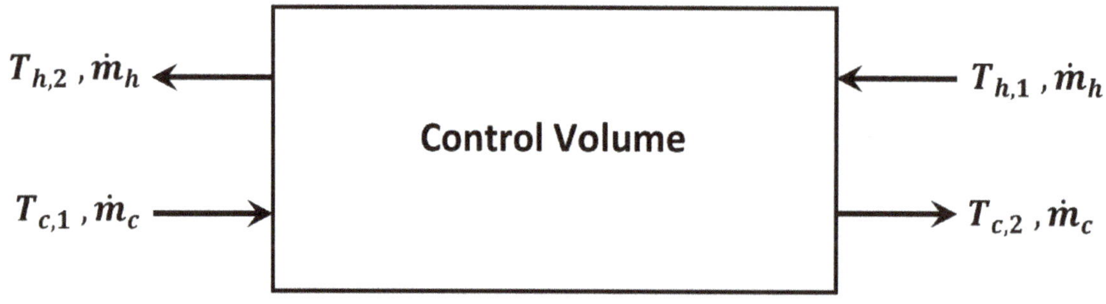

Figure 5.10 Control volume of the heat exchanger.

We can now apply the second law of thermodynamics to the control volume of the heat exchanger (Figure 5.10), assuming steady state and no heat loss:

$$\dot{S}_{G,CV} = \Sigma_{OUTLETS}\ (\dot{m}_i s_i) - \Sigma_{INLETS}\ (\dot{m}_i\, s_i) \geq 0 \tag{5.45}$$

Equation (5.45) when applied to the control volume of Figure 5.10 gives:

$$\dot{S}_{G,CV} = \dot{m}_h (s_2 - s_1)_{hot\ fluid} + \dot{m}_c (s_2 - s_1)_{cold\ fluid} \geq 0 \tag{5.46}$$

Assuming fluids to be incompressible liquids and negligible pressure changes in the flowing streams,

$$(s_2 - s_1)_{hot\ fluid} = c_{p,h} ln\left(\frac{T_{h,2}}{T_{h,1}}\right) \tag{5.47}$$

$$(s_2 - s_1)_{cold\ fluid} = c_{p,c} ln\left(\frac{T_{c,2}}{T_{c,1}}\right) \tag{5.48}$$

Substitution of Equations (5.47) and (5.48) into Equation (5.46), we get:

$$\dot{S}_{G,CV} = \dot{m}_h c_{p,h} ln\left(\frac{T_{h,2}}{T_{h,1}}\right) + \dot{m}_c c_{p,c} ln\left(\frac{T_{c,2}}{T_{c,1}}\right) \geq 0 \tag{5.49}$$

As an example, consider geothermal water as a hot fluid available at 160 °C. Cold water is to be heated from 20 °C to 80 °C using the geothermal water in a double pipe countercurrent heat exchanger. The mass flow rates of hot and cold fluids are 2 kg/s and 1.2 kg/s, respectively. The heat capacities of hot and cold fluids are 4310 J/kg.K and 4180 J/kg.K, respectively. Estimate the rate of entropy generation in the heat exchanger. From Equation (5.44), we get:

$$2 \times 4310(T_{h,2} - 160) + 1.2 \times 4180(80 - 20) = 0 \tag{5.50}$$

Equation (5.50) gives $T_{h,2} = 125.086°C$. From Equation (5.49), we get:

$$\dot{S}_{G,CV} = 2 \times 4310 ln\left(\frac{125.086 + 273.15}{160 + 273.15}\right) + 1.2 \times 4180 ln\left(\frac{80 + 273.15}{20 + 273.15}\right)$$

$$= -724.419 + 934.022 = 209.603\ \frac{\mathrm{J}}{\mathrm{K.s}} \geq 0 \tag{5.51}$$

Thus, the total rate of entropy generation in the double pipe heat exchanger is 209.603 $\mathrm{J/K.s}$. As expected from the second law, $\dot{S}_{G,CV} > 0$ for this irreversible process.

5.6 PERPETUAL MOTION MACHINE OF SECOND KIND

Any device, engine, machine or process that violates the second law of thermodynamics is called a *Perpetual Motion Machine of Second Kind.* According to the second law, all real processes are accompanied by entropy generation in the universe. As an example, Figure 5.11 shows a *Perpetual Motion Machine of Second Kind.* In this cyclic process, there is a continuous production of work while the process is interacting thermally with a single heat reservoir. The transfer of heat from a heat reservoir to the boiler decreases the entropy of the surroundings. Due to the cyclic nature of the process, there is no internal entropy change for the process. Thus, this cyclic process is accompanied by a decrease in the entropy of the universe. This violates the second law. It is impossible for the entropy of the universe to decrease.

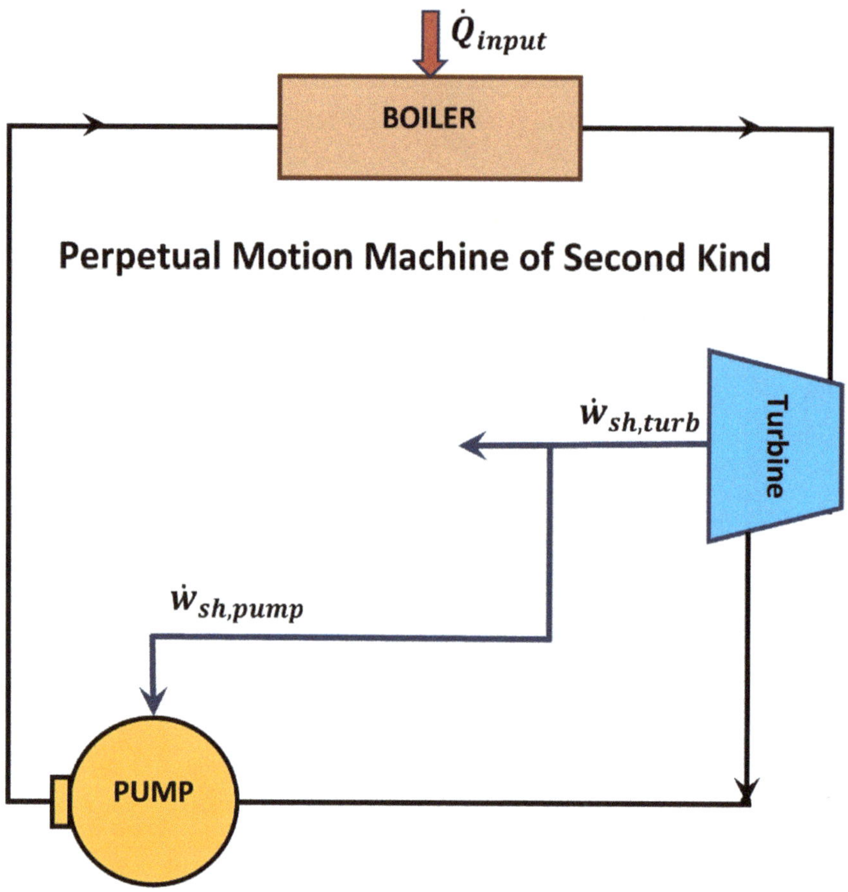

Figure 5.11 Steam power plant as a Perpetual Motion Machine of Second Kind.

NOTATION

c_p	molar constant pressure heat capacity
$c_{p,c}$	molar constant pressure heat capacity of cold fluid
$c_{p,h}$	molar constant pressure heat capacity of hot fluid
g	acceleration due to gravity
h	specific enthalpy
h_s	specific enthalpy at constant entropy
m	mass
$\dot{m}$	mass flow rate
$\dot{m}_c$	mass flow rate of cold fluid
$\dot{m}_h$	mass flow rate of hot fluid
P	pressure
$\dot{Q}$	heat transfer rate
$\dot{Q}_b$	heat transfer rate at the boundary of the system
$\dot{Q}_R$	heat transfer rate from the heat reservoir
s	specific entropy
s_G	entropy generated per unit mass of fluid
$s_{G,CV}$	entropy generated inside control volume per unit mass of fluid
$s_{G,total}$	total entropy generated per unit mass of fluid

S	total entropy
S_{surr}	total entropy of surroundings
$\dot{S}_G$	rate of entropy generation
$\dot{S}_{G,CV}$	rate of entropy generation inside the control volume
$\dot{S}_{G,total}$	rate of total entropy generation
t	time
T	absolute temperature
T_b	absolute temperature at the boundary
T_R	absolute temperature of the heat reservoir
u	specific internal energy
V	velocity of fluid
W_{sh}	shaft work
$W_{sh,min}$	minimum shaft work
$W_{sh,max}$	maximum shaft work
$\dot{W}_{sh}$	rate of shaft work
Z	elevation

Greek Letters

ε_{nozzle}	isentropic nozzle efficiency
$\varepsilon_{turbine}$	isentropic turbine efficiency

FURTHER READING

Pal, R., "Teaching fluid mechanics and thermodynamics simultaneously through pipeline flow experiments", *Fluids* 2019; 4: Article 103, 1-18.
Pal, R., "Teach second law of thermodynamics via analysis of flow through packed beds and consolidated porous media", *Fluids* 2019; 4: Article 116, 1-23.

TUTORIAL PROBLEMS WITH SOLUTION

Problem 1

In a steady flow process, 11.5 kg/s of superheated steam at 175 bar and 450 °C is continuously mixed with 1.65 kg/s of subcooled (compressed) liquid water at 100 bar and 20 °C. The product stream is saturated vapor (steam) at 100 bar. Determine the rate of heat transfer and the rate of entropy generation. The temperature of the surroundings is 298 K.

Solution

The process is shown schematically in the following figure.

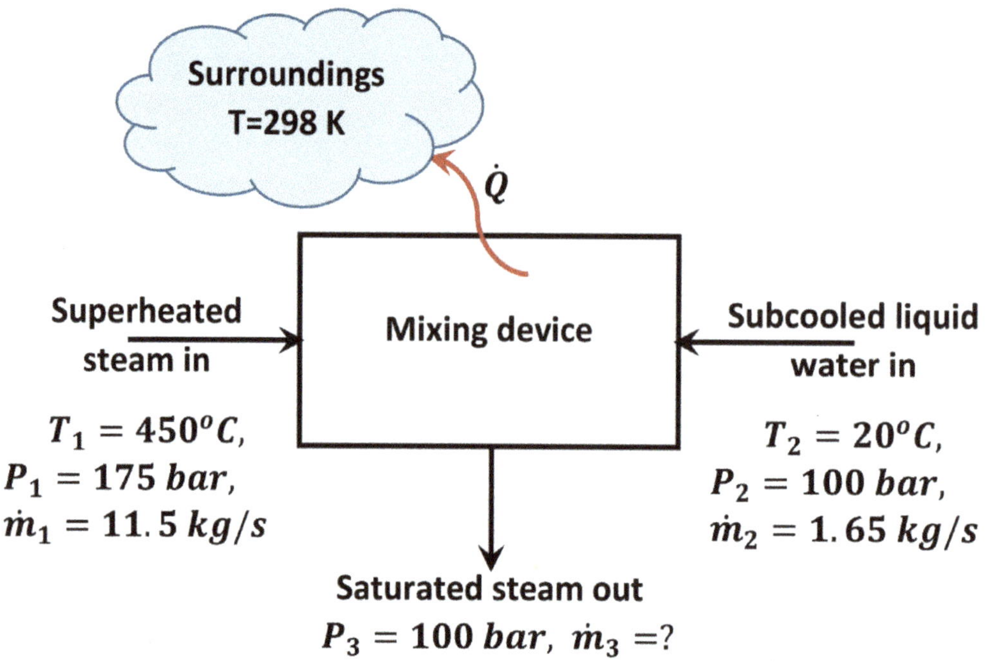

We need specific enthalpies and entropies for the various streams. We obtain this information from the steam tables given in the Appendices M and N.

Superheated steam at 175 bar and 450 °C:

$$h = 3109.7 \text{ kJ/kg} \;;\; s = 6.0184 \text{ kJ/kg.K}$$

Subcooled (compressed) liquid water at 100 bar and 20 °C:

$$h = 93.33 \text{ kJ/kg} \;;\; s = 0.2945 \text{ kJ/kg.K}$$

Saturated steam at 100 bar:

$$h = 2724.7 \text{ kJ/kg} \;;\; s = 5.6149 \text{ kJ/kg.K}$$

Apply first law for open systems:

$$\Sigma_{OUTLETS}\ \dot{m}_i(h_i + \tfrac{1}{2}V_i^2 + gZ_i) - \Sigma_{INLETS}\ \dot{m}_i\ (h_i + \tfrac{1}{2}V_i^2 + gZ_i) = \Sigma\dot{Q} - \dot{W}_{sh}$$

Neglecting kinetic and potential energy changes, and no shaft work:

$$\sum_{OUTLETS} \dot{m}_i h_i - \sum_{INLETS} \dot{m}_i\, h_i = \dot{Q}$$

$$\dot{m}_3 h_3 - (\dot{m}_1 h_1 + \dot{m}_2 h_2) = \dot{Q}$$

Apply mass balance:

$$\dot{m}_3 = \dot{m}_1 + \dot{m}_2 = 11.5 + 1.65 = 13.15 \text{ kg/s}$$

$$\dot{Q} = \dot{m}_3 h_3 - (\dot{m}_1 h_1 + \dot{m}_2 h_2) = 13.15 \times 2724.7 - (11.5 \times 3109.7 + 1.65 \times 93.33) = -85.69 \text{ kW}$$

Thus, heat is rejected to the surroundings at a rate of 85.69 kW.

Apply second law for open systems:

$$\dot{S}_{G,total} = \sum_{OUTLETS} (\dot{m}_i s_i) - \sum_{INLETS} (\dot{m}_i s_i) - \sum \frac{\dot{Q}_i}{T_i} \geq 0$$

$$\dot{S}_{G,total} = \dot{m}_3 s_3 - (\dot{m}_1 s_1 + \dot{m}_2 s_2) - \frac{\dot{Q}}{T_{surr}} = 13.15 \times 5.6149 - (11.5 \times 6.0184 + 1.65 \times 0.2945) - \frac{(-85.69)}{298} = 4.428 \text{ kW/K}.$$

As expected for a natural process, $\dot{S}_{G,total} > 0$

Problem 2

Superheated steam at 35 bars and 450 °C enters the turbine at a velocity of 110 m/s. The exhaust from the turbine is saturated steam at 90 °C and has a velocity of 75 m/s. The turbine develops work of 495 kJ per kg of fluid flowing through the turbine. The average temperature of the surface of the turbine is 468 K. The temperature of the surroundings is 298 K.
Calculate:

a) the amount of heat loss from the turbine per kg of fluid flowing,
b) the amount of entropy production within the turbine per kg of fluid flowing,
c) the total amount of entropy production, internally and externally, per kg of fluid flowing.

Solution

The process is shown schematically in the figure below.

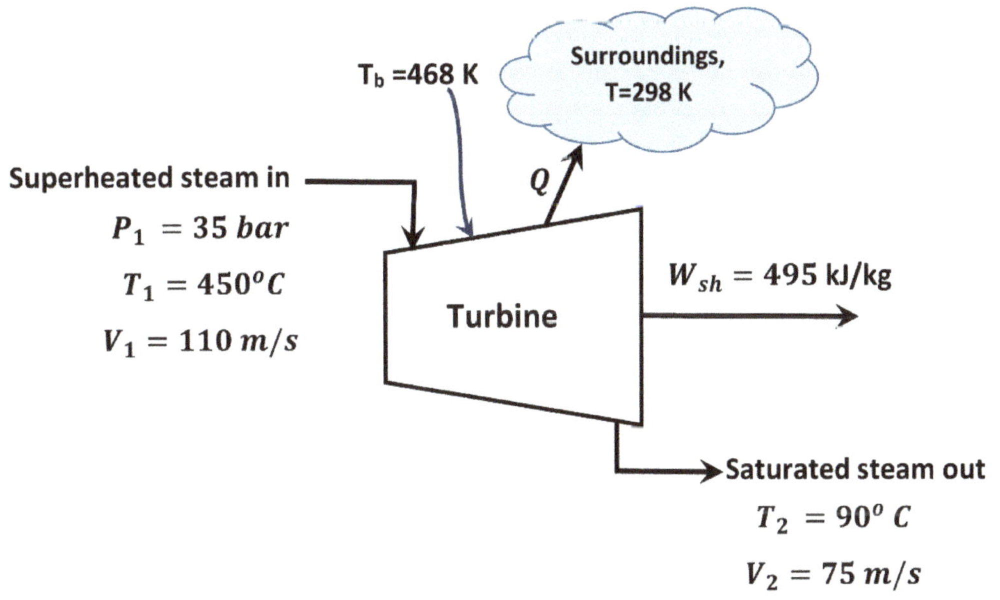

We need specific enthalpies and entropies for the inlet and discharge streams. We obtain this information from the steam tables given in the Appendices M and N.

Superheated steam at 35 bar and 450 °C:

$$h = 3337.2 \text{ kJ/kg} \ ; \ s = 7.0052 \text{ kJ/kg.K}$$

Saturated steam at 90 °C:

$$h = 2660.1 \text{ kJ/kg} \ ; \ s = 7.4799 \text{ kJ/kg.K}$$

Apply first law for open systems:

$$\left[\Delta h + \frac{1}{2}\Delta V^2 + g\Delta Z\right] = Q - W_{sh}$$

Neglecting potential energy changes and re-arranging:

$$Q = W_{sh} + \left[\Delta h + \frac{1}{2}\Delta V^2\right] = 495 + (2660.1 - 3337.2) + \frac{(75^2 - 110^2)}{2\times 10^3} = -185.34 \text{ kJ/kg}$$

Apply entropy balance for control volume:

$$\dot{S}_{G,CV} = \dot{m}(s_2 - s_1) - \frac{\dot{Q}_b}{T_b}$$

On a unit mass basis,

$$s_{G,CV} = (s_2 - s_1) - \frac{Q_b}{T_b} = (7.4799 - 7.0052) - \frac{(-185.34)}{468} = 0.87 \text{ kJ/kg.K}$$

$$s_{G,CV} > 0 \text{, as expected.}$$

Apply second law for open systems:

$$\dot{S}_{G,total} = \dot{m}(s_2 - s_1) - \frac{Q}{T_{surr}} \geq 0$$

On a unit mass basis,

$$s_{G,total} = (s_2 - s_1) - \frac{Q}{T_{surr}} = (7.4799 - 7.0052) - \frac{(-185.34)}{298} = 1.967 \text{ kJ/kg.K}$$

As expected, $S_{G,total} > 0$. Note that $S_{G,total} > S_{G,CV}$ as entropy is generated externally due to irreversible heat transfer between turbine surface and surroundings reservoir.

Problem 3

An ideal gas (air) flows through a converging nozzle in which the diameter is reduced from 150 to 60 mm. The inlet conditions are: $P_1 = 150\ kPa, T_1 = 160^oC, V_1 = 50\ m/s$. Calculate the exit velocity if the exit pressure is $P_2 = 70\ kPa$. Assume isentropic flow. The molecular weight of the gas is 29 g/mol and the heat capacity is $c_p = 7R/2$.

Solution

Refer to the following diagram.

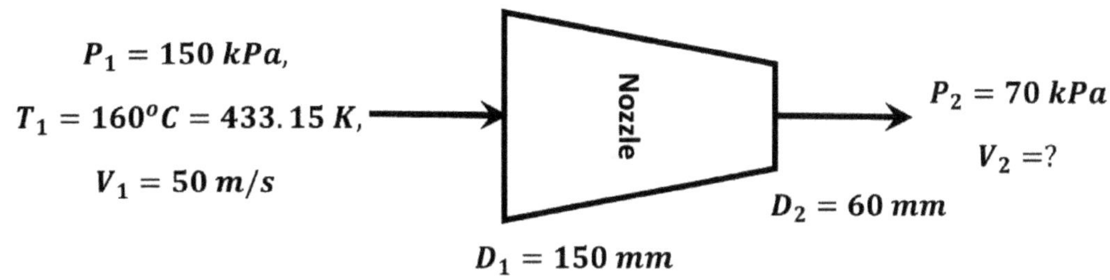

Let us first estimate the exit temperature. The process is isentropic, that is, $\Delta s = 0$

$$\Delta s = c_p ln\left(\frac{T_2}{T_1}\right) - Rln\left(\frac{P_2}{P_1}\right) = 0$$

$$ln\left(\frac{T_2}{T_1}\right) = \frac{R}{c_p} ln\left(\frac{P_2}{P_1}\right) = \frac{8.314}{3.5 \times 8.314} ln\left(\frac{70}{150}\right) = -0.21775$$

$$\frac{T_2}{T_1} = 0.8043\ ;\ T_2 = 348.383\ K$$

From first law for opens systems,

$$h_1 - h_2 = \frac{1}{2}(V_2^2 - V_1^2)$$

$$V_2^2 = 2(h_1 - h_2) + V_1^2$$

For ideal gas,

$$\Delta h = h_2 - h_1 = c_p(T_2 - T_1) = 3.5 \times 8.314 \times (348.383 - 433.15) = -2466.63\ \mathrm{J/mol}$$
$$= \frac{-2466.63}{29} \times 1000 = -85056.38\ \mathrm{J/kg}$$

$$V_2^2 = 2 \times 85056.38 + 50^2 = 172612.75$$

$$V_2 = 415.5\ \mathrm{m/s}$$

Problem 4

An ideal gas expands isentropically through a turbine from $P_1 = 1000\ kPa$ and $T_1 = 800^oC$ to $P_2 = 100\ kPa$. Calculate the work output in kJ/kg. The molecular weight of the gas is 29 g/mol and the heat capacity is $c_p = 7R/2$.

Solution

Refer to the following diagram.

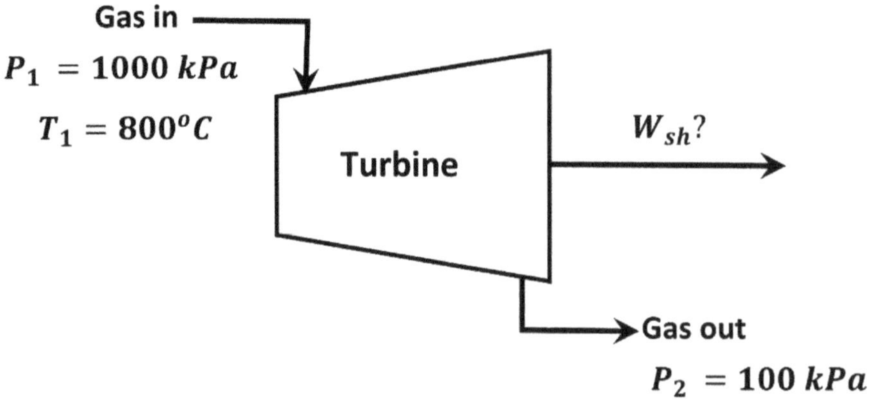

Let us first estimate the exit temperature. The process is isentropic, that is, $\Delta s = 0$

$$\Delta s = c_p ln\left(\frac{T_2}{T_1}\right) - Rln\left(\frac{P_2}{P_1}\right) = 0$$

$$ln\left(\frac{T_2}{T_1}\right) = \frac{R}{c_p} ln\left(\frac{P_2}{P_1}\right) = \frac{8.314}{3.5 \times 8.314} ln\left(\frac{100}{1000}\right) = -0.65788$$

$$\frac{T_2}{T_1} = 0.518\ ;\ T_2 = 555.89\ K$$

From first law for opens systems, neglecting kinetic and potential energy changes, and taking $Q = 0$:

$$W_{sh} = -\Delta h = -c_p(T_2 - T_1) = -3.5 \times 8.314 \times (555.89 - 1073.15) = 15051.75\ \mathrm{J/mol}$$
$$= \frac{15051.75}{29} \times 1000 = 519025.82\ \mathrm{J/kg} = 519.03\ \mathrm{kJ/kg}$$

Problem 5

Steam at 1.2 MPa (12 bars) and 500^oC is expanded in a turbine. The exhaust steam is at 100 kPa. The efficiency of the turbine is 85%. Calculate the work output in kJ/kg. Use steam tables to determine the properties of the steam.

Solution

Refer to the following diagram.

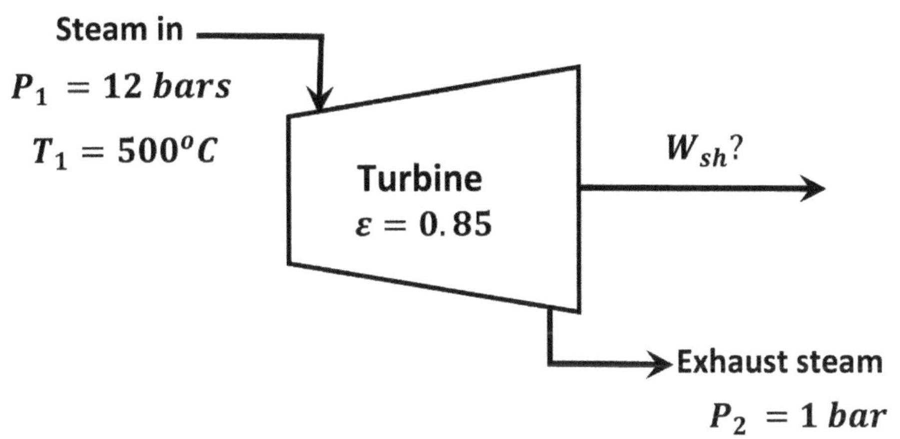

From first law for opens systems, neglecting kinetic and potential energy changes, and taking $Q = 0$:

$$W_{sh} = -\Delta h$$

The efficiency of the turbine is defined as:

$$\varepsilon = \frac{\Delta h}{(\Delta h)_s} = \frac{h_2 - h_1}{h_{2s} - h_1}$$

where Δh is actual enthalpy change and $(\Delta h)_s$ is isentropic enthalpy change. To calculate W_{sh}, we need to calculate actual enthalpy change Δh as:

$$\Delta h = \varepsilon(\Delta h)_s$$

How do we calculate $(\Delta h)_s$?

From steam tables (Appendix N), we find that:

- at 1.2 MPa (12 bars) and 500^oC, steam is superheated with h_1 = 3476.3 kJ/kg and s_1 =7.6759 kJ/kg.K

For isentropic expansion in the turbine, $s_2 = s_1$ = 7.6759 kJ/kg.K.

To calculate $(\Delta h)_s$, we need to find enthalpy h_{2s} at P_2 =1 bar and $s_2 = s_1$ = 7.6759 kJ/kg.K.

From saturated steam tables (Appendix M),

- at P_2 =1 bar , the entropy of saturated vapor s^v = 7.3602 kj/kg.K

Since $s_2 > s^v$, this indicates that the steam is superheated steam at P_2 =1 bar and $s_2 = s_1$ = 7.6759 kJ/kg.K.

From superheated steam tables (Appendix N), at P_2 =1 bar

T	***h* (kJ/kg)**	***s* (kJ/kg.K)**
150^oC	2776.4	7.6134
200^oC	2875.3	7.8343

To determine h_{2s} corresponding to P_2 =1 bar and $s_2 = s_1 = 7.6759$ kJ/kg.K, we need to interpolate the above table as follows:

$$h_{2s} = 2776.4 + \frac{(2875.3-2776.4)}{(7.8343-7.6134)}(7.6759 - 7.6134) = 2804.38 \text{ kJ/kg}$$

Hence,

$$(\Delta h)_s = h_{2s} - h_1 = 2804.38 - 3476.3 = -671.92 \text{ kJ/kg}$$

$$\Delta h = \varepsilon(\Delta h)_s = 0.85 \times (-671.92) = -571.132 \text{ kJ/kg.}$$

Thus, the shaft work produced is

$$W_{sh} = -\Delta h = 571.132 \text{ kJ/kg.}$$

Problem 6

Saturated refrigerant vapor (HFC-134A) at -36°C is compressed isentropically to 5 bars. Assuming negligible changes in kinetic and potential energies, calculate the work required in kJ/kg.

Solution

Refer to the diagram below.

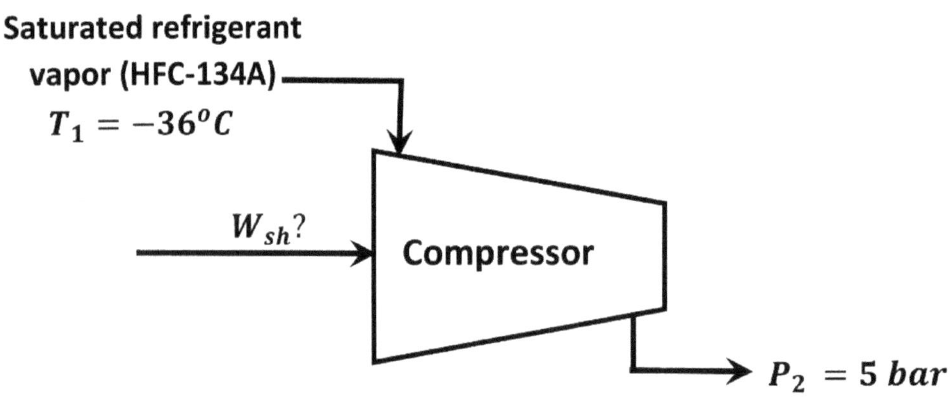

From first law for opens systems, neglecting kinetic and potential energy changes, and taking $Q = 0$:

$$W_{sh} = -\Delta h = h_1 - h_2$$

We need h_1 and h_2 in order to calculate the required work. As the expansion is isentropic, $h_2 = h_{2s}$ (same entropy in states 1 and 2) and $-\Delta h = -(\Delta h)_s = h_1 - h_{2s}$.

From saturated Refrigerant 134A tables given in Appendix K, we find that:

- for saturated refrigerant vapor at $T_1 = -36^oC$, $P_1 = 0.6332$ bar, $h_1 = 225.40$ kJ/kg, $s_1 = 0.9506$ kJ/kg.K

We now need h_{2s} . This is the enthalpy corresponding to P_2 =5 bar and $s_2 = s_1 = 0.9506$ kJ/kg.K. From saturated refrigerant tables (Appendix K),

- at P_2 =5 bar, the entropy of saturated vapor $s^v = 0.9116$ kJ/kg.K

Since $s_2 > s^v$, this indicates that the refrigerant vapor is superheated at P_2 =5 bar and $s_2 = s_1 = 0.9506$ kJ/kg.K.

From superheated refrigerant vapor tables (Appendix L), at P_2 =5 bar

T	***h* (kJ/kg)**	***s* (kJ/kg.K)**
20^oC	260.34	0.9264
30^oC	270.28	0.9597

To determine h_{2s} corresponding to P_2 = 5 bar and $s_2 = s_1 = 0.9506$ kJ/kg.K, we need to interpolate the above table as follows:

$$h_{2s} = 260.34 + \frac{(270.28-260.34)}{(0.9597-0.9264)}(0.9506 - 0.9264) = 267.56 \text{ kJ/kg}$$

Thus, the shaft work required is

$$W_{sh} = -(\Delta h)_s = h_1 - h_{2s} = 225.40 - 267.56 = -42.16 \text{ kJ/kg.}$$

Problem 7

Consider a steady throttling process where an ideal gas at 15 bar and 385 K is throttled to 1 bar. Estimate the final temperature and the change in the entropy of the gas. The heat capacity of the gas $c_p = 7R/2$.

Solution

Throttling is an isenthalpic process. Thus,

$$\Delta h = c_p\Delta T = 0$$

When ideal gas is throttled, the temperature does not change and $T_1 = T_2$. The entropy change is:

$$\Delta s = -Rln\left(\frac{P_2}{P_1}\right) = -8.314 \times ln\left(\frac{1}{15}\right) = 22.51 \text{ J/mol.K}$$

PRACTICE PROBLEMS

Problem 1

Air at 550 K and 1 atm is mixed continuously in a mixing device at a flow rate of 3 mol/s with another air stream at 395 K and 1 atm at a flow rate of 1.5 mol/s. The product air stream leaving the mixing device is at 515 K and 1 atm. The surroundings are at a temperature of 298 K. Determine the rate of entropy generation.

You can assume air to be an ideal gas with a heat capacity of c_p=7R/2.

Problem 2

Three streams of air are mixed continuously in a mixing device. The flow rates and conditions of streams entering the device are specified as follows:

Stream 1: $\dot{n}_1 = 5$ mol/s, $T_1 = 640$ K, $P_1 = 1$ atm

Stream 2: $\dot{n}_2 = 3.5$ mol/s, $T_2 = 440$ K, $P_2 = 1$ atm

Stream 3: $\dot{n}_3 = 4.5$ mol/s, $T_3 = 500$ K, $P_3 = 1$ atm

The product air stream leaving the mixing device is at 550 K and 1 atm. You can assume air to be an ideal gas with a heat capacity $c_p = 7R/2$. The surroundings are at 298 K.

Determine the rate of entropy generation.

Problem 3

Superheated steam at 200 bar (20 MPa) and 450 °C is continuously mixed at a mass flow rate of 15 kg/s with 2.5 kg/s of subcooled (compressed) liquid water at 100 bar (10 MPa) and 40 °C. The product stream is saturated vapor (steam) at 100 bar (10 MPa). Determine the rate of heat transfer and the rate of entropy generation. The temperature of the surroundings is 298 K.

Use steam tables given in the appendix to determine the properties.

Problem 4

Superheated steam flows at steady state through a converging insulated nozzle. The inlet diameter is 4.5 cm. The inlet conditions are: $P_1 = 10\ bars$ (1 MPa) , $T_1 = 300^oC,\ V_1 = 10\,m/s$. Calculate the exit velocity if the exit conditions are: $P_2 = 5\ bars$ (0.5 MPa) , $T_2 = 200^oC$. What is the exit diameter?

Use steam tables given in the appendix to determine the properties.

Problem 5

Consider a converging nozzle in which the diameter is reduced from 125 to 45 mm. Air flows through this nozzle with inlet conditions: $P_1 = 250\ kPa, T_1 = 125^oC, V_1 = 35\ m/s$. Calculate the exit velocity if the exit pressure is $P_2 = 50\ kPa$. Assume isentropic flow. The molecular weight of the gas is 29 g/mol and the heat capacity is $c_p = 7R/2$.

Problem 6

Steam at 3 MPa (30 bars) and 400 °C is expanded in a turbine. The exhaust steam is saturated vapor at 0.6 MPa (6 bars). The power output of the turbine is 2500 kW. What is the mass flow rate of steam through the turbine? What is the rate of entropy generation?

Assume zero heat loss and negligible changes in kinetic and potential energies.

Use steam tables given in the appendix to determine the properties.

Problem 7

Saturated refrigerant vapor (HFC-134A) at -32 °C is compressed to 8 bars. The efficiency of the compressor is 80%. Calculate the work required by the compressor in kJ/kg.

Use refrigerant HFC-134A tables for the properties.

Assume zero heat loss and negligible changes in kinetic and potential energies.

Chapter 6

Thermodynamics of Compressible and Incompressible Flows in Ducts and Pipes

6.1 LEARNING OUTCOMES

Upon completion of this chapter, students will be able to:

- Analyze steady isentropic compressible flow in a duct of varying cross-sectional area
- Explain velocity and pressure variations in subsonic and supersonic isentropic compressible flows in a duct of varying cross-sectional area
- Explain choking phenomenon in isentropic compressible flow in converging-diverging ducts
- Calculate velocity, pressure, temperature, and density variations in isentropic compressible flows
- Analyze steady non-isentropic compressible flows such as Fanno, Rayleigh and isothermal flows. Calculate velocity, pressure, temperature, and density variations in such flows.
- Analyze steady non-isentropic adiabatic and non-adiabatic incompressible flows in constant area pipes. Determine entropy generation in such flows.
- Determine the relative magnitudes of heat transfer and fluid flow irreversibilities in non-adiabatic incompressible flows in pipes.
- Estimate the optimum Reynolds number and optimum pipe diameter in non-adiabatic incompressible flows in pipes using minimization of entropy generation.

6.2 INTRODUCTION

Flow through ducts and pipes is a common occurrence. As noted in Chapter 3, pipelines are used to transport water, oil, gas and many more fluidic substances from one place to another. Pipelines that transport oil and gas form a critical part of the energy industry. In this chapter, we will cover thermodynamics of both compressible and incompressible flows in ducts and pipes.

The words *duct* and *pipe* are often used interchangeably. Ducts are normally made of light materials (aluminum) and are used to distribute air and gases. They can have different shapes of cross-section (round, rectangular, oval, etc.). Pipes, on the other hand, are made from heavier materials (steel) and are used to transport and distribute liquids and gases. They are cylindrical in shape with circular (round) cross-section.

6.3 STEADY ISENTROPIC COMPRESIBLE FLOWS

Compressible flows are important in the design and operation of industrial gas pipelines involving high speed flows. Design and operation of air ducts used for heating and ventilation require a good understanding of compressible flows. Compressible flows are also important in the design and operation of jet engines, rockets and air crafts.

6.3.1 Influence of Duct Area Variation on Pressure and Velocity in Isentropic Compressible Flow

Consider a duct of varying cross-sectional area (see Figure 6.1). A compressible fluid (gas) flows through the duct at steady state. The flow is one-dimensional and the flow properties are uniform at each cross-section. We will assume that the flow is *isentropic* implying that the flow process is adiabatic and reversible (frictionless). There is no shaft work involved and there is no potential energy change.

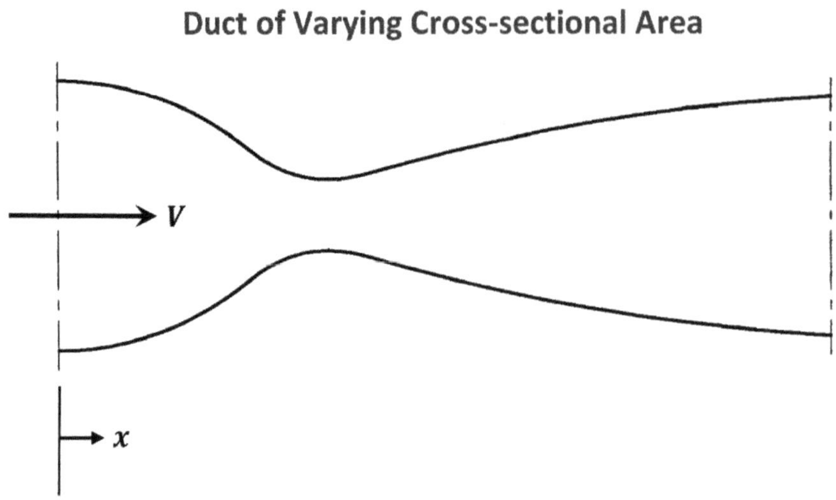

Figure 6.1 Duct with variable cross-sectional area.

The mass flow rate of fluid ($\dot{m}$) through the duct is:

$$\dot{m} = \rho A V \tag{6.1}$$

where ρ is fluid density, A is cross-sectional area of the duct, and V is fluid velocity. Since $\dot{m}$ is constant at steady state, it does not very with distance (x) along the duct. Thus,

$$\frac{d(\rho A V)}{dx} = 0 \tag{6.2}$$

Using the product rule, Equation (6.2) gives:

$$\frac{1}{V}\frac{dV}{dx} + \frac{1}{A}\frac{dA}{dx} + \frac{1}{\rho}\frac{d\rho}{dx} = 0 \tag{6.3}$$

From the first law for control volume, written in a differential form:

$$dh + d(V^2/2) = 0 \tag{6.4}$$

Using the second Tds relation, we obtain the following relation for isentropic process:

$$dh = dP/\rho \tag{6.5}$$

Combining Equations (6.4) and (6.5), we get:

$$dP/\rho + d(V^2/2) = 0 \tag{6.6}$$

Equation (6.6) could be further re-written as:

$$\frac{1}{\rho}\frac{dP}{dx} + V\frac{dV}{dx} = 0 \tag{6.7}$$

Here we make use of the relationship for velocity of sound (c) in a compressible fluid, given as [1]:

$$c = \sqrt{\left(\frac{\partial P}{\partial \rho}\right)_s} \tag{6.8}$$

From Equations (6.7) and (6.8), it follows that:

$$\frac{1}{\rho}\frac{d\rho}{dx} = -\frac{V}{c^2}\frac{dV}{dx} \tag{6.9}$$

Combining mass and energy-based relations, Equations (6.3) and (6.9), yields:

$$\frac{dV}{dx} = \left(\frac{V}{A}\right)\frac{1}{(M^2-1)}\frac{dA}{dx} \tag{6.10}$$

where M is a dimensionless number called *Mach number*, defined as:

$$M = \frac{V}{c} \tag{6.11}$$

From Equations (6.7) and (6.10), we get:

$$\frac{dP}{dx} = -\left(\frac{\rho V^2}{A}\right)\frac{1}{(M^2-1)}\frac{dA}{dx} \tag{6.12}$$

Equations (6.10) and (6.12) are important relations. Equation (6.10) describes the influence of area variation on the fluid velocity and Equation (6.12) describes the influence of area variation on pressure.

Implications of Equations (6.10) and (6.12)

Upon comparison of Equations (6.10) and (6.12), it is clear that velocity and pressure always move in opposite directions. If velocity of the fluid increases, the pressure will decrease and vice versa. Therefore, we need to explore the implications of just one equation, that is, the velocity Equation (6.10). How does the velocity vary when the cross-sectional area of the duct is varied? The variation of velocity with the variation of area depends on the Mach number, M as described below:

Subsonic Flow ($M < 1$): For subsonic flow, $M^2 - 1 < 0$. Under this condition, Equation (6.10) predicts that V and A vary in the opposite directions. As A decreases, velocity increases (pressure decreases) and as A increases, velocity decreases (pressure increases). The variations of velocity with area for subsonic flows are summarized in Figure 6.2. Thus, subsonic flows of compressible fluids behave more like an incompressible fluid flow.

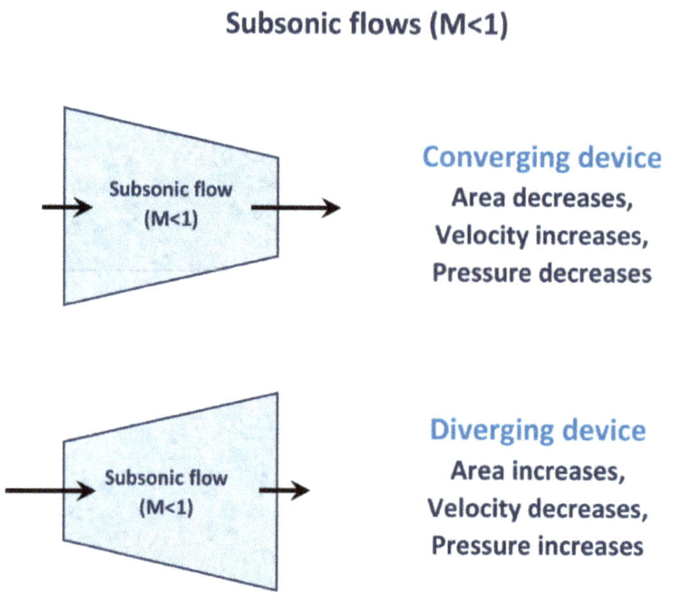

Figure 6.2 Variations of velocity and pressure with area in subsonic flows.

Supersonic Flow ($M > 1$): For supersonic flow, $M^2 - 1 > 0$. Under this condition, Equation (6.10) predicts that V and A vary in the same directions. As A decreases, velocity decreases (pressure increases) and as A increases, velocity increases (pressure decreases). The variations of velocity with area for supersonic flows are summarized in Figure 6.3. Thus, supersonic flows of compressible fluids behave in a manner opposite to incompressible fluid flow.

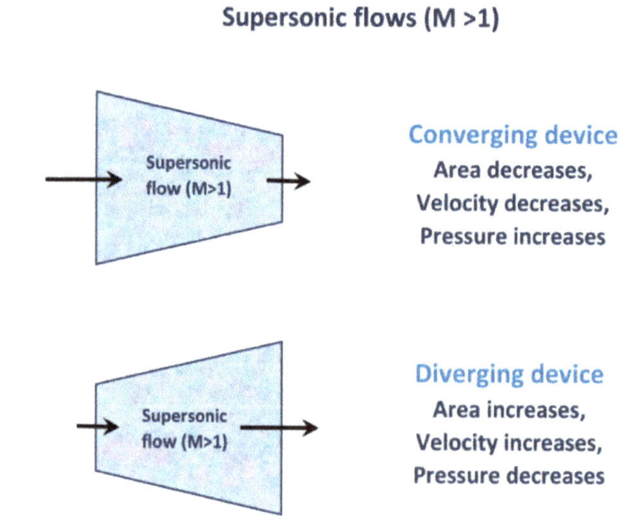

Figure 6.3 Variations of velocity and pressure with area in supersonic flows.

Other Implications of Equation (6.10): a) For subsonic flows ($M^2 - 1 < 0$), velocity is maximum where dA/dx=0, that is, at throat of a converging-diverging device shown in Figure 6.4; b) For supersonic flows ($M^2 - 1 > 0$), velocity is minimum where dA/dx=0, that is, at throat of a converging-diverging device shown in Figure 6.5; and c) Mach number can be unity ($M = 1$) only at the throat of a converging-diverging device where $dA/dx = 0$.

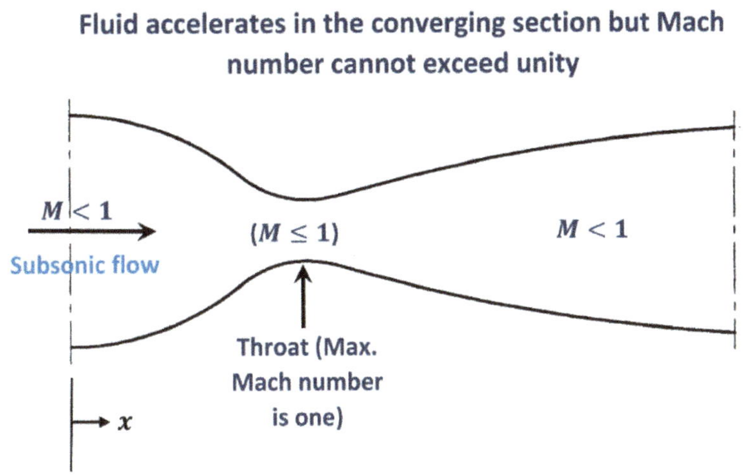

Figure 6.4 Subsonic flow ($M^2 - 1 < 0$). Velocity is maximum at the throat but M cannot exceed unity.

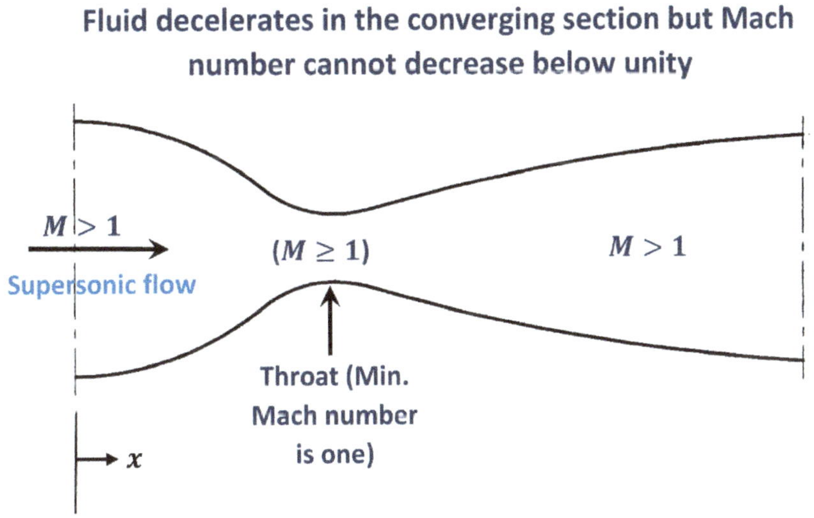

Figure 6.5 Supersonic flow ($M^2 - 1 > 0$). Velocity is minimum at the throat but M cannot decrease below unity.

6.3.2 Choking Phenomenon in Converging-Diverging Ducts

Consider isentropic flow of a compressible fluid (gas) in a converging-diverging (CD) nozzle as shown in Figure 6.6. The nozzle is connected to a large reservoir of stagnant gas at the upstream end. At the downstream end of the CD nozzle, the gas exhausts into an exhaust chamber where the pressure can be controlled and varied. The pressure in the exhaust chamber, denoted as P_b, is often referred to as back-pressure.

When the backpressure P_b is equal to the pressure in the upstream gas reservoir, $P_b = P_o$, then there is no flow through the CD nozzle. If the backpressure is lowered below P_o, then the fluid flows through the nozzle. Figure 6.7 shows the pressure profiles of the nozzle in terms of pressure ratio P/P_o where P is the local static pressure in the nozzle.

When $P_b = P_{b,1}$ (slightly below P_o), flow in the CD nozzle everywhere in the converging and diverging sections and throat is *subsonic* ($M < 1$). However, the pressure decreases and the velocity increases in the converging section. The pressure reaches a minimum value, and the velocity reaches a maximum value at the throat of the nozzle (see Figure 6.7). In the diverging section, the pressure increases and the velocity decreases.

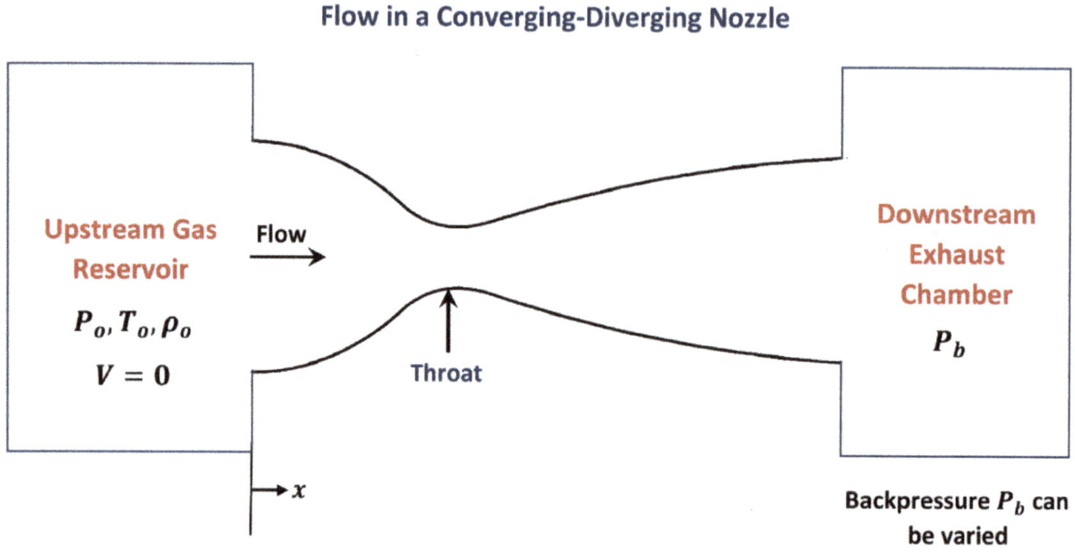

Figure 6.6 Flow of compressible fluid in a converging-diverging nozzle.

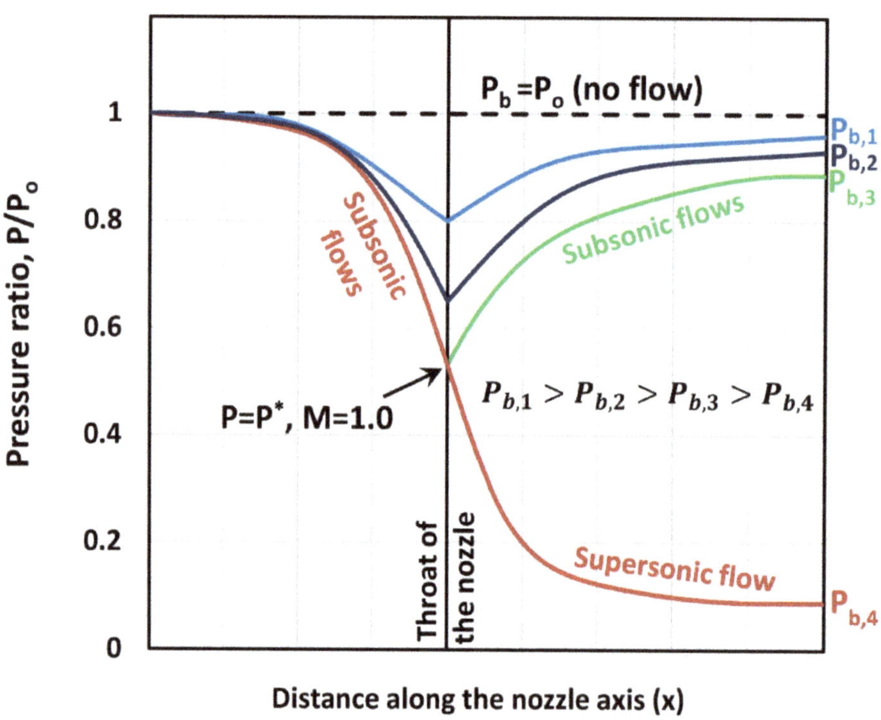

Figure 6.7 Pressure ratio (P/P_o) profiles for isentropic flow of compressible fluid in a converging-diverging nozzle. The backpressure is decreased from P_o to $P_{b,4}$. The nozzle is choked when $P = P^*$ and $M = 1$ at the nozzle throat.

When the backpressure is decreased further to $P_b = P_{b,2}$ (where $P_{b,2} < P_{b,1}$), the mass flow rate through the nozzle increases but the flow everywhere in the nozzle is still *subsonic* ($M < 1$). The velocity everywhere in the nozzle increases and the pressure everywhere decreases as compared with the previous case where $P_b = P_{b,1}$ (see Figure 6.7).

When the backpressure is decreased further from $P_{b,2}$ to $P_{b,3}$ (where $P_{b,3} < P_{b,2}$), the mass flow rate through the nozzle increases but the flow remains subsonic in the converging section. However, at the throat of the nozzle the flow becomes *sonic* ($M = 1$). This condition where the flow becomes sonic at the throat is called *critical condition*. The flow properties at the throat when $M = 1$ are referred to as critical properties and are denoted with * as the superscript. For example, P^* is critical pressure at the throat corresponding to $M = 1$. When the critical condition ($M = 1,\ P = P^*$) is reached at the throat of the CD nozzle, the nozzle becomes *choked*. With further decrease in backpressure, the mass flow rate through the nozzle does not increase anymore. The mass flow rate through the nozzle reaches a maximum value when the critical condition is reached at the throat. Thus, choked flow is a limiting condition where the mass flow rate through the nozzle will not increase with a further reduction in the backpressure for a fixed upstream pressure and temperature.

Once a critical condition is reached at the throat, no change is possible in the converging section where the flow remains subsonic. However, there are two possibilities in the diverging section. The equations governing isentropic flow of compressible fluids give two solutions for the diverging section when the nozzle is choked, depending upon the value of the backpressure. Referring to Figure 6.7, the flow in the diverging section is subsonic when $P_b = P_{b,3}$. When the value of P_b is $P_{b,4}$ (much lower than $P_{b,3}$), the flow in the diverging section becomes *supersonic*. However, the mass flow rate through the nozzle remains the same at $P_{b,3}$ and $P_{b,4}$.

Note that in all cases considered in Figure 6.7, the static pressure in the nozzle at the exit is equal to the backpressure. Thus, there is no expansion or contraction of exhaust plume. The width of the exhaust plume is equal to the diameter of the nozzle at the exit, as shown in Figure 6.8.

Figure 6.9 shows the profiles of Mach number in the nozzle when the critical condition ($M = 1,\ P = P^*$) is reached at the throat. When P_b is $P_{b,3}$, the flow is subsonic ($M < 1$) in the converging section, sonic ($M = 1$) at the throat, and subsonic in the diverging section. When P_b is much lower corresponding to $P_{b,4}$, the flow in the diverging section becomes supersonic ($M > 1$).

Figure 6.10 shows the plot of mass flow rate through the nozzle when the backpressure is decreased. The mass flow rate increases as the backpressure is decreased. However, the mass flow rate through the nozzle becomes constant once the choking occurs. Any further decrease in back pressure does not increase the mass flow rate through the nozzle.

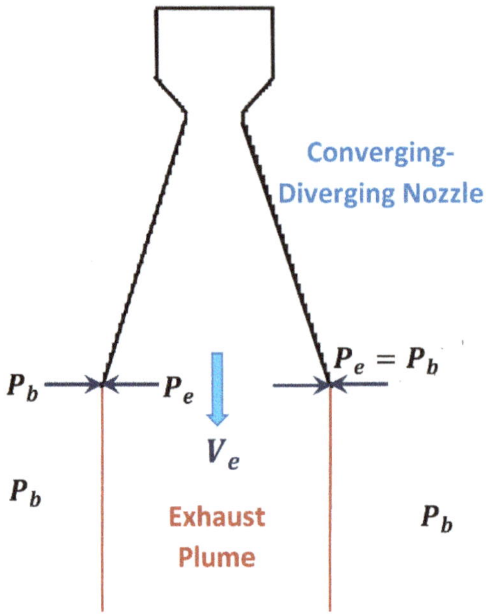

Figure 6.8 Exhaust plume from converging-diverging nozzle under isentropic conditions.

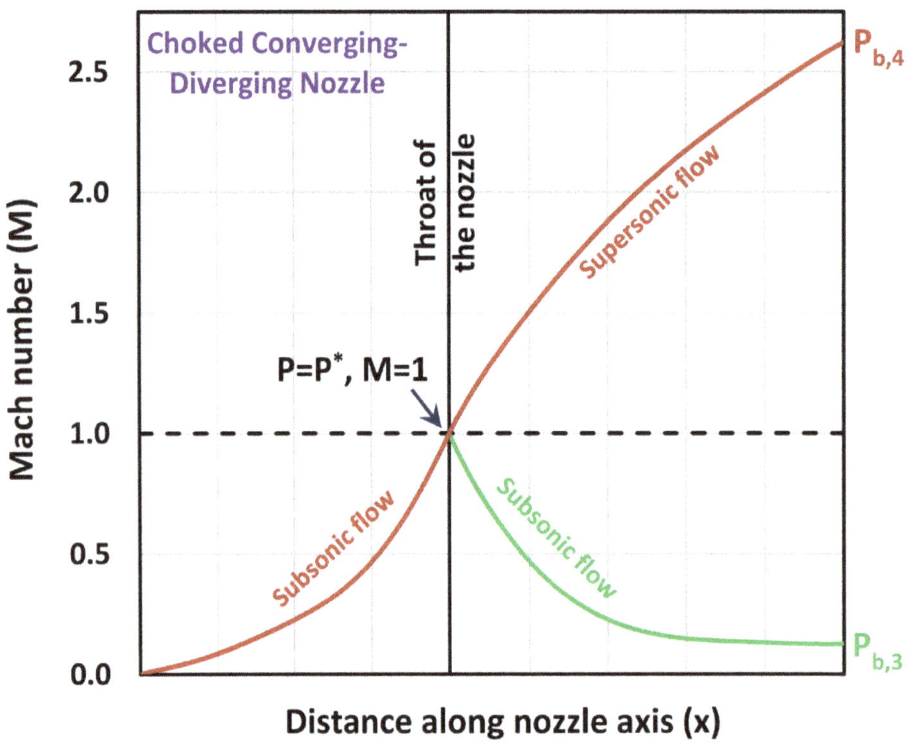

Figure 6.9 Mach number profiles for isentropic flow of compressible fluid in a converging-diverging nozzle. The nozzle is choked, that is, when $P = P^*$ and $M = 1$ at the nozzle throat.

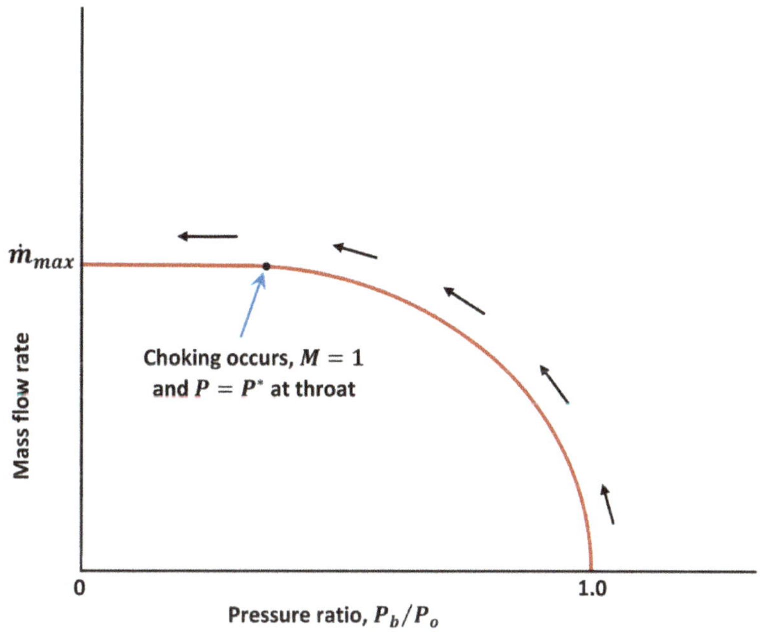

Figure 6.10 Mass flow rate in the nozzle as a function of backpressure ratio, P_b/P_o.

Why Does Choking Occur in Nozzles?

The pressure disturbance in a fluid travels with the speed of sound (sound itself is a pressure disturbance). When backpressure in the downstream exhaust chamber (see Figure 6.6) is decreased below the upstream gas reservoir pressure, a signal (pressure disturbance) is transmitted from the downstream exhaust chamber to the upstream gas reservoir at a sonic speed. The upstream gas reservoir responds to the signal by sending fluid through the nozzle. With further reduction in backpressure, fluid flows from the gas reservoir at a faster rate. However, when the critical condition is reached at the throat, that is, the velocity of the fluid at the throat reaches the sound velocity, the pressure disturbance signal (which travels with the speed of sound) from the downstream exhaust chamber is not able to reach the upstream gas reservoir as the fluid itself is moving in the opposite direction at a sonic speed at the throat. Thus the pressure disturbance signal is washed away and the upstream gas reservoir and converging section of the nozzle are cut off from the backpressure exhaust chamber. Consequently, the nozzle is choked and the mass flow rate through the nozzle becomes constant.

What Happens When $P_{b,4} \leq P_b < P_{b,3}$ and $P_b < P_{b,4}$?

When $P_{b,4} \leq P_b < P_{b,3}$ *and* $P_b < P_{b,4}$ (see Figure 6.7), nothing changes in the converging section and throat. The flow remains subsonic in the converging section and sonic at the throat. The mass flow rate remains constant. However, the flow in the diverging section becomes complicated due to the generation of shocks. Shocks are characterized by sudden changes in the flow regime from supersonic to subsonic accompanied by sharp changes in the properties (pressure, temperature, density). In other words, shocks are irreversible discontinuities with large changes in properties. When P_b is lowered below $P_{b,3}$, the flow becomes supersonic over a certain initial portion of the diverging section and then terminates in a shock within the diverging section. Downstream the shock, the flow in the diverging section remains subsonic and isentropic. With further decrease in P_b the location of the shock moves toward the exit plane and a larger portion of the diverging section is covered by supersonic flow. Interestingly, when $P_b = P_{b,4}$, no shocks are generated, and the flow is supersonic and isentropic over the entire diverging section. This is an ideal condition (design condition) to operate a converging-diverging nozzle when supersonic flows are desired. Only when $P_{b,4} < P_b < P_{b,3}$ and $P_b < P_{b,4}$, shocks are generated. When $P_{b,4} < P_b < P_{b,3}$, the shocks are generated within the nozzle and when $P_b < P_{b,4}$, the shocks occur just outside the nozzle [2].

6.3.3 Further Analysis of Isentropic Compressible Flow in Converging-Diverging Ducts

We begin with Equation (6.6) derived earlier using the first law of thermodynamics and second Tds relation. This equation is integrated with respect to *stagnation state* and corresponding stagnation properties. Stagnation state is an imaginary state defined to be a state where the actual fluid velocity is reduced to zero isentropically. The stagnation properties of interest are: pressure, temperature, and density. If the actual flow is isentropic, then the stagnation properties are constant throughout the flow, and they correspond to the upstream gas reservoir properties (see Figure 6.6). Note that the fluid is stagnant in the upstream gas reservoir. In the following discussion, the stagnation properties are denoted as: P_o , T_o, ρ_o corresponding to stagnation pressure, temperature, and density, respectively.

We assume that the fluid is an ideal gas. For an ideal gas undergoing isentropic (reversible adiabatic) process, the following relation was derived in Chapter 2 (see Equation (2.54)):

$$Pv^{\gamma} = C \tag{6.13}$$

where C is a constant. From Equation (6.13), it follows that:

$$\rho = aP^{1/\gamma} \tag{6.14}$$

where a is another constant ($= C^{-1/\gamma}$). From Equations (6.6) and (6.14), it follows that:

$$dP/aP^{1/\gamma} + d(V^2/2) = 0 \tag{6.15}$$

Upon integration from actual state to stagnation state, we get:

$$-\int_V^0 d(V^2/2) = \frac{1}{a}\int_P^{P_o} \frac{dP}{P^{1/\gamma}} \tag{6.16}$$

Upon evaluation of the integrals, Equation (6.16) gives:

$$\frac{V^2}{2} = -\left(\frac{P}{\rho}\right)\left(\frac{\gamma}{\gamma-1}\right)\left[1-\left(\frac{P_o}{P}\right)^{\frac{\gamma-1}{\gamma}}\right] \tag{6.17}$$

The density of ideal gas can be expressed as:

$$\rho = \frac{P}{RT} \tag{6.18}$$

where R is a specific gas constant, which has a unique value for each gas. The units of R is J/kg.K. From Equations (6.17) and (6.18), it follows that:

$$\frac{V^2}{2} = -(RT)\left(\frac{\gamma}{\gamma-1}\right)\left[1-\left(\frac{P_o}{P}\right)^{\frac{\gamma-1}{\gamma}}\right] \tag{6.19}$$

The velocity of sound in ideal gas is given as:

$$c = \sqrt{\gamma RT} \tag{6.20}$$

Equation (6.20) follows from the defining relation for the velocity of sound in a compressible fluid, that is, Equation (6.8). From Equations (6.19) and (6.20), we get:

$$M^2\left(\frac{\gamma-1}{2}\right) = -\left[1-\left(\frac{P_o}{P}\right)^{\frac{\gamma-1}{\gamma}}\right] \tag{6.21}$$

where M is the Mach number defined in Equation (6.11). Equation (6.21), upon rearrangement, gives:

$$\frac{P}{P_o} = \left[1+\left(\frac{\gamma-1}{2}\right)M^2\right]^{-\frac{\gamma}{\gamma-1}} \tag{6.22}$$

This is an important relation. It allows us to calculate the pressure ratio in isentropic flow in a nozzle at any axial position provided that the Mach number is known.

From Equation (6.14), it can be readily shown that:

$$\frac{\rho}{\rho_o} = \left(\frac{P}{P_o}\right)^{1/\gamma} \tag{6.23}$$

Combining Equations (6.22) and (6.23) gives the density ratio ρ/ρ_o as:

$$\frac{\rho}{\rho_o} = \left[1+\left(\frac{\gamma-1}{2}\right)M^2\right]^{-\frac{1}{\gamma-1}} \tag{6.24}$$

This is another important relation. It allows us to calculate the density ratio in isentropic flow in a nozzle at any axial position provided that the Mach number is known.

Using Equation (6.18), it can be readily shown that:

$$\frac{T}{T_o} = \frac{P/P_o}{\rho/\rho_o} \tag{6.25}$$

Combining Equations (6.22), (6.24) and (6.25) gives:

$$\frac{T}{T_o} = \left[1+\left(\frac{\gamma-1}{2}\right)M^2\right]^{-1} \tag{6.26}$$

This is yet another important relation. It allows us to calculate the temperature ratio in isentropic flow in a nozzle at any axial position provided that the Mach number is known.

Critical Properties

When the flow becomes sonic at the throat, a *critical condition* is reached. The flow properties at the throat when $M = 1$ are referred to as critical properties and are denoted with * as the superscript. Thus,

$$V^* = c^* = \sqrt{\gamma R T^*} \tag{6.27}$$

Upon substitution of $M = 1$ in Equations (6.22), (6.24) and (6.26), we obtain the following expressions for the critical properties:

$$\frac{P^*}{P_o} = \left[\left(\frac{\gamma+1}{2}\right)\right]^{-\frac{\gamma}{\gamma-1}} \tag{6.28}$$

$$\frac{\rho^*}{\rho_o} = \left[\left(\frac{\gamma+1}{2}\right)\right]^{-\frac{1}{\gamma-1}} \tag{6.29}$$

$$\frac{T^*}{T_o} = \left[\left(\frac{\gamma+1}{2}\right)\right]^{-1} \tag{6.30}$$

Thus, the critical properties can be readily calculated from Equations (6.28) to (6.30).

Mach Number Estimation from Area Ratio

The mass flow through the choked nozzle can be expressed as:

$$\dot{m} = \rho A V = \rho^* A^* V^* \tag{6.31}$$

where the * quantities refer to critical properties corresponding to $M = 1$. From Equations (6.27) and (6.31), we get:

$$\frac{A}{A^*} = \left(\frac{1}{M}\right)\left(\frac{\rho^*}{\rho_o}\right)\left(\frac{\rho_o}{\rho}\right)\sqrt{\left(\frac{T^*}{T_o}\right)\left(\frac{T_o}{T}\right)} \tag{6.32}$$

From Equations (6.24), (6.26), (6.29), (6.30), and (6.32), it follows that:

$$\frac{A}{A^*} = \left(\frac{1}{M}\right)\left[\frac{1+\left(\frac{\gamma-1}{2}\right)M^2}{\left(\frac{\gamma+1}{2}\right)}\right]^{\frac{\gamma+1}{2(\gamma-1)}} \tag{6.33}$$

For air $\gamma = 1.4$ and Equation (6.33) becomes:

$$\frac{A}{A^*} = \left(\frac{1}{M}\right)\left[\frac{1+0.2M^2}{1.2}\right]^3 \tag{6.34}$$

Figure 6.11 shows the plot of A/A^* versus M for air. When $A = A^*$, $M = 1$. When $A/A^* > 1$, there are two solutions: one value of M corresponding to subsonic flow and the other value of M corresponding to supersonic flow. Given the area ratio A/A^*, the Mach number M for isentropic flow of air through the nozzle can be determined from either Equation (6.34) or plot shown in Figure 6.11. Note that the critical area A^* where $M = 1$ must be known in order to calculate M at any axial position of the nozzle. Once M is known, all other flow properties (pressure, temperature, density) can be calculated in isentropic flow in a nozzle from the equations derived in section 6.3.3.

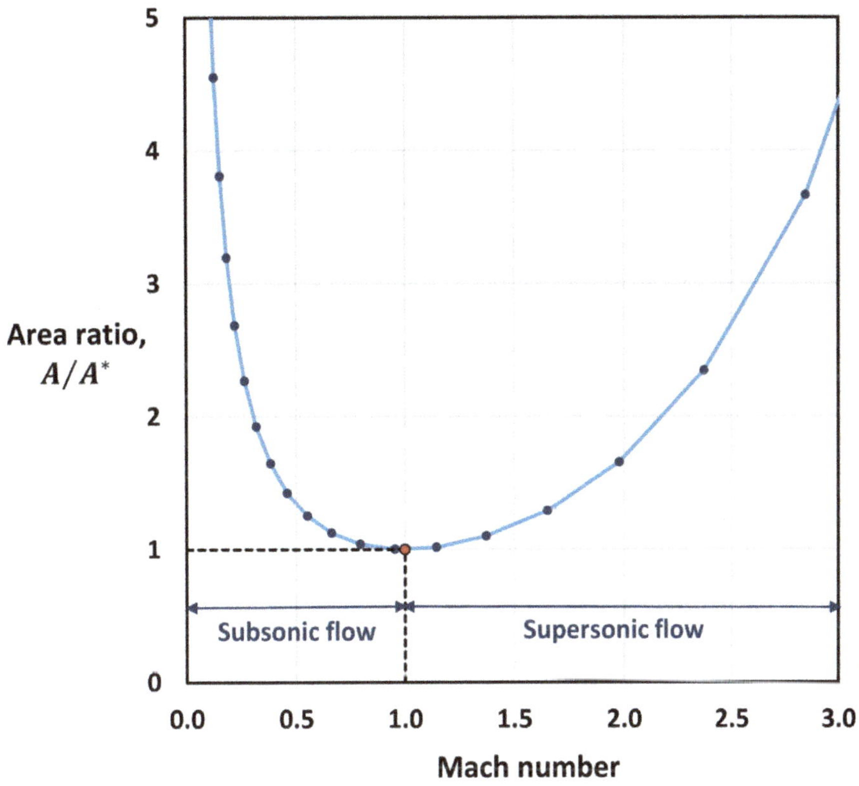

Figure 6.11 Area ratio verses Mach number for isentropic flow of air (ideal gas) in a nozzle of varying area.

Table 6.1 gives M versus A/A^* data generated from Equation (6.34). Instead of solving Equation (6.34) for M, this table could be used to estimate M in solving problems.

Table 6.1 One-dimensional isentropic compressible flow of an ideal gas with $\gamma = 1.4$

M	A/A*	M	A/A*	M	A/A*	M	A/A*
SUBSONIC		0.40	1.590	2.20	2.005	6.4	70.227
0.00	infinity	0.45	1.449	2.30	2.193	6.6	80.323
0.01	57.874	0.50	1.340	2.40	2.403	6.8	91.593
0.02	28.942	0.55	1.255	2.50	2.637	7.0	104.143
0.03	19.301	0.60	1.188	2.60	2.896	7.5	141.841
0.04	14.481	0.65	1.136	2.70	3.183	8.0	190.109
0.05	11.591	0.70	1.094	2.80	3.500	8.5	251.086
0.06	9.666	0.75	1.062	2.90	3.850	9.0	327.189
0.07	8.292	0.80	1.038	3.00	4.235	9.5	421.131
0.08	7.262	0.85	1.021	3.2	5.121	10.0	535.937
0.09	6.461	0.90	1.009	3.4	6.184	10.5	674.963
0.10	5.822	0.95	1.002	3.6	7.450	11.0	841.909
0.11	5.299	1.00	1.00	3.8	8.951	11.5	1040.84
0.12	4.864	SUPERSONIC		4.0	10.719	12.0	1276.21
0.13	4.497	1.10	1.008	4.2	12.792	12.5	1552.87
0.14	4.182	1.20	1.030	4.4	15.210	13.0	1876.07
0.15	3.910	1.30	1.066	4.6	18.018	13.5	2251.53
0.16	3.673	1.40	1.115	4.8	21.264	14.0	2685.38
0.17	3.464	1.50	1.176	5.0	25.000	14.5	3184.25
0.18	3.278	1.60	1.250	5.2	29.283	15.0	3755.24
0.19	3.112	1.70	1.338	5.4	34.175	15.5	4405.97

0.20	2.964	1.80	1.439	5.6	39.740	16.0	5144.55
0.25	2.403	1.90	1.555	5.8	46.050	16.5	5979.66
0.30	2.035	2.00	1.687	6.0	53.180	infinity	infinity
0.35	1.778	2.10	1.837	6.2	61.210		

6.3.4 Isentropic Compressible Flow in Constant Area Duct

Consider a duct of constant cross-sectional area (see Figure 6.12). A compressible fluid (gas) flows through the duct at steady state. The flow is one-dimensional, and the flow properties are uniform at each cross-section. Furthermore, the flow is *isentropic* implying that the flow process is adiabatic and reversible (frictionless). There is no shaft work involved and there is no potential energy change.

Figure 6.12 Duct with constant cross-sectional area.

Equations (6.10) and (6.12) suggest:

$$\frac{dV}{dx} = 0 \tag{6.35}$$

$$\frac{dP}{dx} = 0 \tag{6.36}$$

Thus, velocity and pressure do not vary with axial position in the duct. From Equation (6.4), it follows that enthalpy change is zero:

$$dh = 0 \tag{6.37}$$

Hence, enthalpy is constant. Assuming fluid to be an ideal gas, the temperature of the fluid is constant as well. The Mach number also does not change as velocity of the fluid and velocity of sound are constant. Note that the velocity of sound in a fluid depends on temperature. As pressure and temperature are constant, the density of the fluid is also constant (see Equation (6.18)).

In summary, steady isentropic flow of compressible fluid in a constant area duct does not involve any changes in properties. The flow remains as it enters the duct as there is no mechanism to cause any change. If the flow is subsonic at entrance, it remains subsonic throughout the duct. If the flow at the entrance is sonic, it remains sonic throughout the duct. Likewise, the flow remains supersonic throughout the duct, if the flow is supersonic at the entrance. The Mach number is constant throughout the duct. No choking phenomenon is involved.

6.4 STEADY NON-ISENTROPIC COMPRESSIBLE FLOWS IN CONSTANT AREA DUCTS

6.4.1 Adiabatic Flow with Friction (Fanno Flow)

Adiabatic, steady compressible flow with friction in a constant cross-sectional area duct is called *Fanno flow*. Due to the presence of friction in fluid, entropy generation takes place in such flows. Thus, adiabatic friction flows are non-isentropic irreversible flows. The entropy of the fluid increases in the direction of flow.

For steady adiabatic irreversible flow in a duct of constant area, it can be readily shown [3] that:

$$\frac{dP}{dx} = (T\rho)\left[\frac{1+(\beta V^2/c_p)}{M^2-1}\right]\frac{ds}{dx} \tag{6.38}$$

$$\frac{dV}{dx} = -\left(\frac{T}{V}\right)\left[\frac{M^2+(\beta V^2/c_p)}{M^2-1}\right]\frac{ds}{dx} \tag{6.39}$$

where β is volume expansivity, defined as $\frac{1}{v}\left(\frac{\partial v}{\partial T}\right)_p$, and s is specific entropy.

As entropy increases in the direction of flow ($ds/dx > 0$), the following observations can be made from Equations (6.38) and (6.39): a) In subsonic flow ($M < 1$), $dP/dx < 0$, that is, pressure decreases in the direction of flow, and $dV/dx > 0$, that is, velocity increases in the direction of flow; b) In supersonic flow ($M > 1$), $dP/dx > 0$, that is, pressure increases in the direction of flow, and $dV/dx < 0$, that is, velocity decreases in the direction of flow.

Thus, it can be concluded that the presence of friction in steady compressible flow in a constant area duct causes acceleration of flow in subsonic flows and deceleration of flow in supersonic flows.

Interestingly, the maximum value of Mach number that is obtainable in subsonic flows is unity ($M = 1$) and this value is reached at the end of the duct or pipe. Fanno flow is choked when $M = 1$. We cannot exceed M of unity. When $M = 1$, the entropy of fluid reaches a maximum value ($ds/dx = 0$). Note that if the duct is not long enough, the critical condition of $M = 1$ may not be reached and the flow remains subsonic throughout the duct.

In supersonic flows, the minimum value of Mach number that is obtainable is unity ($M = 1$) and this value is reached at the end of the duct. Fanno flow is choked when M of unity is reached. We cannot go lower than M of unity. Once again, the entropy is maximum at the critical condition of $M = 1$.

Figure 6.13 shows the plot of Fanno line representing the Fanno flow process on a T-s diagram. The entropy increases in the direction of flow (left to right of the diagram) due to irreversible nature of Fanno flow. In subsonic flow, the temperature decreases (velocity increases, pressure decreases) in the direction of flow. In supersonic flow, the temperature increases (velocity decreases, pressure increases) in the direction of flow. At a point of maximum entropy, $M = 1$. Note that subsonic flow cannot become supersonic and vice versa, as that would violate the second law according to which the entropy must increase in irreversible adiabatic process.

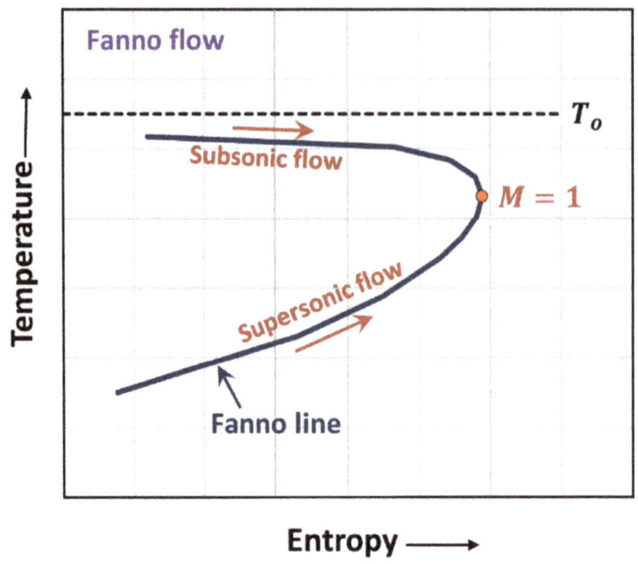

Figure 6.13 Fanno flow process on a T-s diagram. T_o is the constant stagnation temperature in adiabatic flow.

What Happens if the Duct Length is Larger than the Critical Length L^* ($M = 1$ at $x = L^*$)?

Suppose we have subsonic flow at the entrance ($M_1 < 1$). If the actual duct length is larger than the critical length L^* corresponding to the given value of M_1, would the flow accelerate to supersonic velocities? The answer is *no*, since at $M = 1$ the flow is choked to a point of maximum entropy. When $L > L^*$, the flow adjusts itself to a lower mass flow rate. The inlet velocity and Mach number are reduced so that the sonic velocity is still obtained at the exit of the duct.

Suppose we have supersonic flow at the entrance ($M_1 > 1$). If the actual duct length is larger than the critical length L^* corresponding to the given value of M_1, would the flow decelerate to subsonic velocities? When $L > L^*$, the flow will not decelerate to $M < 1$. Instead, it will cause a *normal shock* to occur at such a location that the continuing subsonic flow becomes sonic at the duct exit. Figure 6.14 illustrates the occurrence of normal shock in Fanno flow. Note that shock corresponds to a sudden change in the flow regime from supersonic to subsonic accompanied by sharp changes in P, T and ρ. Also, shocks are accompanied by entropy increase as they are irreversible processes.

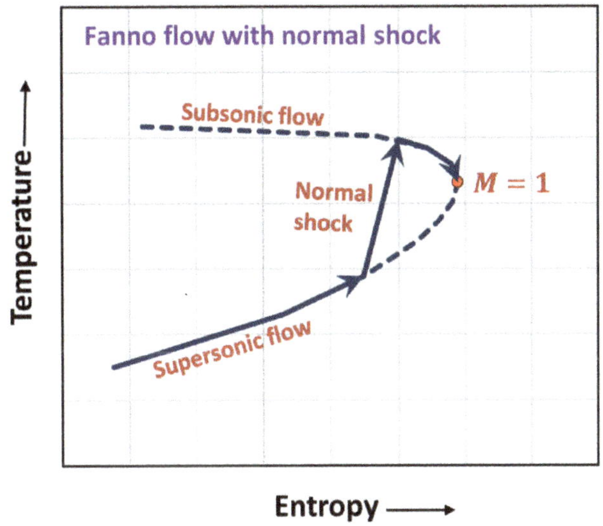

Figure 6.14 Occurrence of normal shock in Fanno flow.

Fanno Flow Equations

Using mass, momentum and energy balances, and ideal gas law, the following equation can be derived for Fanno flow of ideal gas [1]:

$$\frac{1}{\gamma}\left[\frac{1-M^2}{M^2}\right]+\left(\frac{\gamma+1}{2\gamma}\right)ln\left[\frac{\left(\frac{\gamma+1}{2}\right)M^2}{1+\left(\frac{\gamma-1}{2}\right)M^2}\right]=\frac{4f(L^*-L)}{D} \tag{6.40}$$

where f is the average Fanning friction factor in the duct and M is the Mach number at $x = L$ where L is the length measured from the entrance of the duct. This equation allows us to estimate the critical length L^* where $M = 1$, provided that M is known at any position L from the entrance of the duct. If M is known at the entrance itself, then $L = 0$ and the right-hand side term of Equation (6.40) simply becomes $4fL^*/D$.

Equation (6.40) can be used to estimate the Mach number along the axis of the duct. Suppose we want to estimate M_2 at $x = L_2$ and we are given M_1 at $x = L_1$. Before we estimate M_2 at $x = L_2$, note that:

$$\frac{4f(L^*-L_2)}{D}=\frac{4f(L^*-L_1)}{D}-\frac{4f(L_2-L_1)}{D} \tag{6.41}$$

Given M_1 at $x = L_1$, estimate $\frac{4f(L^*-L_1)}{D}$ from Equation (6.40) using $M = M_1$. Calculate $\frac{4f(L^*-L_2)}{D}$ from Equation (6.41). Then estimate $M = M_2$ from Equation (6.40) as the right-hand side of the equation $\frac{4f(L^*-L_2)}{D}$ is now known. Repeat this calculation for different values of x to establish the Mach number values along the direction of flow in the duct.

Table 6.2 gives M versus $4f(L^* - L)/D$ data generated from Equation (6.40) with $\gamma = 1.4$ (air). Instead of solving Equation (6.40) for M, this table could be used to estimate the Mach number in solving problems.

Table 6.2 Data generated from Fanno flow equation (Equation (6.40)) for ideal gas with $\gamma = 1.4$

M	$4f(L^* - L)/D$	M	$4f(L^* - L)/D$	M	$4f(L^* - L)/D$	M	$4f(L^* - L)/D$
0.05	280.020	1.3	0.065	2.55	0.442	3.8	0.616
0.1	66.922	1.35	0.082	2.6	0.453	3.85	0.621
0.15	27.932	1.4	0.100	2.65	0.462	3.9	0.625
0.2	14.533	1.45	0.118	2.7	0.472	3.95	0.629
0.25	8.483	1.5	0.136	2.75	0.481	4	0.633
0.3	5.299	1.55	0.154	2.8	0.490	4.05	0.637
0.35	3.452	1.6	0.172	2.85	0.498	4.1	0.641
0.4	2.308	1.65	0.190	2.9	0.507	4.15	0.645
0.45	1.566	1.7	0.208	2.95	0.514	4.2	0.648
0.5	1.069	1.75	0.225	3	0.522	4.25	0.652
0.55	0.728	1.8	0.242	3.05	0.530	4.3	0.655
0.6	0.491	1.85	0.258	3.1	0.537	4.35	0.658
0.65	0.325	1.9	0.274	3.15	0.544	4.4	0.661
0.7	0.208	1.95	0.290	3.2	0.550	4.45	0.665
0.75	0.127	2	0.305	3.25	0.557	4.5	0.668
0.8	0.072	2.05	0.320	3.3	0.563	4.55	0.671
0.85	0.036	2.1	0.334	3.35	0.569	4.6	0.673
0.9	0.015	2.15	0.348	3.4	0.575	4.65	0.676
0.95	0.003	2.2	0.361	3.45	0.581	4.7	0.679
1	0.000	2.25	0.374	3.5	0.586	4.75	0.682
1.05	0.003	2.3	0.386	3.55	0.592	4.8	0.684
1.1	0.010	2.35	0.398	3.6	0.597	4.85	0.687
1.15	0.021	2.4	0.410	3.65	0.602	4.9	0.689
1.2	0.034	2.45	0.421	3.7	0.607	4.95	0.691
1.25	0.049	2.5	0.432	3.75	0.612	5	0.694

Once M is known at different positions along the duct, one can easily estimate other properties of Fanno flow like T, P, V and ρ. The equations for the prediction of these properties in Fanno flow of ideal gas are described below.

The temperature is given as:

$$\frac{T}{T^*} = \left[\frac{\left(\frac{\gamma+1}{2}\right)}{1+\left(\frac{\gamma-1}{2}\right)M^2}\right] \tag{6.42}$$

where T^* is the critical temperature corresponding to $M = 1$. In order to calculate T/T_1 where T_1 is the temperature at the entrance, we use:

$$\frac{T}{T_1} = \frac{(T/T^*)_M}{(T/T^*)_{M_1}} \tag{6.43}$$

where T corresponds to the location with M and T_1 corresponds to entrance location with $M = M_1$. It should be noted that in Fanno flow of ideal gas, the temperature of the fluid changes along the duct. However, the stagnation temperature throughout the duct remains constant as the flow is adiabatic. In adiabatic flow, the first law of thermodynamics gives:

$$h_o = h + \frac{V^2}{2} = \text{constant} \tag{6.44}$$

where h_o is the stagnation enthalpy. As h_o is constant, it follows that the stagnation temperature T_o is constant as enthalpy is a function of temperature only for ideal gas. Equation (6.44) gives the following expression of the stagnation temperature:

$$T_o = T + \frac{V^2}{2c_p} \tag{6.45}$$

The pressure is given as:

$$\frac{P}{P^*} = \frac{1}{M}\sqrt{\frac{\left(\frac{\gamma+1}{2}\right)}{1+\left(\frac{\gamma-1}{2}\right)M^2}} \tag{6.46}$$

where P^* is the critical pressure corresponding to $M = 1$. In order to calculate P/P_1 where P_1 is the pressure at the entrance, we use:

$$\frac{P}{P_1} = \frac{(P/P^*)_M}{(P/P^*)_{M_1}} \tag{6.47}$$

where P corresponds to the location with M and P_1 corresponds to entrance location with $M = M_1$.

The velocity is given as:

$$\frac{V}{V^*} = M\sqrt{\frac{\left(\frac{\gamma+1}{2}\right)}{1+\left(\frac{\gamma-1}{2}\right)M^2}} \tag{6.48}$$

where V^* is the critical velocity corresponding to $M = 1$. In order to calculate V/V_1 where V_1 is the velocity at the entrance, we use:

$$\frac{V}{V_1} = \frac{(V/V^*)_M}{(V/V^*)_{M_1}} \tag{6.49}$$

where V corresponds to the location with M and V_1 corresponds to entrance location with $M = M_1$.

The density is given as:

$$\frac{\rho}{\rho^*} = \frac{1}{M}\left[\frac{1+\left(\frac{\gamma-1}{2}\right)M^2}{\left(\frac{\gamma+1}{2}\right)}\right] \tag{6.50}$$

where ρ^* is the critical density corresponding to $M = 1$. In order to calculate ρ/ρ_1 where ρ_1 is the density at the entrance, we use:

$$\frac{\rho}{\rho_1} = \frac{(\rho/\rho^*)_M}{(\rho/\rho^*)_{M_1}} \tag{6.51}$$

where ρ corresponds to the location with M and ρ_1 corresponds to entrance location with $M = M_1$

Figure 6.15 shows the plots of T/T^*, P/P^*, V/V^* and ρ/ρ^* versus Mach number. In subsonic Fanno flow of ideal gas (from left to right of the figure), friction causes an increase in fluid velocity, decrease in temperature, decrease in pressure, and decrease in density until the critical condition of $M = 1$ is reached. In supersonic Fanno flow of ideal gas (from right to left of the figure), friction causes a decrease in fluid velocity, increase in temperature, increase in pressure, and increase in density until the critical condition of $M = 1$ is reached.

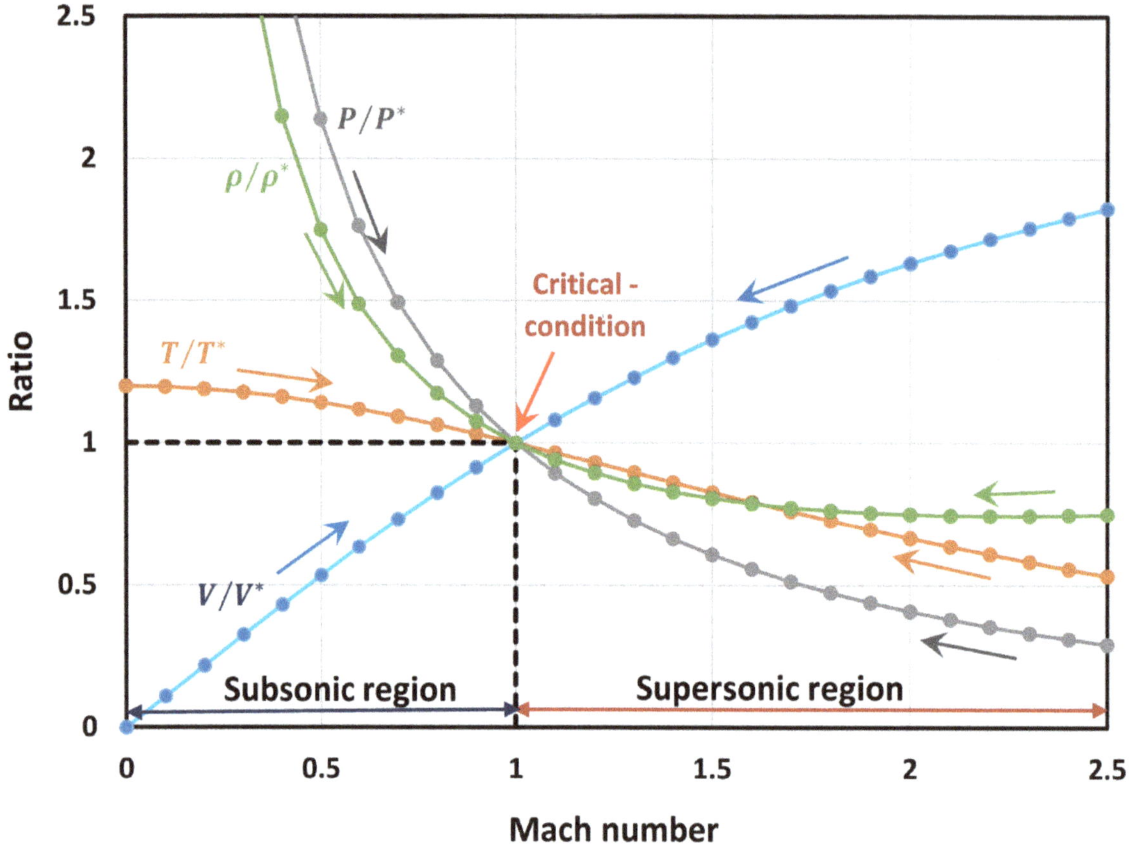

Figure 6.15 Plots of different property ratios in Fanno flow of ideal gas.

The entropy change in Fanno flow of ideal gas can be calculated using the following equation:

$$s - s_1 = c_v ln\left(\frac{T}{T_1}\right) - Rln\left(\frac{\rho}{\rho_1}\right) \tag{6.52}$$

This equation along with temperature and density relations allows the plotting of the Fanno line (see Figure 6.13).

6.4.2 Frictionless Flow with Heat Transfer (Rayleigh Flow)

Steady, frictionless compressible flow in a constant cross-sectional area duct with heat transfer is called *Rayleigh flow*. Due to heat transfer, entropy changes in the direction of flow. The entropy of the fluid increases in the direction of flow when heat is added to the flow. The entropy decreases in the direction of flow when heat is removed from the flow. As the flow is frictionless and hence internally reversible, there is no generation of entropy. However, the flow is non-isentropic due to heat transfer.

Using mass, momentum and energy balances, and ideal gas law, the following equation can be derived for Rayleigh flow of ideal gas [1]:

$$\frac{dV}{V} = \frac{\delta Q}{Tc_p}\left(\frac{1}{1-M^2}\right) \tag{6.53}$$

where δQ is heat transfer per unit mass of fluid. Since the flow is internally reversible (frictionless),

$$Tds = \delta Q \tag{6.54}$$

$$T\frac{ds}{dx} = \frac{\delta Q}{\delta x} \tag{6.55}$$

From Equations (6.53) to (6.55), it follows that:

$$\frac{1}{V}\frac{dV}{dx} = \frac{1}{c_p}\left(\frac{1}{1-M^2}\right)\frac{ds}{dx} \tag{6.56}$$

The following observations can be made from Equation (6.56): a) In subsonic flow ($M < 1$), the velocity increases in the flow direction, if heat is added to the flow ($ds/dx > 0$); b) In subsonic flow ($M < 1$), the velocity decreases in the flow direction, if heat is removed from the flow ($ds/dx < 0$); c) In supersonic flow ($M > 1$), the velocity decreases in the flow direction, if heat is added to the flow ($ds/dx > 0$); d) In supersonic flow ($M > 1$), the velocity increases in the flow direction, if heat is removed from the flow ($ds/dx < 0$).

Figure 6.16 shows the plot of Rayleigh line representing the Rayleigh flow process on a T-s diagram. We are assuming that heat is being added to the flow so entropy increases in the direction of flow (left to right of the diagram). In subsonic flow, the addition of heat causes the fluid temperature to reach a maximum value of T_{max} while speed of the fluid is increasing until the Mach number reaches a value of $1/\sqrt{\gamma}$. The fluid temperature then drops as M approaches the sonic speed ($M = 1$). Once the fluid reaches the critical condition of $M = 1$, it cannot be accelerated further by heating. The flow is choked. If we keep heating the fluid, the flow will adjust itself to lower mass flow rate and the critical state is moved further downstream. In supersonic flow, the addition of heat causes the fluid temperature to increase but the Mach number decreases until M approaches the sonic speed ($M = 1$). Once the fluid reaches the critical condition of $M = 1$, it cannot be decelerated further by heating. The flow is choked. With further heating, the critical state is moved further downstream resulting in the reduction of mass flow rate.

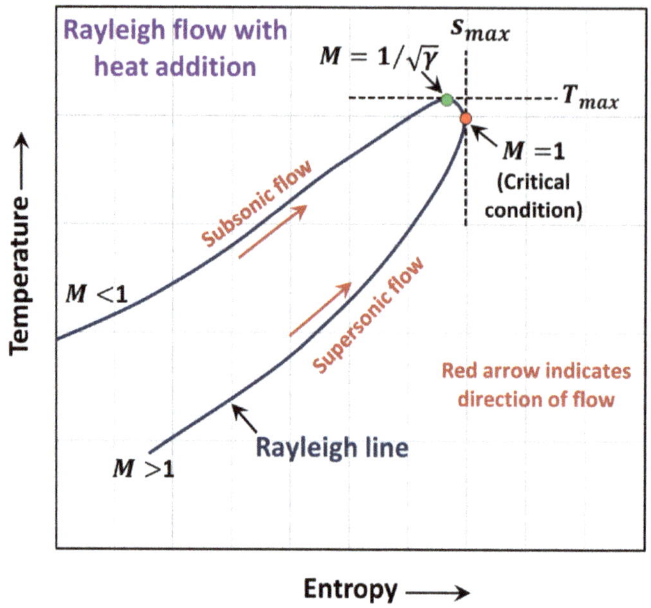

Figure 6.16 Rayleigh flow of ideal gas with heat addition.

Figure 6.17 shows the plot of Rayleigh line when heat is being removed from the flow so entropy decreases in the direction of flow (right to left of the diagram). In subsonic flow, the removal of heat causes deceleration of flow, that is, the Mach number decreases in the direction of flow. The temperature generally decreases in the direction of flow. In supersonic flow, the removal of heat causes acceleration of flow, that is, the Mach number increases in the direction of flow. The temperature decreases in the direction of flow.

As with Fanno flows, normal shocks can occur in Rayleigh flows accompanied by abrupt deceleration of flow from supersonic regime to subsonic regime. Figure 6.18 shows occurrence of normal shock in Rayleigh flow with heat addition.

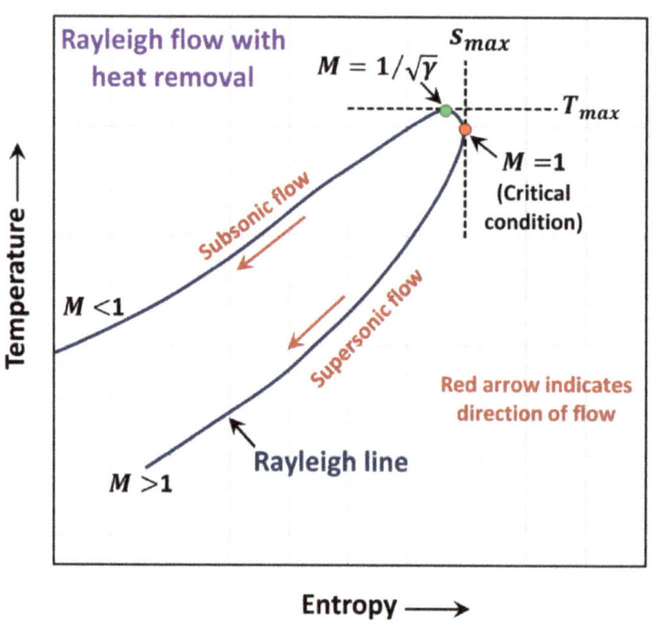

Figure 6.17 Rayleigh flow of ideal gas with heat removal.

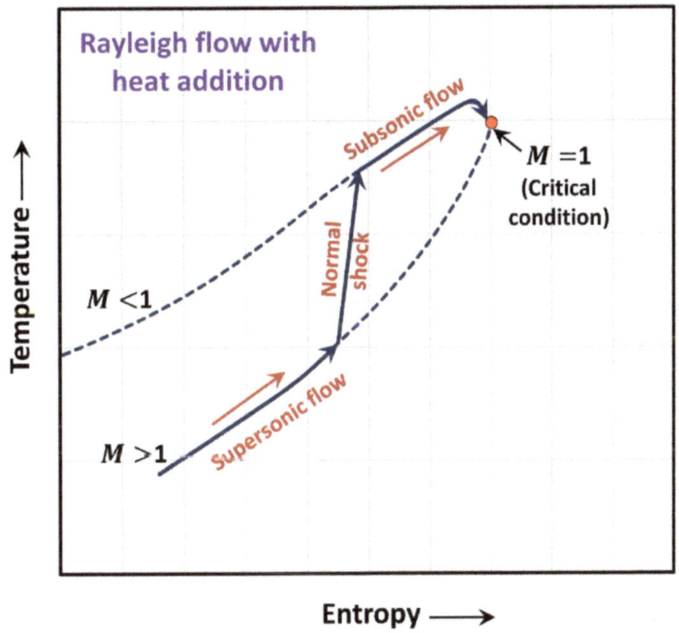

Figure 6.18 Occurrence of normal shock in Rayleigh flow of ideal gas with heat addition.

Rayleigh Flow Equations

The stagnation properties in Rayleigh flow of ideal gas are given as:

$$T_o = T\left[1 + \left(\frac{\gamma-1}{2}\right)M^2\right] \quad (6.57)$$

$$P_o = P\left[1 + \left(\frac{\gamma-1}{2}\right)M^2\right]^{\gamma/(\gamma-1)} \quad (6.58)$$

$$\rho_o = \rho\left[1 + \left(\frac{\gamma-1}{2}\right)M^2\right]^{1/(\gamma-1)} \quad (6.59)$$

Note that stagnation properties are properties that the fluid would have if it were brought to zero velocity isentropically. As Rayleigh flow is non-isentropic flow due to heat transfer, the stagnation properties vary along the duct. For example, the variation of stagnation temperature along the duct can be determined from energy balance as:

$$Q_{1\to2} = c_p\left(T_{o,2} - T_{o,1}\right) \quad (6.60)$$

Thus,

$$T_{o,2} = T_{o,1} + \left(Q_{1\to2}/c_p\right) \quad (6.61)$$

The flow properties in Rayleigh flow are expressed in the form of ratio of actual property value to critical property value at $M = 1$ as:

$$\frac{T}{T^*} = \left[\frac{(1+\gamma)M}{1+\gamma M^2}\right]^2 \quad (6.62)$$

$$\frac{P}{P^*} = \left[\frac{(1+\gamma)}{1+\gamma M^2}\right] \tag{6.63}$$

$$\frac{V}{V^*} = \left[\frac{(1+\gamma)M^2}{1+\gamma M^2}\right] \tag{6.64}$$

$$\frac{\rho}{\rho^*} = \left[\frac{1+\gamma M^2}{(1+\gamma)M^2}\right] \tag{6.65}$$

$$\frac{T_o}{T_o^*} = 2(1+\gamma)M^2\left[\frac{1+\left(\frac{\gamma-1}{2}\right)M^2}{(1+\gamma M^2)^2}\right] \tag{6.66}$$

$$\frac{P_o}{P_o^*} = \left(\frac{(1+\gamma)}{1+\gamma M^2}\right)\left[\left(\frac{2}{1+\gamma}\right)\left\{1+\left(\frac{\gamma-1}{2}\right)M^2\right\}\right]^{\gamma/(\gamma-1)} \tag{6.67}$$

where starred quantities are critical properties, that is, properties corresponding to $M = 1$.

Procedure to Calculate Properties in Rayleigh Flow of Ideal Gas

Given heat transfer per unit mass (Q) and properties at the entrance of the duct V_1, P_1, T_1, we can calculate the properties along the duct as follows:

- Determine M_1 and ρ_1 using $M_1 = V_1/\sqrt{\gamma R T_1}$ and $\rho_1 = P_1/RT_1$
- Determine the critical properties using Equations (6.62) to (6.67) as: $T^* = \frac{T_1}{(T_1/T^*)}$; $P^* = \frac{P_1}{(P_1/P^*)}$; $V^* = \frac{V_1}{(V_1/V^*)}$; $\rho^* = \frac{\rho_1}{(\rho_1/\rho^*)}$; $T_o^* = \frac{T_{o,1}}{(T_{o,1}/T_o^*)}$; $P_o^* = \frac{P_{o,1}}{(P_{o,1}/P_o^*)}$
  Note that $T_{o,1}$ and $P_{o,1}$ can be calculated from Equations (6.57) and (6.58), respectively.
- Using Equation (6.61), calculate $T_{o,2}$ at the new location and hence $T_{o,2}/T_o^*$. The Mach number M_2 corresponding to new location ($T_{o,2}$) can be calculated from Equation (6.66).
- Once M_2 is known, all other properties can be calculated at the new location using Equations (6.62) to (6.65) and (6.67).
- This calculation procedure is repeated for different locations along the duct.

Table 6.3 gives M versus T_o/T_o^* data generated from Equation (6.66) with $\gamma = 1.4$ (air). Instead of solving Equation (6.66) for M, this table could be used to estimate the Mach number in solving problems.

Figure 6.19 shows the plots of T/T^*, P/P^*, V/V^* and ρ/ρ^* versus Mach number in Rayleigh flow with heat addition. In subsonic Rayleigh flow of ideal gas (from left to right of the figure), heat addition causes an increase in fluid velocity, decrease in pressure, and decrease in density until the critical condition of $M = 1$ is reached. The temperature increases with heat addition up until $M = 1/\sqrt{\gamma}$ and then decreases up to the critical condition. In supersonic Rayleigh flow of ideal gas (from right to left of the figure), heat addition causes a decrease in fluid velocity, increase in temperature, increase in pressure, and increase in density until the critical condition of $M = 1$ is reached.

Finally, the entropy change in Rayleigh flow of ideal gas can be calculated using the following equation:

$$s - s_1 = c_v ln\left(\frac{T}{T_1}\right) - Rln\left(\frac{\rho}{\rho_1}\right) \tag{6.68}$$

This equation along with temperature and density relations allows the plotting of the Rayleigh line (see Figures 6.16 and 6.17).

Table 6.3 Data generated from Equation (6.66) for ideal gas with $\gamma = 1.4$

M	T_o/T_o^*	M	T_o/T_o^*	M	T_o/T_o^*	M	T_o/T_o^*
0.05	0.012	1.3	0.958	2.55	0.703	3.8	0.599
0.1	0.047	1.35	0.946	2.6	0.697	3.85	0.596
0.15	0.102	1.4	0.934	2.65	0.691	3.9	0.594
0.2	0.174	1.45	0.922	2.7	0.685	3.95	0.591
0.25	0.257	1.5	0.909	2.75	0.679	4	0.589
0.3	0.347	1.55	0.897	2.8	0.674	4.05	0.587
0.35	0.439	1.6	0.884	2.85	0.669	4.1	0.585
0.4	0.529	1.65	0.872	2.9	0.664	4.15	0.583
0.45	0.614	1.7	0.860	2.95	0.659	4.2	0.581
0.5	0.691	1.75	0.848	3	0.654	4.25	0.579
0.55	0.760	1.8	0.836	3.05	0.649	4.3	0.577
0.6	0.819	1.85	0.825	3.1	0.645	4.35	0.575
0.65	0.868	1.9	0.814	3.15	0.641	4.4	0.573
0.7	0.908	1.95	0.804	3.2	0.637	4.45	0.571
0.75	0.940	2	0.793	3.25	0.633	4.5	0.570
0.8	0.964	2.05	0.784	3.3	0.629	4.55	0.568
0.85	0.981	2.1	0.774	3.35	0.626	4.6	0.567
0.9	0.992	2.15	0.765	3.4	0.622	4.65	0.565
0.95	0.998	2.2	0.756	3.45	0.619	4.7	0.564
1	1.000	2.25	0.748	3.5	0.616	4.75	0.562
1.05	0.998	2.3	0.740	3.55	0.613	4.8	0.561
1.1	0.994	2.35	0.732	3.6	0.610	4.85	0.559
1.15	0.987	2.4	0.724	3.65	0.607	4.9	0.558
1.2	0.979	2.45	0.717	3.7	0.604	4.95	0.557
1.25	0.969	2.5	0.710	3.75	0.601	5	0.556

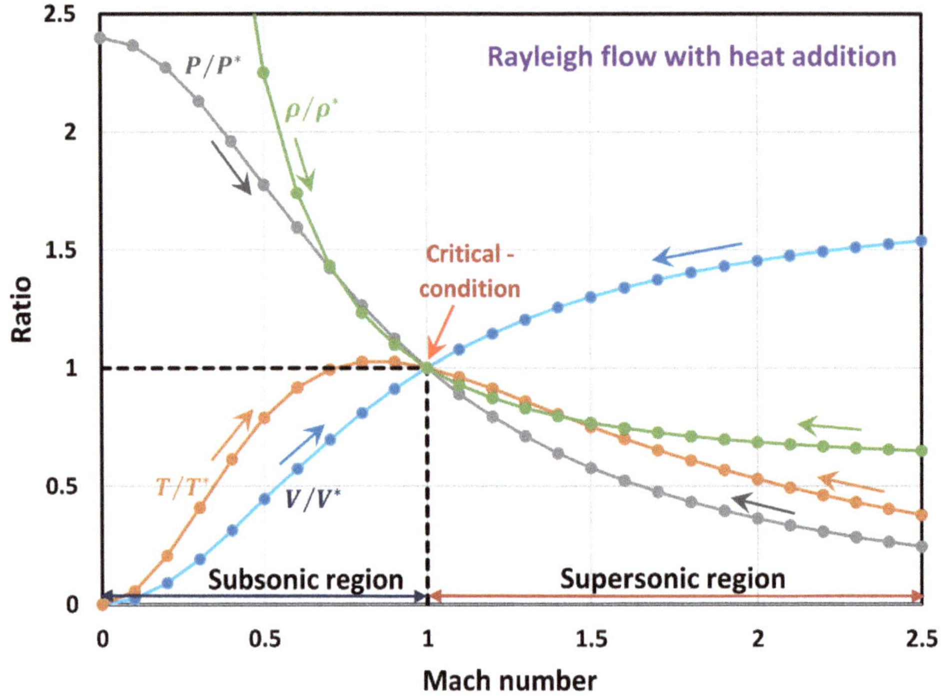

Figure 6.19 Plots of different property ratios in Rayleigh flow of ideal gas with heat addition.

6.4.3 Isothermal Flow with Friction

Isothermal flow normally occurs in long distance pipelines used for the transportation of natural gas. Due to large surface area of the pipe, heat transfer through the walls of the pipe is relatively rapid so that the temperature of the fluid can be treated as approximately constant.

Important features of steady, isothermal flow of compressible fluid (ideal gas) in a constant area duct are as follows:

- The flow is choked at $M = 1/\sqrt{\gamma}$ and not at $M = 1$ as found in Fanno and Rayleigh flows.
- When $M < 1/\sqrt{\gamma}$ (subsonic) at the entrance of the duct, the velocity and Mach number increase along the length of the duct until the flow is choked at $M = 1/\sqrt{\gamma}$. Note that choking occurs at the exit of the duct only if it is long enough that the critical condition of $M = 1/\sqrt{\gamma}$ is reached at the exit. The pressure and density decrease along the duct until the critical condition of $M = 1/\sqrt{\gamma}$ is reached.
- When $M > 1/\sqrt{\gamma}$ (subsonic or supersonic) at the entrance of the duct, the velocity and Mach number decrease along the length of the duct until the flow is choked at $M = 1/\sqrt{\gamma}$. The pressure and density increase along the duct until the critical condition of $M = 1/\sqrt{\gamma}$ is reached. Note that unlike Fanno and Rayleigh flows, supersonic flow in the present isothermal case can come down to subsonic flow.
- Isothermal flow involves both friction and heat transfer effects. For flows with $M < 1/\sqrt{\gamma}$ (subsonic), heat is absorbed by the flow. For flows with $M > 1/\sqrt{\gamma}$ (subsonic or supersonic), heat is rejected by the flow.
- Although the temperature of the fluid is constant, the stagnation temperature varies along the duct due to heat transfer. For flows with $M < 1/\sqrt{\gamma}$, the stagnation temperature increases along the

duct as heat is absorbed. For flows with $M > 1/\sqrt{\gamma}$, the stagnation temperature decreases along the duct as heat is rejected by the flow.

Isothermal Flow Equations

Using mass, momentum and energy balances, and ideal gas law, the following equation can be derived for isothermal flow of ideal gas [4]:

$$\left[\frac{1-\gamma M^2}{\gamma M^2}\right] + ln(\gamma M^2) = \frac{4f(L^*-L)}{D} \tag{6.69}$$

where f is the average Fanning friction factor in the duct and M is the Mach number at $x = L$ where L is the length measured from the entrance of the duct. This equation allows us to estimate the critical length L^* where $M = 1/\sqrt{\gamma}$, provided that M is known at any position L from the entrance of the duct. If M is known at the entrance itself, then $L = 0$ and the right-hand side term of Equation (6.69) simply becomes $4fL^*/D$. Equation (6.69) can be used to estimate the Mach number along the axis of the duct. The approach is similar to that used in Fanno flow (see Equation (6.41). The duct length over which flow passes from Mach number of M_1 to M_2 can be expressed as:

$$\frac{4f(L_2-L_1)}{D} = \frac{4f(L^*-L_1)}{D} - \frac{4f(L^*-L_2)}{D} - \left[\frac{1-\gamma M_1^2}{\gamma M_1^2}\right] - \left[\frac{1-\gamma M_2^2}{\gamma M_2^2}\right] + 2ln\left(\frac{M_1}{M_2}\right) \tag{6.70}$$

Given M_1 at $x = L_1$, one can calculate M_2 at $x = L_2$ using Equation (6.70).

Once M is known at different positions along the duct, one can easily estimate other properties of isothermal flow like P, V and ρ. The equations for the prediction of these properties in isothermal flow of ideal gas are described below.

The pressure and density are given as:

$$\frac{P}{P^*} = \frac{\rho}{\rho^*} = \frac{1}{M\sqrt{\gamma}} \tag{6.71}$$

where P^* and ρ^* are critical pressure and critical density corresponding to $M = 1/\sqrt{\gamma}$.

The velocity is given as:

$$\frac{V}{V^*} = M\sqrt{\gamma} \tag{6.72}$$

where V^* is the critical velocity corresponding to $M = 1/\sqrt{\gamma}$.

It should be noted that although temperature is constant in isothermal flow, the stagnation temperature varies due to heat transfer. The stagnation temperature and pressure in isothermal flow is given as:

$$\frac{T_o}{T_o^*} = \frac{2\gamma}{3\gamma-1}\left[1 + \left(\frac{\gamma-1}{2}\right)M^2\right] \tag{6.73}$$

$$\frac{P_o}{P_o^*} = \frac{1}{M\sqrt{\gamma}}\left[\left(\frac{2\gamma}{3\gamma-1}\right)\left\{1 + \left(\frac{\gamma-1}{2}\right)M^2\right\}\right]^{\gamma/(\gamma-1)} \tag{6.74}$$

Finally note that Equation (6.70) can be re-written in terms of the pressure ratio P_2/P_1 , using the relation $M_2 = M_1 P_1/P_2$, as follows :

$$\frac{4f(L_2-L_1)}{D} = \left[\frac{1-\left(\frac{P_2}{P_1}\right)^2}{\gamma M_1^2}\right] + 2ln\left(\frac{P_2}{P_1}\right) \tag{6.75}$$

Assuming that $L_1 = 0$ (entrance of the duct), where P_1 and M_1 are known, Equation (6.75) can be solved for P_2 over any length L_2 of the pipe. There are two possible solutions of Equation (6.75). Only one solution is acceptable. The other one violates the second law. Note that M_2 cannot exceed $1/\sqrt{\gamma}$. Thus $(P_2/P_1)^2 > \gamma M_1^2$.

Figure 6.20 shows the plots of V/V^*, T_o/T_o^*, T/T^*, P/P^* and ρ/ρ^* versus Mach number. In subsonic isothermal flow of ideal gas with $M < 1/\sqrt{\gamma}$ (from left to right of the figure), the velocity increases, stagnation temperature increases, temperature remains constant, pressure and density decrease in the direction of flow until the critical condition of $M = 1/\sqrt{\gamma}$ is reached. In isothermal flow of ideal gas with $M > 1/\sqrt{\gamma}$ (from right to left of the figure), the velocity decreases, stagnation temperature decreases, temperature remains constant, pressure and density increase in the direction of flow until the critical condition of $M = 1/\sqrt{\gamma}$ is reached. Thus, Mach number tends toward $1/\sqrt{\gamma}$ regardless of the condition at the entrance (subsonic or supersonic).

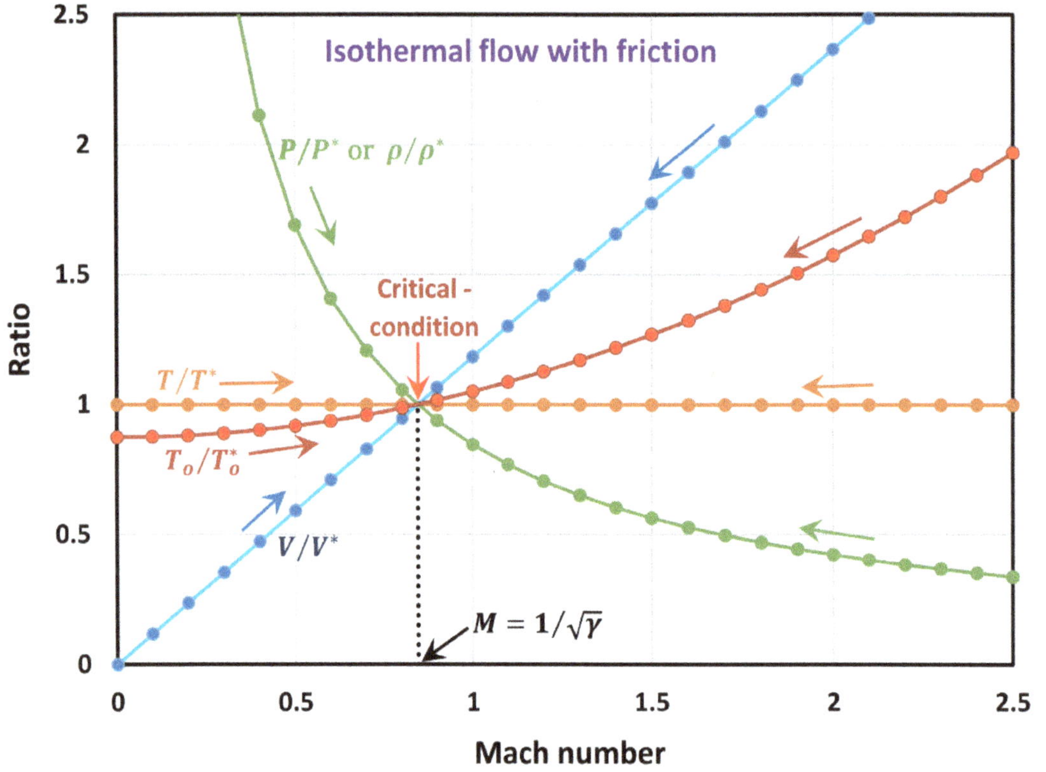

Figure 6.20 Plots of different property ratios in isothermal flow of ideal gas with friction.

6.5 STEADY ADIABATIC AND NON-ADIABATIC INCOMPRESSIBLE FLOWS

Incompressible flows are flows in which the fluid density remains constant. They are very important from a practical point of view. The flow of liquid in pipes can be treated as an incompressible flow as changes in the density of liquid are negligible. In a constant cross-sectional area pipe, the flow of a compressible fluid (gas) experiences velocity changes due to changes in the density of the fluid. However, the velocity remains constant in flow of an incompressible fluid (liquid) in a constant area pipe.

In pipeline flow of liquids (incompressible fluid), dissipation of mechanical energy takes place due to high viscosity of liquids. Viscosity could be interpreted as a measure of internal friction in fluids. The dissipation of mechanical energy due to internal friction in pipeline flow is reflected in the loss of pressure of fluid. The degradation of high quality mechanical energy into frictional heating is simultaneously reflected in the generation of entropy. It is important to quantify entropy generation in flow of viscous incompressible fluids as entropy generation is directly related to mechanical energy dissipation in such flows.

In adiabatic flow of incompressible fluids, the mechanism of entropy generation is the internal friction in fluids. However, in many practical applications the flow of liquids in pipes is not adiabatic, that is, heat is added to or removed from flow. In non-adiabatic flow of liquids, there are two mechanisms of entropy generation, namely internal friction (due to viscosity) and irreversible heat transfer. It is important to quantify entropy generation due to different mechanisms and to minimize entropy generation, in order to improve the thermodynamic efficiency of the pipeline transportation process.

6.5.1 Steady Adiabatic Incompressible Flow

The second law of thermodynamics for control volume is given as:

$$\dot{S}_{G,total} = \Sigma_{OUTLETS}\ (\dot{m}_i s_i) - \Sigma_{INLETS}\ (\dot{m}_i\, s_i) - \Sigma \frac{\dot{Q}_i}{T_i} \geq 0 \tag{6.76}$$

where T_i is the absolute temperature of ith heat reservoir and $\dot{Q}_i$ is the rate of heat transfer between the control volume and ith heat reservoir.

For steady flow in a pipe with no heat transfer, Equation (6.76) reduces to [5]:

$$\dot{S}_{G,total} = \dot{S}_{G,CV} = \dot{m}(\Delta s) > 0 \tag{6.77}$$

There is no entropy generation in the surroundings. All the entropy is generated within the fluid inside the pipe and the rate of entropy generation is the net rate of increase in entropy of the flowing stream.

We can now relate entropy change of the fluid stream to other variables. For pure substances, the relationship between entropy and other state variables is given as:

$$Tds = dh - (dP/\rho) \tag{6.78}$$

where T is the absolute temperature. From the first law of thermodynamics for control volume, the enthalpy change is zero in the absence of heat transfer and shaft work for steady flow in a horizontal pipe of uniform diameter. Consequently, Equation (6.78) reduces to:

$$Tds = -(dP/\rho) \tag{6.79}$$

Assuming incompressible flow and constant temperature, Equation (6.79) upon integration gives:

$$\Delta s = -\frac{\Delta P}{\rho T} \tag{6.80}$$

From Equations (6.77) and (6.80), it follows that:

$$\dot{S}_G = \frac{\dot{m}}{\rho}\left(-\frac{\Delta P}{T}\right) > 0 \tag{6.81}$$

The subscript "*CV*" has been removed from $\dot{S}_{G,CV}$ for the sake of simplicity. We can also express the rate of entropy generation in a pipe on a unit length basis as:

$$\dot{S}'_G = \frac{\dot{m}}{\rho T}\left(-\frac{dP}{dx}\right) \tag{6.82}$$

where $\dot{S}'_G$ is the rate of entropy generation per unit length of the pipe.

Now consider force balance over a differential control volume of length dx as shown in Figure 6.21:

$$\sum F_x = PA - (P + dP)A - \tau_W(\pi D)dx = 0 \tag{6.83}$$

where τ_W is the wall shear stress, D is the pipe internal diameter, and $A = \pi D^2/4$. Equation (6.83) leads to:

$$-(dP/dx) = \frac{\pi D \tau_W}{A} = \frac{4\tau_W}{D} \tag{6.84}$$

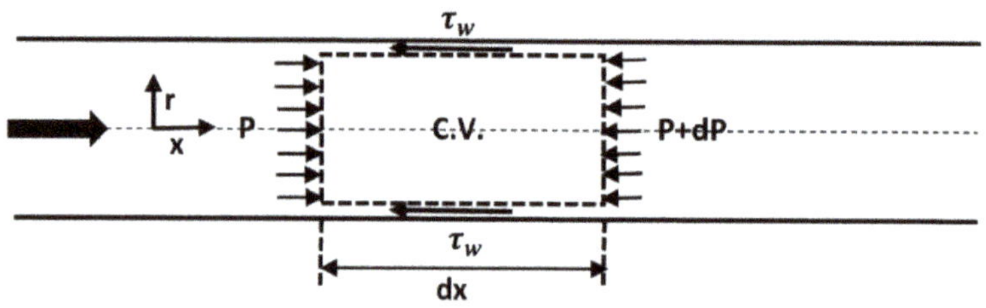

Figure 6.21 Stresses acting on fluid inside the control volume.

The friction factor (f) defined as:

$$f = 2\tau_W/(\rho \bar{V}^2) \tag{6.85}$$

where $\bar{V}$ is the average velocity in the pipe. Upon substitution of τ_W from Equation (6.84) into Equation (6.85), we get

$$(-dP/dx) = \frac{(2\rho\bar{V}^2)}{D} f \tag{6.86}$$

From Equations (6.82) and (6.86), it follows that the rate of entropy generation per unit length of the pipe is [5]:

$$\dot{S}'_G = \left(\frac{\pi}{2T}\right)(\rho D \bar{V}^3) f \tag{6.87}$$

In *laminar* flow of Newtonian fluids ($Re \leq 2100$), friction factor is related to Reynolds number through the following theoretical relationship:

$$f = 16/Re \tag{6.88}$$

where the Reynolds number Re is defined as:

$$Re = \rho D \bar{V}/\mu \tag{6.89}$$

Consequently, Equation (6.87) yields:

$$\dot{S}'_G = \left(\frac{8\pi}{T}\right)\mu\bar{V}^2 \tag{6.90}$$

Thus, entropy generation rate per unit length of pipe in steady laminar flow of a Newtonian fluid is directly proportional to fluid viscosity and square of average velocity in the pipe.

In *turbulent* flow of Newtonian fluids, friction factor is a function of Reynolds number and relative roughness of pipe (ϵ/D). For hydraulically smooth pipes$(\epsilon/D \to 0)$, the friction factor depends only on Re in turbulent regime. The following semi-empirical equation, often referred to as von Karman – Nikuradse equation, describes the $f\ vs.Re$ turbulent behavior of Newtonian fluids in smooth pipes very well:

$$1/\sqrt{f} = 4log_{10}\left(Re\sqrt{f}\right) - 0.40 \tag{6.91}$$

The von Karman – Nikuradse equation is not explicit in friction factor. A number of explicit $f\ vs.Re$ relations are available in the literature. One of the popular ones is the Blasius friction factor equation for turbulent flow of Newtonian fluids in smooth pipes:

$$f = 0.079/Re^{0.25} \tag{6.92}$$

Equations (6.87) and (6.92) give the following expression for entropy generation rate per unit length of pipe [5]:

$$\dot{S}'_G = \left(\frac{0.079\pi}{2T}\right)\mu^{0.25}\rho^{0.75}D^{0.75}\bar{V}^{2.75} \tag{6.93}$$

In turbulent flow, the entropy generation rate per unit length of pipe also depends on pipe diameter and fluid density, in addition to viscosity and fluid velocity dependence. Although the viscosity dependence of $\dot{S}'_G$ in turbulent flow is less severe in comparison with laminar flow, the velocity dependence is stronger in turbulent flow.

Figure 6.22 shows the plots of $\dot{S}'_G\ vs.\ \bar{V}$ on a log-log scale for laminar and turbulent flows generated from Equations (6.90) and (6.93), respectively. The fluid properties used in the equations are: $\mu = 1\ mPa.s$, $\rho = 1000\ kg/m^3$. The temperature used is 298.15 K. A single line of slope 2 is obtained for laminar regime regardless of the pipe diameter. The entropy generation rate per unit length of the pipe increases linearly with the increase in average velocity in the pipe. A family of parallel lines of slope 2.75 is obtained for the turbulent regime. The line shifts upward towards higher entropy generation rate with the increase in the pipe diameter.

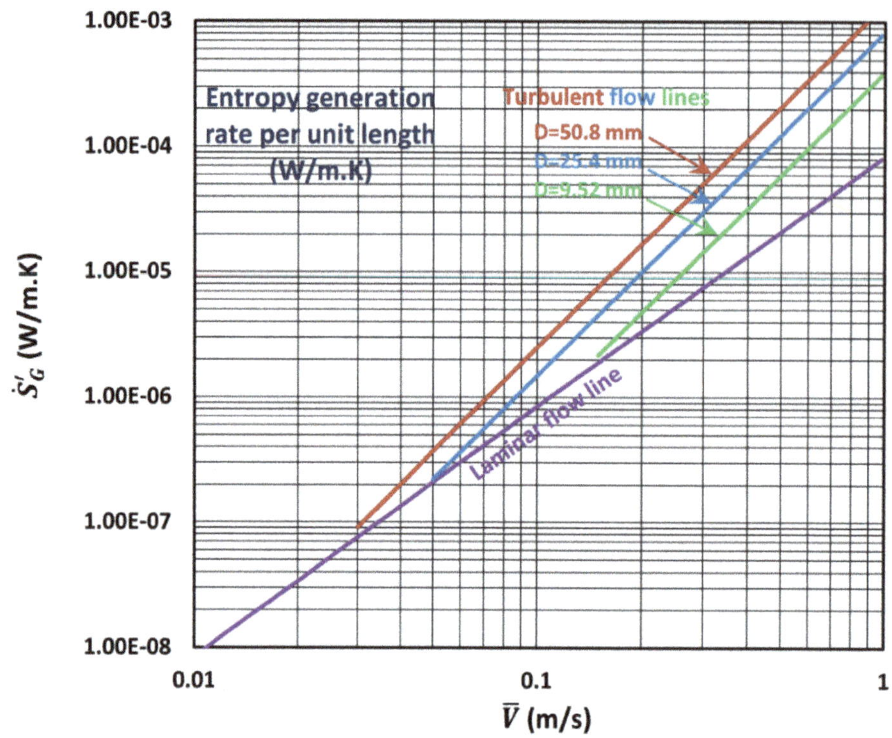

Figure 6.22 Entropy generation rate per unit length of pipe as a function of average velocity in the pipe.

Finally, it should be noted that the rate of mechanical energy dissipation in pipeline flow, per unit length of pipe, is directly proportional to the rate of entropy generation per unit length of pipe and is given as:

$$\dot{F}' = T\dot{S}'_G \tag{6.94}$$

where $\dot{F}'$ is rate of mechanical energy loss per unit length of pipe.

6.5.2 Steady Non-Adiabatic Incompressible Flow

Consider steady flow of an incompressible viscous fluid (liquid) in a constant cross-sectional area pipe with heat transfer at the wall. Let the heat flux $\dot{Q}''$ imposed on the wall be constant. Let the mass flow rate be $\dot{m}$, pipe diameter be D and pipe length be L.

The rate of entropy generation in a differential control volume of thickness dx is:

$$d\dot{S}_G = \dot{m}ds - \frac{\dot{Q}''(\pi D dx)}{T_w} \tag{6.95}$$

Equation (6.95) can be re-written as:

$$\frac{d\dot{S}_G}{dx} = \dot{m}\frac{ds}{dx} - \frac{\dot{Q}''(\pi D)}{T_w} \tag{6.96}$$

From the first law for control volume, in differential form, we get:

$$\dot{m}dh = \dot{Q}''(\pi D dx) \tag{6.97}$$

$$\dot{m}\frac{dh}{dx} = \dot{Q}''(\pi D) \tag{6.98}$$

The second Tds relation is given as:

$$Tds = dh - \frac{dP}{\rho} \tag{6.99}$$

$$T\frac{ds}{dx} = \frac{dh}{dx} - \frac{1}{\rho}\frac{dP}{dx} \tag{6.100}$$

Combining Equations (6.96), (6.98) and (6.100), we get:

$$\frac{d\dot{S}_G}{dx} = \dot{Q}''(\pi D)\left(\frac{1}{T} - \frac{1}{T_w}\right) + \frac{\dot{m}}{\rho T}\left(-\frac{dP}{dx}\right) \tag{6.101}$$

Equation (6.101) could be recast as:

$$\dot{S}_G' = \dot{Q}'\left(\frac{T_w - T}{TT_w}\right) + \frac{\dot{m}}{\rho T}\left(-\frac{dP}{dx}\right) \tag{6.102}$$

where $\dot{Q}'$ $(= \dot{Q}''\pi D)$ is rate of heat transfer per unit length of the pipe. Let the temperature difference $T_w - T$ be ΔT and assuming $\Delta T/T$ is small:

$$\left(\frac{T_w - T}{TT_w}\right) = \frac{\Delta T}{T^2\left(1 + \frac{\Delta T}{T}\right)} \approx \frac{\Delta T}{T^2} \tag{6.103}$$

From Equations (6.102) and (6.103), it follows that:

$$\dot{S}_G' = \dot{Q}'\left(\frac{\Delta T}{T^2}\right) + \frac{\dot{m}}{\rho T}\left(-\frac{dP}{dx}\right) \tag{6.104}$$

The first term on the right hand side of Equation (6.104) gives entropy generation due to irreversible heat transfer and the second term gives entropy generation due to fluid flow irreversibility caused by friction in fluid.

Following Bejan [6], Equation (6.104) could be re-written in terms dimensionless quantities: St and f, where St is Stanton number and f is friction factor. The Stanton number is defined as:

$$St = \frac{\bar{h}}{\rho C_p \bar{V}} \tag{6.105}$$

where $\bar{h}$ is the average heat transfer coefficient and $\bar{V}$ is the average velocity of fluid in the pipe. Note that

$$\dot{Q}'' = \bar{h}\Delta T \tag{6.106}$$

From Equations (6.105) and (6.106), we get:

$$St = \frac{\dot{Q}''/\Delta T}{\rho C_p \bar{V}} = \frac{\dot{Q}' D}{4\dot{m} C_p \Delta T} \tag{6.107}$$

where $\dot{Q}'$ is the rate of heat transfer per unit length of the pipe. Note that $\dot{Q}' = \pi D\dot{Q}''$.

The friction factor, defined in Equation (6.85), can be expressed as:

$$f = \frac{\pi^2 D^5 \rho}{32\dot{m}^2}\left(-\frac{dP}{dx}\right) \tag{6.108}$$

Using Equations (6.107) and (6.108), the entropy generation expression, Equation (6.104), can be re-written as:

$$\dot{S}'_G = \frac{(\dot{Q}')^2 D}{4T^2 \dot{m} C_p St} + \frac{32 f \dot{m}^3}{\pi^2 \rho^2 T D^5} \tag{6.109}$$

Using the following relations,

$$Nu = St\ Re\ Pr \tag{6.110}$$

$$Re = \frac{4\dot{m}}{\pi D \mu} \tag{6.111}$$

Equation (6.109) can be re-written as [6]:

$$\dot{S}'_G = \frac{(\dot{Q}')^2}{\pi T^2 k Nu} + \frac{32 f \dot{m}^3}{\pi^2 \rho^2 T D^5} \tag{6.112}$$

where Nu is Nusselt number, Re is Reynolds number, Pr is Prandtl number, and k is thermal conductivity of fluid. The second term on the right-hand side of Equation (6.112) can be further expressed in terms of Re and Equation (6.112) is modified as:

$$\dot{S}'_G = \frac{(\dot{Q}')^2}{\pi T^2 k Nu} + \frac{\pi^3 \mu^5 f Re^5}{32 \dot{m}^2 \rho^2 T} \tag{6.113}$$

Assuming the flow to be turbulent, Nu increases with Reynolds number Re and fRe^5 also increases with Re. Thus, for given $\dot{Q}'$, $\dot{m}, T$ and working fluid, the heat transfer irreversiblity (first term on right hand side of Equation (6.113)) decreases with the increase in Re whereas the fluid flow irreversibility (second term on right hand side of Equation (6.113)) increases with the increase in Re. As the two terms on the right-hand side of Equation (6.113) vary in opposite directions with the increase in Re, the entropy generation rate $\dot{S}'_G$ goes through a *minimum* value at some value of Re. The Re where $\dot{S}'_G$ is minimum can be determined from the condition that $d\dot{S}'_G/dRe = 0$.

For fully developed turbulent flow in a smooth pipe, the following correlations for Nu and f are applicable:

$$Nu = 0.023 Re^{0.8} Pr^{0.4} \tag{6.114}$$

$$f = 0.079/Re^{0.25} \tag{6.92}$$

Upon substitution of these expressions for Nu and f into Equation (6.113), the following result is obtained:

$$\dot{S}'_G = \frac{13.84(\dot{Q}')^2}{kT^2 Pr^{0.4}} Re^{-0.80} + \frac{0.0446 \mu^5 Re^{4.75}}{\dot{m}^2 \rho^2 T} \tag{6.115}$$

From Equation (6.115), it is clear that with the increase Re (for given $\dot{Q}'$, $\dot{m}, T$ and working fluid), the heat transfer irreversiblity decreases whereas the fluid flow irreversibility increases. For a minimum rate of entropy generation:

$$\frac{d\dot{S}'_G}{dRe} = -\frac{11.072(\dot{Q}')^2}{kT^2 Pr^{0.4}} Re^{-1.80} + \frac{0.2119 \mu^5 Re^{3.75}}{\dot{m}^2 \rho^2 T} = 0 \tag{6.116}$$

Let $Re = Re_{opt}$ where the entropy generation rate is minimum, Equation (6.116) gives:

$$Re_{opt} = 1.851Pr^{-0.072}B_o^{0.36} \tag{6.117}$$

where the parameter B_o is given as:

$$B_o = \frac{\dot{m}\dot{Q}'\rho}{(kT)^{1/2}\mu^{5/2}} \tag{6.118}$$

Note that for given $\dot{Q}'$, $\dot{m}, T$ and working fluid, B_o is fixed.

Equation (6.117) gives the optimum Reynolds number at which the efficiency of the flow process involving heat transfer is maximum from thermodynamics point of view. The magnitude of the combined irreversibilities due to heat transfer and fluid flow are at a minimum at the optimum Reynolds number. Interestingly, an optimum Reynolds number also implies an optimum pipe diameter given as:

$$D_{opt} = \frac{4\dot{m}}{\pi\mu Re_{opt}} = 0.688\left(\frac{\dot{m}}{\mu}\right)Pr^{0.072}B_o^{-0.36} \tag{6.119}$$

For a given $\dot{Q}'$, $\dot{m}, T$ and working fluid, Equation (6.119) gives the optimum pipe diameter where the thermodynamic efficiency of the process is maximum. Thus, thermodynamics plays a very important role in the design of pipelines.

NOTATION

a	constant in Equation (6.14)
A	cross-sectional area
A^*	critical cross-sectional area
B_o	parameter defined in Equation (6.117)
c	speed of sound
C	constant in Equation (6.13)
c^*	critical speed of sound
c_p	specific or molar constant pressure heat capacity
c_v	specific or molar constant volume heat capacity
D	internal diameter of pipe
D_{opt}	optimum pipe diameter
d	differential of exact quantity
f	friction factor
F_x	force component in x-direction
$\dot{F}'$	rate of mechanical energy dissipation per unit length of pipe
h	specific or molar enthalpy
h_o	specific or molar stagnation enthalpy
$\bar{h}$	average heat transfer coefficient
k	thermal conductivity
L	length of pipe
L^*	critical length of pipe
M	Mach number
$\dot{m}$	mass flow rate
Nu	Nusselt number
P	pressure

P_b	backpressure
P_e	pressure at the exit
P_o	stagnation pressure
P^*	critical pressure
Pr	Prandtl number
Q	heat transfer per unit mass of fluid
$\dot{Q}$	heat transfer rate
$\dot{Q}'$	heat transfer rate per unit length
$\dot{Q}''$	heat flux
R	universal gas constant
Re	Reynolds number
Re_{opt}	optimum Reynolds number
s	specific or molar entropy
$\dot{S}_G$	rate of entropy generation
$\dot{S}'_G$	rate of entropy generation per unit length
St	Stanton number
T	temperature
T_o	stagnation temperature
T_w	wall temperature
T^*	critical temperature
V	fluid velocity
V_e	fluid velocity at the exit
$\bar{V}$	average fluid velocity
V^*	critical fluid velocity
x	distance

Greek Letters

β	volume expansivity
γ	heat capacity ratio
Δ	difference operator
δ	differential of inexact quantity
μ	viscosity of fluid
ρ	density of fluid
ρ_o	stagnation density
ρ^*	critical density
τ_w	wall shear stress

REFERENCES

1. Munson, B.R., Young, D.F., Okiishi, T.H., *Fundamentals of Fluid Mechanics*, John Wiley & Sons, New York, 5th edition, 2006.
2. Hibbeler, R.C., *Fluid Mechanics*, Pearson Education, Hoboken, NJ, 2nd edition, 2018.
3. Smith, J.M., Van Ness, H.C., Abbott, M.M., *Introduction to Chemical Engineering Thermodynamics*, McGraw Hill, New York, 6th edition, 2001.
4. Bar-Meir, G., *Fundamentals of Compressible Fluid Mechanics*, Potto project: Genick Bar-Meir, 2006.
5. Pal, R., "Teaching fluid mechanics and thermodynamics simultaneously through pipeline flow experiments", *Fluids* 2019; 4: Article 103, 1-18.
6. Bejan, A., *Entropy Generation Minimization*, CRC Press, Boca Raton, 1996.

TUTORIAL PROBLEMS WITH SOLUTION

Problem 1

A converging nozzle is attached to a large tank of air under pressure. The temperature of air in the tank is 120 °C. When the nozzle is open, the air flows out at a Mach number of 0.5 at the exit. Determine the temperature of the air at the exit? You can assume steady isentropic flow and treat air as ideal gas with $\gamma = 1.4$. Note that the air is at rest inside the tank.

Solution

We are given the stagnation temperature $T_o = 120^oC = 393.15$ K. From Equation (6.26):

$$\frac{T}{T_o} = \left[1 + \left(\frac{\gamma - 1}{2}\right) M^2\right]^{-1} = \left[1 + \left(\frac{1.4 - 1}{2}\right)(0.50)^2\right]^{-1} = 0.9524$$

$$T = 0.9524 \times 393.15 = 374.44$$

There occurs a temperature drop of 393.15-374.44=18.71 K or °C.

Problem 2

Air from outside enters a pipe through a converging section. The pipe diameter is 65 mm. The outside air is at rest at 101.3 kPa and 23 °C. What is the mass flow rate of air in the pipe, if the pressure at entrance of the pipe is 90 kPa. Assume steady isentropic flow and treat air as ideal gas with $\gamma = 1.4$. The gas constant for air is 286.9 J/kg.K.

Solution

$$P_o = 101.3\ kPa,\ T_o = 23 + 273.15 = 296.15 \text{ K}$$

Pressure at the entrance of the pipe = 90 kPa. This is the pressure at the end of the converging section. From Equation (6.22):

$$\frac{P}{P_o} = \left[1 + \left(\frac{\gamma-1}{2}\right) M^2\right]^{-\frac{\gamma}{\gamma-1}}$$

$$\frac{90}{101.3} = \left[1 + \left(\frac{1.4-1}{2}\right) M^2\right]^{-\frac{1.4}{1.4-01}}$$

Solve for M, M=0.4146. Thus, the Mach number is 0.4146 at the entrance of the pipe. The mass flow rate is:

$$\dot{m} = \rho A V$$

$$\frac{\rho}{\rho_o} = \left[1 + \left(\frac{\gamma - 1}{2}\right) M^2\right]^{-\frac{1}{\gamma-1}} = \left[1 + \left(\frac{1.4 - 1}{2}\right)(0.4146)^2\right]^{-\frac{1}{1.4-1}} = 0.919$$

$$\rho_o = P_o/RT_o = 101.3 \times 10^3/(286.9 \times 296.15) = 1.19 \text{ kg/m}^3$$

$$\rho = 1.19 \times 0.919 = 1.094 \text{ kg/m}^3$$

$$\frac{T}{T_o} = \left[1 + \left(\frac{\gamma - 1}{2}\right) M^2\right]^{-1} = \left[1 + \left(\frac{1.4 - 1}{2}\right) (0.4146)^2\right]^{-1} = 0.9668$$

$$T = 0.9668 \times 296.15 = 286.32 \text{ K}$$

$$V = M\sqrt{\gamma RT} = 0.4146\sqrt{1.4 \times 286.9 \times 286.32} = 140.6 \text{ m/s}$$

$$\dot{m} = \rho AV = 1.094 \times \frac{\pi}{4}\left(\frac{65}{1000}\right)^2 \times 140.6 = 0.511 \text{ kg/s}$$

Problem 3

Air enters the converging-diverging nozzle at a speed of 125 m/s with pressure of 378 kPa and temperature of 325 K. The diameter of nozzle at the entrance is 45 mm. What should be the area of the nozzle at the throat so that M=1 at the throat? For the flow to be supersonic in the diverging section, what should be the temperature, pressure and velocity at the end of the diverging section where diameter is 70 mm. Assume flow to be steady isentropic and treat air as ideal gas with $\gamma = 1.4$.

Solution

Refer to the diagram given below.

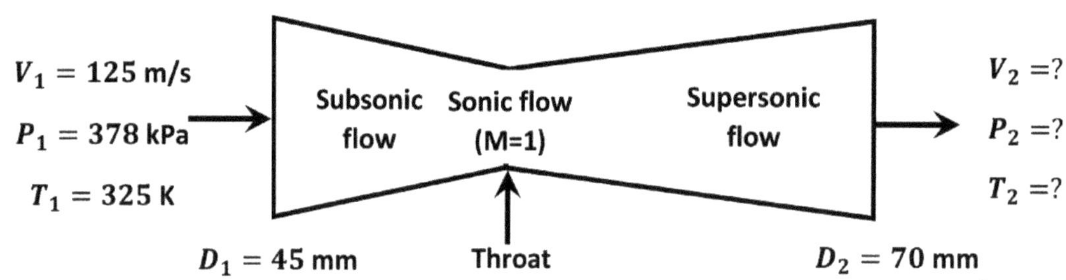

Mach number at the entrance, M_1

$$M_1 = \frac{V_1}{\sqrt{\gamma RT_1}} = \frac{125}{\sqrt{1.4 \times 286.9 \times 325}} = 0.346$$

From Equation (6.34), we get:

$$\frac{A_1}{A^*} = \left(\frac{1}{M}\right)\left[\frac{1 + 0.2M^2}{1.2}\right]^3 = \frac{1}{0.346}\left[\frac{1 + 0.2 \times 0.346^2}{1.2}\right]^3 = 1.7956$$

$$A^* = \frac{A_1}{1.7956} = \frac{\pi(45)^2}{4 \times 1.7956} = 885.74 \text{ mm}^2$$

Thus, the area at the throat of the nozzle for flow to be sonic (M=1) at the throat should be 885.74 mm^2.

At the exit,

$$\frac{A_2}{A^*} = \frac{\pi(70)^2}{4 \times 885.74} = 4.345$$

From Table 6.1,

M	A/A^*
3	4.235
3.2	5.121

By interpolation,

$$M = 3 + \frac{(3.2-3)}{(5.121-4.235)}(4.345-4.235) = 3.025$$

Thus, M=3.025 at the end of the diverging section.

Temperature, pressure, and velocity at the end of the diverging section:

We can calculate temperature (T_2) as follows:

$$\frac{T_2}{T_1} = \frac{T_2/T_o}{T_1/T_o}$$

where

$$\frac{T_1}{T_o} = \left[1+\left(\frac{\gamma-1}{2}\right)M_1^2\right]^{-1} = \left[1+\left(\frac{1.4-1}{2}\right)\times 0.346^2\right]^{-1} = 0.97662$$

$$\frac{T_2}{T_o} = \left[1+\left(\frac{\gamma-1}{2}\right)M_2^2\right]^{-1} = \left[1+\left(\frac{1.4-1}{2}\right)\times 3.025^2\right]^{-1} = 0.35334$$

Thus,

$$\frac{T_2}{T_1} = \frac{T_2/T_o}{T_1/T_o} = \frac{0.35334}{0.97662} = 0.3618$$

$$T_2 = 0.3618 \times 325 = 117.585 \text{ K}$$

The pressure (P_2) can be calculated as follows:

$$\frac{P_2}{P_1} = \frac{P_2/P_o}{P_1/P_o}$$

where

$$\frac{P_1}{P_o} = \left[1+\left(\frac{\gamma-1}{2}\right)M_1^2\right]^{-\frac{\gamma}{\gamma-1}} = \left[1+\left(\frac{1.4-1}{2}\right)\times 0.346^2\right]^{-\frac{1.4}{1.4-1}} = 0.92052$$

$$\frac{P_2}{P_o} = \left[1+\left(\frac{\gamma-1}{2}\right)M_2^2\right]^{-\frac{\gamma}{\gamma-1}} = \left[1+\left(\frac{1.4-1}{2}\right)\times 3.025^2\right]^{-\frac{1.4}{1.4-1}} = 0.02622$$

Thus,

$$\frac{P_2}{P_1} = \frac{P_2/P_o}{P_1/P_o} = \frac{0.02622}{0.92052} = 0.0285$$

$$P_2 = 0.0285 \times 378 = 10.773 \text{ kPa}$$

The velocity is:

$$V_2 = M_2\sqrt{\gamma R T_2} = 3.025 \times \sqrt{1.4 \times 286.9 \times 117.585} = 657.40 \text{ m/s}$$

Problem 4

Air enters a converging nozzle from a large tank at a pressure of 800 kPa and temperature of 315 K. The velocity of air at the entrance is negligible. The cross-sectional area of the nozzle at the exit is 0.0013 m^2.

a) Calculate the Mach number at the exit and the mass flow rate through the nozzle when backpressure is 400kPa.
b) Repeat the calculation in part (a) when backpressure is 650 kPa.

Assume steady isentropic flow and treat air as ideal gas with $\gamma = 1.4$.

Solution

Let us first calculate the critical pressure, P^*. From Equation (6.28), we get:

$$\frac{P^*}{P_o} = \left[\left(\frac{\gamma+1}{2}\right)\right]^{-\frac{\gamma}{\gamma-1}} = \left[\left(\frac{1.4+1}{2}\right)\right]^{-\frac{1.4}{1.4-1}} = 0.5283$$

$$P^* = 0.5283 \times P_o = 0.5283 \times 800 = 422.64 \text{ kPa}$$

For backpressures of 422.64 kPa or less, $M = 1$ at the exit and the nozzle is choked.

a) When $P_b = 400$ kPa, $M = 1$ at the exit and the nozzle is choked.

The mass flow rate is:

$$\dot{m} = \rho A V$$

As the nozzle is choked, the properties are critical properties at the exit. Thus,

$$\dot{m} = \rho A V = \rho^* A V^*$$

$$\frac{\rho^*}{\rho_o} = \left[\left(\frac{\gamma+1}{2}\right)\right]^{-\frac{1}{\gamma-1}} = \left[\left(\frac{1.4+1}{2}\right)\right]^{-\frac{1}{1.4-1}} = 0.6339$$

$$\rho_o = P_o/RT_o = 800 \times 10^3/(286.9 \times 315) = 8.852 \text{ kg/m}^3$$

$$\rho^* = 0.6339 \times 8.852 = 5.611 \text{ kg/m}^3$$

$$\frac{T^*}{T_o} = \left[\left(\frac{\gamma+1}{2}\right)\right]^{-1} = \left[\left(\frac{1.4+1}{2}\right)\right]^{-1} = 0.8333$$

$$T^* = 0.8333 \times 315 = 262.5 \text{ K}$$

$$V_2 = \sqrt{\gamma R T^*} = \sqrt{1.4 \times 286.9 \times 262.5} = 324.71 \text{ m/s}$$

$$\dot{m} = \rho^* A V^* = 5.611 \times 0.0013 \times 324.71 = 2.369 \text{ kg/s}$$

b) When $P_b = 650$ kPa, $P_b > P^*$ and therefore, the flow is subsonic in the nozzle and the exit pressure is $P_b = 650$ kPa.

$$\frac{P_2}{P_o} = \frac{650}{800} = 0.8125$$

$$\frac{P_2}{P_o} = \left[1 + \left(\frac{\gamma - 1}{2}\right) M_2^2\right]^{-\frac{\gamma}{\gamma-1}} = 0.8125$$

Given $\gamma = 1.4$, solve for M_2 :

$$M_2 = 0.5528$$

The mass flow rate is:

$$\dot{m} = \rho AV = \left(\frac{P_2}{RT_2}\right) AV_2$$

We need temperature T_2 to determine $\dot{m}$. The temperature is given by

$$\frac{T_2}{T_o} = \left[1 + \left(\frac{\gamma - 1}{2}\right) M_2^2\right]^{-1} = \left[1 + \left(\frac{1.4 - 1}{2}\right) \times 0.5528^2\right]^{-1} = 0.9424$$

$$T_2 = 0.9424 \times 315 = 296.86 \text{ K}$$

The velocity V_2 is:

$$V_2 = M_2\sqrt{\gamma RT_2} = 0.5528 \times \sqrt{1.4 \times 286.9 \times 296.86} = 190.89 \text{ m/s}$$

Thus, the mass flow rate is:

$$\dot{m} = \left(\frac{P_2}{RT_2}\right) AV_2 = \left(\frac{650\times10^3}{286.9\times296.86}\right) \times 0.0013 \times 190.89 = 1.894 \text{ kg/s}$$

Problem 5

Consider a converging-diverging nozzle with a throat area of 0.0005 m^2 and an exit area of 0.00125 m^2. Air enters the nozzle at a pressure of 600 kPa and temperature of 280 K. The velocity of air at the entrance is negligible. The flow is steady isentropic throughout the nozzle. Calculate:

a) Mass flow rate, exit pressure, exit temperature and exit Mach number when M=0.60 at the throat.
b) Repeat calculations of part (a) when M=1.0 at the throat and the flow is subsonic in the diverging section.
c) Repeat calculations of part (a) when M=1.0 at the throat and the flow is supersonic in the diverging section.

You can treat air as ideal gas with $\gamma = 1.4$.

Solution

Refer to the figure given below.

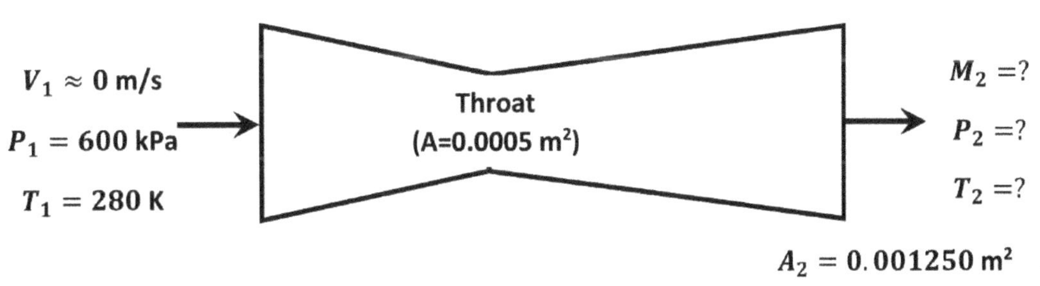

a) $M = 0.60$ at throat with $A_t = 0.0005\ \text{m}^2$. Let us calculate A^* where $M = 1.0$. From Equation (6.34):

$$\frac{A}{A^*} = \left(\frac{1}{M}\right)\left[\frac{1+0.2M^2}{1.2}\right]^3 = \left(\frac{1}{0.60}\right)\left[\frac{1+0.2\times0.6^2}{1.2}\right]^3 = 1.1882$$

$$A^* = \frac{A}{1.1882} = \frac{0.0005}{1.1882} = 4.208\times10^{-4}\ m^2$$

The area ratio at the exit is:

$$\frac{A_2}{A^*} = \frac{0.001250}{4.208\times10^{-4}} = 2.9705$$

The Mach number at the exit, M_2:

From Table 6.1,

M	A/A^*
0.19	3.112
0.20	2.964

By interpolation,

$$M = 0.19 + \frac{(0.20-0.19)}{(2.964-3.112)}(2.9705-3.112) = 0.19956$$

Thus, $M_2 = 0.19956$ at the end of the diverging section. We can now calculate temperature, pressure, and velocity at the exit:

$$\frac{T_2}{T_o} = \left[1+\left(\frac{\gamma-1}{2}\right)M_2^2\right]^{-1} = \left[1+\left(\frac{1.4-1}{2}\right)\times0.19956^2\right]^{-1} = 0.9921$$

$$T_2 = 0.9921\times280 = 277.79\ \text{K}$$

$$\frac{P_2}{P_o} = \left[1+\left(\frac{\gamma-1}{2}\right)M_2^2\right]^{-\frac{\gamma}{\gamma-1}} = \left[1+\left(\frac{1.4-1}{2}\right)\times0.19956^2\right]^{-\frac{1.4}{1.4-1}} = 0.9726$$

$$P_2 = 0.9726\times600 = 583.56\ \text{kPa}$$

Note that $P_o = P_1 = 600$ kPa.

The velocity V_2 is:

$$V_2 = M_2\sqrt{\gamma RT_2} = 0.19956\times\sqrt{1.4\times286.9\times277.79} = 66.66\ \text{m/s}$$

Thus, the mass flow rate is:

$$\dot{m} = \left(\frac{P_2}{RT_2}\right)AV_2 = \left(\frac{583.56\times10^3}{286.9\times277.79}\right)\times0.00125\times66.66 = 0.61\ \text{kg/s}$$

b) $M = 1.0$ at throat. Thus, $A^* = A_t = 0.0005\ \text{m}^2$. The flow is subsonic in the diverging section.

Let us calculate M_2 at the exit. The area ratio at the exit is:

$$\frac{A_2}{A^*} = \frac{0.001250}{0.0005} = 2.5$$

From Table 6.1,

M	A/A^*
0.20	2.964
0.25	2.403

By interpolation,

$$M = 0.20 + \frac{(0.25 - 0.20)}{(2.403 - 2.964)}(2.5 - 2.964) = 0.241$$

Thus, $M_2 = 0.241$ at the end of the diverging section. We can now calculate temperature, pressure, and velocity at the exit:

$$\frac{T_2}{T_o} = \left[1 + \left(\frac{\gamma - 1}{2}\right)M_2^2\right]^{-1} = \left[1 + \left(\frac{1.4 - 1}{2}\right) \times 0.241^2\right]^{-1} = 0.9885$$

$$T_2 = 0.9885 \times 280 = 276.78 \text{ K}$$

$$\frac{P_2}{P_o} = \left[1 + \left(\frac{\gamma - 1}{2}\right)M_2^2\right]^{-\frac{\gamma}{\gamma-1}} = \left[1 + \left(\frac{1.4 - 1}{2}\right) \times 0.241^2\right]^{-\frac{1.4}{1.4-1}} = 0.9604$$

$$P_2 = 0.9604 \times 600 = 576.24 \text{ kPa}$$

Note that $P_o = P_1 = 600$ kPa.

The velocity V_2 is:

$$V_2 = M_2\sqrt{\gamma R T_2} = 0.241 \times \sqrt{1.4 \times 286.9 \times 276.78} = 80.36 \text{ m/s}$$

Thus, the mass flow rate is:

$$\dot{m} = \left(\frac{P_2}{RT_2}\right)AV_2 = \left(\frac{576.24\times10^3}{286.9\times276.78}\right) \times 0.00125 \times 80.36 = 0.73 \text{ kg/s}$$

c) $M = 1.0$ at throat. Thus, $A^* = A_t = 0.0005$ m^2. The flow is supersonic in the diverging section. Let us calculate M_2 at the exit. The area ratio at the exit is:

$$\frac{A_2}{A^*} = \frac{0.001250}{0.0005} = 2.5$$

From Table 6.1,

M	A/A^*
2.4	2.403
2.5	2.637

By interpolation,

$$M = 2.4 + \frac{(2.5 - 2.4)}{(2.637 - 2.403)}(2.5 - 2.403) = 2.441$$

Thus, $M_2 = 2.441$ at the end of the diverging section. We can now calculate temperature, pressure, and velocity at the exit:

$$\frac{T_2}{T_o} = \left[1 + \left(\frac{\gamma - 1}{2}\right) M_2^2\right]^{-1} = \left[1 + \left(\frac{1.4 - 1}{2}\right) \times 2.441^2\right]^{-1} = 0.4563$$

$$T_2 = 0.4563 \times 280 = 127.764 \text{ K}$$

$$\frac{P_2}{P_o} = \left[1 + \left(\frac{\gamma - 1}{2}\right) M_2^2\right]^{-\frac{\gamma}{\gamma-1}} = \left[1 + \left(\frac{1.4 - 1}{2}\right) \times 2.441^2\right]^{-\frac{1.4}{1.4-1}} = 0.0642$$

$$P_2 = 0.0642 \times 600 = 38.52 \text{ kPa}$$

Note that $P_o = P_1 = 600$ kPa.

The velocity V_2 is:

$$V_2 = M_2\sqrt{\gamma R T_2} = 2.441 \times \sqrt{1.4 \times 286.9 \times 127.764} = 552.97 \text{ m/s}$$

Thus, the mass flow rate is:

$$\dot{m} = \left(\frac{P_2}{RT_2}\right) A V_2 = \left(\frac{38.52 \times 10^3}{286.9 \times 127.764}\right) \times 0.00125 \times 552.97 = 0.73 \text{ kg/s}$$

Note that the mass flow rate is the same as in part(b) as the nozzle is choked.

Problem 6

Consider Fanno flow (steady, adiabatic compressible flow with friction) of air in a 25 mm diameter pipe. Air enters the pipe a temperature of 298 K and a velocity of 145 m/s. The average friction factor is 0.014. Determine the critical length L^* of pipe such that the flow becomes sonic at the exit. What is the exit velocity at L^* ? What is the velocity of air at a distance of 0.75 m from the entrance? What is the temperature of air at this location? You can treat air as ideal gas with $\gamma = 1.4$.

Solution

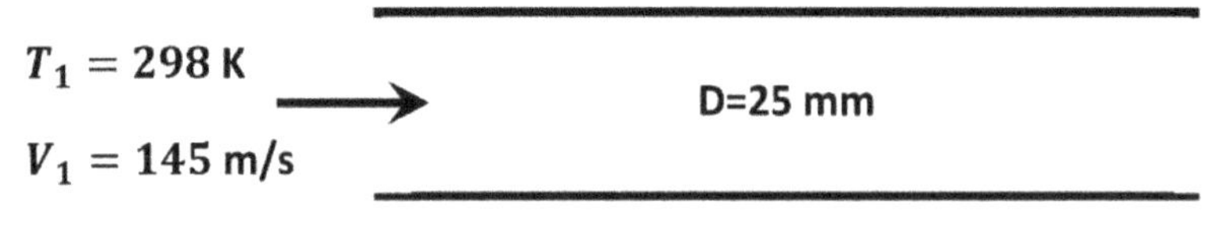

Mach number at the entrance, M_1

$$M_1 = \frac{V_1}{\sqrt{\gamma R T_1}} = \frac{145}{\sqrt{1.4 \times 286.9 \times 298}} = 0.4191$$

From Equation (6.40), we get:

$$\frac{4f(L^* - L_1)}{D} = \frac{1}{\gamma}\left[\frac{1 - M_1^2}{M_1^2}\right] + \left(\frac{\gamma + 1}{2\gamma}\right) ln\left[\frac{\left(\frac{\gamma + 1}{2}\right) M_1^2}{1 + \left(\frac{\gamma - 1}{2}\right) M_1^2}\right]$$

$$= \frac{1}{1.4}\left[\frac{1 - 0.4191^2}{0.4191^2}\right] + \left(\frac{1.4 + 1}{2 \times 1.4}\right) ln\left[\frac{\left(\frac{1.4 + 2}{2}\right) \times 0.4191^2}{1 + \left(\frac{1.4 - 1}{2}\right) \times 0.4191^2}\right]$$

$$= 1.9883$$

With $L_1 = 0$, L^* is as follows:

$$L^* = \frac{1.9883 \times 25 \times 10^{-3}}{4 \times 0.014} = 0.8876 \text{ m}$$

What is exit velocity corresponding to L^*?

From Equation (6.48):

$$\frac{V_1}{V^*} = M_1 \sqrt{\frac{\left(\frac{\gamma + 1}{2}\right)}{1 + \left(\frac{\gamma - 1}{2}\right) M_1^2}} = 0.4191 \sqrt{\frac{\left(\frac{1.4 + 1}{2}\right)}{1 + \left(\frac{1.4 - 1}{2}\right) \times 0.4191^2}} = 0.4512$$

Thus,

$$V^* = \frac{145}{0.4512} = 321.37 \text{ m/s}$$

What is velocity at $L = 0.75$ m?

$$\frac{4f(L^* - L)}{D} = \frac{4 \times 0.014(0.8876 - 0.75)}{25 \times 10^{-3}} = 0.3082$$

Mach number at $L = 0.75$ m?

From Table 6.2:

M	$4f(L^* - L)/D$
0.65	0.325
0.70	0.208

By interpolation,

$$M = 0.65 + \frac{(0.70 - 0.65)}{(0.208 - 0.325)}(0.3082 - 0.325) = 0.6572$$

To calculate the velocity, use Equation (6.48):

$$\frac{V}{V^*} = M \sqrt{\frac{\left(\frac{\gamma + 1}{2}\right)}{1 + \left(\frac{\gamma - 1}{2}\right) M^2}} = 0.6572 \sqrt{\frac{\left(\frac{1.4 + 1}{2}\right)}{1 + \left(\frac{1.4 - 1}{2}\right) \times 0.6572^2}} = 0.6907$$

$$V = 0.6907 \times 321.37 = 221.97 \text{ m/s}$$

Temperature can be calculated as follows:

$$\frac{T}{T_1} = \frac{(T/T^*)_{M=0.6907}}{(T/T^*)_{M=0.4191}}$$

where

$$\frac{T}{T^*} = \left[\frac{\left(\frac{\gamma+1}{2}\right)}{1+\left(\frac{\gamma-1}{2}\right)M^2}\right] = \left[\frac{\left(\frac{1.4+1}{2}\right)}{1+\left(\frac{1.4-1}{2}\right)M^2}\right]$$

$$(T/T^*)_{M=0.6907} = 1.09548; \quad (T/T^*)_{M=0.4191} = 1.15928$$

Using the above values, we get:

$$\frac{T}{T_1} = \frac{(T/T^*)_{M=0.6907}}{(T/T^*)_{M=0.4191}} = \frac{1.09548}{1.15928} = 0.945$$

$$T_1 = 0.945 \times 298 = 281.61 \text{ K}$$

Thus, the velocity has increased from 145 m/s to 221.97 m/s and the temperature has dropped from 298 K to 281.61 K as the air travels 0.75 m downstream the pipe.

Problem 7

Consider steady, compressible flow of air in a pipe of 175 mm. At the entrance, the velocity is 69m/s, the pressure is 113 kPa, and the temperature is 289 K. Heat is added to the flow through the pipe wall at Q' =92 kJ/kg.m.

a) Determine Mach number and density at the entrance.
b) Determine the critical properties $T^*, P^*, V^*, \rho^*, T_o^*$ and P_o^*.
c) Calculate stagnation temperature $T_{o,2}$ and Mach number at $x = 1.9\ m$.
d) Determine the properties T, P, V, ρ, P_o of air at $x = 1.9\ m$.

You can treat air as ideal gas with $\gamma = 1.4$ and c_p=1004 J/kg.K

Solution

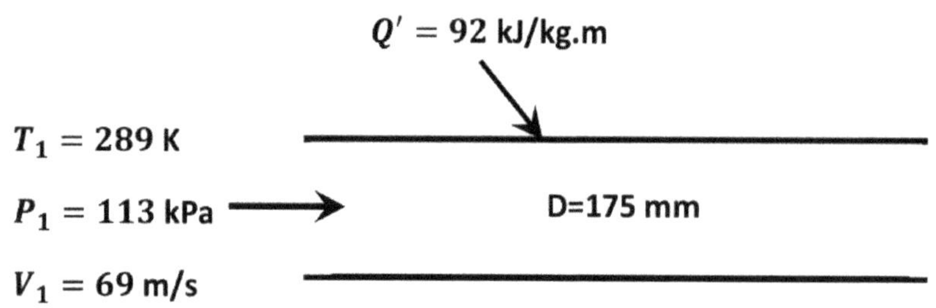

a) Mach number (M_1) and density (ρ_1) at the entrance

$$M_1 = \frac{V_1}{\sqrt{\gamma R T_1}} = \frac{69}{\sqrt{1.4 \times 286.9 \times 289}} = 0.2025$$

$$\rho_1 = \frac{P_1}{RT_1} = \frac{113\times10^3}{286.9\times289} = 1.3629 \text{ kg/m}^3$$

b) Critical properties $T^*, P^*, V^*, \rho^*, T_o^*$ and P_o^*.

$$\frac{T_1}{T^*} = \left[\frac{(1+\gamma)M_1}{1+\gamma M_1^2}\right]^2 = \left[\frac{(1+1.4)\times 0.2025}{1+1.4\times 0.2025^2}\right]^2 = 0.2112 \; ; \; T^* = \frac{T_1}{0.2112} = \frac{289}{0.2112} = 1368.37 \text{ K}$$

$$\frac{P_1}{P^*} = \left[\frac{(1+\gamma)}{1+\gamma M_1^2}\right] = \left[\frac{(1+1.4)}{1+1.4\times 0.2025^2}\right] = 2.2697 \; ; \quad P^* = \frac{P_1}{2.2697} = \frac{113}{2.2697} = 49.786 \text{ kPa}$$

$$\frac{V_1}{V^*} = \left[\frac{(1+\gamma)M_1^2}{1+\gamma M_1^2}\right] = \left[\frac{(1+1.4)\times 0.2025^2}{1+1.4\times 0.2025^2}\right] = 0.09307 \; ; \; V^* = \frac{V_1}{0.09307} = \frac{69}{0.09307} = 741.377 \text{ m/s}$$

$$\frac{\rho_1}{\rho^*} = \left[\frac{1+\gamma M_1^2}{(1+\gamma)M_1^2}\right] = \left[\frac{1+1.4\times 0.2025^2}{(1+1.4)\times 0.2025^2}\right] = 10.7444 \; ; \; \rho^* = \frac{\rho_1}{10.7444} = \frac{1.3629}{10.7444} = 0.12685 \text{ kg/m}^3$$

$$T_{o,1} = T_1\left[1 + \left(\frac{\gamma-1}{2}\right)M_1^2\right] = 289\left[1 + \left(\frac{1.4-1}{2}\right)\times 0.2025^2\right] = 291.37 \text{ K}$$

$$\frac{T_{o,1}}{T_o^*} = 2(1+\gamma)M_1^2\left[\frac{1+\left(\frac{\gamma-1}{2}\right)M_1^2}{\left(1+\gamma M_1^2\right)^2}\right] = 2(1+1.4)\times 0.2025^2\left[\frac{1+\left(\frac{1.4-1}{2}\right)\times 0.2025^2}{(1+1.4\times 0.2025^2)^2}\right] = 0.1775$$

$$T_o^* = \frac{T_{o,1}}{0.1775} = \frac{291.37}{0.1775} = 1641.52 \text{ K}$$

$$P_{o,1} = P_1\left[1 + \left(\frac{\gamma-1}{2}\right)M_1^2\right]^{\gamma/(\gamma-1)} = 113\left[1 + \left(\frac{1.4-1}{2}\right)\times 0.2025^2\right]^{1.4/(1.4-1)} = 116.277 \text{ kPa}$$

$$\frac{P_{o,1}}{P_o^*} = \left(\frac{(1+\gamma)}{1+\gamma M_1^2}\right)\left[\left(\frac{2}{1+\gamma}\right)\left\{1 + \left(\frac{\gamma-1}{2}\right)M_1^2\right\}\right]^{\gamma/(\gamma-1)} = \left(\frac{(1+1.4)}{1+1.4\times\times 0.2025^2}\right)\left[\left(\frac{2}{1+1.4}\right)\left\{1 + \left(\frac{1.4-1}{2}\right)\times 0.2025^2\right\}\right]^{1.4/(1.4-1)} = 1.2338 \; ; \; P_o^* = \frac{P_{o,1}}{1.2338} = \frac{116.277}{1.2338} = 94.64 \text{ kPa}$$

c) Stagnation temperature $T_{o,2}$ and Mach number at $x = 1.9\ m$.

$$T_{o,2} = T_{o,1} + (Q_{1\to 2}/c_p) = T_{o,1} + (Q'L/c_p) = 291.37 + (92\times 10^3 \times 1.9/1004) = 465.47 \text{ K}$$

$$\frac{T_{o,2}}{T_o^*} = \frac{465.47}{1641.52} = 0.2836$$

From Table 6.3:

M	T_o/T_o^*
0.25	0.257
0.30	0.347

By interpolation,

$$M = 0.25 + \frac{(0.30-0.25)}{(0.347-0.257)}(0.2836 - 0.257) = 0.2648$$

d) Properties T, P, V, ρ, P_o of air at $x = 1.9\ m$.

$$\frac{T_2}{T^*} = \left[\frac{(1+\gamma)M_2}{1+\gamma M_2^2}\right]^2 = \left[\frac{(1+1.4)\times 0.2648}{1+1.4\times 0.2648^2}\right]^2 = 0.3349 \; ; \; T_2 = T^*\left(\frac{T_2}{T^*}\right) = 1368.37 \times 0.3349 = 458.27 \text{ K}$$

$$\frac{P_2}{P^*} = \left[\frac{(1+\gamma)}{1+\gamma M_2^2}\right] = \left[\frac{(1+1.4)}{1+1.4\times 0.2648^2}\right] = 2.185 \ ; \ P_2 = P^*\left(\frac{P_2}{P^*}\right) = 49.786 \times 2.185 = 108.78 \text{ kPa}$$

$$\frac{V_2}{V^*} = \left[\frac{(1+\gamma)M_2^2}{1+\gamma M_2^2}\right] = \left[\frac{(1+1.4)\times 0.2648^2}{1+1.4\times 0.2648^2}\right] = 0.1532 \ ; \ V_2 = V^*\left(\frac{V_2}{V^*}\right) = 741.377 \times 0.1532 = 113.58 \text{ m/s}$$

$$\frac{\rho_2}{\rho^*} = \left[\frac{1+\gamma M_2^2}{(1+\gamma)M_2^2}\right] = \left[\frac{1+1.4\times 0.2648^2}{(1+1.4)\times 0.2648^2}\right] = 0.1848; \rho_2 = \rho^*\left(\frac{\rho_2}{\rho^*}\right) = 0.12685 \times 0.1848 = .0234 \text{ kg/m}^3$$

$$\frac{P_{o,2}}{P_o^*} = \left(\frac{(1+\gamma)}{1+\gamma M_2^2}\right)\left[\left(\frac{2}{1+\gamma}\right)\left\{1+\left(\frac{\gamma-1}{2}\right)M_2^2\right\}\right]^{\gamma/(\gamma-1)}$$
$$= \left(\frac{(1+1.4)}{1+1.4\times 0.0.2648^2}\right)\left[\left(\frac{2}{1+1.4}\right)\left\{1 + \left(\frac{1.4-1}{2}\right)\times 0.2648^2\right\}\right]^{1.4/(1.4-1)} = 1.2122$$

$$P_{o,2} = P_o^*\left(\frac{P_{o,2}}{P_o^*}\right) = 94.64 \times 1.2122 = 114.723 \text{ kPa}$$

Thus, the Mach number increases from 0.2025 to 0.2648, the velocity increases from 69 m/s to 113.58 m/s, the temperature increases from 289 K to 458.267 K, the pressure drops from 113 kPa to 108.78 kPa and the density decreases from 1.3629 kg/m^3 to 0.0234 kg/m^3.

Problem 8

Consider steady, compressible flow of natural gas (methane) in a long smooth pipeline with a diameter of 1.8 ft. At the entrance, the gas velocity is 10 m/s, the pressure is 650 kPa, and the temperature is 293 K. The length of the pipe is 15 km. The friction factor is constant, equal to 0.0026. Estimate the pressure at the exit of the pipe. You can assume that the flow is isothermal (293 K). You can treat natural gas as ideal gas with $\gamma = 1.32$. The gas constant is 518.28 J/kg.K.

Solution

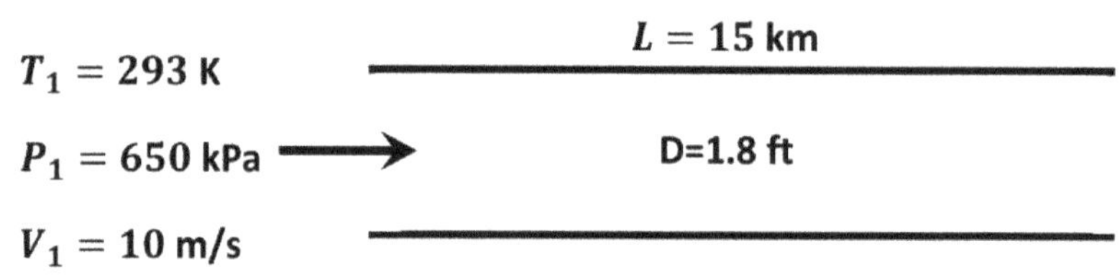

Isothermal flow,

$$M_1 = \frac{V_1}{\sqrt{\gamma R T_1}} = \frac{10}{\sqrt{1.32 \times 518.28 \times 293}} = 0.02234$$

From Equation (6.69):

$$\frac{4f(L^*)}{D} = \left[\frac{1-\gamma M^2}{\gamma M^2}\right] + ln(\gamma M^2) = \left[\frac{1-1.32\times 0.02234^2}{1.32\times 0.02234^2}\right] + ln(1.32 \times 0.02234^2) = 1509.633$$

$$L^* = \frac{1509.633 \times 1.8 \times 0.3048}{4 \times 0.0026} = 79638.947 \text{ m} = 79.64 \text{ km}$$

The corresponding critical Mach number is:

$$M^* = \frac{1}{\sqrt{\gamma}} = \frac{1}{\sqrt{1.32}} = 0.87$$

Since $L \ll L^*$, the flow is obviously subsonic throughout the pipe.

To estimate P_2, use Equation (6.75):

$$\frac{4f(L_2 - L_1)}{D} = \left[\frac{1 - \left(\frac{P_2}{P_1}\right)^2}{\gamma M_1^2}\right] + 2ln\left(\frac{P_2}{P_1}\right)$$

$$\frac{4f(L_2 - L_1)}{D} = \frac{4 \times 0.0026 \times (15 - 0) \times 1000}{1.8 \times 0.3048} = 284.34$$

Thus,

$$284.34 = \left[\frac{1 - \left(\frac{P_2}{P_1}\right)^2}{1.32 \times 0.02234^2}\right] + 2ln\left(\frac{P_2}{P_1}\right)$$

Upon simplification,

$$-ln\left(\frac{P_2}{P_1}\right) + 758.98\left[\left(\frac{P_2}{P_1}\right)^2 - 1\right] + 142.17 = 0$$

The solution is:

$$\frac{P_2}{P_1} = 0.9014 \;\; ; \;\; P_2 = 0.9014 \times 650 = 585.91 \text{ kPa.}$$

PRATICE PROBLEMS

Problem 1
A pipe is connected to a large reservoir of air. Air enters the pipe at a velocity of 135 m/s. The pressure and temperature of air at the entrance are: 270 kPa and 75 °C. Determine the stagnation pressure and temperature of air. Assume flow to be isentropic. Air behaves as an ideal gas with $\gamma = 1.4$.

Problem 2
Air from a large reservoir enters a pipe with 1.75 inch-diameter through a converging nozzle. The exit diameter of the nozzle and the pipe diameter are the same. What should be the pressure at the entrance of the pipe (end of converging nozzle) so that the mass flow rate of air through the pipe is maximum? What is this mass flow rate? Assume isentropic flow. Air behaves as an ideal gas with $\gamma = 1.4$.

Problem 3
A converging nozzle is attached to a large tank of nitrogen gas. The pressure of the gas inside the tank is 471 kPa and the temperature is 998 K. The exit diameter of the nozzle is 250 mm. If the pressure of gas at the exit of the nozzle is 271 kPa, what is the mass flow rate of nitrogen? What is the pressure of the gas at the exit of the nozzle if the mass flow rate of nitrogen through the nozzle is maximum? Assume isentropic flow. Nitrogen behaves as an ideal gas with $\gamma = 1.4$. The gas constant for nitrogen is 296.8 J/kg.K.

Problem 4

A converging-diverging nozzle is attached to a large reservoir of air. The pressure of air in the reservoir is 298 kPa. The diameter of the nozzle at the throat is 45 mm. The diameter at the end of the diverging section of the nozzle is 92 mm. The nozzle is choked. Assuming isentropic flow throughout the nozzle,

a) Calculate the exit pressure for the flow to be subsonic throughout the nozzle.
b) Calculate the exit pressure for the flow to be supersonic throughout the diverging section.

Air is an ideal gas with $\gamma = 1.4$.

Problem 5

A converging-diverging nozzle is attached to a large tank of air. The diameter of the nozzle at the throat is 38 mm. The air in the tank is at 515 kPa and 25 °C. What is the maximum mass flow rate of air through the nozzle?

Assuming isentropic flow throughout the nozzle. Air is an ideal gas with $\gamma = 1.4$.

Problem 6

Consider Fanno flow of air in a 75 mm diameter pipe. The Mach number at the entrance is 0.40. The average friction factor in the pipe is 0.008.

a) Calculate the Mach number at the exit of the pipe, if the pipe length is 1.5 m.
b) What will be the Mach number at the entrance of the pipe, if the pipe length is 7 m?

Assume air to be an ideal gas with $\gamma = 1.4$.

Problem 7

Air from a large reservoir is drawn into a 80 mm-diameter pipe through a converging section. The flow in this converging section is isentropic. The pressure at the entrance of the pipe is 73 kPa. The temperature and pressure of air in the reservoir are: 300 K and 120 kPa. The flow in the pipe is steady adiabatic. The pipe has an average friction factor of 0.007.

a) Determine the Mach number and velocity at the entrance of the pipe.
b) What is the mass flow rate of air in the pipe?
c) What is the critical length of the pipe?
d) What is the Mach number at a location L=0.85 m? Also, estimate pressure and velocity at this location.

Assume air to be an ideal gas with $\gamma = 1.4$.

Problem 8

Consider steady, frictionless compressible flow of air in a pipe. The conditions at the entrance are: T_1=256 K, P_1=313 kPa, and M_1=0.40. The pipe is heated at $Q' = 90$ kJ/kg.m.

e) Determine velocity and density at the entrance.
f) Determine the critical properties $T^*, P^*, V^*, \rho^*, T_o^*$ and P_o^*.
g) Calculate stagnation temperature $T_{o,2}$ and Mach number at $x = 2.5\ m$.
h) Determine the properties T, P, V, ρ, P_o of air at $x = 2.5\ m$.

You can treat air as ideal gas with $\gamma = 1.4$ and c_p=1004 J/kg.K

Problem 9

Natural gas, which can be treated as pure methane, is flowing in a 2.5 ft diameter pipeline at steady state. The length of the pipeline is 12 km. The gas enters the pipeline at 25 m/s. The conditions at the entrance are: P_1= 715 kPa and T_1=300 K. The average friction factor for the pipe is 0.0027. Estimate the pressure at the exit of the pipe, assuming isothermal flow. You can treat natural gas as ideal gas with $\gamma = 1.32$. The gas constant is 518.28 J/kg.K.

Problem 10

The speed of sound in a fluid, denoted as c, is given as:

$$c = \sqrt{\left(\frac{\partial P}{\partial \rho}\right)_s}$$

Using this equation, show that the speed of sound in an ideal gas is:

$$c = \sqrt{\gamma RT}$$

Chapter 7

Thermodynamic Properties of Pure Fluids

7.1 LEARNING OUTCOMES

Upon completion of this chapter, students will be able to:

- Estimate compressibility factor and hence molar volume of real fluids (gases and liquids) using the generalized compressibility factor correlations such as the Lee/Kesler correlation
- Estimate compressibility factor and hence molar volume of real fluids using the cubic equations of state (EOS) such as van der Waals, Redlich-Kwong, Soave-Redlich-Kwong, and Peng-Robinson EOS
- Estimate residual enthalpy and entropy of real fluids using the generalized correlations such as the Lee/Kesler correlation
- Estimate residual enthalpy and entropy of real fluids using the cubic EOS such as van der Waals, Redlich-Kwong, Soave-Redlich-Kwong, and Peng-Robinson EOS
- Calculate enthalpy and entropy changes for real fluids
- Estimate latent heat and entropy of vaporization of pure substances
- Estimate absolute values of enthalpy and entropy of real fluids with respect to a reference state
- Estimate extensive thermodynamic properties (volume, enthalpy, entropy) of two-phase vapor/liquid systems.

7.2 INTRODUCTION

The knowledge of the thermodynamic properties (volume, enthalpy and entropy) of fluids is required in the design and operation of industrial processes. For example, the application of the first law of thermodynamics to an industrial process requires knowledge of enthalpy changes. The second law requires knowledge of entropy changes.

In this chapter, the estimation of the thermodynamic properties of pure fluids (gases and liquids) is covered. The properties of interest are compressibility factor (hence volume), enthalpy and entropy. Note that compressibility factor estimation is required not only in the calculation of volume of fluids but also in the calculation of enthalpy and entropy of fluids.

7.3 COMPRESSIBILITY FACTOR

The compressibility factor of a fluid (gas or liquid) is defined as:

$$Z = \frac{Pv}{RT} \tag{7.1}$$

where v is molar volume of the fluid. If Z is known, the molar volume of the fluid can be calculated as:

$$v = \frac{ZRT}{P} \tag{7.2}$$

Note that the compressibility factor is simply the ratio of actual volume of the fluid to the volume of ideal gas at the same temperature and pressure:

$$Z = \frac{v(T,P)}{v^{ig}(T,P)} \tag{7.3}$$

7.3.1 Ideal Gas

An ideal gas is one in which 1) the molecules are point molecules, that is, they do not occupy any space and 2) the molecules do not interact with each other. An ideal gas obeys the following equation of state:

$$Pv = RT \tag{7.4}$$

The compressibility factor of an ideal gas is unity, that is:

$$Z^{ig} = \frac{Pv^{ig}}{RT} = 1 \tag{7.5}$$

Note that the ideal gas model fails to describe phase transitions such as condensation of vapor into liquid. The condensation of vapor into liquid can only happen if there are attractive forces between the molecules and the molecules are of finite size (not point molecules).

7.3.2 Real Gases

In real gases, the molecules interact with each other and the molecules are of finite size. Consequently, the compressibility factor may deviate positively or negatively from unity depending on the nature of the intermolecular forces (attractive or repulsive) between the molecules of the gas. For real gases, the following statements are generally true:

- $Z = 1$ in the limit $P \to 0$.
- The deviation of Z from ideal gas value of unity is the greatest near the critical point.
- Z generally approaches unity, regardless of the pressure, when the temperatures are very high, that is, $T \gg T_c$ where T_c is the critical temperature.
- Z generally approaches unity, regardless of the temperature, when $P \ll P_c$ where P_c is the critical pressure.
- Z deviates from unity at low temperatures or high pressures where intermolecular forces and molecular size become important.
- $Z < 1$ when intermolecular forces are attractive in nature resulting in the actual volume of a gas to be less than the ideal gas at the same temperature and pressure.
- $Z > 1$ when intermolecular forces are repulsive in nature resulting in the actual volume of a gas to be more than the ideal gas at the same temperature and pressure.

The compressibility factor of a real gas depends on temperature, pressure and nature of material:

$$Z = Z(T, P, nature\ of\ material) \tag{7.6}$$

The nature of material is taken care of by critical properties. Thus, Z is expressed in terms of the reduced temperature (T_r) and reduced pressure (P_r) as:

$$Z = Z(T_r, P_r) \tag{7.7}$$

where

$$T_r = \frac{T}{T_c} \tag{7.8}$$

$$P_r = \frac{P}{P_c} \tag{7.9}$$

Such compressibility factor correlations are called *generalized* compressibility correlations as they are valid for all gases. Different gases, when compared at the same reduced temperature T_r and reduced pressure P_r, should have the same compressibility factor.

Figure 7.1 shows the typical generalized compressibility factor chart for a pure substance.
As noted earlier, the deviation of Z from ideal gas value of unity is the greatest near the critical point. Also at high T_r, Z approaches unity regardless of the pressure. At low P_r, Z approaches unity regardless of the temperature.

Figure 7.2 shows the generalized compressibility factor chart for air based on experimental data. Once again, note that: a) at low pressures, Z approaches unity regardless of the temperature; b) at high T_r, Z approaches unity regardless of the pressure; and c) the deviation of Z from ideal gas value of unity is the greatest near the critical point.

The correlations of the form of Equation (7.7) are referred to as *two-parameter* generalized compressibility factor correlations. It has been found that appreciable improvement occurs in the correlation and prediction of compressibility factor if a third parameter, characteristic of molecular structure, is introduced in the correlation. This third parameter is the so-called *acentric factor*, denoted as ω.

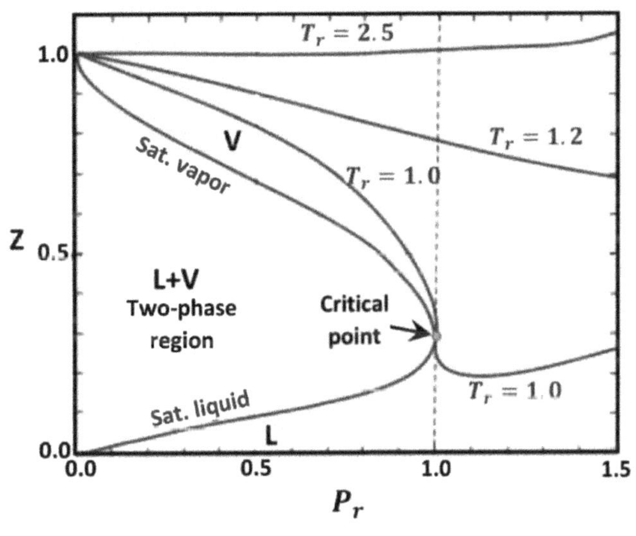

Figure 7.1 Typical generalized compressibility factor chart for a pure substance.

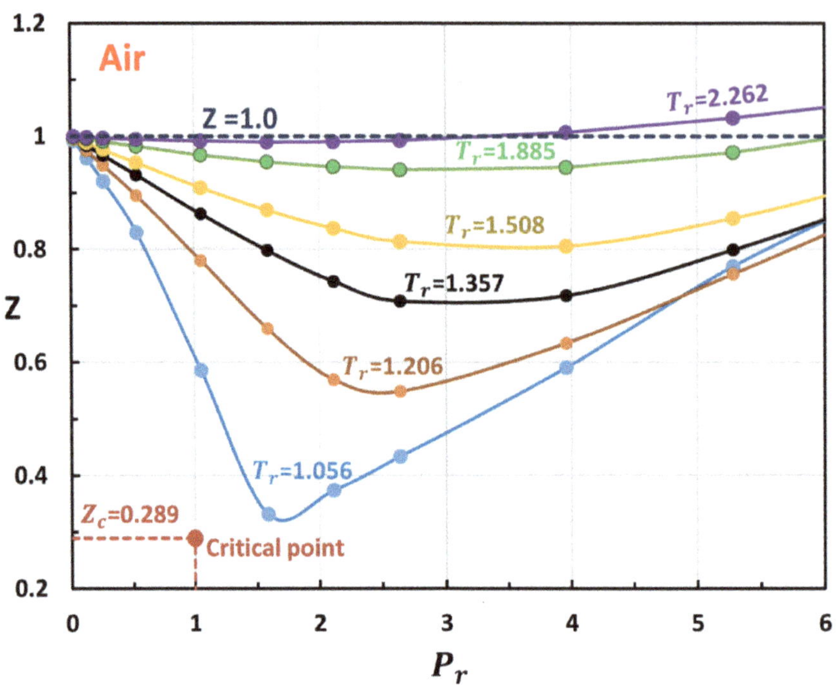

Figure 7.2 Generalized compressibility factor chart for air.

Thus,

$$Z = Z(T_r, P_r, \omega) \tag{7.10}$$

The acentric factor depends on the shape of the molecules. For spherical molecules, $\omega = 0$. According to the *three-parameter* generalized compressibility factor correlation of the form of Equation (7.10), different fluids having the same value of ω, when compared at the same values of T_r and P_r, should have the same compressibility factor.

An example of a three-parameter generalized compressibility factor correlation is the Pitzer generalized virial coefficient correlation, given as:

$$Z = 1 + B^o \frac{P_r}{T_r} + \omega B^1 \frac{P_r}{T_r} \tag{7.11}$$

where

$$B^o = 0.083 - \frac{0.422}{T_r^{1.6}} \tag{7.12}$$

$$B^1 = 0.139 - \frac{0.172}{T_r^{4.2}} \tag{7.13}$$

Given the critical properties and acentric factor, Z can be estimated from Equation (7.11) for any gas at specified temperature and pressure. The Pitzer correlation is based on a truncated form of the virial expansion of Z. It is valid for gases only at low to moderate pressures where Z varies linearly with pressure.

7.3.3 Liquids

There are a number of different correlations available in the literature to predict the molar volumes of saturated liquids [1]. The following equation proposed by Rackett is used extensively:

$$v_r = \frac{v}{v_c} = Z_c^{(1-T_r)^{0.2857}} \tag{7.14}$$

where v is molar volume of saturated liquid at T_r, v_r is reduced volume, v_c and Z_c are critical volume and critical compressibility factor, respectively.

7.3.4 Lee/Kesler Correlation for Gases and Liquids

The Lee/Kesler generalized compressibility factor correlation is a three-parameter correlation given as:

$$Z = Z^o + \omega Z^1 \tag{7.15}$$

where Z^o and Z^1 are functions of T_r and P_r:

$$Z^o = Z^o(T_r, P_r) \tag{7.16}$$

$$Z^1 = Z^1(T_r, P_r) \tag{7.17}$$

The values of Z^o and Z^1 are given in table forms in Appendix J over broad ranges of P_r and T_r.

The Lee/Kesler correlation is applicable to both gases and liquids. However, it is most suitable for nonpolar and slightly polar fluids. It gives significant errors when applied to highly polar fluids such as water and ammonia.

7.3.5 Cubic Equations of State for Gases and Liquids

Van der Waals proposed the first cubic EOS (equation of state) in 1873 using simple molecular arguments. Let us recall the ideal gas law which assumes that the molecules are non-interacting, point (size-less) molecules:

$$P = \frac{RT}{v} \tag{7.18}$$

The major drawback of the ideal gas model is that it fails to describe phase transition from vapor to liquid. Another unrealistic implication of the ideal gas model is that the gas can be compressed to zero volume at very high pressures ($P \to \infty$). Van der Waals argued that the molecules of a gas are not point molecules. The molecules have a finite size. He introduced a term " b" in the ideal gas law to account for the finite size of the molecules as:

$$P = \frac{RT}{v-b} \tag{7.19}$$

The term "b" is the excluded volume of the molecules per mole. The volume available for a mole of gas for *free motion* is $v - b$ (not v). The excluded volume could be interpreted as a volume of one mole of tightly packed molecules. Thus, it is impossible to compress a real gas to zero volume due to repulsion between molecules upon contact.

Van der Waals also considered long range attractive forces between the molecules. Due to the presence of attractive forces between the molecules, the pressure exerted by the molecules on the walls of the container is less. Therefore, he modified Equation (7.19) further as:

$$P = \frac{RT}{v-b} - \frac{a}{v^2} \tag{7.20}$$

This is the *celebrated van der Waals EOS* where the term "a" takes into account the attraction between the molecules and the term "b" takes into account the finite size of the molecules. The vdW (van der Waals) EOS allows phase transition from gas phase to liquid phase and is valid for both gases and liquids. Van der Waals won the 1910 *Nobel Prize* in physics for the development of this EOS.

The vdW (van der Waals) EOS could be re-written as:

$$v^3 - \left(b + \frac{RT}{P}\right)v^2 + \frac{av}{P} - \frac{ab}{P} = 0 \tag{7.21}$$

The highest power of v in Equation (7.21) is 3 and hence the name cubic EOS.

The constants "a" and "b" in vdW EOS can be determined by noting that the critical isotherm passes through a horizontal inflection at the critical point (see Figure 7.3). Consequently, the following mathematical conditions are applicable at the critical point:

$$\left(\frac{\partial P}{\partial v}\right)_T = 0\,;\ \left(\frac{\partial^2 P}{\partial v^2}\right)_T = 0 \tag{7.22}$$

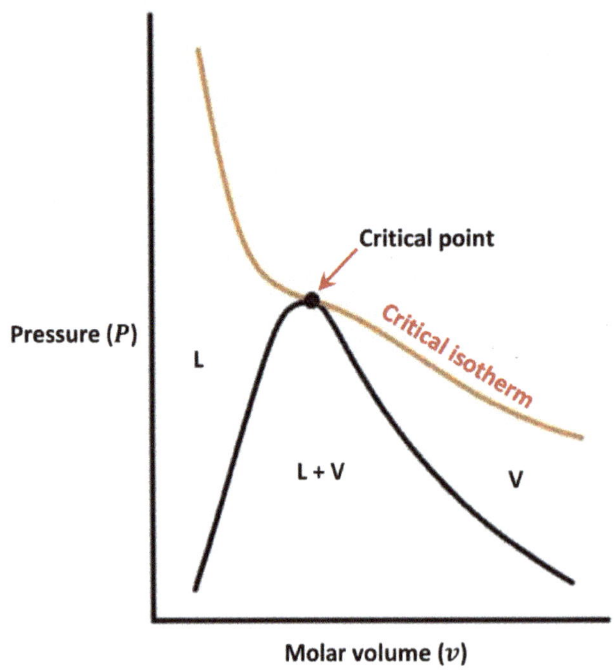

Figure 7.3 Critical isotherm corresponding to $T = T_c$.

The application of the conditions given in Equation (7.22) to vdW EOS gives:

$$\left(\frac{\partial P}{\partial v}\right)_T = \frac{2a}{v_c^3} - \frac{RT_c}{(v_c - b)^2} = 0 \tag{7.23}$$

$$\left(\frac{\partial^2 P}{\partial v^2}\right)_T = -\frac{6a}{v_c^4} + \frac{2RT_c}{(v_c - b)^3} = 0 \tag{7.24}$$

The critical isotherm itself is given as:

$$P_c = \frac{RT_c}{v_c - b} - \frac{a}{v_c^2} \tag{7.25}$$

From Equations (7.23) to (7.25), it follows that:

$$a = \frac{27}{64}\frac{R^2T_c^2}{P_c} \tag{7.26}$$

$$b = \frac{1}{8}\frac{RT_c}{P_c} \tag{7.27}$$

$$v_c = \frac{3}{8}\frac{RT_c}{P_c} \tag{7.28}$$

We can further express the compressibility factor at critical point as:

$$Z_c = \frac{P_c v_c}{RT_c} = \frac{3}{8} = 0.375 \tag{7.29}$$

Van der Waals EOS was a great improvement over the ideal gas law. Unlike ideal gas law, vdW EOS is applicable to both liquids and gases. However, the vdW EOS is rarely used in engineering applications as it is not very accurate. A number of more accurate cubic equations of state have been proposed in the literature. They are modifications of the vdW EOS. Among them, the most popular ones are: Redlich-Kwong (RK) EOS, Soave-Redlich-Kwong (SRK) EOS and Peng-Robinson (PR) EOS. They are summarized in Table 7.1.

Table 7.1 Cubic equations of state

Name	EOS	a	b	α
vdW	$P = \dfrac{RT}{v-b} - \dfrac{a\alpha}{v^2}$	$0.421875\left(\dfrac{R^2T_c^2}{P_c}\right)$	$0.125\dfrac{RT_c}{P_c}$	1
RK	$P = \dfrac{RT}{v-b} - \dfrac{a\alpha}{v(v+b)}$	$0.42748\left(\dfrac{R^2T_c^2}{P_c}\right)$	$0.08664\dfrac{RT_c}{P_c}$	$\dfrac{1}{\sqrt{T_r}}$
SRK	$P = \dfrac{RT}{v-b} - \dfrac{a\alpha_{SRK}}{v(v+b)}$	$0.42748\left(\dfrac{R^2T_c^2}{P_c}\right)$	$0.08664\dfrac{RT_c}{P_c}$	α_{SRK}
PR	$P = \dfrac{RT}{v-b} - \dfrac{a\alpha_{PR}}{\{v+(1-\sqrt{2})b\}\{v+(1+\sqrt{2})b\}}$	$0.45724\left(\dfrac{R^2T_c^2}{P_c}\right)$	$0.07779\dfrac{RT_c}{P_c}$	α_{PR}
$\alpha_{SRK} = \left[1 + (0.480 + 1.574\omega - 0.176\omega^2)(1-\sqrt{T_r})\right]^2$ $\alpha_{PR} = \left[1 + (0.37464 + 1.5422\omega - 0.26992\omega^2)(1-\sqrt{T_r})\right]^2$				

The cubic equations of state could be written in a general form as:

$$P = \frac{RT}{v-b} - \frac{a'}{\{v+\varepsilon b\}\{v+\sigma b\}} \tag{7.30}$$

where

$$a' = a\alpha = \psi\left(\frac{R^2T_c^2}{P_c}\right)\alpha(T_r,\omega)\ ; \quad b = \Omega\frac{RT_c}{P_c} \tag{7.31}$$

For a given EOS, $\varepsilon, \sigma, \psi, \Omega$ are pure numbers (same for all substances) whereas α is, in general, a function of T_r and ω. Table 7.2 summarizes the values of various parameters for different cubic equations of state. The critical compressibility factor (Z_c) values obtained for different EOS are also listed in the table.

Table 7.2 Parameters for different cubic equations of state

Name	ε	σ	ψ	Ω	$\alpha(T_r, \omega)$	Z_c
vdW	0	0	0.421875	0.125	1	0.375
RK	0	1	0.42748	0.08664	$1/\sqrt{T_r}$	1/3
SRK	0	1	0.42748	0.08664	α_{SRK}	1/3
PR	$1-\sqrt{2}$	$1+\sqrt{2}$	0.45724	0.07779	α_{PR}	0.3074
$\alpha_{SRK} = \left[1 + (0.480 + 1.574\omega - 0.176\omega^2)(1 - \sqrt{T_r})\right]^2$ $\alpha_{PR} = \left[1 + (0.37464 + 1.5422\omega - 0.26992\omega^2)(1 - \sqrt{T_r})\right]^2$						

Estimation of Volume from Cubic EOS

Cubic EOS can be solved analytically for volume. It has three volume roots. One is guaranteed to be real but the other two may be real or complex. Only real and positive roots (greater than constant b) are physically meaningful. When $T > T_c$, the EOS has only one real and positive root. When $T = T_c$ and $P = P_c$, there are three real and positive roots, all equal to v_c. For $T < T_c$, the EOS may have one or three real and positive roots depending on the value of the pressure.

Although the cubic equations of state can be solved analytically, they are normally solved using iterative procedures in industrial applications. The commonly used iterative method to solve cubic EOS is the *Newton-Raphson method*. Let us re-write the general form of the cubic EOS, Equation (7.30), as follows:

$$f(v) = P - \frac{RT}{v-b} + \frac{a'}{\{v+\varepsilon b\}\{v+\sigma b\}} = 0 \tag{7.32}$$

According to the Newton-Raphson method, the iterative scheme is as follows:

$$v_{i+1} = v_i - \frac{f(v_i)}{f'(v_i)} \tag{7.33}$$

where $f(v)$ is given by Equation (7.32) and $f'(v)$ is the derivative of $f(v)$ given as:

$$f'(v) = \frac{RT}{(v-b)^2} - \frac{a'}{(v+b\varepsilon)(v+\sigma b)^2} - \frac{a'}{(v+\sigma b)(v+\varepsilon b)^2} \tag{7.34}$$

The iterative scheme, Equation (7.33), gives improved value of the root (v_{i+1}) over the previous value of the root (v_i). The iteration process is terminated when the deviation between two successive values is within the tolerance level, $[|v_{i+1} - v_i|/v_i] \times 100 \leq t$ where t is the percent tolerance level.

The iterative scheme requires an initial guess of the root. For gases, the initial guess v_0 is taken as ideal gas value ($= RT/P$). For liquids, $v_0 > b$. We cannot take $v_o = b$ as $f(v)$ and $f'(v)$ diverge when $v_o = b$. The convergence problems are most likely avoided if v_o is chosen to be close to b, $1.05b < v_o < 1.5b$.

Estimation of Compressibility Factor from Cubic EOS

Although the compressibility factor can be calculated readily once the volume is estimated, it can also be calculated directly using an iterative procedure. The general cubic EOS, Equation (7.30), can be re-arranged as:

$$\frac{Pv}{RT} = 1 + \frac{bP}{RT} - \frac{a'(v-b)}{RT\{v+\varepsilon b\}\{v+\sigma b\}} \tag{7.35}$$

This equation can be recast in the following form:

$$Z = 1 + \beta - q\beta\frac{(Z-\beta)}{(Z+\varepsilon\beta)(Z+\sigma\beta)} \tag{7.36}$$

where

$$\beta = \frac{bP}{RT} \tag{7.36a}$$

$$q = \frac{a'}{bRT} \tag{7.36b}$$

We can re-arrange Equation (7.36) as:

$$f(Z) = Z - 1 - \beta + q\beta\frac{(Z-\beta)}{(Z+\varepsilon\beta)(Z+\sigma\beta)} = 0 \tag{7.37}$$

We now use the *Newton-Raphson iterative method* to find the root of Equation (7.37). The iterative scheme is:

$$Z_{i+1} = Z_i - \frac{f(Z_i)}{f'(Z_i)} \tag{7.38}$$

where $f'(Z)$ is the derivative of $f(Z)$, given as:

$$f'(Z) = 1 + q\beta\left[\frac{(Z+\varepsilon\beta)(Z+\sigma\beta)-(Z-\beta)(2Z+\beta\varepsilon+\beta\sigma)}{(Z+\varepsilon\beta)^2(Z+\sigma\beta)^2}\right] \tag{7.39}$$

The iterative scheme, Equation (7.38), gives improved value of the root (Z_{i+1}) over the previous value of the root (Z_i). The iteration process is terminated when the deviation between two successive values is within the tolerance level, $[|Z_{i+1} - Z_i|/Z_i] \times 100 \leq t$ where t is the percent tolerance level. The iterative scheme requires an initial guess of the root. For gases, the initial guess Z_o is taken as unity (ideal gas value). For liquids, $Z_o = \beta$.

It is important to realize that there is no guarantee that the Newton-Raphson iterative process will always converge to the desired root. To avoid convergence problems, Smith et al. [2] suggest the usage of Equation (7.36) in the following iterative form to obtain the *gas* root:

$$Z_{i+1} = 1 + \beta - q\beta\frac{(Z_i-\beta)}{(Z_i+\varepsilon\beta)(Z_i+\sigma\beta)} \tag{7.40}$$

The gas root can be obtained by successive substitution of Z_i starting with the initial guess of $Z_0 = 1$. For the *liquid* root, Smith et al. [2] suggest that Equation (7.36) be re-arranged as:

$$Z = \beta + (Z + \varepsilon\beta)(Z + \sigma\beta)\left(\frac{1+\beta-Z}{q\beta}\right) \tag{7.41a}$$

Equation (7.41a) gives the following iterative scheme for the *liquid* root:

$$Z_{i+1} = \beta + (Z_i + \varepsilon\beta)(Z_i + \sigma\beta)\left(\frac{1+\beta-Z_i}{q\beta}\right) \tag{7.41b}$$

where the initial guess for the liquid root is $Z_o = \beta$.

7.4 ENTHALPY AND ENTROPY CHANGES

7.4.1 Ideal Gas

The specific (molar) enthalpy and entropy changes for ideal gas are given as follows:

$$\Delta h^{ig} = \int_{T_1}^{T_2} c_p^{ig}\, dT \tag{7.42}$$

$$\Delta s^{ig} = \int_{T_1}^{T_2} \frac{c_p^{ig}}{T} dT - R ln\left(\frac{P_2}{P_1}\right) \tag{7.43}$$

where c_p^{ig} is the molar heat capacity of gas in ideal gas state. The heat capacity c_p^{ig} is generally a function of temperature and can be expressed as:

$$\frac{c_p^{ig}}{R} = A + BT + CT^2 + DT^{-2} \tag{7.44}$$

where A, B, C, D are constants for a given substance and R is the universal gas constant.

Upon substitution of the expression for c_p^{ig} given by Equation (7.44) into Equations (7.42) and (7.43) and carrying out the integration, we get:

$$\Delta h^{ig} = c_{pm,h}^{ig} \Delta T \tag{7.45}$$

$$\Delta s^{ig} = c_{pm,s}^{ig} ln\left(\frac{T_2}{T_1}\right) - R ln\left(\frac{P_2}{P_1}\right) \tag{7.46}$$

where

$$\frac{c_{pm,h}^{ig}}{R} = \frac{\int_{T_1}^{T_2} \frac{c_p^{ig}}{R} dT}{(T_2 - T_1)} = A + BT_{am} + \frac{C}{3}[4T_{am}^2 - T_1T_2] + \frac{D}{T_1T_2} \tag{7.47}$$

$$\frac{c_{pm,s}^{ig}}{R} = \frac{\int_{T_1}^{T_2} \frac{c_p^{ig}}{RT} dT}{ln\left(\frac{T_2}{T_1}\right)} = A + BT_{lm} + CT_{am}T_{lm} + \frac{DT_{am}T_{lm}}{(T_1T_2)^2} \tag{7.48}$$

T_{am} is the arithmetic-mean temperature and T_{lm} is the log-mean temperature, defined as:

$$T_{am} = \frac{T_1 + T_2}{2} \tag{7.49}$$

$$T_{lm} = \frac{T_2 - T_1}{ln\left(\frac{T_2}{T_1}\right)} \tag{7.50}$$

Note that the mean heat capacity expressions for enthalpy calculation, $c_{pm,h}^{ig}$ given by Equation (7.47), and entropy calculation, $c_{pm,s}^{ig}$ given by Equation (7.48), are different.

7.4.2 Real Gases and Liquids

The calculation of enthalpy and entropy changes of fluids, especially real gases, makes use of the concept of *residual property*. For any extensive thermodynamic property (e) on a unit mole basis, the corresponding molar residual property (e^R) is defined as:

$$e^R(T, P) = e(T, P) - e^{ig}(T, P) \tag{7.51}$$

where e is the molar value of the extensive property at given T and P, e^{ig} is the molar value of the same property in ideal gas state at the same T and P, and e^R is the molar value of residual property at the same

T and P. Note that e, e^{ig}, and e^R are all molar quantities and are compared at the same T and P. Also, the superscript of e^R should not be confused with the universal gas constant. The superscript here refers to "residual".

The residual property is a measure of departure from ideal gas behavior. As compressibility factor is also a measure of departure from ideal gas behavior, it follows that the residual property is linked to the compressibility factor. As an example, consider the molar residual volume (v^R) of a fluid, defined as:

$$v^R = v - v^{ig} \tag{7.52}$$

Substituting the expressions for v and v^{ig} into Equation (7.52), we obtain:

$$v^R = \frac{ZRT}{P} - \frac{RT}{P} = (Z-1)\frac{RT}{P} \tag{7.53}$$

Thus, v^R is related to compressibility factor Z. If the material itself is ideal gas, $Z = 1$ and $v^R = 0$. Like v^R is related to Z, the residual enthalpy (h^R) and residual entropy (s^R) are also related to the compressibility factor.

It can be readily shown that the residual enthalpy is related to compressibility factor as follows:

$$\frac{h^R}{RT} = -T\int_0^P \left(\frac{\partial Z}{\partial T}\right)_P \frac{dP}{P} \tag{7.54}$$

Likewise, the residual entropy is related to compressibility factor as:

$$\frac{s^R}{R} = -T\int_0^P \left(\frac{\partial Z}{\partial T}\right)_P \frac{dP}{P} - \int_0^P (Z-1)\frac{dP}{P} \tag{7.55}$$

The derivations of Equations (7.54) and (7.55) are given in the Appendix D.

We can re-write Equations (7.54) and (7.55) in terms of reduced variables as:

$$\frac{h^R}{RT_c} = -T_r^2\int_0^{P_r} \left(\frac{\partial Z}{\partial T_r}\right)_{P_r} \frac{dP_r}{P_r} \tag{7.56}$$

$$\frac{s^R}{R} = -T_r\int_0^{P_r} \left(\frac{\partial Z}{\partial T_r}\right)_{P_r} \frac{dP_r}{P_r} - \int_0^{P_r} (Z-1)\frac{dP_r}{P_r} \tag{7.57}$$

7.4.3 Estimation of Residual Enthalpy and Entropy

The residual enthalpy and entropy of fluids (real gases and liquids) are estimated from Equations (7.56) and (7.57). The compressibility factor information is needed to evaluate the integrals in Equations (7.56) and (7.57). Note that the concept of residual property, although applicable to both real gases and liquids, is generally used for real gases only. In real gas calculations, the residual property terms ($h^R$and s^R) are generally small as compared with ideal gas terms ($h^{ig}$and s^{ig}). Thus, the residual property of real gas is basically a small correction term to ideal gas calculation. However, in the case of liquids, the residual properties ($h^R$and s^R) are large and dominant and ideal gas terms ($h^{ig}$and s^{ig}) are small. This defeats the purpose of working with a residual property as a measure of small correction to an ideal gas value. Therefore, the use of residual properties is restricted to prediction of only real gas (not liquid) properties.

Lee/Kesler Correlation

Lee/Kesler correlation in the form of Equations (7.15) to (7.17) gives compressibility factor of real gases and liquids. Upon substitution of Z from Equation (7.15) into the residual enthalpy expression, Equation (7.56), we get:

$$\frac{h^R}{RT_c} = \frac{(h^R)^o}{RT_c} + \omega \frac{(h^R)^1}{RT_c} \tag{7.58}$$

where

$$\frac{(h^R)^o}{RT_c} = -T_r^2 \int_0^{P_r} \left(\frac{\partial Z^o}{\partial T_r}\right)_{P_r} \frac{dP_r}{P_r} \tag{7.59}$$

$$\frac{(h^R)^1}{RT_c} = -T_r^2 \int_0^{P_r} \left(\frac{\partial Z^1}{\partial T_r}\right)_{P_r} \frac{dP_r}{P_r} \tag{7.60}$$

Upon substitution of Z from Equation (7.15) into the residual entropy expression, Equation (7.57), we get:

$$\frac{s^R}{R} = \frac{(s^R)^o}{R} + \omega \frac{(s^R)^1}{R} \tag{7.61}$$

$$\frac{(s^R)^o}{R} = -\int_0^{P_r} \left[T_r \left(\frac{\partial Z^o}{\partial T_r}\right)_{P_r} + Z^o - 1\right] \frac{dP_r}{P_r} \tag{7.62}$$

$$\frac{(s^R)^1}{R} = -\int_0^{P_r} \left[T_r \left(\frac{\partial Z^1}{\partial T_r}\right)_{P_r} + Z^1\right] \frac{dP_r}{P_r} \tag{7.63}$$

Equations (7.58) and (7.61) are to be used to estimate residual enthalpy and entropy. Z^o and Z^1 are functions of T_r and P_r. The integrals in Equations (7.59) and (7.60) and in Equations (7.62) and (7.63) are evaluated using compressibility factor data either numerically or graphically to obtain the quantities $\frac{(h^R)^o}{RT_c}$, $\frac{(h^R)^1}{RT_c}$, $\frac{(s^R)^o}{R}$ and $\frac{(s^R)^1}{R}$ as functions of T_r and P_r. These quantities are given in form of tables in Appendix J. Given T_r and P_r, one can determine $\frac{(h^R)^o}{RT_c}$, $\frac{(h^R)^1}{RT_c}$, $\frac{(s^R)^o}{R}$ and $\frac{(s^R)^1}{R}$ from the tables, and then use these quantities in Equations (7.58) and (7.61) to estimate residual enthalpy and entropy.

Pitzer Generalized Virial Coefficient Correlation

Pitzer generalized virial coefficient correlation, Equation (7.11), is valid for real gases only at low to moderate pressures where Z varies linearly with pressure. It can be used directly in Equations (7.56) and (7.57) to develop the expressions for residual enthalpy and entropy. Using the Pitzer correlation, it can be readily shown that:

$$\frac{h^R}{RT_c} = P_r\left[B^o - T_r\frac{dB^o}{dT_r} + \omega\left(B^1 - T_r\frac{dB^1}{dT_r}\right)\right] \tag{7.64}$$

$$\frac{s^R}{R} = P_r\left[\frac{dB^o}{dT_r} + \omega\frac{dB^1}{dT_r}\right] \tag{7.65}$$

where

$$\frac{dB^o}{dT_r} = \frac{0.675}{T_r^{2.6}} \tag{7.66}$$

$$\frac{dB^1}{dT_r} = \frac{0.722}{T_r^{5.2}} \tag{7.67}$$

Thus, residual enthalpy and entropy can be calculated from Equations (7.64) and (7.65) at any given T_r and P_r .

Cubic Equations of State

Cubic equations of state are applicable to both real gases and liquids. The compressibility factor can be determined from a cubic EOS using the Newton-Raphson iterative procedure as described in section 7.3.5.

The following residual property equations can be derived readily [2] from the general form of the cubic EOS, that is, Equation (7.30) or (7.36):

$$\frac{h^R}{RT} = Z - 1 + \left[\frac{dln\alpha}{dlnT_r} - 1\right] qI \tag{7.68}$$

$$\frac{s^R}{R} = ln(Z - \beta) + \left[\frac{dln\alpha}{dlnT_r}\right] qI \tag{7.69}$$

where

$$I = \frac{1}{\sigma - \varepsilon} ln\left[\frac{Z+\sigma\beta}{Z+\varepsilon\beta}\right] \quad \text{when } \varepsilon \neq \sigma$$
$$= \beta/(Z + \varepsilon\beta) \quad \text{when } \varepsilon = \sigma \tag{7.70}$$

The quantities ε and σ are pure numbers dependent on the type of EOS (see Table 7.2). The parameters β, q and α have been defined earlier (see Equations (7.36a)-(7.36b) and Table 7.2).

Note that once Z is estimated using the iterative procedure, no more iteration is required to calculate the residual enthalpy and entropy.

7.4.4 Calculation of Enthalpy and Entropy Changes for Real Gas

Figure 7.4 shows a computational path for the calculation of enthalpy and entropy changes for a real gas in going from state 1 (T_1, P_1) to state 2 (T_2, P_2). As enthalpy and entropy are state properties, the changes are the same regardless of the path that is followed. Thus, the actual path indicated by dashed green line is replaced by a three-step computational path shown in Figure 7.4: Step 1 where real gas at (T_1, P_1) is transformed to ideal gas at (T_1, P_1) by a fictitious process, Step 2 where ideal gas undergoes a change in state from (T_1, P_1) to (T_2, P_2), and Step 3 where ideal gas at (T_2, P_2) is transformed to real gas at (T_2, P_2) by a fictitious process. Thus, the enthalpy and entropy changes for a real gas in going from state 1 (T_1, P_1) to state 2 (T_2, P_2) are:

$$\Delta h = -h_1^R + \Delta h^{ig} + h_2^R \tag{7.71}$$

$$\Delta s = -s_1^R + \Delta s^{ig} + s_2^R \tag{7.72}$$

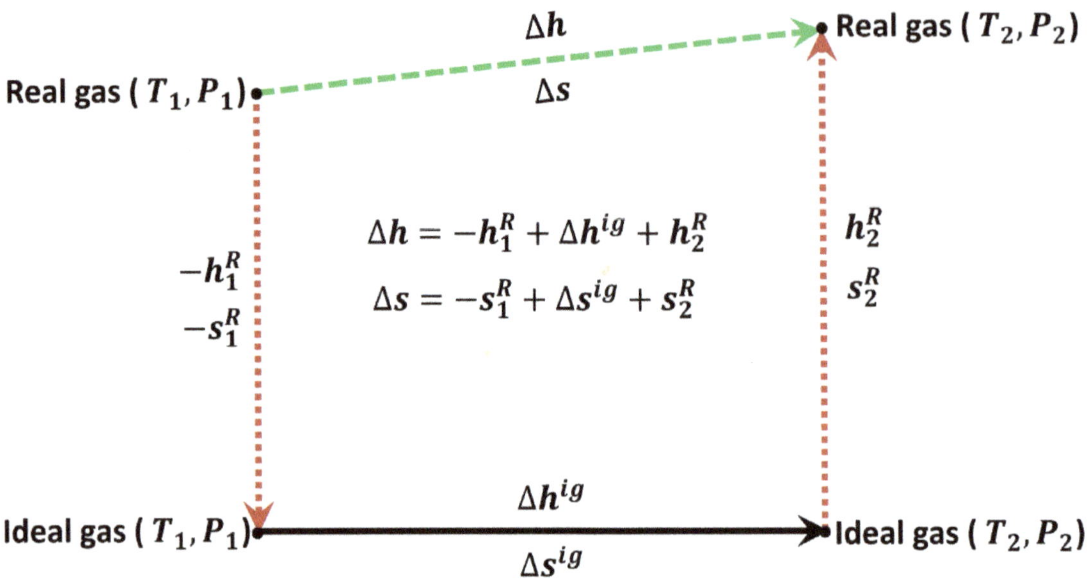

Figure 7.4 Computational path for the calculation of enthalpy and entropy changes for a real gas.

7.5 ABSOLUTE VALUES OF ENTHALPY AND ENTROPY

It is impossible to calculate the absolute value of enthalpy. We can only measure changes in enthalpy. It is possible to calculate absolute value of entropy using the third law of thermodynamics. However, in engineering applications we need only changes in enthalpy and entropy rather than the absolute values of enthalpy and entropy. Therefore, the absolute values of enthalpy and entropy are calculated relative to some arbitrary reference state. The changes in enthalpy and entropy do not depend on the reference state as the reference values cancel out. Obviously, one must use the same source of thermodynamic data (table or chart with a fixed reference state) to calculate the changes.

Reference state of a material is any convenient state, specified arbitrarily. The temperature, pressure, and actual phase of the material are specified in the reference state. The values are assigned to enthalpy and entropy arbitrarily in the reference state. Typically, enthalpy and entropy are set equal to zero in the reference state. As an example, the reference state for steam tables is chosen to be saturated liquid water at $T = 0.01^oC$ and $P = 0.611\ kPa$ (triple point of water where ice, water, and steam all co-exist in thermodynamic equilibrium). The enthalpy and entropy are set equal to zero for saturated liquid water at triple point.

7.5.1 Estimation of Absolute Enthalpy and Entropy with Respect to Different Reference States

Figure 7.5 shows a computational path for the calculation of absolute values of enthalpy and entropy for a real gas at T and P with respect to a given reference state. In the reference state, the temperature and pressure are specified as T_o and P_o. The pressure in the reference state is low enough for the actual gas to behave as an ideal gas. The enthalpy and entropy are set equal to zero in the reference state.

Based on the computational path shown in Figure 7.5, the absolute values of enthalpy and entropy of real gas at T and P are given as:

$$h = \Delta h^{ig} + h^R \tag{7.73}$$

$$s = \Delta s^{ig} + s^R \tag{7.74}$$

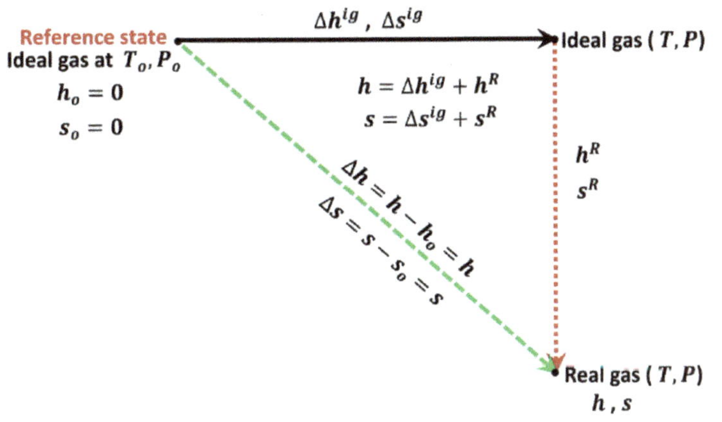

Figure 7.5 Calculation of absolute values of enthalpy and entropy for a real gas with respect to ideal gas as a reference state.

Figure 7.6 shows a computational path for the calculation of absolute values of enthalpy and entropy for a subcooled liquid at T and P with respect to a given reference state. In the reference state, the temperature and pressure are specified as T_o and P_o. The pressure in the reference state is low enough for the actual gas to behave as an ideal gas. The enthalpy and entropy are set equal to zero in the reference state.

Based on the computational path shown in Figure 7.6, the absolute values of enthalpy and entropy of a subcooled liquid at T and P are given as:

$$h = \Delta h^{ig} + h^R - \Delta h^{vap} + \Delta h^{liq} \tag{7.75}$$

$$s = \Delta s^{ig} + s^R - \Delta s^{vap} + \Delta s^{liq} \tag{7.76}$$

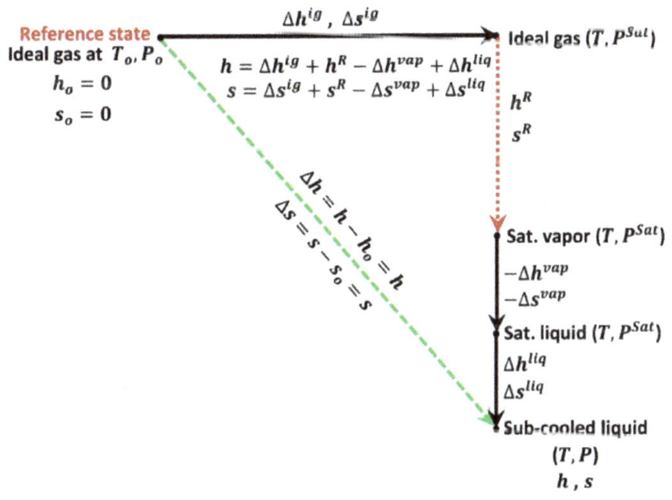

Figure 7.6 Calculation of absolute values of enthalpy and entropy for a subcooled liquid with respect to ideal gas as a reference state.

Figure 7.7 shows a computational path for the calculation of absolute values of enthalpy and entropy for a real gas at T and P with respect to a given reference state. In the reference state, the temperature and pressure are specified as T_o and P_o. The material in the reference state is subcooled liquid. The enthalpy and entropy are set equal to zero in the reference state.

Based on the computational path shown in Figure 7.7, the absolute values of enthalpy and entropy of a real gas at T and P are given as:

$$h = \Delta h^{liq} + \Delta h^{vap} - h_1^R + \Delta h^{ig} + h_2^R \tag{7.77}$$

$$s = \Delta s^{liq} + \Delta s^{vap} - s_1^R + \Delta s^{ig} + s_2^R \tag{7.78}$$

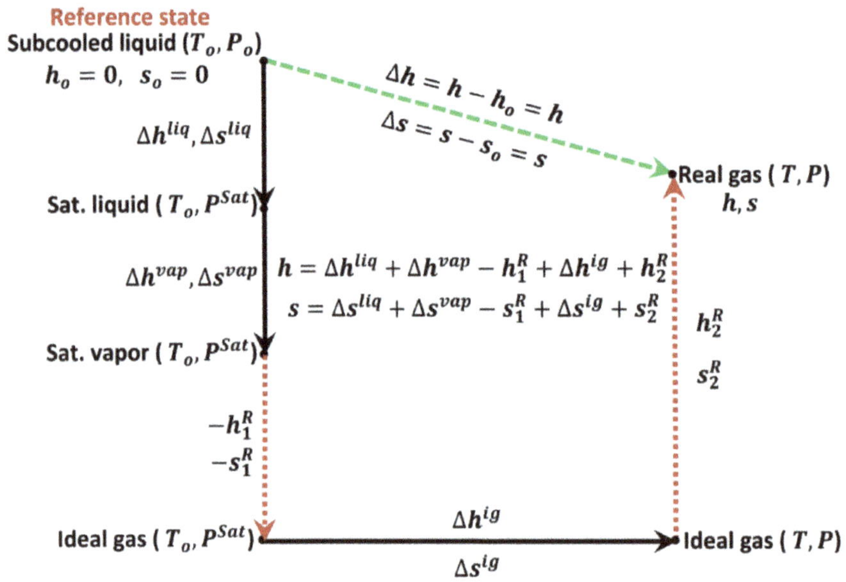

Figure 7.7 Calculation of absolute values of enthalpy and entropy for a real gas with respect to subcooled liquid as a reference state.

7.5.2 Estimation of Latent Heat and Entropy of Vaporization (Δh^{vap} and Δs^{vap})

In the calculation of absolute values of enthalpy and entropy of substances, the latent heat of vaporization and entropy of vaporization may be required (see Figures 7.6 and 7.7). The latent heat of vaporization and entropy of vaporization are related to each other as:

$$\Delta s^{vap} = \frac{\Delta h^{vap}}{T} \tag{7.79}$$

Rough estimate of the Δs^{vap} of pure liquid at normal boiling point is given by the *Trouton's rule:*

$$\frac{\Delta s_n^{vap}}{R} \approx 10.5 \tag{7.80}$$

where Δs_n^{vap} is entropy of vaporization at normal boiling point. From Equations (7.79) and (7.80):

$$\frac{\Delta h_n^{vap}}{RT_n} \approx 10.5 \tag{7.81}$$

where Δh_n^{vap} is latent heat of vaporization at normal boiling point T_n.

The latent heat of vaporization and the entropy of vaporization are functions of temperature. Figures 7.8 and 7.9 show the plots of Δh^{vap} and Δs^{vap} versus temperature for water. They decrease with the increase in temperature. At critical point ($T_c = 373.95^oC = 647.1\ K$ for water), Δh^{vap} and Δs^{vap} both become zero. At the critical point, the liquid and vapor phases are indistinguishable.

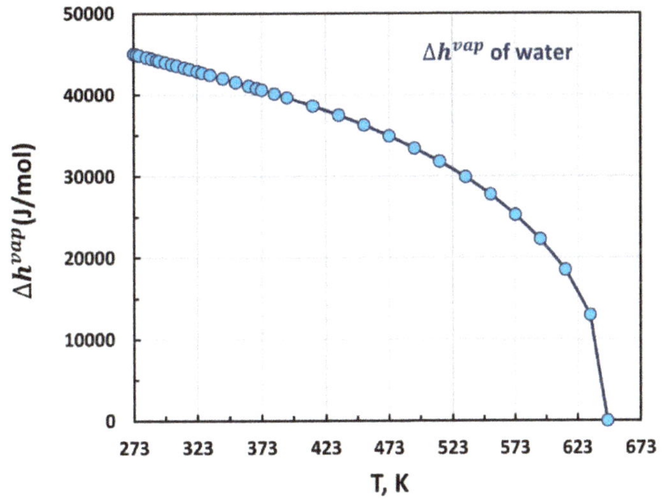

Figure 7.8 Variation of latent heat of vaporization of water with temperature.

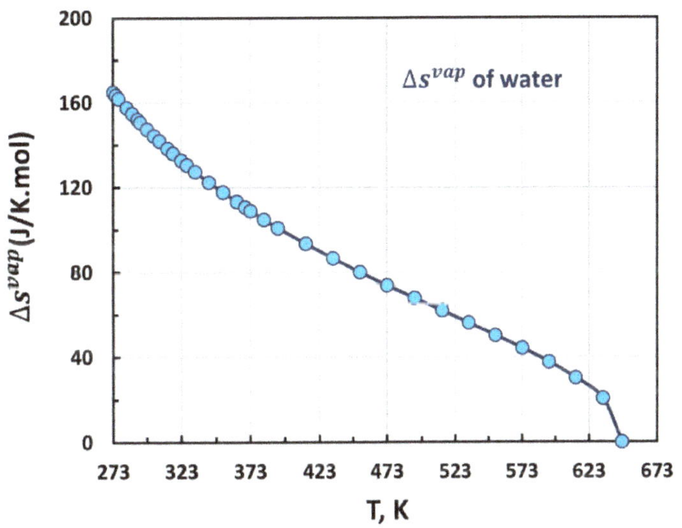

Figure 7.9 Variation of entropy of vaporization of water with temperature.

Latent Heat of Vaporization from Clapeyron Equation

The Clapeyron equation is given as follows:

$$\Delta h^{vap} = T\Delta v^{vap}\frac{dP^{sat}}{dT} \tag{7.82}$$

The latent heat of vaporization can be calculated from Equation (7.82) provided that the vapor pressure versus temperature data are available. The vapor pressure versus temperature behavior is usually described by the Antoine equation:

$$lnP^{sat} = A - \frac{B}{T+C} \tag{7.83}$$

where A, B and C are constants for a given material. From Equations (7.82) and (7.83), it follows that:

$$\Delta h^{vap} = T\Delta v^{vap} P^{sat} \frac{B}{(T+C)^2} \tag{7.84}$$

If P^{sat} is low, one can treat the vapor phase as an ideal gas. Thus,

$$\Delta v^{vap} \approx v^{vap} = \frac{RT}{P^{sat}} \tag{7.85}$$

From Equations (7.84) and (7.85):

$$\Delta h^{vap} = RT^2 \frac{B}{(T+C)^2} \tag{7.86}$$

Riedel-Watson Empirical Method of Estimating Latent Heat of Vaporization

The Riedel-Watson empirical method of estimating Δh^{vap} involves two steps: Δh^{vap} is first estimated at normal boiling point using the Riedel equation. Using the Δh^{vap} value at normal boiling point, Δh^{vap} is estimated at a given temperature from the Watson equation:

$$\frac{\Delta h_n^{vap}}{RT_n} = \frac{1.092(\ln P_c - 1.013)}{0.930 - T_{rn}} \tag{7.87}$$

$$\frac{\Delta h_2^{vap}}{\Delta h_1^{vap}} = \left(\frac{1 - T_{r2}}{1 - T_{r1}}\right)^{0.38} \tag{7.88}$$

Equation (7.87) is the Riedel equation and Equation (7.88) is the Watson equation. In the Riedel equation, P_c is the critical pressure in bars and T_{rn} is the reduced normal boiling point T_n/T_c.

7.5.3 Estimation of Enthalpy and Entropy Changes for Liquids (Δh^{liq} *and* Δs^{liq})

In the calculation of absolute values of enthalpy and entropy of substances, the enthalpy and entropy changes of liquid phase may be required (see Figures 7.6 and 7.7). The changes in enthalpy and entropy for pure single-phase substances can be expressed as:

$$dh = c_p dT + v(1 - \beta T)dP \tag{7.89}$$

$$ds = c_p \frac{dT}{T} - \beta v dP \tag{7.90}$$

where β is volume expansivity, defined as:

$$\beta = \frac{1}{v}\left(\frac{\partial v}{\partial T}\right)_P \tag{7.91}$$

Assuming liquids to be *incompressible* ($\beta = 0$), Equations (7.89) and (7.90) reduce to:

$$dh = c_p dT + v dP \tag{7.92}$$

$$ds = c_p \frac{dT}{T} \tag{7.93}$$

Upon integration:

$$\Delta h^{liq} = c_p^{liq}(T_2 - T_1) + v^{liq}(P_2 - P_1) \tag{7.94}$$

$$\Delta s^{liq} = c_p^{liq} ln\left(\frac{T_2}{T_1}\right) \tag{7.95}$$

We have assumed that c_p^{liq} and v^{liq} are constant.

7.6 THERMODYNAMIC PROPERTIES OF TWO-PHASE VAPOR/LIQUID SYSTEMS

Thermodynamic properties of two-phase vapor/liquid systems are required in the analysis and design of power cycles and refrigeration cycles. In the preceding sections, we learnt how to estimate enthalpy and entropy of single-phase systems (gases and liquids). In this section, we will discuss how to calculate enthalpy and entropy of two-phase vapor/liquid systems at equilibrium.

7.6.1 Pressure-Volume Phase Diagram of a Pure Substance

Figure 7.10 shows the $P - v$ phase diagram of a pure substance such as water. It shows two-phase dome (vapor-liquid mixture region) in the middle, single-phase subcooled liquid region on the left and single-phase superheated vapor region on the right. At the top of the two-phase dome is the critical point of the substance. A constant temperature line (isotherm) is also shown at one temperature. To explain the $P - v$ phase diagram further, let us follow the isotherm from point A to point D. At point A, the substance is in subcooled liquid state. As pressure is decreased at constant temperature, the specific or molar volume of the substance increases. The substance remains in subcooled liquid state until point B is reached. At point B, the substance is in saturated liquid state and the pressure is vapor pressure of the substance at given T. With further expansion of the substance, the pressure and temperature remain constant but some of the liquid phase is converted into vapor phase and we have a two-phase vapor-liquid mixture. The vapor content of the two-phase mixture continues to increase as we go from point B to point C. At point C, all the liquid is vaporized, and the substance is in single-phase saturated vapor state. With further expansion, the pressure decreases at the same T and the vapor becomes superheated vapor. At point D, the substance is in superheated vapor phase.

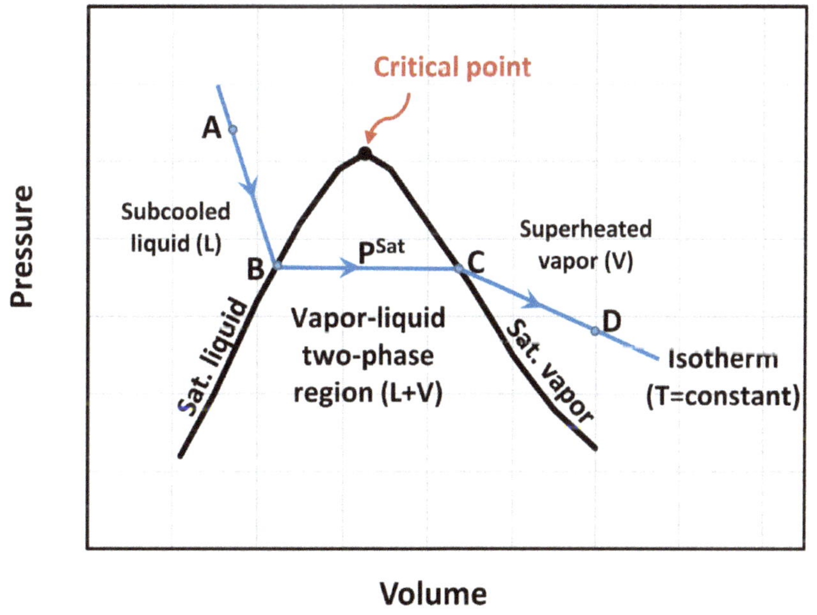

Figure 7.10 Pressure-volume phase diagram of a pure substance.

7.6.2 Quality, Enthalpy and Entropy of Two-Phase Vapor-Liquid Mixtures

Consider a vapor/liquid two-phase system at equilibrium, as shown in Figure 7.11. Let m be the total mass of the two-phase mixture. Thus,

$$m = m^l + m^v \tag{7.96}$$

where m^l is liquid mass and m^v is vapor mass. The mass fractions of the vapor and liquid phases are:

$$x^v = \frac{m^v}{m}\ ; \qquad x^l = \frac{m^l}{m} \tag{7.97}$$

Also,

$$x^l + x^v = 1 \tag{7.98}$$

The mass fraction of vapor in a vapor-liquid mixture (x^v) is also called *quality* of the vapor-liquid mixture.

The total enthalpy of a two-phase mixture is:

$$H = (enthalpy\ of\ vapor) + (enthalpy\ of\ liquid) \tag{7.99}$$

$$H = m^v h^v + m^l h^l \tag{7.100}$$

where h^v and h^l are specific enthalpies of saturated vapor and liquid phases, respectively.

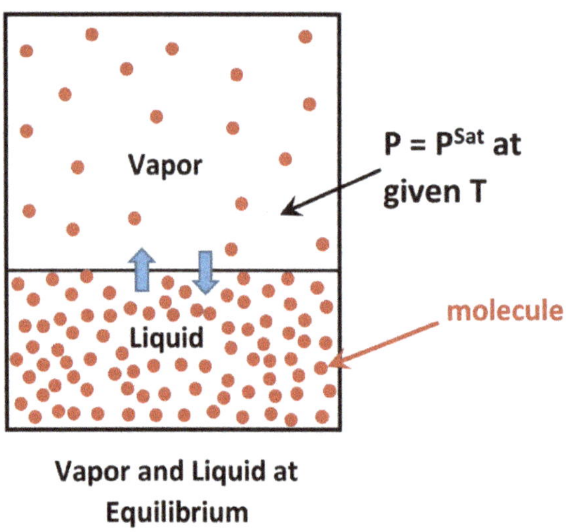

Figure 7.11 Vapor and liquid of a pure substance in equilibrium at a given temperature.

The specific enthalpy of a two-phase mixture is:

$$h = \frac{H}{m} = x^v h^v + x^l h^l \tag{7.101}$$

Using Equation (7.98), Equation (7.101) can be re-written as:

$$h = x^v h^v + (1 - x^v) h^l \tag{7.102}$$

This equation can be used to calculate the specific enthalpy of a two-phase vapor/liquid system provided that the specific enthalpies of saturated vapor and liquid phases are known.

In a similar manner, one can derive the following relationship for the calculation of specific entropy of a two-phase vapor/liquid system:

$$s = x^v s^v + (1 - x^v) s^l \tag{7.103}$$

NOTATION

a	parameter in cubic equation of state
a'	parameter in cubic equation of state, define din Equation (7.31)
A	coefficient in heat capacity expression, Equation (7.44), and Antoine equation, Equation (7.83)
b	parameter in cubic equation of state
B	coefficient in heat capacity expression, Equation (7.44), and Antoine equation, Equation (7.83)
B^o	parameter given by Equation (7.12)
B^1	parameter given by Equation (7.13)
C	coefficient in heat capacity expression, Equation (7.44), and Antoine equation, Equation (7.83)
c_p	specific or molar constant-pressure heat capacity
c_p^{ig}	molar constant-pressure heat capacity of gas in ideal gas state
c_p^{liq}	specific or molar constant-pressure heat capacity of liquid
$c_{pm,h}^{ig}$	mean molar constant-pressure heat capacity of gas in ideal gas state for enthalpy calculation
$c_{pm,s}^{ig}$	mean molar constant-pressure heat capacity of gas in ideal gas state for entropy calculation
D	coefficient in heat capacity expression, Equation (7.44)
e	molar value of an extensive property
e^{ig}	molar value of extensive property in ideal gas state
e^R	molar value of residual property
h	molar or specific enthalpy
H	total enthalpy
h^{ig}	molar enthalpy of gas in ideal gas state
h^l	molar or specific enthalpy of liquid
h^R	molar residual enthalpy
$(h^R)^o$	residual-enthalpy function defined in Equation (7.59)
$(h^R)^1$	residual-enthalpy function defined in Equation (7.60)
h^v	molar or specific enthalpy of vapor
I	quantity defined in Equation (7.70)
m	mass
m^l	mass of liquid phase
m^v	mass of vapor phase
P	pressure
P_c	critical pressure
P_r	reduced pressure
P^{sat}	saturation pressure or vapor pressure
q	parameter defined in Equation (7.36b)

R	universal gas constant
s	molar or specific entropy
s^{ig}	molar entropy of gas in ideal gas state
s^{l}	molar or specific entropy of liquid
s^{R}	molar residual entropy
s^{v}	molar or specific entropy of vapor
$(s^{R})^{o}$	residual-entropy function defined in Equation (7.62)
$(s^{R})^{1}$	residual-entropy function defined in Equation (7.63)
T	temperature
T_c	critical temperature
T_n	normal boiling point temperature
T_r	reduced temperature
T_{am}	arithmetic mean temperature
T_{lm}	log-mean temperature
v	molar or specific volume
v_c	critical molar volume
v_r	reduced molar volume
v^{ig}	molar volume of gas in ideal gas state
v^{liq}	molar or specific volume of liquid
v^{R}	molar residual volume
x^{l}	mass fraction of liquid in two-phase liquid-vapor mixture
x^{v}	mass fraction of vapor in two-phase liquid-vapor mixture
Z	compressibility factor
Z_c	critical compressibility factor
Z^{o}	compressibility factor function in Lee/Kesler generalized compressibility factor correlation, Equation (7.15)
Z^{1}	compressibility factor function in Lee/Kesler generalized compressibility factor correlation, Equation (7.15)

Greek Letters

α	parameter in cubic equations of state, see Table 7.1
α_{PR}	parameter in PR equation of state, see Table 7.1
α_{SRK}	parameter in SRK equation of state, see Table 7.1
β	parameter in cubic equation of state, defined in Equation (7.36a)
Δ	difference operator
Δh^{liq}	specific enthalpy change of liquid
Δh^{vap}	latent heat or enthalpy change of vaporization
Δh_n^{vap}	latent heat or enthalpy change of vaporization at normal boiling point
Δs^{liq}	specific enthalpy change of liquid
Δs^{vap}	entropy change of vaporization
Δs_n^{vap}	entropy change of vaporization at normal boiling point
ε	parameter in cubic equation of state, Equation (7.30), see Table 7.2
σ	parameter in cubic equation of state, Equation (7.30), see Table 7.2
ψ	parameter in cubic equation of state, Equation (7.31), see Table 7.2
Ω	parameter in cubic equation of state, Equation (7.31), see Table 7.2
ω	acentric factor

REFERENCES

1. Al-Rubaiee, A.H.A., Prediction and correlation of saturated liquid molar volume at any temperature (including normal boiling point) for pure components and mixtures, M.Sc. thesis, chemical engineering, Nahrain

college of engineering, Baghdad, Iraq, 2009.
2. Smith, J.M., Van Ness, H.C., Abbott, M.M., *Introduction to Chemical Engineering Thermodynamics*, McGraw Hill, New York, 6th edition, 2001.

TUTORIAL PROBLEMS WITH SOLUTION

Problem 1

Estimate compressibility factor and molar volume of ethanol vapor at 415 °C and 53 bar using the following:

a) Ideal gas law
b) Pitzer generalized virial coefficient correlation
c) Lee/Kesler correlation

Solution

a) $Z = 1$ for ideal gas.

$$v^{ig} = \frac{RT}{P} = \frac{8.314\times(415+273.15)}{53\times10^5} = 1.08\times10^{-3}\ \text{m}^3/\text{mol} = 1080\ \text{cm}^3/\text{mol}$$

b) Pitzer generalized virial coefficient correlation:

$$Z = 1 + B^o\frac{P_r}{T_r} + \omega B^1\frac{P_r}{T_r}$$

where

$$B^o = 0.083 - \frac{0.422}{T_r^{1.6}}$$

$$B^1 = 0.139 - \frac{0.172}{T_r^{4.2}}$$

From table in Appendix F:

For ethanol, $\omega = 0.645, T_c = 513.9$ K, $P_c = 61.48$ bar

$$T_r = \frac{T}{T_c} = \frac{415+273.15}{513.9} = 1.34\ ;\ P_r = \frac{P}{P_c} = \frac{53}{61.48} = 0.862$$

$$B^o = 0.083 - \frac{0.422}{T_r^{1.6}} = 0.083 - \frac{0.422}{1.34^{1.6}} = -0.1812$$

$$B^1 = 0.139 - \frac{0.172}{T_r^{4.2}} = 0.139 - \frac{0.172}{1.34^{4.2}} = 0.0887$$

$$Z = 1 + (-0.1812)\times\frac{0.862}{1.34} + 0.645\times0.0887\times\frac{0.862}{1.34} = 0.9202$$

$$v = \frac{ZRT}{P} = \frac{0.9202\times8.314\times(415+273.15)}{53\times10^5} = 0.993\times10^{-4}\ \text{m}^3/\text{mol} = 993\ \text{cm}^3/\text{mol}$$

c) Lee/Kesler correlation:

$$Z = Z^o + \omega Z^1$$

where Z^o and Z^1 are functions of T_r and P_r given in Table form.

$$T_r = 1.34\ ;\ P_r = 0.862$$

From Lee/Kesler table in Appendix J, Z^o values are:

$P_r \rightarrow$ / $T_r \downarrow$	0.80	1.0
1.30	0.8764	0.8438
1.40	0.9062	0.8827

We will do double linear interpolation to obtain Z^o at $T_r = 1.34\ ;\ P_r = 0.862$

T_r	Z^o at $P_r = 0.862$ by linear interpolation
1.30	$Z^o = 0.8764 + \frac{(0.8438 - 0.8764)}{(1.0 - 0.80)}(0.862 - 0.80) = 0.8663$
1.40	$Z^o = 0.9062 + \frac{(0.8827 - 0.9062)}{(1.0 - 0.80)}(0.862 - 0.80) = 0.8989$

Z^o at $T_r = 1.34$ ($P_r = 0.862$) by linear interpolation

$$Z^o = 0.8663 + \frac{(0.8989 - 0.8663)}{(1.4 - 1.30)}(1.34 - 1.30) = 0.8793$$

Thus, $Z^o = 0.8793$ at $T_r = 1.34\ ;\ P_r = 0.862$

From Lee/Kesler table in Appendix J, Z^1 values are:

$P_r \rightarrow$ / $T_r \downarrow$	0.80	1.0
1.30	0.0612	0.0819
1.40	0.0661	0.0857

We will do double linear interpolation to obtain Z^1 at $T_r = 1.34\ ;\ P_r = 0.862$

T_r	Z^1 at $P_r = 0.862$ by linear interpolation
1.30	$Z^1 = 0.0612 + \frac{(0.0819 - 0.0612)}{(1.0 - 0.80)}(0.862 - 0.80) = 0.0676$
1.40	$Z^1 = 0.0661 + \frac{(0.0857 - 0.0661)}{(1.0 - 0.80)}(0.862 - 0.80) = 0.0722$

Z^1 at $T_r = 1.34$ ($P_r = 0.862$) by linear interpolation

$$Z^1 = 0.0676 + \frac{(0.0722 - 0.0676)}{(1.4 - 1.30)}(1.34 - 1.30) = 0.0694$$

Thus, $Z^1 = 0.0694$ at $T_r = 1.34\ ;\ P_r = 0.862$

The compressibility factor is:

$$Z = Z^o + \omega Z^1 = 0.8793 + 0.645 \times 0.0694 = 0.924$$

The molar volume is:

$$v = \frac{ZRT}{P} = \frac{0.924 \times 8.314 \times (415 + 273.15)}{53 \times 10^5} = 0.997 \times 10^{-3}\ \mathrm{m^3/mol} = 997\ \mathrm{cm^3/mol}$$

Problem 2

Ammonia gas is stored in a 50-liter vessel at a constant temperature of 58 °C. The mass of ammonia is 0.65 kg. Calculate the pressure of ammonia gas using

a) Ideal gas law
b) Cubic equations of state

From table in Appendix F:
Molecular weight = 17.031, $\omega = 0.253$, $T_c = 405.7$ K, $P_c = 112.8$ bar

Solution

a) Ideal gas law

$$n = m/M = 0.65 \times 1000/17.031 = 38.166 \text{ moles}$$

$$P = \frac{nRT}{V} = \frac{38.166 \times 8.314 \times 331.15}{50 \times 10^{-3}} = 2.102 \times 10^6 \text{ Pa} = 21.02 \text{ bar}$$

b) Cubic equations of state

$$P = \frac{RT}{v - b} - \frac{a'}{\{v + \varepsilon b\}\{v + \sigma b\}}$$

where

$$a' = \psi\left(\frac{R^2 T_c^2}{P_c}\right)\alpha \ ; \quad b - \Omega \frac{RT_c}{P_c}$$

Given that: $\omega = 0.253$, $T_c = 405.7$ K, $P_c = 112.8$ bar

$$v = \frac{V}{n} = \frac{50 \times 10^{-3}}{38.166} = 1.31 \times 10^{-3}\ \mathrm{m^3/mol}\ ; \quad T_r = \frac{T}{T_c} = \frac{58 + 273.15}{405.7} = 0.8162$$

Summary of parameters for different cubic equations of state and the resulting pressure values

Name	ε	σ	ψ	Ω	α	a'	b	P
vdW	0	0	0.421875	0.125	1	0.425505	3.74E-05	1915444 Pa = 19.154 bar
RK	0	1	0.42748	0.08664	$1/\sqrt{T_r}$ $= 1.106883$	0.477242	2.59E-05	1871364 Pa =18.714 bar
SRK	0	1	0.42748	0.08664	α_{SRK} $= 1.174439$	0.506369	2.59E-05	1854721 Pa = 18.547 bar
PR	$1-\sqrt{2}$	$1+\sqrt{2}$	0.45724	0.07779	α_{PR} $= 1.149578$	0.530156	2.33E-05	1841231 Pa = 18.412 bar
$\alpha_{SRK} = \left[1 + (0.480 + 1.574\omega - 0.176\omega^2)(1 - \sqrt{T_r})\right]^2$ $\alpha_{PR} = \left[1 + (0.37464 + 1.5422\omega - 0.26992\omega^2)(1 - \sqrt{T_r})\right]^2$								

Parameters a' and b are in SI units

Comments: The pressure value calculated from the ideal gas law is significantly larger than those calculated from the cubic equations of state. Among the cubic equations of state, vdW gives the largest value of pressure. The pressure values obtained by RK, SRK, and PR equations of state are similar.

Problem 3

Estimate the molar volumes of n-butane in saturated-vapor and saturated-liquid states using cubic equations of state at 350 K where vapor pressure is 9.46 bar. Compare the estimated values with the experimental values given as: $v^g = 2482\ \mathrm{cm^3/mol}$ and $v^l = 115\ \mathrm{cm^3/mol}$. Use Newton-Raphson iterative method to solve the problem.

Solution

The general form of the cubic EOS can be expressed as:

$$f(v) = P - \frac{RT}{v-b} + \frac{a'}{\{v+\varepsilon b\}\{v+\sigma b\}} = 0 \tag{7.32}$$

where $a' = \psi\left(\frac{R^2T_c^2}{P_c}\right)\alpha\ ;\quad b = \Omega\frac{RT_c}{P_c}$

According to the *Newton-Raphson* method, the iterative scheme is as follows:

$$v_{i+1} = v_i - \frac{f(v_i)}{f'(v_i)} \tag{7.33}$$

where $f'(v)$ is the derivative of $f(v)$ given as:

$$f'(v) = \frac{RT}{(v-b)^2} - \frac{a'}{(v+\varepsilon b)(v+\sigma b)^2} - \frac{a'}{(v+\sigma b)(v+\varepsilon b)^2} \tag{7.34}$$

Newton-Raphson iterative procedure:

- Make an initial guess of the molar volume v
- Calculate $f(v)$ from Equation (7.32)
- Calculate $f'(v)$ from Equation (7.34)
- Calculate improved value of v from Equation (7.33)

For n-butane, the properties table in Appendix F gives: $\omega = 0.20$, $T_c = 425.1$ K, $P_c = 37.96$ bar
The reduced temperature (T_r) is: $T_r = T/T_c = 350/425.1 = 0.823336$.

Summary of parameters for different cubic equations of state

Name	ε	σ	ψ	Ω	α	a'	b
vdW	0	0	0.421875	0.125	1	1.3882249	0.000116382
RK	0	1	0.42748	0.08664	$1/\sqrt{T_r} =$ 1.102076	1.5502558	8.06665E-05
SRK	0	1	0.42748	0.08664	$\alpha_{SRK} =$ 1.151251	1.6194285	8.06665E-05
PR	$1-\sqrt{2}$	$1+\sqrt{2}$	0.45724	0.07779	$\alpha_{PR} =$	1.6978072	7.24267E-05

					1.128413		
$\alpha_{SRK} = \left[1 + (0.480 + 1.574\omega - 0.176\omega^2)(1 - \sqrt{T_r})\right]^2$ $\alpha_{PR} = \left[1 + (0.37464 + 1.5422\omega - 0.26992\omega^2)(1 - \sqrt{T_r})\right]^2$							

Parameters a' and b are in SI units

Iterations for molar volume of gas (initial guess value is molar volume of ideal gas):

vdW		RK		SRK		PR	
v_i	v_{i+1}	v_i	v_{i+1}	v_i	v_{i+1}	v_i	v_{i+1}
0.003076	0.002614	0.003076	0.002471	0.003076	0.002426	0.003076	0.002380
0.002614	0.002665	0.002471	0.002552	0.002426	0.002517	0.002380	0.002482
0.002665	0.002666	0.002552	0.002554	0.002517	0.002519	0.002482	0.002485
0.002666	0.002666	0.002554	0.002554	0.002519	0.002519	0.002485	0.002485
0.002666	0.002666	0.002554	0.002554	0.002519	0.002519	0.002485	0.002485
Final vdW value $= 0.002666\ \text{m}^3/\text{mol}$ $= 2666\ \text{cm}^3/\text{mol}$		Final RK value $= 0.002554\ \text{m}^3/\text{mol}$ $= 2554\ \text{cm}^3/\text{mol}$		Final SRK value $= 0.002519\ \text{m}^3/\text{mol}$ $= 2519\ \text{cm}^3/\text{mol}$		Final PR value $= 0.002485\ \text{m}^3/\text{mol}$ $= 2485\ \text{cm}^3/\text{mol}$	

Comment: The convergence occurs within 3-4 iterations. The experimental value is $v^g = 2482\ \text{cm}^3/\text{mol}$. The value predicted by the PR equation of state is very close to the experimental value. The largest error occurs in the case of the vdW equation of state.

Iterations for molar volume of liquid (initial guess value is $\mathbf{1.1} \times \boldsymbol{b}$):

vdW		RK		SRK		PR	
v_i	v_{i+1}	v_i	v_{i+1}	v_i	v_{i+1}	v_i	v_{i+1}
0.000128	0.000136	0.000089	0.000095	0.000089	0.000095	0.000080	0.000085
0.000136	0.000147	0.000095	0.000103	0.000095	0.000103	0.000085	0.000092
0.000147	0.000161	0.000103	0.000114	0.000103	0.000112	0.000092	0.000100
0.000161	0.000174	0.000114	0.000124	0.000112	0.000121	0.000100	0.000107
0.000174	0.000185	0.000124	0.000131	0.000121	0.000126	0.000107	0.000111
0.000185	0.000190	0.000131	0.000133	0.000126	0.000128	0.000111	0.000113
0.000190	0.000191	0.000133	0.000133	0.000128	0.000128	0.000113	0.000113
0.000191	0.000191	0.000133	0.000133	0.000128	0.000128	0.000113	0.000113
Final vdW value $= 0.000191\ \text{m}^3/\text{mol}$ $= 191\ \text{cm}^3/\text{mol}$		Final RK value $= 0.000133\ \text{m}^3/\text{mol}$ $= 133\ \text{cm}^3/\text{mol}$		Final SRK value $= 0.000128\ \text{m}^3/\text{mol}$ $= 128\ \text{cm}^3/\text{mol}$		Final PR value $= 0.000113\ \text{m}^3/\text{mol}$ $= 113\ \text{cm}^3/\text{mol}$	

Comment: The convergence occurs within 6-8 iterations. The experimental value is $v^l = 115\ \text{cm}^3/\text{mol}$. The value predicted by the PR equation of state is very close to the experimental value. The largest error occurs in the case of the vdW equation of state.

Problem 4

Re-do problem 3 using cubic equations of state in compressibility factor form, that is, first estimate the compressibility factor from cubic equations of state using the Newton-Raphson iterative method and then calculate the molar volumes of n-butane in saturated-vapor and saturated-liquid states at 350 K and 9.46 bar.

Solution

The general form of the cubic equation of state in terms of compressibility factor is:

$$f(Z) = Z - 1 - \beta + q\beta \frac{(Z-\beta)}{(Z+\varepsilon\beta)(Z+\sigma\beta)} = 0 \qquad (7.37)$$

where $\beta = bP/RT$, $q = a'/bRT$, $a' = \psi\left(\frac{R^2T_c^2}{P_c}\right)\alpha$, and $b = \Omega\frac{RT_c}{P_c}$. Note that β and q could also be expressed as: $\beta = \Omega P_r/T_r$ and $q = \psi\alpha/\Omega T_r$.

The *Newton-Raphson* iterative scheme is:

$$Z_{i+1} = Z_i - \frac{f(Z_i)}{f'(Z_i)} \qquad (7.38)$$

where $f'(Z)$ is the derivative of $f(Z)$, given as:

$$f'(Z) = 1 + q\beta\left[\frac{(Z+\varepsilon\beta)(Z+\sigma\beta)-(Z-\beta)(2Z+\beta\varepsilon+\beta\sigma)}{(Z+\varepsilon\beta)^2(Z+\sigma\beta)^2}\right] \qquad (7.39)$$

Newton-Raphson iterative procedure:

- Make an initial guess of the molar volume Z
- Calculate $f(Z)$ from Equation (7.37)
- Calculate $f'(Z)$ from Equation (7.39)
- Calculate improved value of Z from Equation (7.38)

For n-butane, the properties table in Appendix F gives: $\omega = 0.20$, $T_c = 425.1$ K, $P_c = 37.96$ bar
The reduced temperature (T_r) is: $T_r = T/T_c = 350/425.1 = 0.823336$. The reduced pressure ($P_r$) is : $P_r = P/P_c = 9.46/37.96 = 0.2492$

Summary of parameters for different cubic equations of state

Name	ε	σ	ψ	Ω	α	a'	b	β	q
vdW	0	0	0.421875	0.125	1	1.3882249	0.000116382	0.0378354	4.09917857
RK	0	1	0.42748	0.08664	$1/\sqrt{T_r} =$ 1.102076	1.5502558	8.06665E-05	0.0262245	6.6043784

SRK	0	1	0.42748	0.08664	$\alpha_{SRK} =$ 1.151251	1.6194285	8.06665E-05	0.0262245	6.89906672
PR	1-$\sqrt{2}$	1+$\sqrt{2}$	0.45724	0.07779	$\alpha_{PR} =$ 1.128413	1.6978072	7.24267E-05	0.0235457	8.05585463
$\alpha_{SRK} = \left[1 + (0.480 + 1.574\omega - 0.176\omega^2)\left(1 - \sqrt{T_r}\right)\right]^2$ $\alpha_{PR} = \left[1 + (0.37464 + 1.5422\omega - 0.26992\omega^2)\left(1 - \sqrt{T_r}\right)\right]^2$									

Parameters a' and b are in SI units

Iterations for compressibility factor of gas (initial guess value of Z is 1):

vdW		RK		SRK		PR	
Z_i	Z_{i+1}	Z_i	Z_{i+1}	Z_i	Z_{i+1}	Z_i	Z_{i+1}
1.00000	0.86816	1.00000	0.83640	1.00000	0.82629	1.00000	0.81626
0.86816	0.86667	0.83640	0.83044	0.82629	0.81905	0.81626	0.80804
0.86667	0.86670	0.83044	0.83043	0.81905	0.81903	0.80804	0.80802
0.86670	0.86670	0.83043	0.83043	0.81903	0.81903	0.80802	0.80802
0.86670	0.86670	0.83043	0.83043	0.81903	0.81903	0.80802	0.80802
Final vdW value of Z $= 0.86670$		Final RK value of Z $= 0.83043$		Final SRK value of Z $= 0.81903$		Final PR value of Z $= 0.80802$	
$v = ZRT/P$ $= 2.666 \times 10^{-3}\ \mathrm{m^3/mol}$ $= 2666\ \mathrm{cm^3/mol}$		$v = ZRT/P$ $= 2.554 \times 10^{-3}\ \mathrm{m^3/mol}$ $= 2554\ \mathrm{cm^3/mol}$		$v = ZRT/P$ $= 2.519 \times 10^{-3}\ \mathrm{m^3/mol}$ $= 2519\ \mathrm{cm^3/mol}$		$v = ZRT/P$ $= 2.485 \times 10^{-3}\ \mathrm{m^3/mol}$ $= 2485\ \mathrm{cm^3/mol}$	

Comment: The convergence occurs within 3-4 iterations. The values of the volumes estimated from the compressibility factor approach are the same as those obtained in problem 3 (as expected).

Iterations for compressibility factor of liquid (initial guess value of Z is β):

vdW		RK		SRK		PR	
Z_i	Z_{i+1}	Z_i	Z_{i+1}	Z_i	Z_{i+1}	Z_i	Z_{i+1}
0.03784	0.02852	0.02622	0.03410	0.02622	0.03377	0.02355	0.02936
0.02852	0.01383	0.03410	0.04011	0.03377	0.03916	0.02936	0.03408
0.01383	-0.01146	0.04011	0.04286	0.03916	0.04129	0.03408	0.03625
-0.01146	-0.06169	0.04286	0.04331	0.04129	0.04155	0.03625	0.03659
-0.06169	-0.19139	0.04331	0.04332	0.04155	0.04155	0.03659	0.03660
-0.19139	-0.87158	0.04332	0.04332			0.03660	0.03660
-0.87158	1.76098						
1.76098	0.90905						
0.90905	0.86630						

0.86630	0.86671						
0.86671	0.86670						
0.86670	0.86670						
Does not converge to the liquid root. Z values even become negative. Finally converges to *gas root.*		Final RK value of Z $= 0.04332$ $v = ZRT/P$ $= 1.333 \times 10^{-4}\ \text{m}^3/\text{mol}$ $= 133.3\ \text{cm}^3/\text{mol}$		Final RK value of Z $= 0.04155$ $v = ZRT/P$ $= 1.278 \times 10^{-4}\ \text{m}^3/\text{mol}$ $= 127.8\ \text{cm}^3/\text{mol}$		Final RK value of Z $= 0.03660$ $v = ZRT/P$ $= 1.126 \times 10^{-4}\ \text{m}^3/\text{mol}$ $= 112.6\ \text{cm}^3/\text{mol}$	

Comment: The iterative process does not converge to the liquid root in the case of the vdW equation of state. The intermediate values of Z even become negative. The convergence finally occurs to the gas root. In all other cases (RK, SRK, PR), convergence occurs to liquid root within 4-5 iterations. The values of the molar volumes of liquid estimated from the compressibility factor approach are the same as those obtained in problem 3.

To obtain the *liquid root* using vdW equation of state, the following iterative scheme suggested by Smith et al. is applied:

$$Z_{i+1} = \beta + (Z_i + \varepsilon\beta)(Z_i + \sigma\beta)\left(\frac{1+\beta-Z_i}{q\beta}\right) \tag{7.41b}$$

where the initial guess for the liquid root is $Z_o = \beta$. This scheme converges to the liquid root as shown below:

vdW					
Z_i	Z_{i+1}	Z_i	Z_{i+1}	Z_i	Z_{i+1}
0.03784	0.04707	0.06145	0.06161	0.06206	0.06207
0.04707	0.05199	0.06161	0.06173	0.06207	0.06207
0.05199	0.05501	0.06173	0.06182	0.06207	0.06208
0.05501	0.05701	0.06182	0.06188	0.06208	0.06208
0.05701	0.05839	0.06188	0.06193	0.06208	0.06208
0.05839	0.05937	0.06193	0.06197	0.06208	0.06208
0.05937	0.06007	0.06197	0.06200	0.06208	0.06208
0.06007	0.06059	0.06200	0.06202	0.06208	0.06208
0.06059	0.06096	0.06202	0.06204	0.06208	0.06209
0.06096	0.06124	0.06204	0.06205	0.06209	0.06209
0.06124	0.06145	0.06205	0.06206	0.06209	0.06209
Final vdW value of Z $= 0.06209$ $v = ZRT/P = 1.91 \times 10^{-4}\ \text{m}^3/\text{mol} = 191\ \text{cm}^3/\text{mol}$					

Problem 5

Consider superheated steam at 300 °C and 0.4 MPa (4 bar). Calculate the residual properties (volume, enthalpy, and entropy) using Pitzer generalized virial coefficient correlation. From table in Appendix F, $\omega = 0.345$, $T_c = 647.1$ K, $P_c = 220.55$ bar

Solution

$$T = 300 + 273.15 = 573.15 \text{ K}; T_r = \frac{T}{T_c} = \frac{573.15}{647.1} = 0.88572 \,;\, P_r = \frac{P}{P_c} = \frac{4}{220.55} = 0.01814$$

Pitzer generalized virial coefficient correlation for compressibility factor:

$$Z = 1 + B^o \frac{P_r}{T_r} + \omega B^1 \frac{P_r}{T_r} = 1 + (-4294)\left(\frac{0.01814}{0.88572}\right) + 0.345 \times (-0.1473)\left(\frac{0.01814}{0.88572}\right) = 0.9902$$

Note that:

$$B^o = 0.083 - \frac{0.422}{T_r^{1.6}} = 0.083 - \frac{0.422}{0.88572^{1.6}} = -0.4294$$

$$B^1 = 0.139 - \frac{0.172}{T_r^{4.2}} = 0.139 - \frac{0.172}{0.88572^{4.2}} = -0.1473$$

Thus,

$$v^R = (Z-1)\frac{RT}{P} = (0.9902 - 1)\frac{8.314 \times 573.15}{4 \times 10^5} = -116.7 \text{ cm}^3\text{/mol}$$

Pitzer generalized virial coefficient correlation for residual enthalpy:

$$\frac{h^R}{RT_c} = P_r\left[B^o - T_r\frac{dB^o}{dT_r} + \omega\left(B^1 - T_r\frac{dB^1}{dT_r}\right)\right] = 0.01814\Big[-0.4294 - 0.88572\frac{dB^o}{dT_r} + 0.345(-0.1473 - 0.88572)\frac{dB^1}{dT_r}\Big]$$

where

$$\frac{dB^o}{dT_r} = \frac{0.675}{T_r^{2.6}} = \frac{0.675}{0.88572^{2.6}} = 0.9254$$

$$\frac{dB^1}{dT_r} = \frac{0.722}{T_r^{5.2}} = \frac{0.722}{0.88572^{5.2}} = 1.357$$

Thus,

$$\frac{h^R}{RT_c} = -0.0311; \qquad h^R = -0.0311(8.314 \times 647.1) = -167.84 \text{ J/mol}$$

The residual entropy is:

$$\frac{s^R}{R} = P_r\left[\frac{dB^o}{dT_r} + \omega\frac{dB^1}{dT_r}\right] = 0.01814[0.9254 + 0.345 \times 1.357] = -0.0253$$

$$s^R = -0.0253 \times 8.314 = -0.21 \text{ J/mol.K}$$

Problem 6

Calculate residual enthalpy and residual entropy of superheated steam at 300 °C and 4 MPa (40 bar) using the Lee/Kesler correlation. From table in Appendix F, $\omega = 0.345$, $T_c = 647.1$ K, $P_c = 220.55$ bar.

Solution

$$T = 300 + 273.15 = 573.15 \text{ K} \,;\, T_r = \frac{T}{T_c} = \frac{573.15}{647.1} = 0.88572 \,;\, P_r = \frac{P}{P_c} = \frac{40}{220.55} = 0.1814$$

$$\frac{h^R}{RT_c} = \frac{(h^R)^o}{RT_c} + \omega\frac{(h^R)^1}{RT_c} \;; \qquad \frac{s^R}{R} = \frac{(s^R)^o}{R} + \omega\frac{(s^R)^1}{R}$$

We obtain the values of $\frac{(h^R)^o}{RT_c}$, $\frac{(h^R)^1}{RT_c}$, $\frac{(s^R)^o}{R}$ and $\frac{(s^R)^1}{R}$ from Lee/Kesler tables given in Appendix J.

We first calculate the *residual enthalpy*.

From Lee/Kesler table in Appendix J, $\frac{(h^R)^o}{RT_c}$ values are:

$P_r \rightarrow$ / $T_r \downarrow$	**0.1**	**0.2**
0.85	-0.141	-0.300
0.90	-0.126	-0.264

We use double linear interpolation to estimate $\frac{(h^R)^o}{RT_c}$ at $T_r = 0.88572$ and $P_r = 0.1814$.

T_r	$\frac{(h^R)^o}{RT_c}$ at $P_r = 0.1814$ by linear interpolation
0.85	$\frac{(h^R)^o}{RT_c} = -0.141 + \frac{(-0.300 - (-0.141))}{(0.2 - 0.1)}(0.1814 - 0.1) = -0.2704$
0.90	$\frac{(h^R)^o}{RT_c} = -0.126 + \frac{(-0.264 - (-0.126))}{(0.2 - 0.1)}(0.1814 - 0.1) = -0.2383$

$\frac{(h^R)^o}{RT_c}$ at $T_r = 0.88572$ ($P_r = 0.1814$) by linear interpolation

$$\frac{(h^R)^o}{RT_c} = -0.2704 + \frac{(-0.2383 - (-0.2704))}{(0.9 - 0.85)}(0.88572 - 0.85) = -0.2475$$

Thus, $\frac{(h^R)^o}{RT_c} = -0.2475$ at $T_r = 0.88572$; $P_r = 0.1814$

Now we use double linear interpolation to estimate $\frac{(h^R)^1}{RT_c}$ at $T_r = 0.88572$ and $P_r = 0.1814$.

T_r	$\frac{(h^R)^1}{RT_c}$ at $P_r = 0.1814$ by linear interpolation
0.85	$\frac{(h^R)^1}{RT_c} = -0.182 + \frac{(-0.401-(-0.182))}{(0.2-0.1)}(0.1814-0.1) = -0.3603$
0.90	$\frac{(h^R)^1}{RT_c} = -0.144 + \frac{(-0.308-(-0.144))}{(0.2-0.1)}(0.1814-0.1) = -0.2775$

$\frac{(h^R)^1}{RT_c}$ at $T_r = 0.88572$ ($P_r = 0.1814$) by linear interpolation

$$\frac{(h^R)^1}{RT_c} = -0.3603 + \frac{(-0.2775-(-0.3603))}{(0.9-0.85)}(0.88572-0.85) = -0.3011$$

Thus, $\frac{(h^R)^1}{RT_c} = -0.3011$ at $T_r = 0.88572$; $P_r = 0.1814$

$$\frac{h^R}{RT_c} = \frac{(h^R)^o}{RT_c} + \omega\frac{(h^R)^1}{RT_c} = -0.2475 + 0.345 \times (-0.3011) = -0.3514$$

Hence, the *residual enthalpy* is:

$$h^R = -0.3514(RT_c) = -0.3514 \times 8.314 \times 647.1 = -1890.53 \text{ J/mol}$$

Now we calculate the *residual entropy*.

From Lee/Kesler table in Appendix J, $\frac{(s^R)^o}{R}$ values are:

$P_r \rightarrow$ / $T_r \downarrow$	0.1	0.2
0.85	-0.111	-0.239
0.90	-0.094	-0.199

We use double linear interpolation to estimate $\frac{(s^R)^o}{R}$ at $T_r = 0.88572$ and $P_r = 0.1814$.

T_r	$\frac{(s^R)^o}{R}$ at $P_r = 0.1814$ by linear interpolation
0.85	$\frac{(s^R)^o}{R} = -0.111 + \frac{(-0.239-(-0.111))}{(0.2-0.1)}(0.1814-0.1) = -0.2152$
0.90	$\frac{(s^R)^o}{R} = -0.094 + \frac{(-0.199-(-0.094))}{(0.2-0.1)}(0.1814-0.1) = -0.1795$

$\frac{(s^R)^o}{R}$ at $T_r = 0.88572$ ($P_r = 0.1814$) by linear interpolation

$$\frac{(s^R)^o}{R} = -0.2152 + \frac{(-0.1795-(-0.2152))}{(0.9-0.85)}(0.88572-0.85) = -0.1897$$

From Lee/Kesler table in Appendix J, $\frac{(s^R)^1}{R}$ values are:

$P_r \rightarrow$ / $T_r \downarrow$	0.1	0.2
0.85	-0.183	-0.400
0.90	-0.140	-0.301

We use double linear interpolation to estimate $\frac{(s^R)^1}{R}$ at $T_r = 0.88572$ and $P_r = 0.1814$.

T_r	$\frac{(s^R)^1}{R}$ at $P_r = 0.1814$ by linear interpolation
0.85	$\frac{(s^R)^1}{R} = -0.183 + \frac{(-0.40-(-0.183))}{(0.2-0.1)}(0.1814-0.1) = -0.3596$
0.90	$\frac{(s^R)^1}{R} = -0.140 + \frac{(-0.301-(-0.140))}{(0.2-0.1)}(0.1814-0.1) = -0.27105$

$\frac{(s^R)^o}{R}$ at $T_r = 0.88572$ ($P_r = 0.1814$) by linear interpolation

$$\frac{(s^R)^o}{R} = -0.3596 + \frac{(-0.27105-(-0.3596))}{(0.9-0.85)}(0.88572-0.85) = -0.2963$$

Thus, $\frac{(s^R)^1}{R} = -0.2963$ at $T_r = 0.88572$; $P_r = 0.1814$

$$\frac{s^R}{R} = \frac{(s^R)^o}{R} + \omega\frac{(s^R)^1}{R} = -0.1897 + 0.345 \times (-0.2963) = -0.2919$$

Hence, the *residual entropy* is:

$$s^R = -0.2919R = -0.2919 \times 8.314 = -2.427 \text{ J/mol.K}$$

Problem 7

Estimate the residual properties (enthalpy, and entropy) on n-butane at $T = 463$ K and 47 bar using the RK, SRK, and PR cubic equations of state. From table in Appendix F, $\omega = 0.20$, $T_c = 425.1$ K, $P_c = 37.96$ bar

Solution

We first estimate the compressibility factor from the cubic equations of state. The general form of the cubic equation of state in terms of compressibility factor is:

$$f(Z) = Z - 1 - \beta + q\beta\frac{(Z-\beta)}{(Z+\varepsilon\beta)(Z+\sigma\beta)} = 0 \quad (7.37)$$

where $\beta = bP/RT$, $q = a'/bRT$, $a' = \psi\left(\frac{R^2T_c^2}{P_c}\right)\alpha$, and $b = \Omega\frac{RT_c}{P_c}$. Note that β and q could also be expressed as: $\beta = \Omega P_r/T_r$ and $q = \psi\alpha/\Omega T_r$.

The *Newton-Raphson* iterative scheme is:

$$Z_{i+1} = Z_i - \frac{f(Z_i)}{f'(Z_i)} \quad (7.38)$$

where $f'(Z)$ is the derivative of $f(Z)$, given as:

$$f'(Z) = 1 + q\beta\left[\frac{(Z+\varepsilon\beta)(Z+\sigma\beta)-(Z-\beta)(2Z+\beta\varepsilon+\beta\sigma)}{(Z+\varepsilon\beta)^2(Z+\sigma\beta)^2}\right] \quad (7.39)$$

For n-butane, the properties table in Appendix F gives: $\omega = 0.20$, $T_c = 425.1$ K, $P_c = 37.96$ bar
The reduced temperature (T_r) is: $T_r = T/T_c = 463/425.1 = 1.0892$. The reduced pressure ($P_r$) is: $P_r = P/P_c = 47/37.96 = 1.2381$

Summary of parameters for different cubic equations of state

Name	ε	σ	ψ	Ω	α	a'	b	β	q
RK	0	1	0.42748	0.08664	$1/\sqrt{T_r} =$ 0.958198	1.3478666	8.06665E-05	0.0984918	4.34072755
SRK	0	1	0.42748	0.08664	$\alpha_{SRK} =$ 0.93244725	1.3116444	8.06665E-05	0.0984918	4.22407619
PR	$1-\sqrt{2}$	$1+\sqrt{2}$	0.45724	0.07779	$\alpha_{PR} =$ 0.942202	1.4176345	7.24267E-05	0.0884312	5.0848081
$\alpha_{SRK} = \left[1 + (0.480 + 1.574\omega - 0.176\omega^2)(1 - \sqrt{T_r})\right]^2$ $\alpha_{PR} = \left[1 + (0.37464 + 1.5422\omega - 0.26992\omega^2)(1 - \sqrt{T_r})\right]^2$									

Parameters a' and b are in SI units

Iterations for compressibility factor of gas (initial guess value of Z is 1):

RK		SRK		PR	
Z_i	Z_{i+1}	Z_i	Z_{i+1}	Z_i	Z_{i+1}
1.00000	0.64896	1.00000	0.66559	1.00000	0.64169
0.64896	0.57543	0.66559	0.60523	0.64169	0.57522
0.57543	0.56872	0.60523	0.60144	0.57522	0.57061
0.56872	0.56866	0.60144	0.60143	0.57061	0.57058
0.56866	0.56866	0.60143	0.60143	0.57058	0.57058
Final RK value of Z = 0.56866		Final SRK value of Z = 0.60143		Final PR value of Z = 0.57058	

Comment: The convergence occurs within 4-5 iterations.

Now that the compressibility factor is known, we can readily calculate the residual properties from the following equations:

$$\frac{h^R}{RT} = Z - 1 + \left[\frac{dln\alpha}{dlnT_r} - 1\right] qI \quad (7.68)$$

$$\frac{s^R}{R} = ln(Z - \beta) + \left[\frac{dln\alpha}{dlnT_r}\right] qI \quad (7.69)$$

where

$$I = \frac{1}{\sigma-\varepsilon} ln\left[\frac{Z+\sigma\beta}{Z+\varepsilon\beta}\right] \quad \text{when } \varepsilon \neq \sigma$$
$$= \beta/(Z+\varepsilon\beta) \quad \text{when } \varepsilon = \sigma \tag{7.70}$$

In order to calculate h^R and s^R, we need the values of the derivative $\left(\frac{dln\alpha}{dlnT_r}\right)$ and the parameter I for different equations of state (RK, SRK, and PR). The expressions are given below for different equations of state:

Name	Expressions for the derivative $\left(\frac{dln\alpha}{dlnT_r}\right)$	Parameter I
RK	$\alpha = 1/\sqrt{T_r}$; $\frac{dln\alpha}{dlnT_r} = -\frac{1}{2}$	$I = ln[(Z+\beta)/Z]$
SRK	$\alpha = \left[1 + (0.480 + 1.574\omega - 0.176\omega^2)(1-\sqrt{T_r})\right]^2$ $\frac{dln\alpha}{dlnT_r} = \frac{T_r}{\alpha}\left(\frac{d\alpha}{dT_r}\right) = -\frac{1}{\alpha\sqrt{T_r}} f(\omega)[1 + f(\omega)(1-\sqrt{T_r})]$ where $f(\omega) = (0.480 + 1.574\omega - 0.176\omega^2)$	$I = ln[(Z+\beta)/Z]$
PR	$\alpha = \left[1 + (0.37464 + 1.5422\omega - 0.26992\omega^2)(1-\sqrt{T_r})\right]^2$ $\frac{dln\alpha}{dlnT_r} = \frac{T_r}{\alpha}\left(\frac{d\alpha}{dT_r}\right) = -\frac{1}{\alpha\sqrt{T_r}} f(\omega)[1 + f(\omega)(1-\sqrt{T_r})]$ where $f(\omega) = (0.37464 + 1.5422\omega - 0.26992\omega^2)$	$I = \frac{1}{2\sqrt{2}} ln\left[\frac{Z+\sigma\beta}{Z+\varepsilon\beta}\right]$ where $\sigma = 1+\sqrt{2}$ $\varepsilon = 1-\sqrt{2}$

Given T_r (= 1.0892), P_r (= 1.2381), ω (= 0.2), and Z values, the quantities β, q, I and derivative $dln\alpha/dlnT_r$, and hence the residual properties, can be easily calculated. They are summarized below:

Name	Z	β	q	$dln\alpha/dlnT_r$	I	h^R, J/mol	s^R, J/mol.K
RK	0.56866	0.0984918	4.34072755	-0.5	0.159735	-5663.94	-9.15660
SRK	0.60143	0.0984918	4.22407619	-0.76069	0.151659	-5876.07	-9.76560
PR	0.57058	0.0884312	5.0848081	-0.663645	0.135834	-6076.19	-9.87600

Problem 8

Consider compression of propane gas at 5 bar and 45 °C to 110 bar and 213 °C. Calculate the changes in enthalpy and entropy for this compression process. Use PR equation of state to calculate the residual properties.

From table in Appendix F, $\omega = 0.152$, $T_c = 369.8$ K, $P_c = 42.48$ bar

The heat capacity for propane is:

$$\frac{c_p^{ig}}{R} = A + BT + CT^2 + DT^{-2}$$

where $A = 1.213$, $B = 28.785 \times 10^{-3}$, $C = -8.824 \times 10^{-6}$, $D = 0$

Solution

The figure given below shows the computational path for the calculation of enthalpy and entropy changes for a real gas in going from state 1 (T_1, P_1) to state 2 (T_2, P_2). Thus, the enthalpy and entropy changes for a real gas in going from state 1 (T_1, P_1) to state 2 (T_2, P_2) are:

$$\Delta h = -h_1^R + \Delta h^{ig} + h_2^R$$

$$\Delta s = -s_1^R + \Delta s^{ig} + s_2^R$$

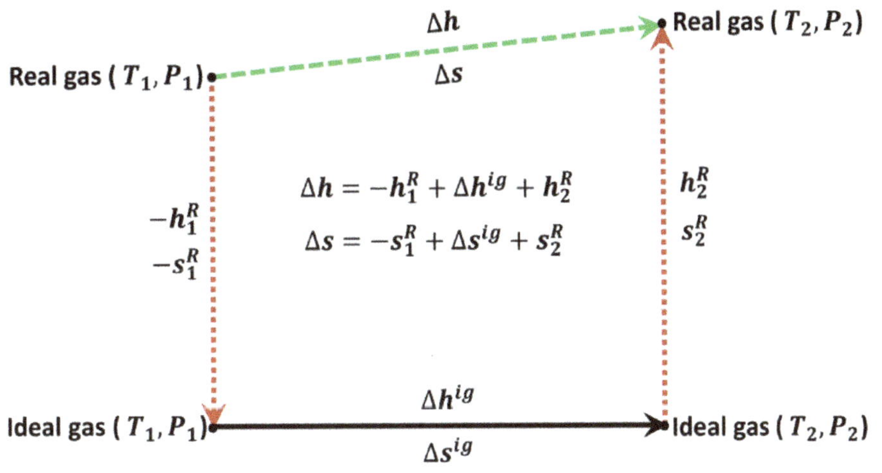

Calculation of h_1^R and s_1^R:

First calculate compressibility factor using the Newton-Raphson iterative process for PR equation of state, see Equations (7.37) - (7.39). For propane: $\omega = 0.152$, $T_c = 369.8$ K, $P_c = 42.48$ bar. The reduced temperature (T_r) is : $T_r = T/T_c = 318.15/369.8 = 0.86033$. The reduced pressure ($P_r$) is : $P_r = P/P_c = 5/42.48 = 0.1177$.

Summary of parameters for PR cubic equation of state

Name	ε	σ	ψ	Ω	α	a'	b	β	q
PR	1-√2	1+√2	0.45724	0.0777 9	$\alpha_{PR} =$ 1.089269	1.108277 7	5.6301E- 05	0.010642 5	7.4420136 7
$\alpha_{PR} = \left[1 + (0.37464 + 1.5422\omega - 0.26992\omega^2)(1 - \sqrt{T_r})\right]^2$									

Newton-Raphson iterations for compressibility factor of gas (initial guess value of Z is 1):

Z_i	Z_{i+1}
1.00000	0.92860
0.92860	0.92817
0.92817	0.92817

Thus, Z=0.92817

Once Z is known, we can easily calculate the residual properties using Equations (7.68) - (7.70). The calculations are summarized below:

Z	β	q	$dln\alpha/dlnT_r$	I	h_1^R, J/mol	s_1^R, J/mol.K
0.92817	0.0106425	7.44201367	-0.6227112	0.011337	-552.137	-1.15242

Calculation of Δh^{ig} *and* Δs^{ig}*:*

$$\Delta h^{ig} = c_{pm,h}^{ig} \Delta T \tag{7.45}$$

$$\Delta s^{ig} = c_{pm,s}^{ig} ln\left(\frac{T_2}{T_1}\right) - Rln\left(\frac{P_2}{P_1}\right) \tag{7.46}$$

where

$$\frac{c_{pm,h}^{ig}}{R} = A + BT_{am} + \frac{C}{3}[4T_{am}^2 - T_1T_2] + \frac{D}{T_1T_2} \tag{7.47}$$

$$\frac{c_{pm,s}^{ig}}{R} = A + BT_{lm} + CT_{am}T_{lm} + \frac{DT_{am}T_{lm}}{(T_1T_2)^2} \tag{7.48}$$

T_{am} is the arithmetic-mean temperature and T_{lm} is the log-mean temperatures, defined as:

$$T_{am} = \frac{T_1+T_2}{2} \tag{7.49}$$

$$T_{lm} = \frac{T_2-T_1}{ln\left(\frac{T_2}{T_1}\right)} \tag{7.50}$$

For propane, $A = 1.213$, $B = 28.785 \times 10^{-3}$, $C = -8.824 \times 10^{-6}$, $D = 0$. Given that $T_1 = 318.15$ K, $T_2 = 486.15$ K, $P_1 = 5$ bar, and $P_2 = 110$ bar. The calculations are summarized below:

T_{am}	T_{lm}	$\frac{c_{pm,h}^{ig}}{R}$	$\frac{c_{pm,s}^{ig}}{R}$	Δh^{ig}, J/mol	Δs^{ig}, J/mol.K
402.15	396.2318	11.34108	11.21248	15840.67	13.82604

Calculation of h_2^R *and* s_2^R*:*

Calculate compressibility factor using the Newton-Raphson iterative process for PR equation of state. The reduced temperature (T_r) is : $T_r = T/T_c = 486.15/369.8 = 1.31463$. The reduced pressure ($P_r$) is : $P_r = P/P_c = 110/42.48 = 2.58945$.

Summary of parameters for PR cubic equation of state

Name	ε	σ	ψ	Ω	α	a'	b	β	q
PR	1-√2	1+√2	0.45724	0.07779	$\alpha_{PR} =$ 0.831093	0.8455968	5.6301 E-05	0.1532246	3.71592397
$\alpha_{PR} = \left[1 + (0.37464 + 1.5422\omega - 0.26992\omega^2)\left(1 - \sqrt{T_r}\right)\right]^2$									

Newton-Raphson iterations for compressibility factor of gas (initial guess value of Z is 1):

Z_i	Z_{i+1}
1.00000	0.71028
0.71028	0.69439
0.69439	0.69435
0.69435	0.69435

Thus, Z=0.69435.

Once Z is known, the residual properties are calculated using Equations (7.68) - (7.70). The calculations are summarized below:

Z	β	q	$dln\alpha/dlnT_r$	I	h_2^R, J/mol	s_2^R, J/mol.K
0.69435	0.1532246	3.71592397	-0.5767134	0.184881	-5613.55	-8.39970

Calculation of Δh *and* Δs*:*

$$\Delta h = -h_1^R + \Delta h^{ig} + h_2^R = -(-552.137) + 15840.67 + (-5613.55) = 10779.26 \text{ J/mol}$$

$$\Delta s = -s_1^R + \Delta s^{ig} + s_2^R = -(-1.15242) + 13.82604 + (-8.3997) = 6.57876 \text{ J/mol.K}$$

Problem 9

A substance is in superheated vapor state at 183 °C and 57 bar. What are the enthalpy and entropy of this superheated vapor assuming that the enthalpy and entropy are zero for saturated liquid at -6.25 °C and P^{sat} = 101.325 kPa. Use SRK equation of state to calculate the residual properties.
Given that:
$\omega = 0.191$, $T_c = 420$ K, $P_c = 40.43$ bar, $T_n = 266.9$ K.

$$\frac{c_p^{ig}}{R} = A + BT + CT^2 + DT^{-2}$$

where $A = 1.967$, $B = 31.63 \times 10^{-3}$, $C = -9.837 \times 10^{-6}$, $D = 0$

Solution

The computational path is shown as follows.

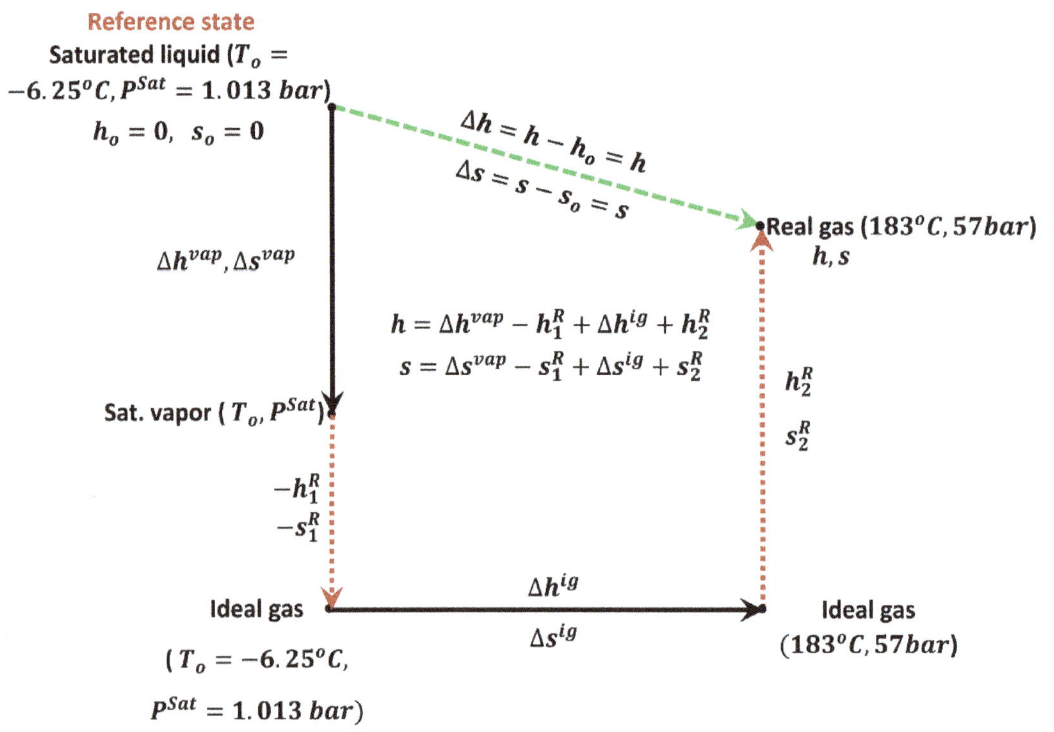

Following the computational path shown in the figure, the enthalpy and entropy of superheated vapor (real gas) at 183 °C and 57 bar can be calculated as:

$$h = \Delta h^{vap} - h_1^R + \Delta h^{ig} + h_2^R$$

$$s = \Delta s^{vap} - s_1^R + \Delta s^{ig} + s_2^R$$

Calculation of Δh^{vap} and Δs^{vap}:

We can use the following Riedel equation to estimate Δh_n^{vap}:

$$\frac{\Delta h_n^{vap}}{RT_n} = \frac{1.092(lnP_c - 1.013)}{0.930 - T_{rn}} \tag{7.87}$$

Note that vaporization calculation is needed at -6.25^oC. This is the normal boiling point (P^{sat} = 1 atm = 1.013 bar).

$$\frac{\Delta h_n^{vap}}{RT_n} = \frac{1.092(lnP_c - 1.013)}{0.930 - T_{rn}} = \frac{1.092(ln40.43 - 1.013)}{0.930 - 0.635476} = 9.96095$$

$$\Delta h_n^{vap} = 9.96095 \times 8.314 \times 266.9 = 22103.41 \text{ J/mol}$$

$$\Delta s_n^{vap} = \frac{\Delta h_n^{vap}}{T_n} = \frac{22103.41}{266.9} = 82.8153 \quad \text{J/mol.K}$$

Thus, $\Delta h^{vap} = 22103.41$ J/mol; $\Delta s^{vap} = 82.8153$ J/mol.K

Calculation of h_1^R and s_1^R :

First calculate compressibility factor using the Newton-Raphson iterative process for SRK equation of state, see Equations (7.37) - (7.39). Given: $\omega = 0.191$, $T_c = 420$ K, $P_c = 40.43$ bar. The reduced temperature (T_r) is: $T_r = T/T_c = 266.9/420 = 0.635476$. The reduced pressure ($P_r$) is: $P_r = P/P_c = 1.01325/40.43 = 0.025062$.

Summary of parameters for SRK cubic equation of state

Name	ε	σ	ψ	Ω	α	a'	b	β	q
SRK	0	1	0.42748	0.08664	$\alpha_{SRK} =$ 1.338731257	1.7259336	7.48297E-05	0.0034169	10.3942098
$\alpha_{SRK} = \left[1 + (0.480 + 1.574\omega - 0.176\omega^2)\left(1 - \sqrt{T_r}\right)\right]^2$									

Newton-Raphson iterations for compressibility factor of gas (initial guess value of Z is 1):

Z_i	Z_{i+1}
1.00000	0.96699
0.96699	0.96695
0.96695	0.96695

Thus, Z= 0.96695

Once Z is known, we can easily calculate the residual properties using Equations (7.68) - (7.70). The calculations are summarized below:

Z	β	q	$dln\alpha/dlnT_r$	I	h_1^R, J/mol	s_1^R, J/mol.K
0.96695	0.0034169	10.3942098	-0.83939	0.003527	-222.991	-0.56473

Calculation of Δh^{ig} and Δs^{ig} :

Use Equations (7.45) – (7.50) to estimate enthalpy and entropy changes for ideal gas. For the given gas, $A = 1.967$, $B = 31.63 \times 10^{-3}$, $C = -9.837 \times 10^{-6}$, $D = 0$. Given that $T_1 = 266.9$ K, $T_2 = 456.15$ K, $P_1 = 1.01325$ bar, and $P_2 = 57$ bar. The calculations are summarized below:

T_{am}	T_{lm}	$\frac{c_{pm,h}^{ig}}{R}$	$\frac{c_{pm,s}^{ig}}{R}$	Δh^{ig}, J/mol	Δs^{ig}, J/mol.K
361.525	353.1128	12.08698	11.88018	19017.95	19.43202

Calculation of h_2^R and s_2^R :

Calculate compressibility factor using the Newton-Raphson iterative process for SRK equation of state. The reduced temperature (T_r) is: $T_r = T/T_c = 456.15/420 = 1.086071$. The reduced pressure ($P_r$) is : $P_r = P/P_c = 57/40.43 = 1.40984$.

Summary of parameters for SRK cubic equation of state

Name	ε	σ	ψ	Ω	α	a'	b	β	q
SRK	0	1	0.42748	0.08664	$\alpha_{PR} =$ 0.93580247	1.2064654	7.48297E-05	0.1124686	4.25131372
$\alpha_{PR} = \left[1 + (0.37464 + 1.5422\omega - 0.26992\omega^2)(1 - \sqrt{T_r})\right]^2$									

Newton-Raphson iterations for compressibility factor of gas (initial guess value of Z is 1):

Z_i	Z_{i+1}
1.00000	0.61869
0.61869	0.52899
0.52899	0.52005
0.52005	0.51997
0.51997	0.51997

Thus, Z=0.51997

Once Z is known, the residual properties are calculated using Equations (7.68) - (7.70). The calculations are summarized below:

Z	β	q	$dln\alpha/dlnT_r$	I	h_2^R, J/mol	s_2^R, J/mol.K
0.51997	0.1124686	4.25131372	-0.76796	0.195812	-7402.01	-12.77867

Calculation of h and s:

$$h = \Delta h^{vap} - h_1^R + \Delta h^{ig} + h_2^R = 22103.41 - (-222.991) + 19017.95 + (-7402.01) = 33942.34 \text{ J/mol}$$

$$s = \Delta s^{vap} - s_1^R + \Delta s^{ig} + s_2^R = 82.8153 - (-0.56473) + 19.43202 + (-12.77867) = 19.43202 \text{ J/mol.K}$$

Problem 10

Liquid-vapor mixture of refrigerant HFC-134A is in equilibrium at 10 bars. The mixture consists of equal volumes of liquid and vapor. The total volume of the mixture is 1 m^3.

a) What is the quality of the two-phase liquid-vapor mixture?
b) What is the total mass of the two-phase mixture?
c) What are the total enthalpy and entropy of the mixture?

From table in Appendix K,

$$P^{sat} = 10 \text{ bar}, T^{sat} = 39.39\ °\text{C}$$

Property	Saturated liquid	Saturated vapor
v	0.8695 cm^3/g	20.2 cm^3/g
h	105.29 kJ/kg	267.97 kJ/kg
s	0.3838 kJ/kg.K	0.9043 kJ/kg.K

Solution

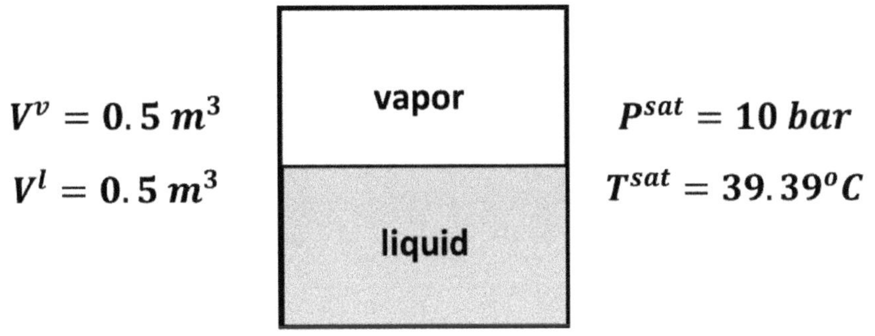

Vapor-liquid mixture of refrigerant HFC-134A

a) Volume of vapor $= mx^v v^v$

Volume of liquid $= m(1 - x^v)v^l$

Since the volumes of vapor and liquid are equal,

$$mx^v v^v = m(1 - x^v)v^l$$

$$x^v v^v = (1 - x^v)v^l$$

$$x^v = \frac{v^l}{v^v + v^l} = \frac{0.8695}{20.2 + 0.8695} = 0.04127$$

b) Volume of vapor $= mx^v v^v = 0.5$

$$m = \frac{0.5}{x^v v^v} = \frac{0.5 \times 10^3}{0.04127 \times 20.2} = 599.77 \text{ kg}$$

c) Total enthalpy

$$H = m[x^v h^v + (1 - x^v)h^l] = 599.7[0.04127(267.97) + (1 - 0.04127)(105.29)] = 67{,}176.5 \text{ kJ}$$

Total entropy

$$S = m[x^v s^v + (1 - x^v)s^l] = 599.7[0.04127(0.9043) + (1 - 0.04127)(0.3838)] = 243.08 \text{ kJ/K.}$$

PRACTICE PROBLEMS

Problem 1

Estimate compressibility factor and molar volume of propane gas at 50 °C and 10 bar using the following:

a) Ideal gas law
b) Pitzer generalized virial coefficient correlation

Problem 2

Estimate compressibility factor and molar volume of n-butane gas at 130 °C and 20 bar using the following:

a) Ideal gas law
b) Pitzer generalized virial coefficient correlation
c) Lee/Kesler correlation

Problem 3

a) Estimate the volume occupied by 1 kg of sulfur dioxide gas at 100 °C and 25 bar. Use Lee/Kesler correlation.
b) What is the mass of sulfur dioxide gas contained in a 1 m^3 vessel at 80 °C and 15 bar. Use Pitzer generalized virial coefficient correlation.

Problem 4

Ethylene gas is stored in a 0.2 m^3 vessel at a constant temperature of 25°C. The mass of the gas is 35 kg. Calculate the pressure of the gas using

a) Ideal gas law
b) Cubic equations of state

Problem 5

Estimate the molar volumes of chlorine in saturated-vapor and saturated-liquid states using cubic equations of state at 353.15 K where the vapor pressure is 27.43 bar.

Problem 6

Estimate the residual properties (v^R, h^R, s^R) of superheated steam at 300 °C and 2 MPa (20 bar) using the steam tables. You can treat superheated steam at 300 °C and a low pressure of 0.01 MPa as ideal gas. Compare the values obtained from steam table data with those estimated from Pitzer generalized virial coefficient correlation.

Problem 7

Estimate the residual properties (v^R, h^R, s^R) of methane gas at 200 K and 110 bar using Lee/Kesler correlation.

Problem 8

Re-do problem 7 using RK, SRK, and PR cubic equations of state.

Problem 9

Consider expansion of ethylene gas at 50 bar and 200 °C to 10 bar and 30 °C. Calculate the changes in enthalpy and entropy for this expansion. Use SRK equation of state to calculate the residual properties.

Problem 10

Estimate the molar enthalpy and entropy of isobutene gas at 120 °C and 25 bar. You can set enthalpy and entropy equal to zero for isobutane in ideal gas state at 1 bar and 0 °C. Use the PR equation of state to calculate the residual properties.

Problem 11

Estimate the molar enthalpy and entropy of 1,3-butadiene vapor at 450 K and 20 bar. The reference state is saturated liquid 1,3-butadiene at 268.7 K and P^{sat} =101.325 kPa. The enthalpy and entropy in the reference state are zero. Use the PR equation of state to calculate the residual properties.

Problem 12

Consider wet steam at 100 °C. The specific volume of the wet steam is 500 cm^3/g. Determine the quality, specific enthalpy, and specific entropy of this wet steam.

Problem 13

A stream of some gas enters a turbine at 250 °C and 50 bar and undergoes isentropic expansion to 5 bar. Calculate the work produced by the turbine (a) assuming that the gas behaves like an ideal gas, and b) using the Pitzer generalized virial coefficient correlation to estimate the residual properties.
Given that: $c_p = 50$ J/mol.K , $\omega = 0.10$, $T_c = 285$ K, $P_c = 55$ bar.

Chapter 8

Power Cycles

8.1 LEARNING OUTCOMES

Upon completion of this chapter, students will be able to:

- Appreciate the importance of second law in the development of power cycles
- Describe Carnot cycle based steam power plant and problems associated with it
- Derive the expression for thermal efficiency of a Carnot cycle based steam power plant and identify the factors which affect the thermal efficiency of the cycle
- Describe the Rankine cycle based steam power plant and discuss the factors which affect the thermal efficiency of the cycle
- Describe reheat, regenerative, and combined reheat/regenerative cycles
- Describe supercritical Rankine cycle and supercritical reheat cycle
- Solve problems dealing with steam power plants based on Carnot cycle, Rankine cycle with and without reheat and regeneration, supercritical Rankine cycles with and without reheat and regeneration.

8.2 INTRODUCTION

Power cycles are heat-to-work conversion cyclic processes in which heat is absorbed by the process from a heat source and converted into work. According to the second law of thermodynamics, it is impossible for a power cycle to convert the heat absorbed completely into work. Only a part of the heat absorbed by the process can be converted into work. Some of the heat absorbed by the process must be rejected to a heat sink. Figure 8.1 shows the scheme on which the power cycles are based. Heat is absorbed from the heat source by the working fluid of the cyclic process, part of it is converted into work and part of it is rejected into the heat sink. The thermal efficiency of a power cycle, defined as the ratio of work produced to heat input, is always less than one. If the working substance that undergoes the power cycle is gas and remains gas throughout the cyclic process, the power cycle is referred to as *gas* power cycle. When the working substance undergoes phase changes from liquid to vapor and vice versa during different stages of the cyclic process, the power cycle is called *vapor* power cycle.

This chapter is devoted to the analysis of steam (vapor) power cycles, also referred to as steam power plants, where the working substance (H_2O) is alternately vaporized and condensed. Steam power plants are used extensively for the generation of electric power at a large scale.

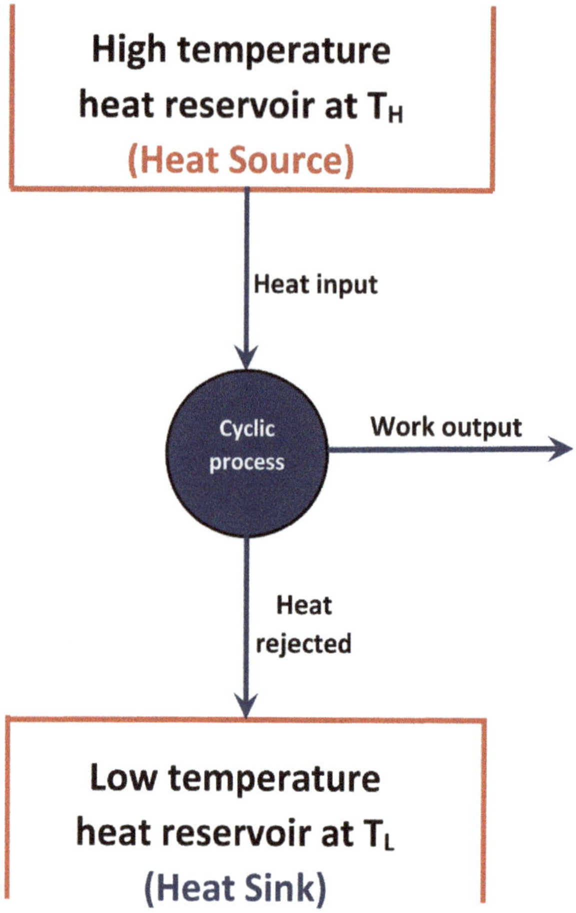

Figure 8.1 Conversion of heat into work by a power cycle.

8.3 CARNOT CYCLE BASED STEAM POWER PLANT

The simplest vapor power cycle is the Carnot cycle. Figure 8.2 shows a Carnot cycle on a P versus v diagram for H_2O. This Carnot cycle is the basis for a simple steam power plant. The working substance (H_2O) undergoes the Carnot cyclic process to produce work from heat. The Carnot cycle $1 \rightarrow 2 \rightarrow 3 \rightarrow 4 \rightarrow 1$ operates reversibly and involves four steps: two adiabatic steps and two isothermal steps. Along step $1 \rightarrow 2$, the working substance undergoes reversible adiabatic compression where work is absorbed by the process. Step $2 \rightarrow 3$ is a reversible isothermal vaporization process where heat is absorbed by the process at high temperature T_H. Step $3 \rightarrow 4$ is a reversible adiabatic expansion process where work is produced by the process. Step $4 \rightarrow 1$ is a reversible isothermal condensation process where heat is rejected by the process at low temperature T_L. Note that in state 1, the working fluid is a two-phase steam-water mixture. In state 2, the working fluid is saturated liquid water. State 3 consists of saturated steam and state 4 is a two-phase steam-water mixture.

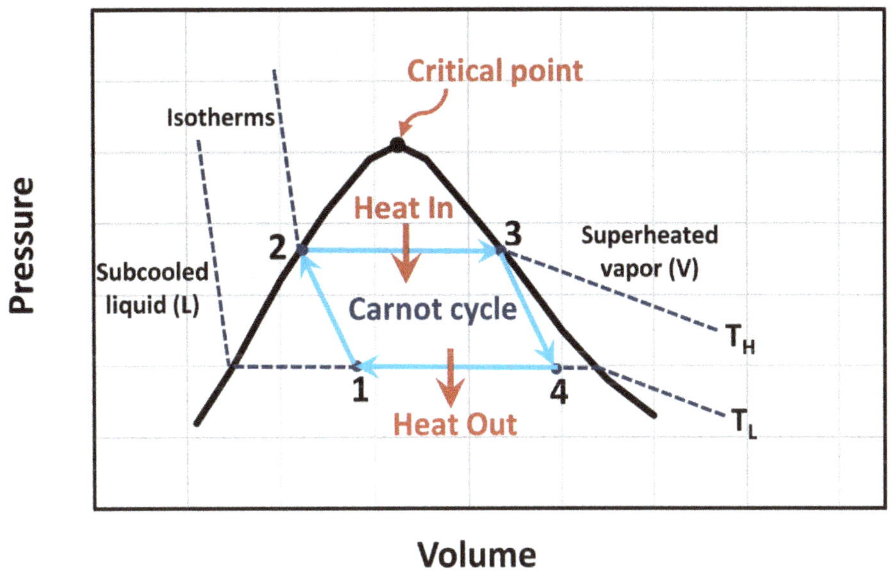

Figure 8.2 Carnot cycle as the basis for a simple steam power plant.

Figure 8.3 shows the equipment diagram for a simple steam power plant based on the Carnot cycle. The working fluid in state 1 (wet steam) is pumped through a pump and enters the boiler as high pressure saturated liquid water (state 2). In the boiler, the working fluid absorbs heat from the heat source (combustion of fuel) to generate steam (state 3, saturated steam). The saturated steam at high pressure and high temperature is expanded in the turbine to produce work. The discharge from the turbine (state 4, wet steam at low pressure) is passed to the condenser where heat is rejected to a heat sink (cold water) and condensation of steam to liquid water takes place. The vapor-liquid mixture from the condenser (state 1, wet steam) is fed back to the pump. Referring to Figure 8.2, step 1 → 2 occurs in a pump, step 2 → 3 occurs in a boiler, step 3 → 4 occurs in a turbine, and step 4 → 1 occurs in a condenser. Note that the work required by the pump along step 1 → 2 is much smaller than the work produced in the turbine along step 3 → 4. There is a net production of work in a steam power plant, as expected.

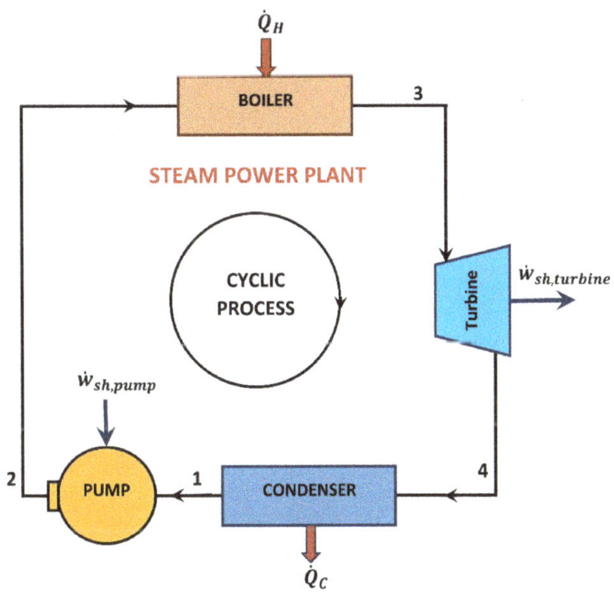

Figure 8.3 Equipment diagram for a simple steam power plant.

When analyzing power cycles, it is often preferred to draw the cycle on a temperature (T) versus entropy (s) diagram (instead of P versus v diagram). Figure 8.4 shows a Carnot cycle on T versus s diagram for substance H_2O. Steps $1 \rightarrow 2$ and $3 \rightarrow 4$ are reversible adiabatic processes. No entropy change occurs in reversible adiabatic processes. Such *isentropic* processes are simply vertical lines on a T versus s diagram.

The expression for the thermal efficiency of Carnot cycle based steam power plant can be derived readily using the T versus s diagram. The thermal efficiency can be expressed as:

$$\eta = \frac{W_{cycle}}{Heat\ input} = \frac{Q_{cycle}}{Q_{2\rightarrow 3}} = 1 + \frac{Q_{4\rightarrow 1}}{Q_{2\rightarrow 3}} \tag{8.1}$$

As steps $2 \rightarrow 3$ and $4 \rightarrow 1$ are reversible isothermal processes,

$$Q_{2\rightarrow 3} = T_H(s_3 - s_2) \tag{8.2}$$

$$Q_{4\rightarrow 1} = T_L(s_1 - s_4) \tag{8.3}$$

From Equations (8.1) to (8.3), we get:

$$\eta = 1 + \left(\frac{T_L}{T_H}\right)\left(\frac{s_1 - s_4}{s_3 - s_2}\right) \tag{8.4}$$

From Figure 8.4, $s_1 = s_2$ and $s_3 = s_4$. Consequently, Equation (8.4) reduces to:

$$\eta = 1 - \left(\frac{T_L}{T_H}\right) \tag{8.5}$$

Thus, the thermal efficiency of a steam power plant based on Carnot cycle depends only on the temperatures in the boiler and condenser.

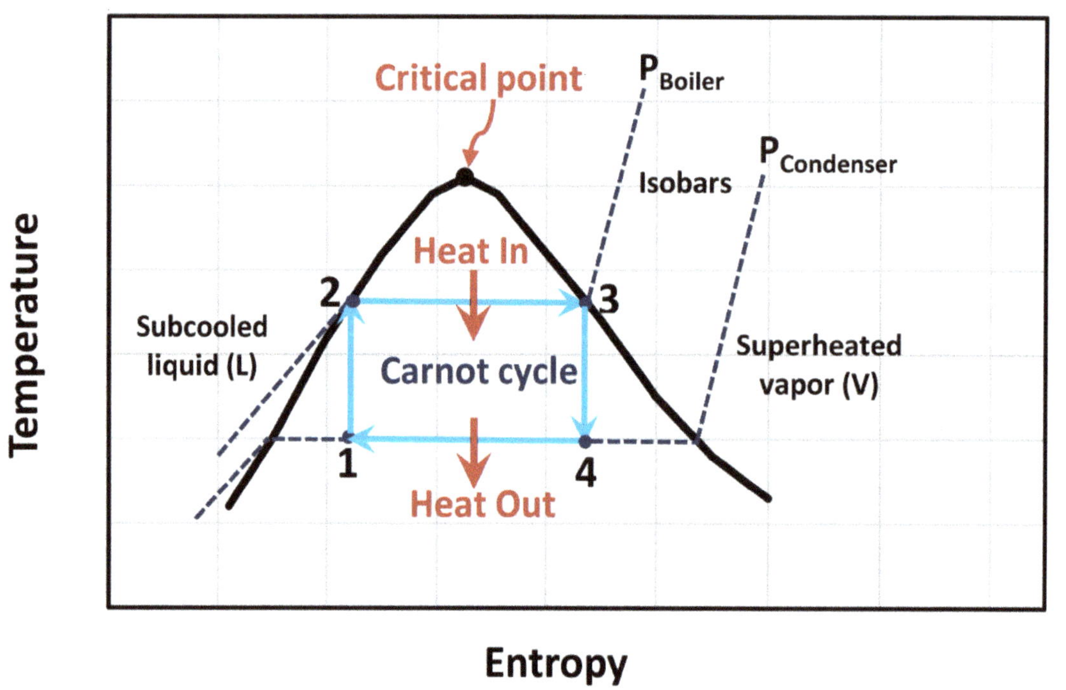

Figure 8.4 Carnot cycle based steam power plant on a T versus s diagram.

How to Improve the Thermal Efficiency of a Simple Steam Power Plant?

- The thermal efficiency can be increased by increasing the boiler temperature T_H. The boiler pressure is related to T_H. It is simply the vapor pressure of water corresponding to T_H, that is, $P_{Boiler} = P^{Sat}$ corresponding to T_H. Thus, the boiler temperature can be increased by increasing the boiler pressure. The higher the boiler pressure, the higher the thermal efficiency. However, it should be noted that increasing the boiler pressure raises the capital (equipment) cost of the plant as high pressure equipment requires heavier construction of the boiler and related equipment.

- The thermal efficiency of a steam power plant can be increased by decreasing the condenser temperature T_L. As condenser pressure is related to T_L ($P_{Condenser} = P^{Sat}$ corresponding to T_L), the condenser temperature can be decreased by lowering $P_{Condenser}$. The lower the condenser pressure, the higher the thermal efficiency of the plant. Note that the lowest condenser pressure we can have is P^{Sat} corresponding to the ambient temperature as heat can be rejected to the surroundings only if T_L is larger than the ambient temperature. P^{Sat} is about 2.3 kPa (much lower than the atmospheric pressure of 101.32 kPa) at 20 °C. Thus, condenser can be operated at a very low pressure.

Problems with Carnot Cycle Based Steam Power Plant

Practical steam plants are rarely based on Carnot cycle. The main problems are in steps 1 → 2 and 3 → 4 of the cycle (see Figure 8.5) as explained below.

- Step 1 → 2 requires pumping of a two-phase mixture of vapor and liquid. Pumps work very well with liquids. The presence of gas/vapor in the liquid is problematic. It deteriorates the pressure characteristics of the pump causing sharp losses in head and flow or even total cessation of pump flow.

- Step 3 → 4 is also problematic. In this step, the turbine takes in saturated steam and discharges wet steam (mixture of vapor and liquid) with high liquid content. The liquid content is high enough to cause severe erosions problems for the turbine. The liquid content of the turbine exhaust should be no more than 10% to avoid serious erosion problems.

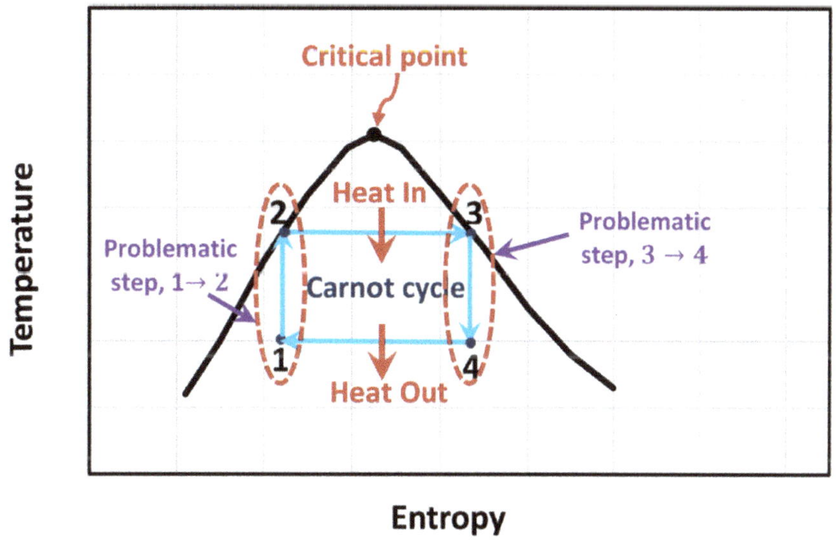

Figure 8.5 Problematic steps in the operation of steam power plant based on Carnot cycle.

8.4 IMPROVED STEAM POWER PLANT- RANKINE CYCLE

To overcome the problems associated with Carnot cycle, the actual steam power plant is based on the *Rankine cycle*. Figure 8.6 shows the Rankine cycle. Like Carnot cycle, the Rankine cycle consists of four reversible steps: two constant pressure steps (heating in boiler and cooling in condenser) and two isentropic steps (compression in pump and expansion in turbine). The four steps are further described as follows:

- Step 1 → 2 Isentropic (reversible adiabatic) compression of liquid in a pump to produce sub-cooled liquid at point 2.
- Step 2 → 3 Isobaric heat addition in a boiler to produce superheated steam at point 3.
- Step 3 → 4 Isentropic (reversible adiabatic) expansion in a turbine to produce wet steam at point 4.
- Step 4 → 1 Isobaric heat rejection in a condenser to produce saturated liquid at point 1.

The Rankine cycle differs from the Carnot cycle of Figure 8.5 in two major ways. First, the isobaric cooling step 4 → 1 results in complete condensation of vapor and the fluid to be pumped is 100% liquid (saturated liquid at point 1), and second, the isobaric heating step 2 → 3 is carried out well beyond the saturated vapor point to produce superheated steam. The expansion of superheated steam in a turbine (step 3 → 4) produces wet steam with moisture content not too large.

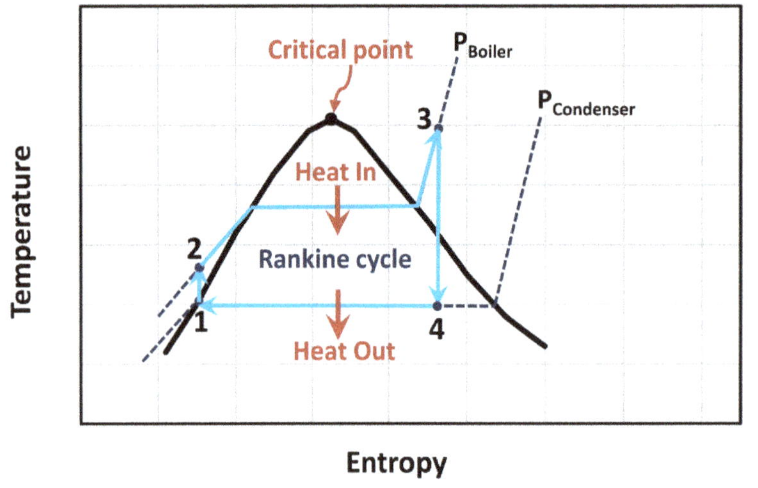

Figure 8.6 The Rankine cycle.

The Rankine cycle shown in Figure 8.6 is a reversible (ideal) cycle. In actual practice, the work producing step (3 → 4) and the work-requiring step (1 → 2) are irreversible and hence non-isentropic. The entropy increases in steps 3 → 4 and 1 → 2. The actual power cycle is shown in Figure 8.7. The ideal Rankine cycle is 1 → 2′ → 3 → 4′ → 1 and the actual Rankine cycle is 1 → 2 → 3 → 4 → 1.

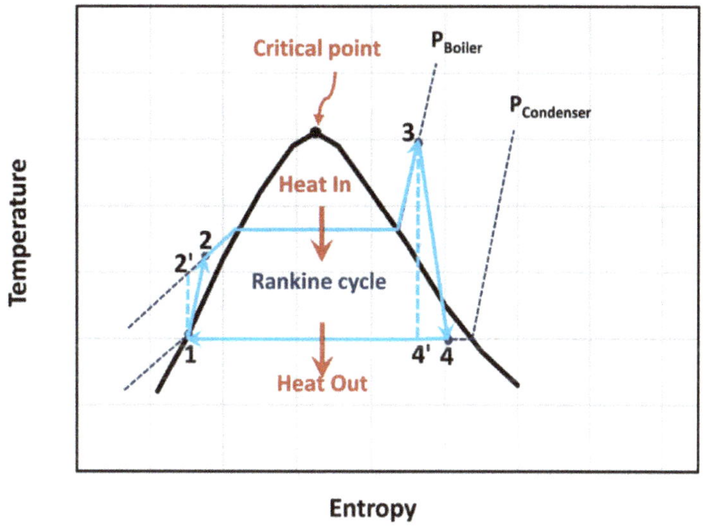

Figure 8.7 Practical Rankine cycle.

Thermal Efficiency of Ideal Rankine Cycle

Referring to Figure 8.6, the thermal efficiency of ideal Rankine cycle is:

$$\eta = \frac{Q_{2\to3}+Q_{4\to1}}{Q_{2\to3}} = 1 + \frac{Q_{4\to1}}{Q_{2\to3}} = 1 + \frac{T_L(s_1-s_4)}{Q_{2\to3}} \tag{8.6}$$

The heat input $Q_{2\to3}$ can be expressed as:

$$Q_{2\to3} = \int_2^3 Tds = \bar{T}_{Boiler}(s_3 - s_2) \tag{8.7}$$

Substitution of $Q_{2\to3}$ from Equation (8.7) into Equation (8.6) gives:

$$\eta = 1 + \frac{T_L(s_1-s_4)}{\bar{T}_{Boiler}(s_3-s_2)} = 1 - \frac{T_L}{\bar{T}_{Boiler}} \tag{8.8}$$

Note that $s_3 = s_4$ and $s_1 = s_2$. Thus, the thermal efficiency of Rankine cycle increases with the increase in average boiler temperature (average temperature over which heat is added in step $2 \to 3$) and decrease in condenser temperature.

Effect of Boiler Temperature on Thermal Efficiency

The thermal efficiency of the Rankine cycle increases with the increase in the average temperature of the boiler (see Equation (8.8)). We can increase the average boiler temperature at constant pressure by increasing the degree of superheating of the steam leaving the boiler, as shown in Figure 8.8. The average boiler temperature over the heat addition process $2 \to 3'$ in the boiler is clearly higher in comparison with the average temperature over the process $2 \to 3$.

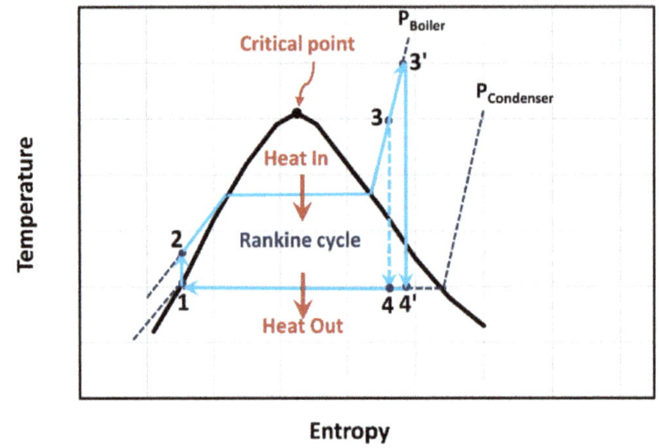

Figure 8.8 Effect of degree of superheating of steam leaving the boiler on the Rankine cycle.

The other advantage of raising the steam temperature leaving the boiler and entering the turbine is that the quality of steam discharged from the turbine increases. Note that the quality of steam in state 4′ is higher than that in state 4. Higher quality implies less moisture content (less water droplets) and hence less erosion of turbine blades. It is a common practice to maintain steam quality of at least 90% in the turbine exhaust steam. However, it should be pointed out that the temperature of the steam leaving the boiler cannot be increased indefinitely. There are metallurgical limitations (for example, meltdown) imposed by the materials used in the construction of the boiler. In practice, power plants are rarely operated at temperatures higher than 600 °C or so.

Effect of Boiler Pressure on Thermal Efficiency

Figure 8.9 shows the effect of boiler pressure on the Rankine cycle while keeping the maximum temperature (temperature of the superheated steam leaving the boiler) constant. With the increase in boiler pressure from P_1 to P_2 the average temperature in the boiler increases. Note that the average temperature in the boiler at P_1 is over step $2 \rightarrow 3$ and the average temperature at P_2 is over step $2' \rightarrow 3'$. Thus, the thermal efficiency of the Rankine cycle increases with the increase in pressure (see Equation (8.8)). However, there is one disadvantage of raising the boiler pressure, that is, the quality of the exhaust steam from the turbine decreases with the increase in pressure. Referring to Figure 8.9, the quality of steam at point 4′ is lower than the quality at point 4.

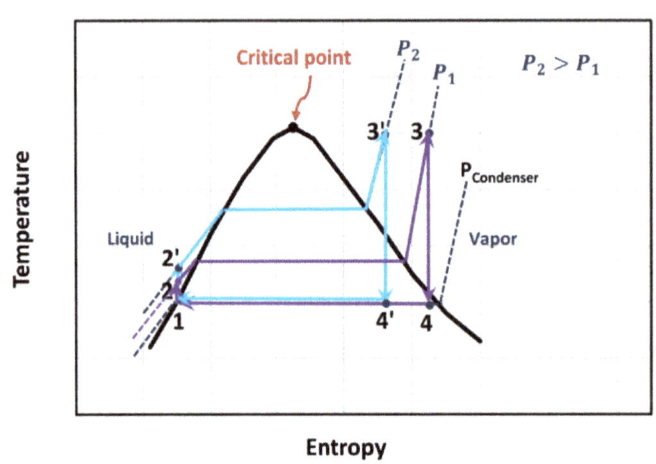

Figure 8.9 Effect of boiler pressure on the Rankine cycle.

8.5 REHEAT CYCLE

In the preceding section, it was observed that operating the Rankine cycle at high boiler pressure favors high thermal efficiency. However, high boiler pressures result in high moisture (water droplets) in the steam leaving the turbine. One can, in principle, increase the temperature of the steam entering the turbine to improve the quality of the steam leaving the turbine. But most metals used in the construction of equipment (boiler, turbine, etc.) cannot withstand temperatures above a certain maximum value (T_{max}), about 600 °C. So how do we increase the steam quality leaving the turbine without increasing the temperature above T_{max}? It is here that the *reheat cycle* comes in handy.

Figure 8.10 shows the reheat cycle. The isentropic expansion of steam leaving the boiler at point 3 is now carried out in two steps: step 3 → 4 and step 5 → 6. These expansions are carried out in two separate turbines (or in a two-stage turbine). After expansion in step 3 → 4, the exhaust from the turbine is reheated at constant pressure P' along step 4 → 5, there by raising the temperature of the steam to point 5. The expansion of superheated steam at point 5 is then carried out along step 5 → 6 in a second turbine. The exhaust from the second turbine is high quality steam with moisture content well below 10%. Note that if the superheated steam at point 3 is expanded in a single step 3 → 4′, the exhaust from the turbine will have an unacceptable amount of moisture (water droplets). The reheating of steam along step 4 → 5 improves the quality of exhaust steam from state 4′ to 6.

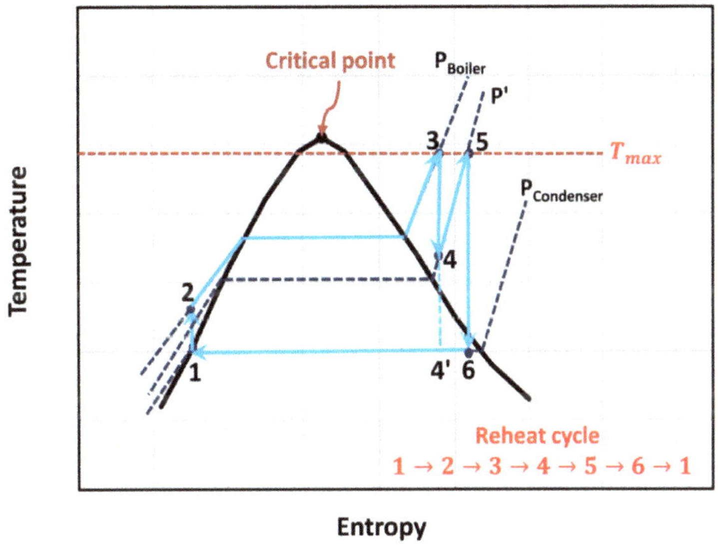

Figure 8.10 The Reheat cycle.

The processes shown in the reheat cycle (Figure 8.10) are reflected in the equipment diagram of the reheat cycle shown in Figure 8.11. Expansion in step 3 → 4 is carried out in turbine I. The discharge from turbine I is passed through the heating coil in the boiler (step 4 → 5) to reheat the steam to state 5. Expansion step 5 → 6 is carried out in turbine II. The steam discharged from turbine II in state 6 is of high quality with acceptable moisture content (<10%).

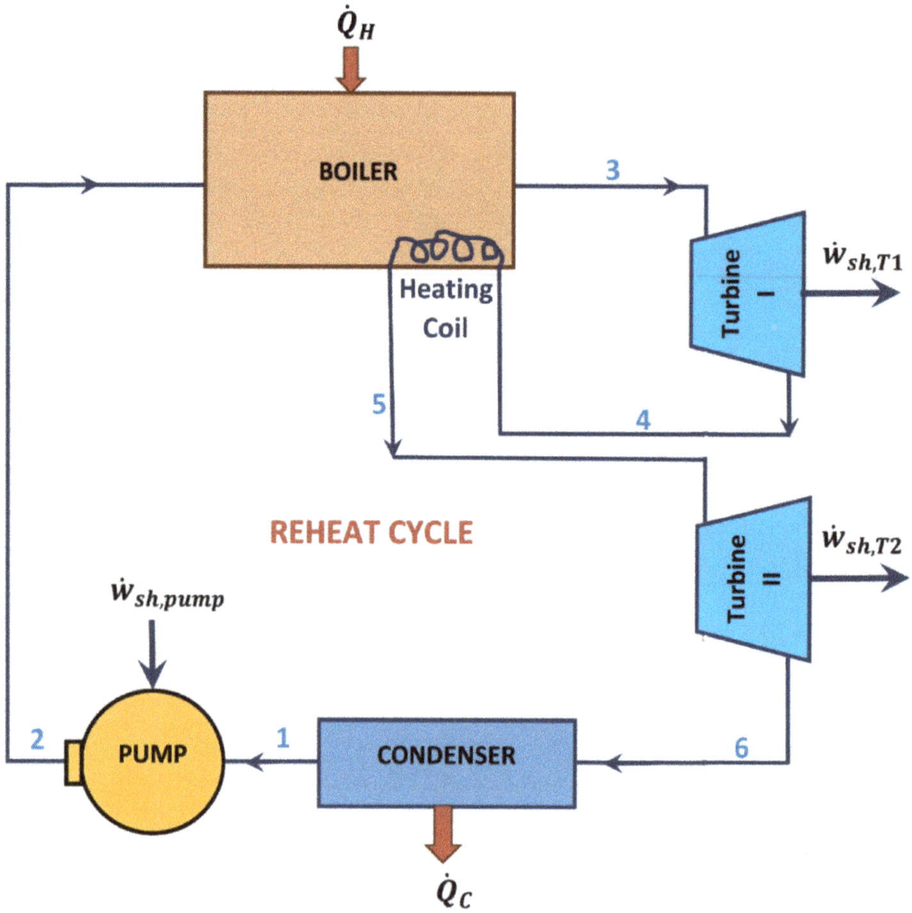

Figure 8.11 Equipment diagram of the reheat cycle.

8.6 REGENERATIVE CYCLE

According to Equation (8.8), the thermal efficiency of a Rankine cycle can be increased by increasing the average temperature at which heat is added to the boiler. In a regenerative cycle, the average temperature in the boiler is increased by preheating the water before it enters the boiler. Note that the water entering the boiler is often referred to as *feedwater*. To preheat the feedwater, some superheated steam is extracted from the turbine and mixed with the water stream from the condenser in a feedwater heater (mixing device) before pumping it to the boiler.

Figure 8.12 shows the equipment diagram for the regenerative cycle. The superheated steam leaving the boiler in state 5 is expanded in turbine I where some work is produced but the exhaust steam from turbine I is still a superheated steam given by state 6. A fraction of the superheated steam leaving turbine I in state 6 is diverted to a feedwater heater. The rest of the superheated steam leaving turbine I is expanded in turbine II to state 7 (wet steam consisting of water-vapor mixture). The wet steam in state 7 is fully condensed in a condenser to saturated liquid water in state 1. The water from the condenser is pumped through pump I (condensate pump) to a mixing device (feedwater heater) where superheated steam in state 6 extracted from turbine I and water in state 2 (subcooled liquid water) are mixed at constant pressure (discharge pressure of turbine I). The direct mixing of subcooled water and superheated steam results in a saturated liquid water. This saturated liquid water is state 3 is pumped through pump II (feedwater pump) to the boiler in state 4 (subcooled liquid water at boiler pressure).

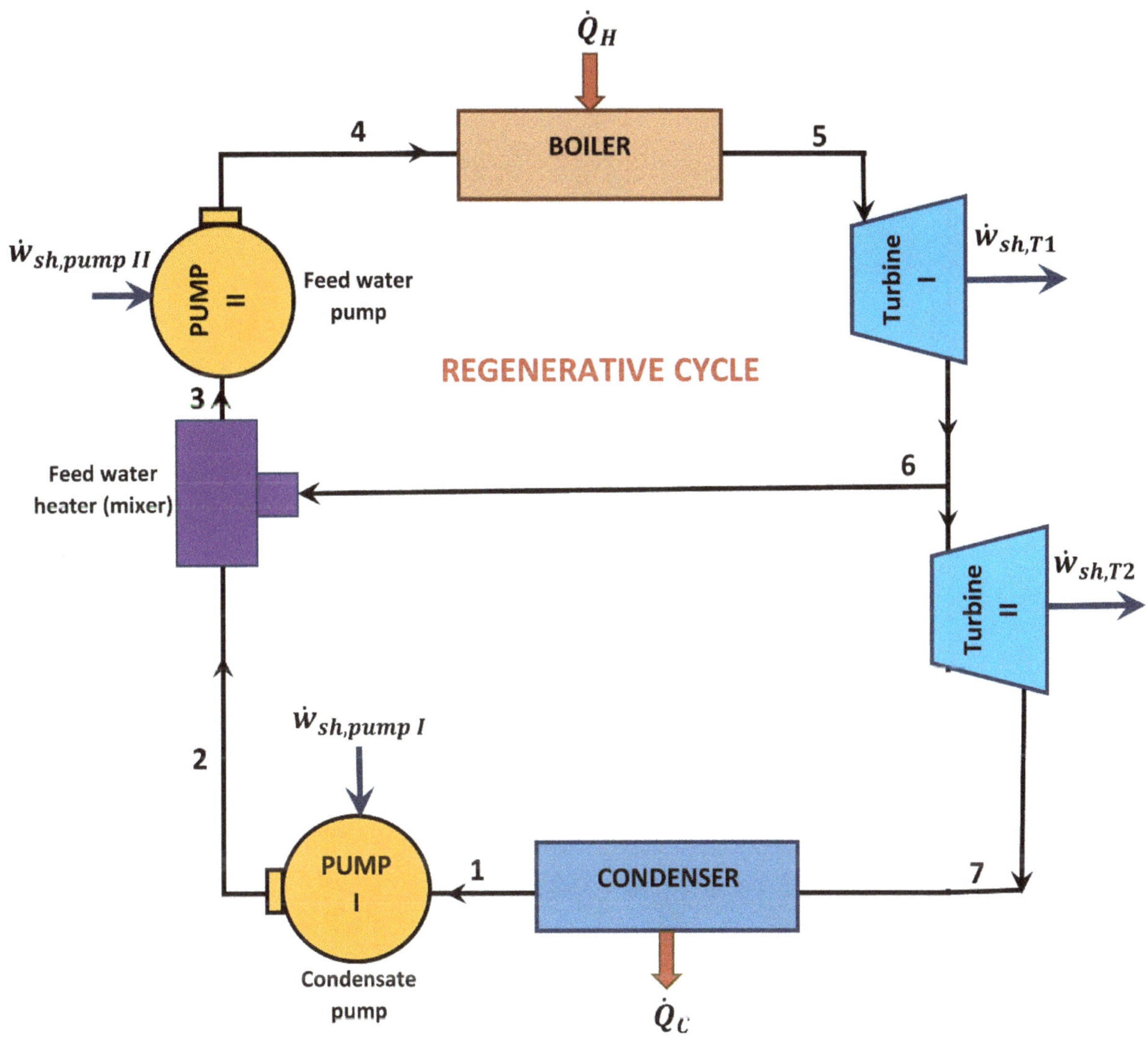

Figure 8.12 Equipment diagram of the regenerative cycle.

Figure 8.13 shows the regenerative cycle on a T-s diagram. Heat is added to the boiler along step $4 \rightarrow 5$. The superheated steam in state 5 is expanded along step $5 \rightarrow 6$ in turbine I. Part of the superheated steam discharged from turbine I is diverted to feedwater heater. The remaining steam is expanded along step $6 \rightarrow 7$ in turbine II. The discharge steam from turbine II in state 7 is fully condensed to saturated liquid water along step $7 \rightarrow 1$. The saturated liquid water in state 1 is pumped through a condensate pump to subcooled state 2. The subcooled liquid water in state 2 is mixed with superheated steam extracted from turbine I in state 6 in a mixing device at constant pressure to produce saturated liquid water in state 3. The saturated liquid water emerging from a mixer in state 3 is pumped through a feedwater pump to the boiler in subcooled liquid state 4.

It is important to realize that if there was no feedwater heating and the saturated liquid water from the condenser in state 1 was pumped directly to the boiler in state 2, the average temperature over which heat would be added in the boiler will be lower. Comparing the average temperature over step $4 \rightarrow 5$ (with feedwater heater) with the average temperature over step $2' \rightarrow 5$ (no feedwater heater), the average temperature over step $4 \rightarrow 5$ is significantly higher. Thus, the thermal efficiency of the cycle is improved due to feedwater heating.

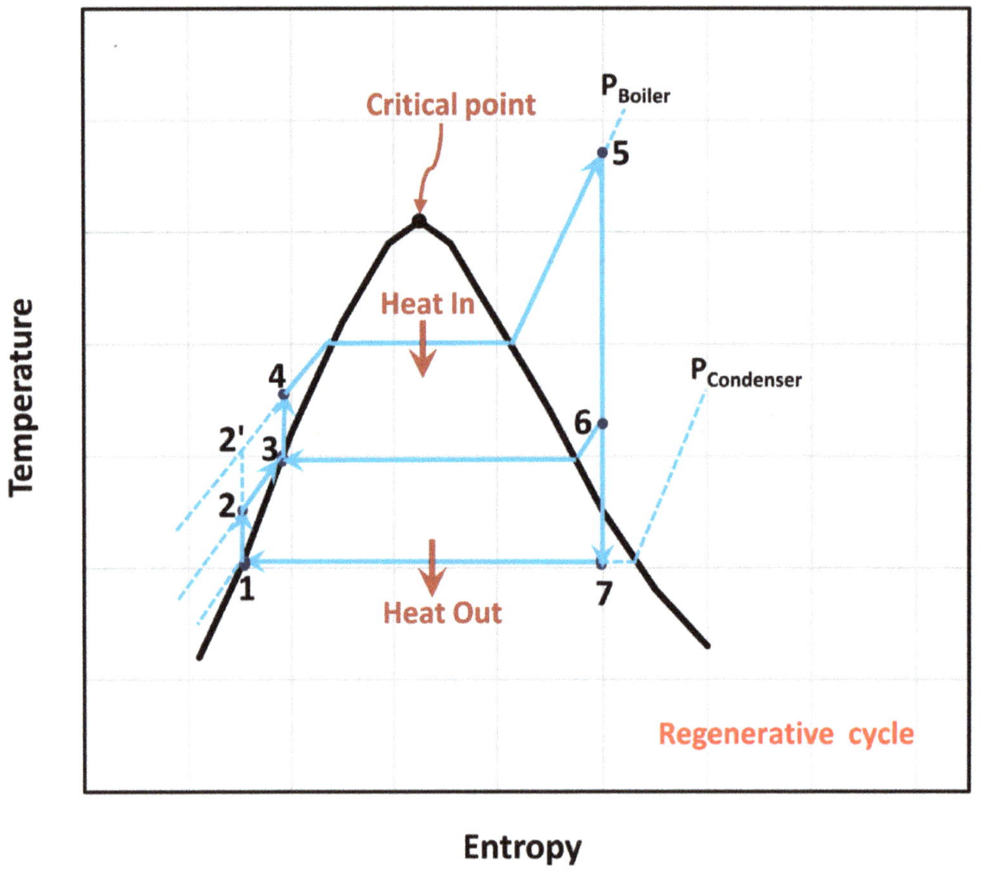

Figure 8.13 The Regenerative cycle.

8.7 COMBINED REHEAT/REGENERATIVE CYCLE

In the regenerative cycle (Figure 8.13), the moisture content (water droplets) of the exhaust steam from the second turbine (state 7, Figure 8.13) may be large enough (>10%) to cause erosion problems in the turbine. To avoid the moisture problem, it is not uncommon to combine the reheat and regenerative cycles.

Figure 8.14 shows the combined reheat/regenerative cycle. Now three turbines (or a three-stage turbine) are used to expand the superheated steam generated in the boiler. Superheated steam in state 5 is expanded in turbine I along isentropic step 5 → 6. The superheated exhaust steam in state 6 is reheated at constant pressure along step 6 → 7 by passing it through the boiler. The superheated steam in state 7 is expanded in turbine II along isentropic step 7 → 8. A portion of the steam discharged from turbine II in state 8 is diverted to a feedwater heater. The remaining steam from turbine II is expanded from state 8 to state 9 in turbine III. The exhaust steam from turbine III in state 9 is of high quality with moisture content well below 10%. The remaining portion of the cycle is the same as that of the regenerative cycle (Figure 8.13).

Figure 8.15 shows the equipment diagram for the combined reheat/regenerative cycle. It reflects all the steps shown in Figure 8.14.

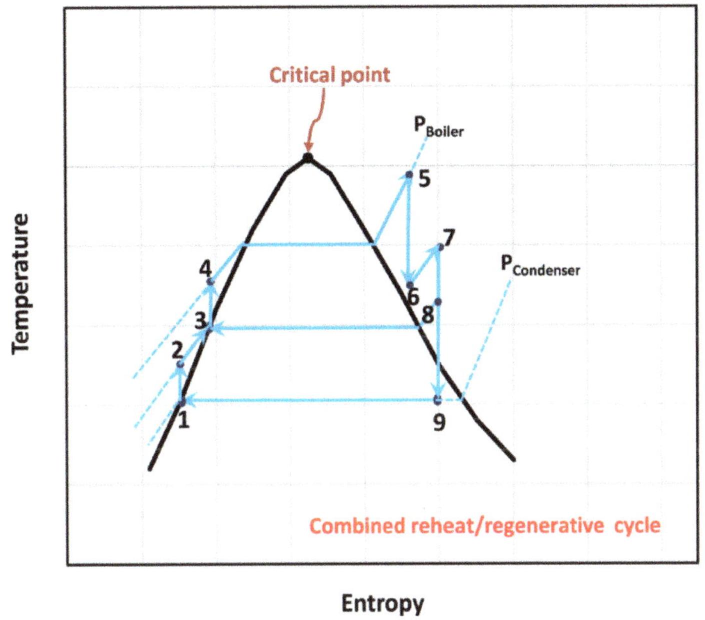

Figure 8.14 The combined reheat/regenerative cycle.

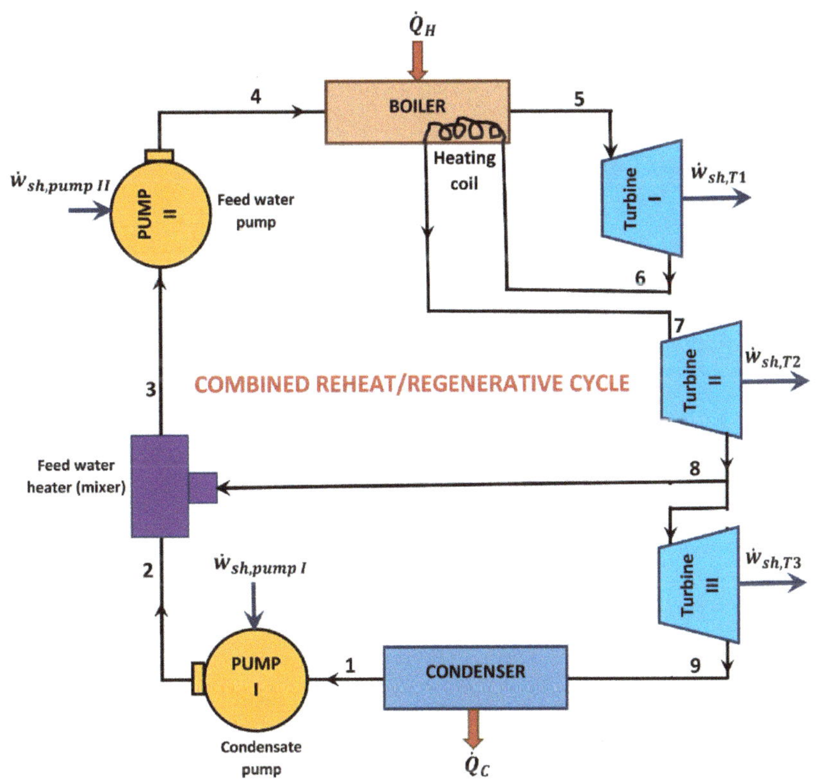

Figure 8.15 Equipment diagram of the combined reheat/regenerative cycle.

8.8 SUPERCRITICAL RANKINE CYCLE

As already pointed out, the thermal efficiency of a Rankine cycle can be increased by increasing the average temperature at which heat is added in the boiler. One way to increase the average temperature of heat addition is to operate the boiler at pressures higher than the critical pressure. The critical properties of water are: $T_c = 374.14\ ^oC$ and $P_c = 22.09\ MPa$.

Figure 8.16 shows the supercritical Rankine cycle where the boiler is operated at a pressure significantly above the critical pressure of 22.09 MPa. The heat is added to the boiler along step 2 → 3. Now no phase change (vaporization) occurs during the heat addition step.

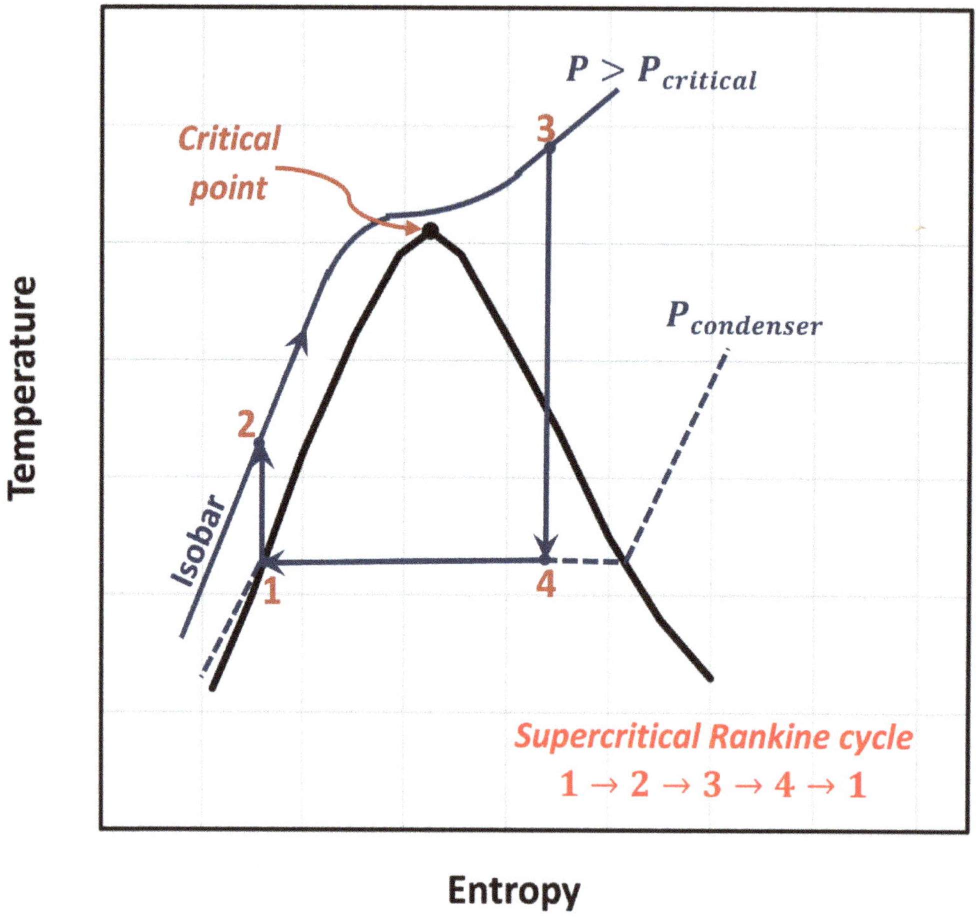

Figure 8.16 The supercritical Rankine cycle.

As the quality of the exhaust steam from the turbine in state 4 may not be high enough to avoid moisture/erosion problems in the turbine, the supercritical reheat cycle, shown in Figure 8.17, is preferred.

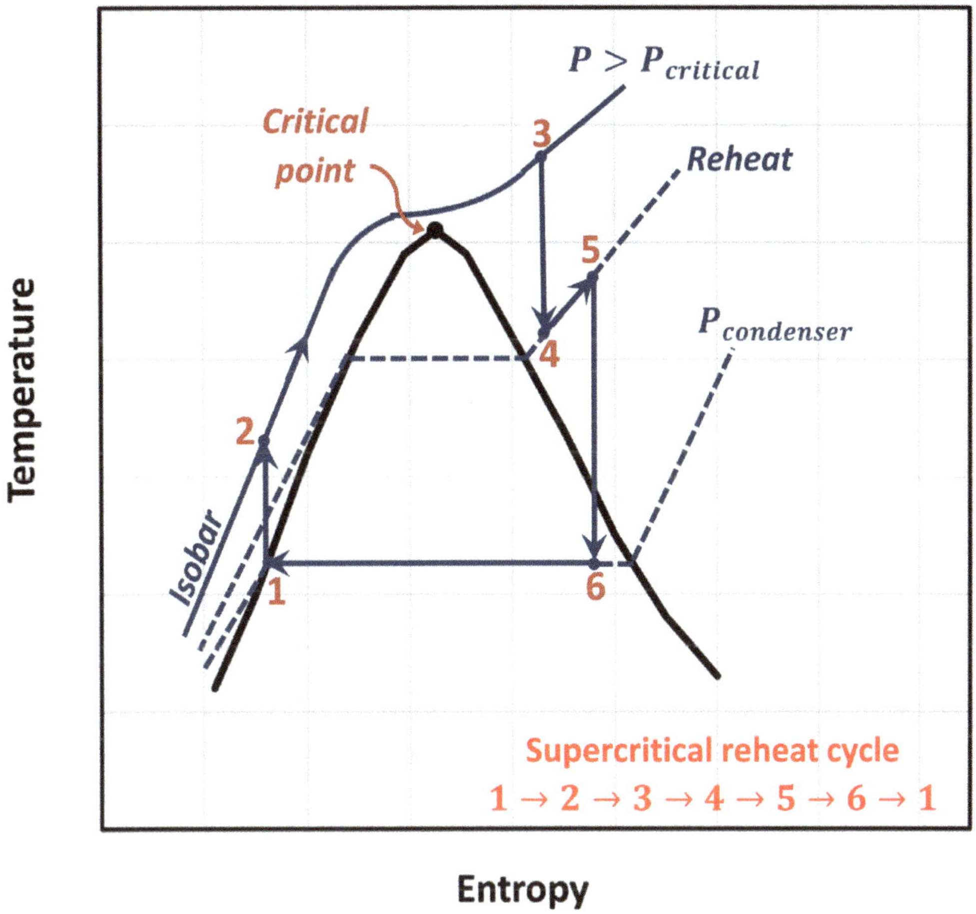

Figure 8.17 The supercritical reheat cycle.

One major disadvantage of a supercritical Rankine cycle is that it requires heavier construction and more expensive materials of construction to withstand high stresses at elevated temperatures and pressures. Thus, the savings associated with the increase in thermal efficiency of the plant must justify the increased costs of the equipment.

NOTATION

h	specific enthalpy
h^L	specific enthalpy of liquid
h^v	specific enthalpy of vapor
$\dot{m}$	mass flow rate
P	pressure
P_c	critical pressure
$P_{Condenser}$	pressure in the condenser
P_{Boiler}	pressure in the boiler
P^{sat}	saturation pressure or vapor pressure
Q	heat transfer per unit mass of fluid
Q_{cycle}	net heat transfer per unit mass of fluid in a cyclic process
$\dot{Q}_C$	heat transfer rate in the condenser

$\dot{Q}_H$	heat transfer rate in the boiler
s	specific entropy
s^L	specific entropy of liquid
s^v	specific entropy of vapor
T	temperature
T_c	critical temperature
T_H	temperature of high temperature heat reservoir
T_L	temperature of low temperature heat reservoir
$\bar{T}$	average temperature
v	specific volume
v^v	specific volume of vapor
v^L	specific volume of liquid
W_{cycle}	net work transfer per unit mass of fluid in a cyclic process
$\dot{W}_{sh}$	rate of shaft work
x^v	mass fraction of vapor in a two-phase liquid-vapor mixture

Greek Letters

η	thermal efficiency of a cyclic process

TUTORIAL PROBLEMS WITH SOLUTION

Problem 1

Consider a steam power plant operating on a Carnot cycle. Saturated liquid water enters the boiler at a pressure of 6 MPa (60 bar). The saturated steam leaves the boiler at 6 MPa and enters the turbine. The wet steam is discharged from the turbine at 0.006 MPa (0.06 bar). Determine:

a) the boiler temperature
b) the condenser temperature
c) the quality of the exhaust steam from the turbine
d) the quality of the steam leaving the condenser
e) the heat transfer in the boiler per unit mass of fluid
f) the heat transfer in the condenser per unit mass of fluid
g) the thermal efficiency of the cycle

Solution

Refer to figure given below.

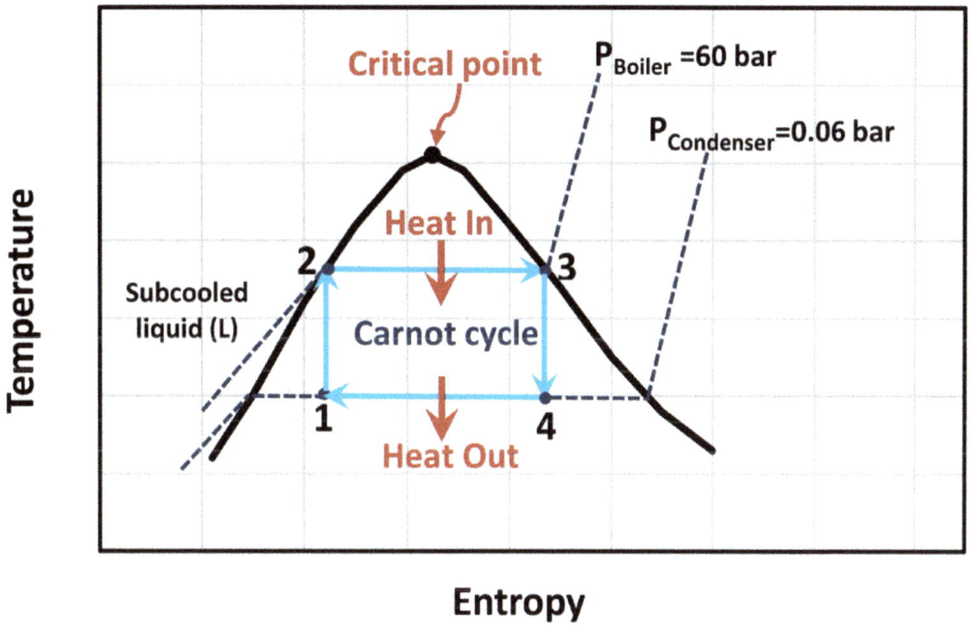

From saturated steam tables (Appendix M):

P	T^{sat}	$h^L(kJ/kg)$	$h^v(kJ/kg)$	$s^L(kJ/kg.K)$	$s^v(kJ/kg.K)$
6 MPa	275.6 °C	1213.3	2784.3	3.0273	5.8900
0.006 MPa	36.2 °C	151.5	2567.4	0.5208	8.3312

a) Boiler temperature is 275.6 °C.

b) Condenser temperature is 36.2 °C.

c) Quality at point 4:

$$s_3^v = 5.890 = s_4 = x_4^v s_4^v + (1 - x_4^v) s_4^L = x_4^v(8.3312) + (1 - x_4^v) \times 0.5208$$

$$x_4^v = (5.89 - 0.5208)/(8.3312 - 0.5208) = 0.6874$$

d) Quality at point 1:

$$s_2^L = 3.0273 = s_1 = x_1^v s_1^v + (1 - x_1^v) s_1^L = x_1^v(8.3312) + (1 - x_1^v) \times 0.5208$$

$$x_1^v = (3.0273 - 0.5208)/(8.3312 - 0.5208) = 0.3209$$

e) Heat transfer in boiler

$$Q_{23} = h_3 - h_2 = 2784.3 - 1213.3 = 1571 \text{ kJ/kg}$$

f) Heat transfer in condenser

$$Q_{41} = h_1 - h_4$$

$$h_1 = x_1^v h_1^v + (1 - x_1^v) h_1^L = 0.3209 \times 2567.4 + (1 - 0.3209) \times 151.5 = 926.762 \text{ kJ/kg}$$

$$h_4 = x_4^v h_4^v + (1 - x_4^v) h_4^L = 0.6874 \times 2567.4 + (1 - 0.6874) \times 151.5 = 1812.19 \text{ kJ/kg}$$

$$Q_{41} = h_1 - h_4 = 926.762 - 1812.19 = -885.428 \text{ kJ/kg}$$

g) Thermal efficiency of cycle

$$\eta = \frac{W_{cycle}}{Q_{Boiler}} = \frac{Q_{cycle}}{Q_{Boiler}} = \frac{Q_{Boiler} + Q_{condenser}}{Q_{Boiler}} = \frac{1571 + (-885.428)}{1571} = 0.4364$$

Let us now verify the result of part (g) using the Carnot formula for thermal efficiency as:

$$\eta = 1 - \frac{T_L}{T_H} = 1 - \frac{36.2 + 273.15}{275.6 + 273.15} = 0.4363$$

Problem 2

Consider a steam power plant operating on an ideal Rankine cycle. The boiler pressure is 4 MPa (40 bar) and the condenser pressure is 0.008 MPa (8 kPa). The temperature of the superheated steam leaving the boiler and entering the turbine is 450 °C. Estimate the thermal efficiency of the cycle.

Solution

Refer to figure given below.

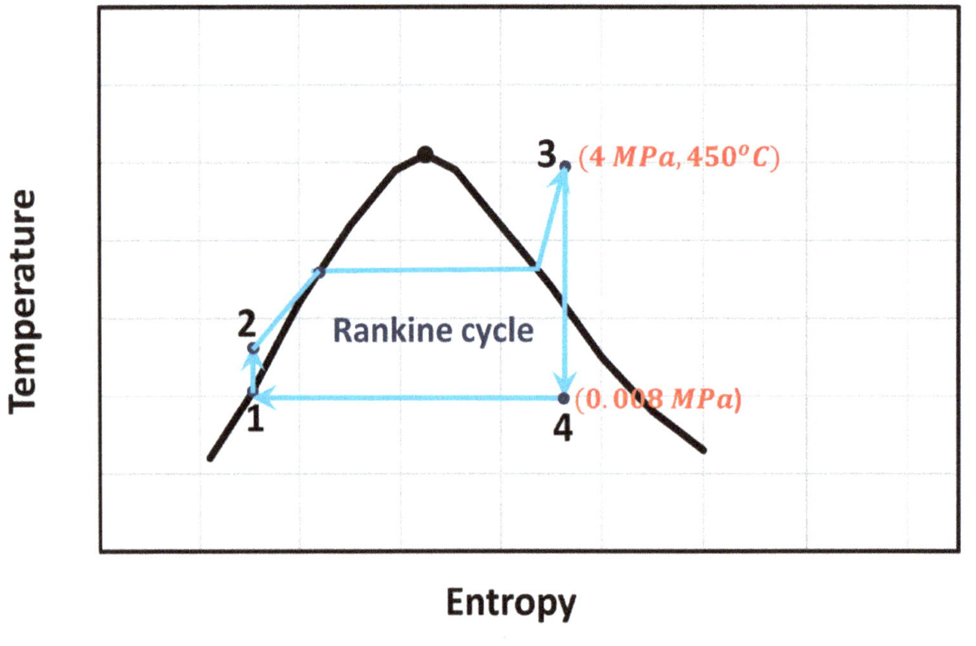

Superheated steam in state 3 (4 MPa, 450 °C):

From superheated steam tables (Appendix N),

$h_3 = 3330.3$ kJ/kg ; $s_3 = 6.9363$ kJ/kg.K

Wet steam in state 4 (0.008 MPa):

From saturated steam tables (Appendix M),

$v^L(m^3/kg)$	$v^v(m^3/kg)$	$h^L(kJ/kg)$	$h^v(kJ/kg)$	$s^L(kJ/kg.K)$	$s^v(kJ/kg.K)$
0.001008	18.10	173.9	2577	0.5924	8.2295

The quality of steam in state 4:

$$s_4 = s_3 = 6.9363 = x_4^v s_4^v + (1 - x_4^v)s_4^L = x_4^v(8.2295) + (1 - x_4^v)(0.5924)$$

$$x_4^v = \frac{6.9363 - 0.5924}{8.2295 - 0.5924} = 0.8307$$

$$h_4 = x_4^v h_4^v + (1 - x_4^v)h_4^L = 0.8307 \times 2577 + (1 - 0.8307) \times 173.9 = 2170.16 \text{ kJ/kg}$$

$$h_1 = 173.9 \text{ kJ/kg}$$

Pump Work:

$$W_{sh,pump} = -\frac{(P_2 - P_1)}{\rho} = -v(P_2 - P_1) = -0.001008(4 - 0.008) \times 10^3 = -4.024 \text{ kJ/kg}$$

Note that we have assumed that the fluid (liquid water) is incompressible, and $v = v_1$.

Since,

$$W_{sh,pump} = -\Delta h = h_1 - h_2 = -4.024 \text{ kJ/kg}$$

$$h_1 = 173.9 \text{ kJ/kg}$$

$$h_2 = 173.9 + 4.024 = 177.924 \text{ kJ/kg}$$

Turbine Work:

$$W_{sh,turbine} = -\Delta h = h_3 - h_4 = 3330.3 - 2170.16 = 1160.14 \text{ kJ/kg}$$

Heat Transfer in Boiler:

$$Q_{boiler} = \Delta h = h_3 - h_2 = 3330.3 - 177.924 = 3152.376 \text{ kJ/kg}$$

Thermal Efficiency:

$$\eta = \frac{W_{sh,turbine} + W_{sh,pump}}{Q_{Boiler}} = \frac{1160.14 + (-4.024)}{3152.376} = 0.367$$

Thus, the thermal efficiency of the given Rankine cycle is 36.7%.

Problem 3

Re-do problem 2 by increasing the boiler pressure to 6 MPa while keeping the temperature of the super-heated steam and the condenser pressure the same as in problem 2. Does the thermal efficiency of the cycle increase by increasing the boiler pressure? If yes, what is the percent increase in thermal efficiency by increasing the boiler pressure from 4 MPa to 6 MPa?

Solution

Superheated steam in state 3 (6 MPa, 450 °C):

From superheated steam tables (Appendix N),

$h_3 = 3301.8$ J/kg ; $s_3 = 6.7193$ kJ/kg.K

Wet steam in state 4 (0.008 MPa):

From saturated steam tables (Appendix M),

$v^L(m^3/kg)$	$v^v(m^3/kg)$	$h^L(kJ/kg)$	$h^v(kJ/kg)$	$s^L(kJ/kg.K)$	$s^v(kJ/kg.K)$
0.001008	18.10	173.9	2577	0.5924	8.2295

The quality of steam in state 4:

$$s_4 = s_3 = 6.7193 = x_4^v s_4^v + (1 - x_4^v)s_4^L = x_4^v(8.2295) + (1 - x_4^v)(0.5924)$$

$$x_4^v = \frac{6.7193 - 0.5924}{8.2295 - 0.5924} = 0.8023$$

$$h_4 = x_4^v h_4^v + (1 - x_4^v)h_4^L = 0.8023 \times 2577 + (1 - 0.8023) \times 173.9 = 2101.91 \text{ kJ/kg}$$

$$h_1 = 173.9 \text{ kJ/kg}$$

Pump Work:

$$W_{sh,pump} = -\frac{(P_2 - P_1)}{\rho} = -v(P_2 - P_1) = -0.001008(6 - 0.008) \times 10^3 = -6.04 \text{ kJ/kg}$$

Note that we have assumed that the fluid (liquid water) is incompressible, and $v = v_1$.

Since,

$$W_{sh,pump} = -\Delta h = h_1 - h_2 = -6.04 \text{ kJ/kg}$$

$$h_1 = 173.9 \text{ kJ/kg}$$

$$h_2 = 173.9 + 6.04 = 179.94 \text{ kJ/kg}$$

Turbine Work:

$$W_{sh,turbine} = -\Delta h = h_3 - h_4 = 3301.8 - 2101.91 = 1199.89 \text{ kJ/kg}$$

Heat Transfer in Boiler:

$$Q_{boiler} = \Delta h = h_3 - h_2 = 3301.8 - 179.94 = 3121.86 \text{ kJ/kg}$$

Thermal Efficiency:

$$\eta = \frac{W_{sh,turbine} + W_{sh,pump}}{Q_{Boiler}} = \frac{1199.89 + (-6.04)}{3121.86} = 0.382$$

Thus, the thermal efficiency of the Rankine cycle is increased from 36.7% to 38.2% by increasing the boiler pressure from 4 MPa to 6 MPa. The percent increase in efficiency is 4.1. However, the disadvantage of increasing the boiler pressure is that the quality of the exhaust steam from the turbine decreases from 0.8307 to 0.8023.

Problem 4

In problem 2, if we increase the temperature of the superheated steam from 450 °C to 550 °C while keeping the boiler pressure at 4 MPa and the condenser pressure at 0.008 MPa, does the thermal efficiency of the cycle increase? If yes, by what percentage?

Solution

Superheated steam in state 3 (4 MPa, 550 °C):

From superheated steam tables (Appendix N), at 4 MPa:

Temperature	$h(kJ/kg)$	$s(kJ/kg.K)$
500 °C	3445.3	7.0901
600 °C	3674.4	7.3688

By linear interpolation, the enthalpy and entropy at 4 MPa, 550 °C are:

$$h_3 = 3445.3 + \left(\frac{3674.4-3445.3}{600-500}\right) \times (550 - 500) = 3559.85 \text{ kJ/kg}$$

$$s_3 = 7.0901 + \left(\frac{7.3688-7.0901}{600-500}\right) \times (550 - 500) = 7.22945 \text{ kJ/kg.K}$$

Wet steam in state 4 (0.008 MPa):

From saturated steam tables (Appendix M),

$v^L(m^3/kg)$	$v^v(m^3/kg)$	$h^L(kJ/kg)$	$h^v(kJ/kg)$	$s^L(kJ/kg.K)$	$s^v(kJ/kg.K)$
0.001008	18.10	173.9	2577	0.5924	8.2295

The quality of steam in state 4:

$$s_4 = s_3 = 7.22945 = x_4^v s_4^v + (1 - x_4^v)s_4^L = x_4^v(8.2295) + (1 - x_4^v)(0.5924)$$

$$x_4^v = \frac{7.22945 - 0.5924}{8.2295 - 0.5924} = 0.8691$$

$$h_4 = x_4^v h_4^v + (1 - x_4^v)h_4^L = 0.8691 \times 2577 + (1 - 0.8691) \times 173.9 = 2262.43 \text{ kJ/kg}$$

$$h_1 = 173.9 \text{ kJ/kg}$$

Pump Work:

$$W_{sh,pump} = -\frac{(P_2-P_1)}{\rho} = -v(P_2 - P_1) = -0.001008(4 - 0.008) \times 10^3 = -4.024 \text{ kJ/kg}$$

Note that we have assumed that the fluid (liquid water) is incompressible, and $v = v_1$.

Since,

$$W_{sh,pump} = -\Delta h = h_1 - h_2 = -4.024 \text{ kJ/kg}$$

$$h_1 = 173.9 \text{ kJ/kg}$$

$$h_2 = 173.9 + 4.024 = 177.924 \text{ kJ/kg}$$

Turbine Work:

$$W_{sh,turbine} = -\Delta h = h_3 - h_4 = 3559.85 - 2262.43 = 1297.42 \text{ kJ/kg}$$

Heat Transfer in Boiler:

$$Q_{boiler} = \Delta h = h_3 - h_2 = 3559.85 - 177.924 = 3381.926 \text{ kJ/kg}$$

Thermal Efficiency:

$$\eta = \frac{W_{sh,turbine} + W_{sh,pump}}{Q_{Boiler}} = \frac{1297.42 + (-4.024)}{3381.926} = 0.382$$

Thus, the thermal efficiency of the Rankine cycle is increased from 36.7% to 38.2% by increasing the temperature of the superheated steam from 450 °C to 550 °C while keeping the boiler pressure at 4 MPa. The percent increase in efficiency is 4.1. The additional advantage of increasing the temperature of the superheated steam is that the quality of the exhaust steam from the turbine increases from 0.8307 to 0.8691.

Problem 5

In problem 2, if we now decrease the condenser pressure to 0.003 MPa while keeping the boiler pressure at 4 MPa and the temperature of the superheated steam at 450 °C, will it increase the thermal efficiency? If yes, by what percentage?

Solution

Superheated steam in state 3 (4 MPa, 450 °C):

From superheated steam tables (Appendix N),

$h_3 = 3330.3$ kJ/kg; $s_3 = 6.9363$ kJ/kg.K

Wet steam in state 4 (0.003 MPa):

From saturated steam tables (Appendix M),

$v^L(m^3/kg)$	$h^L(kJ/kg)$	$h^v(kJ/kg)$	$s^L(kJ/kg.K)$	$s^v(kJ/kg.K)$
0.001003	101	2545.5	0.3544	8.5784

The quality of steam in state 4:

$$s_4 = s_3 = 6.9363 = x_4^v s_4^v + (1 - x_4^v) s_4^L = x_4^v(8.5784) + (1 - x_4^v)(0.3544)$$

$$x_4^v = \frac{6.9363 - 0.3544}{8.5784 - 0.3544} = 0.8003$$

$$h_4 = x_4^v h_4^v + (1 - x_4^v) h_4^L = 0.8003 \times 2545.5 + (1 - 0.8003) \times 101 = 2057.33 \text{ kJ/kg}$$

$$h_1 = 101 \text{ kJ/kg}$$

Pump Work:

$$W_{sh,pump} = -\frac{(P_2 - P_1)}{\rho} = -v(P_2 - P_1) = -0.001003(4 - 0.003) \times 10^3 = -4.009 \text{ kJ/kg}$$

Note that we have assumed that the fluid (liquid water) is incompressible, and $v = v_1$.

Since,

$$W_{sh,pump} = -\Delta h = h_1 - h_2 = -4.009 \text{ kJ/kg}$$

$$h_1 = 101 \text{ kJ/kg}$$

$$h_2 = 101 + 4.009 = 105.009 \text{ kJ/kg}$$

Turbine Work:

$$W_{sh,turbine} = -\Delta h = h_3 - h_4 = 3330.3 - 2057.33 = 1272.97 \text{ kJ/kg}$$

Heat Transfer in Boiler:

$$Q_{boiler} = \Delta h = h_3 - h_2 = 3330.3 - 105.009 = 3225.291 \text{ kJ/kg}$$

Thermal Efficiency:

$$\eta = \frac{W_{sh,turbine} + W_{sh,pump}}{Q_{Boiler}} = \frac{1272.97 + (-4.009)}{3225.291} = 0.393$$

Thus, the thermal efficiency of the Rankine cycle is increased from 36.7% to 39.3% by decreasing the condenser pressure 0.008 MPa to 0.003 MPa while keeping the boiler pressure at 4 MPa and superheated steam temperature of 450 °C. The percent increase in efficiency is 7.1. However, the quality of the exhaust steam from the turbine decreases from 0.8307 to 0.8003.

Problem 6

Consider a steam power plant operating on a practical Rankine cycle. The boiler pressure is 12.5 MPa (125 bar) and the condenser pressure is 0.008 MPa (8 kPa). The temperature of the superheated steam leaving the boiler and entering the turbine is 550 °C. The efficiency of the turbine is 0.76 and the pump efficiency is 0.72. The plant is producing power at a rate of 73,000 kW.

Calculate:

a) The mass flow rate of fluid
b) The rate of heat addition in the boiler
c) The rate of heat rejection in the condenser
d) The thermal efficiency of the plant

Solution

Refer to the figure below

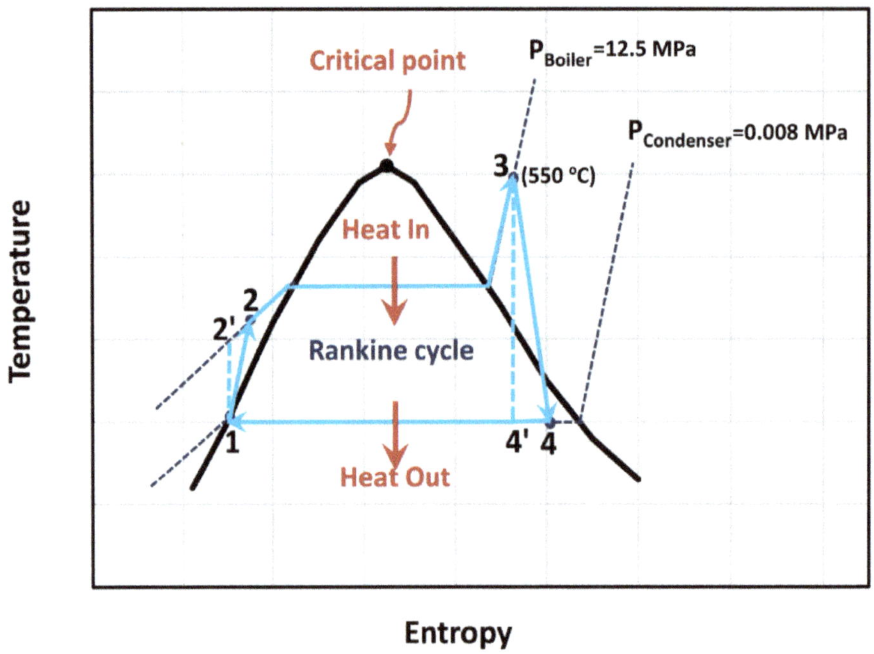

Superheated steam in state 3 (12.5 MPa, 550 °C):

From superheated steam tables (Appendix N),

$h_3 = 3475.2$ kJ/kg; $s_3 = 6.6290$ kJ/kg.K

Wet steam at 0.008 MPa:

From saturated steam tables (Appendix M),

$v^L(m^3/kg)$	$h^L(kJ/kg)$	$h^v(kJ/kg)$	$s^L(kJ/kg.K)$	$s^v(kJ/kg.K)$
0.001008	173.9	2577	0.5924	8.2295

The quality of steam in state 4′:

$$s_{4'} = s_3 = 6.6290 = x^v_{4'}s^v_{4'} + (1 - x^v_{4'})s^L_{4'} = x^v_{4'}(8.2295) + (1 - x^v_{4'})(0.5924)$$

$$x^v_{4'} = \frac{6.6290 - 0.5924}{8.2295 - 0.5924} = 0.7904$$

$$h_{4'} = x^v_{4'}h^v_{4'} + (1 - x^v_{4'})h^L_{4'} = 0.7904 \times 2577 + (1 - 0.7904) \times 173.9 = 2073.31 \text{ kJ/kg}$$

Turbine Work:

$$W_{sh,turbine} = -(\Delta h)_{actual} = -\varepsilon(\Delta h)_{isentropic} = -0.76(h_{4'} - h_3) = -0.76(2073.31 - 3475.2) = 1065.44 \text{ kJ/kg}$$

$$h_4 - h_3 = -1065.44\ ;\ h_4 = 3475.2 - 1065.44 = 2409.76 \text{ kJ/kg}$$

Pump Work:

$$W_{sh,pump} = -\frac{(P_2-P_1)}{\rho} = -v(P_2 - P_1) = -0.001008(12.5 - 0.008) \times 10^3 = -12.592 \text{ kJ/kg}$$

This is isentropic work. The actual work required is higher.

$$W_{sh,pump} = -(\Delta h)_{actual} = -(\Delta h)_{isentropic}/\varepsilon = -12.592/0.72 = -17.49 \text{ kJ/kg}$$

$$h_1 - h_2 = -17.49$$

$$h_1 = 173.9 \text{ kJ/kg}$$

$$h_2 = 173.9 + 17.49 = 191.39 \text{ kJ/kg}$$

Net Work:

The net work produced by the plant is:

$$W_{sh,net} = W_{sh,turbine} + W_{sh,pump} = 1065.44 - 17.49 = 1047.95 \text{ kJ/kg}$$

a) The mass flow rate of fluid:

$$Power\ ouput = 73000 = \dot{m} \times 1047.95;\ \dot{m} = 73000/1047.95 = 69.66 \text{ kg/s}$$

b) The rate of heat addition in the boiler:

$$Q_{boiler} = \Delta h = h_3 - h_2 = 3475.2 - 191.39 = 3283.81 \text{ kJ/kg}$$

$$\dot{Q}_{boiler} = 69.66 \times 3283.81 = 228.75 \times 10^3 \text{ kJ/s}$$

c) The rate of heat rejection in the condenser:

$$Q_{condenser} = \Delta h = h_1 - h_4 = 173.9 - 2409.76 = -2235.86 \text{ kJ/kg}$$

$$\dot{Q}_{condenser} = -69.66 \times 2235.86 = -155.75 \times 10^3 \text{ kJ/s}$$

d) The thermal efficiency of the plant

$$\eta = \frac{W_{sh,turbine} + W_{sh,pump}}{Q_{Boiler}} = \frac{1065.44 - 17.49}{3283.81} = 0.319$$

Problem 7

Consider a steam power plant operating on a reheat cycle (see figure below). Superheated steam enters the first turbine at 4 MPa (40 bar) and 600 °C. The steam discharged from the first turbine is reheated at a pressure of 0.5 MPa to 600 °C and then expanded in the second turbine to 4 kPa. Determine the thermal efficiency of the reheat cycle.

Solution

Refer to the figure below

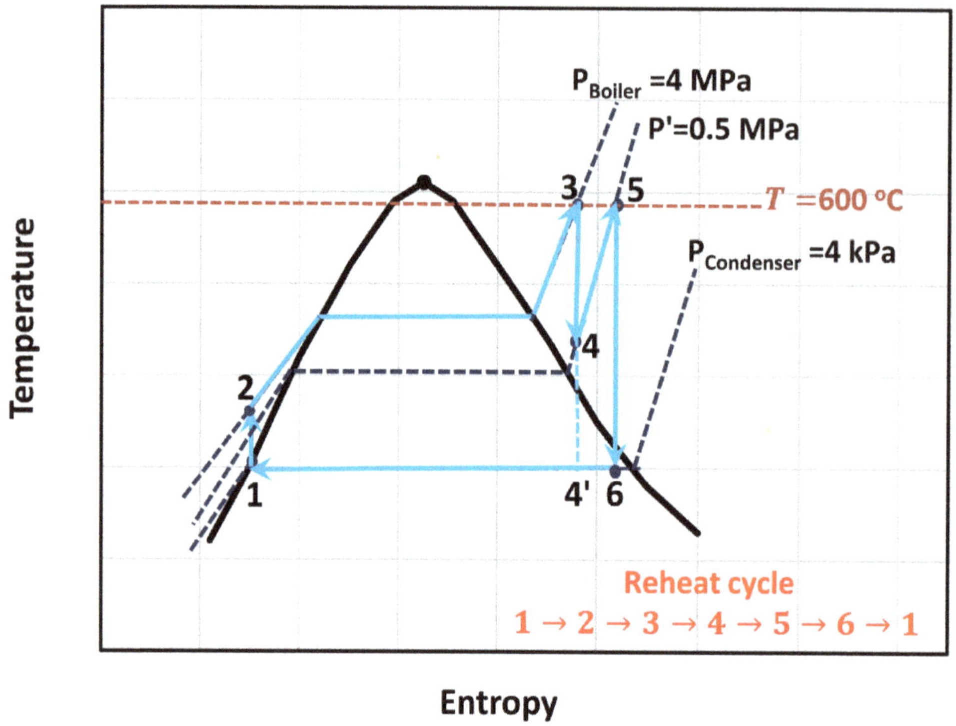

Superheated steam in state 3 (4 MPa, 600 °C):

From superheated steam tables (Appendix N),

$h_3 = 3674.4$ kJ/kg; $s_3 = 7.3688$ kJ/kg.K

Properties of steam at point 4 (0.5 MPa)?

$P_4 = 0.5\ MPa$ and $s_4 = s_3 = 7.3688$ kJ/kg.K . From superheated steam tables, at $P = 0.5\ MPa$

h (kJ/kg)	s (kJ/kg.K)
2960.7	7.2709
3064.2	7.4599

By linear interpolation:

$$h_4 = 2960.7 + \left(\frac{3064.2-2960.7}{7.4599-7.2709}\right) \times (7.3688 - 7.2709) = 3014.312 \text{ kJ/kg}$$

Properties of steam at point 5 (0.5 MPa, 600 °C)?

From superheated steam tables (Appendix N),

$h_5 = 3701.7$ kJ/kg; $s_5 = 8.3522$ kJ/kg.K

Properties of wet steam at point 6 (0.004 MPa)?

From saturated steam tables (Appendix M),

$v^L(m^3/kg)$	$h^L(kJ/kg)$	$h^v(kJ/kg)$	$s^L(kJ/kg.K)$	$s^v(kJ/kg.K)$
0.001004	121.4	2554.4	0.4225	8.4754

The quality of steam in state 6:

$$s_6 = s_5 = 8.3522 \ = x_6^v s_6^v + (1 - x_6^v) s_6^L = x_6^v(8.4754) + (1 - x_6^v)(0.4225)$$

$$x_4^v = \frac{8.3522 - 0.4225}{8.4754 - 0.4225} = 0.9847$$

$$h_6 = x_6^v h_6^v + (1 - x_6^v) h_6^L = 0.9847 \times 2554.4 + (1 - 0.9847) \times 121.4 = 2517.175 \text{ kJ/kg}$$

Pump Work:

$$W_{sh,pump} = -\frac{(P_2 - P_1)}{\rho} = -v(P_2 - P_1) = -0.001004(4 - 0.004) \times 10^3 = -4.012 \text{ kJ/kg}$$

$$h_1 - h_2 = -4.012$$

$$h_1 = 121.4 \text{ kJ/kg}$$

$$h_2 = 121.4 + 4.012 = 125.412 \text{ kJ/kg}$$

Turbine Work:

$$W_{sh,turbine,1} = -\Delta h = (h_3 - h_4) = 3674.4 - 3014.312 = 660.088 \text{ kJ/kg}$$

$$W_{sh,turbine,2} = -\Delta h = (h_5 - h_6) = 3701.7 - 2517.175 = 1184.525 \text{ kJ/kg}$$

Boiler heat input:

$$Q_{23} = \Delta h = h_3 - h_2 = 3674.4 - 125.412 = 3548.988 \text{ kJ/kg}$$

$$Q_{45} = \Delta h = h_5 - h_4 = 3701.7 - 3014.312 = 687.388 \text{ kJ/kg}$$

Thermal efficiency of the plant:

$$\eta = \frac{W_{sh,net}}{Q_{Boiler,total}} = \frac{660.088 + 1184.525 - 4.012}{3548.988 + 687.388} = 0.434$$

Problem 8

A steam power plant is operating on a regenerative cycle (see Figure 8.13). The boiler pressure is 4 MPa and the temperature of superheated steam leaving the boiler is 450°C. A portion of the steam discharged from the first turbine is diverted to the feed-water heater at a pressure of 400 kPa. The condenser pressure is 8 kPa. Determine the thermal efficiency of the regenerative cycle. Compare the efficiency of this cycle with the efficiency of problem 2 cycle. As a basis for calculations, you can take the mass flow rate of steam leaving the boiler equal to 1 kg/s.

Solution

Refer to the figures below.

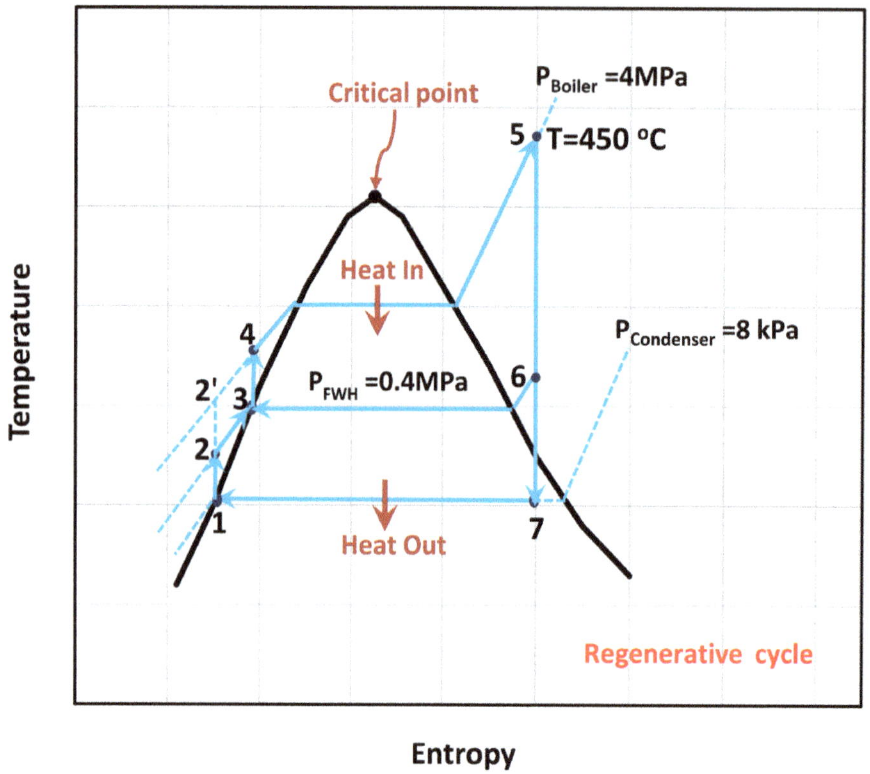
Critical point
PBoiler =4MPa
5
T=450 °C
Heat In
4
PCondenser =8 kPa
2'
3
PFWH =0.4MPa
6
2
1
Heat Out
7
Regenerative cycle
Temperature
Entropy

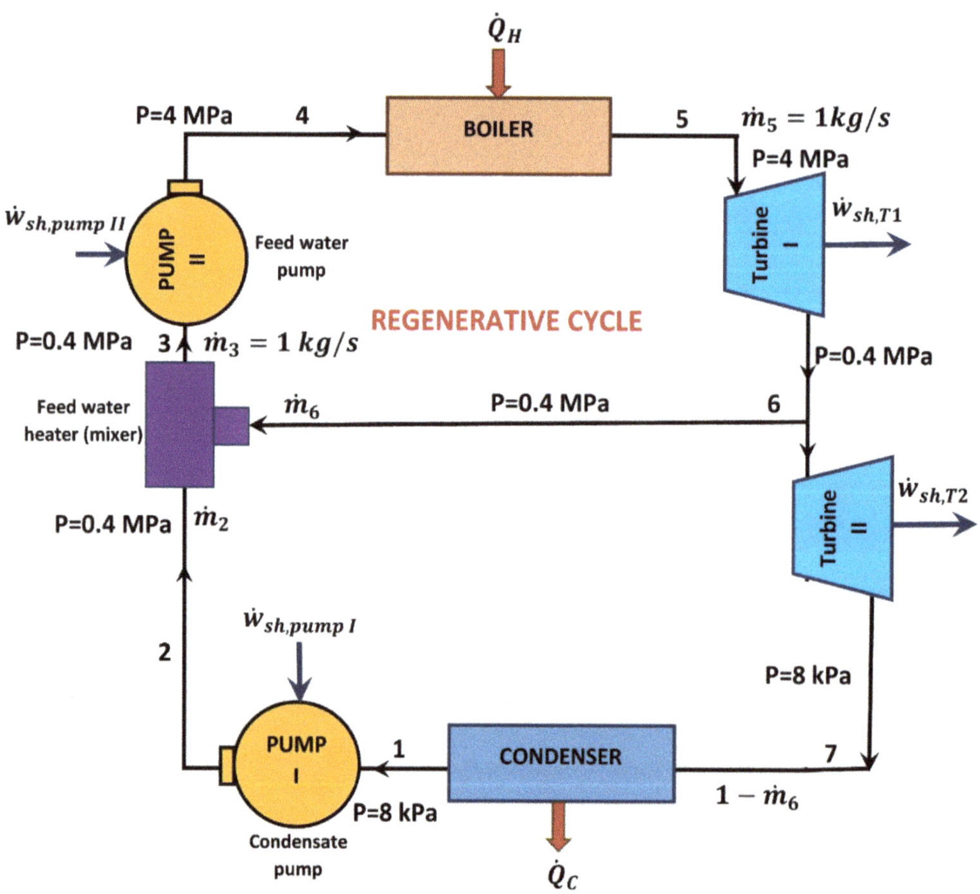
$\dot{Q}_H$
P=4 MPa
4
BOILER
5
$\dot{m}_5 = 1kg/s$
P=4 MPa
$\dot{W}_{sh,pump\ II}$
PUMP II
Feed water pump
Turbine I
$\dot{W}_{sh,T1}$
REGENERATIVE CYCLE
P=0.4 MPa
3
$\dot{m}_3 = 1\ kg/s$
P=0.4 MPa
Feed water heater (mixer)
$\dot{m}_6$
P=0.4 MPa
6
Turbine II
$\dot{W}_{sh,T2}$
P=0.4 MPa
$\dot{m}_2$
$\dot{W}_{sh,pump\ I}$
2
P=8 kPa
PUMP I
1
CONDENSER
7
P=8 kPa
$1 - \dot{m}_6$
Condensate pump
$\dot{Q}_C$

Superheated steam in state 5 (4 MPa, 450 °C):

From superheated steam tables (Appendix N),

$h_5 = 3330.3$ kJ/kg; $s_5 = 6.9363$ kJ/kg.K

Superheated steam in state 6 (0.4 MPa):

$s_6 = s_5 = 6.9363$ kJ/kg.K , $P = 0.4$ MPa

From superheated steam tables (Appendix N), at $P = 0.4\ MPa$

h (kJ/kg)	s (kJ/kg.K)
2752.8	6.9299
2860.5	7.1706

By linear interpolation:

$$h_6 = 2752.8 + \left(\frac{2860.5-2752.8}{7.1706-6.9299}\right) \times (6.9363 - 6.9299) = 2755.66 \text{ kJ/kg}$$

Properties of wet steam at point 7 (0.008 MPa)?

From saturated steam tables (Appendix M),

$v^L(m^3/kg)$	$v^v(m^3/kg)$	$h^L(kJ/kg)$	$h^v(kJ/kg)$	$s^L(kJ/kg.K)$	$s^v(kJ/kg.K)$
0.001008	18.10	173.9	2577	0.5924	8.2295

The quality of steam in state 7:

$$s_7 = s_6 = s_5 = 6.9363 = x_7^v s_7^v + (1 - x_7^v)s_7^L = x_7^v(8.2295) + (1 - x_7^v)(0.5924)$$

$$x_7^v = \frac{6.9363 - 0.5924}{8.2295 - 0.5924} = 0.8307$$

$$h_7 = x_7^v h_7^v + (1 - x_7^v)h_7^L = 0.8307 \times 2577 + (1 - 0.8307) \times 173.9 = 2170.16 \text{ kJ/kg}$$

Pump Work:

$$W_{sh,pump,1} = -\frac{(P_2-P_1)}{\rho} = -v(P_2 - P_1) = -0.001008(400 - 8) = -0.3951 \text{ kJ/kg}$$

$$h_1 - h_2 = -0.3951$$

$$h_1 = 173.9 \text{ kJ/kg}$$

$$h_2 = 173.9 + 0.3951 = 174.295 \text{ kJ/kg}$$

$$W_{sh,pump,2} = -\frac{(P_4-P_3)}{\rho} = -v(P_4 - P_3) = -0.001084(4 - 0.4) \times 10^3 = -3.9024 \text{ kJ/kg}$$

$$h_3 - h_4 = -3.9024$$

From saturated steam tables, at P=0.4 MPa, saturated liquid enthalpy is $h_3 = 604.7$ kJ/kg. Also note that $v = 0.001084$ m^3/kg for saturated liquid. Thus,

$$h_4 = 604.7 + 3.9024 = 608.6024 \text{ kJ/kg}$$

Mass and energy balances around feed water heater:

The mass and energy balances around the feed water heater are needed to calculate the mass flow rate $\dot{m}_6$.

Mass balance gives: $$\dot{m}_2 + \dot{m}_6 = \dot{m}_3 = 1$$

Energy balance gives: $$\dot{m}_3 h_3 = \dot{m}_2 h_2 + \dot{m}_6 h_6$$

From these equations, we get:

$$\dot{m}_6 = \frac{h_3 - h_2}{h_6 - h_2} = \frac{604.7 - 174.295}{2755.66 - 174.295} = 0.1667 \text{ kg/s}; \; \dot{m}_2 = 1 - \dot{m}_6 = 0.8333 \text{ kg/s}$$

Net work produced by the cyclic process:

$$\dot{W}_{\text{sh,pump,1}} = -\dot{m}_2 \times 0.3951 = -0.8333 \times 0.3951 = -0.3292 \ \text{kW}$$

$$\dot{W}_{\text{sh,pump,2}} = -\dot{m}_3 \times 3.9024 = -1 \times 3.9024 = -3.9024 \ \text{kW}$$

$$\dot{W}_{\text{sh,turbine,1}} = -\dot{m}_5 \times (h_6 - h_5) = -1 \times (2755.66 - 3330.3) = 574.64 \ \text{kW}$$

$$\dot{W}_{\text{sh,turbine,2}} = -(1 - \dot{m}_6) \times (h_7 - h_6) = -0.8333 \times (2170.16 - 2755.66) = 487.897 \text{ kW}$$

$$\text{Net work produced} = (-0.3292) + (-3.9024) + 574.64 + 487.897 = 1058.3 \ \text{kW}$$

Heat input to the boiler:

$$\dot{Q}_{boiler} = \dot{m}_5 (h_5 - h_4) = 1 \times (3330.3 - 608.6024) = 2721.698 \ \text{kW}$$

Thermal efficiency of the cyclic process:

$$\eta = \frac{W_{sh,net}}{Q_{Boiler}} = \frac{1058.3}{2721.698} = 0.389$$

Comparison of efficiency of this regenerative cycle with the efficiency of problem 2 cycle:

The regenerative cycle has a higher efficiency (38.9%) as compared with the efficiency of problem 2 cycle (36.7%). The increase in efficiency is 6%. Note that the boiler and condenser pressures are the same in these cycles and that the temperature of the superheated steam leaving the boiler is also the same. Thus, feed water heating is advantageous as it increases the thermal efficiency of the cycle.

Problem 9

Consider a supercritical reheat/regeneration cycle based steam power plant. The boiler pressure is 35 MPa and the temperature of superheated steam leaving the boiler is 600°C. The boiler steam is expanded in a turbine (first turbine) to 4 MPa. A portion of the steam discharged from the first turbine at 4 MPa is diverted to the feed-water heater. The remaining portion of steam is reheated at a constant pressure of 4 MPa to 500 °C. This steam is expanded in the second turbine to a condenser pressure of 20 kPa. Determine the thermal efficiency of this supercritical reheat/regeneration cycle. As a basis for calculations, you can take the mass flow rate of steam leaving the boiler equal to 1 kg/s.

Solution

Refer to the figures below.

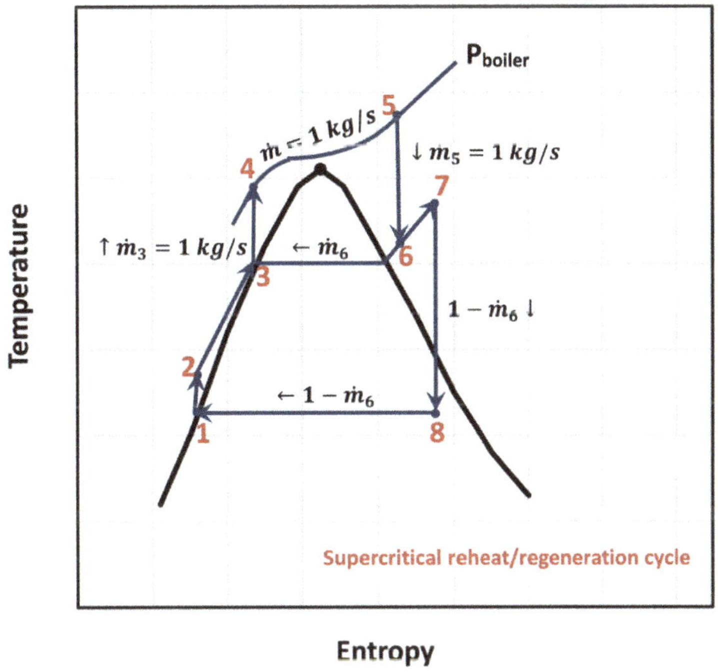

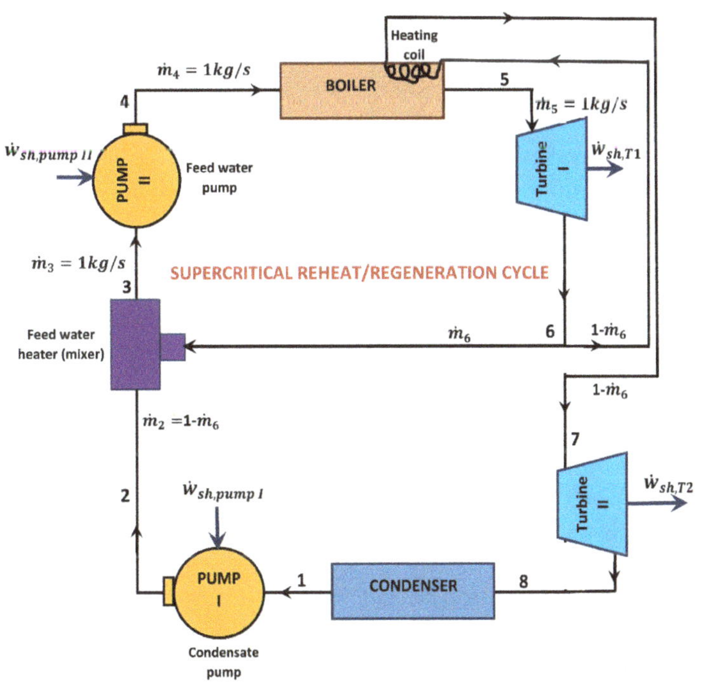

Superheated steam in state 5 (35 MPa, 600 °C):

From superheated steam tables (Appendix N),

$h_5 = 3395.5$ kJ/kg; $s_5 = 6.1179$ kJ/kg.K

Superheated steam in state 6 (4 MPa):

$s_6 = s_5 = 6.1179$ kJ/kg.K , $P = 4$ MPa

From superheated steam tables (Appendix N), at $P = 4\ MPa$

h (kJ/kg)	*s* (kJ/kg.K)
2801.4	6.0709
2886.2	6.2285

By linear interpolation:

$$h_6 = 2801.4 + \left(\frac{2886.2-2802.4}{6.2285-6.0709}\right) \times (6.1179 - 6.0709) = 2826.69 \text{ kJ/kg}$$

Superheated steam in state 7 (4 MPa, 500°C):

From superheated steam tables (Appendix N),

$h_7 = 3445.3$ kJ/kg; $s_7 = 7.0901$ kJ/kg.K

Properties of wet steam at point 8 (0.02 MPa)?

From saturated steam tables (Appendix M),

$v^L(m^3/kg)$	$h^L(kJ/kg)$	$h^v(kJ/kg)$	$s^L(kJ/kg.K)$	$s^v(kJ/kg.K)$
0.001017	251.4	2609.7	0.8319	7.9093

The quality of steam in state 8:

$$s_8 = s_7 = 7.0901 = x_8^v s_8^v + (1 - x_8^v)s_8^L = x_8^v(7.9093) + (1 - x_8^v)(0.8319)$$

$$x_8^v = \frac{7.0901 - 0.8319}{7.9093 - 0.8319} = 0.8843$$

$$h_8 = x_8^v h_8^v + (1 - x_8^v)h_8^L = 0.8843 \times 2609.7 + (1 - 0.8843) \times 251.4 = 2336.84 \text{ kJ/kg}$$

Properties of saturated liquid in state 1 (0.02 MPa) and state 3 (4 MPa)?

$$h_1 = h_1^L = 251.4 \text{ kJ/kg};\ v_1 = v_1^L = 0.001017\ (m^3/kg)$$

$$h_3 = h_3^L = 1087.3 \text{ kJ/kg};\ v_3 = v_3^L = 0.001252\ (m^3/kg)$$

Pump Work:

$$W_{sh,pump,1} = -\frac{(P_2-P_1)}{\rho_1} = -v_1(P_2 - P_1) = -0.001017 \times (4 - 0.02) \times 10^3 = -4.048 \text{ kJ/kg}$$

$$h_1 - h_2 = -4.048$$

$$h_1 = 251.4 \text{ kJ/kg}$$

$$h_2 = 251.4 + 4.048 = 255.448 \text{ kJ/kg}$$

$$W_{sh,pump,2} = -\frac{(P_4-P_3)}{\rho_3} = -v_3(P_4 - P_3) = -0.001252 \times (35 - 4) \times 10^3 = -38.812 \text{ kJ/kg}$$

$$h_3 - h_4 = -38.812$$

$$h_4 = 1087.3 + 38.812 = 1126.112 \text{ kJ/kg}$$

Mass and energy balances around feed water heater:

The mass and energy balances around the feed water heater are needed to calculate the mass flow rate $\dot{m}_6$.

Mass balance gives: $\dot{m}_2 + \dot{m}_6 = \dot{m}_3 = 1$

Energy balance gives: $\dot{m}_3 h_3 = \dot{m}_2 h_2 + \dot{m}_6 h_6$

From these equations, we get:

$$\dot{m}_6 = \frac{h_3-h_2}{h_6-h_2} = \frac{1087.3-255.448}{2826.69-255.448} = 0.3235 \text{ kg/s}; \quad \dot{m}_2 = 1 - \dot{m}_6 = 0.6765 \text{ kg/s}$$

Net work produced by the cyclic process:

$$\dot{W}_{sh,pump,1} = -\dot{m}_2 \times 4.048 = -0.6765 \times 4.048 = -2.7385 \text{ kW}$$

$$\dot{W}_{sh,pump,2} = -\dot{m}_3 \times 38.812 = -1 \times 38.812 = -38.812 \text{ kW}$$

$$\dot{W}_{sh,turbine,1} = -\dot{m}_5 \times (h_6 - h_5) = -1 \times (2826.69 - 3395.5) = 568.81 \text{ kW}$$

$$\dot{W}_{sh,turbine,2} = -(1 - \dot{m}_6) \times (h_8 - h_7) = -0.6765 \times (2336.84 - 3445.3) = 749.87 \text{ kW}$$

$$\text{Net work produced} = (-2.7385) + (-38.812) + 568.81 + 749.87 = 1277.13 \text{ kW}$$

Heat input to the boiler:

$$\dot{Q}_{45} = \dot{m}_5 (h_5 - h_4) = 1 \times (3395.5 - 1126.112) = 2269.388 \text{ kW}$$

$$\dot{Q}_{67} = (1 - \dot{m}_6)(h_7 - h_6) = 0.6765 \times (3445.3 - 2826.69) = 418.49 \text{ kW}$$

Thermal efficiency of the cyclic process:

$$\eta = \frac{W_{sh,net}}{Q_{Boiler}} = \frac{-2.7385 - 38.812 + 568.81 + 749.87}{2269.388 + 418.49} = \frac{1277.13}{2687.878} = 0.475$$

PRACTICE PROBLEMS

Problem 1

Consider a steam power plant operating on a Carnot cycle. The circulation rate of fluid in the cycle is 10 kg/s. Given that the boiler temperature is 240 °C and the condenser temperature is 20°C, determine the following:

a) the boiler pressure
b) the condenser pressure
c) the quality of the exhaust steam from the turbine
d) the quality of the steam leaving the condenser
e) the rate of heat transfer in the boiler
f) the rate of heat transfer in the condenser
g) the rate of shaft work output of the turbine
h) the rate of shaft work required by the pump
i) the thermal efficiency of the cycle

Problem 2

Consider a steam power plant operating on an ideal Rankine cycle. The boiler pressure is 7 MPa (70 bar) and the condenser pressure is 0.012 MPa (12 kPa). The temperature of the superheated steam entering the turbine is 500 °C. Determine the following:

a) the quality of the exhaust steam from the turbine
b) the heat transfer in the boiler per unit mass of fluid
c) the heat transfer in the condenser per unit mass of fluid
d) the pump shaft work per unit mass of fluid
e) the turbine shaft work per unit mass of fluid
f) the thermal efficiency of the cycle

Problem 3

A steam power plant is developed to use agricultural and forest solid waste as fuel. The solid waste is burnt in a fluidized bed to generate heat needed for the boiler. The maximum temperature in the boiler is limited to 350 °C. The boiler pressure is 1 MPa. The condenser of the plant operates at 0.06 MPa. Assuming ideal Rankine cycle, determine the following:

a) the quality of the exhaust steam from the turbine
b) the heat transfer in the boiler per unit mass of fluid
c) the heat transfer in the condenser per unit mass of fluid
d) the pump shaft work per unit mass of fluid
e) the turbine shaft work per unit mass of fluid
f) the thermal efficiency of the cycle

Problem 4

A steam power plant operating on a practical Rankine cycle has a power rating of 67,000 kW. The superheated steam leaving the boiler is at a pressure of 9 MPa (90 bar) and a temperature of 450 °C. The condenser pressure is 0.006 MPa (6 kPa). The efficiency of the turbine is 0.79 and the pump efficiency is 0.74. Determine:

a) the mass flow rate of fluid circulating in the cycle loop
b) the rate of heat addition in the boiler
c) the rate of heat rejection in the condenser
d) the thermal efficiency of the cycle

Problem 5

Consider a steam power plant operating on a reheat cycle. Superheated steam enters the first turbine at 5 MPa (50 bar) and 600 °C. The steam discharged from the first turbine is reheated at a pressure of 0.6 MPa to 600 °C and then expanded in the second turbine to 4 kPa. Determine:

a) the quality of the steam exhausted from the second turbine
b) the net work produced per unit mass of fluid
c) the thermal efficiency of the cycle

What is the quality of the exhaust steam if reheating is not used, that is, the superheated steam at 5 MPa and 600 °C is expanded in a single turbine to a condenser pressure of 4 kPa?

Problem 6

Consider a steam power plant operating on a reheat/regeneration cycle. Superheated steam at 5 MPa (50 bar) and 600 °C is expanded in the first turbine to 0.6 MPa. A portion of the steam discharged from the first turbine at 0.6 MPa is diverted to the feed-water heater. The remaining portion of steam is reheated at a constant pressure of 0.6 MPa to 600 °C. This reheated steam is expanded in the second turbine to a condenser pressure of 4 kPa.

Determine the thermal efficiency of the cycle. As a basis for calculations, you can take the mass flow rate of steam leaving the boiler equal to 1 kg/s.

Compare the efficiency of this reheat/regeneration cycle with the reheat cycle of problem 5.

Problem 7

Consider a steam power plant operating on a reheat cycle with two reheat stages. Superheated steam leaving the boiler at 9 MPa (90 bar) and 550 °C is expanded in the first turbine to 1.6 MPa. The exhaust steam from the first turbine is reheated at a constant pressure of 1.6 MPa to 550 °C and then expanded in the second turbine to 0.2 MPa. The exhaust steam from the second turbine is once again reheated at a constant pressure of 0.2 MPa to 550 °C and then expanded in the third turbine to a condenser pressure of 0.06 MPa. Determine the thermal efficiency of this reheat cycle.

Problem 8

Consider a steam power plant based on supercritical Rankine cycle. The boiler pressure is 40 MPa and the temperature of superheated steam leaving the boiler is 600°C. The boiler steam is expanded in a turbine to a condenser pressure of 10 kPa. The mass flow rate of steam leaving the boiler is 100 kg/s. Determine:

a) the power output of the plant
b) the thermal efficiency of the plant

Problem 9

The supercritical steam power plant of problem 8 is modified by adding a feed water heater. The superheated steam at 40 MPa and 600°C is expanded in the first turbine to 5 MPa. A portion of the steam at 5 MPa is diverted to the feed-water heater. The remaining portion of the steam is expanded in the second turbine to a condenser pressure of 10 kPa. The mass flow rate of steam leaving the boiler is 100 kg/s. Determine:

a) the quality of the steam discharged from the second turbine
b) the power output of the plant
c) the thermal efficiency of the plant

Problem 10

The supercritical steam power plant of problem 9 is further modified by adding a reheat stage. The superheated steam at 40 MPa and 600°C is expanded in the first turbine to 5 MPa. A portion of the steam at 5 MPa is diverted to the feed-water heater. The remaining portion of the steam is reheated at 5 MPa to 600°C before expanding in the second turbine to a condenser pressure of 10 kPa. The mass flow rate of steam leaving the boiler is 100 kg/s. Determine:

a) the quality of the steam discharged from the second turbine

b) the power output of the plant
c) the thermal efficiency of the plant

Chapter 9

Internal Combustion Engines

9.1 LEARNING OUTCOMES

Upon completion of this chapter, students will be able to:

- Explain the difference between steam power plant type heat engine and internal combustion engine
- Describe the Otto engine and the air-standard Otto cycle
- Develop the expression for thermal efficiency of air-standard Otto cycle
- Describe the diesel engine and the air-standard diesel cycle
- Develop the expression for thermal efficiency of air-standard diesel cycle
- Compare the performances of Otto and diesel cycles
- Describe the gas turbine engine and the Brayton cycle
- Develop the expression for thermal efficiency of Brayton cycle
- Describe the working of rocket and jet engines
- Solve problems dealing with Otto, diesel, and Brayton cycles
- Calculate the thrust developed in rocket and jet engines

9.2 INTRODUCTION

In the previous chapter, we studied a steam power plant. The working substance in a steam power plant is an inert (non-reactive) medium which is circulated in a closed loop continuously. The working fluid undergoes a cyclic process involving vaporization (heat absorption), expansion, condensation (heat rejection), and compression, all in such a way as to generate power. The combustion of a fuel is carried out externally to the working medium and heat is transferred from the burning fuel to the working fluid through the walls of the boiler. Likewise, heat is rejected by the working fluid to the surroundings through the walls of the condenser. Thus, a steam power plant involves heat transfer surfaces through which heat is absorbed or rejected by the working medium.

Like a steam power plant, *internal combustion engines* are devices which convert heat into work or mechanical energy. However, the heat in internal combustion engines is generated within the working medium itself by burning the fuel present in the working medium. Strictly speaking, the working substance in an internal combustion engine does not undergo a cyclic process although work/mechanical energy is produced on a continuous basis. Internal combustion devices are *one-pass (open cycle)* devices in that the working fluid (air) enters the device and is mixed with a fuel, undergoes combustion and leaves the device as exhaust gases. However, internal combustion engines are often modelled as cyclic devices as an approximation. It is important to note that an internal combustion engine generally involves a piston-cylinder arrangement, unlike a steam power plant which involves an interconnected series of different components (boiler, turbine, condenser, and pump). Thus, they are light and compact in structure.

In this chapter, we will describe and analyze the following internal combustion engines: Otto engine, Diesel engine, gas-turbine engine, rocket engine, and jet engine.

9.3 THE OTTO ENGINE

The Otto engine is the most commonly used internal combustion engine. It is used extensively in automobiles (cars). It is a 4-stroke reciprocating piston/cylinder type internal combustion engine using *spark ignition*.

Figure 9.1 shows the working of the Otto engine. It begins with the *intake stroke* during which the piston moves downward drawing a fuel (gasoline)/air mixture through an open valve. The pressure during the intake stroke is constant. During the *second stroke*, the values are closed, and the fuel/air mixture is compressed adiabatically. At the end of compression, the mixture is ignited using a *spark plug*. The combustion occurs rapidly such that the volume of the fluid mixture remains nearly constant, but the temperature and pressure rise substantially. In the *third stroke (power stroke)*, the gas mixture expands adiabatically, and work is produced. The piston then executes the *fourth (exhaust) stroke*. The exhaust valve is opened, and the combustion gases are purged from the cylinder to the atmosphere.

Figure 9.2 shows the plot of pressure versus volume behavior of gas contained in the piston/cylinder assembly. Path $0 \rightarrow 1$ corresponds to first (fuel/air intake) stroke. Path $1 \rightarrow 2 \rightarrow 3$ corresponds to second (compression) stroke. Path $3 \rightarrow 4 \rightarrow 1$ corresponds to third (expansion) stroke, and path $1 \rightarrow 0$ corresponds to the fourth (exhaust) stroke. Note that during path $2 \rightarrow 3$ of the second stroke, the volume of the gas is constant, but the temperature and pressure rise substantially. Also, during path $4 \rightarrow 1$ of the third stroke, the exhaust valve is opened, and the pressure falls rapidly to atmospheric pressure at constant volume.

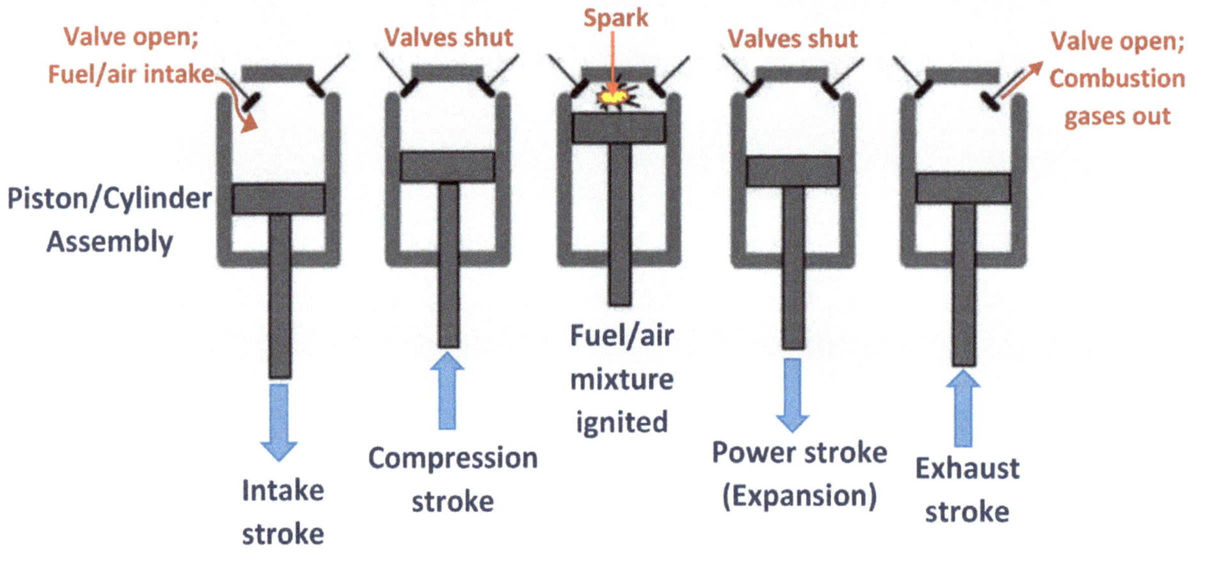

Figure 9.1 Working of the Otto engine.

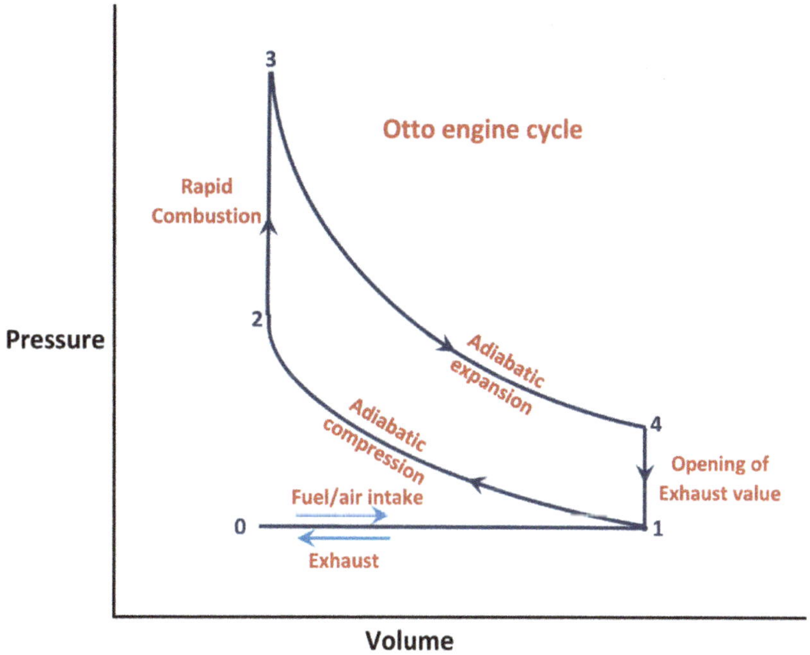

Figure 9.2 Pressure versus volume behavior of gas in Otto engine.

Thermal Efficiency of the Otto Engine

In order to estimate the thermal efficiency of the Otto engine, the Otto engine is modelled as the idealized Otto cycle, shown in Figure 9.3. The idealized Otto cycle is also referred to as air-standard Otto cycle. In the air-standard Otto cycle, a closed system consisting of air as the working substance undergoing the cyclic process. The cycle $1 \rightarrow 2 \rightarrow 3 \rightarrow 4 \rightarrow 1$ consists of reversible adiabatic compression of air along path $1 \rightarrow 2$, reversible isochoric heating along path $2 \rightarrow 3$, reversible adiabatic expansion along path $3 \rightarrow 4$, and finally reversible isochoric cooling along path $4 \rightarrow 1$.

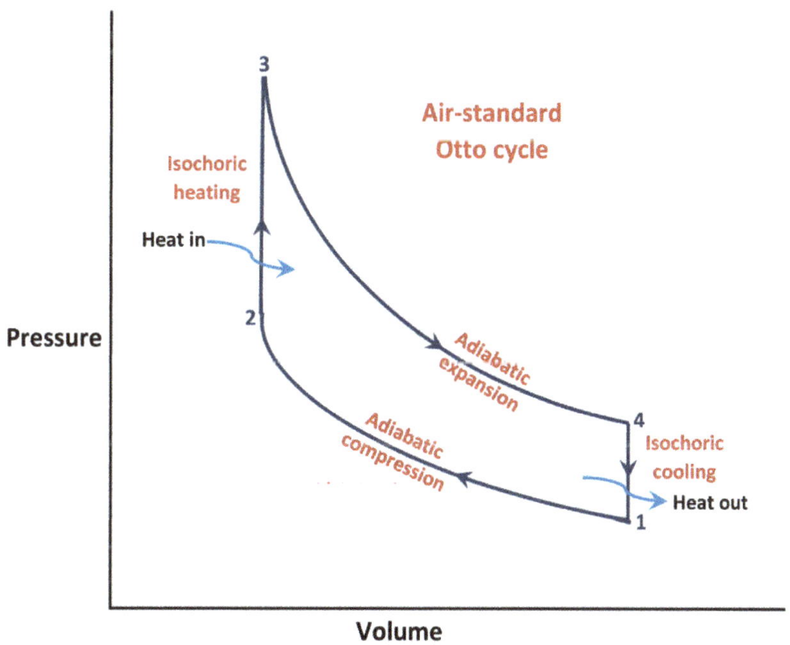

Figure 9.3 Air-standard Otto cycle.

The thermal efficiency of any cyclic process is defined as:

$$\eta = \frac{W_{cycle}}{Heat\ input} = \frac{Q_{cycle}}{Heat\ input} \tag{9.1}$$

For the air-standard Otto cycle of Figure 9.3,

$$\eta = \frac{Q_{2\to 3}+Q_{4\to 1}}{Q_{2\to 3}} = 1 + \frac{Q_{4\to 1}}{Q_{2\to 3}} \tag{9.2}$$

Using 1 mol of air as the basis for calculation and assuming air is an ideal gas,

$$Q_{2\to 3} = c_v(T_3 - T_2) \tag{9.3}$$

$$Q_{4\to 1} = c_v(T_1 - T_4) \tag{9.4}$$

From Equations (9.2) to (9.3), we get:

$$\eta = 1 - \left(\frac{T_4 - T_1}{T_3 - T_2}\right) \tag{9.5}$$

Using ideal gas law,

$$T_1 = \frac{P_1 v_1}{R}\ ;\ T_2 = \frac{P_2 v_2}{R}\ ; T_3 = \frac{P_3 v_2}{R}\ ; T_4 = \frac{P_4 v_1}{R} \tag{9.6}$$

Substituting the expressions for temperature into Equation (9.5),

$$\eta = 1 - r\left(\frac{P_4 - P_1}{P_3 - P_2}\right) \tag{9.7}$$

where r is the compression ratio, defined as v_1/v_2. For reversible adiabatic processes:

$$P_4 v_1^{\gamma} = P_3 v_2^{\gamma} \tag{9.8}$$

$$P_1 v_1^{\gamma} = P_2 v_2^{\gamma} \tag{9.9}$$

Dividing Equation (9.8) by Equation (9.9),

$$\frac{P_4}{P_1} = \frac{P_3}{P_2} \tag{9.10}$$

From Equations (9.7) and (9.10), we get:

$$\eta = 1 - r\left(\frac{P_1}{P_2}\right) \tag{9.11}$$

Equation (9.9) gives

$$\frac{P_1}{P_2} = \left(\frac{v_2}{v_1}\right)^{\gamma} = \left(\frac{1}{r}\right)^{\gamma} \tag{9.12}$$

Substitution of the pressure ratio P_1/P_2 from Equation (9.12) into Equation (9.11) yields:

$$\eta = 1 - \left(\frac{1}{r}\right)^{\gamma - 1} \tag{9.13}$$

Thus, the thermal efficiency of the Otto cycle/engine depends on the compression ratio r $(= v_1/v_2)$. The thermal efficiency increases with the increase in r.

Figure 9.4 shows the plot of η versus r for air with $\gamma = 1.4$. The efficiency increases rapidly with the increase in r at low values of r, but at high values of r, η increases less rapidly. Thus, it is advantageous to operate the Otto engine at high compression ratios. The compression ratio is close to 10 for typical automobiles.

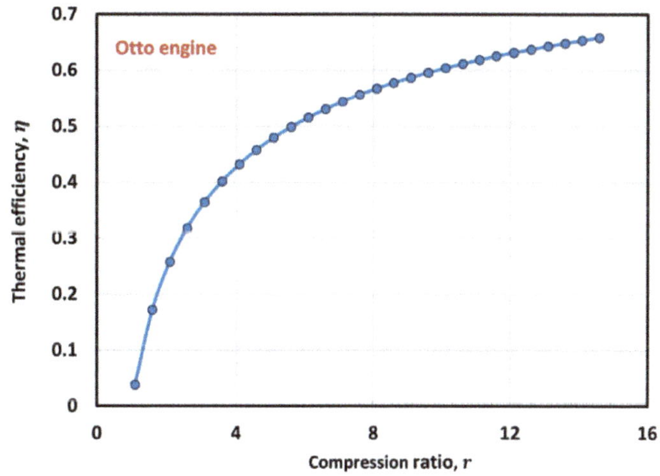

Figure 9.4 Thermal efficiency of the air-standard Otto cycle.

9.4 THE DIESEL ENGINE

The diesel engine is similar to the Otto engine except that the compression ratio is now much larger (larger than 15). Consequently, the temperature of air at the end of compression exceeds the ignition temperature of the diesel fuel and no external spark is needed to ignite the fuel. The diesel fuel is simply injected into the compressed air and the combustion is initiated spontaneously. The diesel engine is also referred to as *compression-ignition* engine as ignition occurs spontaneously without any spark plug. Also note that the pressure is nearly constant during the combustion process as the fuel is injected slowly to the compressed air.

Figure 9.5 shows the working of the diesel engine. The first stroke is the intake stroke during which fresh air is drawn into the cylinder. The pressure is constant during the intake stroke. The second stroke involves adiabatic compression of gas. At the end of air compression, the fuel (diesel) is injected slowly into the cylinder. Due to high temperature of the air, the combustion of the fuel occurs spontaneously at constant pressure. The third stroke is the power stroke during which the gas mixture expands adiabatically, and work is produced. In the final fourth stroke, the exhaust value is opened, and the combustion gases are purged from the cylinder.

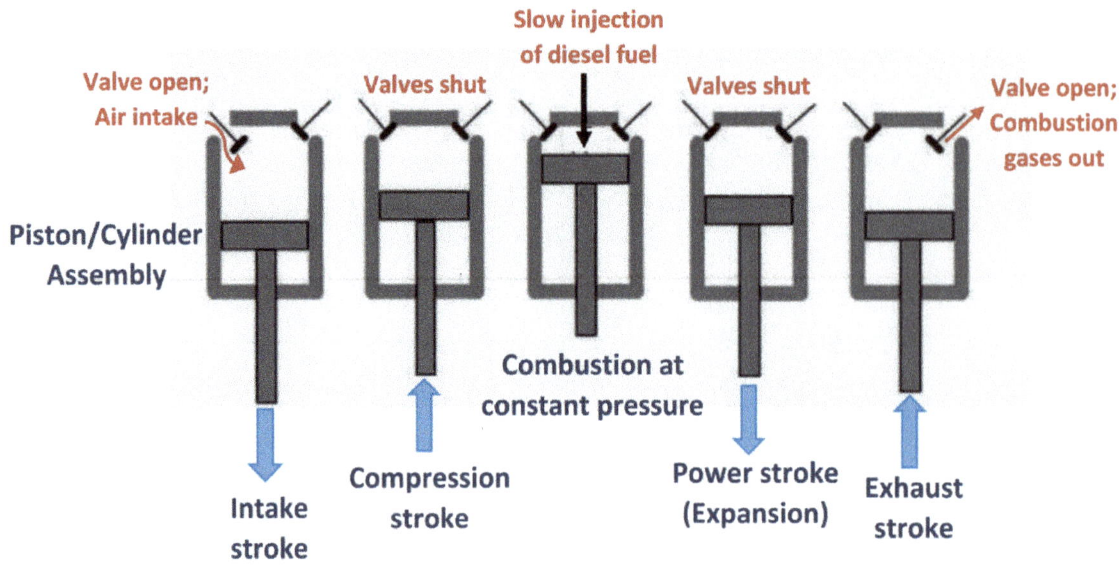

Figure 9.5 Working of the diesel engine.

Figure 9.6 shows the plot of pressure versus volume behavior of the gas in the piston/cylinder assembly of the diesel engine. The P versus v behavior is similar to that of the Otto engine cycle (see Figure 9.2) except that the process $2 \rightarrow 3$ involving combustion of the fuel is now a constant pressure process. In the Otto engine, the combustion process $2 \rightarrow 3$ is a constant volume process.

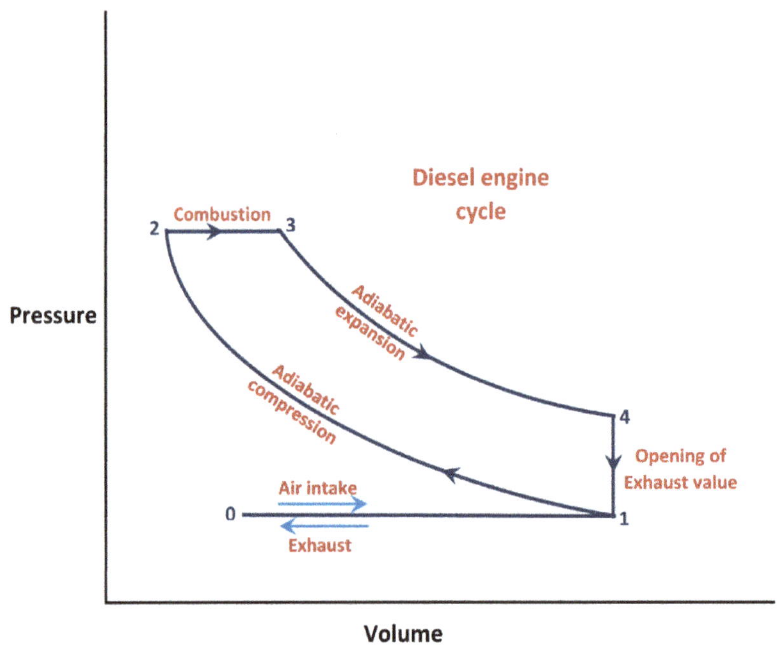

Figure 9.6 Pressure versus volume behavior of gas in the diesel engine.

Thermal Efficiency of the Diesel Engine

To estimate the thermal efficiency of the diesel engine, an air-standard diesel cycle, shown in Figure 9.7, is used as a model of the actual diesel engine cycle. The working substance is air that undergoes the cyclic process shown in Figure 9.7. The cycle $1 \rightarrow 2 \rightarrow 3 \rightarrow 4 \rightarrow 1$ consists of reversible adiabatic compression of air along path $1 \rightarrow 2$, reversible isobaric heating along path $2 \rightarrow 3$, reversible adiabatic expansion along path $3 \rightarrow 4$, and finally reversible isochoric cooling along path $4 \rightarrow 1$.

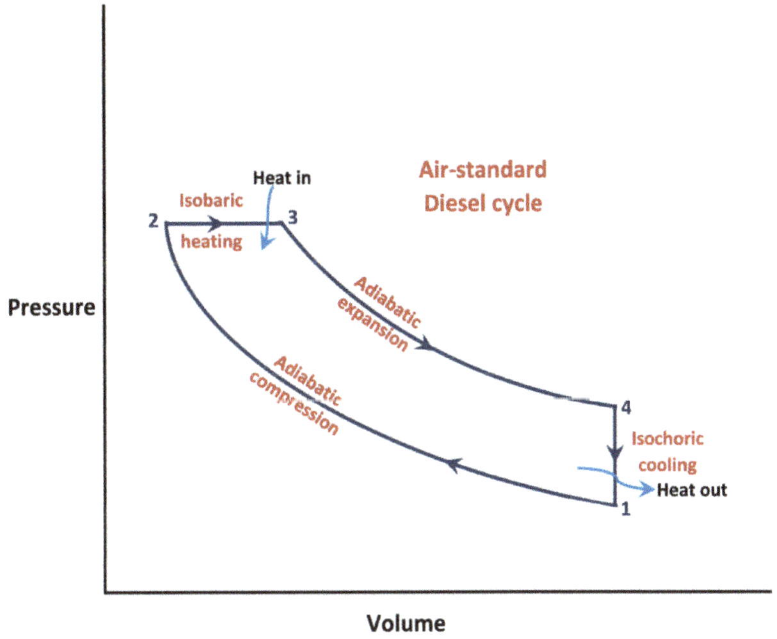

Figure 9.7 Air-standard diesel cycle.

The thermal efficiency of the given cyclic process can be expressed as:

$$\eta = \frac{Q_{2\rightarrow3}+Q_{4\rightarrow1}}{Q_{2\rightarrow3}} = 1 + \frac{Q_{4\rightarrow1}}{Q_{2\rightarrow3}} \tag{9.14}$$

Using 1 mol of air as the basis for calculation and assuming air is an ideal gas,

$$Q_{2\rightarrow3} = c_p(T_3 - T_2) \tag{9.15}$$

$$Q_{4\rightarrow1} = c_v(T_1 - T_4) \tag{9.16}$$

From Equations (9.14) to (9.16), we get:

$$\eta = 1 - \left(\frac{1}{\gamma}\right)\left(\frac{T_4-T_1}{T_3-T_2}\right) \tag{9.17}$$

For reversible adiabatic processes involving an ideal gas of constant heat capacity ratio γ:

$$T_4 v_4^{\gamma-1} = T_3 v_3^{\gamma-1} \tag{9.18}$$

$$T_1 v_1^{\gamma-1} = T_2 v_2^{\gamma-1} \tag{9.19}$$

From Equations (9.18) and (9.19), we get:

$$\frac{T_4}{T_3} = \left(\frac{v_3}{v_4}\right)^{\gamma-1} \; ; \; \frac{T_1}{T_2} = \left(\frac{v_2}{v_1}\right)^{\gamma-1} \tag{9.20}$$

Let the compression ratio r be defined as v_1/v_2 and the expansion ratio r_e be defined as v_4/v_3. Re-writing Equation (9.20) in terms of r and r_e gives:

$$\frac{T_4}{T_3} = \left(\frac{1}{r_e}\right)^{\gamma-1} \; ; \; \frac{T_1}{T_2} = \left(\frac{1}{r}\right)^{\gamma-1} \tag{9.21}$$

From Equations (9.17) and (9.21), it follows that:

$$\eta = 1 - \left(\frac{1}{\gamma}\right)\left[\frac{T_3\left(\frac{1}{r_e}\right)^{\gamma-1} - T_2\left(\frac{1}{r}\right)^{\gamma-1}}{T_3 - T_2}\right] \tag{9.22}$$

To simplify Equation (9.22) further, we make use of the ideal gas law as:

$$\frac{T_2}{T_3} = \frac{P_2 v_2}{P_3 v_3} = \frac{v_2}{v_3} = \frac{v_2/v_1}{v_3/v_1} = \frac{r_e}{r} \tag{9.23}$$

Note that $P_2 = P_3$. Using Equation (9.23), the temperatures in Equation (9.22) can be eliminated to get the following expression for the thermal efficiency of air-standard diesel cycle:

$$\eta = 1 - \left(\frac{1}{\gamma}\right)\left[\frac{\left(\frac{1}{r_e}\right)^{\gamma-1} - \left(\frac{r_e}{r}\right)\left(\frac{1}{r}\right)^{\gamma-1}}{1 - \left(\frac{r_e}{r}\right)}\right] \tag{9.24}$$

Upon re-arrangement, Equation (9.24) could be re-written as:

$$\eta = 1 - \left(\frac{1}{\gamma}\right)\left[\frac{(1/r_e)^{\gamma} - (1/r)^{\gamma}}{(1/r_e) - (1/r)}\right] \tag{9.25}$$

Thus, the thermal efficiency of the diesel engine depends on both expansion and compression ratios.

The thermal efficiency of a diesel engine could also be expressed in terms of the compression ratio r and so-called *cut-off ratio* r_c, defined as:

$$r_c = \frac{v_3}{v_2} \tag{9.26}$$

It can be readily shown that:

$$r = r_c r_e \tag{9.27}$$

Upon substitution of $r_e = r/r_c$ into the expression for η, we get:

$$\eta = 1 - \left(\frac{1}{r}\right)^{\gamma-1}\left[\frac{r_c^{\gamma} - 1}{\gamma(r_c - 1)}\right] \tag{9.28}$$

Note that the thermal efficiency of the diesel cycle differs from the Otto cycle only by the term in the square brackets. For $r_c > 1$, the square bracket term is greater than unity. Hence for the same compression ratio r, the thermal efficiency of the diesel engine is less than that of the Otto engine. This can be seen more clearly in Figure 9.8 generated from Equation (9.28) using $\gamma = 1.4$. The thermal efficiency decreases with the increase in cut-off ratio r_c for any given compression ratio r. Note that $r_c = 1$ for the Otto cycle.

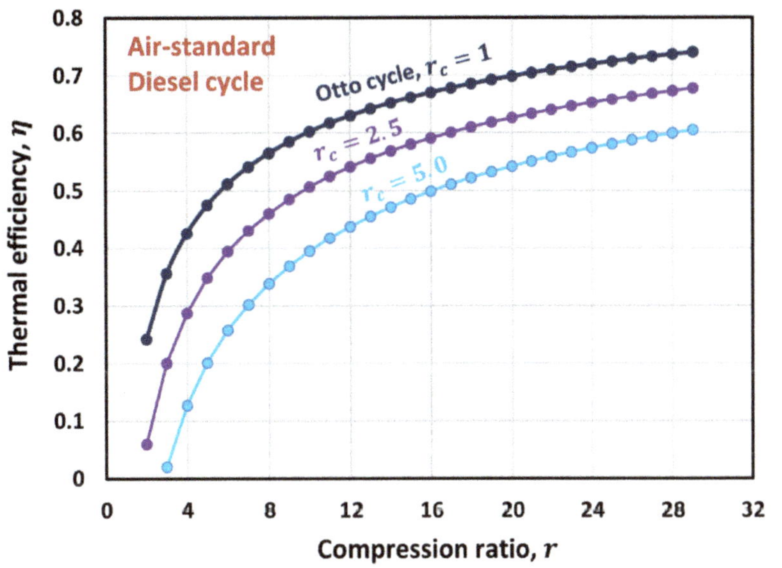

Figure 9.8 Thermal efficiency of air-standard diesel cycle.

9.5 THE GAS TURBINE ENGINE

The gas turbine engine is an open cycle (one-pass) internal combustion engine that produces work from heat. Gas turbine engines are used extensively in industrial applications for the generation of electric power. Some of their advantages over a steam power plant are: a) they are small in size, weigh less, and have low cost; b) they are mobile; and c) they do not need any cooling water.

Figure 9.9 shows a schematic diagram of the gas turbine engine. Fresh air from the atmosphere enters the compressor. The compressed air passes through a combustion chamber where fuel is burned at constant pressure. The high pressure, high temperature combustion gas leaving the combustion chamber is expanded in a turbine to produce work. Part of the work produced in a turbine is used to run the compressor. The combustion gas from the turbine is discharged to the surroundings.

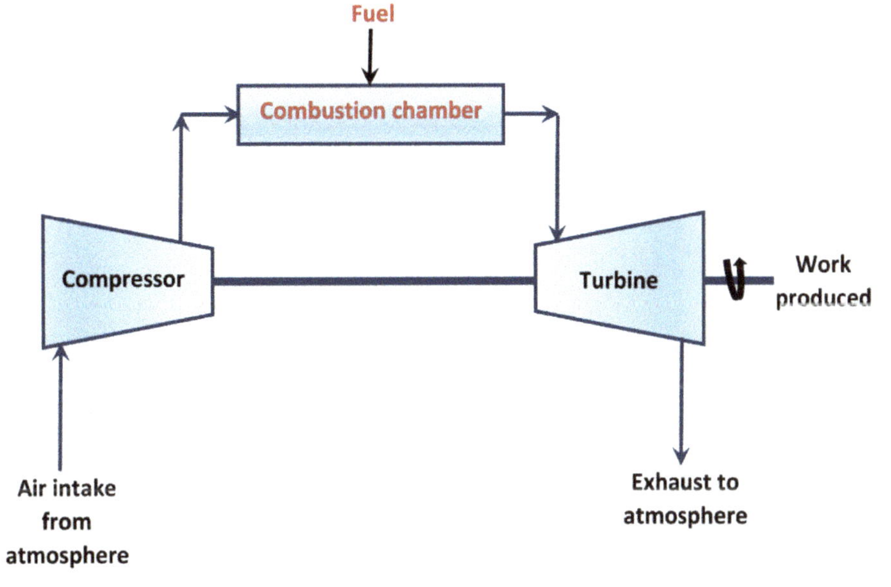

Figure 9.9 Schematic of a gas turbine engine.

Thermal Efficiency of the Gas Turbine Engine

In order to estimate the thermal efficiency of a gas turbine engine, the process is modelled as a cyclic process shown in Figure 9.10. The working substance is assumed to be air that undergoes the cyclic process. The cycle $1 \rightarrow 2 \rightarrow 3 \rightarrow 4 \rightarrow 1$ consists of reversible adiabatic compression of air in a compressor along path $1 \rightarrow 2$, reversible isobaric heating along path $2 \rightarrow 3$ representing combustion process in the combustion chamber, reversible adiabatic expansion in a turbine along path $3 \rightarrow 4$, and finally reversible isobaric cooling along path $4 \rightarrow 1$ representing discharge of combustion gas from the turbine to the atmosphere. The cyclic process shown in Figure 9.10 is called the *Brayton cycle*.

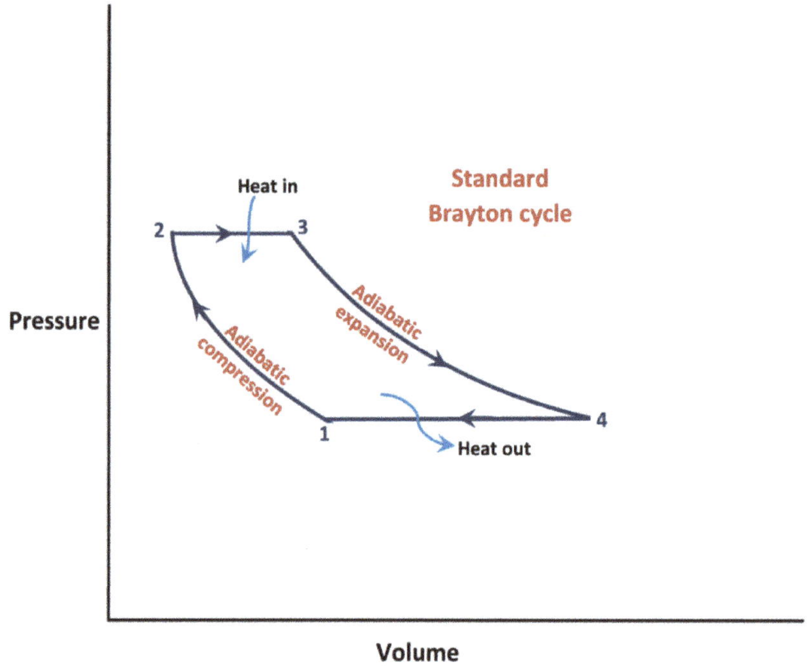

Figure 9.10 Air-standard Brayton cycle.

The thermal efficiency of the given cyclic process can be expressed as:

$$\eta = \frac{W_{3\rightarrow4} + W_{1\rightarrow2}}{Q_{2\rightarrow3}} \tag{9.29}$$

Assuming air to be an ideal gas, the work and heat terms (on a molar basis) can be calculated using the first law for flow processes as:

$$W_{1\rightarrow2} = -\Delta h_{1\rightarrow2} = c_p(T_1 - T_2) \tag{9.30}$$

$$W_{3\rightarrow4} = -\Delta h_{3\rightarrow4} = c_p(T_3 - T_4) \tag{9.31}$$

$$Q_{2\rightarrow3} = \Delta h_{2\rightarrow3} = c_p(T_3 - T_2) \tag{9.32}$$

Substituting the expressions for $W_{1\rightarrow2}$, $W_{3\rightarrow4}$, and $Q_{2\rightarrow3}$ into Equation (9.29), we get:

$$\eta = 1 - \left(\frac{T_4 - T_1}{T_3 - T_2}\right) \tag{9.33}$$

For reversible adiabatic processes involving an ideal gas of constant heat capacity ratio γ:

$$\frac{T_1}{T_2} = \left(\frac{P_1}{P_2}\right)^{(\gamma-1)/\gamma} \quad ; \quad \frac{T_4}{T_3} = \left(\frac{P_4}{P_3}\right)^{(\gamma-1)/\gamma} = \left(\frac{P_1}{P_2}\right)^{(\gamma-1)/\gamma} \tag{9.34}$$

Note that $P_4 = P_1$ and $P_3 = P_2$. Elimination of temperature from Equation (9.33) using Equation (9.34) gives:

$$\eta = 1 - \frac{1}{\left(\frac{P_2}{P_1}\right)^{(\gamma-1)/\gamma}} \tag{9.35}$$

As the pressure ratio P_2/P_1 across the compressor increases, the thermal efficiency of the gas turbine engine increases. Figure 9.11 shows the plot of η versus pressure ratio P_2/P_1.

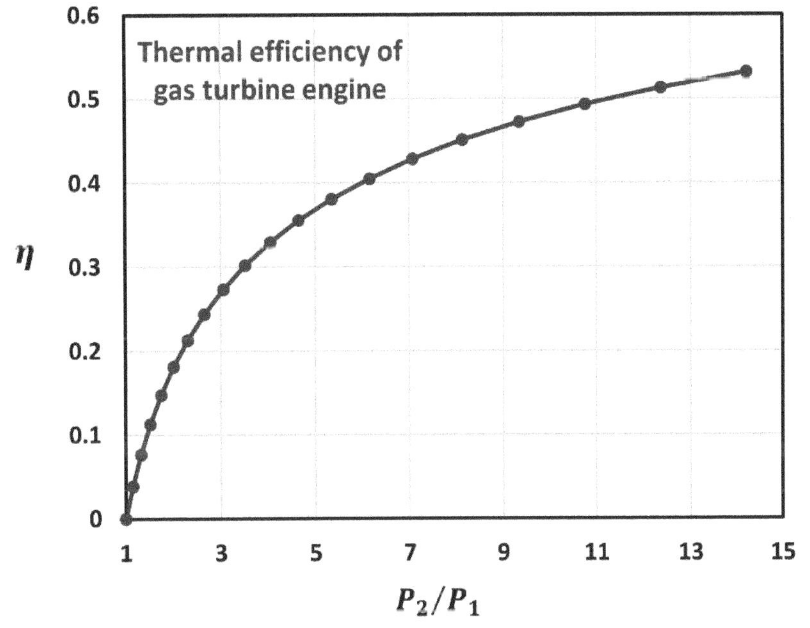

Figure 9.11 Thermal efficiency of a gas turbine engine as a function of compressor pressure ratio.

It should be noted that a large portion of the work produced in the turbine is consumed by the compressor itself. This limits the thermal efficiency of the gas turbine engine. The ratio of compressor work to turbine work in gas turbine engines is typically in the range of 50 to 80%. However, in a steam power plant, the work required to drive the pump is much lower, typically 1 to 2 % of the work produced in the turbine. The average specific volume of gas flowing through the compressor is many times larger than the specific volume of liquid flowing through the pump. Thus, for the same rise in pressure, the compressor requires a much larger work input per unit mass of fluid as compared with the pump.

9.6 THE ROCKET ENGINE

In a rocket engine, the fuel (assumed liquid such as kerosene or liquid hydrogen) and the oxidizing agent (liquid oxygen) are pumped into the combustion chamber and burned. The combustion takes place at a constant pressure. The high pressure, high temperature combustion gases are expanded in converging-diverging nozzle, as shown in Figure 9.12. The exhaust gases leaving the nozzle have a high velocity relative to the rocket. Thus, heat is converted into kinetic energy of gases in a rocket engine. The discharge of high velocity gas from the nozzle provides forward *thrust* to the rocket in the direction of motion.

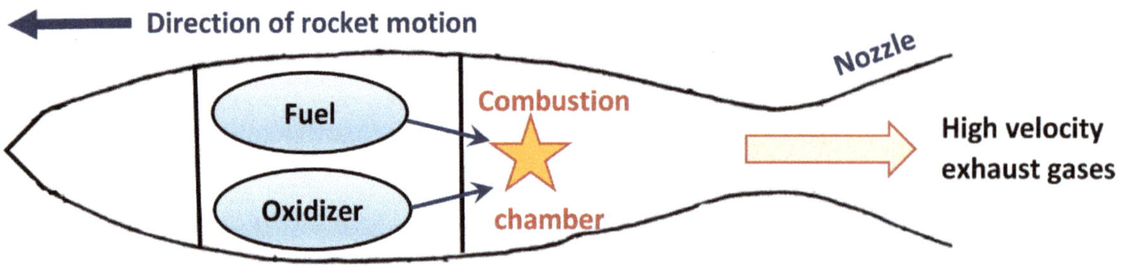

Figure 9.12 The liquid-fuel rocket engine.

Estimation of Thrust of a Rocket Engine

Consider the motion of a rocket in gravity free space. At time t, the rocket is moving with a velocity $\vec{V}$ in the x-direction (indicated by unit vector $\hat{\imath}$) as observed by a stationary observer on earth. In time $t + \Delta t$, the rocket is moving with a higher velocity $\vec{V} + \Delta\vec{V}$. During the period Δt, the rocket has ejected gas of mass Δm at a velocity $\vec{U}_R$ relative to the rocket. The ejected mass is moving at a velocity of $\vec{U}$ with respect to a stationary observer on earth. Figure 9.13 shows the motion of the rocket at times t and $t + \Delta t$. Note that:

$$\vec{U}_R = \vec{U} - \left(\vec{V} + \Delta\vec{V}\right) = -U_R\hat{\imath} \tag{9.36}$$

where U_R is the magnitude of the relative velocity of the ejected mass. The mass is ejected in the direction opposite to the direction of motion of the rocket.

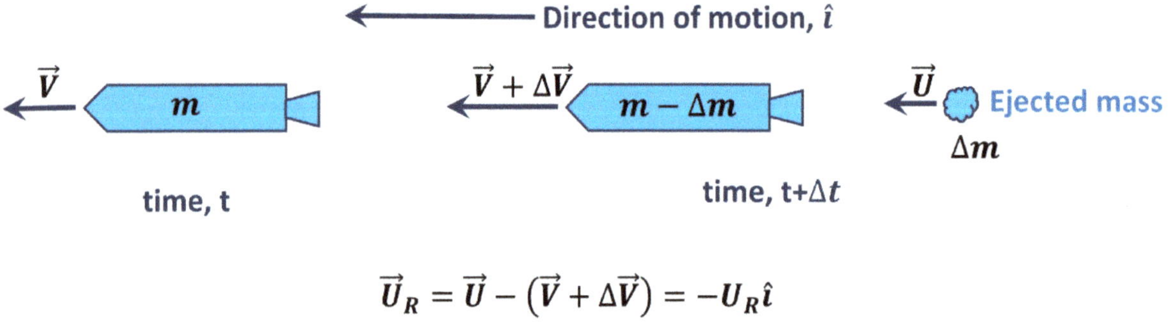

Figure 9.13 Motion of a rocket at times t and $t + \Delta t$.

As there is no external force acting on the rocket, the principle of *conservation of linear momentum* gives:

$$\frac{d\vec{M}}{dt} = 0 \tag{9.37}$$

where $\vec{M}$ is momentum, defined as mass times velocity. From Equation (9.37), it follows that:

$$\frac{\vec{M}_{t+\Delta t} - \vec{M}_t}{\Delta t} = 0 \tag{9.38}$$

$$\frac{[(m-\Delta m)(\vec{V}+\Delta\vec{V})+(\Delta m)(\vec{U})]-[m\vec{V}]}{\Delta t}=0 \tag{9.39}$$

Upon simplification, Equation (9.39) gives:

$$m\left(\frac{\Delta\vec{V}}{\Delta t}\right)=\left(\frac{\Delta m}{\Delta t}\right)[(\vec{V}+\Delta\vec{V})-(\vec{U})] \tag{9.40}$$

In the limit $\Delta t \to 0$,

$$m\left(\frac{d\vec{V}}{dt}\right)=q[(\vec{V}+\Delta\vec{V})-(\vec{U})] \tag{9.41}$$

where q is rate of mass ejection (positive quantity). Using Equation (9.36), Equation (9.41) can be re-written as:

$$m\left(\frac{d\vec{V}}{dt}\right)=q[U_R\hat{\imath}] \tag{9.42}$$

Thus, the rocket accelerates in the forward direction indicated by $\hat{\imath}$ (see Figure 9.13) as U_R and q are positive quantities. Equivalently, a thrust force $(m\,d\vec{V}/dt)$ acts on the rocket in the forward direction, that is, the direction of motion of the rocket. We can re-write Equation (9.42) as:

$$F=qU_R \tag{9.43}$$

where F is the forward thrust on the rocket. Thus, faster the speed of the exhaust gas (relative to the rocket) and greater the rate of mass ejection from the rocket, larger is the forward thrust of the rocket in the direction of motion.

9.7 THE JET ENGINE

Jet engines are used extensively to power the airplanes. Figure 9.14 shows a schematic diagram of the jet engine. Fresh air enters the engine at the front and passes through a diffuser before being further compressed by a compressor. The compressed air enters the combustion chamber where fuel is injected and burned. The hot combustion gas at high pressure is passed through a turbine to produce work just enough to run the compressor. The gas from the turbine is expanded in a nozzle to increase its kinetic energy. The discharge of high velocity exhaust gas from the nozzle provides the forward thrust to the aircraft.

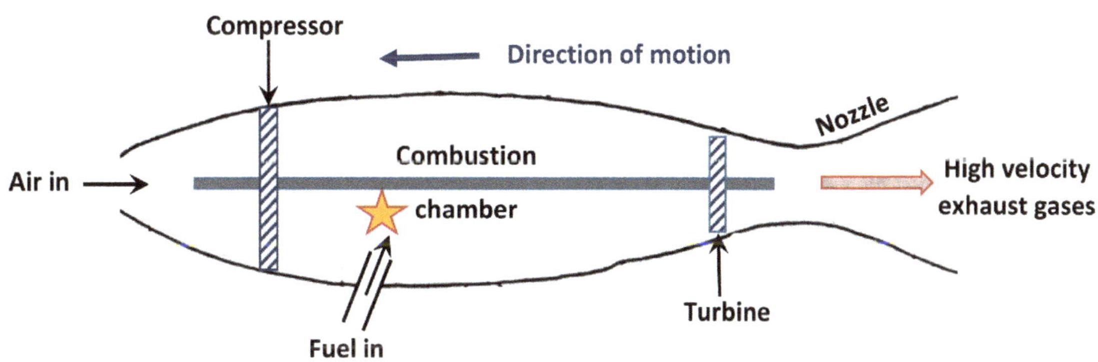

Figure 9.14 The jet engine.

The forward thrust of the jet engine F is given as:

$$F = \dot{m}\left(U_{R,e} - U_{R,i}\right) \tag{9.43}$$

where $\dot{m}$ is the mass flow rate of air through the engine, $U_{R,e}$ is the magnitude of the exhaust velocity relative to the engine ($U_{R,e}$ is a positive number), and $U_{R,i}$ is the magnitude of the relative velocity of air entering the engine ($U_{R,i}$ is a positive number). Note that $U_{R,i}$ is simply the magnitude of the aircraft velocity relative to a stationary observer on earth. Equation (9.43) can be readily derived using the principle of conservation of linear momentum. Obviously $U_{R,e}$ must be larger than $U_{R,i}$ for a net forward thrust on the aircraft.

NOTATION

c_p	specific or molar constant-pressure heat capacity
c_v	specific or molar constant-volume heat capacity
F	forward thrust force
h	specific or molar enthalpy
$\hat{\imath}$	unit vector
m	mass
$\vec{M}$	momentum vector
$\dot{m}$	mass flow rate
P	pressure
q	rate of mass ejection
Q	heat transfer per unit mole or mass
$\dot{Q}$	heat transfer rate
Q_{cycle}	net heat transfer in a cyclic process
R	specific or universal gas constant
r	compression ratio
r_c	cut-off ratio
r_e	expansion ratio
t	time
T	temperature
u	specific or molar internal energy
$\vec{U}$	velocity vector of ejected mass
U_R	speed of exhaust gas relative to the rocket
$\vec{U}_R$	velocity vector of ejected mass relative to the rocket
$U_{R,e}$	speed of exhaust gas relative to jet engine
$U_{R,i}$	speed of air entering relative to jet engine
v	specific or molar volume
V	velocity
$\vec{V}$	velocity vector
W	work per unit mole or mass
$\dot{W}$	rate of work
W_{cycle}	net work in a cyclic process
W_{net}	net work in a cyclic process
W_{sh}	shaft work

Greek Letters

γ	heat capacity ratio
Δ	difference operator
η	thermal efficiency of a cyclic process

TUTORIAL PROBLEMS WITH SOLUTION

Problem 1

An air-standard Otto cycle (see Figure 9.3) is used to model a 4-stroke reciprocating piston/cylinder internal combustion engine using spark ignition. The compression ratio is 12. The pressure and temperature at the beginning of compression are: $P_1 = 120$ kPa and $T_1 = 293$ K. The heat addition per unit mass of fluid (air) is 1250 kJ/kg. Determine the following:

a) the temperature and pressure at the end of each process of the Otto cycle
b) the net work produced per unit mass of fluid
c) the thermal efficiency of the cycle
d) the thermal efficiency of a Carnot cycle operating between maximum and minimum temperatures

You can assume air to be an ideal gas with constant heat capacities: $c_p = (7/2)R$, $c_v = (5/2)R$, $\gamma = 1.4$. The specific gas constant R of air is 286.9 J/kg.K

Solution

Refer to Figure 9.3.

a) Given: $P_1 = 120$ kPa, $T_1 = 293$ K, $r = 12$, and $Q_{23} = 1250$ kJ/kg

Process 1 → 2: Reversible adiabatic compression

$$\frac{P_2}{P_1} = \left(\frac{v_1}{v_2}\right)^{\gamma} = (r)^{\gamma} = (12)^{1.4} = 32.423\ ;\ \ P_2 = 32.423 \times 120 = 3890.76 \text{ kPa}$$

$$\frac{T_2}{T_1} = \left(\frac{v_1}{v_2}\right)^{\gamma-1} = (12)^{0.4} = 2.702\ ; T_2 = 2.702 \times 293 = 791.686 \text{ K}$$

Process 2 → 3: Reversible isochoric heating

$$Q_{23} = \Delta u_{23} = c_v(T_3 - T_2)$$

$$T_3 = T_2 + \frac{Q_{23}}{c_v} = 791.686 + \frac{1250}{2.5 \times 286.9} = 2534.45 \text{ K}$$

$$P_3 = \frac{T_3}{T_2} P_2 = \frac{2534.45}{791.686} \times 3890.76 = 12455.62 \text{ kPa}$$

Process 3 → 4: Reversible adiabatic expansion

$$\frac{P_4}{P_3} = \left(\frac{v_3}{v_4}\right)^{\gamma} = \left(\frac{1}{r}\right)^{\gamma} = \left(\frac{1}{12}\right)^{1.4} = 0.030842\ ;\ \ P_4 = 0.030842 \times 12455.62 = 384.16 \text{ kPa}$$

$$\frac{T_4}{T_3} = \left(\frac{v_3}{v_4}\right)^{\gamma-1} = \left(\frac{1}{r}\right)^{0.4} = \left(\frac{1}{12}\right)^{0.4} = 0.37011\ ; T_4 = 0.37011 \times 2534.45 = 938.02 \text{ K}$$

b) the net work produced per unit mass of fluid

$$W_{12} = -\Delta u_{12} = -c_v(T_2 - T_1) = -2.5 \times 286.9 \times (791.686 - 293) = -357682.53 \text{ J/kg} = -357.683 \text{ kJ/kg}$$

$$W_{23} = 0$$

$$W_{34} = -\Delta u_{34} = -c_v(T_4 - T_3) = -2.5 \times 286.9 \times (938.02 - 2534.45) = -1145039.42 \text{ J/kg} = 1145.039 \text{ kJ/kg}$$

$$W_{41} = 0$$

$$W_{net} = W_{12} + W_{23} + W_{34} + W_{41} = -357.683 + 1145.039 = 787.356 \text{ kJ/kg}$$

c) the thermal efficiency of the cycle

$$\eta = \frac{W_{net}}{Q_{23}} = \frac{787.356}{1250} = 0.6299$$

We can also apply the formula:

$$\eta = 1 - \left(\frac{1}{r}\right)^{\gamma-1} = 1 - \left(\frac{1}{12}\right)^{0.4} = 0.6299$$

d) the Carnot cycle efficiency

$$\eta = 1 - \frac{T_L}{T_H} = 1 - \frac{293}{2534.45} = 0.884$$

Problem 2

An air-standard diesel cycle (see Figure 9.7) is used to model a 4-stroke reciprocating piston/cylinder internal combustion engine using compression ignition. The compression ratio is 20 and the cut-off ratio is 3. The pressure and temperature at the beginning of compression are: $P_1 = 110$ kPa and $T_1 = 293$ K. Determine the following:

a) the temperature and pressure at the end of each process of the cycle
b) the net work produced per unit mass of fluid
c) heat input to the process
d) the thermal efficiency of the cycle
e) compare the efficiency of the cycle with the efficiency of an Otto cycle with the same compression ratio

You can assume air to be an ideal gas with constant heat capacities: $c_p = (7/2)R$, $c_v = (5/2)R$, $\gamma = 1.4$

Solution

Refer to Figure 9.7.

a) Given: $P_1 = 110$ kPa, $T_1 = 293$ K, $r = 20$, and $r_c = 3$

Process 1 → 2: Reversible adiabatic compression

$$\frac{P_2}{P_1} = \left(\frac{v_1}{v_2}\right)^{\gamma} = (r)^{\gamma} = (20)^{1.4} = 66.289\ ; \quad P_2 = 66.289 \times 110 = 7291.8 \text{ kPa}$$

$$\frac{T_2}{T_1} = \left(\frac{v_1}{v_2}\right)^{\gamma-1} = (20)^{0.4} = 3.314 \; ; T_2 = 3.314 \times 293 = 971 \text{ K}$$

Process 2 → 3: Reversible isobaric heating

$$P_3 = P_2 = 7291.8 \text{ kPa}$$

$$T_3 = \left(\frac{v_3}{v_2}\right) T_2 = 3 \times 971 = 2913 \text{ K}$$

Process 3 → 4: Reversible adiabatic expansion

$$\frac{P_4}{P_3} = \left(\frac{v_3}{v_4}\right)^{\gamma} = \left(\frac{1}{r_e}\right)^{\gamma} = \left(\frac{r_c}{r}\right)^{1.4} = \left(\frac{3}{20}\right)^{1.4} = 0.0702 \; ; \quad P_4 = 0.0702 \times 7291.8 = 511.88 \text{ kPa}$$

$$\frac{T_4}{T_3} = \left(\frac{v_3}{v_4}\right)^{\gamma-1} = \left(\frac{3}{20}\right)^{0.4} = 0.4682 \; ; T_4 = 0.4682 \times 2913 = 1363.87 \text{ K}$$

b) the net work produced per unit mass of fluid

$$W_{12} = -\Delta u_{12} = -c_v(T_2 - T_1) = -2.5 \times 286.9 \times (971 - 293) = -486295.5 \text{ J/kg} = -486.296 \text{ kJ/kg}$$

$$W_{23} = P_2(v_3 - v_2) = R(T_3 - T_2) = 286.9(2913 - 971) = 557159.8 \text{ J/kg} = 557.16 \text{ kJ/kg}$$

$$W_{34} = -\Delta u_{34} = -c_v(T_4 - T_3) = -2.5 \times 286.9 \times (1363.87 - 2913) = -1111113.5 \text{ J/kg} = 1111.11 \text{ kJ/kg}$$

$$W_{41} = 0$$

$$W_{net} = W_{12} + W_{23} + W_{34} + W_{41} = -486.296 + 557.16 + 1111.11 = 1181.97 \text{ kJ/kg}$$

c) heat input to the process

$$Q_{23} = \Delta h_{23} = c_p(T_3 - T_2) = 3.5 \times 286.9 \times (2913 - 971) = 1950059.3 \text{ J/kg} = 1950.06 \text{ kJ/kg}$$

d) the thermal efficiency of the cycle

$$\eta = \frac{W_{net}}{Q_{23}} = \frac{1181.97}{1950.06} = 0.606$$

We can also apply the formula:

$$\eta = 1 - \left(\frac{1}{r}\right)^{\gamma-1} \left[\frac{r_c^{\gamma} - 1}{\gamma(r_c - 1)}\right] = 1 - \left(\frac{1}{20}\right)^{0.4} \left[\frac{3^{1.4} - 1}{1.4(3 - 1)}\right] = 0.606$$

e) the efficiency of an Otto cycle with the same compression ratio

$$\eta = 1 - \left(\frac{1}{r}\right)^{\gamma-1} = 1 - \left(\frac{1}{20}\right)^{0.4} = 0.6983$$

Thus, for the same compression ratio r, the thermal efficiency of the diesel engine is less than that of the Otto engine.

Problem 3

An air-standard Brayton cycle (see Figure 9.10) is used to model a gas turbine engine (see Figure 9.9). Air at 293 K and 101.3 kPa enters the compressor at a volumetric rate of 7 m^3/s. The pressure ratio across the compressor is 12. The temperature of air leaving the combustion chamber is 1295 K.
Determine the following:

a) the net power produced
b) the rate of heat addition in the combustion chamber
c) the thermal efficiency of the cycle

You can assume air to be an ideal gas with constant heat capacities: $c_p = (7/2)R$, $c_v = (5/2)R$, $\gamma = 1.4$

Solution

Refer to Figure 9.10.

Given: $P_1 = 101.3$ kPa, $T_1 = 293$ K, volumetric flow rate of air is 7 m^3/s, $P_2/P_1 = 12$, $T_3 = 1295$ K.

$$\dot{m} = \frac{P_1}{RT_1}(air\ flow\ rate) = \frac{101.3 \times 10^3}{286.9 \times 293}(7) = 8.435 \text{ kg/s}$$

$$\frac{P_2}{P_1} = \left(\frac{v_1}{v_2}\right)^{\gamma} = (r)^{\gamma} = (20)^{1.4} = 66.289\ ;\quad P_2 = 66.289 \times 110 = 729.8 \text{ kPa}$$

$$\frac{T_2}{T_1} = \left(\frac{P_2}{P_1}\right)^{(\gamma-1)/\gamma} = (12)^{(0.4/1.4)} = 2.034\ ; T_2 = 2.034 \times 293 = 595.96 \text{ K}$$

$$\frac{T_4}{T_3} = \left(\frac{P_1}{P_2}\right)^{(\gamma-1)/\gamma} = \left(\frac{1}{12}\right)^{(0.4/1.4)} = 0.4917\ ; T_4 = 0.4917 \times 1295 = 636.75 \text{ K}$$

a) the net power produced

$$W_{12} = -\Delta h_{12} = c_p(T_1 - T_2) = 3.5 \times 286.9 \times (293 - 595.96) = -304217.28 \text{ J/kg} = -304.217 \text{ kJ/kg}$$

$$\dot{W}_{12} = -304.217 \times \dot{m} = -304.217 \times 8.435 = -2566.07 \text{ kW}$$

$$W_{34} = -\Delta h_{34} = c_p(T_3 - T_4) = 3.5 \times 286.9 \times (1295 - 636.75) = 660981.74 \text{ J/kg} = 660.982 \text{ kJ/kg}$$

$$\dot{W}_{34} = 660.982 \times \dot{m} = 660.982 \times 8.435 = 5575.38 \text{ kW}$$

$$\dot{W}_{net} = \dot{W}_{12} + \dot{W}_{34} = -2566.07 + 5575.38 = 3009.31 \text{ kW}$$

b) the rate of heat addition in the combustion chamber

$$Q_{23} = \Delta h_{23} = c_p(T_3 - T_2) = 3.5 \times 286.9 \times (1295 - 595.96) = 701941.06 \text{ J/kg} = 701.941 \text{ kJ/kg}$$

$$\dot{Q}_{23} = 701.941 \times \dot{m} = 701.941 \times 8.435 = 5920.87 \text{ kW}$$

c) the thermal efficiency of the cycle

$$\eta = \frac{\dot{W}_{net}}{\dot{Q}_{23}} = \frac{3009.31}{5920.87} = 0.5083$$

We can also apply the formula:

$$\eta = 1 - \frac{1}{\left(\frac{P_2}{P_1}\right)^{(\gamma-1)/\gamma}} = 1 - \frac{1}{(12)^{(1.4-1)/1.4}} = 0.5083$$

Problem 4

Air enters a turbojet engine at 10 kPa and 215 K and an inlet velocity of 290 m/s. The mass flow rate of air is 25 kg/s. The compression ratio is 12 and the temperature at the turbine inlet is 950 °C. Calculate the following:

a) the temperature and pressure at the end of each unit (diffuser, compressor, combustor, turbine, and nozzle)
b) the velocity of gas at the nozzle exit
c) the thrust produced by the engine
d) the rate of fuel consumption. The heating value of the fuel is 8100 kJ/kg.

You can assume air to be an ideal gas with constant heat capacities: $c_p = (7/2)R$, $c_v = (5/2)R$, $\gamma = 1.4$. The flow in the diffuser, compressor, turbine, and nozzle is isentropic. The work produced by the turbine is equal to the work required by the compressor.

Solution

Refer to the figure below.

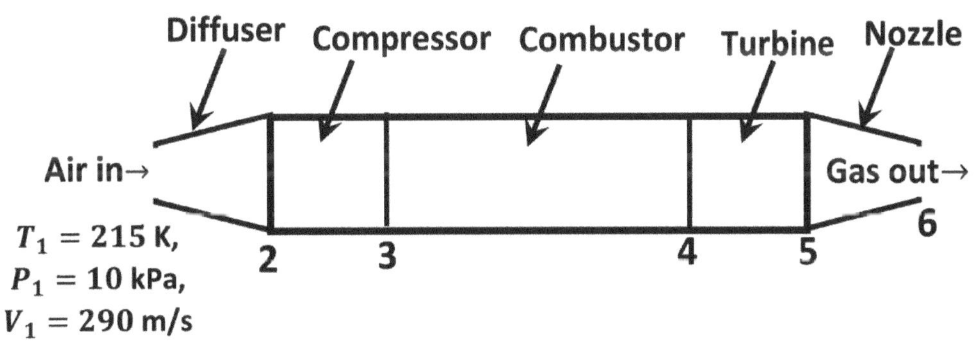

a) the temperature and pressure at the end of each unit (diffuser, compressor, combustor, turbine, and nozzle)

Diffuser:

Apply the first law for open systems to diffuser (no heat transfer, no shaft work, no potential energy change),

$$h_2 + \frac{V_2^2}{2} = h_1 + \frac{V_1^2}{2}$$

Assuming the velocity at the end of diffuser to be negligible ($V_2 \approx 0$),

$$h_2 - h_1 = \frac{V_1^2}{2} = c_p(T_2 - T_1)$$

$$T_2 = T_1 + \frac{V_1^2}{2C_p} = 215 + \frac{290^2}{2\times3.5\times286.9} = 256.88\text{ K}$$

Note that the gas constant of air is 286.9 J/kg.K.

Assuming the process to be isentropic,

$$\frac{P_2}{P_1} = \left(\frac{T_2}{T_1}\right)^{\gamma/(\gamma-1)} = \left(\frac{256.88}{215}\right)^{1.4/(1.4-1)} = 1.864\ ; \quad P_2 = 1.864 \times 10 = 18.64\text{ kPa}$$

Compressor:

$$T_2 = 256.88\text{ K},\ P_2 = 18.64\text{ kPa, Compression ratio=12}$$

$$\frac{P_3}{P_2} = 12\ ;\ P_3 = 12 \times 18.64 = 223.68\text{ kPa}$$

Assuming the process to be isentropic,

$$\frac{T_3}{T_2} = \left(\frac{P_3}{P_2}\right)^{(\gamma-1)/\gamma} = (12)^{(1.4-1)/1.4} = 2.034\ ; \quad T_3 = 2.034 \times 256.88 = 522.49\text{ K}$$

Combustor:

$$P_4 = P_3 = 223.68\text{ kPa},\ T_4 = 950 + 273.15 = 1223.15\text{ K}$$

Turbine:

The work produced by the turbine is equal to the work required by the compressor:

$$W_{sh,turbine} = -W_{sh,compressor}\ ;\ h_4 - h_5 = h_3 - h_2\ ; T_4 - T_5 = T_3 - T_2$$

$$T_5 = T_4 - T_3 + T_2 = 1223.15 - 522.49 + 256.88 = 957.54\text{ K}$$

Assuming the process to be isentropic,

$$\frac{P_5}{P_4} = \left(\frac{T_5}{T_4}\right)^{\gamma/(\gamma-1)} = \left(\frac{957.54}{1223.15}\right)^{1.4/(1.4-1)} = 0.4245\ ; \quad P_5 = 0.4245 \times 223.68 = 94.95\text{ kPa}$$

Nozzle:

$$P_6 = P_1 = 10\text{ kPa}$$

$$\frac{T_6}{T_5} = \left(\frac{P_6}{P_5}\right)^{(\gamma-1)/\gamma} = \left(\frac{10}{94.95}\right)^{(1.4-1)/1.4} = 0.5257\ ; \quad T_6 = 0.5257 \times 957.54 = 503.38\text{ K}$$

b) the velocity of gas at the nozzle exit

Apply the first law for open systems to the whole turbojet engine between inlet and exit (no heat transfer, no shaft work, no potential energy change),

$$h_6 + \frac{V_6^2}{2} = h_1 + \frac{V_1^2}{2}$$

$$h_6 - h_1 = c_p(T_6 - T_1)$$

$$V_6^2 = V_1^2 + 2c_p(T_6 - T_1) = (290)^2 + 2 \times 3.5 \times 286.9 \times (503.38 - 215) = 663253.554$$

$$V_6 = 814.40 \text{ m/s}$$

c) the thrust produced by the engine

$$F = \dot{m}(V_6 - V_1) = 25 \times (814.40 - 290) = 13110 \text{ N} = 13.11 \text{ kN}$$

d) the rate of fuel consumption. The heating value of the fuel is 8100 kJ/kg.

$$\dot{Q}_{combustor} = \dot{m}c_p(T_4 - T_3) = 25 \times 3.5 \times 286.9 \times (1223.15 - 522.49) = 17589193.48 \text{ W} = 17589.19 \text{ kW}$$

$$\text{Fuel consumption: } \dot{m}_{fuel} \times 8100 = 17589.19 \;;\; \dot{m}_{fuel} = 2.17 \text{ kg/s}$$

Problem 5

A rocket engine ejects mass at a rate of 25 kg/s with an exhaust velocity of 2,900 m/s.

a) Calculate the thrust produced by the engine assuming the exit pressure of gas is equal to the ambient pressure.
b) If the initial mass of the rocket is 25,000 kg, what is the change in velocity of the rocket in 5 minutes.

Solution

Given: $q = 25$ kg/s, $U_R = 2900$ m/s

a) Thrust produced

$$F = qU_R = 25 \times 2900 = 72500 \text{ N} = 72.5 \text{ kN}$$

b) change in velocity of the rocket in 5 minutes.

From conservation of momentum,

$$m\left(\frac{d\vec{V}}{dt}\right) = q[U_R\hat{\imath}] \tag{9.42}$$

The mass of the rocket decreases at a rate of q:

$$-\frac{dm}{dt} = q$$

From the above equations, we get:

$$m\left(\frac{dV}{dt}\right) = -U_R\frac{dm}{dt}$$

Upon re-arrangement and integration:

$$\int_V^{V+\Delta V} dV = -U_R\int_{m_o}^{m}\frac{dm}{m}$$

Thus,

$$\Delta V = U_R ln\left(\frac{m_o}{m}\right) = 2900 \times ln\left(\frac{25000}{25000-25\times5\times60}\right) = 1034.36 \text{ m/s}$$

Note that the mass m of the rocket after 5 minutes is $(25000 - 25 \times 5 \times 60) = 17500$ kg.

PRATICE PROBLEMS

Problem 1

Consider an air-standard Otto cycle as a model for a 4-stroke reciprocating piston/cylinder internal combustion engine using spark ignition. The compression ratio is 7.5. The pressure and temperature at the beginning of compression are: $P_1 = 101.3$ kPa and $T_1 = 298$ K. The net work output of the cycle per unit mass of fluid (air) is 900 kJ/kg. Determine the following:

a) the temperature and pressure at the end of each process of the Otto cycle
b) the heat addition per unit mass of fluid
c) the thermal efficiency of the cycle
d) the thermal efficiency of a Carnot cycle operating between maximum and minimum temperatures

You can assume air to be an ideal gas with constant heat capacities: $c_p = (7/2)R$, $c_v = (5/2)R$, $\gamma = 1.4$

Problem 2

Consider an air-standard diesel cycle (see Figure 9.7) as a model for a 4-stroke reciprocating piston/cylinder internal combustion engine using compression ignition. The cycle absorbs 1152 J/mol of heat at constant pressure (step 2 → 3 of Figure 9.7). The pressure and temperature at the start of the compression process are: $P_1 = 105$ kPa and $T_1 = 298$ K. The pressure at the end of the compression process is 5.3 bar. Determine the following:

a) the temperature, pressure, and molar volume at the end of each process of the cycle
b) the compression ratio, cut-off ratio, and expansion ratio
c) the net work produced per mole of fluid
d) the thermal efficiency of the cycle

You can assume air to be an ideal gas with constant heat capacities: $c_p = (7/2)R$, $c_v = (5/2)R$, $\gamma = 1.4$

Problem 3

A gas turbine engine (see Figure 9.9) is modeled as an air-standard Brayton cycle (see Figure 9.10). Air at 298 K and 110 kPa enters the compressor at a certain mass flow rate. The air leaves the compressor at a pressure of 1000 kPa. The temperature of air leaving the combustion chamber is 1455 K. The power output of the cycle is 720 kW.

Determine the following:

a) the thermal efficiency of the cycle
b) the rate of heat addition in the combustion chamber
c) the mass flow rate of air

You can assume air to be an ideal gas with constant heat capacities: $c_p = (7/2)R$, $c_v = (5/2)R$, $\gamma = 1.4$

Problem 4

Air enters a turbojet engine at 100 kPa and 298 K and an inlet velocity of 225 m/s. The mass flow rate of air is 35 kg/s. The compression ratio is 9.2 and the temperature at the turbine inlet is 1025 °C. Calculate the following:

a) the temperature and pressure at the end of each unit (diffuser, compressor, combustor, turbine, and nozzle)
b) the velocity of gas at the nozzle exit
c) the thrust produced by the engine
d) the rate of fuel consumption. The heating value of the fuel is 7800 kJ/kg.

You can assume air to be an ideal gas with constant heat capacities: $c_p = (7/2)R$, $c_v = (5/2)R$, $\gamma = 1.4$. The flow in the diffuser, compressor, turbine, and nozzle is isentropic. The work produced by the turbine is equal to the work required by the compressor.

Problem 5

A rocket engine ejects mass at a rate of 28 kg/s. The thrust produced by the engine is 100 kN. Determine the exhaust velocity of gas. If the change in the velocity of the rocket is 850 m/s in 3.5 minutes, what is the initial mass of the rocket?

Chapter 10

Refrigeration Cycles

10.1 LEARNING OUTCOMES

Upon completion of this chapter, students will be able to:

- Appreciate the importance of second law in the development of refrigeration cycles
- Describe Carnot vapor-compression refrigeration cycle and problems associated with it
- Derive the expression for coefficient-of-performance (COP) of a Carnot vapor-compression refrigeration cycle
- Describe the practical modified - Carnot refrigeration cycle and derive the expression for its COP
- Describe the vapor-compression refrigeration cycle using throttling valve for expansion. Derive the expression for its COP
- Explain the importance of cascading of vapor-compression refrigeration cycles
- Describe absorption refrigeration cycle. Explain the difference between vapor-compression and absorption refrigeration cycles
- Describe heat pump based on Carnot vapor-compression cycle and derive the expression for its COP.
- Describe the vapor-compression heat pump cycle using throttling valve for expansion. Derive the expression for its COP
- Describe air conditioner
- Solve problems dealing with refrigeration cycles, heat pumps and air conditioners.

10.2 INTRODUCTION

The maintenance of a temperature well below that of the surroundings in a selected region of space is referred to as *refrigeration*. As the surroundings are at high temperature heat continuously leaks into the cold region (e.g., cold storage warehouse) through the boundary wall, as shown schematically in Figure 10.1. Heat may also enter the cold region with warm air entering through door openings. The continuous removal of heat that leaks into the cold region from the high temperature surroundings is required in order to maintain the cold region temperature at a low temperature T_L.

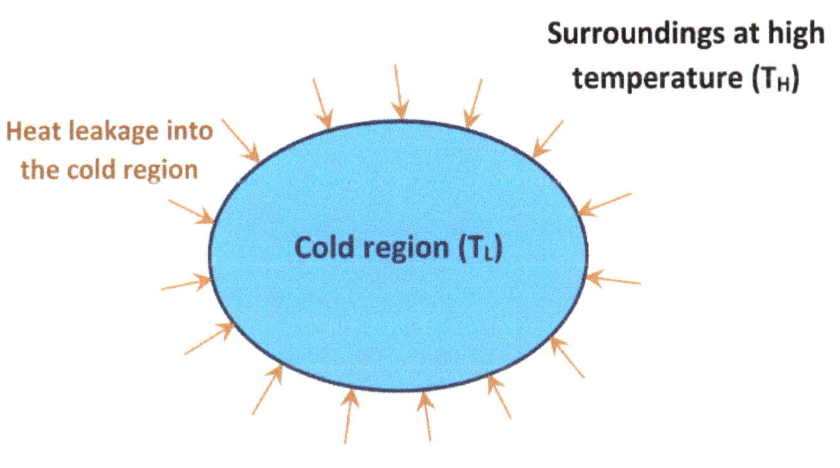

Figure 10.1 Cold region at T_L surrounded by high temperature surroundings at T_H.

According to the second law of thermodynamics, it is impossible to develop a cyclic process whose sole effect is the transfer of heat from a low temperature region to a high temperature region. Practical experience indicates that we need to supply mechanical energy to the cyclic process in order to achieve the transfer of heat from low temperature region to a high temperature region. Figure 10.2 shows the scheme on which the refrigeration cycles are based. Heat is absorbed from the low temperature region by the working fluid of the cyclic process and transferred to a high temperature region with the help of work input to the process.

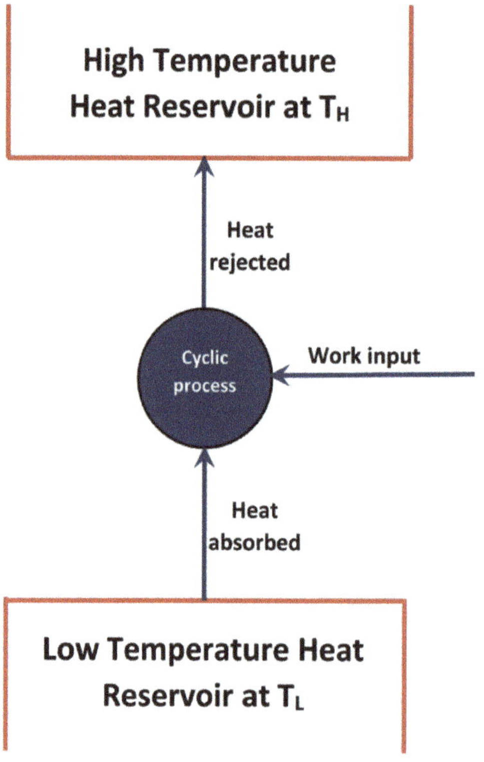

Figure 10.2 Transfer of heat from cold region to hot region by a cyclic process.

In this chapter, we will cover vapor-compression refrigeration cycles, absorption refrigeration cycle, heat pump cycle, and air-conditioning cycle. In all these cycles, heat is transferred from a low temperature region to a high temperature region.

10.3 VAPOR-COMPRESSION REFRIGERATION CYCLES

In a vapor-compression refrigeration cycle, the refrigerant is alternately vaporized and condensed to transfer heat from cold region to hot region. The mechanical energy is supplied to the process with the help of a compressor.

10.3.1 Carnot Refrigeration Cycle

The simplest possible vapor-compression refrigeration cycle is the Carnot cycle, shown in Figure 10.3. The Carnot refrigeration cycle is a reversible cycle involving four steps:

- Step $1 \rightarrow 2$: isentropic (reversible adiabatic) compression where work is absorbed by the process.
- Step $2 \rightarrow 3$: reversible isothermal condensation process where heat is rejected by the process to the high temperature surroundings at T_H.
- Step $3 \rightarrow 4$: isentropic (reversible adiabatic) expansion where work is produced by the process.

- Step 4 → 1 : reversible isothermal evaporation process where heat is absorbed by the process from cold region at T_L.

Note that the Carnot refrigeration cycle is a net work absorbing cycle, that is, work required in the compression process (step 1 → 2) is much larger than the work produced in the expansion process (step 3 → 4).

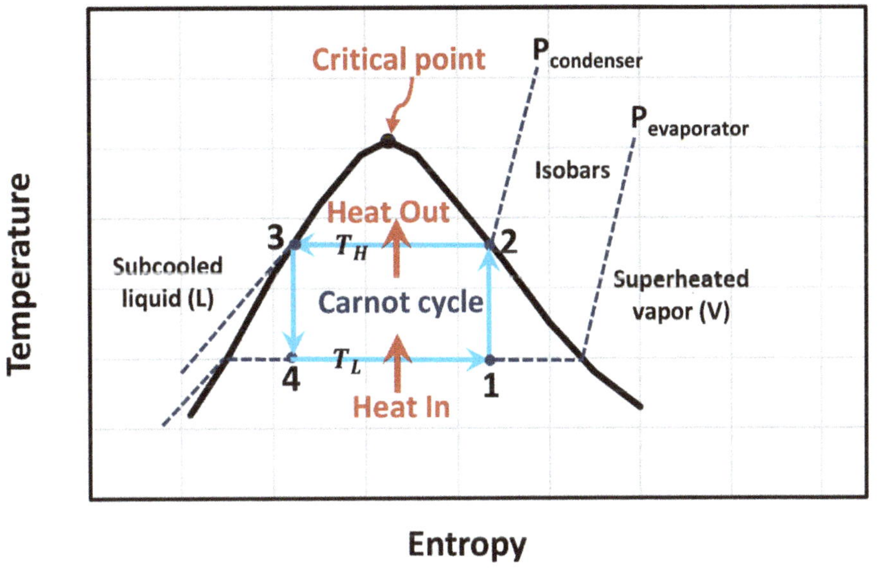

Figure 10.3 Carnot refrigeration cycle.

Figure 10.4 shows the equipment diagram of a refrigeration plant based on Carnot vapor-compression refrigeration cycle. The refrigerant in state 1 (two-phase liquid-vapor mixture at T_L and $P_{evaporator}$) is compressed adiabatically to state 2 (saturated vapor at T_H and $P_{condenser}$) by the compressor. The refrigerant in state 2 passes through a condenser where it undergoes phase change from 100% saturated vapor to 100% saturated liquid as a result of heat rejection to the warm region at T_H. The temperature and pressure remain constant in the condenser. The refrigerant in state 3 (saturated liquid at T_H and $P_{condenser}$) is expanded adiabatically in a turbine to state 4 (two-phase liquid-vapor mixture at T_L and $P_{evaporator}$). The refrigerant then passes through the evaporator where some of the liquid refrigerant is vaporized as a result of heat absorption from the cold region at T_L. The temperature and pressure remain constant in the evaporator. Thus heat is continuously absorbed from the cold region and rejected into the warm region at the expense of mechanical energy. As mentioned earlier, the cycle requires a net mechanical energy input.

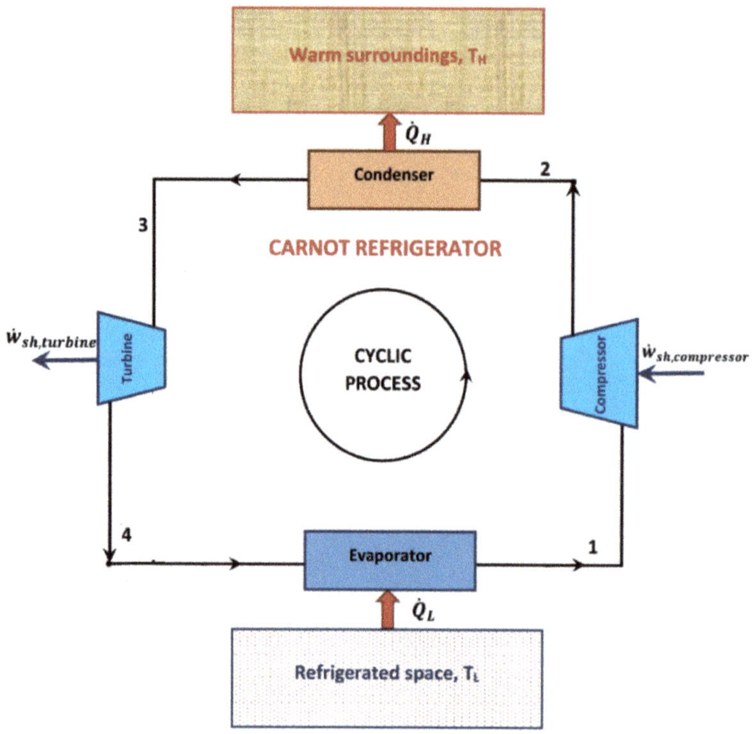

Figure 10.4 Equipment diagram of a refrigeration plant based on Carnot vapor-compression refrigeration cycle.

The performance of a refrigeration cycle is measured in terms of the "*Coefficient of Performance (COP)*", defined as:

$$\omega = \frac{Heat\ removed\ from\ refrigerated\ space}{|W_{cycle}|} = \frac{Q_{4\to1}}{-(Q_{cycle})} = \frac{Q_{4\to1}}{-(Q_{2\to3}+Q_{4\to1})} \quad (10.1)$$

where $|W_{cycle}|$ is the modulus (positive value) of W_{cycle} . As steps $2 \to 3$ and $4 \to 1$ are reversible isothermal processes (see Figure 10.3),

$$Q_{2\to3} = T_H(s_3 - s_2) \quad (10.2)$$

$$Q_{4\to1} = T_L(s_1 - s_4) \quad (10.3)$$

From Equations (10.1) to (10.3), we get:

$$\omega = \frac{T_L(s_1-s_4)}{-[T_H(s_3-s_2)+T_L(s_1-s_4)]} \quad (10.4)$$

From Figure 10.3, $s_1 = s_2$ and $s_3 = s_4$. Consequently, Equation (10.4) reduces to:

$$\omega = \left(\frac{T_L}{T_H-T_L}\right) \quad (10.5)$$

Thus, the COP of a Carnot refrigeration cycle depends only on the temperatures in the evaporator and condenser. This is the maximum possible ω for any refrigerator operating between two given temperatures T_L and T_H. The COP of a refrigeration cycle is usually greater than 1. Higher the COP, lower is the operating cost of the refrigerator.

Figure 10.5 shows the variation of Carnot refrigerator COP with the increase in evaporator temperature T_L for different values of condenser temperature T_H. The COP increases with the increase in T_L for any given T_H. The COP decreases as T_H increases.

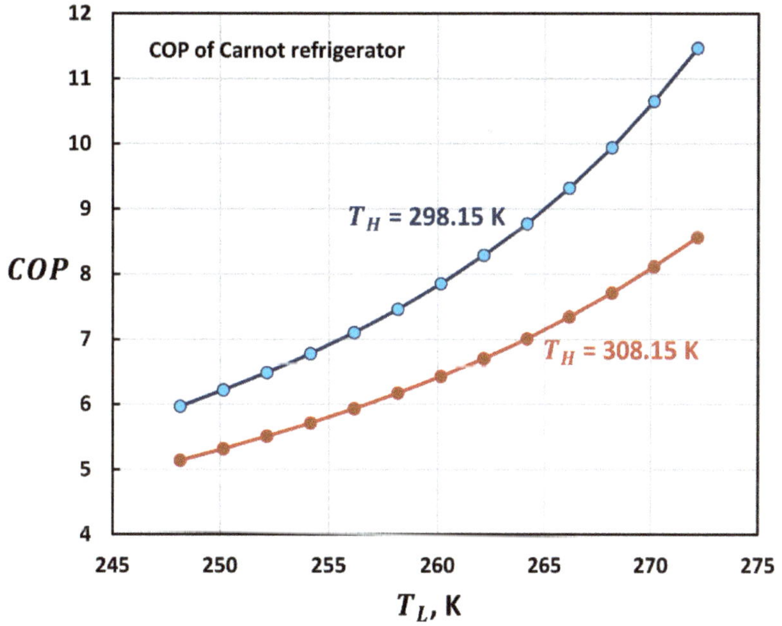

Figure 10.5 Variation of Carnot refrigerator COP with cold region temperature T_L for different values of hot region temperature T_H.

Problem with Carnot Refrigeration Cycle

Practical refrigeration processes are rarely based on the Carnot cycle. The main problem is in step 1 → 2, that is, compression of vapor-liquid mixture (see Figure 10.6). Compressors are generally designed for *dry compression* (vapor without any liquid). Compression of vapor-liquid mixture, known as *wet compression*, damages the compressor due to the presence of liquid droplets in the vapor.

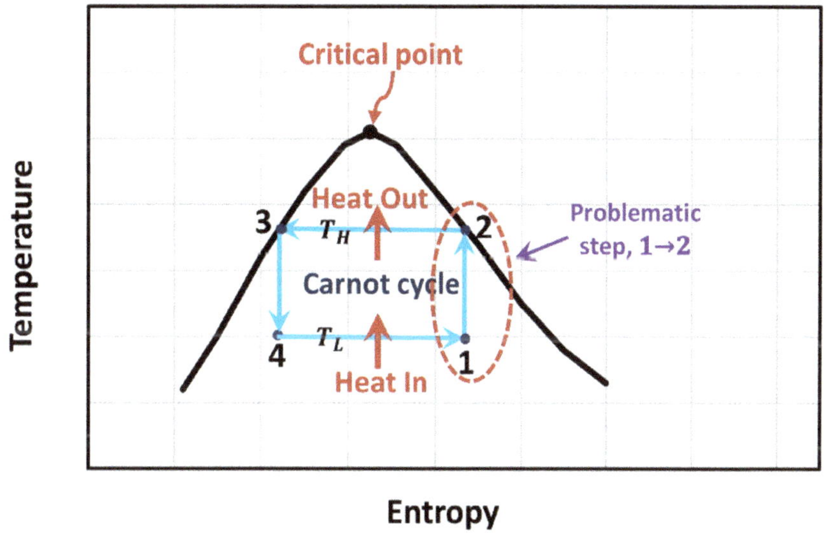

Figure 10.6 Problematic step in the operation of Carnot refrigeration cycle.

10.3.2 Modified-Carnot Refrigeration Cycle

Figure 10.7 shows the modified - Carnot refrigeration cycle. To overcome the problem associated with the Carnot refrigeration cycle, the isobaric heat absorption step 4 → 1 is carried out up to the saturated vapor point 1. The compression of dry vapor (step 1 → 2) is carried out easily in a compressor. Upon isentropic compression of vapor (step 1 → 2), a superheated vapor is produced. The heat rejection is then carried out along isobar (step 2 → 3) corresponding to condenser pressure.

The COP of a modified - Carnot refrigeration cycle can be expressed as:

$$\omega = \left(\frac{T_L}{\bar{T}_H - T_L}\right) \tag{10.6}$$

where $\bar{T}_H$ is the average temperature over step 2 → 3. As $\bar{T}_H$ is higher than T_H, the COP of a modified - Carnot refrigerator is expected to be lower than that of the Carnot refrigeration cycle operating between T_H and T_L.

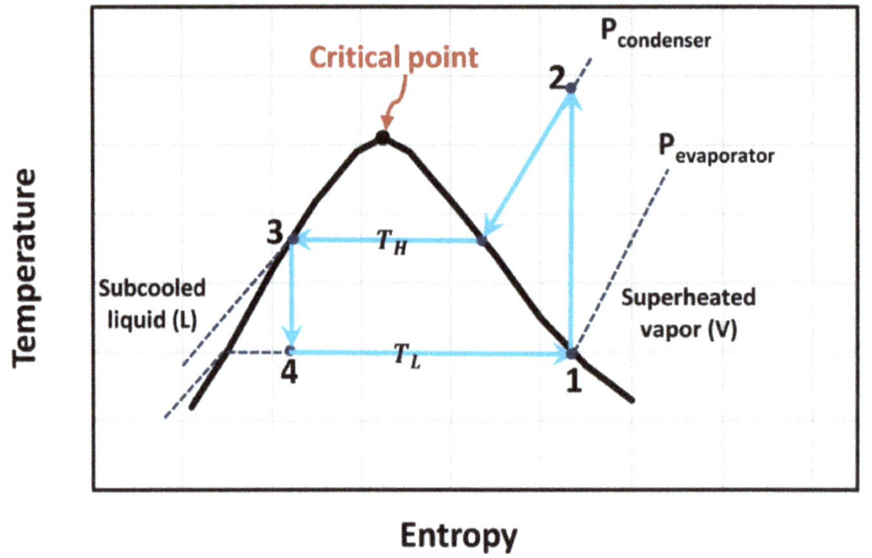

Figure 10.7 The modified – Carnot refrigeration cycle.

The modified – Carnot refrigeration cycle shown in Figure 10.7 is a reversible (ideal) cycle. In practice, steps 1 → 2 and 3 → 4 are irreversible adiabatic processes generating entropy. The practical version of the modified – Carnot refrigeration cycle is shown in Figure 10.8 as 1 → 2 → 3 → 4 → 1. The ideal cycle is 1 → 2′ → 3 → 4′ → 1. The entropy increases in steps 1 → 2 and 3 → 4.

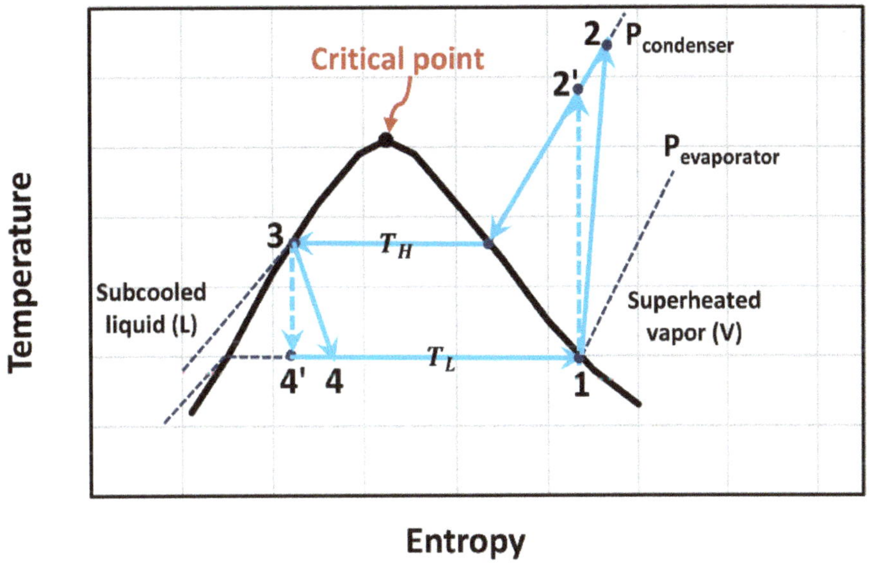

Figure 10.8 Practical modified – Carnot refrigeration cycle.

Let us now determine the COP of the practical refrigeration cycle 1 → 2 → 3 → 4 → 1 of Figure 10.8. By definition,

$$\omega = \frac{Heat\ removed\ from\ refrigerated\ space}{|W_{cycle}|} = \frac{Q_{4\to1}}{-(Q_{cycle})} = \frac{Q_{4\to1}}{-(Q_{2\to3}+Q_{4\to1})} \tag{10.1}$$

From first law for open systems:

$$Q_{4\to1} = h_1 - h_4 \tag{10.7}$$

$$Q_{2\to3} = h_3 - h_2 \tag{10.8}$$

From Equations (10.1), (10.7), (10.8), it follows that:

$$\omega = \frac{h_1 - h_4}{(h_2-h_3)-(h_1-h_4)} \tag{10.9}$$

10.3.3 Vapor-Compression Refrigeration Cycle Using Throttling Valve for Expansion

The modified-Carnot refrigeration cycle utilizes turbine to extract work in the expansion step 3 → 4. The work produced in the expansion of liquid refrigerant (step 3 → 4) is usually small as the specific volume of liquid is small. Note that the shaft work produced in a turbine (step 3 → 4) is given as:

$$W_{sh} = -\int_3^4 vdP = \int_{P_{evaporator}}^{P_{condenser}} vdP \tag{10.10}$$

Equation (10.10) follows from the mechanical energy equation (Equation (3.26)) neglecting kinetic and potential energy changes and friction loss.

Thus, installation of a turbine to extract work from high pressure liquid refrigerant is not justified economically. In practice, the turbine is replaced by a simple *throttling valve* to save initial capital and maintenance costs.

Figure 10.9 shows the equipment diagram of a vapor-compression refrigeration cycle where turbine is replaced by a throttling valve. The T-s diagram of the cycle is shown in Figure 10.10. Note that the compression process (step 1 → 2) is irreversible adiabatic and hence non-isentropic. The expansion of refrigerant in a throttling valve (step 3 → 4) is also irreversible adiabatic and hence non-isentropic. Consequently, the entropy increases in steps 1 → 2 and 3 → 4. However, the enthalpy does not change in the throttling process as the process is *isenthalpic* and therefore, the specific enthalpy of refrigerant in state 4 is equal to the specific enthalpy in state 3:

$$h_4 = h_3 \tag{10.11}$$

Upon substitution of Equation (10.11) into Equation (10.9), the expression for the COP of the refrigeration cycle becomes:

$$\omega = \frac{h_1 - h_4}{h_2 - h_1} \tag{10.12}$$

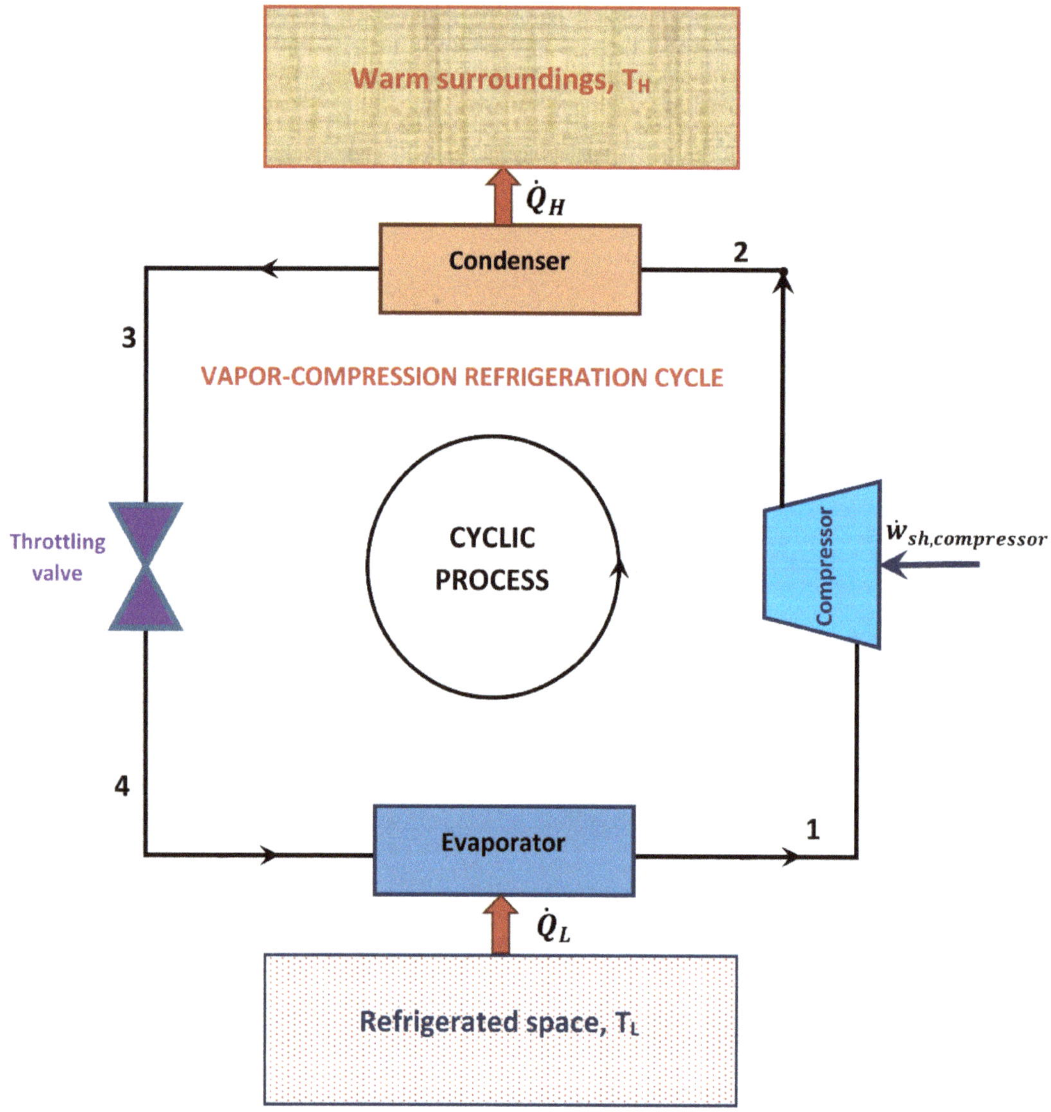

Figure 10.9 Equipment diagram of a vapor-compression refrigeration cycle using throttling valve for expansion.

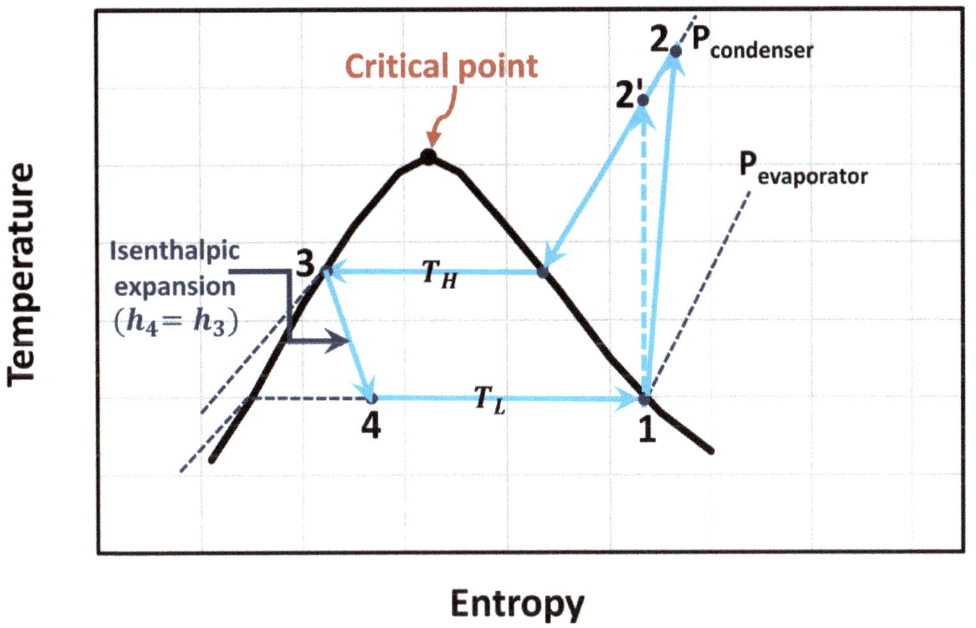

Figure 10.10 Vapor-compression refrigeration cycle using throttling valve for expansion.

10.4 CASCADING OF VAPOR-COMPRESSION REFRIGERATION CYCLES

For refrigeration applications requiring ultra-low temperatures (for example, T_L lower than -50^oC) and high refrigeration effect, conventional single vapor-compression refrigeration cycles are not effective. Even if the desired ultra-low temperature can be reached using a single refrigeration cycle, the COP will be very low. Furthermore, the ratio of condenser pressure to evaporator pressure would be too high for the compressor to function efficiently. In such applications, *cascading* of vapor-compression refrigeration cycles is done. Using a series of vapor-compression refrigeration cycles, one can achieve ultra-low refrigeration temperatures with good refrigeration effect and high COP.

Figure 10.11 shows the equipment diagram for a *cascade* of two vapor-compression refrigeration cycles. The two refrigeration cycles are arranged in series with a countercurrent flow heat exchanger as a link between the two cycles. The energy rejected by the low temperature cycle (cycle A) during cooling/condensation of refrigerant superheated vapor leaving the compressor of cycle A is utilized to evaporate the refrigerant of the high temperature cycle (cycle B) using the countercurrent flow heat exchanger. The net effect of the cascade is the removal of heat from the low temperature region (refrigerated space) at T_L and rejection of heat in the high temperature surroundings at T_H.

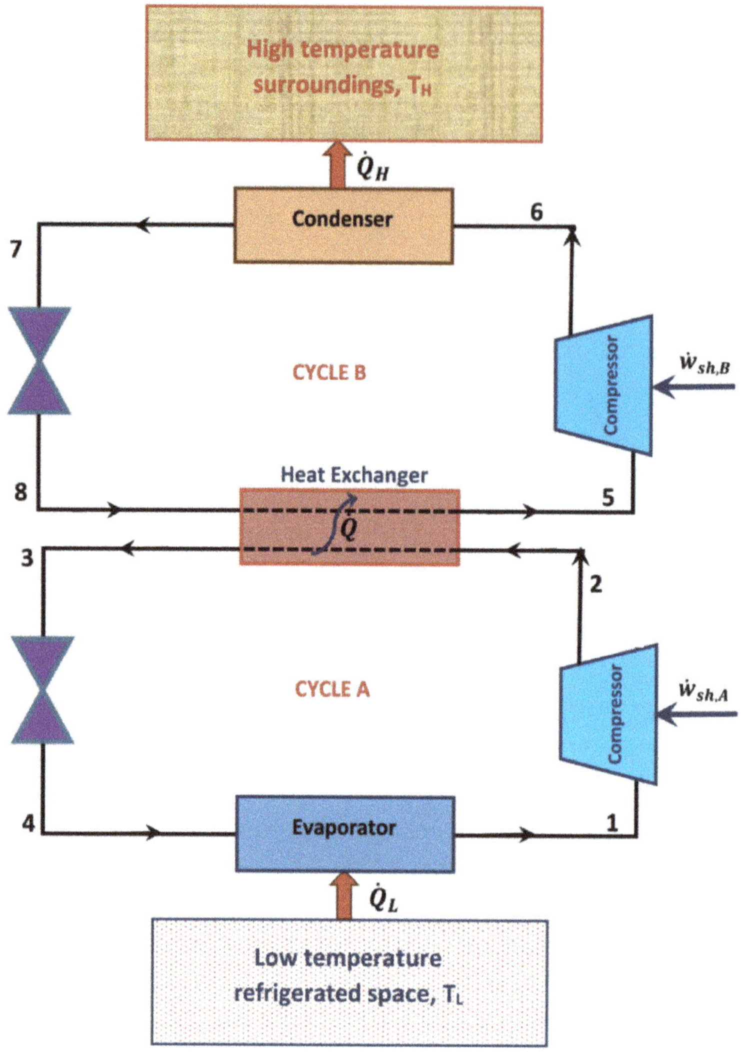

Figure 10.11 Cascade of two vapor-compression refrigeration cycles.

Figure 10.12 shows the T-s diagram of the cascade of two vapor-compression refrigeration cycles. The low temperature cycle (cycle A) corresponds to cycle 1 → 2 → 3 → 4 → 1 marked in light blue color. The high temperature cycle (cycle B) corresponds to cycle 5 → 6 → 7 → 8 → 5 marked in purple color.

The COP of the refrigeration process is:

$$\omega = \frac{Heat\ removal\ from\ refrigerated\ space}{Net\ work\ input} \tag{10.13}$$

Upon comparison of a single vapor-compression refrigeration cycle (1 → 2′ → 7 → 4′ → 1) with a cascade of two refrigeration cycles in series (cycles A and B, Figure 10.12), the COP of the cascade is higher due to two reasons: 1) the amount of heat removal from the refrigerated region (that is, refrigeration capacity) is increased as shown by the hatched rectangle area under the line 4 → 4′ in Figure 10.12, and 2) the amount of compressor work required to achieve the refrigeration is reduced as shown by the hatched area 2 → 2′ → 6 → 5 → 2 in Figure 10.12. An increase in refrigeration capacity and a decrease in net work input both favor the COP (see Equation (10.13)).

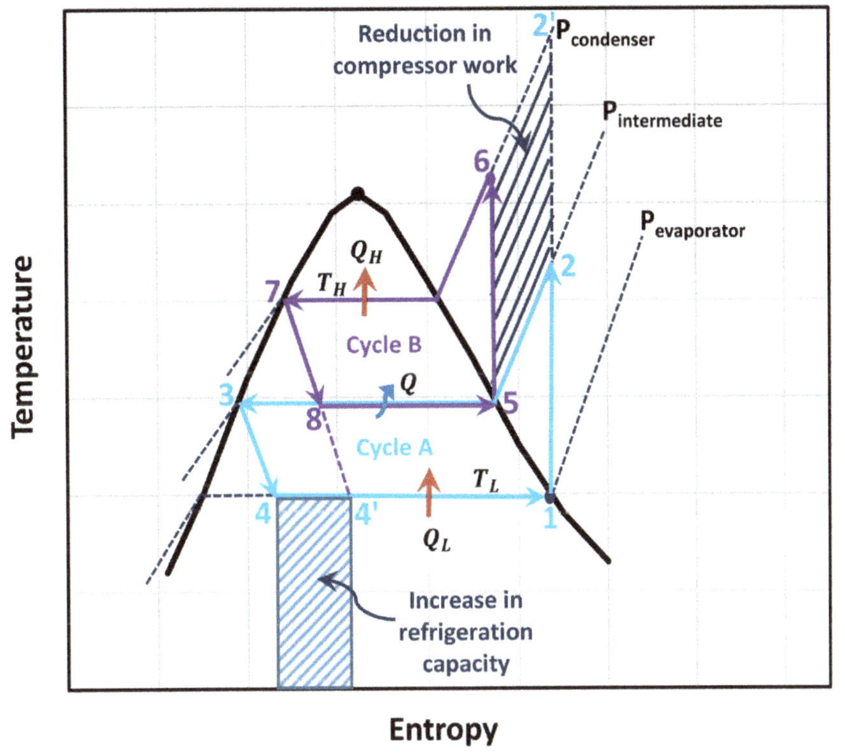

Figure 10.12 T-s diagram of cascade of two vapor-compression refrigeration cycles.

In Figure 10.12, it is assumed that the same refrigerant is used in the two cycles. One can use different refrigerants in the two cycles. A refrigerant with high boiling point can be used in the high temperature cycle (cycle B) and a refrigerant with low boiling point can be used in the low temperature cycle (cycle A). For example, ammonia (normal boiling point of -33.34 °C) can be used in the high temperature cycle and carbon dioxide (normal boiling point of -57°C) can be used in the low temperature cycle.

The number of refrigeration cycles can be increased to more than two if extremely low temperatures are required in the refrigerated region. For example, Figure 10.13 shows a cascade with three vapor-compression refrigeration cycles. As we go from higher temperature cycle to lower temperature cycle, the refrigerant with lower boiling point should be used.

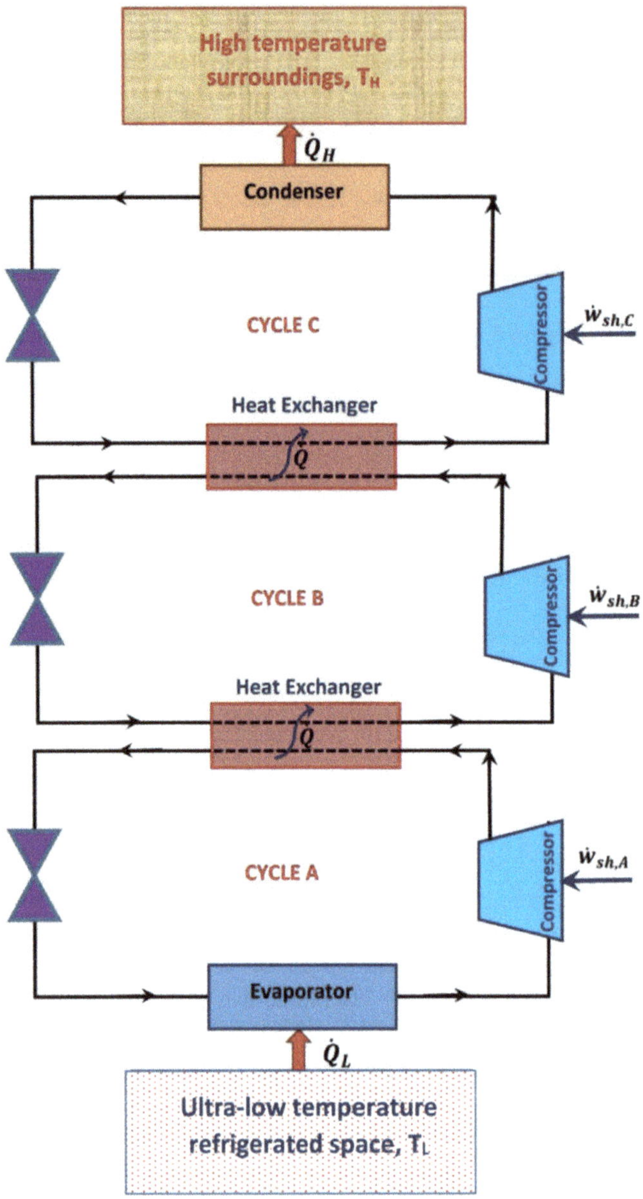

Figure 10.13 Cascade of three vapor-compression refrigeration cycles.

10.5 ABSORPTION REFRIGERATION CYCLE

The main difference between a vapor-compression and an absorption refrigeration cycle is the method of compression of refrigerant. In a vapor-compression refrigeration cycle, the refrigerant vapor from the evaporator is compressed to high pressure using a mechanical device, namely the compressor (see Figure 10.9). In a vapor-absorption cycle, the compressor is not used. The compressor is replaced by a different set up that utilizes heat as a source of energy to drive the refrigeration cycle. Thus, the absorption refrigeration cycle is *heat-operated* cycle as the energy required to operate the cycle comes from heat whereas the vapor-compression refrigeration cycle is *work-operated* cycle as the energy required to operate the cycle comes from mechanical work.

Figure 10.14 shows the absorption refrigeration cycle. The section to the left of the dashed line in Figure 10.14 is the same as in a vapor-compression refrigeration cycle (see Figure 10.9). Only the section to the right of Figure 10.14 is different. Instead of using a mechanical compressor, we now have several different pieces of equipment to achieve the function of a compressor. The saturated low-pressure refrigerant vapor (for example, ammonia vapor) from the evaporator enters the absorber where it is absorbed by the liquid absorbent (for example, water). The formation of the refrigerant solution is exothermic in nature and heat is generated. As the amount of refrigerant absorbed increases with the decrease in temperature, it is important that the solution in the absorber is cooled to keep the temperature as low as possible. Thus, heat is rejected from the absorber to the surroundings. The strong refrigerated solution is pumped to the generator at condenser pressure. The refrigerant liquid solution is preheated in a heat exchanger before it enters the generator. In the generator, the refrigerant solution is boiled off using heat from a high temperature heat source. The high pressure high temperature refrigerant vapor passes on to the condenser. The weak refrigerant solution from the generator is returned to the absorber for recharging. On its way to the absorber, the refrigerant solution is cooled in a heat exchanger and its pressure is reduced through a valve. The only work input to the process is through the pump. However, the work required to pump a liquid solution is small as the specific volume of liquid is very small in comparison with gases. For a majority of applications of vapor absorption refrigeration cycle, ammonia is used as a refrigerant and water is used as an absorbent.

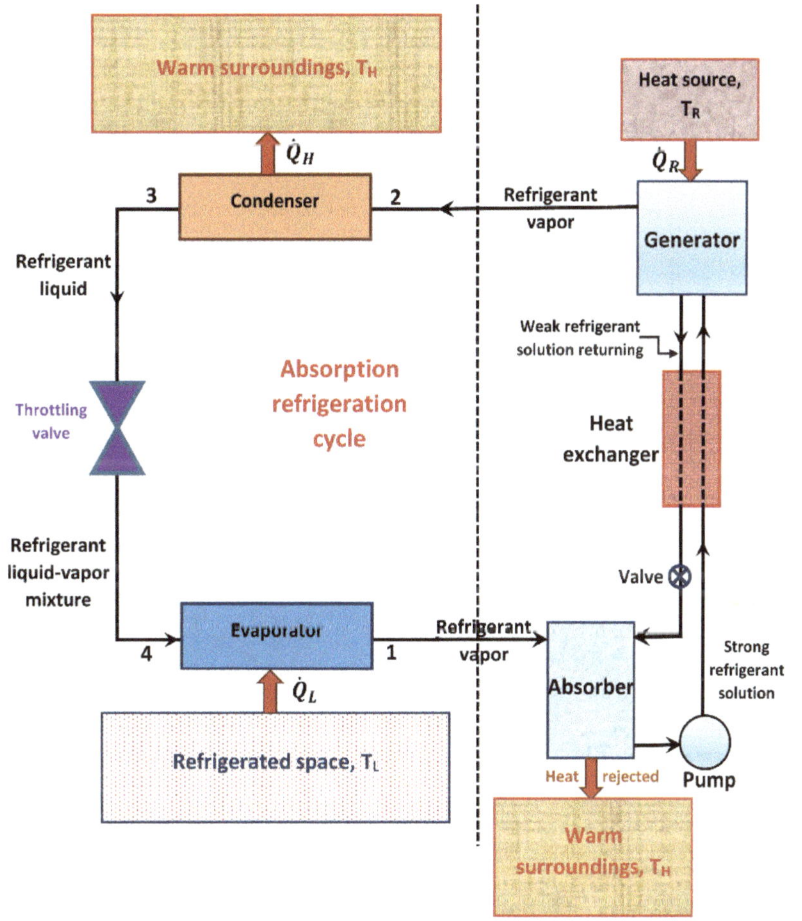

Figure 10.14 Absorption refrigeration cycle.

The absorption refrigeration cycle shown in Figure 10.14 can be modelled as a combination of a refrigerator and heat engine, as shown in Figure 10.15. The section to the right of the dashed line in Figure 10.14 is equivalent to a heat engine which absorbs heat from a heat source at T_R, converts part of the heat absorbed into work and rejects the remaining heat to the surroundings at T_H. The section to the left of the dashed line in Figure 10.14 is equivalent to a refrigerator driven by work produced by the heat engine.

From a practical point of view, it is of interest to know how much heat is required from the heat source at T_R to remove one unit of heat from the refrigerated region at T_L. Thus, estimation of $\dot{Q}_R/\dot{Q}_L$ is important. We can estimate the *minimum* $\dot{Q}_R/\dot{Q}_L$ using the model shown in Figure 10.15 by assuming that the refrigerator and heat engine are reversible cyclic devices.

For a reversible refrigerator, the COP is given as:

$$\omega = \frac{\dot{Q}_L}{|\dot{W}|} = \frac{T_L}{T_H - T_L} \tag{10.14}$$

For a reversible heat engine, the thermal efficiency is given as:

$$\eta = \frac{|\dot{W}|}{\dot{Q}_R} = 1 - \frac{T_H}{T_R} \tag{10.15}$$

Eliminating $|\dot{W}|$ from Equations (10.14) and (10.15) gives:

$$\frac{\dot{Q}_R}{\dot{Q}_L} = \left(\frac{T_R}{T_R - T_H}\right)\left(\frac{T_H - T_L}{T_L}\right) \tag{10.16}$$

This is the *minimum* value of $\dot{Q}_R/\dot{Q}_L$. For the actual absorption refrigerator, the value of $\dot{Q}_R/\dot{Q}_L$ would be larger due to irreversibilities in the process.

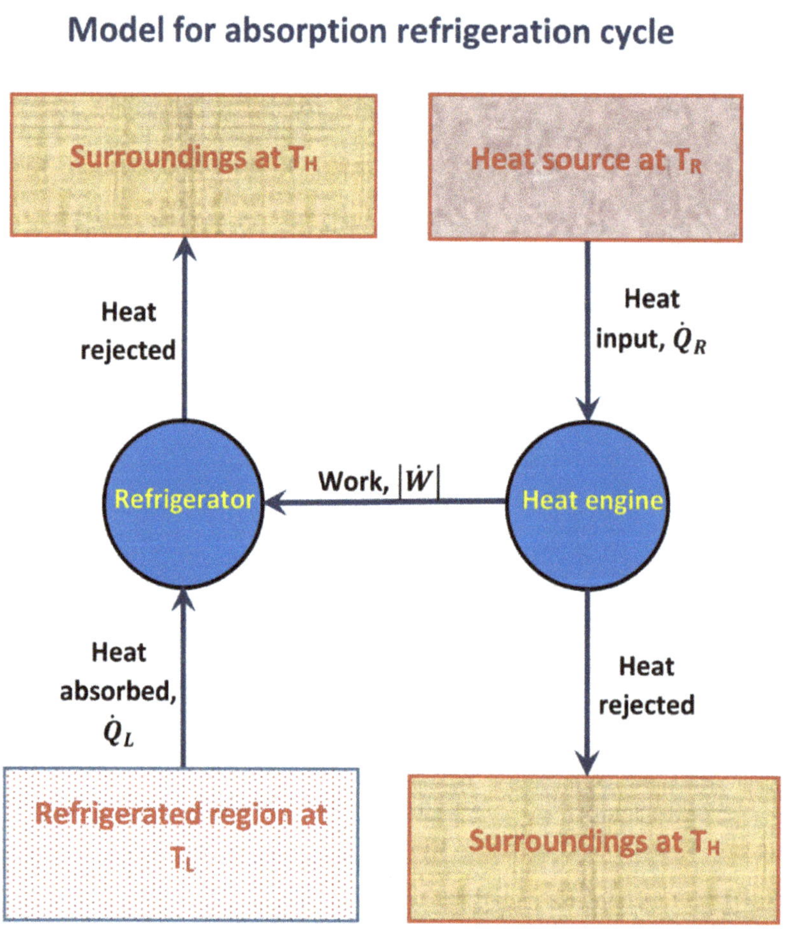

Figure 10.15 Combination of refrigerator and heat engine as a model for the absorption refrigeration cycle.

10.6 HEAT PUMP

Heat pumps are cyclic devices used to heat a house or building during the winter (cold weather). They remove heat from a cold region (outside the house) and transfer it to the high temperature region (inside the house). Thus, they are identical to refrigerators except that their objective is different. In refrigerators the desired objective is refrigeration, that is, removal of heat from a low temperature region, whereas in heat pumps the desired objective is to transfer heat to a high temperature region (inside the house).

10.6.1 Carnot Heat Pump

A Carnot heat pump is shown schematically in Figure 10.16. It is based on a Carnot vapor-compression refrigeration cycle, shown in Figure 10.3. The equipment diagram of a Carnot heat pump is the same as that of a Carnot refrigerator shown in Figure 10.4.

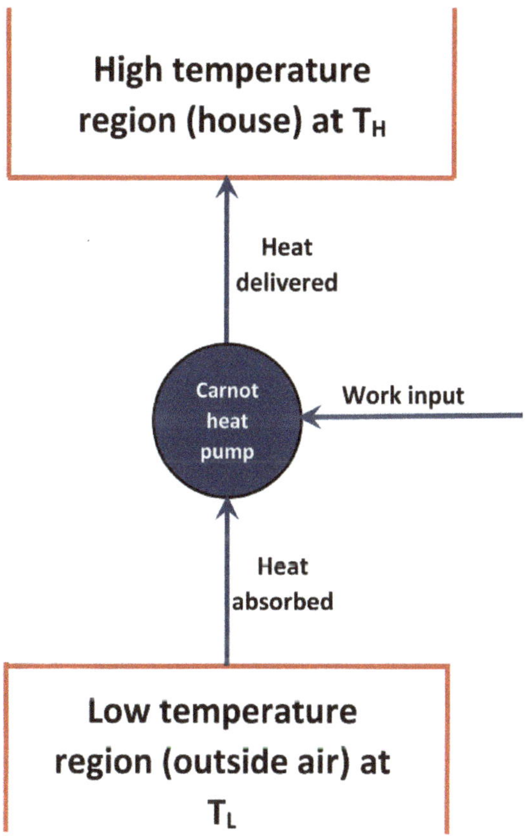

Figure 10.16 Carnot heat pump.

The COP of a heat pump is defined as:

$$\omega = \frac{Heat\ tranferred\ to\ high\ temperature\ region}{Net\ work\ input} \tag{10.17}$$

Referring to Figure 10.3, the COP of a Carnot heat pump can be expressed as:

$$\omega = \frac{Q_{2\to3}}{(Q_{2\to3}+Q_{4\to1})} = \frac{T_H(s_3-s_2)}{[T_H(s_3-s_2)+T_L(s_1-s_4)]} \tag{10.18}$$

From Figure 10.3, $s_1 = s_2$ and $s_3 = s_4$. Consequently, Equation (10.18) reduces to:

$$\omega = \left(\frac{T_H}{T_H - T_L}\right) \tag{10.19}$$

This is the maximum possible ω for any heat pump operating between two given temperatures T_L and T_H. Note that this expression for the COP of a heat pump is different from that of the COP of a refrigerator. In the COP expression for a refrigerator (see Equation (10.5)), the temperature T_L is in the numerator whereas in Equation (10.19), temperature T_H is in the numerator.

10.6.2 Practical Heat Pumps

The Carnot heat pump utilizes a turbine to produce work in the expansion step (3 → 4, Figure 10.3). As the work produced in the turbine is small, the practical heat pumps use a throttling valve to carry out the expansion. The practical vapor-compression heat pump cycle is shown in Figure 10.17 below. Note that this heat pump cycle is identical to the vapor-compression refrigeration cycle of Figure 10.9.

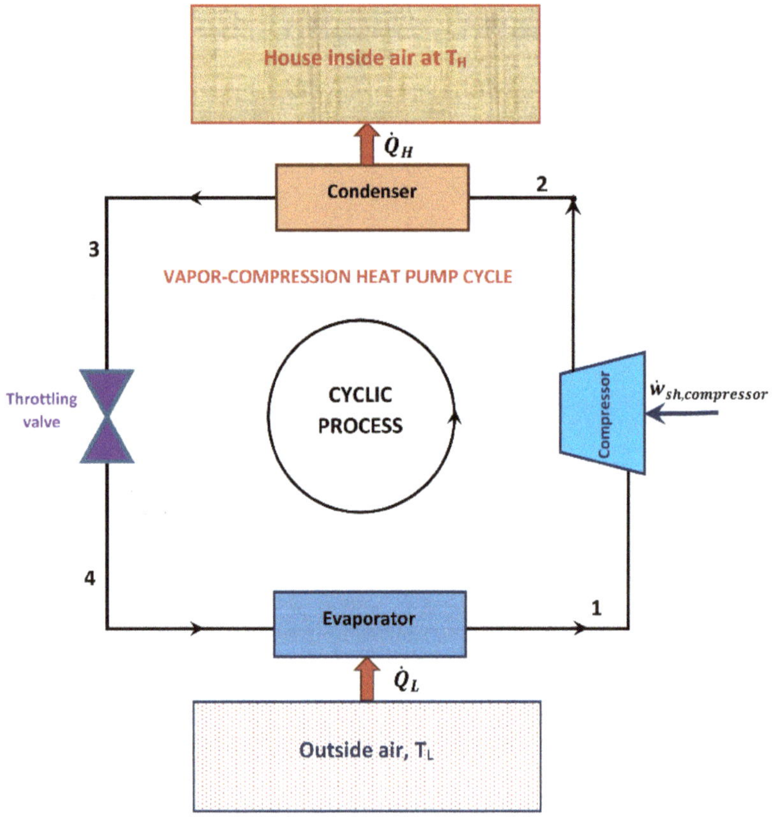

Figure 10.17 Practical heat pump cycle.

Using the definition of COP of a heat pump given in Equation (10.17) and the T-s diagram of Figure 10.10, the COP of the practical vapor-compression heat pump cycle of Figure 10.17 can be expressed as:

$$\omega = \frac{Q_{2\rightarrow 3}}{(Q_{2\rightarrow 3} + Q_{4\rightarrow 1})} = \frac{(h_3 - h_2)}{[(h_3 - h_2) + (h_1 - h_4)]} \tag{10.20}$$

As the throttling process is isenthalpic, $h_3 = h_4$. Consequently, Equation (10.20) reduces to:

$$\omega = \frac{(h_2 - h_3)}{(h_2 - h_1)} \tag{10.21}$$

10.7 AIR CONDITIONER

The purpose of an air conditioner is to cool the air inside the house by absorbing heat from the inside air at low temperature T_L and rejecting it to the outside air at high temperature T_H. The working principle of an air conditioner is the same as that of a refrigerator. Figure 10.18 shows the air-conditioning cycle. It is identical to the vapor-compression refrigeration cycle of Figure 10.9.

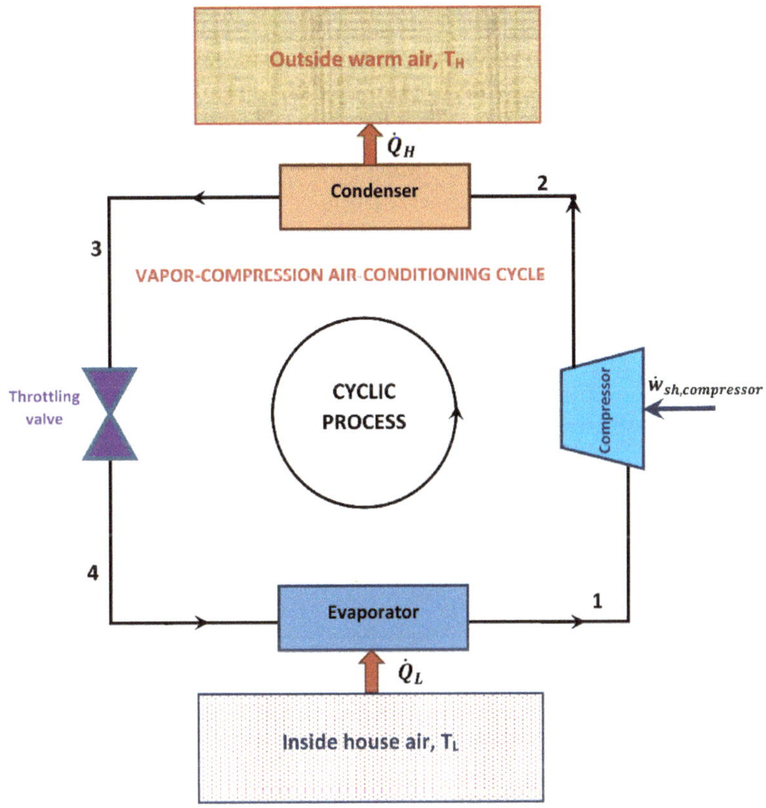

Figure 10.18 Air-conditioning cycle.

As the desired objective of an air-conditioning cycle is the same as that of the refrigeration cycle, that is, to remove heat from the low temperature region, the COP expression for the refrigeration cycle, Equation (10.12), is also applicable to the air-conditioning cycle of Figure 10.18.

Finally, it should be noted that the same device (for example, a heat pump) can be used to heat or cool the house by simply switching the direction of the flow of refrigerant with the help of a reversing valve installed in the flow loop. This is possible as the equipment for a heat pump and air conditioner is the same. Figure 10.19 shows a simplified equipment flow chart of heat pump-cum-air conditioner device. The device acts as a heat pump when the refrigerant flow direction is clockwise. In this case, the inside heat exchanger acts as a condenser and the outside heat exchanger acts as an evaporator. The condenser releases heat to the inside air of the house. When the flow direction of the refrigerant is reversed with the help of a reversing valve, that is, the flow is made anti-clockwise, the same device becomes an air-conditioner. Now the inside heat exchanger acts as an evaporator and the outside heat exchanger acts as a condenser. Heat in this case is removed from the inside air of the house.

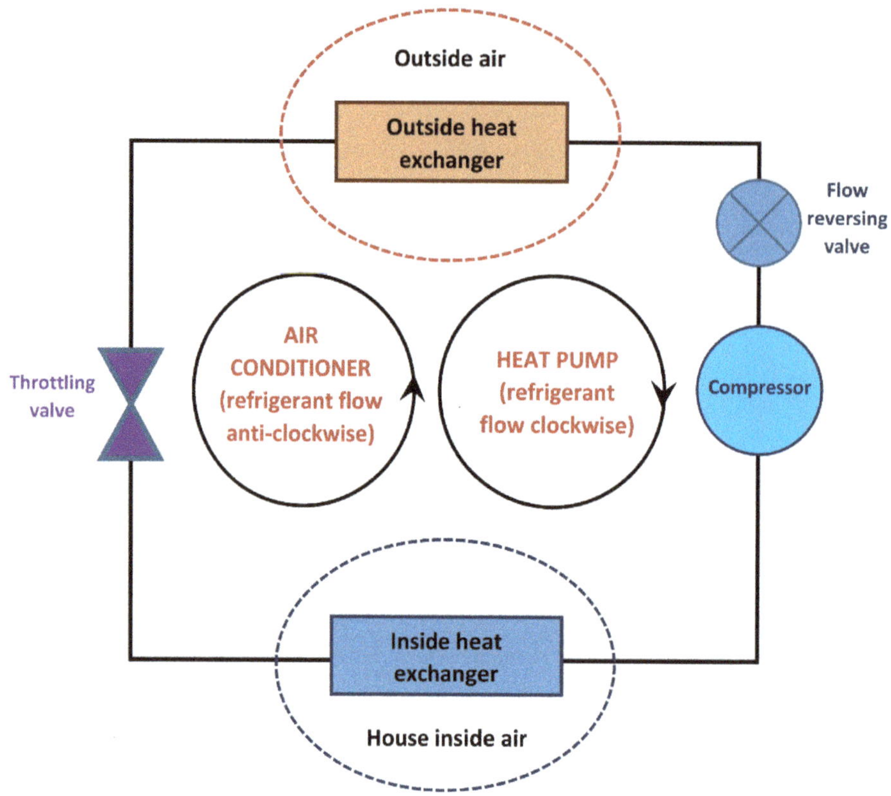

Figure 10.19 Heat pump-cum-air conditioner.

NOTATION

COP	coefficient of performance
h	specific enthalpy
h^L	specific enthalpy of liquid
h^v	specific enthalpy of vapor
$\dot{m}$	mass flow rate
P	pressure
$P_{condenser}$	pressure in the condenser
$P_{evaporator}$	pressure in the evaporator
Q	heat transfer per unit mass
Q_{cycle}	net heat transfer in a cyclic process
$\dot{Q}$	heat transfer rate
$\dot{Q}_{condenser}$	heat transfer rate in the condenser
$\dot{Q}_{evaporator}$	heat transfer rate in the evaporator
$\dot{Q}_H$	heat transfer rate at high temperature
$\dot{Q}_L$	heat transfer rate at low temperature
$\dot{Q}_R$	heat transfer rate from heat reservoir
s	specific entropy
s^L	specific entropy of liquid
s^v	specific entropy of vapor
T	temperature
T_H	temperature of high temperature region

T_L	temperature of low temperature region
T_R	temperature of heat reservoir
$\bar{T}$	average temperature
W	work on a unit mass basis
W_{cycle}	net work in a cyclic process
W_{sh}	shaft work on a unit mass basis
$W_{sh,cycle}$	net shaft work in a cyclic process
$\dot{W}_{sh}$	rate of shaft work
x^v	mass fraction of vapor in two-phase liquid-vapor mixture

Greek Letters

ω	coefficient of performance

TUTORIAL PROBLEMS WITH SOLUTION

Problem 1

A vapor-compression refrigeration cycle using a throttling valve for expansion requires 2.3 kW of power to run the compressor. The rate of heat removal from the refrigerated region is 5.5 kW.
Calculate the following:

a) The COP of the refrigeration cycle
b) The rate of heat rejection in the condenser
c) If the condensation temperature is 35 °C, what is the minimum temperature that can be maintained in the refrigerated region.

Solution

Given: $\dot{W}_{sh,compressor} = -2.3$ kW ; $\dot{Q}_{evaporator} = 5.5$ kW

a) The COP of the refrigeration cycle

$$\omega = \frac{Heat\ removed\ from\ refrigerated\ space}{|W_{cycle}|} = \frac{5.5}{2.3} = 2.39$$

b) The rate of heat rejection in the condenser

From energy balance,

$$\dot{Q}_{evaporator} + \dot{Q}_{condenser} = \dot{W}_{sh,compressor}$$

$$\dot{Q}_{condenser} = (-2.3) - 5.5 = -7.8 \text{ kW}$$

c) If the condensation temperature is 35 °C, what is the minimum temperature that can be maintained in the refrigerated region.

$$\omega = \left(\frac{T_L}{T_H - T_L}\right) = 2.39 \quad ; \; T_L = 2.39(308.15 - T_L)\,; \qquad T_L = 2.39 \times 308.15/3.39 = 217.25 \text{ K}$$

Thus, the minimum temperature that can be maintained in the refrigerated region is 217.25-273.15=-55.9 °C.

Problem 2

A Carnot vapor-compression refrigeration cycle (see Figure 10.3) is using HFC-134A as a refrigerant. The condenser is operating at a temperature of 30 °C and the evaporator is operating at -12 °C.
Calculate the following:

a) The work requirement of the compressor per unit mass of refrigerant
b) The work produced in the turbine per unit mass of refrigerant
c) The heat transfer per unit mass of refrigerant in the evaporator
d) The COP of the refrigeration cycle

Solution

Refer to Figure 10.3.

Let us first determine the relevant properties in different states of the cycle:

State 2 (condenser inlet/compressor outlet):

In state 2, the refrigerant is saturated vapor at 30 °C. From saturated refrigerant tables (Appendix K), at a temperature of 30 °C:

$h_2 = 263.50$ kJ/kg ; $s_2 = 0.9070$ kJ/kg.K

State 3 (turbine inlet/condenser outlet):

In state 3, the refrigerant is saturated liquid at 30 °C. From saturated refrigerant tables (Appendix K), at a temperature of 30 °C:

$h_3 = 91.49$ kJ/kg ; $s_3 = 0.3396$ kJ/kg.K

State 4 (evaporator inlet/turbine outlet):

In state 4, the refrigerant is a vapor-liquid mixture at -12 °C. From saturated refrigerant tables (Appendix K), at a temperature of -12 °C:

$h^L = 34.39$ kJ/kg ; $h^v = 240.15$ kJ/kg ; $s^L = 0.1388$ kJ/kg.K ; $s^v = 0.9267$ kJ/kg.K

In state 4, $s_4 = s_3 = 0.3396 = x_4^v(0.9267) + (1 - x_4^v) \times 0.1388$. Thus,

$$x_4^v = \frac{0.3396 - 0.1388}{0.9267 - 0.1388} = 0.2549$$

$$h_4 = x_4^v h_4^v + (1 - x_4^v) \times h_4^L = 0.2549 \times 240.15 + (1 - 0.2549) \times 34.39 = 86.838 \text{ kJ/kg}$$

State 1 (compressor inlet/evaporator outlet):

In state 1, the refrigerant is a vapor-liquid mixture at -12 °C. From saturated refrigerant tables (Appendix K), at a temperature of -12 °C:

$h^L = 34.39$ kJ/kg ; $h^v = 240.15$ kJ/kg ; $s^L = 0.1388$ kJ/kg.K ; $s^v = 0.9267$ kJ/kg.K

In state 1, $s_1 = s_2 = 0.9070 = x_1^v(0.9267) + (1 - x_1^v) \times 0.1388$. Thus,

$$x_1^v = \frac{0.9070 - 0.1388}{0.9267 - 0.1388} = 0.975$$

$$h_1 = x_1^v h_1^v + (1 - x_1^v) \times h_1^L = 0.975 \times 240.15 + (1 - 0.975) \times 34.39 = 235.006 \text{ kJ/kg}$$

a) The work requirement of the compressor per unit mass of refrigerant

$$W_{sh,compressor} = -\Delta h = -(h_2 - h_1) = -(263.50 - 235.006) = -28.494 \text{ kJ/kg}$$

b) The work produced in the turbine per unit mass of refrigerant

$$W_{sh,turbine} = -\Delta h = -(h_4 - h_3) = -(86.838 - 91.49) = 4.652 \text{ kJ/kg}$$

c) The heat transfer per unit mass of refrigerant in the evaporator

$$Q_{evaporator} = \Delta h = (h_1 - h_4) = (235.006 - 86.838) = 148.168 \text{ kJ/kg}$$

d) The COP of the refrigeration cycle

$$\omega = \frac{Q_{evaporator}}{|W_{sh,cycle}|} = \frac{148.168}{28.494 - 4.652} = 6.21$$

We can also apply the formula:

$$\omega = \left(\frac{T_L}{T_H - T_L}\right) = \frac{(-12 + 273.15)}{30 - (-12)} = 6.21$$

Problem 3

An ideal vapor-compression refrigeration cycle (see figure) is using HFC-134A as a refrigerant. The condenser is operating at a pressure of 10 bar. The evaporator pressure is 2 bar. The mass flow rate of the refrigerant is 0.1 kg/s.

Calculate the following:

a) The quality of vapor-liquid mixture leaving the throttling device
b) The refrigeration rate in kW
c) The power requirement of the compressor in kW
d) The COP of the refrigeration cycle

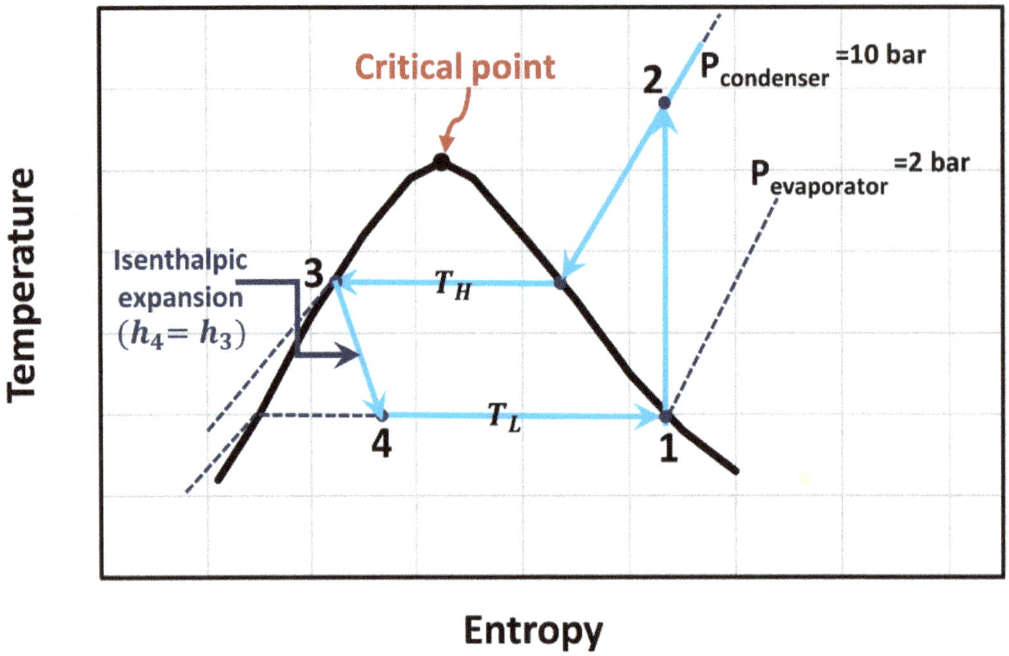

Solution

Refer to the figure given above.

We determine the relevant properties in different states of the cycle:

State 1 (compressor inlet/evaporator outlet):

In state 1, the refrigerant is saturated vapor at 2 bar. From saturated refrigerant tables (Appendix K), at a pressure of 2 bar ($T^{sat} = -10.09$ °C):

$h^L = 36.84$ kJ/kg ; $h^v = 241.30$ kJ/kg ; $s^L = 0.1481$ kJ/kg.K ; $s^v = 0.9253$ kJ/kg.K
$h_1 = 241.30$ kJ/kg ; $s_1 = 0.9253$ kJ/kg.K

State 2 (condenser inlet/compressor outlet):

In state 2, the refrigerant is superheated vapor at 10 bar. Also, $s_2 = s_1 = 0.9253$ kJ/kg.K

From superheated refrigerant tables (Appendix L), at a pressure of 10 bar:

h (kJ/kg)	*s* (kJ/kg.K)
268.68	0.9066
280.19	0.9428

By interpolation:

$$h_2 = 268.68 + \frac{(280.19-268.68)}{(0.9428-0.9066)} \times (0.9253 - 0.9066) = 274.626 \text{ kJ/kg}$$

State 3 (Throttling valve inlet/condenser outlet):

In state 3, the refrigerant is saturated liquid at 10 bar. From saturated refrigerant tables (Appendix K), at a pressure of 10 bar ($T^{sat} = 39.39$ °C):

$h^L = 105.29$ kJ/kg ; $h^v = 267.97$ kJ/kg ; $s^L = 0.3838$ kJ/kg.K ; $s^v = 0.9043$ kJ/kg.K

Thus,

$h_3 = 105.29$ kJ/kg ; $s_3 = 0.3838$ kJ/kg.K

State 4 (evaporator inlet/throttling valve outlet):

$h_4 = h_3 = 105.29\ \text{kJ/kg}$ (throttling process is isenthalpic)

a) The quality of vapor-liquid mixture leaving the throttling device

In state 4, the refrigerant is a vapor-liquid mixture at 2 bar. From saturated refrigerant tables (Appendix K), at a pressure of 2 bar:

$h^L = 36.84$ kJ/kg ; $h^v = 241.30$ kJ/kg ; $s^L = 0.1481$ kJ/kg.K ; $s^v = 0.9253$ kJ/kg.K

Since $h_4 = h_3 = 105.29\ \text{kJ/kg}$, it follows that:

$105.29 = x_4^v(241.30) + (1 - x_4^v) \times 36.84$. Thus,

$$x_4^v = \frac{105.29 - 36.84}{241.30 - 36.84} = 0.3348$$

b) The refrigeration rate in kW

$$\dot{Q}_{evaporator} = \dot{m}\Delta h = \dot{m}(h_1 - h_4) = 0.1 \times (241.30 - 105.29) = 13.601 \text{ kW}$$

We can also express the refrigeration rate in units of *tons* of refrigeration (TR) as:

$$\dot{Q}_{evaporator} = \frac{13.601}{3.517} = 3.867 \text{ TR (note that 1 ton = 3.517 kW)}$$

c) The power requirement of the compressor in kW

$$\dot{W}_{sh,compressor} = -\dot{m}\Delta h = -\dot{m}(h_2 - h_1) = 0.1 \times (274.626 - 241.30) = -3.3326 \text{ kW}$$

d) The COP of the refrigeration cycle

e)

$$\omega = \frac{\dot{Q}_{evaporator}}{|\dot{W}_{sh,cycle}|} = \frac{13.601}{3.3326} = 4.081$$

Note that the COP of a Carnot refrigerator operating between the same condensation and evaporation temperatures is expected to be higher, as shown below:

$$\omega = \left(\frac{T_L}{T_H - T_L}\right) = \frac{(-10.09 + 273.15)}{39.39 - (-10.09)} = 5.316$$

Problem 4

Re-consider problem 3 assuming that the compressor is 73 percent efficient (see figure below). Calculate the power requirement of the compressor in this case. What is the COP of the cycle?

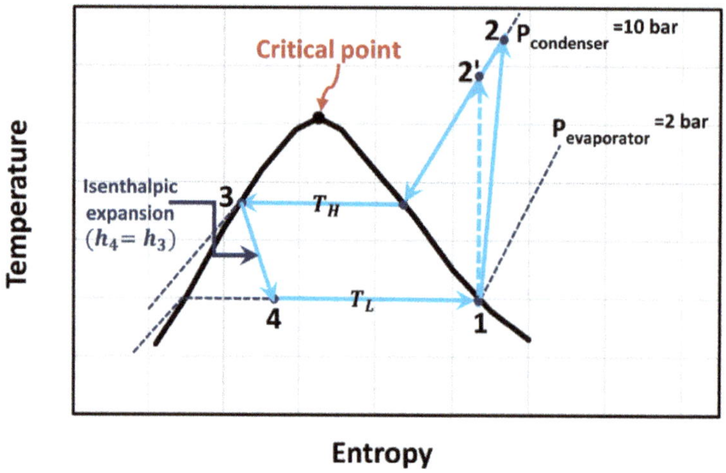

Solution

In comparison to problem 3, the compressor work requirement and the COP of the cycle are expected to be different due to compressor inefficiency. The compressor work requirement is expected to increase, and the COP of the cycle is expected to be lower. The rate of refrigeration ($\dot{Q}_{evaporator}$) is expected to remain the same.

State 1: $h_1 = 241.30$ kJ/kg ; $s_1 = 0.9253$ kJ/kg.K

State 2': $s_2' = s_1 = 0.9253$ kJ/kg.K, $P = 10$ bar, $\eta = 0.73$
$h_2' = 274.626$ kJ/kg (see problem 3 solution)

$$\eta = \frac{(\Delta h)_{isentropic}}{(\Delta h)_{actual}} \text{ ; } (\Delta h)_{actual} = h_2 - h_1 = \frac{h_2' - h_1}{\eta} = \frac{274.626 - 241.30}{0.73} = 45.652 \text{ kJ/kg}$$

$$h_2 = 45.652 + h_1 = 45.652 + 241.30 = 286.952 \text{ kJ/kg}$$

Power requirement of the compressor:

$$\dot{W}_{sh,compressor} = -\dot{m}\Delta h = -\dot{m}(h_2 - h_1) = -0.1 \times (286.952 - 241.30) = -4.5652 \text{ kW}$$

COP of the cycle

$$\omega = \frac{\dot{Q}_{evaporator}}{|\dot{W}_{sh,cycle}|} = \frac{13.601}{4.5652} = 2.979$$

As expected, the power requirement of the compressor is increased from 3.3326 to 4.5652 kW. The COP is decreased from 4.081 to 2.979.

Problem 5

A cascade of two vapor-compression refrigeration cycles (see Figures 10.11 and 10.12) operates between condenser pressure of 1.2 MPa and evaporator pressure of 120 kPa. The same refrigerant (HFC-134A) is used in the two cycles. The mass flow rate of refrigerant in the low pressure cycle is 1 kg/s.
Calculate the following:

a) The mass flow rate of refrigerant in the high pressure cycle
b) The refrigeration rate in kJ/s
c) The COP of the cascade

You can take the intermediate pressure at which heat is exchanged between the two cycles as the geometric mean of P_c and P_e : $P_i = \sqrt{P_c P_e}$ where P_c is the pressure of the condenser and P_e is the pressure of the evaporator.

Solution

Refer to the figure below.

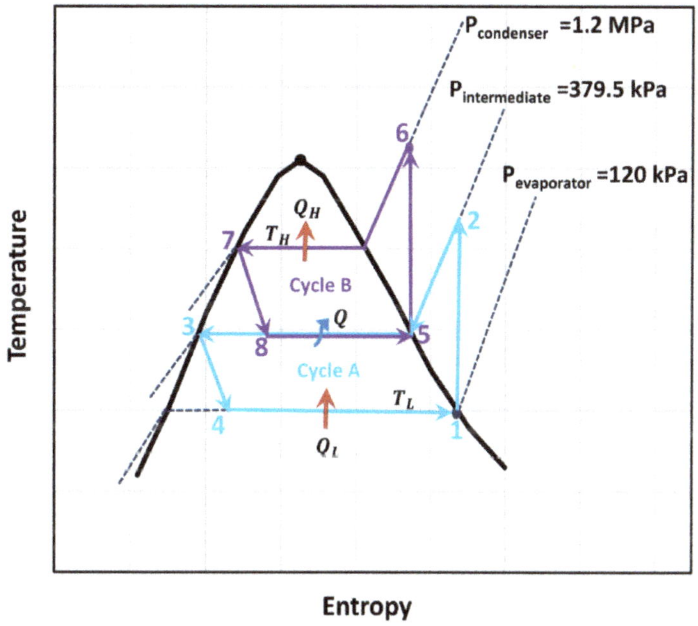

The intermediate pressure at which heat is exchanged between the two cycles is taken as the geometric mean of P_c and P_e :

$$P_i = \sqrt{P_c P_e} = \sqrt{1.2 \times 1000 \times 120} = 379.5 \text{ kPa}$$

We now determine the relevant properties in different states of the cascade.

State 1:

$P_e = 120\ kPa$. From saturated refrigerant table (Appendix K),
$h^L = 21.32$ kJ/kg ; $h^v = 233.86$ kJ/kg ; $s^L = 0.0879$ kJ/kg.K ; $s^v = 0.9354$ kJ/kg.K

As the refrigerant is in saturated vapor state,
$h_1 = 233.86$ kJ/kg ; $s_1 = 0.9354$ kJ/kg.K

State 2:

$P = 379.5\ kPa = 0.3795\ MPa$; $s_2 = s_1 = 0.9354$ kJ/kg.K

From superheated refrigerant table (Appendix L),
At $T = 10^oC$,

$P(MPa)$	h (kJ/kg)	s (kJ/kg.K)
0.4	253.35	0.9182
0.32	255.65	0.9427

By interpolation, we find h and s at 10^oC and $0.3795\ MPa$:

$$h = 255.65 + \frac{(253.35-255.65)}{(0.4-0.32)} \times (0.3795 - 0.32) = 253.939 \text{ kJ/kg}$$

$$s = 0.9427 + \frac{(0.9182-0.9427)}{(0.4-0.32)} \times (0.3795 - 0.32) = 0.9245 \text{ kJ/kg}$$

At $T = 20^oC$,

$P(MPa)$	h (kJ/kg)	s (kJ/kg.K)
0.4	262.96	0.9515
0.32	264.95	0.9749

By interpolation, we find h and s at 20^oC and $0.3795\ MPa$:

$$h = 264.95 + \frac{(262.96-264.95)}{(0.4-0.32)} \times (0.3795 - 0.32) = 263.47 \text{ kJ/kg}$$

$$s = 0.9749 + \frac{(0.9515-0.9749)}{(0.4-0.32)} \times (0.3795 - 0.32) = 0.9575 \text{ kJ/kg}$$

Thus, we have at $P = 0.3795\ MPa$:

T	h (kJ/kg)	s (kJ/kg.K)
10^oC	253.939	0.9245
20^oC	263.47	0.9575

We require h corresponding to $P = 0.3795\ MPa$ and $s_2 = s_1 = 0.9354$ kJ/kg.K

Interpolation of the data given in the above table, we get:

$$h_2 = 253.939 + \frac{(263.47-253.939)}{(0.9575-0.9245)} \times (0.9354 - 0.9245) = 257.087 \text{ kJ/kg}$$

State 3:

Saturated liquid at $P = 0.3795\ MPa$. From saturated refrigerant table (Appendix K),

$P(MPa)$	h (kJ/kg)
0.36	57.82
0.4	62.00

By interpolation, we find h at 0.3795 MPa:

$$h_3 = 57.82 + \frac{(62.0-57.82)}{(0.4-0.36)} \times (0.3795 - 0.36) = 59.86 \text{ kJ/kg}$$

State 4:

$$h_4 = h_3 = 59.86 \text{ kJ/kg}$$

State 5:

Saturated vapor at $P = 0.3795\ MPa$. From saturated refrigerant table (Appendix K),

$P(MPa)$	h^v (kJ/kg)	s^v (kJ/kg.K)
0.36	250.58	0.9160
0.4	252.32	0.9145

By interpolation,

$$h_5 = 250.58 + \frac{(252.32-250.58)}{(0.4-0.36)} \times (0.3795 - 0.36) = 251.428 \text{ kJ/kg}$$

$$s_5 = 0.9160 + \frac{(0.9145-0.9160)}{(0.4-0.36)} \times (0.3795 - 0.36) = 0.9153 \text{ kJ/kg}$$

State 6:

Superheated vapor at $P = 1.2\ MPa$; $s_6 = s_5 = 0.9153$ kJ/kg.K.

From superheated refrigerant table (Appendix L), at $P = 1.2\ MPa$

h (kJ/kg)	s (kJ/kg.K)
270.99	0.9023
275.52	0.9164

By interpolation,

$$h_6 = 270.99 + \frac{(275.52-270.99)}{(0.9164-0.9023)} \times (0.9153 - 0.9023) = 275.167 \text{ kJ/kg}$$

State 7:

Saturated liquid at $P = 1.2\ MPa$. From saturated refrigerant table,
$h_7 = 115.76$ kJ/kg

State 8:

$h_8 = h_7 = 115.76$ kJ/kg

Now that all enthalpies are known, we can do the required calculations.

a) The mass flow rate of refrigerant in the high pressure cycle

Energy balance over the intermediate heat exchanger gives:

$$\dot{m}_B(h_5 - h_8) = \dot{m}_A(h_2 - h_3)$$

where $\dot{m}_A$ is the mass flow rate of refrigerant in the low pressure cycle (cycle A, see figure) and $\dot{m}_B$ is the mass flow rate of refrigerant in the high pressure cycle (cycle B, see figure).

$$\dot{m}_A = 1 \text{ kg/s} \; ; \; \dot{m}_B = \dot{m}_A \frac{(h_2 - h_3)}{(h_5 - h_8)} = 1 \times \frac{(257.087 - 59.86)}{(251.428 - 115.76)} = 1.4996 \text{ kg/s}$$

b) The refrigeration rate in kJ/s

$$\dot{Q}_L = \dot{Q}_{41} = \dot{m}_A(h_1 - h_4) = 1 \times (233.86 - 59.86) = 174 \text{ kW} = 174/3.517 = 49.47 \text{ TR}$$

c) The COP of the cascade

$$\omega = \frac{\dot{Q}_L}{|\dot{W}_{sh,compressor\ A} + \dot{W}_{sh,compressor\ B}|}$$

$$\dot{W}_{sh,compressor\ A} = -\dot{m}_A(h_2 - h_1) = -1 \times (257.087 - 233.86) = -23.227 \text{ kW}$$

$$\dot{W}_{sh,compressor\ B} = -\dot{m}_B(h_6 - h_5) = -1.4996 \times (275.167 - 251.428) = -35.599 \text{ kW}$$

$$\omega = \frac{\dot{Q}_L}{|\dot{W}_{sh,compressor\ A} + \dot{W}_{sh,compressor\ B}|} = \frac{174}{|-23.227 - 35.599|} = \frac{174}{58.826} = 2.958$$

Problem 6

Consider an absorption refrigeration cycle (see Figure 10.14). The refrigerated region is maintained at a temperature of -18°C ($T_L = 255.15$ K). The heat is provided to the process by combustion of natural gas. The heat is available at a temperature of 250 °C ($T_R = 523.15$ K). The surroundings are at a temperature of 26 °C ($T_H = 299.15$ K). Determine the minimum amount of heat that the heat source at T_R must supply (Q_R) in order to remove one unit of heat (Q_L) from the refrigerated region.

Solution

Based on the model system for absorption refrigeration shown in Figure 10.15, the minimum value of Q_R/Q_L is given by Equation (10.16) as:

$$\frac{Q_R}{Q_L} = \left(\frac{T_R}{T_R - T_H}\right)\left(\frac{T_H - T_L}{T_L}\right) = \left(\frac{523.15}{523.15 - 299.15}\right)\left(\frac{299.15 - 255.15}{255.15}\right) = 0.4027 \qquad (10.16)$$

Thus, the minimum amount of heat that the heat source at T_R should supply per unit heat removed from the refrigerated region is 0.4027. For an actual absorption refrigerator, the value of heat to be supplied is generally much larger.

Problem 7

An ideal vapor-compression heat pump cycle (see figure below) using HFC-134A refrigerant provides 260 kW to heat a home. The operating pressures of condenser and evaporator are 12 and 2 bar, respectively.

Calculate the following:

a) The mass flow rate of refrigerant
b) The power requirement of a compressor
c) The COP of the heat pump cycle

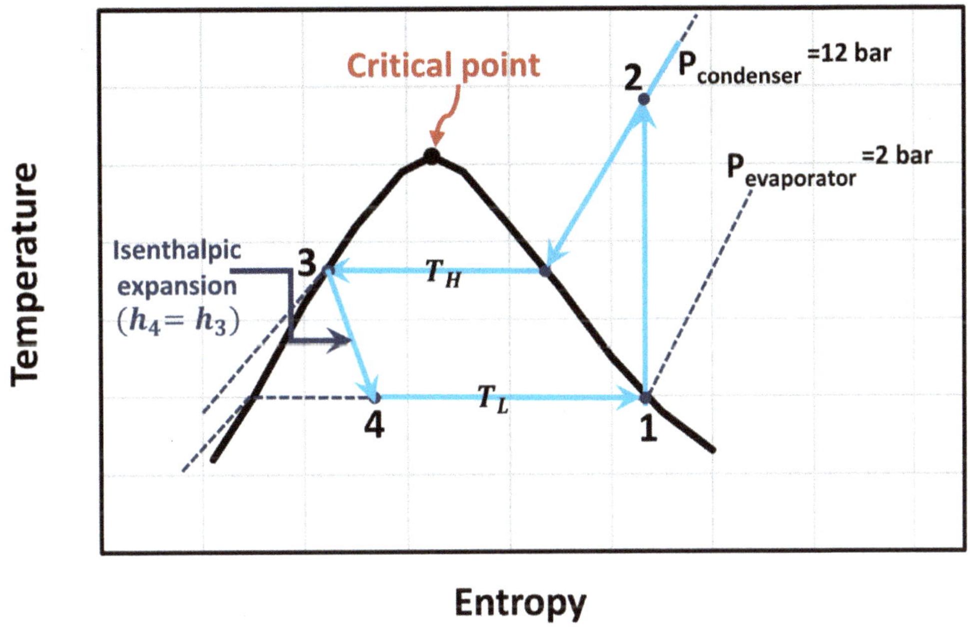

Solution

Refer to the figure above.

We first determine the relevant properties in different states of the cycle.

State 1:

$P = 2\ bar$. From saturated refrigerant table (Appendix K),
$h^L = 36.84$ kJ/kg ; $h^v = 241.30$ kJ/kg ; $s^L = 0.1481$ kJ/kg.K ; $s^v = 0.9253$ kJ/kg.K

As the refrigerant is in saturated vapor state,
$h_1 = 241.30$ kJ/kg ; $s_1 = 0.9253$ kJ/kg.K

State 2:

$P = 12\ bar$; $s_2 = s_1 = 0.9253$ kJ/kg.K

From superheated refrigerant table (Appendix L),

At $P = 12\ bar$,

h (kJ/kg)	s (kJ/kg.K)
275.52	0.9164
287.44	0.9527

By interpolation,

$$h_2 = 275.52 + \frac{(287.44-275.52)}{(0.9527-0.9164)} \times (0.9253 - 0.9164) = 278.44 \text{ kJ/kg}$$

State 3:

Saturated liquid at $P = 12\ bar$. From saturated refrigerant table (Appendix K),
$h_3 = 115.76$ kJ/kg

State 4:

$h_4 = h_3 = 115.76$ kJ/kg

a) The mass flow rate of refrigerant

$$\dot{Q}_H = \dot{Q}_{23} = \dot{m}(h_3 - h_2)\ \ ;\ \ \dot{m} = \dot{Q}_H/(h_3 - h_2) = -260/(115.76 - 278.44) = 1.598 \text{ kg/s}$$

b) The power requirement of a compressor

$$\dot{W}_{sh,compressor} = -\dot{m}(h_2 - h_1) = -1.598 \times (278.44 - 241.30) = -59.35 \text{ kW}$$

c) The COP of the heat pump cycle

$$\omega = \frac{(h_2-h_3)}{(h_2-h_1)} = \frac{278.44-115.76}{278.44-241.30} = 4.38 \tag{10.21}$$

We can also use,

$$\omega = \frac{|\dot{Q}_H|}{|\dot{W}_{sh,compressor}|} = \frac{260}{59.35} = 4.38$$

PRACTICE PROBLEMS

Problem 1

A Carnot vapor-compression refrigeration cycle is using HFC-134A as a refrigerant. The evaporator temperature is -4 °C. Saturated vapor enters the condenser and exits the condenser as saturated liquid at 34 °C. The circulation rate of refrigerant in the cycle flow loop is 0.075 kg/s.
Determine the following:

a) The operating pressures of the condenser and evaporator
b) The quality of vapor-liquid mixture entering the compressor
c) The quality of vapor-liquid mixture at the exit of the turbine
d) The refrigeration rate in tons
e) The net power input to the cycle
f) The COP of the refrigeration cycle

Problem 2

An ideal vapor-compression refrigeration cycle is operating at steady state with HFC-134A as a refrigerant. The saturated vapor enters the compressor at -12 °C and leaves the compressor as superheated vapor at 8 bar. The condenser is operating at a pressure of 8 bar. The mass flow rate of refrigerant is 0.12 kg/s.
Calculate the following:

a) The quality of vapor-liquid mixture leaving the throttling device

b) The refrigeration capacity in kW and tons
c) The compressor power requirement in kW
d) The COP of the refrigeration cycle

What is the COP of a Carnot vapor-compression refrigeration cycle operating between the same condensation and evaporation temperatures?

Problem 3

A vapor-compression refrigeration cycle of Figure 10.10 is operating at steady state with water as a refrigerant. The evaporator is operating at 0.8 kPa and the condenser is operating at a pressure of 6 kPa. The compressor is 81% efficient. The refrigeration capacity of the cycle is 1000 kW.
Calculate the following:

a) The mass flow rate of refrigerant
b) The rate of heat transfer in the condenser
c) The power requirement of a compressor
d) The COP of the cycle

Compare the COP of the cycle with the COP of a Carnot refrigeration cycle operating with the same condenser and evaporator pressures.

Problem 4

A cascade of two ideal vapor-compression refrigeration cycles (see Figures 10.11 and 10.12) operates between condenser pressure of 1.2 MPa and evaporator pressure of 120 kPa. The intermediate pressure where heat is exchanged between the two cycles is 400 kPa. The same refrigerant (HFC-134A) is used in the two cycles. The refrigeration produced by the cascade is 8 tons.
Calculate the following:

a) The mass flow rate of refrigerant in the low pressure cycle
b) The mass flow rate of refrigerant in the high pressure cycle
c) The total work requirement of the cascade in kW
d) The COP of the cascade

Problem 5

As absorption refrigeration cycle is modelled as a combination of heat engine and refrigerator, as shown in Figure 10.15. The work produced by the heat engine is used by the refrigerator in extracting heat from a refrigerated region. The refrigerated region is maintained at a temperature of -5°C. The temperature of the heat source of the engine is 215 °C. The heat engine and the refrigerator both discard heat to the surroundings at 27 °C.
Calculate:

a) The rate of heat absorbed by the heat engine from the heat source, assuming both heat engine and refrigerator are ideal (Carnot devices)
b) The rate of heat absorbed by the heat engine from the heat source, if the COP of the refrigerator is 55% of the Carnot refrigerator and the thermal efficiency of the heat engine is 55% of the Carnot heat engine.

Problem 6

An ideal vapor-compression heat pump cycle using HFC-134A refrigerant provides 100 MJ/h to heat a building. The operating pressures of condenser and evaporator are 9 and 1.8 bar, respectively.
Calculate the following:

a) The mass flow rate of refrigerant
b) The volumetric flow rate of refrigerant entering the compressor
c) The power requirement of a compressor
d) The COP of the heat pump cycle.

Problem 7

The air conditioner of an automobile is using an ideal vapor-compression cycle with HFC-134A as the refrigerant. The operating pressures of condenser and evaporator are 3.6 and 12 bar, respectively. Calculate the following:

a) The work requirement of the compressor for each kJ of cooling provided by the air conditioner
b) The COP of the air-conditioning cycle.

Chapter 11

Phase Equilibrium

11.1 LEARNING OUTCOMES

Upon completion of this chapter, students will be able to:

- Define phase equilibrium and state the thermodynamic criterion for phase equilibrium in terms of chemical potential
- Define the partial molar property of a species in a mixture or solution
- State the Gibbs theorem for ideal gas mixture
- Describe the important properties of ideal gas mixtures
- State the summability relation for mixtures and solutions
- Define ideal liquid solution. Describe the important properties of ideal liquid solutions
- Derive Raoult's law for ideal systems
- Define bubble and dew points (pressure and temperature) of multicomponent systems
- Perform bubble and dew points calculations for ideal multicomponent systems using Raoult's law
- Construct P-x-y, T-x-y, and y-x diagrams for ideal binary systems using the Raoult's law
- Perform flash calculations for ideal multicomponent systems using the Rachford-Rice flash equation

11.2 INTRODUCTION

A closed system consisting of two or more phases is said to be in *phase equilibrium* when the compositions and amounts of the phases present in the system are no longer changing with time, that is, they are time-invariant and there is no further tendency towards any changes in the compositions and amounts of various phases.

The thermodynamic criterion for phase equilibrium of a multicomponent chemical system is that the *chemical potential* of each species is the same in all the phases present in the system. That is, the system is free of any chemical potential gradients. Also, it is understood that the temperature and pressure are uniform throughout the whole system. Thus:

$$\mu_i\,(phase\ 1) = \mu_i\,(phase\ 2) = \mu_i\,(phase\ 3) = \mu_i\,(phase\ 4) = \cdots \tag{11.1}$$

where μ_i is the chemical potential of species i.

When the chemical potential of any species i is different in different phases, the transfer of that species takes place from a phase with higher chemical potential to a phase with lower chemical potential. It should also be noted that it is the difference in chemical potentials of a species between different phases that drives the movement of any species from one phase to another. The species always moves in the direction of decreasing chemical potential, just like current flows in the direction of decreasing voltage potential.

In this chapter we will restrict our discussion to phase equilibrium of multicomponent vapor-liquid system.

11.3 PHASE EQUILIBRIUM OF MULTICOMPONENT VAPOR-LIQUID SYSTEM

Consider a closed system consisting of multicomponent vapor and liquid phases, as shown in Figure 11.1. The temperature and pressure are uniform throughout the system.

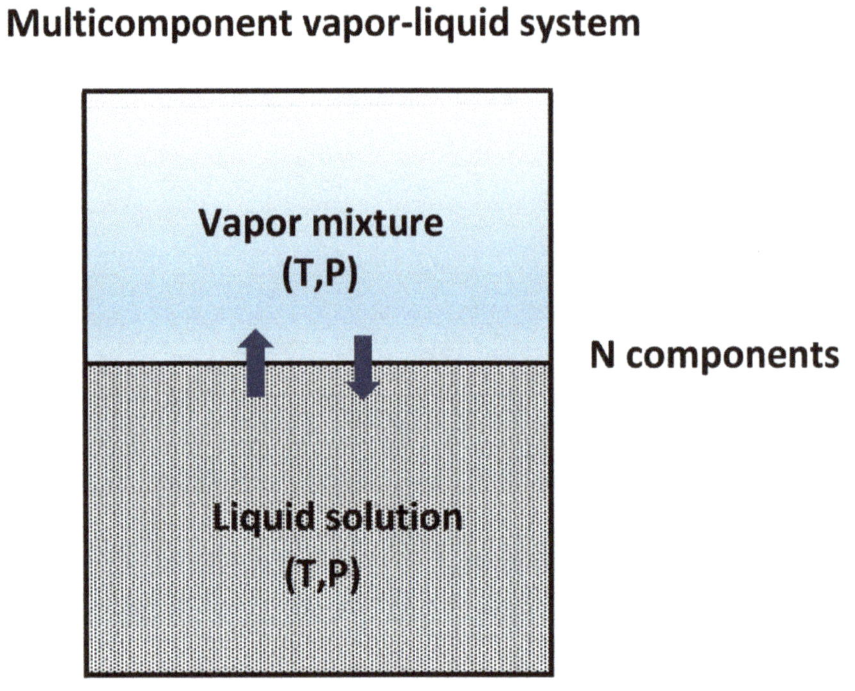

Figure 11.1 Closed system containing multicomponent vapor and liquid phases.

For this system to be in *phase equilibrium*, the following thermodynamic criterion must be satisfied:

$$\mu_i^V(T,P) = \mu_i^L(T,P) \tag{11.2}$$

where μ_i^V is the chemical potential of species i in the vapor phase and μ_i^L is the chemical potential in the liquid phase. This equality of chemical potentials must hold for every species present in the system.

The system is not in phase equilibrium if

$$\mu_i^V(T,P) \neq \mu_i^L(T,P) \tag{11.3}$$

The species i will vaporize from liquid phase to vapor phase if

$$\mu_i^L(T,P) > \mu_i^V(T,P) \tag{11.4}$$

The species i will condense from vapor phase to liquid phase if

$$\mu_i^V(T,P) > \mu_i^L(T,P) \tag{11.5}$$

The species will continue to move from one phase to another, and hence the compositions and amounts of vapor and liquid phases will continue to change, until the equality of chemical potentials is established.

11.3.1 Some Definitions and Basic Concepts

Gibbs Free Energy

Gibbs free energy, denoted by G, is a thermodynamic extensive property defined as:

$$G = H - TS \tag{11.6}$$

Upon taking differential of G,

$$dG = dH - TdS - SdT \tag{11.7}$$

The TdS relation is

$$TdS = dH - VdP \tag{11.8}$$

From Equations (11.7) and (11.8), we get:

$$dG = -SdT + VdP \tag{11.9}$$

This is an important relation. It is applicable to pure substances or mixtures of *constant composition*. We can also write,

$$G = G(T, P) \tag{11.10}$$

Hence,

$$dG = \left(\frac{\partial G}{\partial T}\right)_P dT + \left(\frac{\partial G}{\partial P}\right)_T dP \tag{11.11}$$

Upon comparison of Equations (11.9) and (11.11), we get:

$$\left(\frac{\partial G}{\partial T}\right)_P = -S \tag{11.12}$$

$$\left(\frac{\partial G}{\partial P}\right)_T = V \tag{11.13}$$

For systems of *variable composition,* such as multicomponent vapor phase or liquid phase *open* to mass transfer of species,

$$G = G(T, P, n_1, n_2, n_3, \ldots n_N) \tag{11.14}$$

where n_1, n_2, n_3, etc. are the number of moles of different species (total of N species) present in the system. Upon differentiation of Equation (11.14), we obtain:

$$dG = \left(\frac{\partial G}{\partial T}\right)_{P,n_i} dT + \left(\frac{\partial G}{\partial P}\right)_{T,n_i} dP + \sum_{i=1}^{N} \left(\frac{\partial G}{\partial n_i}\right)_{P,T,n_j} dn_i \tag{11.15}$$

where the subscript n_i in the partial derivatives of G indicates that the moles of all species are held constant and the subscript n_j in the partial derivative indicates that the moles of all species other than the species i are held constant. As the temperature and pressure partial derivatives of G (partial derivatives in first two terms on the right hand side of Equation (11.15)) are constant composition derivatives, they can be expressed as:

$$\left(\frac{\partial G}{\partial T}\right)_{P,n_i} = -S \tag{11.16}$$

$$\left(\frac{\partial G}{\partial P}\right)_{T,n_i} = V \tag{11.17}$$

Hence Equation (11.15) can be re-written as:

$$dG = -SdT + VdP + \sum_{i=1}^{N}\left(\frac{\partial G}{\partial n_i}\right)_{P,T,n_j} dn_i \tag{11.18}$$

The *chemical potential* of any species i is defined as:

$$\mu_i = \left(\frac{\partial G}{\partial n_i}\right)_{P,T,n_j} \tag{11.19}$$

Therefore, Equation (11.18) becomes

$$dG = -SdT + VdP + \sum_{i=1}^{N}\mu_i dn_i \tag{11.20}$$

This is an important relation. It is applicable to mixtures of *variable composition.*

Chemical Potential

Equation (11.19) defines the chemical potential. The chemical potential is also called *partial molar Gibbs free energy* as it is the partial derivative of G with respect to number of moles of species. The chemical potential represents the change in total Gibbs free energy of a system due to addition of a differential amount of a species at constant T and P. It tells us how sensitive is the Gibbs free energy of a system of variable composition to changes in the number of moles of a species. Let us consider a species i . A high chemical potential of a species i means G is highly sensitive to changes in the moles of species i. Likewise, a low chemical potential means G is less sensitive to changes in the moles of species i. If one mole of species i moves from a region of high chemical potential to a region of low chemical potential, the decrease in Gibbs free energy is more than the increase in Gibbs free energy. Thus, the overall Gibbs free energy of the system decreases when a species moves in the direction of decreasing chemical potential. This is consistent with the criterion of thermodynamic equilibrium of a system at constant T and P. The criterion is given as:

$$(dG)_{T,P} \leq 0 \tag{11.21}$$

The equality in Equation (11.21) is applicable at equilibrium and the inequality is valid when the system is not at equilibrium. According to the criterion of thermodynamic equilibrium (Equation (11.21)), the Gibbs free energy decreases as we approach the equilibrium, and at equilibrium the Gibbs free energy is minimum.

For pure (single-component) substances, the chemical potential is simply the molar Gibbs free energy of the substance:

$$\mu = \left(\frac{\partial G}{\partial n}\right)_{P,T} = \left(\frac{\partial (ng)}{\partial n}\right)_{P,T} = g \tag{11.22}$$

where g is the molar Gibbs free energy which is constant at constant T and P.

11.3.2 Criterion for Phase Equilibrium

Consider a closed system consisting of multicomponent vapor and liquid phases at *equilibrium.* The temperature and pressure are uniform throughout the system. As the overall closed system is at equilibrium,

$$(dG)_{T,P} = 0 \tag{11.23}$$

Since G is an extensive property,

$$G = G^L + G^V \tag{11.24}$$

Thus, at equilibrium:

$$dG = dG^L + dG^V = 0 \tag{11.25}$$

Since liquid and vapor phases are open systems (species can move from liquid phase to vapor phase and vice versa), we can apply Equation (11.20) to individual phases. Thus, Equations (11.20) and (11.25) give the following result for a multicomponent vapor and liquid phases at equilibrium at constant T and P:

$$dG = \sum_{i=1}^{N} \mu_i^L dn_i^L + \mu_i^V dn_i^V = 0 \tag{11.26}$$

Now imagine small-small (that is, differential) amounts of different species are transferred from one phase to another (say from liquid phase to vapor phase). This is a virtual (imaginary) transfer of species as the system is in equilibrium and no spontaneous changes can occur. The mass balance for any species i gives:

$$dn_i^V = -dn_i^L \tag{11.27}$$

From Equations (11.26) and (11.27), it follows that:

$$dG = \sum_{i=1}^{N} (\mu_i^L - \mu_i^V) dn_i^L = 0 \tag{11.28}$$

Since dn_i^L are independent and arbitrary, the only way Equation (11.28) can be satisfied is when:

$$\mu_i^L = \mu_i^V \tag{11.29}$$

Thus, vapor and liquid phases at the same T and P are in equilibrium with each other when the chemical potential of each species is the same in both phases.

11.4 VAPOR-LIQUID PHASE EQUILIBRIUM IN IDEAL SYSTEMS

11.4.1 Ideal Gas Mixtures

An ideal gas mixture is a mixture of ideal gases. The molecules of different species are point molecules and they do not interact with each other. Each component has the same temperature and occupies the full volume of the mixture. However, each component has its own partial pressure (P_i) given as:

$$P_i = \left(\frac{n_i}{V}\right) RT \tag{11.30}$$

where n_i is the number of moles of species i. As the ideal gas law is applicable to the mixture,

$$P = \left(\frac{n}{V}\right) RT \tag{11.31}$$

where n is the total number of moles of the mixture. Dividing Equation (11.30) by Equation (11.31), we get:

$$\frac{P_i}{P} = \frac{n_i}{n} = y_i \tag{11.32}$$

$$P_i = y_i P \tag{11.33}$$

where y_i is the mole fraction of species i in the mixture. Thus, partial pressure of a species i in an ideal gas mixture is equal to the mole fraction of species i multiplied by the total pressure. Also, the sum of partial pressures is equal to the total pressure:

$$\sum P_i = \sum y_i P = P \sum y_i = P \tag{11.34}$$

Furthermore, the partial molar volume of species i in an ideal gas mixture is equal to the pure species molar volume at the mixture temperature and pressure:

$$\bar{v}_i^{ig}(T,P) = v_i^{ig}(T,P) \tag{11.35}$$

where $\bar{v}_i^{ig}$ is the partial molar volume of a species i in the ideal gas mixture and v_i^{ig} is the molar volume of pure ideal gas species i, both compared at the same T and P.

Note that by definition,

$$\bar{v}_i^{ig}(T,P) = \left(\frac{\partial V}{\partial n_i}\right)_{T,P,n_j} = \left(\frac{\partial(nRT/P)}{\partial n_i}\right)_{T,P,n_j} = \frac{RT}{P}\left(\frac{\partial n}{\partial n_i}\right)_{n_j} \tag{11.36}$$

The total number of moles is: $n = n_i + \sum_j n_j$ where n_j is the number of moles of all species other than the species i. Hence the partial derivative $(\partial n/\partial n_i)_{n_j} = 1$. Consequently, Equation (11.36) reduces to:

$$\bar{v}_i^{ig}(T,P) = \frac{RT}{P} = v_i^{ig}(T,P) \tag{11.37}$$

Gibbs Theorem

An ideal gas mixture can be defined through the *Gibbs theorem* which states that "Any partial molar property (other than volume) of a species in an ideal gas mixture is equal to the corresponding molar property of the species as a pure ideal gas at a temperature of the mixture but at a pressure equal to its partial pressure in the mixture".

Let M be any extensive property (say Gibbs free energy, G) of a mixture. Let $\bar{m}_i$ be the corresponding partial molar property of species i in the mixture, defined as:

$$\bar{m}_i = \left(\frac{\partial M}{\partial n_i}\right)_{T,P,n_j} \tag{11.38}$$

According to the Gibbs theorem:

$$\bar{m}_i^{ig}(T,P) = m_i^{ig}(T,P_i) \tag{11.39}$$

where $\bar{m}_i^{ig}$ is the partial molar property of a species i in the ideal gas mixture at mixture T and P and m_i^{ig} is the molar property of pure ideal gas species i at mixture T and partial pressure of species i, P_i.

Examples of Equation (11.39) are as follows:

$$\bar{g}_i^{ig}(T,P) = g_i^{ig}(T,P_i) \tag{11.40}$$

$$\bar{s}_i^{ig}(T,P) = s_i^{ig}(T,P_i) \tag{11.41}$$

$$\bar{h}_i^{ig}(T,P) = h_i^{ig}(T,P_i) \tag{11.42}$$

where g, s, and h are molar Gibbs free energy, molar entropy, and molar enthalpy, respectively. It should also be noted that Equation (11.40) could be re-cast in terms of chemical potentials as:

$$\mu_i^{ig}(T,P) = \mu_{i,pure}^{ig}(T,P_i) \tag{11.43}$$

where $\mu_i^{ig}(T,P)$ is the chemical potential of species i in an ideal gas mixture at T and P and $\mu_{i,pure}^{ig}(T,P_i)$ is the chemical potential of pure ideal gas species i at T and P_i.

Important Properties of Ideal Gas Mixtures

- The volume change upon mixing of ideal gases to form an ideal gas mixture at constant T and P is zero. That is,

$$\Delta v_{mixing}^{ig} = 0 \tag{11.44}$$

- The enthalpy change upon mixing of ideal gases to form an ideal gas mixture at constant at T and P is zero. That is,

$$\Delta h_{mixing}^{ig} = 0 \tag{11.45}$$

- The entropy change upon mixing of ideal gases to form an ideal gas mixture at constant at T and P is given as:

$$\Delta s_{mixing}^{ig} = -R\sum y_i \ln y_i > 0 \tag{11.46}$$

- The Gibbs free change upon mixing of ideal gases to form an ideal gas mixture at constant at T and P is given as:

$$\Delta g_{mixing}^{ig} = RT\sum y_i \ln y_i < 0 \tag{11.47}$$

One can easily prove the above relations using the Gibbs theorem (or Equation (11.37)) and the following *summability relation*:

$$m = \sum y_i \bar{m}_i \tag{11.48}$$

where m is the molar property of the mixture at given T and P and $\bar{m}_i$ is the corresponding partial molar property of species i in the mixture, defined in Equation (11.38), at the same T and P as that of mixture. Note that this summability relation is valid for both ideal and non-ideal gas mixtures. In the case of ideal gas mixtures, we can use the Gibbs theorem (Equation (11.39)) to express the summability relation as:

$$m^{ig} = \sum y_i m_i^{ig}(T,P_i) \tag{11.49}$$

where m^{ig} is the molar property of the ideal gas mixture at given T and P, and $m_i^{ig}(T, P_i)$ is the corresponding molar property of pure ideal gas species i at mixture T and partial pressure of species i, P_i. Note that Equation (11.49) is valid for all properties other than volume.

As an example, consider the proof of Equation (11.44). The volume change upon mixing of ideal gases at constant T and P can be expressed as:

$$\Delta v_{mixing}^{ig} = v^{ig} - \sum y_i v_i^{ig} \quad (11.50)$$

where v^{ig} is the molar volume of ideal gas mixture and v_i^{ig} is the molar volume of pure ideal gas species i at the same T and P as that of mixture. Using the summability relation, Equation (11.48), we can re-write Equation (11.50) as:

$$\Delta v_{mixing}^{ig} = \sum y_i \bar{v}_i^{ig} - \sum y_i v_i^{ig} \quad (11.51)$$

Using Equation (11.37), Equation (11.51) becomes:

$$\Delta v_{mixing}^{ig} = \sum y_i v_i^{ig} - \sum y_i v_i^{ig} = 0 \quad (11.52)$$

Thus volume change upon mixing of ideal gases to form an ideal gas mixture at constant T and P is zero.

Let us consider another proof, that of Equation (11.46). The entropy change upon mixing of ideal gases at constant T and P can be expressed as:

$$\Delta s_{mixing}^{ig} = s^{ig} - \sum y_i s_i^{ig} \quad (11.53)$$

where s^{ig} is the molar entropy of ideal gas mixture and s_i^{ig} is the molar entropy of pure ideal gas species i at the same T and P as that of mixture. For pure ideal gas species i, we can also write:

$$s_i^{ig}(T, P) = s_i^{ig}(T, P_i) - R ln\left(\frac{P}{P_i}\right) \quad (11.54)$$

This equation simply reflects the effect of pressure on the entropy of pure ideal gas at constant temperature. Using Equation (11.49), we can write:

$$s^{ig} = \sum y_i s_i^{ig}(T, P_i) = \sum y_i s_i^{ig}(T, P) + R \sum y_i ln\left(\frac{1}{y_i}\right) \quad (11.55)$$

Note that $P_i = y_i P$. From Equations (11.53) and (11.55), we get:

$$\Delta s_{mixing}^{ig} = R \sum y_i \, ln\left(\frac{1}{y_i}\right) = -R \sum y_i \, ln(y_i) > 0 \quad (11.56)$$

Thus, entropy of mixing of ideal gases to form an ideal gas mixture at constant T and P is always positive, as expected from the second law of thermodynamics.

11.4.2 Ideal Liquid Solutions

When the chemical species of a liquid solution are chemically and structurally very similar so that the attractions between like and unlike molecules are the same, the liquid solution behaves as an ideal solution. For example, liquid solutions of hexane and heptane are ideal as both species are linear hydrocarbons with a difference of a single $-CH_2$ group. Likewise, benzene and toluene form ideal liquid solutions as they are chemically and structurally very similar. Some examples of non-ideal liquid solutions are: acetone and ethanol, chloroform and benzene, methyl alcohol and water, ethanol and water. The species in these solutions are chemically and structurally dissimilar.

Mathematically, we can define an ideal liquid solution as one which obeys the following relation:

$$\bar{g}_i^{id} = g_i + RTln(x_i) \tag{11.57}$$

where $\bar{g}_i^{id}$ is the partial molar Gibbs free energy of species i in the solution, g_i is the molar Gibbs free energy of pure liquid species i at the same T and P as that of solution, and x_i is the mole fraction of species i in the solution. Note that $\bar{g}_i^{id}$ is the same as chemical potential of species i in the ideal solution ($\bar{g}_i^{id} = \mu_i^{id}$).

Like the gas mixtures (ideal or non-ideal), the liquid solutions also follow the summability relation:

$$m = \sum x_i \bar{m}_i \tag{11.58}$$

where m is the molar property of the solution at given T and P and $\bar{m}_i$ is the corresponding partial molar property of species i in the solution at the same T and P as that of solution. For example,

$$g = \sum x_i \bar{g}_i \tag{11.59}$$

where g is the molar Gibbs free energy of the solution at given T and P and $\bar{g}_i$ is the corresponding partial molar Gibbs free energy of species i in the solution at the same T and P as that of solution.

Important Properties of Ideal Liquid Solutions

- The volume change upon mixing of liquids to form an ideal liquid solution at constant T and P is zero. That is,

$$\Delta v_{mixing}^{id} = 0 \tag{11.60}$$

- The enthalpy change upon mixing of liquids to form an ideal liquid solution at constant at T and P is zero. That is,

$$\Delta h_{mixing}^{id} = 0 \tag{11.61}$$

- The entropy change upon mixing of liquids to form an ideal liquid solution at constant at T and P is given as:

$$\Delta s_{mixing}^{id} = -R\sum x_i \, lnx_i > 0 \tag{11.62}$$

- The Gibbs free change upon mixing of liquids to form an ideal liquid solution at constant at T and P is given as:

$$\Delta g_{mixing}^{id} = RT\sum x_i \, lnx_i < 0 \tag{11.63}$$

As an example, consider the proof of Equation (11.63). The Gibbs free energy change upon mixing of liquids at constant T and P can be expressed as:

$$\Delta g_{mixing}^{id} = g^{id} - \sum x_i g_i \tag{11.64}$$

where g^{id} is the molar Gibbs free energy of ideal liquid solution and g_i is the molar Gibbs free energy of pure liquid species i at the same T and P as that of solution. Using the summability relation, Equation (11.59), we can re-write Equation (11.64) as:

$$\Delta g_{mixing}^{id} = \sum x_i \bar{g}_i^{id} - \sum x_i g_i \tag{11.65}$$

Making use of Equation (11.57), Equation (11.65) simplifies to Equation (11.63).

11.4.3 Raoult's Law

Consider a closed system consisting of multicomponent vapor and liquid phases, as shown in Figure 11.1. The temperature and pressure are uniform throughout the system. When this system is in *equilibrium*, the compositions of vapor and liquid phases are related to each other. Assuming that the liquid solution is an ideal liquid solution, and the vapor mixture is an ideal gas mixture, the relationship between vapor and liquid compositions is given as follows:

$$y_i P = x_i P_i^{sat} \tag{11.66}$$

This is the celebrated *Raoult's law*, where y_i is the mole fraction of species i in the vapor mixture, x_i is the mole fraction of species i in the liquid solution, P is the total pressure, and P_i^{sat} is the vapor pressure of pure species i at the temperature of the system. The Raoult's law can be used to perform different types of vapor-liquid equilibrium calculations including bubble point, dew point, and flash calculations.

11.4.4 Derivation of Raoult's Law

Once again, consider a closed system consisting of multicomponent vapor and liquid phases, as shown in Figure 11.1. The temperature and pressure are uniform throughout the system. The system is in equilibrium. Assume that the vapor phase is an ideal gas mixture, and the liquid solution is an ideal liquid solution. As the system is in equilibrium, we can apply the criterion of phase equilibrium given in Equation (11.29) as:

$$\mu_{i,solu}^{id}(T,P) = \mu_{i,mix}^{ig}(T,P) \tag{11.67}$$

where $\mu_{i,solu}^{id}(T,P)$ is the chemical potential of species i in the liquid phase (ideal liquid solution) and $\mu_{i,mix}^{ig}(T,P)$ is the chemical potential of species i in the vapor phase (ideal gas mixture).

Expansion of $\mu_{i,mix}^{ig}(T,P)$ term of Equation (11.67)

From Gibbs theorem, Equation (11.43):

$$\mu_{i,mix}^{ig}(T,P) = \mu_{i,pure}^{ig}(T,P_i) \tag{11.68}$$

where $\mu_{i,pure}^{ig}(T,P_i)$ is the chemical potential of pure ideal gas species i at system T and partial pressure P_i.

For pure substances, chemical potential and molar Gibbs free energy are the same. Hence, we can use Equation (11.9) in the following form

$$d\mu = -s dT + v dP \tag{11.69}$$

For ideal gas at constant temperature, Equation (11.69) becomes:

$$d\mu^{ig} = RTdlnP \tag{11.70}$$

Integration of Equation (11.69) at constant temperature gives,

$$\mu_{i,pure}^{ig}(T,P) = \mu_{i,pure}^{ig}(T,P_i) + RTln\left(\frac{P}{P_i}\right) \tag{11.71}$$

Since $P_i = y_i$P, Equation (11.71) can be re-written as:

$$\mu_{i,pure}^{ig}(T,P_i) = \mu_{i,pure}^{ig}(T,P) + RTlny_i \tag{11.72}$$

Consequently, Equation (11.68) becomes:

$$\mu_{i,mix}^{ig}(T,P) = \mu_{i,pure}^{ig}(T,P) + RTlny_i \tag{11.73}$$

Expansion of $\mu_{i,solu}^{id}(T,P)$ ***term of Equation (11.67)***

The defining relation for ideal liquid solution, Equation (11.57), can be expressed in terms of chemical potentials as:

$$\mu_{i,solu}^{id}(T,P) = \mu_{i,pure}^{L}(T,P) + RTln(x_i) \tag{11.74}$$

Raoult's Law

Substitute the expressions for $\mu_{i,mix}^{ig}(T,P)$ and $\mu_{i,solu}^{id}(T,P)$ in the criterion of phase equilibrium, Equation (11.67), to obtain:

$$\mu_{i,pure}^{ig}(T,P) + RTlny_i = \mu_{i,pure}^{L}(T,P) + RTln(x_i) \tag{11.75}$$

Upon re-arrangement, Equation (11.75) gives:

$$\mu_{i,pure}^{ig}(T,P) - \mu_{i,pure}^{L}(T,P) = RTln\left(\frac{x_i}{y_i}\right) \tag{11.76}$$

Integration of Equation (11.70) from P_i^{sat} to P at constant temperature gives,

$$\mu_{i,pure}^{ig}(T,P) = \mu_{i,pure}^{ig}(T,P_i^{sat}) + RTln\left(\frac{P}{P_i^{sat}}\right) \tag{11.77}$$

where P_i^{sat} is vapor pressure of pure species i at temperature T. From Equations (11.76) and (11.77), we obtain:

$$\mu_{i,pure}^{ig}(T,P_i^{sat}) + RTln\left(\frac{P}{P_i^{sat}}\right) - \mu_{i,pure}^{L}(T,P) = RTln\left(\frac{x_i}{y_i}\right) \tag{11.78}$$

Neglecting the effect of pressure on the chemical potential of pure liquid,

$$\mu_{i,pure}^{L}(T,P) \approx \mu_{i,pure}^{L}(T,P_i^{sat}) \tag{11.79}$$

From Equations (11.78) and (11.79), we get:

$$\mu_{i,pure}^{ig}(T, P_i^{sat}) - \mu_{i,pure}^{L}(T, P_i^{sat}) = RTln\left(\frac{x_i P_i^{sat}}{y_i P}\right) \tag{11.80}$$

From vapor-liquid equilibrium condition of pure species i,

$$\mu_{i,pure}^{ig}(T, P_i^{sat}) = \mu_{i,pure}^{L}(T, P_i^{sat}) \tag{11.81}$$

Consequently, Equation (11.80) simplifies to

$$RTln\left(\frac{x_i P_i^{sat}}{y_i P}\right) = 0 \tag{11.82}$$

Equation (11.82) gives the Raoult's law:

$$y_i P = x_i P_i^{sat} \tag{11.66}$$

The Raoult's law could also be expressed as:

$$y_i = K_i x_i \tag{11.83}$$

$$K_i = P_i^{sat}/P \tag{11.84}$$

where K_i is called the phase equilibrium constant or distribution coefficient of a species i. K_i is a function of both temperature and pressure.

The applications of the Raoult's law are treated in the following section.

11.5 VAPOR-LIQUID EQUILIBRIUM CALCULATIONS USING RAOULT'S LAW

In the design of distillation columns, bubble point and dew point calculations are required. The construction of P-x-y and T-x-y phase diagrams, and y-x equilibrium curve are also needed in practical applications. The bubble point/dew point calculations and the construction of P-x-y, T-x-y, and y-x diagrams can be performed readily using the Raoult's law. The Raoult's law could also be used to perform flash calculations needed in the design of flash separators. The basic assumption in all these calculations is that the vapor mixture and liquid solution are ideal in nature.

11.5.1 Bubble Point and Dew Point Calculations

A liquid solution at saturation condition is said to be at its bubble point. With small changes in temperature or pressure, the liquid solution at bubble point begins to boil. A vapor mixture at saturation condition is said to be at its dew point. With small changes in temperature or pressure, the vapor mixture at dew point begins to condense.

Pure Substances

Consider a subcooled liquid at a fixed temperature T_o initially in state 1 as shown on a P-V-T diagram in Figure 11.2. If the pressure of this subcooled liquid is decreased from state 1 along an isotherm, state 2 is reached where the pressure is equal to the vapor pressure (P^{Sat}) of the liquid at T_o . In state 2, the liquid is saturated and begins to form bubbles. The pressure where the first bubble appears at a fixed temperature of T_o is P^{Sat} . This pressure is called *bubble point pressure (BUBLP)*. Also note that the pressure does not change until the entire liquid is vaporized or boiled off.

Let us now follow the path from state 4 to state 3 along an isotherm (fixed temperature of T_o) as shown in Figure 11.2. We have superheated vapor in state 4. With the increase in pressure of this superheated vapor, we reach state 3 where the vapor becomes saturated vapor. In saturated vapor state 3, the vapor begins to condense. The pressure where the first liquid droplet (also called dew) is formed at a fixed temperature of T_o is P^{Sat}. This pressure is called *dew point pressure (DEWP)*. Also note that the pressure does not change until the entire vapor is condensed.

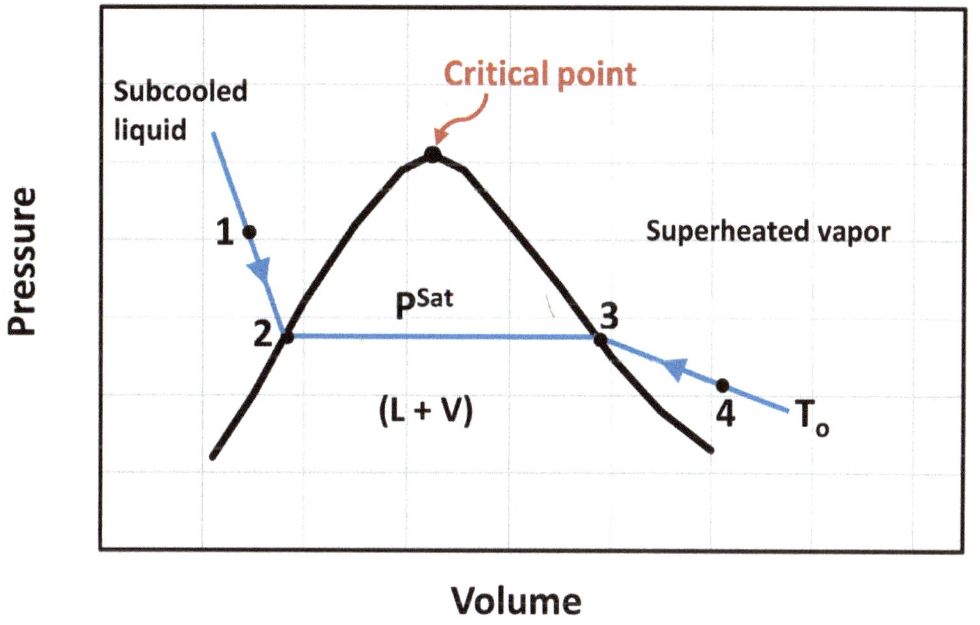

Figure 11.2 Boiling of pure liquid upon decrease of pressure to $P^{Sat}$and condensation of pure vapor upon increase of pressure to P^{Sat}at constant temperature.

Thus, the bubble point pressure and dew point pressure at any given temperature are the same for pure (single-component) substances, that is, $BUBLP = DEWP = P^{Sat}$. This implies that boiling and condensation of pure substances at any given temperature occur at a single pressure corresponding to vapor pressure P^{Sat} at that temperature.

We now consider a different experiment where pressure is fixed, and temperature is varied. Referring to Figure 11.3, we heat a subcooled liquid in state 1 at constant pressure P_o until state 2 is reached. In state 2, the liquid is saturated, and it begins to boil. The temperature where the first bubble is formed at a fixed pressure of P_o is T^{Sat}. This temperature is called *bubble point temperature (BUBLT)*. The temperature remains constant until the entire liquid is vaporized. What happens when we cool a superheated vapor in state 4 at constant pressure? Upon isobaric cooling, the vapor becomes saturated in state 3 and it begins to condense. The temperature where the first liquid droplet (dew) is formed at a fixed temperature of P_o is T^{Sat}. This temperature is called *dew point temperature (DEWT)*. The temperature remains constant until the entire vapor is condensed.

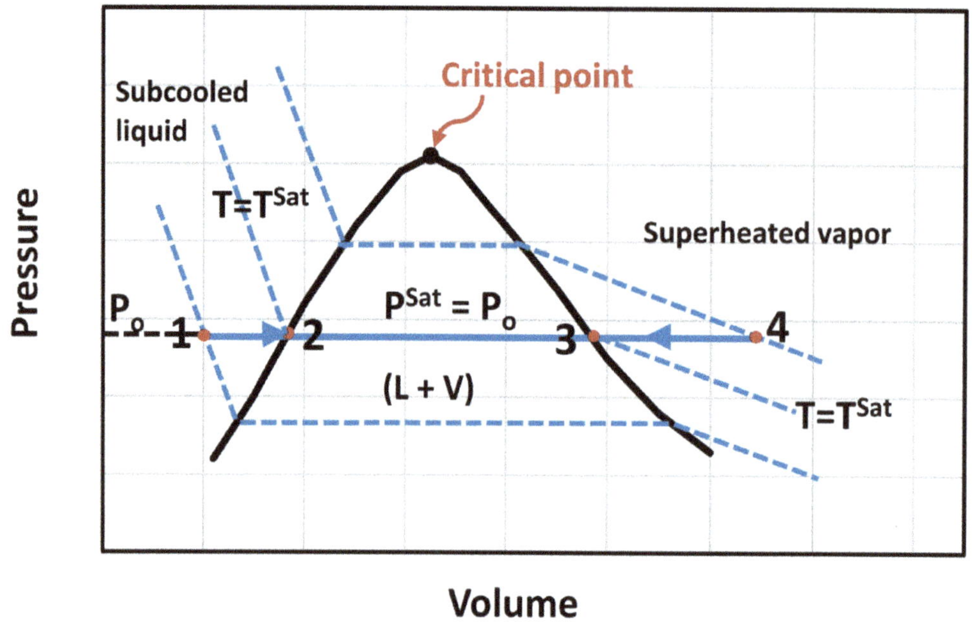

Figure 11.3 Boiling of pure liquid upon increase of temperature to $T^{Sat}$and condensation of pure vapor upon decrease of temperature to T^{Sat}at constant pressure.

Thus, we have observed that the bubble point temperature is the same as the dew point temperature at any given pressure for pure (single-component) substances, that is, $BUBLT = DEWT = T^{Sat}$. The boiling and condensation of pure substances at any given pressure occur at a single temperature corresponding to T^{Sat} at that pressure.

Mixtures of Two or More Components

In the case of systems with two or more components, boiling and condensation do not occur at a single pressure at a fixed temperature. That is, *BUBLP* and *DEWP* are not the same. The boiling and condensation take place over a range of pressures at any given temperature. For the same overall composition of the system, *BUBLP* > *DEWP* at a fixed temperature. Likewise, the boiling and condensation do not occur at a single temperature at a fixed pressure. The *BUBLT* and *DEWT* are different and boiling and condensation occur over a range of temperatures. For the same overall composition of the system *DEWT* > *BUBLT* at a fixed pressure. Furthermore, the bubble points (*BUBLP* and *BUBPT*) and dew points (*DEWP* and *DEWT*) vary with the system composition.

As an example, consider a binary system consisting of two chemical species. Figure 11.4 shows the P-x-y phase diagram of a binary system at a fixed temperature. Starting from a subcooled liquid solution of known composition in state 1 (Figure 11.4), we decrease the pressure of the system along the path $1 \rightarrow 2 \rightarrow 3$. At point 2, the liquid solution begins to boil. The pressure corresponding to state 2 where the first bubble appears is the bubble point pressure *(BUBLP)* of the solution. In order to continue boiling of the liquid solution into the vapor phase, the pressure has to be decreased further from state 2 to state 3. In state 3, the last liquid droplet is vaporized. The pressure corresponding to state 3 is *DEWP*. Note that this boiling behavior is different from what was observed in the case of pure single-component system where boiling occurred at a constant pressure. In the case of multi-component systems, boiling at a fixed temperature occurs over a range of pressures $DEWP \leq P \leq BUBLP$. Also note that *DEWP* where the first liquid droplet (dew) is formed is the same pressure where the last droplet is vaporized. If we begin from superheated state 4 and increase the pressure (keeping temperature fixed) along the path $4 \rightarrow 3 \rightarrow 2$, the first liquid droplet is formed in state 3 and to continue condensation further we need to increase the pressure further until the entire vapor is condensed in state 2. Thus, condensation at a fixed temperature occurs over a range of pressures $DEWP \leq P \leq BUBLP$.

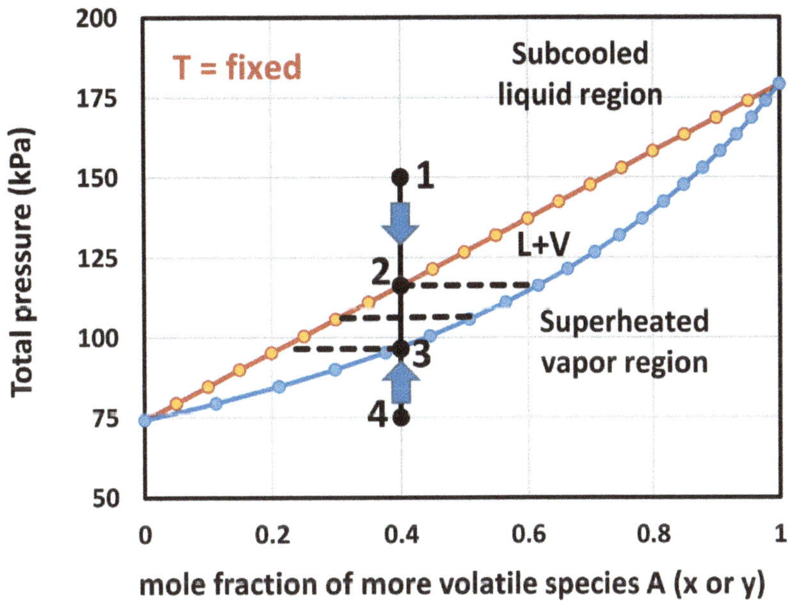

Figure 11.4 P-x-y phase diagram of a binary system at a fixed temperature.

In the preceding discussion, the temperature was held constant, and pressure was varied. Now we will keep the pressure fixed and vary the temperature of a binary system. Thus, we need T-x-y phase diagram of a binary system at a fixed pressure to define bubble and dew points. Figure 11.5 shows a T-x-y diagram for a binary system consisting of species "*A*" and "*B*" with species "*A*" being more volatile ($T_A^{Sat} \ll T_B^{Sat}$). Starting from a subcooled liquid solution of known composition in state 1 (Figure 11.5) we heat the solution isobarically along the path $1 \rightarrow 2 \rightarrow 3$. At point 2, the liquid solution begins to boil. The temperature

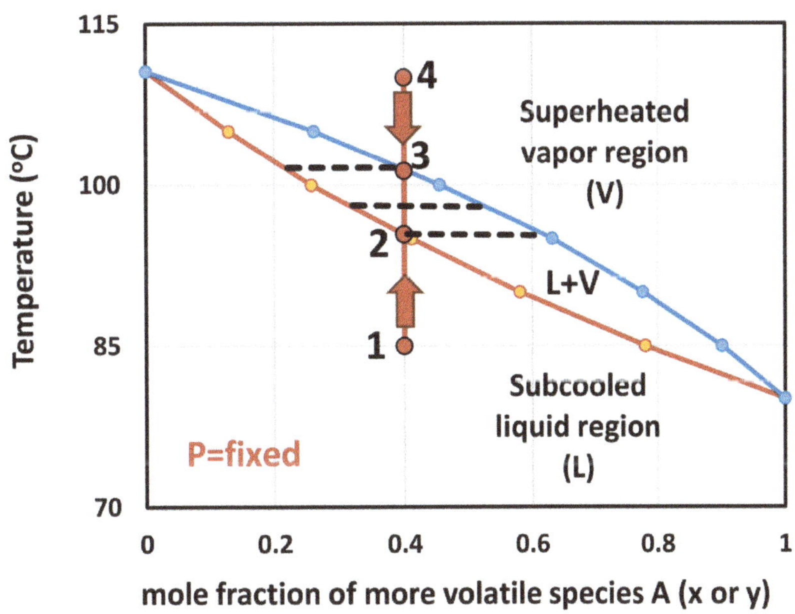

Figure 11.5 T-x-y phase diagram of a binary system at a fixed pressure.

corresponding to state 2 where the first bubble appears is the bubble point temperature *(BUBLT)* of the solution. In order to continue boiling of the liquid solution into the vapor phase, the temperature has to be increased further from state 2 to state 3. In state 3, the last liquid droplet is vaporized. The temperature corresponding to state 3 is *DEWT*. Thus, unlike single-component systems, boiling at a fixed pressure occurs over a range of temperatures $BUBLT \leq T \leq DEWT$ in multi-component systems. If we begin from superheated state 4 and cool the vapor by decreasing the temperature (keeping pressure fixed) along the path $4 \rightarrow 3 \rightarrow 2$, the first liquid droplet is formed in state 3 and to continue condensation further we need to decrease the temperature further until the entire vapor is condensed in state 2. Thus condensation, like boiling, at a fixed pressure occurs over a range of temperatures $BUBLT \leq T \leq DEWT$.

Different Classes of Bubble and Dew Points Calculations

There are four types of bubble and dew points calculations. They are as follows:

- *BUBLP:* Given liquid composition and temperature, calculate the pressure where boiling begins and the composition of the first bubble formed. We can assume that the liquid is initially in sub-cooled liquid state. If we reduce the pressure (keeping temperature fixed), at what pressure would boiling begin?
- *DEWP:* Given vapor composition and temperature, calculate the pressure where condensation begins and the composition of the first droplet (dew) formed. We can assume that the vapor is initially in superheated vapor state. If we increase the pressure (keeping temperature fixed), at what pressure would condensation begin?
- *BUBLT:* Given liquid composition and pressure, calculate the temperature where boiling begins and the composition of the first bubble formed. We can assume that the liquid is initially in sub-cooled liquid state. If we increase the temperature (keeping pressure fixed), at what temperature would boiling begin?
- *DEWT:* Given vapor composition and pressure, calculate the temperature where condensation begins and the composition of the first droplet (dew) formed. We can assume that the vapor is initially in superheated vapor state. If we decrease the temperature (keeping pressure fixed), at what temperature would condensation begin?

We will assume that the Raoult law is applicable, that is, the liquid solution and vapor mixture are both ideal in nature.

BUBLP Calculation: Given $\{x_i\}$ and T, calculate P and $\{y_i\}$

From Raoult's law:

$$y_i = x_i \frac{P_i^{Sat}}{P} \qquad (11.85)$$

As $\sum y_i = 1$, Equation (11.85) gives the bubble point pressure as:

$$P = \sum x_i P_i^{Sat} \qquad (11.86)$$

The liquid composition is specified, and the vapor pressures of different species can be calculated from the Antoine equation, given below, as the temperature is specified:

$$lnP^{sat} = A - \frac{B}{T+C} \qquad (11.87)$$

where A, B, C are constants for a given substance. These constants are available in the literature for a large number of substances. Once the bubble point pressure is known, the composition of the first bubble can be calculated from Equation (11.85).

***DEWP Calculation: Given $\{y_i\}$ and T, calculate P* and $\{x_i\}$**

From Raoult's law:

$$x_i = y_i \frac{P}{P_i^{Sat}} \tag{11.88}$$

As $\sum x_i = 1$, Equation (11.88) gives the dew point pressure as:

$$P = \frac{1}{\sum y_i / P_i^{Sat}} \tag{11.89}$$

The vapor composition is specified, and the vapor pressures of different species can be calculated from the Antoine equation, Equation (11.87), as the temperature is specified. Once the dew point pressure is known, the composition of the first liquid droplet (dew) can be calculated from Equation (11.88).

***BUBLT Calculation: Given $\{x_i\}$ and P, calculate T* and $\{y_i\}$**

When temperature is unknown, an iterative procedure is required. The iterative procedure can be developed as follows.

Equation (11.86) could be re-written as:

$$P = P_r^{Sat} \sum x_i P_i^{Sat} / P_r^{Sat} \tag{11.90}$$

where species "*r*" is the "reference species", selected somewhat arbitrarily.

Equation (11.90) can be re-arranged as:

$$P_r^{Sat} = P / \sum x_i \alpha_{ir} \tag{11.91}$$

where α_{ir} is the relative volatility of species i with respect to species r, defined as $\alpha_{ir} = P_i^{sat} / P_r^{sat}$. From the Antoine equations for species "*i*" and "*r*", one can write:

$$ln\alpha_{ir} = A_i - A_r + \frac{B_r}{T+C_r} - \frac{B_i}{T+C_i} \tag{11.92}$$

Upon re-arrangement of the Antoine equation for the reference species "*r*", we get:

$$T = \frac{B_r}{A_r - lnP_r^{Sat}} - C_r \tag{11.93}$$

where $A, B, and\ C$ are coefficients in the Antoine equations for species "*i*" and "*r*". The reference species "*r*" is fixed whereas "*i*" refers to any species in a multi-component system.

The iterative procedure for the estimation of T is shown in Figure 11.6. The initial guess of temperature can be made as follows:

$$T_{guess} = \sum x_i T_i^{Sat} \tag{11.94}$$

where T_i^{Sat} is the boiling point of species i estimated from the Antoine equation at the given total pressure P. Equation (11.94) is not a valid equation, and it is to be used only for initial guess purposes. Following the procedure shown in Figure 11.6, an improved value of temperature is calculated from Equation (11.93). If the calculated temperature is within the set tolerance level, the iteration process is stopped otherwise the calculated value of temperature is taken as the new guess value and the iterative process is continued. The percent error between two successive iterations can be calculated and compared with the percent tolerance level ε as follows:

$$\%\, Error = \left|\frac{T_{k+1}-T_k}{T_k}\right| \times 100 \leq \varepsilon \tag{11.95}$$

Once the temperature is estimated, one can use Equation (11.85) to calculate the vapor composition.

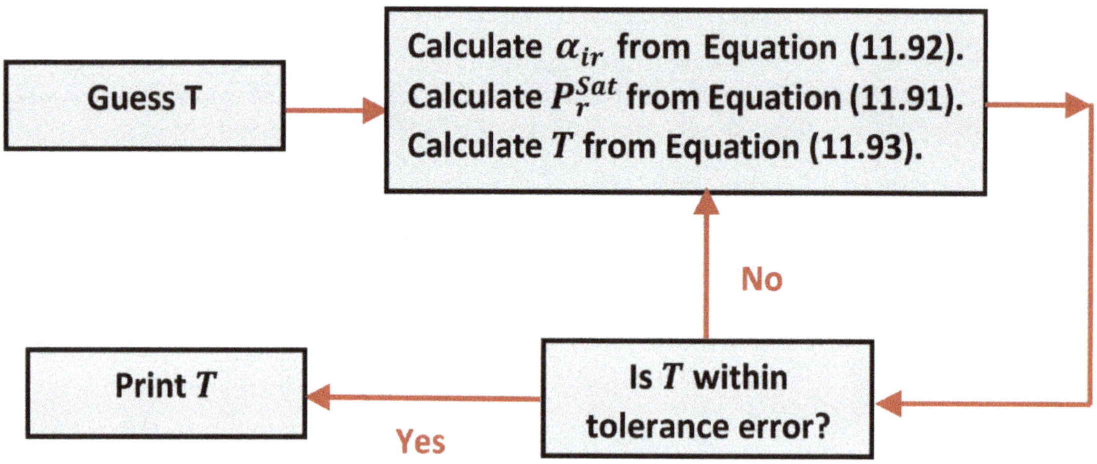

Figure 11.6 Iterative procedure for the calculation of *BUBLT*.

DEWT Calculation: Given $\{y_i\}$ *and* P*, calculate* T *and* $\{x_i\}$

Once again, an iterative procedure is required as temperature is unknown. The iterative procedure is similar to that used in *BUBLT* calculation. Equation (11.89) can be re-written as:

$$P = \frac{P_r^{Sat}}{P_r^{Sat} \sum y_i / P_i^{Sat}} \tag{11.96}$$

Using the definition of relative volatility, Equation (11.96) can be re-arranged as:

$$P_r^{Sat} = P \sum y_i / \alpha_{ir} \tag{11.97}$$

Equations (11.92) and (11.93) are applicable to *DEWT* calculation. Thus, the iteration scheme for *DEWT* calculation is as shown in Figure 11.7.

The initial guess of temperature can be made as follows:

$$T_{guess} = \sum y_i\, T_i^{Sat} \tag{11.98}$$

where T_i^{Sat} is the boiling point of species i estimated from the Antoine equation at the given total pressure P. Following the procedure shown in Figure 11.7, an improved value of temperature is calculated from Equation (11.93). If the calculated temperature is within the set tolerance level (see Equation (11.95)), the iteration process is stopped otherwise the calculated value of temperature is taken as the new guess value and the iterative process is continued. Once the temperature is estimated, one can use Equation (11.88) to calculate the liquid (dew) composition.

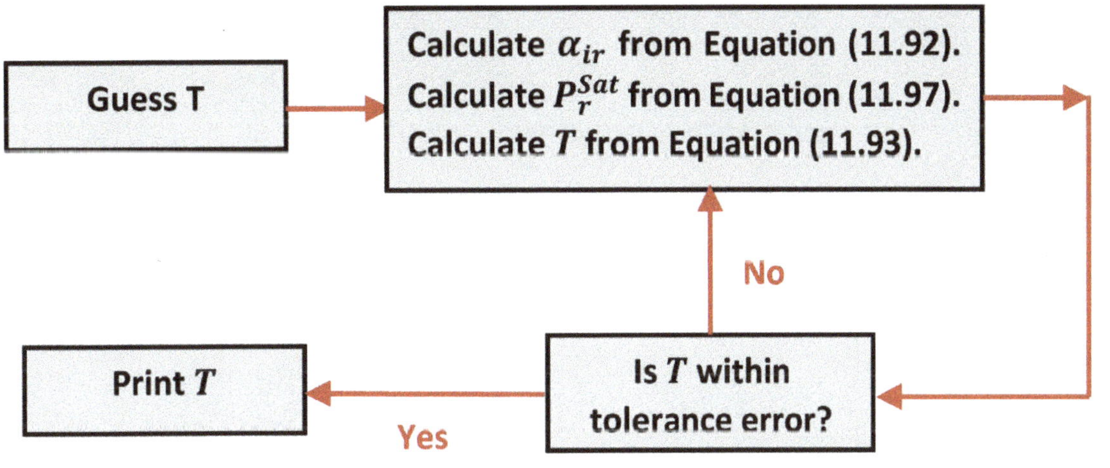

Figure 11.7 Iterative procedure for the calculation of *DEWT*.

11.5.2 Construction of P-x-y and T-x-y Phase Diagrams and y-x Equilibrium Curve

The phase diagrams (P-x-y and T-x-y) and equilibrium y-x curve for binary systems can be constructed using the Raoult law assuming that the liquid solution and vapor mixture are ideal in nature.

Figure 11.4 shows a P-x-y diagram (temperature fixed) of a binary system of species "A" and "B" where species "A" is more volatile. The P-x-y diagram gives the compositions of liquid and vapor phases in equilibrium with each other at different pressures at a fixed temperature. If the liquid solution and vapor mixture are assumed to be ideal, the compositions of the vapor mixture and liquid solution are related by the Raoult law. Thus, the Raoult law can be used to construct the P-x-y phase diagram and the corresponding y-x equilibrium curve.

Figure 11.5 shows the T-x-y diagram (pressure fixed) of a binary system of species "A" and "B" where species "A" is more volatile. The T-x-y diagram gives the compositions of liquid and vapor phases in equilibrium with each other at different temperatures at a fixed pressure. Assuming ideal liquid solution and vapor mixture, the Raoult law can be used to construct the T-x-y phase diagram and the corresponding y-x equilibrium curve.

For a binary system of species "A" and "B" (A being more volatile), the governing equations are as follows:

$$P_A = x_A P_A^{Sat} \tag{11.99}$$

$$P_B = x_B P_B^{Sat} = (1 - x_A) P_B^{Sat} \tag{11.100}$$

$$P = P_A + P_B = x_A P_A^{Sat} + (1 - x_A) P_B^{Sat} \tag{11.101}$$

Solving for x_A , Equation (11.101) gives:

$$x_A = \left[\frac{P-P_B^{Sat}}{P_A^{Sat}-P_B^{Sat}}\right] \tag{11.102}$$

The vapor composition y_A can be calculated as:

$$y_A = \frac{x_A P_A^{Sat}}{P} = \left(\frac{P_A^{Sat}}{P}\right)\left[\frac{P-P_B^{Sat}}{P_A^{Sat}-P_B^{Sat}}\right] \tag{11.103}$$

The phase diagrams (P-x-y, T-x-y) and the y-x equilibrium curve can now be constructed using Equations (11.102) and (11.103) along with the Antoine equations for species "A" and "B".

To construct the P-x-y diagram at a given temperature, the vapor pressures P_A^{Sat} and P_B^{Sat} are calculated at the specified temperature using the Antoine equations. Then a number of P values are selected over the pressure range $P_B^{Sat} < P < P_A^{Sat}$ (say 10 values equally spaced between $P_B^{Sat}$and P_A^{Sat}). The liquid and vapor compositions (x_A and y_A) corresponding to any chosen P are calculated using Equations (11.102) and (11.103). The plots of P vs x_A (bubble point curve) and P vs y_A (dew point curve) give the P-x-y phase diagram. The plot of y_A vs x_A gives the equilibrium curve at constant temperature.

To construct the T-x-y diagram at a given pressure, we need to calculate T_A^{Sat} and T_B^{Sat} corresponding to the specified pressure using the Antoine equations. Then a number of T values are selected over the temperature range $T_A^{Sat} < T < T_B^{Sat}$ (say 10 values equally spaced between $T_A^{Sat}$and T_B^{Sat}). The vapor pressures of species A and B ($P_A^{Sat}$and P_B^{Sat}) are calculated at the selected values of T using the Antoine equations. The liquid and vapor compositions (x_A and y_A) corresponding to any chosen T are calculated using Equations (11.102) and (11.103). The plots of T vs x_A (bubble point curve) and T vs y_A (dew point curve) give the T-x-y phase diagram. The plot of y_A vs x_A gives the equilibrium curve at constant pressure. The isobaric y_A vs x_A equilibrium curve is used extensively in the design of the distillation columns.

11.5.3 Flash Calculations

Flash distillation is a single-stage continuous distillation process shown schematically in Figure 11.8. Liquid feed is pumped through a heater to raise its temperature. Liquid at high temperature and pressure is passed through a throttling valve where a sharp reduction in pressure occurs. The throttling valve is located right at the entry into a flash drum. A sharp reduction in pressure of a hot liquid causes flashing (rapid vaporization) in the flash drum. The vapor enriched in more volatile components is taken off as overhead product while the liquid lean in more volatile components (enriched in less volatile components) is withdrawn from the bottom of the drum. The vapor and liquid leaving the flash drum are at equilibrium with each other.

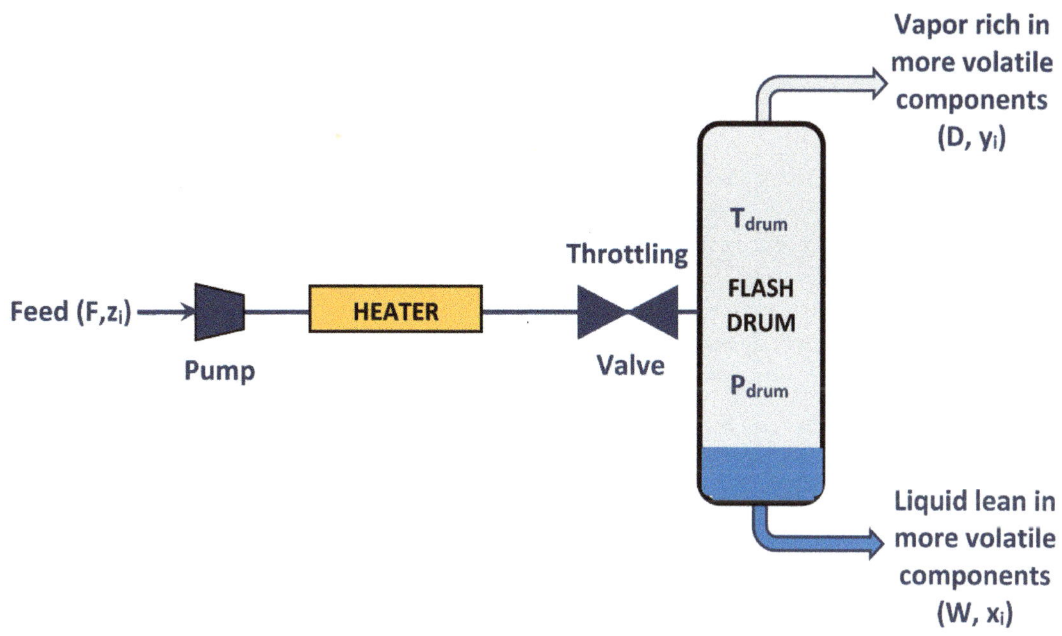

Figure 11.8 Flash distillation process.

Consider flash distillation of multicomponent feed. Given the feed flow rate (F, mol/s), feed composition (set of $\{z_i\}$ values), flash drum temperature and pressure, calculate the vapor-distillate flow rate (D, mol/s), distillate composition (set of $\{y_i\}$ values), bottoms-liquid flow rate (W, mol/s) and bottoms composition (set of $\{x_i\}$ values).

The component mass balance on species "i" gives:

$$z_i F = y_i D + x_i W \tag{11.104}$$

Using overall mass balance and the equilibrium relation (Raoult's law) $y_i = K_i x_i$, we can re-write Equation (11.104) as:

$$y_i = \frac{K_i z_i}{1+\psi(K_i-1)} \tag{11.105}$$

where $\psi = D/F$ (fraction of feed that goes to the distillate product). Alternatively, we can write:

$$x_i = \frac{z_i}{1+\psi(K_i-1)} \tag{11.106}$$

Because $\sum y_i = 1$ and $\sum x_i = 1$, Equations (11.105) and (11.106) yield:

$$\sum \frac{K_i z_i}{1+\psi(K_i-1)} = 1 \tag{11.107}$$

$$\sum \frac{z_i}{1+\psi(K_i-1)} = 1 \tag{11.108}$$

Assuming that the Raoult's law is applicable, the distribution coefficient $K_i(T,P)$ of a chemical species is given as P_i^{sat}/P. The vapor pressure of pure species i at the given flash temperature can be determined from the Antoine equation. The feed composition is given. The only unknown in Equations (11.107) and (11.108) is ψ. Either of these two equations (Equation (11.107) or (11.108)) can be used to estimate ψ by trial. Unfortunately, Equations (11.107) and (11.108) do not have good convergence properties. However, the combination of Equations (11.107) and (11.108), that is, $\sum(y_i - x_i) = 0$, is found to have good convergence properties. Thus:

$$\sum \frac{(K_i-1)z_i}{1+\psi(K_i-1)} = 0 \tag{11.109}$$

Equation (11.109) has excellent convergence properties and can be used to estimate the unknown ψ. Equation (11.109) is often referred to as the *Rachford-Rice flash equation.*

One can use the Newton-Raphson method (also called Newton method) to find the root of the Rachford-Rice equation. Equation (11.109) can be re-written as:

$$f(\psi) = \sum \frac{(K_i-1)z_i}{1+\psi(K_i-1)} = 0 \tag{11.110}$$

The Newton-Raphson method is illustrated schematically in Figure 11.9. It requires an initial guess of the root. Let x_o be the initial guess of the root of $f(x) = 0$. In order to estimate an improved value of the root, draw a tangent to the curve $y = f(x)$ at $(x_o, f(x_o))$. The intersection of the tangent line with the x-axis gives the improved estimate x_1. The successive values of x generated in this manner converge to the root of $f(x) = 0$.

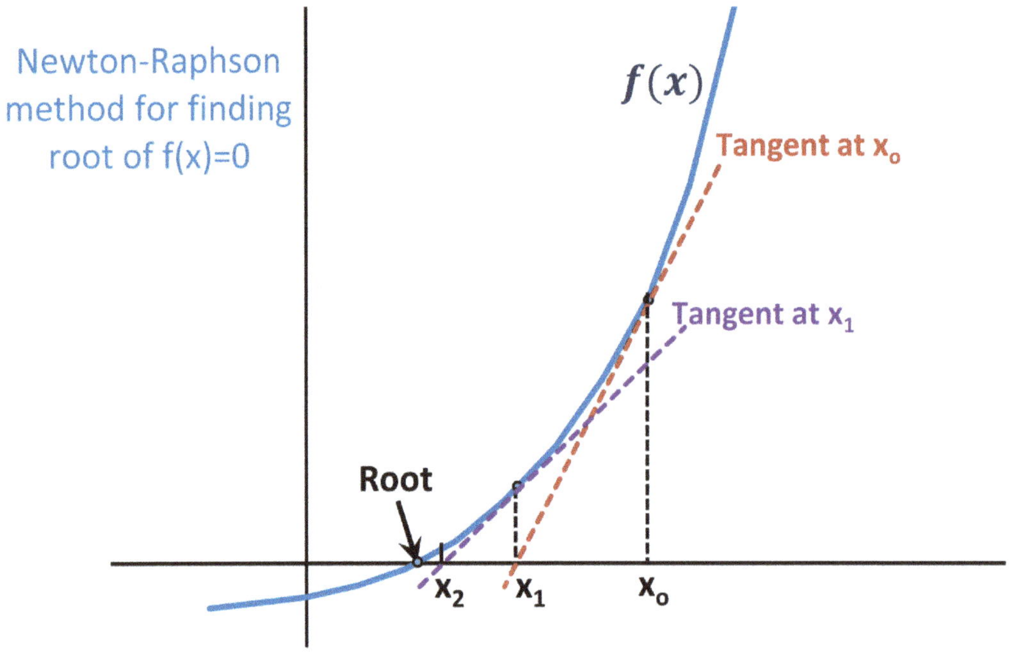

Figure 11.9 Newton-Raphson method for finding root of $f(x) = 0$.

The formula for estimating the improved value of the root (x_{k+1}) over the previous value of the root (x_k) can be derived by expressing the slope of the tangent line as:

$$slope = f'(x_k) = \frac{f(x_k)-0}{(x_k - x_{k+1})} \tag{11.111}$$

Thus,

$$x_{k+1} = x_k - \frac{f(x_k)}{f'(x_k)} \tag{11.112}$$

The derivative of $f(\psi)$ in Rachford-Rice equation, Equation (11.110), is given as:

$$f'(\psi) = -\sum \frac{(K_i-1)^2 z_i}{[1+\psi(K_i-1)]^2} \tag{11.113}$$

Hence, we can write:

$$\psi_{k+1} = \psi_k - \frac{f(\psi_k)}{f'(\psi_k)} \tag{11.114}$$

Equation (11.114), along with Equation (11.113), can be used to estimate ψ by successive substitution. The range of ψ is $0 \leq \psi \leq 1$. The initial guess could be taken as $\psi_o = 0.50$. The iteration process can be terminated when the deviation between two successive values is within the tolerance level, $[|(\psi_{k+1} - \psi_k)|/\psi_k] \times 100 \leq \varepsilon$.

Once ψ is known, the vapor composition can be calculated from Equation (11.105) and the liquid composition can be calculated from the equilibrium relation: $x_i = y_i/K_i$. The vapor-distillate flow rate is $D = \psi F$ and the bottoms-liquid flow rate is $W = (1-\psi)F$.

NOTATION

A	Antoine equation coefficient
A_r	Antoine equation coefficient for reference species
B	Antoine equation coefficient
B_r	Antoine equation coefficient for reference species
$BUBLP$	bubble point pressure
$BUBLT$	bubble point temperature
C	Antoine equation coefficient
C_r	Antoine equation coefficient for reference species
D	vapor distillate molar flow rate
$DEWP$	dew point pressure
$DEWT$	dew point temperature
F	feed molar flow rate
f	function
f'	derivative of function f
g	molar Gibbs free energy of pure substance, mixture or solution
g_i	molar Gibbs free energy of pure species i
$\bar{g}_i$	partial molar Gibbs free energy of species i in a mixture or solution
g^{id}	molar Gibbs free energy of ideal solution
$\bar{g}_i^{id}$	partial molar Gibbs free energy of species i in ideal solution
g^{ig}	molar Gibbs free energy of ideal gas mixture
g_i^{ig}	molar Gibbs free energy of pure species i in ideal gas state
$\bar{g}_i^{ig}$	partial molar Gibbs free energy of species i in ideal gas mixture
G	total Gibbs free energy of pure substance, mixture or solution
G^L	total Gibbs free energy of liquid phase
G^V	total Gibbs energy of vapor phase
h	molar enthalpy of pure substance, mixture or solution
h_i	molar enthalpy of pure species i
$\bar{h}_i$	partial molar enthalpy of species i in a mixture or solution
h^{id}	molar enthalpy of ideal solution
$\bar{h}_i^{id}$	partial molar enthalpy of species i in ideal solution

h^{ig}	molar enthalpy of ideal gas mixture
h_i^{ig}	molar enthalpy of pure species i in ideal gas state
$\bar{h}_i^{ig}$	partial molar enthalpy of species i in ideal gas mixture
H	total enthalpy of pure substance, mixture or solution
K_i	distribution coefficient of species i
m	molar property of mixture or solution
M	any extensive property
m_i	molar property of pure species i
$\bar{m}_i$	partial molar property of species i in mixture or solution
m^{id}	molar property of ideal solution
$\bar{m}_i^{id}$	partial molar property of species i in ideal solution
m^{ig}	molar property of ideal gas mixture
m_i^{ig}	molar property of pure species i in ideal gas state
$\bar{m}_i^{ig}$	partial molar property of species i in ideal gas mixture
n	total number of moles
n_i	moles of species i
N	total number of components or species in the system
P	pressure
P_i	partial pressure of species i
P^{sat}	saturation or vapor pressure
P_i^{sat}	saturation or vapor pressure of pure species i
P_r^{sat}	saturation or vapor pressure of reference species r
R	universal gas constant
S	total entropy
s	molar entropy of pure substance, mixture or solution
s_i	molar entropy of pure species i
$\bar{s}_i$	partial molar entropy of species i in a mixture or solution
s^{id}	molar entropy of ideal solution
$\bar{s}_i^{id}$	partial molar entropy of species i in ideal solution
s^{ig}	molar entropy of ideal gas mixture
s_i^{ig}	molar entropy of pure species i in ideal gas state
$\bar{s}_i^{ig}$	partial molar entropy of species i in ideal gas mixture
T	temperature
T^{sat}	saturation or vapor pressure
T_i^{sat}	saturation or vapor pressure of pure species i
V	total volume
v	molar volume of pure substance, mixture or solution
v_i	molar volume of pure species i
$\bar{v}_i$	partial molar volume of species i in a mixture or solution
v^{id}	molar volume of ideal solution
$\bar{v}_i^{id}$	partial molar volume of species i in ideal solution
v^{ig}	molar volume of ideal gas mixture
v_i^{ig}	molar volume of pure species i in ideal gas state
$\bar{v}_i^{ig}$	partial molar volume of species i in ideal gas mixture
W	molar flow rate of bottoms product
x	mole fraction in liquid phase
x_i	mole fraction of species i in liquid phase
y	mole fraction in vapor phase

y_i	mole fraction of species i in vapor phase
z_i	mole fraction of species i in the feed

Greek Letters

α_{ir}	relative volatility of species i with respect to reference species r
Δ	difference operator
Δg_{mixing}^{ig}	molar Gibbs free change upon mixing of ideal gases
Δg_{mixing}^{id}	molar Gibbs free change upon mixing of liquids to form an ideal liquid solution
Δh_{mixing}^{ig}	molar enthalpy change upon mixing of ideal gases
Δh_{mixing}^{id}	molar enthalpy change upon mixing of liquids to form an ideal liquid solution
Δs_{mixing}^{ig}	molar entropy change upon mixing of ideal gases
Δs_{mixing}^{id}	molar entropy change upon mixing of liquids to form an ideal liquid solution
Δv_{mixing}^{ig}	molar volume change upon mixing of ideal gases
Δv_{mixing}^{id}	molar volume change upon mixing of liquids to form an ideal liquid solution
μ	chemical potential of pure substance
μ_i	chemical potential of species i in a mixture or solution
μ^{ig}	chemical potential of pure substance in ideal gas state
μ_i^L	chemical potential of species i in a liquid solution
μ_i^V	chemical potential of species i in a vapor mixture
$\mu_{i,mix}^{ig}$	chemical potential of species i in an ideal gas mixture
$\mu_{i,pure}^{ig}$	chemical potential of pure species i in ideal gas state
$\mu_{i,pure}^{L}$	chemical potential of pure species i in liquid state
$\mu_{i,solu}^{id}$	chemical potential of species i in an ideal solution
ψ	fraction of feed that goes to distillate product in flash distillation

TUTORIAL PROBLEMS WITH SOLUTION

Problem 1

Prove that the volume change upon mixing of liquids to form an ideal liquid solution at constant T and P is zero. That is,

$$\Delta v_{mixing}^{id} = 0$$

Solution

For a pure substance (liquid or gas):

$$dg = -sdT + vdP$$

For any liquid species i in pure state,

$$dg_i = -s_i dT + v_i dP$$

In analogy to the above equation, we can write the following equation in terms of partial molar properties for a constant composition liquid solution:

$$d\bar{g}_i = -\bar{s}_i dT + \bar{v}_i dP$$

Thus,

$$\left(\frac{\partial \bar{g}_i}{\partial P}\right)_{T,x} = \bar{v}_i$$

For an ideal liquid solution,

$$\left(\frac{\partial \bar{g}_i^{id}}{\partial P}\right)_{T,x} = \bar{v}_i^{id}$$

By definition of ideal liquid solution, $\bar{g}_i^{id} = g_i + RTln(x_i)$. Therefore,

$$\left(\frac{\partial \bar{g}_i^{id}}{\partial P}\right)_{T,x} = \left(\frac{\partial g_i}{\partial P}\right)_{T,x}$$

For a pure liquid,

$$\left(\frac{\partial g_i}{\partial P}\right)_{T,x} = v_i$$

Thus,

$$\bar{v}_i^{id} = \left(\frac{\partial \bar{g}_i^{id}}{\partial P}\right)_{T,x} = \left(\frac{\partial g_i}{\partial P}\right)_{T,x} = v_i$$

This indicates that the partial molar volume of any species in an ideal liquid solution is equal to the molar volume of the pure liquid species at the same T and P.

The volume change upon mixing of pure liquids is:

$$\Delta v_{mixing}^{id} = v^{id} - \sum x_i v_i$$

where v^{id} is the molar volume of ideal solution. Upon using the summability relation for v^{id}, and $\bar{v}_i^{id} = v_i$, we get:

$$\Delta v_{mixing}^{id} = v^{id} - \sum x_i v_i = \sum x_i \bar{v}_i^{id} - \sum x_i v_i = \sum x_i v_i - \sum x_i v_i = 0$$

Hence, the volume change upon mixing of liquids to form an ideal liquid solution at constant T and P is zero.

Problem 2

Prove that the entropy change upon mixing of liquids to form an ideal liquid solution at constant at T and P is given as:

$$\Delta s_{mixing}^{id} = -R \sum x_i \, lnx_i > 0$$

Solution

For a constant composition liquid solution:

$$d\bar{g}_i = -\bar{s}_i dT + \bar{v}_i dP$$

Thus,

$$\left(\frac{\partial \bar{g}_i}{\partial T}\right)_{P,x} = \bar{s}_i$$

For an ideal liquid solution,

$$\left(\frac{\partial \bar{g}_i^{id}}{\partial T}\right)_{P,x} = -\bar{s}_i^{id}$$

By definition of ideal liquid solution, $\bar{g}_i^{id} = g_i + RTln(x_i)$. Therefore,

$$\left(\frac{\partial \bar{g}_i^{id}}{\partial T}\right)_{P,x} = \left(\frac{\partial g_i}{\partial T}\right)_{P,x} + Rln(x_i)$$

For a pure liquid,

$$\left(\frac{\partial g_i}{\partial T}\right)_{P,x} = -s_i$$

Thus,

$$\left(\frac{\partial \bar{g}_i^{id}}{\partial T}\right)_{P,x} = -\bar{s}_i^{id} = \left(\frac{\partial g_i}{\partial T}\right)_{P,x} + Rln(x_i) = -s_i + Rln(x_i)$$

$$\bar{s}_i^{id} = s_i - Rln(x_i)$$

The entropy change upon mixing of liquids at constant *T* and *P* is:

$$\Delta s_{mixing}^{id} = s^{id} - \sum x_i s_i$$

where s^{id} is the molar entropy of ideal liquid solution and s_i is the molar entropy of pure liquid species i at the same *T* and *P* as that of solution. Using the summability relation,

$$s^{id} = \sum x_i \bar{s}_i^{id} = \sum x_i s_i - R\sum x_i \, ln(x_i)$$

Thus,

$$\Delta s_{mixing}^{id} = s^{id} - \sum x_i s_i = \left[\sum x_i s_i - R\sum x_i \, ln(x_i)\right] - \sum x_i s_i = -R\sum x_i \, ln(x_i)$$

Note that $\Delta s_{mixing}^{id} > 0$ as $ln(x_i)$ terms are all negative due to $x_i < 1$.

Problem 3

A vapor mixture consisting of 3 components (A, B and C) is in equilibrium with a liquid solution having the following composition: $x_A = 0.15,\ x_B = 0.40,\ and\ x_C = 0.45$. The temperature is 80 °C. The vapor pressures of the pure species are given by the following Antoine equations:

$$lnP_A^{Sat} = 14.5463 - \frac{2940.46}{T+237.22}$$

$$lnP_B^{Sat} = 14.2724 - \frac{2945.47}{T+224}$$

$$lnP_C^{Sat} = 14.2043 - \frac{2972.64}{T+209}$$

where T is in °C and the vapor pressures are in kPa. Using Raoult's law, determine P and vapor composition.

Solution

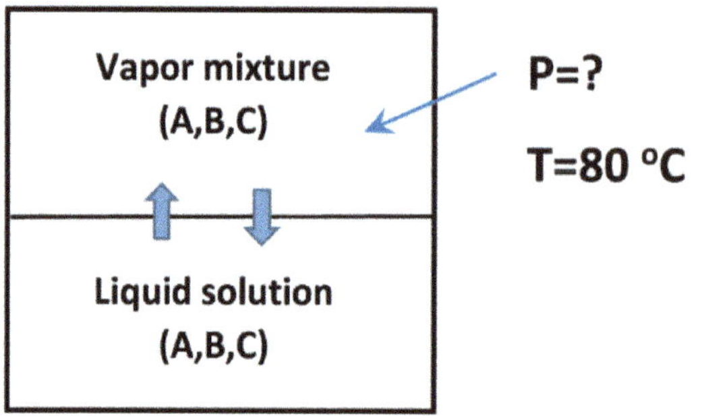

At 80 °C, the vapor pressures of pure species are calculated from the given Antoine equations as:

$$P_A^{Sat} = 195.75\ kPa$$
$$P_B^{Sat} = 97.842\ kPa$$
$$P_C^{Sat} = 50.32\ kPa$$

From Raoult's law:

$$P = x_A P_A^{Sat} + x_B P_B^{Sat} + x_C P_C^{Sat}$$

$$P = 0.15(195.75) + 0.40(97.842) + 0.45(50.32) = 91.143\ kPa$$

Now that P is known, vapor composition can be calculated from the Raoult law as:

$$y_A = \frac{x_A P_A^{Sat}}{P} = \frac{0.15(195.75)}{91.143} = 0.322$$

$$y_B = \frac{x_B P_B^{Sat}}{P} = \frac{0.40(97.842)}{91.143} = 0.429$$

$$y_C = \frac{x_C P_C^{Sat}}{P} = \frac{0.45(50.325)}{91.143} = 0.248$$

Problem 4

Consider a 3-component system species 1, 2 and 3. For this system, the vapor pressures of the pure species are given by the Antoine equations as:

$$lnP_1^{Sat} = 13.7819 - \frac{2726.81}{T+217.572}$$
$$lnP_2^{Sat} = 13.932 - \frac{3056.96}{T+217.625}$$
$$lnP_3^{Sat} = 13.9726 - \frac{3259.93}{T+212.30}$$

where T is in °C and the vapor pressures are in kPa. Species 1 is the most volatile species.
Assuming that the system is ideal in nature and follows the Raoult law, do the following calculations:

a) BUBLP calculation. Given that $T = 120^oC,\ x_1 = 0.25, x_2 = 0.35, x_3 = 0.40$
b) DEWP calculation. Given that $T = 110^oC,\ y_1 = 0.40, y_2 = 0.35, y_3 = 0.25$
c) BUBLT calculation. Given that $P = 95\ kPa,\ x_1 = 0.45, x_2 = 0.20, x_3 = 0.35$

d) DEWT calculation. Given that $P = 100\ kPa,\ y_1 = 0.50, y_2 = 0.30, y_3 = 0.20$
Use species 1 as the reference species for BUBLT and DEWT calculations.

Solution

a) BUBLP calculation. Given that $T = 120^oC,\ x_1 = 0.25, x_2 = 0.35, x_3 = 0.40$

$$P = \sum x_i P_i^{Sat}$$

Species	Ai	Bi	Ci	T(°C)	P_i^{Sat}(kPa)
1	13.7819	2726.81	217.572	120	300.1211
2	13.932	3056.96	217.625	120	131.3275
3	13.9726	3259.93	212.3	120	64.2254

$$P = x_1P_1^{Sat} + x_2P_2^{Sat} + x_3P_3^{Sat} = 0.25 \times 300.1211 + 0.35 \times 131.3275 + 0.40 \times 64.2254 = 146.6851 \text{ kPa}$$

Composition of first vapor bubble:

$$y_1 = x_1P_1^{Sat}/P = 0.25 \times 300.1211/146.6851 = 0.5115$$
$$y_2 = x_2P_2^{Sat}/P = 0.35 \times 131.3275/146.6851 = 0.3134$$
$$y_3 = x_3P_3^{Sat}/P = 0.45 \times 64.2254/146.6851 = 0.1751$$

b) DEWP calculation. Given that $T = 110^oC,\ y_1 = 0.40, y_2 = 0.35, y_3 = 0.25$

$$P = 1/\sum y_i/P_i^{Sat}$$

Species	Ai	Bi	Ci	T(°C)	P_i^{Sat}(kPa)
1	13.7819	2726.81	217.572	110	234.5321
2	13.932	3056.96	217.625	110	99.617
3	13.9726	3259.93	212.3	110	47.3713

$$P = 1/\left(y_1/P_1^{Sat} + y_2/P_2^{Sat} + y_3/P_3^{Sat}\right) = 1/[(0.40/234.5321) + (0.35/99.617) + (0.25/47.3713)] = 95.2705 \text{ kPa}$$

Composition of the first liquid drop:

$$x_1 = y_1P/P_1^{Sat} = 0.40 \times 95.2705/234.5321 = 0.1625$$
$$x_2 = y_2P/P_2^{Sat} = 0.35 \times 95.2705/99.617 = 0.3347$$
$$x_3 = y_3P/P_3^{Sat} = 0.25 \times 95.2705/47.3713 = 0.5028$$

c) BUBLT calculation. Given that $P = 95\ kPa,\ x_1 = 0.45, x_2 = 0.20, x_3 = 0.35$

We are given liquid composition and P, we are required to determine the bubble point T and initial bubble composition. We follow the trial-and-error procedure discussed in section 11.5.1.

Initial guess of bubble point temperature:

$$T_{guess} = \sum x_i T_i^{Sat}$$

At P = 95 kPa, Antoine equations give: $T_1^{Sat} = 77.9203^oC, T_2^{Sat} = 108.3421^oC,\ T_3^{Sat} = 133.8117^oC.$ Thus, $T_{guess} = 103.5667^oC.$

At $T = 103.5667^oC,$ calculate α_{ir} from Equation (11.92) with species 1 as the reference species: $\alpha_{11} = 1.0, \alpha_{21} = 0.4163,\ \alpha_{31} = 0.1942$. From Equation (11.91):

$$P_1^{Sat} = P/\sum x_i\,\alpha_{ir}$$

$$= 95/(0.45 \times 1 + 0.20 \times 0.4163 + 0.35 \times 0.1942) = 158.0095 \text{ kPa}$$

Using Equation (11.93), we obtain:

$$T = \frac{B_1}{A_1 - lnP_1^{Sat}} - C_1 = 95.163^oC$$

Since T of 95.163^oC is significantly off from the assumed value of 103.5667^oC, we need another iteration.

At $T = 95.163^oC, \alpha_{11} = 1.0, \alpha_{21} = 0.405,\ \alpha_{31} = 0.1840.$ Using Equation (11.91):

$$P_1^{Sat} = P/\sum x_i\,\alpha_{ir}$$

$$= 95/(0.45 \times 1 + 0.20 \times 0.405 + 0.35 \times 0.1840) = 159.56 \text{ kPa}$$

$$T = \frac{B_1}{A_1 - lnP_1^{Sat}} - C_1 = 95.513^oC$$

This value of T is very close to the assumed value of 95.163^oC. Thus T=95.513^oC.

Composition of vapor bubble:

$$y_1 = x_1P_1^{Sat}/P = 0.45 \times 159.5454/95 = 0.7557$$
$$y_2 = x_2P_2^{Sat}/P = 0.20 \times 64.6873/95 = 0.1362$$
$$y_3 = x_3P_3^{Sat}/P = 0.35 \times 29.426/95 = 0.1084$$

Note that at T=95.513^oC, the vapor pressures of pure species are determined from the Antoine equations.

d) DEWT calculation. Given that $P = 100\ kPa,\ y_1 = 0.50, y_2 = 0.30, y_3 = 0.20$

We are given vapor composition and P, we are required to determine T and dew (first liquid drop) composition. We follow the trial-and-error procedure discussed in section 11.5.1.

Initial guess of dew point temperature:

$$T_{guess} = \sum y_i\,T_i^{Sat}$$

At P = 100 kPa, Antoine equations give: $T_1^{Sat} = 79.572^oC, T_2^{Sat} = 110.135^oC,\ T_3^{Sat} = 135.707^oC.$

Thus, $T_{guess} = 99.9678^oC.$

At $T = 99.9678^oC$, Equation (11.92) gives $\alpha_{11} = 1.0, \alpha_{21} = 0.4115,\ \alpha_{31} = 0.1898$. Using Equation (11.97):

$$P_1^{Sat} = P\sum y_i/\alpha_{i3} = 1 \times (y_1/\alpha_{13} + y_2/\alpha_{23} + y_3/\alpha_{33} + y_4/\alpha_{43}) = 228.26 \text{ kPa}$$

Using Equation (11.93), we obtain:

$$T = \frac{B_1}{A_1 - lnP_1^{Sat}} - C_1 = 108.936^oC$$

Since T of 108.936^oC is significantly off from the assumed value of 99.9678^oC, we need another iteration.

At $T = 108.936^oC,\ \alpha_{11} = 1.0, \alpha_{21} = 0.4234,\ \alpha_{31} = 0.2007$. Using Equation (11.97):

$$P_1^{Sat} = P\sum y_i/\alpha_{i3} = 1 \times (y_1/\alpha_{13} + y_2/\alpha_{23} + y_3/\alpha_{33} + y_4/\alpha_{43}) = 220.514 \text{ kPa}$$

Using Equation (11.93), we obtain:

$$T = \frac{B_1}{A_1 - lnP_1^{Sat}} - C_1 = 107.5925^oC$$

This value of T is very close to the assumed value of 108.936^oC. Thus T=107.5925^oC.

Composition of liquid droplet:

$$x_1 = y_1P/P_1^{Sat} = 0.50 \times 100/220.514 = 0.2267$$
$$x_2 = y_2P/P_2^{Sat} = 0.30 \times 100/92.9684 = 0.3227$$
$$x_3 = y_3P/P_3^{Sat} = 0.20 \times 100/43.8991 = 0.4556$$

Note that at T=107.5925^oC, the vapor pressures of pure species are determined from the Antoine equations.

Problem 5

Consider a binary system of species 1 and 2 at a total pressure of 101.32 kPa. Assuming that the system is ideal in nature and follows the Raoult law, construct the T-x-y and y-x diagrams for the system at 101.32 kPa. For this system, the vapor pressures of the pure species are given by the Antoine equations as:

$$lnP_1^{Sat} = 13.7819 - \frac{2726.81}{T+217.572}$$
$$lnP_2^{Sat} = 13.932 - \frac{3056.96}{T+217.625}$$

where T is in °C and the vapor pressures are in kPa. Species 1 is more volatile.

Solution

The boiling points of pure species 1 and 2 at 101.32 kPa are calculated from the given Antoine equations as: $T_1^{Sat} = 80^oC$ and $T_2^{Sat} = 110.6^oC$. Table given below shows at different temperatures the values of vapor pressures of pure species and mole fractions of more volatile component (species 1) in the vapor and liquid phases at equilibrium with each other. The Antoine equations are used to calculate the vapor pressures. Equations (11.102) and (11.103) are used to calculate the mole fractions in liquid and vapor phases at equilibrium.

T (°C)	$P_1^{Sat}(kPa)$	$P_2^{Sat}(kPa)$	x	y
80	101.32	38.89	1.00	1.00
83	111.02	43.09	0.86	0.94
86	121.44	47.64	0.73	0.87
89	132.60	52.57	0.61	0.80
92	144.53	57.91	0.50	0.71
95	157.28	63.66	0.40	0.62
98	170.88	69.86	0.31	0.53
101	185.37	76.53	0.23	0.42
104	200.78	83.70	0.15	0.30
107	217.15	91.38	0.08	0.17
110	234.52	99.61	0.01	0.03
110.6	238.11	101.32	0.00	0.00

The T-x-y and y-x diagrams for the system at 101.32 kPa are developed using the data shown in the table. Note that x and y refer to mole fractions of more volatile species (species 1). The diagrams are shown in the figures given below.

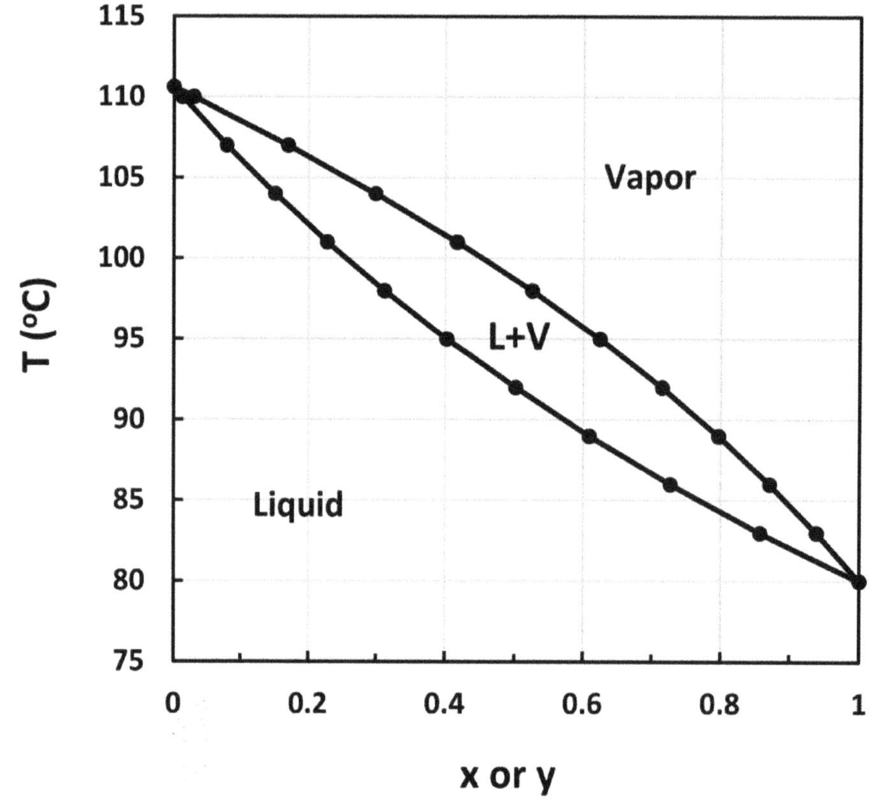

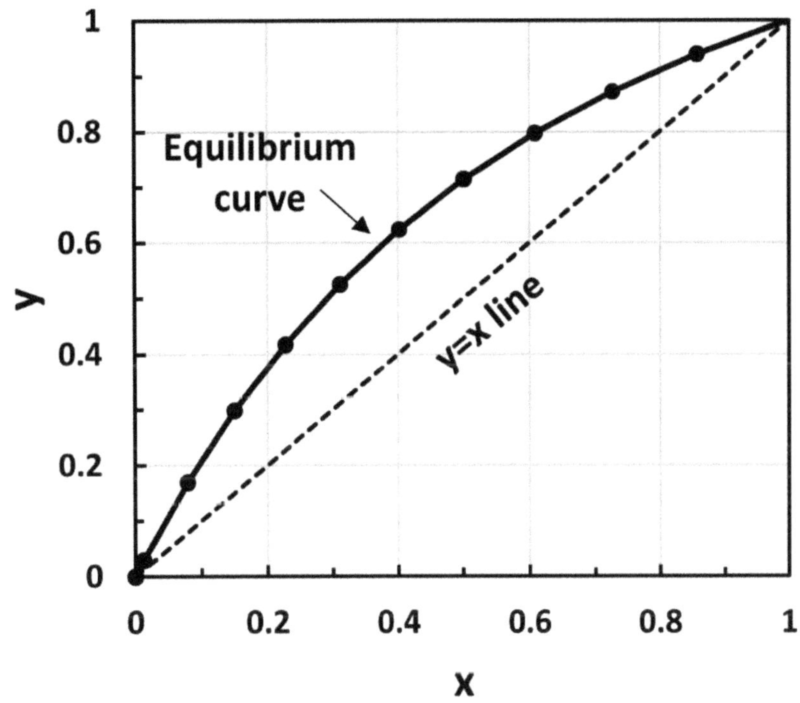

Problem 6

A flash chamber operating at 80 °C and 110 kPa is separating 100 kmol/h of a feed that is 1 3-component mixture of species 1, 2 and 3. The feed composition is z_1=0.42, z_2 = 0.38 and z_3=0.20. Would flashing occur at given temperature (80 °C) and pressure (110 kPa))? If yes, find the product compositions and flow rates.
You can assume that Raoult's law is applicable. The vapor pressures of the pure species at 80 °C are as follows: $P_1^{Sat} = 195.75\ kPa$, $P_2^{Sat} = 97.84\ kPa$, $P_3^{Sat} = 50.32\ kPa$.

Solution

The problem deals with PT-flash calculation of multicomponent systems described in section 11.5.3. For flashing to occur, $P_{dew} < P < P_{bubble}$. The bubble and dew pressures are as follows:

$$P_{bubble} = 0.42(P_1^{Sat}) + 0.38(P_2^{Sat}) + 0.20(P_3^{Sat}) = 129.46\ kPa$$

$$P_{dew} = 1/[(0.42/P_1^{Sat}) + (0.38/P_2^{Sat}) + (0.20/P_3^{Sat})] = 99.96\ kPa$$

As the pressure in the chamber (110 kPa) falls in the range $P_{dew} < P < P_{bubble}$, flashing will occur at the given chamber conditions.

We now need to solve the Rachford-Rice equation, Equation (11.109), using the Newton-Raphson method, given that: z_1=0.42, z_2 = 0.38, z_3=0.20, $K_1 = P_1^{sat}/P = 1.7795$, $K_2 = P_2^{sat}/P = 0.8895$, $K_3 = P_3^{sat}/P = 0.4575$

The Rachford-Rice equation in the present case can be written as:

$$f(\psi) = \sum \frac{(K_i - 1)z_i}{1 + \psi(K_i - 1)} = 0$$

$$f(\psi) = \frac{0.32729}{1+\psi(0.7795)} - \frac{0.04199}{1-\psi(0.1105)} - \frac{0.1085}{1-\psi(0.5425)} = 0$$

The derivative of $f(\psi)$ is as follows:

$$f'(\psi) = -\sum \frac{(K_i-1)^2 z_i}{[1+\psi(K_i-1)]^2}$$

$$f'(\psi) = -\left[\frac{0.2552}{\left(1+\psi(0.7795)\right)^2} + \frac{4.64\times10^{-3}}{\left(1-\psi(0.1105)\right)^2} + \frac{0.05886}{\left(1-\psi(0.5425)\right)^2}\right]$$

Equation (11.114) gives the iteration scheme to solve for ψ as:

$$\psi_{k+1} = \psi_k - \frac{f(\psi_k)}{f'(\psi_k)}$$

The following table gives the iterations starting with the initial guess of ψ_o=0.50

ψ_k	ψ_{k+1}
0.50	0.669938
0.669938	0.666925
0.666925	0.666922
0.666922	0.666922

Thus, the solution is ψ=0.666922≈0.667. Given that F=100 kmol/h, D=ψF=66.7kmol/h, and W=33.3 kmol/h.

The vapor composition follows from Equation (11.105):

$$y_i = \frac{K_i z_i}{1+\psi(K_i-1)}$$

The liquid composition follows from the equilibrium relation:

$$x_i = y_i/K_i$$

The values of y_i and x_i are summarized below.

species	y_i	x_i
1	0.4917	0.2763
2	0.3649	0.4102
3	0.1434	0.3135

PRACTICE PROBLEMS

Problem 1

Prove that the enthalpy change upon mixing of ideal gases to form an ideal gas mixture at constant at T and P is zero. That is,

$$\Delta h^{ig}_{mixing} = 0$$

Problem 2

Prove that the The Gibbs free change upon mixing of ideal gases to form an ideal gas mixture at constant at T and P is given as:

$$\Delta g_{mixing}^{ig} = RT \sum y_i \, ln y_i < 0$$

Problem 3

Prove that the enthalpy change upon mixing of liquids to form an ideal liquid solution at constant at T and P is zero. That is,

$$\Delta h_{mixing}^{id} = 0$$

Problem 4

A liquid solution of benzene and toluene is placed in a piston and cylinder assembly. The benzene amount is 6 moles and the toluene amount is 4 moles. The temperature is fixed at 298 K. The vapor pressures of pure benzene and pure toluene at 298 K are 96.4 mmHg and 28.9 mmHg, respectively. The pressure above the liquid solution is initially 760 mmHg. If the pressure above the solution is reduced from 760 mmHg, at what pressure would the first vapor bubble appear? What is the composition of this vapor bubble? You can assume that the liquid solution and vapor mixture are ideal in nature.

Problem 5

A 4-component vapor mixture containing 25 mole% species 1, 25 mole% species 2, 25 mole% species 3, and 25 mole% species 4 has a total pressure of 1 bar. For this system, the vapor pressures of the pure species are given by the Antoine equation:

$$lnP_i^{Sat} = A_i - \frac{B_i}{T+C_i}$$

where T is in K and the vapor pressures are in bars. The Antoine coefficients are given in the following table:

Species	1	2	3	4
A_i	9.2131	9.1325	9.2164	9.2535
B_i	2477.07	2766.63	2697.55	2911.32
C_i	-39.94	-50.50	-48.78	-56.51

a) Determine the dew point temperature of this vapor.
b) What is the composition of the first drop of liquid?

You can assume that the vapor mixture and liquid solution are ideal in nature. In your calculations, use species 3 as the reference species.

Problem 6

Consider a 4-component liquid solution containing 25 mole% species 1, 25 mole% species 2, 25 mole% species 3, and 25 mole% species 4 is at a pressure of 1 bar. For this system, the vapor pressures of the pure species are given by the Antoine equation:

$$lnP_i^{Sat} = A_i - \frac{B_i}{T+C_i}$$

where T is in K and the vapor pressures are in bars. The Antoine coefficients are given in the following table:

Species	1	2	3	4
A_i	9.2131	9.1325	9.2164	9.2535
B_i	2477.07	2766.63	2697.55	2911.32
C_i	-39.94	-50.50	-48.78	-56.51

a) Determine the bubble point temperature of this liquid solution.
b) What is the composition of the first vapor bubble that is formed?

You can assume that the liquid solution and vapor mixture are ideal in nature. In your calculations, use species 3 as the reference species.

Problem 7

Consider a binary system of species 1 and 2 at a total pressure of 101.32 kPa. Assuming that the system is ideal in nature and follows the Raoult law, construct the T-x-y and y-x diagrams for the system at 101.32 kPa. For this system, the vapor pressures of the pure species are given by the Antoine equations as:

$$lnP_1^{Sat} = 13.8193 - \frac{2696.04}{T+224.317}$$

$$lnP_2^{Sat} = 13.9346 - \frac{3123.13}{T+209.635}$$

where T is in °C and the vapor pressures are in kPa. Species 1 is more volatile.

When a liquid solution containing 25 mole% species 1 is heated at a constant pressure of 101.32 kPa, at what temperature would the first vapor bubble form and what is its composition?

Problem 8

A 4-component feed mixture of species 1 to 4 with composition z_1=0.25, z_2 = 0.15, z_3=0.20 and z_4=0.40 undergoes flash distillation at 50 °C and 200 kPa. First verify that flashing will occur at the given conditions. What is the fraction of feed that goes to the vapor distillate product and what are the product compositions?

You can assume that Raoult's law is applicable. The vapor pressures of the pure species at 50 °C are as follows: $P_1^{Sat} = 1400\ kPa,\ P_2^{Sat} = 480\ kPa,\ P_3^{Sat} = 160\ kPa,\ P_4^{Sat} = 60\ kPa$

Chapter 12

Vapor-Liquid Equilibrium in Non-Ideal Systems

12.1 LEARNING OUTCOMES

Upon completion of this chapter, students will be able to:

- State the generalized Raoult's law valid for ideal and non-ideal multicomponent vapor-liquid systems
- Define fugacity and fugacity coefficient for real gases and liquids
- Estimate fugacity and fugacity coefficient for real gases and liquids using generalized correlations and cubic equations of state
- Define fugacity and fugacity coefficient of a species in a real gas mixture
- Estimate fugacity and fugacity coefficient of a species in a real gas mixture using virial expansion and cubic equations of state
- Define activity coefficient of a species in a non-ideal liquid solution
- State the Lewis-Randall rule
- Discuss different models for activity coefficients for binary and multicomponent systems
- Derive fugacity criterion for phase equilibrium of multicomponent vapor-liquid system
- Derive the generalized Raoult's law
- Perform bubble and dew points calculations for non-ideal systems using generalized Raoult's law
- Perform flash calculation for non-ideal systems using the Rachford-Rice flash equation

12.2 INTRODUCTION

In the preceding chapter, the following form of the Raoult's law was derived:

$$y_i P = x_i P_i^{sat} \tag{12.1}$$

The assumptions involved in the derivation were as follows:

- The liquid and vapor phases are in equilibrium with each other.
- The liquid solution is ideal.
- The vapor mixture is ideal.
- The effect of pressure on the chemical potential of pure liquid is negligible.

In real vapor-liquid systems, some of these assumptions are generally not valid. While the vapor mixtures can often be approximated as ideal gas mixtures if the pressures are not high, the liquid solutions are rarely ideal in nature.

A very general form of the Raoult's law is:

$$y_i \hat{\varphi}_i P = x_i \gamma_i f_i^L \tag{12.2}$$

where $\hat{\varphi}_i$ is the fugacity coefficient of species i in the vapor mixture, γ_i is the activity coefficient of species i in the liquid solution, and f_i^L is the fugacity of pure liquid species i at the system temperature and pressure. This form of the Raoult's law is free of any restrictions except that the vapor phase and liquid phase are assumed to be in equilibrium. We will refer to Equation (12.2) as the *generalized Raoult's law*.

In this chapter, we will focus on real (non-ideal) multicomponent vapor-liquid systems and use the generalized Raoult's law for VLE (vapor-liquid equilibrium) calculations. As the generalized Raoult's law contains new terms such as fugacity, fugacity coefficient, and activity coefficient, these new terms need to be defined and discussed. We also need to know how to evaluate these quantities before we can apply the generalized Raoult's law.

12.3 FUGACITY AND FUGACITY COEFFICIENT OF PURE GASES

For pure ideal gas, the differential of molar Gibbs free energy can be expressed as:

$$dg^{ig} = -s^{ig}dT + v^{ig}dP \tag{12.3}$$

At constant T,

$$dg^{ig} = RTdlnP \tag{12.4}$$

This equation is valid only for an ideal gas. For a real gas, one can write an analogous expression for dg as:

$$dg = RTdlnf \tag{12.5}$$

where pressure P is replaced by fugacity f. Equation (12.5) defines the fugacity of a real gas. It has the units of pressure. For an ideal gas, the fugacity is equal to pressure: $f^{ig} = P$. Thus, the fugacity of a real gas may be interpreted as an effective pressure.

The fugacity coefficient (φ) of a gas is defined as:

$$\varphi = \frac{f}{P} \tag{12.6}$$

For ideal gas, $\varphi^{ig} = 1$. For real gases, φ deviates from unity. Thus, the fugacity coefficient is a measure of deviation of real gas behavior from ideal gas. As compressibility factor and residual properties are also measures of deviation of real gas behavior from ideal gas, the fugacity coefficient must be related to compressibility factor and residual properties.

Upon subtraction of Equation (12.4) from Equation (12.5),

$$d\left(g - g^{ig}\right) = dg^R = RTdln\varphi \tag{12.7}$$

where g^R is the residual molar Gibbs free energy. Upon integration of Equation (12.7) at constant T, we get:

$$g^R = RTln\varphi + C(T) \tag{12.8}$$

where $C(T)$ is the integration constant, function of T in general. Since $g^R = 0$ for ideal gas and $\varphi^{ig} = 1$, the integration constant $C(T) = 0$. Consequently,

$$\frac{g^R}{RT} = ln\varphi \tag{12.9}$$

The residual molar Gibbs free energy can be expressed in terms of residual molar enthalpy and residual molar entropy as:

$$\frac{g^R}{RT} = \frac{h^R}{RT} - \frac{s^R}{R} \tag{12.10}$$

Recall from Chapter 7,

$$\frac{h^R}{RT_c} = -T_r^2 \int_0^{P_r} \left(\frac{\partial Z}{\partial T_r}\right)_{P_r} \frac{dP_r}{P_r} \tag{12.11}$$

$$\frac{s^R}{R} = -T_r \int_0^{P_r} \left(\frac{\partial Z}{\partial T_r}\right)_{P_r} \frac{dP_r}{P_r} - \int_0^{P_r} (Z-1) \frac{dP_r}{P_r} \tag{12.12}$$

From Equations (12.9) to (12.12),

$$ln\varphi = \frac{g^R}{RT} = \int_0^{P_r} (Z-1) \frac{dP_r}{P_r} \tag{12.13}$$

Thus, compressibility factor data are needed in order to calculate the fugacity coefficient and hence fugacity of real gases.

12.3.1 Generalized Correlations for the Fugacity Coefficient of Gases

The Lee/Kesler generalized compressibility correlation is a three-parameter correlation given as:

$$Z = Z^o + \omega Z^1 \tag{12.14}$$

where Z^o and Z^1 are functions of T_r and P_r :

$$Z^o = Z^o(T_r, P_r) \tag{12.15}$$

$$Z^1 = Z^1(T_r, P_r) \tag{12.16}$$

The values of Z^o and Z^1 are given in a table forms in Appendix J over broad ranges of P_r and T_r.

Upon substitution of Equation (12.14) into Equation (12.13) and integrating, it can be readily shown that:

$$ln\varphi = ln\varphi^o + \omega ln\varphi^1 \tag{12.17}$$

where

$$ln\varphi^o = \int_0^{P_r} (Z^o - 1) \frac{dP_r}{P_r} \tag{12.18}$$

$$ln\varphi^1 = \int_0^{P_r} Z^1 \frac{dP_r}{P_r} \tag{12.19}$$

Equation (12.17) could also be expressed as:

$$\varphi = \varphi^o (\varphi^1)^\omega \tag{12.20}$$

where φ^o and φ^1 are functions of T_r and P_r :

$$\varphi^o = \varphi^o(T_r, P_r) \tag{12.21}$$

$$\varphi^1 = \varphi^1(T_r, P_r) \tag{12.22}$$

The values of φ^o and φ^1 are given in a table form in Appendix J over broad ranges of P_r and T_r.

Although the Lee/Kesler fugacity coefficient correlation, Equation (12.17) or Equation (12.20), is applicable over broad ranges of P_r and T_r, its main disadvantage is that it involves the usage of tables for φ^o and φ^1. Thus, it is not suitable for computer calculations.

However, at *low to moderate pressures* one can use the Pitzer generalized virial coefficient correlation for compressibility factor to derive the expression for fugacity coefficient for gases. The Pitzer correlation is given as:

$$Z = 1 + B^o \frac{P_r}{T_r} + \omega B^1 \frac{P_r}{T_r} \tag{12.23}$$

where

$$B^o = 0.083 - \frac{0.422}{T_r^{1.6}} \tag{12.24}$$

$$B^1 = 0.139 - \frac{0.172}{T_r^{4.2}} \tag{12.25}$$

Upon substitution of Equation (12.23) into Equation (12.13) and carrying out the integration, the following expression for fugacity coefficient is obtained:

$$ln\varphi = \left(\frac{P_r}{T_r}\right)(B^o + \omega B^1) \tag{12.26}$$

where B^o and B^1 are given in Equations (12.24) and (12.25), respectively. Thus, the fugacity coefficient for gases at low to moderate pressures can be computed using the Pitzer fugacity coefficient correlation, Equation (12.26).

12.3.2 Cubic Equations of State for the Fugacity Coefficient of Gases

Recall from Chapter 7, the general form of the cubic EOS, in terms of the compressibility factor, is expressed as:

$$Z = 1 + \beta - q\beta \frac{(Z-\beta)}{(Z+\varepsilon\beta)(Z+\sigma\beta)} \tag{12.27}$$

where $\beta = bP/RT$, $q = a'/bRT$, $b = \Omega RT_c/P_c$, and $a' = \psi(R^2T_c^2/P_c)\alpha(T_r, \omega)$. Table 12.1 gives the values of the parameters for different cubic EOS.

Table 12.1 Parameters for different cubic equations of state

Name	ϵ	σ	ψ	Ω	$\alpha(T_r, \omega)$	Z_c
vdW	0	0	0.421875	0.125	1	0.375
RK	0	1	0.42748	0.08664	$1/\sqrt{T_r}$	1/3
SRK	0	1	0.42748	0.08664	α_{SRK}	1/3
PR	$1-\sqrt{2}$	$1+\sqrt{2}$	0.45724	0.45724	α_{PR}	0.3074
$\alpha_{SRK} = \left[1 + (0.480 + 1.574\omega - 0.176\omega^2)(1 - \sqrt{T_r})\right]^2$ $\alpha_{PR} = \left[1 + (0.37464 + 1.5422\omega - 0.26992\omega^2)(1 - \sqrt{T_r})\right]^2$						

The following equation for fugacity coefficient can be derived readily [1] from Equation (12.13) using the compressibility factor cubic EOS, that is, Equation (12.27):

$$ln\varphi = Z - 1 - ln(Z - \beta) - qI \tag{12.28}$$

where

$$I = \frac{1}{\sigma - \varepsilon} ln \left[\frac{Z + \sigma\beta}{Z + \varepsilon\beta}\right] \quad \text{when } \varepsilon \neq \sigma \tag{12.29}$$

$$= \beta/(Z + \varepsilon\beta) \quad \text{when } \varepsilon = \sigma \tag{12.30}$$

The compressibility factor can be determined from a cubic EOS using the Newton-Raphson iterative procedure as described in section 7.3.5. Once Z is estimated using the iterative procedure, no more iteration is required to calculate the fugacity coefficient from Equation (12.28).

12.4 FUGACITY AND FUGACITY COEFFICIENT OF PURE LIQUIDS

The defining equations for fugacity and fugacity coefficient of liquids are the same as those for gases. Thus,

$$dg^L = RTdlnf^L \tag{12.31}$$

Equation (12.31) defines the fugacity of a pure liquid. It has the units of pressure. The fugacity coefficient (φ^L) of a liquid is defined as:

$$\varphi^L = \frac{f^L}{P} \tag{12.32}$$

12.4.1 Criterion for Phase Equilibrium between Pure Liquid and its Vapor

Consider pure liquid and its vapor at equilibrium at temperature T and pressure P^{sat}. Since the phases are at equilibrium,

$$\mu^V = \mu^L \tag{12.33}$$

For pure substances, $\mu = g$. Re-writing Equation (12.33) in terms of molar Gibbs free energy:

$$g^V = g^L \tag{12.34}$$

Upon integrating Equation (12.31) from liquid phase to vapor phase, we get:

$$g^V - g^L = RTln\left(\frac{f^V}{f^L}\right) \tag{12.35}$$

Using the chemical potential equilibrium criterion, Equation (12.33), we get a new criterion of phase equilibrium in terms of fugacities:

$$f^V = f^L \tag{12.36}$$

Thus, pure liquid and its vapor are in phase equilibrium when their fugacities are equal. We can further rewrite Equation (12.36) as:

$$f^V(T, P^{sat}) = f^L(T, P^{sat}) = f^{sat} \tag{12.37}$$

where f^{sat} is fugacity of either saturated vapor or saturated liquid.

Since $f = \varphi P$, we can express the phase equilibrium criterion (Equation (12.37)) in terms of the fugacity coefficient as follows:

$$\varphi^V(T, P^{sat})P^{sat} = \varphi^L(T, P^{sat})P^{sat} = \varphi^{sat}P^{sat} \tag{12.38}$$

As P^{sat} cancels out,

$$\varphi^V(T, P^{sat}) = \varphi^L(T, P^{sat}) = \varphi^{sat} \tag{12.39}$$

This is another criterion for phase equilibrium between pure liquid and its vapor that the fugacity coefficients of liquid and vapor phases are the same at equilibrium.

12.4.2 Calculation of Fugacity and Fugacity Coefficient of Liquids

Although the fugacity coefficient, and hence fugacity, of pure liquids can be calculated directly from Lee/Kesler correlation and the cubic EOS, it is preferred to use the following two-step process to calculate the fugacity and fugacity coefficient of pure liquids.

In the first step, the fugacity of liquid f^L at T and P^{sat} is obtained from the gas phase calculation of f^V at T and P^{sat}. Note that $f^L(T, P^{sat}) = f^V(T, P^{sat})$. Once $f^L(T, P^{sat})$ is known through gas phase calculation, $f^L(T, P)$ can be calculated by taking into account the effect of pressure on the fugacity of liquid as discussed below.

For pure substances,

$$dg = -sdT + vdP \tag{12.40}$$

At constant temperature,

$$dg^L = v^L dP \tag{12.41}$$

From Equations (12.31) and (12.41),

$$RTdlnf^L = v^L dP \tag{12.42}$$

Upon integration of Equation (12.42) from P^{sat} to P, we obtain:

$$ln\left(\frac{f^L}{f^{sat}}\right) = \frac{v^L}{RT}(P - P^{sat}) \tag{12.43}$$

Equation (12.43) could be re-written as:

$$f^L(T, P) = f^{sat} exp\left[\frac{v^L(P - P^{sat})}{RT}\right] \tag{12.44}$$

As

$$f^{sat} = \varphi^{sat} P^{sat} \tag{12.45}$$

Equation (12.44) becomes:

$$f^L(T, P) = \varphi^{sat} P^{sat} exp\left[\frac{v^L(P - P^{sat})}{RT}\right] \tag{12.46}$$

where φ^{sat} is the fugacity coefficient of either saturated vapor or saturated liquid at T and P^{sat}. Thus, the fugacity of liquid at any T and P can be calculated from Equation (12.46). It requires the calculation of vapor or gas phase fugacity coefficient at saturation condition corresponding to T and P^{sat}.

12.5 FUGACITY AND FUGACITY COEFFICIENT OF GAS MIXTURES

Recall from Chapter 11 (Equation (11.73) of Chapter 11),

$$\mu_{i,mix}^{ig}(T,P) = \mu_{i,pure}^{ig}(T,P) + RTlny_i \tag{12.47}$$

That is, the chemical potential of species i in an ideal gas mixture at temperature T and pressure P is equal to the chemical potential of pure ideal gas species i at T and P plus the term $RTlny_i$.

The differentiation of Equation (12.47) gives:

$$d\mu_{i,mix}^{ig}(T,P) = d\mu_{i,pure}^{ig}(T,P) + RTdlny_i \tag{12.48}$$

For pure ideal gas species i,

$$d\mu_{i,pure}^{ig}(T,P) = RTdlnP \tag{12.49}$$

Combining Equations (12.48) and (12.49),

$$d\mu_{i,mix}^{ig}(T,P) = RTdln(y_iP) \tag{12.50}$$

This equation is valid only for an *ideal gas mixture*. For a *real gas mixture*, one can write an analogous expression for $d\mu$ as:

$$d\mu_{i,mix} = RTdln\hat{f}_i \tag{12.51}$$

where $\hat{f}_i$ is the fugacity of species i in a real gas mixture at T and P. Equation (12.51) defines the fugacity of species i in a real gas mixture. It has the units of pressure. In an ideal gas mixture, the fugacity of species i is equal to *partial pressure*: $\hat{f}_i^{ig} = y_iP$. Thus, the fugacity of species i in a real gas mixture may be interpreted as an effective partial pressure.

The fugacity coefficient ($\hat{\varphi}_i$) of a species i in a real gas mixture is defined as:

$$\hat{\varphi}_i = \frac{\hat{f}_i}{y_iP} \tag{12.52}$$

For an ideal gas mixture,

$$\hat{\varphi}_i^{ig} = \frac{\hat{f}_i^{ig}}{y_iP} = \frac{y_iP}{y_iP} = 1 \tag{12.53}$$

12.5.1 Calculation of Fugacity Coefficient of a Species in a Gas Mixture Using Virial Expansion of Compressibility Factor

The two-term virial expansion of compressibility factor of a pure gas in terms of pressure can be expressed as:

$$Z = 1 + \frac{BP}{RT} \tag{12.54}$$

This is the simplest form of virial expansion with only 1st order correction in Z of ideal gas. It is valid at low to moderate pressures.

The fugacity coefficient of a pure substance is (see Equation (12.13)):

$$ln\varphi = \int_0^P (Z-1)\frac{dP}{P} \tag{12.55}$$

Upon substitution of Z from Equation (12.54) into Equation (12.55), we obtain:

$$ln\varphi = \frac{BP}{RT} \tag{12.56}$$

The virial coefficient B of pure gas is given as:

$$B = \frac{RT_c}{P_c}(B^o + \omega B^1) \tag{12.57}$$

where B^o and B^1 are functions of T_r, given by Equations (12.24) and (12.25). Equation (12.56) can be re-written in terms of reduced variables as:

$$ln\varphi = \left(\frac{BP_c}{RT_c}\right)\left(\frac{P_r}{T_r}\right) \tag{12.58}$$

Equation (12.58), in combination with Equation (12.57), is the same as Equation (12.26).

The fugacity coefficient relation for pure gas (Equation (12.56)), developed on the basis of two-term virial expansion of compressibility factor in P, can be extended to real gas mixtures at low to moderate pressures [1] as follows:

$$ln\hat{\varphi}_k = \frac{P}{RT}\left[B_{kk} + \frac{1}{2}\sum_i\sum_j y_i y_j\left(2\delta_{ik} - \delta_{ij}\right)\right] \tag{12.59}$$

where y's are mole fractions of the species, and δ_{ik} and δ_{ij} are defined below:

$$\delta_{ik} = 2B_{ik} - B_{ii} - B_{kk} \;;\; \delta_{ij} = 2B_{ij} - B_{ii} - B_{jj} \tag{12.60}$$

Note that the dummy indices i and j in Equation (12.59) are to run over all species. Also,

$$\delta_{ii} = 0 \;;\; \delta_{jj} = 0\,;\; \delta_{ki} = \delta_{ik} \tag{12.61}$$

The coefficients B's are given as:

$$B_{ij} = \left(\frac{RT_{cij}}{P_{cij}}\right)\left(B^o + \omega_{ij}B^1\right) \tag{12.62}$$

where

$$\omega_{ij} = \frac{\omega_i + \omega_j}{2} \tag{12.63}$$

$$T_{cij} = \left(T_{ci}T_{cj}\right)^{1/2} \tag{12.64}$$

$$P_{cij} = \frac{Z_{cij}RT_{cij}}{v_{cij}} \tag{12.65}$$

$$Z_{cij} = \frac{Z_{ci} + Z_{cj}}{2} \tag{12.66}$$

$$v_{cij} = \left(\frac{v_{ci}^{1/3} + v_{cj}^{1/3}}{2}\right)^3 \tag{12.67}$$

Note that when $i = j$, the quantities B_{ii}, ω_{ij}, T_{cij}, P_{cij}, Z_{cij}, v_{cij} all become pure species i quantities.

As an example, consider a binary gas mixture. For species 1, Equation (12.59) gives:

$$ln\hat{\varphi}_1 = \frac{P}{RT}\left[B_{11} + \frac{1}{2}\sum_i \sum_j y_i y_j (2\delta_{i1} - \delta_{ij})\right]$$

$$= \frac{P}{RT}\left[B_{11} + \frac{1}{2}\sum_j \{y_1 y_j (2\delta_{11} - \delta_{1j}) + y_2 y_j (2\delta_{21} - \delta_{2j})\}\right]$$

$$= \frac{P}{RT}\left[B_{11} + \frac{1}{2}\sum_j y_1 y_j (2\delta_{11} - \delta_{1j}) + \frac{1}{2}\sum_j y_2 y_j (2\delta_{21} - \delta_{2j})\right]$$

$$= \frac{P}{RT}\left[B_{11} + \frac{1}{2}\{y_1 y_1 (2\delta_{11} - \delta_{11}) + y_1 y_2 (2\delta_{11} - \delta_{12})\} + \frac{1}{2}\{y_2 y_1 (2\delta_{21} - \delta_{21}) + y_2 y_2 (2\delta_{21} - \delta_{22})\}\right] \quad (12.68)$$

Since $\delta_{11} = 0,\ \delta_{22} = 0, \delta_{12} = \delta_{21}$, Equation (12.68) simplifies to:

$$ln\hat{\varphi}_1 = \frac{P}{RT}\left[B_{11} + \frac{1}{2}\{y_1 y_2 (-\delta_{12})\} + \frac{1}{2}\{y_2 y_1 (\delta_{21}) + y_2 y_2 (2\delta_{21})\}\right]$$

$$= \frac{P}{RT}\left[B_{11} + \frac{\delta_{12}}{2}\{-y_1 y_2 + y_2 y_1 + 2y_2 y_2\}\right] \quad (12.69)$$

Upon simplification, Equation (12.69) gives:

$$ln\hat{\varphi}_1 = \frac{P}{RT}[B_{11} + y_2^2 \delta_{12}] \quad (12.70)$$

In a similar manner, one can show that for species 2:

$$ln\hat{\varphi}_2 - \frac{P}{RT}[B_{22} + y_1^2 \delta_{12}] \quad (12.71)$$

where

$$\delta_{12} = 2B_{12} - B_{11} - B_{22} \quad (12.72)$$

Also note that:

$$B_{11} = B_1 = \left(\frac{RT_{c1}}{P_{c1}}\right)(B^o + \omega_1 B^1) \quad (12.73)$$

$$B_{22} = B_2 = \left(\frac{RT_{c2}}{P_{c2}}\right)(B^o + \omega_2 B^1) \quad (12.74)$$

$$B_{12} = \left(\frac{RT_{c12}}{P_{c12}}\right)(B^o + \omega_{12} B^1) \quad (12.75)$$

where B^o and B^1 are functions of reduced temperature, given in Equations (12.24) and (12.25), respectively.

12.5.2 Calculation of Fugacity Coefficient of a Species in a Gas Mixture Using Cubic EOS

The fugacity coefficient of a species in a real gas mixture can be determined using the cubic equations of state. Two steps are involved in the estimation of the fugacity coefficient. In the first step, the compressibility factor of the gas mixture is estimated using the cubic EOS. In the second step, the fugacity coefficient is calculated from the fugacity coefficient equation expressed in terms of the mixture compressibility factor.

Step 1: Estimation of Compressibility Factor of Gas Mixture

The cubic equation of state for mixtures has exactly the same form as that of pure species (see Equation (12.27) with Table 12.1). However, the material parameters are now mixture parameters. Thus,

$$Z = 1 + \beta_m - q_m\beta_m \frac{(Z-\beta_m)}{(Z+\varepsilon\beta_m)(Z+\sigma\beta_m)} \tag{12.76}$$

where

$$\beta_m = \frac{b_m P}{RT} \tag{12.77}$$

$$q_m = \frac{a'_m}{b_m RT} \tag{12.78}$$

$$b_m = \sum y_i\, b_i \tag{12.79}$$

$$a'_m = \sum_i \sum_j y_i y_j a'_{ij} \tag{12.80}$$

$$a'_{ij} = \left(a'_i a'_j\right)^{1/2} \tag{12.81}$$

For pure species:

$$a'_i = \psi\left(\frac{R^2T_{ci}^2}{P_{ci}}\right)\alpha_i(T_{ri}, \omega_i) \tag{12.82}$$

$$b_i = \Omega\left(\frac{RT_{ci}}{P_{ci}}\right) \tag{12.83}$$

The calculation procedure for the compressibility factor of gas mixture is as follows:

- From the given values of T_{ci}, P_{ci}, and ω_i for *pure species*, calculate T_{ri}, $\alpha_i(T_{ri}, \omega_i)$, a'_i, and b_i
- Calculate the mixture parameters b_m, a'_m, β_m, and q_m
- Solve Equation (12.76) for mixture Z using the Newton-Raphson iterative procedure as described in section 7.3.5.

Step 2: Estimation of Fugacity Coefficient of Species in a Gas Mixture

The fugacity coefficient of species in a gas mixture can be calculated from the following equation [1]:

$$ln\hat{\varphi}_i = \frac{b_i}{b_m}(Z-1) - ln(Z-\beta_m) - \bar{q}_i I \tag{12.84}$$

where

$$I = \frac{1}{\sigma-\varepsilon} ln\left[\frac{Z+\sigma\beta_m}{Z+\varepsilon\beta_m}\right] \quad \text{when } \varepsilon \neq \sigma \tag{12.85}$$

$$= \beta_m/(Z+\varepsilon\beta_m) \quad \text{when } \varepsilon = \sigma \tag{12.86}$$

$$\bar{q}_i = q_m\left(1 + \frac{\bar{a}'_i}{a'_m} - \frac{b_i}{b_m}\right) \tag{12.87}$$

$$\bar{a}_i' = \left[\frac{\partial(na_m')}{\partial n_i}\right]_{T,n_j} \quad (12.88)$$

12.6 FUGACITY AND ACTIVITY COEFFICIENT OF LIQUID SOLUTIONS

The defining equation for liquid solution fugacity is the same as that of gas mixture, that is, Equation (12.51). Thus, for liquid solution:

$$d\mu_{i,solu} = RTdln\hat{f}_i \quad (12.89)$$

where $\hat{f}_i$ is the fugacity of species i in the liquid solution at T and P. Upon integration of Equation (12.89) from pure liquid species i to species i in liquid solution, we get:

$$\mu_{i,solu} - \mu_{i,pure} = RTln\frac{\hat{f}_i}{f_i^L} \quad (12.90)$$

Recall from Chapter 11 (see Equation (11.57)), the following relation that defines an *ideal liquid solution*:

$$\mu_{i,solu}^{id} = \mu_{i,pure} + RTln(x_i) \quad (12.91)$$

Upon subtracting Equation (12.91) from Equation (12.90):

$$\mu_{i,solu} - \mu_{i,solu}^{id} = RTln\left(\frac{\hat{f}_i}{x_i f_i^L}\right) \quad (12.92)$$

The left hand side of Equation (12.92) is the *excess chemical potential* μ_i^E. We can now define the *activity coefficient* of species i in the liquid solution at T and P as:

$$\mu_i^E = \mu_{i,solu} - \mu_{i,solu}^{id} = RTln(\gamma_i) \quad (12.93)$$

where γ_i is the *activity coefficient* of species i in the liquid solution. Upon comparison of Equations (12.92) and (12.93), we find:

$$\gamma_i = \frac{\hat{f}_i}{x_i f_i^L} \quad (12.94)$$

Thus, the activity coefficient of species i in the liquid solution at T and P is the fugacity of species i in the liquid solution at T and P divided by the product of mole fraction of species i in the liquid solution and fugacity of pure liquid species i at T and P.

12.6.1 Lewis-Randall Rule

Referring to Equation (12.93), the excess chemical potential of any species is zero in an ideal liquid solution. Thus, the activity coefficient of any species i in an ideal liquid solution is unity:

$$\gamma_i^{id} = 1 \quad (12.95)$$

Consequently, Equation (12.94) gives:

$$\hat{f}_i^{id} = x_i f_i^L \quad (12.96)$$

Equation (12.96) represents the *Lewis/Randall rule*. The fugacity of species i in an ideal liquid solution at T and P is equal to the product of mole fraction of species i in the liquid solution and fugacity of pure liquid species i at T and P.

12.6.2 Models for Activity Coefficient

Binary Solutions

The popular models for activity coefficients are: Margules model, Van laar model, and Wilson model. They are given below.

Margules model:

$$ln\gamma_1 = x_2^2[A_{12} + 2(A_{21} - A_{12})x_1] \qquad (12.97)$$

$$ln\gamma_2 = x_1^2[A_{21} + 2(A_{12} - A_{21})x_2] \qquad (12.98)$$

where A_{12} and A_{21} are constants for the binary system. However, they depend on the temperature of the system

Van laar model:

$$ln\gamma_1 = A'_{12}\left[1 + \frac{A'_{12}x_1}{A'_{21}x_2}\right]^{-2} \qquad (12.99)$$

$$ln\gamma_2 = A'_{21}\left[1 + \frac{A'_{21}x_2}{A'_{12}x_1}\right]^{-2} \qquad (12.100)$$

where A'_{12} and A'_{21} are constants for the binary system. However, they depend on the temperature of the system

Wilson model:

$$ln\gamma_1 = -ln(x_1 + x_2\Omega_{12}) + x_2\left[\frac{\Omega_{12}}{x_1+x_2\Omega_{12}} - \frac{\Omega_{21}}{x_2+x_1\Omega_{21}}\right] \qquad (12.101)$$

$$ln\gamma_2 = -ln(x_2 + x_1\Omega_{21}) - x_1\left[\frac{\Omega_{12}}{x_1+x_2\Omega_{12}} - \frac{\Omega_{21}}{x_2+x_1\Omega_{21}}\right] \qquad (12.102)$$

where Ω_{12} and Ω_{21} are constants for the binary system. However, they depend on the temperature of the system. The temperature-dependence of Ω_{12} and Ω_{21} is further expressed as:

$$\Omega_{12} = \frac{v_2}{v_1}exp\left(-\frac{a_{12}}{RT}\right) \qquad (12.103)$$

$$\Omega_{21} = \frac{v_1}{v_2}exp\left(-\frac{a_{21}}{RT}\right) \qquad (12.104)$$

where a_{12} and a_{21} are constants independent of temperature and composition. They depend only on the chemical nature of the binary system. v_1 and v_2 are molar volumes of species 1 and 2.

Multicomponent Solutions

The Wilson model can be readily extended to multicomponent systems as follows [1]:

$$ln\gamma_i = -ln\left(\sum_j x_j \Omega_{ij}\right) - \sum_k \frac{x_k \Omega_{ki}}{\sum_j x_j \Omega_{kj}} \qquad (12.105)$$

where $\Omega_{ij} = 1$ for $i = j$. Also note that the summations are over all the species present in the solution.

The temperature-dependence of Ω_{ij} is as follows:

$$\Omega_{ij} = \frac{v_j}{v_i} exp\left(-\frac{a_{ij}}{RT}\right) \qquad (i \neq j) \qquad (12.106)$$

where a_{ij} is a constant independent of solution composition and temperature.

12.7 DERIVATION OF GENERALIZED RAOULT'S LAW

Consider real (non-ideal) multicomponent vapor-liquid system at equilibrium. The system being at equilibrium satisfies the following phase equilibrium criterion:

$$\mu_i^V(T,P) = \mu_i^L(T,P) \qquad (12.107)$$

The defining equation for fugacity of a species in vapor mixture or liquid solution is:

$$d\mu_i = RTdln\hat{f}_i \qquad (12.108)$$

Upon integration of Equation (12.108) for species i from liquid phase to vapor phase, we get:

$$\mu_i^V(T,P) - \mu_i^L(T,P) = RTdln\left(\frac{\hat{f}_i^V}{\hat{f}_i^L}\right) \qquad (12.109)$$

From Equations (12.107) and (12.109), it follows that at equilibrium:

$$\hat{f}_i^V(T,P) = \hat{f}_i^L(T,P) \qquad (12.110)$$

This is the fugacity criterion of phase equilibrium. When liquid solution and vapor mixture are at phase equilibrium, the fugacities of each species are the same in liquid and vapor phases.

From Equation (12.52),

$$\hat{f}_i^V = \hat{\varphi}_i^V y_i P \qquad (12.111)$$

From Equation (12.94),

$$\hat{f}_i^L = \gamma_i x_i f_i^L \qquad (12.112)$$

Substitute the fugacity expressions, Equations (12.111) to (12.112), into the fugacity criterion of phase equilibrium, Equation (12.110), we get:

$$y_i \hat{\varphi}_i^V P = \gamma_i x_i f_i^L \qquad (12.113)$$

This is the generalized Raoult's law valid for ideal and non-ideal vapor-liquid systems at equilibrium. The generalized Raoult's law could also be expressed as:

$$y_i \Phi_i P = \gamma_i x_i P_i^{sat} \qquad (12.114)$$

where

$$\Phi_i = \left(\frac{\hat{\varphi}_i^V}{\varphi_i^{sat}}\right) exp\left[-\frac{v_i^L(P-P_i^{sat})}{RT}\right] \tag{12.115}$$

The exponential term in Equation (12.115), referred to as Poynting factor, is usually very close to unity at low to moderate pressures. Therefore, Equation (12.115) can be often simplified to:

$$\Phi_i = \left(\frac{\hat{\varphi}_i^V}{\varphi_i^{sat}}\right) \tag{12.116}$$

12.8 APPLICATIONS OF GENERALIZED RAOULT'S LAW

As noted in Chapter 11, bubble point, dew point, and flash calculations are required in the design of distillation columns and flash separators. In Chapter 11, these calculations were performed using the ideal version of the Raoult's law (Equation (12.1)). Here we will extend the same calculations to non-ideal vapor-liquid systems using the generalized Raoult's law (Equation (12.114)).

Different Classes of Bubble and Dew Points Calculations

Recall from Chapter 11, there are four types of bubble point and dew point calculations. They are as follows:

- *BUBLP:* Given liquid composition and temperature, calculate the pressure where boiling begins and the composition of the first bubble formed.
- *DEWP:* Given vapor composition and temperature, calculate the pressure where condensation begins and the composition of the first droplet (dew) formed.
- *BUBLT:* Given liquid composition and pressure, calculate the temperature where boiling begins and the composition of the first bubble formed.
- *DEWT:* Given vapor composition and pressure, calculate the temperature where condensation begins and the composition of the first droplet (dew) formed.

The calculations will require the knowledge of Φ_i's (Equation (12.115)), the activity coefficients of species in the liquid phase (γ_i), and the vapor pressures of pure species (P_i^{sat}). We can express the predictive equations for Φ_i, γ_i, and P_i^{sat} for a multicomponent vapor-liquid system of N species in the general forms as follows:

$$\Phi_i = F(T, P, y_1, y_2, \dots\dots, y_{N-1}) \tag{12.117}$$

$$\gamma_i = F(T, x_1, x_2, \dots\dots, x_{N-1}) \tag{12.118}$$

$$P_i^{sat} = F(T) \tag{12.119}$$

where F indicates "function of". Thus, Φ_i is a function of $T, P,$ and vapor composition, γ_i is a function of T and liquid composition, and P_i^{sat} is a function of T. We are assuming that the effect of P on the activity coefficient is negligible.

BUBLP Calculation: Given $\{x_i\}$ and T, calculate P **and** $\{y_i\}$

From generalized Raoult's law:

$$y_i = \frac{x_i\gamma_i P_i^{sat}}{\Phi_i P} \tag{12.120}$$

As $\sum y_i = 1$, Equation (12.120) gives the bubble point pressure as:

$$P = \sum_i \frac{x_i \gamma_i P_i^{sat}}{\Phi_i} \quad (12.121)$$

The liquid composition is specified, and the vapor pressures of different species can be calculated from the Antoine equation, given below, as the temperature is specified:

$$lnP_i^{sat} = A_i - \frac{B_i}{T + C_i} \quad (12.122)$$

where A_i, B_i, C_i are constants for a given substance. These constants are available in the literature for a large number of substances.

An iteration scheme is required to calculate the bubble point pressure and bubble composition. The iteration scheme is shown in Figure 12.1.

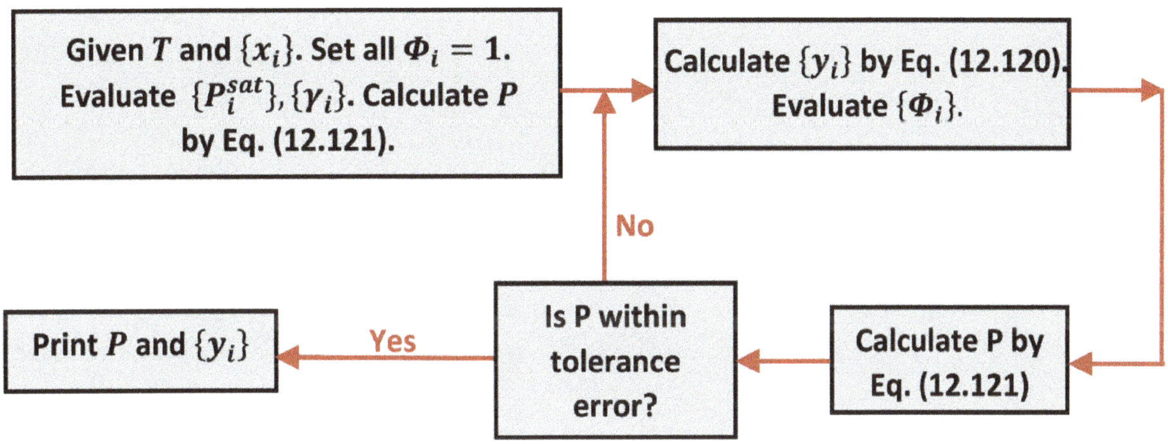

Figure 12.1 Iteration scheme for the calculation of BUBLP.

DEWP Calculation: Given $\{y_i\}$ and T, calculate P and $\{x_i\}$

From generalized Raoult's law:

$$x_i = \frac{y_i \Phi_i P}{\gamma_i P_i^{Sat}} \quad (12.123)$$

As $\sum x_i = 1$, Equation (12.123) gives the dew point pressure as:

$$P = \frac{1}{\sum y_i \Phi_i / \gamma_i P_i^{Sat}} \quad (12.124)$$

The vapor composition is specified, and the vapor pressures of different species can be calculated from the Antoine equation, Equation (12.122), as the temperature is specified.

An iteration scheme is required to calculate the dew point pressure and the composition of the first liquid droplet (dew). The iteration scheme is shown in Figure 12.2. Note that the sum of the calculated values of

x_i may not be unity. Therefore, they are normalized in the iterative procedure (see Figure 12.2) by setting $x_i = x_i/\sum_i x_i$.

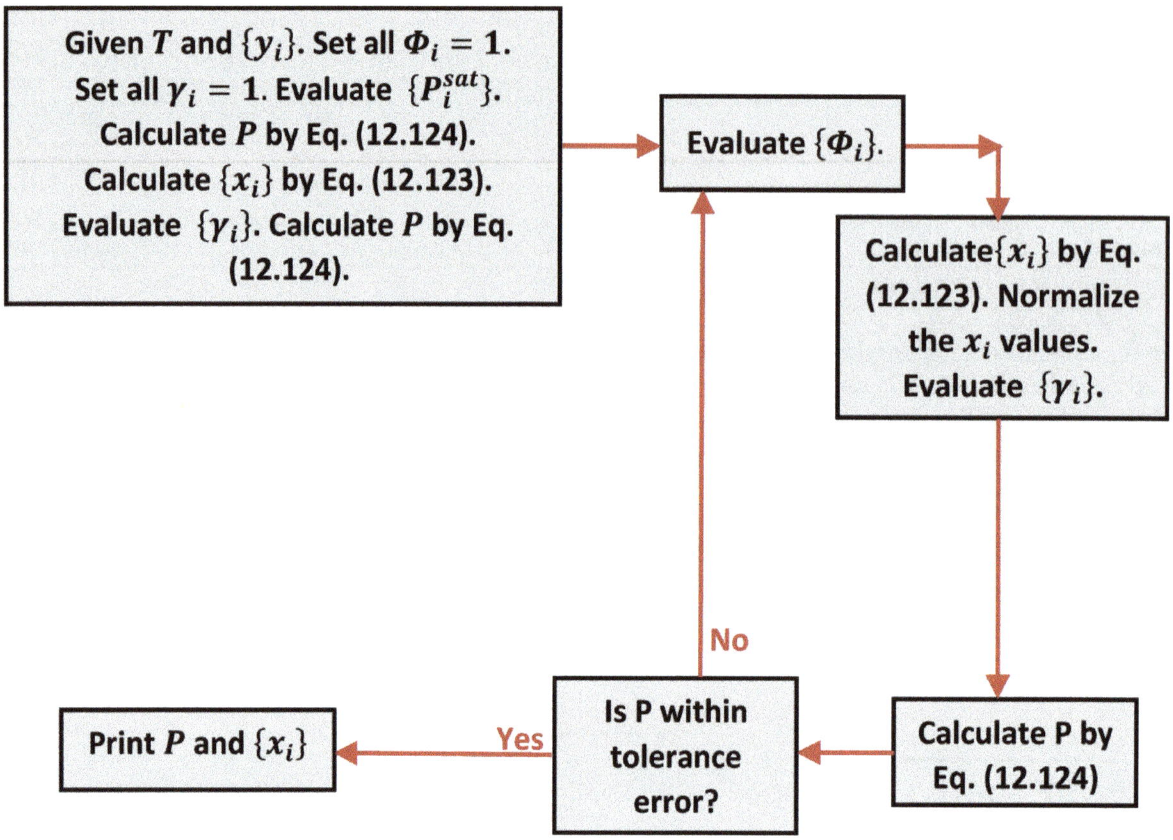

Figure 12.2 Iteration scheme for the calculation of DEWP.

BUBLT Calculation: Given $\{x_i\}$ and P, calculate T and $\{y_i\}$

Now the temperature is unknown. We first make an initial estimate of temperature as follows:

$$T = \sum x_i T_i^{Sat} \tag{12.125}$$

where T_i^{Sat} is the boiling point of species i estimated from the Antoine equation at the given total pressure P, that is:

$$T_i^{Sat} = \frac{B_i}{A_i - lnP} - C_i \tag{12.126}$$

Note that Equation (12.126) is not a valid equation, and it is to be used only for the purpose of initial guess of temperature. Strictly speaking, P should be P_i^{sat} in Equation (12.126).

The equations needed for the iterative procedure to solve for BUBLT are developed as follows.

Multiply and divide Equation (12.121) with P_r^{sat} on the right hand side:

$$P = P_r^{sat} \sum_i \frac{x_i \gamma_i P_i^{sat}}{\Phi_i P_r^{sat}} \tag{12.127}$$

where species "*r*" is the "reference species", selected somewhat arbitrarily. Equation (12.127) can be re-arranged as:

$$P_r^{Sat} = \frac{P}{\sum_i (x_i \gamma_i / \Phi_i)(P_i^{sat}/P_r^{Sat})} \tag{12.128}$$

Upon re-arrangement of the Antoine equation for the reference species "*r*", we get:

$$T = \frac{B_r}{A_r - \ln P_r^{Sat}} - C_r \tag{12.129}$$

This completes the development of equations needed for the iterative procedure to solve for BUBLT. The iterative procedure for the calculation of BUBLT in shown in Figure 12.3.

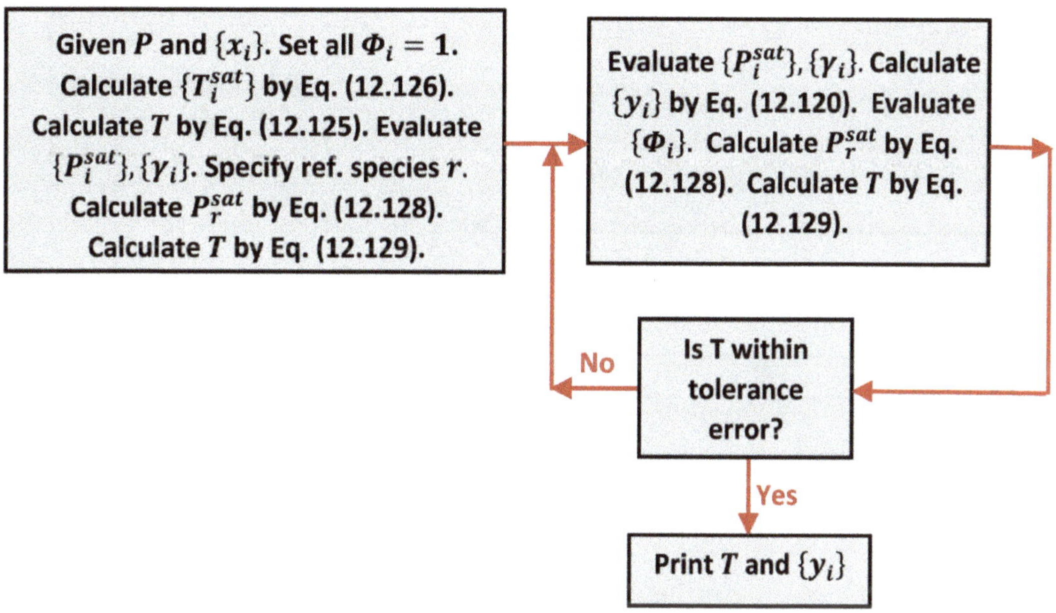

Figure 12.3 Iteration scheme for the calculation of BUBLT.

DEWT Calculation: Given $\{y_i\}$ and P, calculate T and $\{x_i\}$

Once again, the temperature is unknown. We first make an initial estimate of temperature as follows:

$$T = \sum y_i T_i^{Sat} \tag{12.130}$$

where T_i^{Sat} is the boiling point of species i estimated from the Antoine equation at the given total pressure P, that is, Equation (12.126).

The equations needed for the iterative procedure to solve for DEWT are developed as follows.

Multiply and divide Equation (12.124) with P_r^{sat} on the right hand side:

$$P = \frac{P_r^{sat}}{P_r^{sat} \sum y_i \Phi_i / \gamma_i P_i^{sat}} \quad (12.131)$$

where species "r" is the "reference species", selected somewhat arbitrarily. Equation (12.131) can be re-arranged as:

$$P_r^{Sat} = P \sum_i (y_i \Phi_i / \gamma_i)(P_r^{sat} / P_i^{Sat}) \quad (12.132)$$

This completes the development of equations needed for the iterative procedure to solve for BUBLT. The iterative procedure for the calculation of DEWT in shown in Figure 12.4.

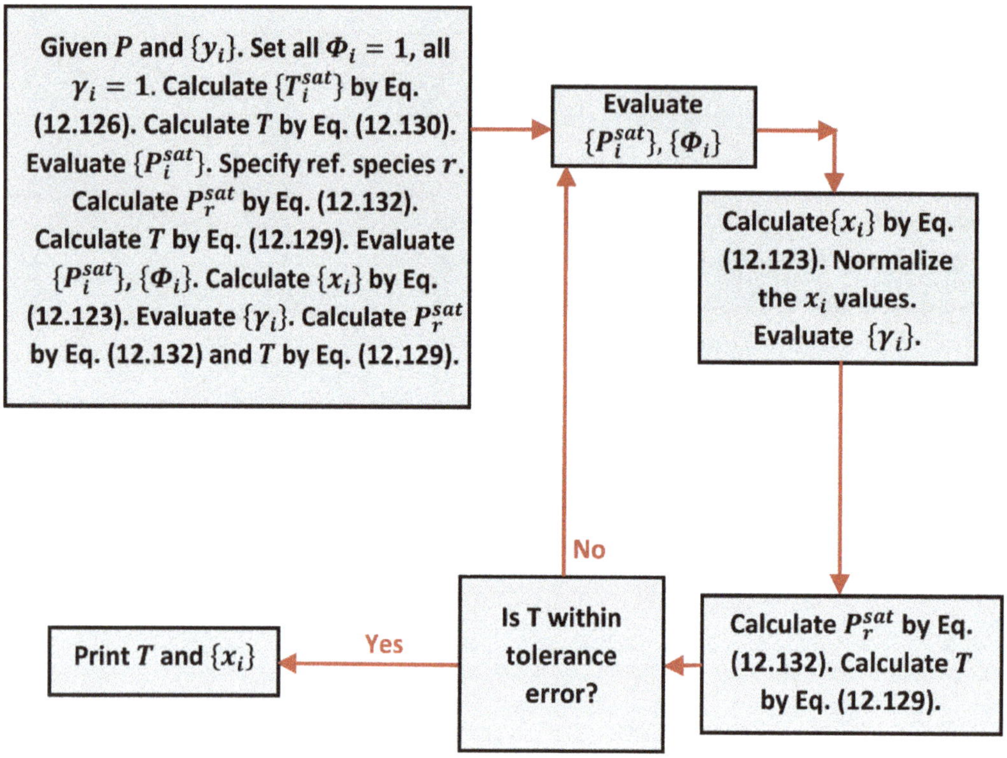

Figure 12.4 Iteration scheme for the calculation of DEWT.

Flash Calculation

Consider flash distillation of multicomponent feed. Given the feed flow rate (F, mol/s), feed composition (set of $\{z_i\}$ values), flash drum temperature and pressure, calculate the vapor-distillate flow rate (D, mol/s), distillate composition (set of $\{y_i\}$ values), bottoms-liquid flow rate (W, mol/s) and bottoms composition (set of $\{x_i\}$ values). This problem was solved in Chapter 11 assuming that the liquid and vapor phases are ideal in nature. Here we will solve this problem for non-ideal liquid and vapor phases.

Recall from Chapter 11, we need to solve for $\psi = D/F$ (fraction of feed that goes to the distillate product) using the *Rachford-Rice flash equation*:

$$\sum \frac{(K_i - 1) z_i}{1 + \psi (K_i - 1)} = 0 \quad (12.133)$$

The distribution coefficients K_i's for non-ideal vapor-liquid system are given as:

$$K_i = \frac{y_i}{x_i} = \frac{\gamma_i P_i^{sat}}{\Phi_i P} \tag{12.134}$$

In the case of ideal vapor-liquid system, the distribution coefficient K_i of a chemical species is given as P_i^{sat}/P. It is a function of just T and P. However, for non-ideal vapor-liquid systems, K_i is a complicated function of T, P, liquid composition $\{x_i\}$, and vapor composition $\{y_i\}$.

As noted in Chapter 11, one can use the Newton-Raphson iterative method to find the root of the Rachford-Rice equation. Equation (12.133) can be re-written as:

$$f(\psi) = \sum \frac{(K_i-1)z_i}{1+\psi(K_i-1)} = 0 \tag{12.135}$$

The derivative of $f(\psi)$ is as follows:

$$f'(\psi) = -\sum \frac{(K_i-1)^2 z_i}{[1+\psi(K_i-1)]^2} \tag{12.136}$$

The Newton-Raphson iterative formula to solve for ψ is given as:

$$\psi_{k+1} = \psi_k - \frac{f(\psi_k)}{f'(\psi_k)} \tag{12.137}$$

where ψ_{k+1} is the improved value of the root over the previous value of the root ψ_k. Equation (12.137) can be used to estimate ψ by successive substitution. Once ψ is known, the vapor and liquid compositions can be calculated from the following equations:

$$y_i = \frac{K_i z_i}{1+\psi(K_i-1)} \tag{12.138}$$

$$x_i = \frac{y_i}{K_i} = \frac{z_i}{1+\psi(K_i-1)} \tag{12.139}$$

The vapor-distillate flow rate is $D = \psi F$ and the bottoms-liquid flow rate is $W = (1-\psi)F$.

Figure 12.5 shows the iteration scheme for flash calculation. The feed composition $\{z_i\}$, feed flow rate (F, mol/s), flash T and P are given. DEWP and BUBLP calculations are performed at given T to determine if flashing would indeed occur at given T and P. For flashing to occur, the flash P must fall in between the dew point and bubble point pressures, that is, $P_{dew} < P < P_{bubl}$.

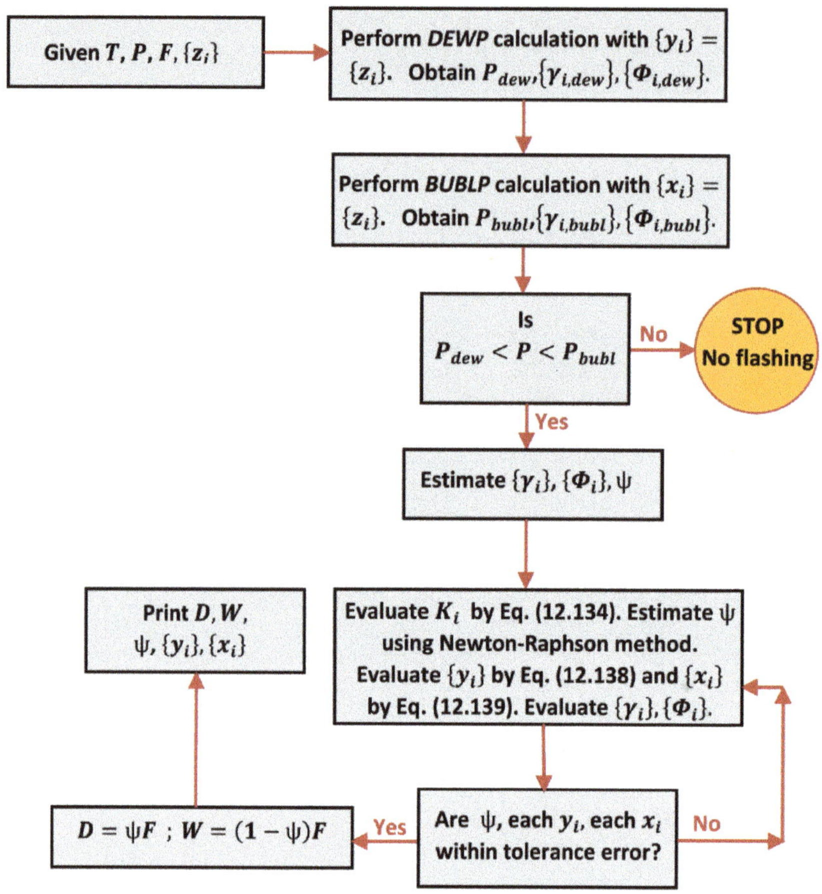

Figure 12.5 Iteration scheme for flash calculation.

The DEWP and BUBLP calculations also generate the following quantities: $\{\gamma_{i,dew}\}$, $\{\Phi_{i,dew}\}$, $\{\gamma_{i,dew}\}$, and $\{\Phi_{i,dew}\}$. For the flash calculation, the initial estimates of γ_i and Φ_i are made using linear interpolation of dew point and bubble point values, as shown below [1]:

$$\gamma_i = \gamma_{i,dew} + \left(\frac{P - P_{dew}}{P_{bubl} - P_{dew}}\right)\left(\gamma_{i,bubl} - \gamma_{i,dew}\right) \tag{12.140}$$

$$\Phi_i = \Phi_{i,dew} + \left(\frac{P - P_{dew}}{P_{bubl} - P_{dew}}\right)\left(\Phi_{i,bubl} - \Phi_{i,dew}\right) \tag{12.141}$$

Likewise, the initial estimate of ψ to be used in the Newton-Raphson method is made using the linear interpolation as follows:

$$\psi = \psi_{dew} + \left(\frac{P - P_{dew}}{P_{bubl} - P_{dew}}\right)\left(\psi_{bubl} - \psi_{dew}\right) \tag{12.142}$$

As $\psi_{dew} = 1$ and $\psi_{bubl} = 0$, Equation (12.142) becomes:

$$\psi = \frac{P_{bubl} - P}{P_{bubl} - P_{dew}} \tag{12.143}$$

Using the initial estimates of γ_i and Φ_i, the distribution coefficients K_i's are calculated from Equation (12.134). Note that P_i^{sat}'s are determined from the Antoine equations. Using the K_i's and initial estimate of ψ, the improved value of ψ is obtained from the Newton-Raphson iterative method. The vapor composition is calculated from Equation (12.138) and the liquid composition is calculated from Equation (12.139). Once the vapor and liquid compositions are known, the activity coefficients γ_i's and Φ_i's can be evaluated. The internal loop iteration (see Figure 12.5) is continued until ψ , each y_i, and each x_i are within the set tolerance error.

NOTATION

A_i	Antoine coefficient for species i
A_r	Antoine coefficient for species r
A_{12}	Margules constant, Equations (12.97) and (12.98)
A_{21}	Margules constant, Equations (12.97) and (12.98)
A'_{12}	Van Laar constant, Equations (12.99) and (12.100)
A'_{21}	Van Laar constant, Equations (12.99) and (12.100)
a'	parameter in cubic equation of state for pure substances
a'_i	parameter in cubic equation of state for pure species i, Equation (12.82)
a'_j	parameter in cubic equation of state for pure species j, Equation (12.81)
a'_m	parameter in cubic equation of state for mixtures, Equation (12.80)
a_{12}	constant in Wilson model for binary systems, Equation (12.103)
a_{21}	constant in Wilson model for binary systems, Equation (12.104)
a_{ij}	constant in Wilson model for multicomponent systems, Equation (12.106)
a'_{ij}	parameter in cubic equation of state for mixtures, Equations (12.80) and (12.81)
$\bar{a}'_i$	parameter defined in Equation (12.88)
B	virial coefficient for pure gas
B_i	Antoine coefficient for species i
B_r	Antoine coefficient for species r
B_{ii}, B_{jj}, B_{kk}	virial coefficients of pure species i, j, k
B_{ij}, B_{ik}	virial cross cefficients
B^o	quantity given in Equation (12.24)
B^1	quantity given in Equation (12.25)
b	parameter in cubic equation of state for pure substances
b_i	parameter in cubic equation of state for pure species i
b_m	parameter in parameter in cubic equation of state for mixtures
C	integration constant, Equation (12.8)
C_i	Antoine coefficient for species i
C_r	Antoine coefficient for species r
F	Feed flow rate
f	fugacity of pure substance or some function
f'	derivative of function f
f^{ig}	fugacity of pure ideal gas
f^L	fugacity of pure liquid
f^V	fugacity of pure vapor
f^{sat}	fugacity of pure substance at saturation conditions
f_i^L	fugacity of pure liquid species i
$\hat{f}_i$	fugacity of species i in mixture or solution
$\hat{f}_i^{id}$	fugacity of species i in ideal solution
$\hat{f}_i^{ig}$	fugacity of species i in ideal gas mixture
$\hat{f}_i^L$	fugacity of species i in liquid solution

$\hat{f}_i^V$	fugacity of species i in vapor mixture
g	molar Gibbs free energy of pure substance
g^{ig}	molar Gibbs free energy of pure substance in ideal gas state
g^L	molar Gibbs free energy of pure liquid
g^R	residual molar Gibbs free energy of pure substance
g^V	molar Gibbs free energy of pure vapor
h^R	molar residual enthalpy
I	quantity defined in Equations (12.29) and (12.85)
K_i	distribution coefficient defined in Equation (12.134)
P	pressure
P_{bubl}	bubble point pressure
P_c	critical pressure
P_{cij}	critical pressure of mixture of species i and j
P_{dew}	dew point pressure
P_r	reduced pressure
P^{sat}	saturation or vapor pressure
P_i^{sat}	saturation or vapor pressure of species i
P_r^{sat}	saturation or vapor pressure of reference species
q	parameter in cubic EOS for pure substances, see Equation (12.27)
q_m	parameter in cubic EOS for mixtures, see Equation (12.78)
$\bar{q}_i$	quantity defined in Equation (12.87)
R	universal gas constant
s	molar or specific entropy
s^{ig}	molar entropy of ideal gas
s^R	molar residual entropy
T	temperature
T_c	critical temperature
T_{ci}	critical temperature of species i
T_{cj}	critical temperature of species j
T_{cij}	critical temperature of mixture of species i and j
T_r	reduced temperature
T_i^{sat}	saturation temperature of species i
v	molar volume of pure substance
v_c	critical molar volume of pure substance
v_{ci}	critical molar volume of species i
v_{cj}	critical molar volume of species j
v_{cij}	critical molar volume of mixture of species i and j
v^{ig}	molar volume of ideal gas
v^L	molar volume of liquid
W	molar flow rate of bottoms product
x_i	mole fraction of species i in liquid solution
y_i	mole fraction of species i in vapor mixture
z_i	mole fraction of species i in feed mixture
Z	compressibility factor
Z_c	critical compressibility factor
Z_{ci}	critical compressibility factor of species i
Z_{cj}	critical compressibility factor of species j
Z_{cij}	critical compressibility factor of mixture of species i and j
Z^o	compressibility factor function in Lee/Kesler correlation, see Equation (12.15)
Z^1	compressibility factor function in Lee/Kesler correlation, see Equation (12.16)

Greek Letters

α	parameter in cubic EOS, see Table 12.1
β	parameter in cubic EOS, see Equation (12.27)
β_m	parameter in cubic EOS for mixtures, see Equation (12.77)
γ_i	activity coefficient of species i
$\gamma_{i,bubl}$	activity coefficient of species i at bubble point
$\gamma_{i,dew}$	activity coefficient of species i at dew point
γ_i^{id}	activity coefficient of species i in ideal solution
δ_{ii}	parameter defined in Equation (12.61)
δ_{ij}	parameter defined in Equation (12.60)
δ_{jj}	parameter defined in Equation (12.61)
δ_{ik}	parameter defined in Equation (12.60)
δ_{ki}	parameter defined in Equation (12.61)
ε	parameter in cubic EOS, see Table 12.1
μ	chemical potential of pure substance
μ_i	chemical potential of species i in mixture or solution
$\mu_{i,mix}$	chemical potential of species i in mixture
$\mu_{i,pure}$	chemical potential of pure species i
$\mu_{i,solu}$	chemical potential of species i in liquid solution
μ^L	chemical potential of pure substance in liquid state
μ^V	chemical potential of pure substance in vapor state
μ_i^E	excess chemical potential of species i
μ_i^L	chemical potential of species i in liquid solution
μ_i^V	chemical potential of species i in vapor mixture
$\mu_{i,solu}^{id}$	chemical potential of species i in ideal liquid solution
$\mu_{i,mix}^{ig}$	chemical potential of species i in ideal gas mixture
$\mu_{i,pure}^{ig}$	chemical potential of pure ideal gas species i
σ	parameter in cubic EOS, see Table 12.1
Φ_i	parameter in generalized Raoult's law, define din Equation (12.115)
$\Phi_{i,bubl}$	Φ_i at bubble point
$\Phi_{i,dew}$	Φ_i at dew point
φ	fugacity coefficient of pure substance
φ^o	function in Lee/Kesler fugacity coefficient correlation, see Equation (12.21)
φ^1	function in Lee/Kesler fugacity coefficient correlation, see Equation (12.22)
φ^L	fugacity coefficient of pure liquid
φ^{sat}	fugacity coefficient of pure substance at saturation condition
φ^V	fugacity coefficient of pure vapor
φ^{ig}	fugacity coefficient of pure ideal gas
$\hat{\varphi}_i$	fugacity coefficient of species i in mixture or solution
$\hat{\varphi}_i^{ig}$	fugacity coefficient of species i in ideal gas mixture
$\hat{\varphi}_i^L$	fugacity coefficient of species i in liquid solution
$\hat{\varphi}_i^V$	fugacity coefficient of species i in vapor mixture
ψ	fraction of feed that goes to the distillate product or parameter in cubic EOS, see Table 12.1
Ω	parameter in cubic EOS, see Table 12.1
Ω_{12}	parameter in Wilson model, see Equation (12.103)
Ω_{21}	parameter in Wilson model, see Equation (12.104)
ω	acentric factor of pure substance

ω_i	acentric factor of pure species i
ω_j	acentric factor of pure species j
ω_{ij}	acentric factor of mixture of species i and j
ψ_{bubl}	fraction of feed that goes to the distillate product at bubble point
ψ_{dew}	fraction of feed that goes to the distillate product at dew point

REFERENCES

1. Smith, J.M., Van Ness, H.C., Abbott, M.M., *Introduction to Chemical Engineering Thermodynamics*, McGraw Hill, New York, 6th edition, 2001.

TUTORIAL PROBLEMS WITH SOLUTION

Problem 1

a) Estimate the fugacity of nitrogen gas at 300 K and 15 bar. Use the Pitzer correlation. The critical properties are: $T_c = 126.2$ K, $P_c = 34$ bar, $\omega = 0.038$.
b) Estimate the fugacity of nitrogen gas at 150 K and 100 bar. Use the Lee/Kesler correlation.

Solution

a) The Pitzer correlation for the fugacity coefficient is given as:

$$ln\varphi = \left(\frac{P_r}{T_r}\right)(B^o + \omega B^1)$$

where B^o and B^1 are given as:

$$B^o = 0.083 - \frac{0.422}{T_r^{1.6}} \quad ; \quad B^1 = 0.139 - \frac{0.172}{T_r^{4.2}}$$

$$T_r = T/T_c = 300/126.2 = 2.3772\,; P_r = P/P_c = 15/34 = 0.4412;\, B^o = -0.022586\,; B^1 = 0.13447.$$

$$ln\varphi = \left(\frac{P_r}{T_r}\right)(B^o + \omega B^1) = \left(\frac{0.4412}{2.3772}\right)(-0.022586 + 0.038 \times 0.13447) = -3.2435 \times 10^{-3}$$

$$\varphi = 0.99676\,; f = \varphi P = 0.99676 \times 15 = 14.9514 \text{ bar}$$

b) The Lee/Kesler correlation for the fugacity coefficient is given as:

$$ln\varphi = ln\varphi^o + \omega ln\varphi^1 \quad ; \quad \varphi = \varphi^o(\varphi^1)^\omega$$

$$T_r = T/T_c = 150/126.2 = 1.1886\,; P_r = P/P_c = 100/34 = 2.9412$$

From Lee/Kesler table in the Appendix J, φ^o values are:

$P_r \rightarrow$ / $T_r \downarrow$	2.0	3.0
1.15	0.6026	0.4875
1.20	0.6546	0.5420

We will do double linear interpolation to obtain φ^o at $T_r = 1.1886$; $P_r = 2.9412$

T_r	φ^o at $P_r = 2.9412$ by linear interpolation
1.15	$\varphi^o = 0.6026 + \dfrac{(0.4875 - 0.6026)}{(3-2)}(2.9412 - 2) = 0.49427$
1.20	$\varphi^o = 0.6546 + \dfrac{(0.5420 - 0.6546)}{(3-2)}(2.9412 - 2) = 0.54862$

Another linear interpolation to obtain φ^o at $T_r = 1.1886$ ($P_r = 2.9412$) :

$$\varphi^o = 0.49427 + \frac{(0.54862 - 0.49427)}{(1.2 - 1.15)}(1.1886 - 1.15) = 0.53623$$

Thus, $\varphi^o = 0.53623$ at $T_r = 1.1886$; $P_r = 2.9412$

From Lee/Kesler table in the Appendix J, φ^1 values are:

$P_r \rightarrow$ / $T_r \downarrow$	**2.0**	**3.0**
1.15	1.1376	1.1858
1.20	1.1588	1.2388

We will do double linear interpolation to obtain φ^1 at $T_r = 1.1886$; $P_r = 2.9412$

T_r	φ^1 at $P_r = 2.9412$ by linear interpolation
1.15	$\varphi^1 = 1.1376 + \dfrac{(1.1858 - 1.1376)}{(3-2)}(2.9412 - 2) = 1.18297$
1.20	$\varphi^1 = 1.1588 + \dfrac{(1.2388 - 1.1588)}{(3-2)}(2.9412 - 2) = 1.2341$

Another linear interpolation to obtain φ^1 at $T_r = 1.1886$ ($P_r = 2.9412$) :

φ^1 at $T_r = 1.1886$ ($P_r = 2.9412$) by linear interpolation

$$\varphi^1 = 1.18297 + \frac{(1.2341 - 1.18297)}{(1.2 - 1.15)}(1.1886 - 1.15) = 1.22244$$

Thus, $\varphi^1 = 1.22244$ at $T_r = 1.1886$; $P_r = 2.9412$

The fugacity coefficient is:

$$\varphi = \varphi^o(\varphi^1)^\omega = 0.53623 \times (1.22244)^{0.038} = 0.5403$$

The fugacity is: $f = \varphi P = 0.5403 \times 100 = 54.03$ bar

Problem 2

Estimate the fugacity of nitrogen gas at 150 K and 100 bar. Use the Peng-Robinson equation of state. The critical properties are: $T_c = 126.2$ K, $P_c = 34$ bar, $\omega = 0.038$.

Solution

The fugacity coefficient is given as:

$$ln\varphi = Z - 1 - ln(Z - \beta) - qI$$

where

$$I = \frac{1}{\sigma - \varepsilon} ln\left[\frac{Z + \sigma\beta}{Z + \varepsilon\beta}\right] \quad \text{when } \varepsilon \neq \sigma$$

$$= \beta/(Z + \varepsilon\beta) \quad \text{when } \varepsilon = \sigma$$

We need the compressibility factor value to calculate the fugacity coefficient. The compressibility factor is determined from the Peng-Robinson cubic EOS using the Newton-Raphson iterative procedure. The reduced temperature (T_r) is: $T_r = T/T_c = 150/126.2 = 1.1886$. The reduced pressure ($P_r$) is : $P_r = P/P_c = 100/34 = 2.9412$.

Summary of parameters for PR cubic equation of state

Name	ϵ	σ	ψ	Ω	α	a'	b	β	q
PR	1-$\sqrt{2}$	1+$\sqrt{2}$	0.45724	0.07779	$\alpha_{PR} =$ 0.923413	0.1367099	2.40057 E-05	0.1924921	4.5665119
$\alpha_{PR} = \left[1 + (0.37464 + 1.5422\omega - 0.26992\omega^2)(1 - \sqrt{T_r})\right]^2$									

Newton-Raphson iterations for compressibility factor of gas (initial guess value of Z is 1):

Z_i	Z_{i+1}
1.00000	0.53622
0.53622	0.53009
0.53009	0.53013
0.53013	0.53013

Thus, Z=0.53013. The fugacity coefficient and fugacity are calculated as follows:

$$I = \frac{1}{\sigma - \varepsilon} ln\left[\frac{Z + \sigma\beta}{Z + \varepsilon\beta}\right] = 0.2801777\,; \quad \varphi = \exp[Z - 1 - ln(Z - \beta) - qI] = 0.515033\,;: \; f = \varphi P = 0.515033 \times 100 = 51.5033 \text{ bar.}$$

The values of fugacity coefficient and fugacity calculated from the PR equation of state are close to the ones obtained from the Lee/Kesler correlation.

Problem 3

Estimate the fugacity of liquid methanol at 64.70 °C and 100 bar. At 64.70 °C, the vapor pressure of methanol is 1.013 bar. Repeat the calculation at 64.70 °C and 275 bar. Given that: $\omega = 0.564$, $T_c = 512.6$ K, $P_c = 80.97$ bar, $Z_c = 0.224$, $v_c = 118\ \text{cm}^3/\text{mol}$.

Solution

The fugacity of liquid at any *T* and *P* is given as:

$$f^L(T, P) = \varphi^{sat} P^{sat} exp\left[\frac{v^L(P - P^{sat})}{RT}\right] \tag{12.46}$$

where φ^{sat} is the fugacity coefficient of saturated vapor (or saturated liquid) at T and P^{sat}. Let us first estimate $\varphi^{sat}(T, P^{sat})$ using the Pitzer correlation:

$$ln\varphi = \left(\frac{P_r}{T_r}\right)(B^o + \omega B^1)$$

where B^o and B^1 are given as:

$$B^o = 0.083 - \frac{0.422}{T_r^{1.6}} \quad ; \quad B^1 = 0.139 - \frac{0.172}{T_r^{4.2}}$$

$$T_r = T/T_c = (64.7 + 273.15)/512.6 = 0.6591 \,;\, P_r = P/P_c = 1.013/80.97 = 0.01251;\, B^o = -0.739224 \,;\, B^1 = -0.851679.$$

$$ln\varphi = \left(\frac{P_r}{T_r}\right)(B^o + \omega B^1) = \left(\frac{0.01251}{0.6591}\right)(-0.739224 - 0.564 \times 0.851679) = -0.02315$$

Thus, $\varphi^{sat} = 0.9771$.

We can estimate v^L from the Rackett equation:

$$v_r = \frac{v}{v_c} = Z_c^{(1-T_r)^{0.2857}} = (0.224)^{(1-0.6591)^{0.2057}} = 0.33284$$

$$v = 0.33284 \times 118 = 39.275 \text{ cm}^3/\text{mol}$$

The liquid fugacity at $T = 64.7 + 273.15 = 337.85$ K and $P = 100$ bar:

$$f^L = \varphi^{sat} P^{sat} exp\left[\frac{v^L(P - P^{sat})}{RT}\right] = 0.9771 \times 1.013 \times exp\left[\frac{39.275 \times (100 - 1.013)}{83.14 \times 337.85}\right] = 1.1367 \text{ bar}$$

The liquid fugacity at $T = 64.7 + 273.15 = 337.85$ K and $P = 275$ bar:

$$f^L = \varphi^{sat} P^{sat} exp\left[\frac{v^L(P - P^{sat})}{RT}\right] = 0.9771 \times 1.013 \times exp\left[\frac{39.275 \times (275 - 1.013)}{83.14 \times 337.85}\right] = 1.452 \text{ bar}$$

Note that the fugacity of liquid methanol is close to P^{sat} (=1.013 bar) at the given temperature. If P^{sat} at given temperature is small enough that the saturated vapor can be treated as an ideal gas, $\varphi^{sat} = 1$. In addition, if the poynting factor (exponential term in f^L expression) is close to unity, $f^L = P^{sat}$.

Problem 4

Consider an equimolar gas mixture of two species 1 and 2. The mixture is at $T = 168$ °C and $P = 37$ bar. Estimate the fugacity coefficients of species 1 and 2 in the given gas mixture using Equations (12.70) to (12.75). The critical properties of the pure species 1 and 2 are given as:

species	T_c, K	P_c, bar	Z_c	ω	v_c, cm^3/mol
1	282.3	50.40	0.281	0.087	131
2	365.6	46.65	0.289	0.140	188.4

Solution

Equations (12.70) to (12.75) are used to estimate the fugacity coefficients. The values of the relevant quantities are summarized in the following table. For cross-coefficients, Equations (12.63) – (12.67) are used.

Species	T_c, K	ω	T_r	B^o	B^1	Z_c	v_c	P_c(bar)	B_{ij}
1	282.3	0.087	1.562699	-0.12359	0.112621	0.281	131	50.40	$B_{11} = -52.9915$
2	365.6	0.140	1.206647	-0.22945	0.060857	0.289	188.4	46.65	$B_{22} = -143.954$
Cross values	$T_{c12} =$ 321.2614	$\omega_{12} =$ 0.1135	1.373181	-0.17107	0.093598	$Z_{c12} =$ 0.285	$v_{c12} =$ 157.9661	$P_{c12} =$ 48.18917	$B_{12} = -88.9285$

v_c and B_{ij} are in units of cm³/mol; $\delta_{12} = 2B_{12} - B_{11} - B_{22} = 19.08879$ cm³/mol

$$ln\hat{\varphi}_1 = \frac{P}{RT}[B_{11} + y_2^2\delta_{12}] = -0.04864; \quad \hat{\varphi}_1 = 0.95252$$

$$ln\hat{\varphi}_2 = \frac{P}{RT}[B_{22} + y_1^2\delta_{12}] = -0.14041; \quad \hat{\varphi}_2 = 0.869004$$

Problem 5

Consider a binary liquid solution of acetone (species 1) and methanol (species 2) at 50 °C. Plot γ_1 and γ_2 as functions of liquid composition x_1. Use Wilson model to calculate γ_1 and γ_2. Determine the limiting values of γ_1 and γ_2: $\gamma_1\ (x_1 = 0), \gamma_1\ (x_1 = 1.0), \gamma_2\ (x_1 = 0)$, and $\gamma_2\ (x_1 = 1.0)$.

Given that:

$v_1 = 74.05$ cm³/mol ; $v_2 = 40.73$ cm³/mol ; $a_{12} = -161.88$ cal/mol ; $a_{21} = 583.11$ cal/mol

Solution

The activity coefficients are calculated using the Wilson model (Equations (12.101) - (12.104)). The data generated from the Wilson model are plotted as shown in the following figure.

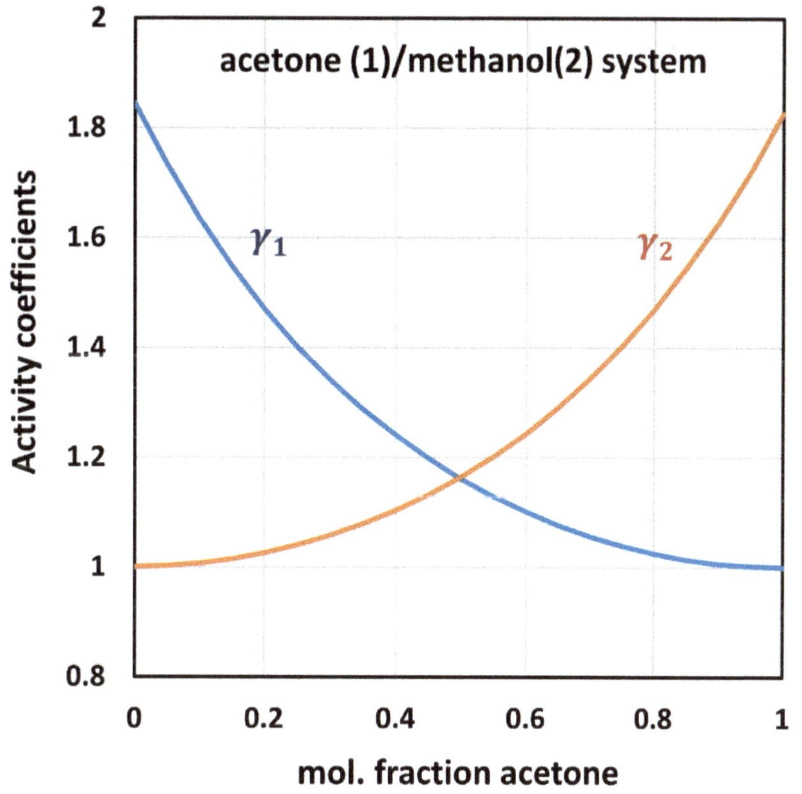

The limiting values of the activity coefficients are: $\gamma_1\,(x_1 = 0) = 1.844998$, $\gamma_1\,(x_1 = 1.0) = 1$, $\gamma_2\,(x_1 = 0) = 1$, $\gamma_2\,(x_1 = 1.0) = 1.826873$.

Problem 6

Consider a binary system of methanol (species 1) and methyl acetate (species 2). The vapor pressures of the pure components are given by the following Antoine equations:

$$log_{10}P_1^{sat} = 7.20277 - \frac{1580.08}{T - 33.65}$$

$$log_{10}P_2^{sat} = 6.18621 - \frac{1156.43}{T - 53.46}$$

where P is in kPa and T is in K. The activity coefficients of the components in liquid solution are given by the Wilson model (Equations (12.101) - (12.104)). The following data are given regarding the Wilson model:

$v_1 = 40.73$ cm^3/mol ; $v_2 = 79.84$ cm^3/mol ; $a_{12} = 813.18$ cal/mol ; $a_{21} = -31.19$ cal/mol

For methanol (1)/methyl acetate (2) system,

Chemical	ω	T_c, K	P_c, bar	Z_c	v_c, cm^3/mol
methanol (1)	0.564	512.6	80.97	0.224	118
methyl acetate (2)	0.331	506.6	47.50	0.257	228

Do the following:

a) Given T=330 K, $x_1 = 0.35$, $x_2 = 0.65$, estimate the bubble point pressure and composition of bubble

b) Given T=330 K, $y_1 = 0.50, y_2 = 0.50$, estimate the dew point pressure and composition of liquid drop
c) Given P=110 kPa, $x_1 = 0.65, x_2 = 0.35$, estimate the bubble point temperature and composition of bubble
d) Given P=110 kPa, $y_1 = 0.37, y_2 = 0.63$, estimate the dew point temperature and composition of liquid drop

The vapor mixture is non-ideal.

Solution

a) Given T=330 K, $x_1 = 0.35, x_2 = 0.65$, estimate the bubble point pressure and composition of bubble

This is *BUBLP* calculation. We follow the iteration scheme shown in Figure 12.1.

- Vapor pressures from Antoine equations:

$$P_1^{sat} = 74.2962 \text{ kPa} ; \ P_2^{sat} = 101.0248 \text{ kPa}$$

- Activity coefficients from the Wilson model:

$$\Omega_{12} = 0.567172 \ ; \Omega_{21} = 0.534998 ; \gamma_1 = 1.534188 \ ; \gamma_2 = 1.133216$$

- Estimation of pressure

$$P = \sum_i \frac{x_i \gamma_i P_i^{sat}}{\Phi_i} = x_1 \gamma_1 P_1^{sat} + x_2 \gamma_2 P_2^{sat} = 114.3084 \text{ kPa}$$

Note that Φ_i's are assumed to be 1 here.

- Calculation of vapor composition

$$y_1 = \frac{x_1 \gamma_1 P_1^{sat}}{P} = 0.349 ; \qquad y_2 = \frac{x_2 \gamma_2 P_2^{sat}}{P} = 0.651$$

- Fugacity Coefficients of vapor phase using Equations (12.70) - (12.75)

The values of the relevant quantities are summarized in the following table. For cross-coefficients, Equations (12.63) – (12.67) are used.

Species	T_c, **K**	ω	T_r	B^o	B^1	Z_c	v_c	P_c**(bar)**	B_{ij}
1	512.6	0.564	0.643777	-0.77076	-0.95455	0.224	118	80.97	$B_{11} = -689.044$
2	506.6	0.331	0.651402	-0.75483	-0.90179	0.257	228	47.5	$B_{22} = -933.989$
Cross values	$T_{c12} =$ 509.5912	$\omega_{12} =$ 0.4475	0.647578	-0.76276	-0.92785	Z_{c12} =0.2405	v_{c12} =167.01	P_{c12} =61.01	$B_{12} = -818.015$

v_c and B_{ij} are in units of cm³/mol; $\delta_{12} = 2B_{12} - B_{11} - B_{22} = -12.9974$ cm³/mol

$$ln\hat{\varphi}_1 = \frac{P}{RT}[B_{11} + y_2^2\delta_{12}] = -0.02894\ ;\ \ \hat{\varphi}_1 = 0.971477$$

$$ln\hat{\varphi}_2 = \frac{P}{RT}[B_{22} + y_1^2\delta_{12}] = -0.03898\ ;\ \ \hat{\varphi}_2 = 0.961771$$

Note that $P = 114.3084$ kPa and $T = 330$ K

For pure species, φ_i^{sat}'s are calculated using Equation (12.56). Thus,

$$\varphi_1^{sat} = exp\left(\frac{B_{11}P_1^{sat}}{RT}\right) = 0.981514\ ;\ \varphi_2^{sat} = exp\left(\frac{B_{22}P_2^{sat}}{RT}\right) = 0.966194$$

$$\Phi_1 = \left(\frac{\hat{\varphi}_1^V}{\varphi_1^{sat}}\right) = \frac{0.971477}{0.981514} = 0.989774\ ;\ \ \Phi_2 = \left(\frac{\hat{\varphi}_2^V}{\varphi_2^{sat}}\right) = \frac{0.961771}{0.966194} = 0.995422$$

- Estimation of improved pressure

$$P = \sum_i \frac{x_i\gamma_i P_i^{sat}}{\Phi_i} = \frac{x_1\gamma_1 P_1^{sat}}{\Phi_1} + \frac{x_2\gamma_2 P_2^{sat}}{\Phi_2} = \frac{0.35\times 1.534188\ \times 74.2962}{0.989774} + \frac{0.65\times 1.133216\times 101.0248}{0.995422} = 115.063 \text{ kPa}$$

This value of pressure is very close to the previous value of 114.3084 kPa. Thus, there is no need for further iteration.

- The bubble point pressure is 115.063 kPa. The vapor composition is as follows:

$$y_1 = \frac{x_1\gamma_1 P_1^{sat}}{\Phi_1 P} = \frac{0.35 \times 1.534188 \times 74.2962}{0.989774 \times 115.063} = 0.3503$$

$$y_2 = \frac{x_2\gamma_2 P_2^{sat}}{\Phi_2 P} = \frac{0.65 \times 1.133216 \times 101.0248}{0.995422 \times 115.063} = 0.6497$$

Note: We are assuming that Φ_i's do not change with slight change in pressure from 114.3084 kPa to 115.063 kPa.

- It is of interest to compare the bubble point pressure of 115.063 kPa and the corresponding bubble composition with *ideal system* values shown below:

$$P = \sum x_i\, P_i^{sat} = 0.35 \times 74.2962 + 0.65 \times 101.0248 = 91.6698 \text{ kPa}$$

$$y_1 = \frac{x_1 P_1^{sat}}{P} = \frac{0.35 \times 74.2962}{91.6698} = 0.2837$$

$$y_2 = \frac{x_2 P_2^{sat}}{P} = \frac{0.65 \times 101.0248}{91.6698} = 0.7163$$

Clearly, the ideal system calculations of bubble point pressure and vapor composition are significantly off from the non-ideal calculations.

b) Given T=330 K, $y_1 = 0.50, y_2 = 0.50$, estimate the dew point pressure and composition of liquid drop

This is *DEWP* calculation. We follow the iteration scheme shown in Figure 12.2.

- Vapor pressures from Antoine equations:

$$P_1^{sat} = 74.2962 \text{ kPa} ;\ P_2^{sat} = 101.0248 \text{ kPa}$$

- Estimation of pressure

$$P = \frac{1}{\Sigma y_i \Phi_i/\gamma_i P_i^{sat}} = \frac{1}{\frac{y_1}{P_1^{sat}} + \frac{y_2}{P_2^{sat}}} = 85.623 \text{ kPa}$$

Note that Φ_i's and $\gamma_i' s$ are assumed to be 1 here.

- Calculation of liquid composition

$$x_1 = \frac{y_1 P}{P_1^{sat}} = 0.576227 ;\quad x_2 = \frac{y_2 P}{P_2^{sat}} = 0.423772$$

- Activity coefficients from the Wilson model:

$$\Omega_{12} = 0.567172 \ ;\Omega_{21} = 0.534998 ;\gamma_1 = 1.205954 ;\gamma_2 = 1.394853$$

- Estimation of improved pressure

$$P = \frac{1}{\Sigma y_i \Phi_i/\gamma_i P_i^{sat}} = \frac{1}{\frac{y_1}{\gamma_1 P_1^{sat}} + \frac{y_2}{\gamma_2 P_2^{sat}}} = \frac{1}{\frac{0.5}{1.205954 \times 74.2962} + \frac{0.5}{1.394853 \times 101.0248}} = 109.544 \text{ kPa}$$

- Fugacity Coefficients of vapor phase using Equations (12.70) - (12.75)

The values of the relevant quantities are summarized in the following table. For cross-coefficients, Equations (12.63) – (12.67) are used.

Species	T_c**, K**	ω	T_r	B^o	B^1	Z_c	v_c	P_c**(bar)**	B_{ij}
1	512.6	0.564	0.643777	-0.77076	-0.95455	0.224	118	80.97	$B_{11} =$ -689.044
2	506.6	0.331	0.651402	-0.75483	-0.90179	0.257	228	47.5	$B_{22} =$ -933.989
Cross values	$T_{c12} =$ 509.5912	$\omega_{12} =$ 0.4475	0.647578	-0.76276	-0.92785	Z_{c12} =0.2405	v_{c12} =167.01	P_{c12} =61.01	$B_{12} =$ -818.015

v_c and B_{ij} are in units of cm³/mol; $\delta_{12} = 2B_{12} - B_{11} - B_{22} = -12.9974$ cm³/mol

$$ln\hat{\varphi}_1 = \frac{P}{RT}[B_{11} + y_2^2 \delta_{12}] = -0.027641 ;\quad \hat{\varphi}_1 = 0.972737$$

$$ln\hat{\varphi}_2 = \frac{P}{RT}[B_{22} + y_1^2 \delta_{12}] = -0.037421 ;\quad \hat{\varphi}_2 = 0.963271$$

Note that $P = 109.544$ kPa and $T = 330$ K

For pure species, φ_i^{sat}'s are calculated using Equation (12.56). Thus,

$$\varphi_1^{sat} = exp\left(\frac{B_{11}P_1^{sat}}{RT}\right) = 0.981514\ ;\ \varphi_2^{sat} = exp\left(\frac{B_{22}P_2^{sat}}{RT}\right) = 0.966194$$

$$\Phi_1 = \left(\frac{\hat{\varphi}_1^V}{\varphi_1^{sat}}\right) = \frac{0.972737}{0.981514} = 0.99106\ ;\ \Phi_2 = \left(\frac{\hat{\varphi}_2^V}{\varphi_2^{sat}}\right) = \frac{0.963271}{0.966194} = 0.99697$$

- Calculation of liquid composition

$$x_1 = \frac{y_1\Phi_1 P}{\gamma_1 P_1^{sat}} = \frac{0.5 \times 0.99106 \times 111.3717}{1.205954 \times 74.2962} = 0.615953$$

$$x_2 = \frac{y_2\Phi_2 P}{\gamma_2 P_2^{sat}} = \frac{0.5\times0.99697\times111.3717}{1.394853\times101.0248} = 0.393977\ ;\ \sum x_i = 1.00993$$

Normalize x values. Thus, $x_1 = \frac{0.615953}{1.00993} = 0.6099\ ;\ x_2 = \frac{0.393977}{1.00993} = 0.3901$

- Evaluate activity coefficients from the Wilson model

$$x_1 = 0.6099 \text{ and } x_2 = 0.3901\ ;\ \gamma_1 = 1.173297;\ \gamma_2 = 1.451795$$

- Estimation of improved pressure

$$P = \frac{1}{\sum y_i\,\Phi_i/\gamma_i P_i^{Sat}} = \frac{1}{\frac{y_1\Phi_1}{\gamma_1 P_1^{sat}} + \frac{y_2\Phi_2}{\gamma_2 P_2^{sat}}} = \frac{1}{\frac{0.5\times0.99106}{1.173297\times74.2962} + \frac{0.5\times0.99697}{1.451795\times101.0248}} = 110.092 \text{ kPa}$$

This value is very close to the previous value of 109.544 kPa. Hence no more iteration is needed.

- The dew point pressure is 110.092 kPa. The liquid composition is as follows:

$$x_1 = \frac{y_1\Phi_1 P}{\gamma_1 P_1^{sat}} = \frac{0.5 \times 0.99106 \times 110.092}{1.173297 \times 74.2962} = 0.6258$$

$$x_2 = \frac{y_2\Phi_2 P}{\gamma_2 P_2^{sat}} = \frac{0.5 \times 0.99697 \times 110.092}{1.451795 \times 101.0248} = 0.3742$$

Note: We are assuming that Φ_i's do not change with slight change in pressure from 109.544 kPa to 110.092 kPa.

c) Given P=110 kPa, $x_1 = 0.65$, $x_2 = 0.35$, estimate the bubble point temperature and co-position of bubble

This is *BUBLT* calculation. We follow the iteration scheme shown in Figure 12.3.

- T_i^{sat}'s from Antoine equations at P=110 kPa:

$$T_1^{sat} = 339.7853 \text{ K}\ ;\ T_2^{sat} = 332.4663 \text{ K};$$

Initial guess of Temperature: $T = \sum x_i T_i^{Sat} = 337.2237$ K

- Vapor pressures from Antoine equations:

$$P_1^{sat} = 99.5038 \text{ kPa}\,;\; P_2^{sat} = 129.087 \text{ kPa}$$

- Activity coefficients from the Wilson model:

$$\Omega_{12} = 0.582441\,;\, \Omega_{21} = 0.534453\,;\, \gamma_1 = 1.137139\,;\, \gamma_2 = 1.514599$$

Vapor pressures of reference species (species 1):

$$P_1^{Sat} = \frac{P}{x_1\gamma_1 + (x_2\gamma_2)(P_2^{sat}/P_1^{Sat})} = \frac{110}{(0.65\times1.137139) + (0.35\times1.514599\times(129.087/99.5038)} = 77.0926 \text{ kPa}$$

- Improved value of temperature

$$T = \frac{B_1}{A_1 - \log P_1^{Sat}} - C_1 = \frac{1580.08}{7.20277 - \log(77.0926)} - (-33.65) = 330.895 \text{ K}$$

- Vapor pressures from Antoine equations:

$$P_1^{sat} = 77.09396 \text{ kPa}\,;\; P_2^{sat} = 104.2121 \text{ kPa}$$

- Activity coefficients from the Wilson model:

$$\Omega_{12} = 0.569078\,;\, \Omega_{21} = 0.534929\,;\, \gamma_1 = 1.138482\,;\, \gamma_2 = 1.526101$$

- Calculation of vapor composition

$$y_1 = \frac{x_1\gamma_1 P_1^{sat}}{\Phi_1 P} = \frac{0.65 \times 1.138482 \times 77.09396}{1 \times 110} = 0.51864$$

$$y_2 = \frac{x_2\gamma_2 P_2^{sat}}{\Phi_2 P} = \frac{0.35 \times 1.526101 \times 104.2121}{1 \times 110} = 0.50603$$

$$\sum y_i = 0.51864 + 0.50603 = 1.02467$$

Normalize y values: $y_1 = \frac{0.51864}{1.02467} = 0.5062\,;\, y_2 = \frac{0.50603}{1.02467} = 0.4938$

- Fugacity Coefficients of vapor phase using Equations (12.70) - (12.75)

The values of the relevant quantities are summarized in the following table. For cross-coefficients, Equations (12.63) – (12.67) are used.

Species	T_c, K	ω	T_r	B^o	B^1	Z_c	v_c	P_c(bar)	B_{ij}
1	512.6	0.564	0.645523	- 0.76707	- 0.94218	0.224	118	80.97	$B_{11} = -$ 683.42
2	506.6	0.331	0.653168	-0.7512	- 0.89002	0.257	228	47.5	$B_{22} = -$ 927.322
Cross values	$T_{c12} =$ 509.5912	$\omega_{12} =$ 0.4475	0.649334	-0.7591	- 0.91578	Z_{c12} =0.2405	v_{c12} =167.01	P_{c12} =61.01	$B_{12} = -$ 811.726

v_c and B_{ij} are in units of cm³/mol; $\delta_{12} = 2B_{12} - B_{11} - B_{22} = -12.7008$ cm³/mol

$$ln\hat{\varphi}_1 = \frac{P}{RT}[B_{11} + y_2^2\delta_{12}] = -0.02745 \;;\; \hat{\varphi}_1 = 0.972923$$

$$ln\hat{\varphi}_2 = \frac{P}{RT}[B_{22} + y_1^2\delta_{12}] = -0.037209 \;;\; \hat{\varphi}_2 = 0.963475$$

Note that $P = 110$ kPa and $T = 330.895$ K

For pure species, φ_i^{sat}'s are calculated using Equation (12.56). Thus,

$$\varphi_1^{sat} = exp\left(\frac{B_{11}P_1^{sat}}{RT}\right) = 0.98103 \;;\; \varphi_2^{sat} = exp\left(\frac{B_{22}P_2^{sat}}{RT}\right) = 0.965482$$

$$\Phi_1 = \left(\frac{\hat{\varphi}_1^V}{\varphi_1^{sat}}\right) = \frac{0.972923}{0.98103} = 0.99174;\; \Phi_2 = \left(\frac{\hat{\varphi}_2^V}{\varphi_2^{sat}}\right) = \frac{0.963475}{0.965482} = 0.99792$$

- Vapor pressures of reference species (species 1):

$$P_1^{Sat} = \frac{P}{(x_1\gamma_1/\Phi_1)+(x_2\gamma_2/\Phi_2)(P_2^{sat}/P_1^{Sat})} = \frac{110}{\left(\frac{0.65\times1.138482}{0.99174}\right)+\left(\frac{0.35\times1.526101}{0.99792}\right)\left(\frac{104.2121}{77.09396}\right)} = 74.845 \text{ kPa}$$

- Improved value of temperature

$$T = \frac{B_1}{A_1 - logP_1^{Sat}} - C_1 = \frac{1580.08}{7.20277 - \log(74.845)} \quad (-33.65) = 330.18 \text{ K}$$

This value of temperature is very close to the previous value of 330.895 K. Hence no more iteration is needed.

- Thus, the bubble point temperature is 330.18 K and the vapor composition is as follows:

$$y_1 = \frac{x_1\gamma_1P_1^{sat}}{\Phi_1P} = \frac{0.65 \times 1.138634 \times 74.85191}{0.99174 \times 110} = 0.5078$$

$$y_2 = \frac{x_2\gamma_2P_2^{sat}}{\Phi_2P} = \frac{0.35 \times 1.527424 \times 101.6591}{0.99792 \times 110} = 0.4951$$

$$\sum y_i = 0.5078 + 0.4951 = 1.0029$$

Normalize y values: $y_1 = \frac{0.5078}{1.0029} = 0.5063$; $y_2 = \frac{0.4951}{1.0029} = 0.4937$

Note that γ_i and P_i^{sat} are calculated at 330.18 K.

d) Given P=110 kPa, $y_1 = 0.37, y_2 = 0.63$, estimate the dew point temperature and composition of liquid drop

This is *DEWT* calculation. We follow the iteration scheme shown in Figure 12.4.

- T_i^{sat}'s from Antoine equations at P=110 kPa:

$$T_1^{sat} = 339.7853 \text{ K}; \quad T_2^{sat} = 332.4663 \text{ K};$$

$$\text{Initial guess of Temperature: } T = \sum y_i T_i^{Sat} = 335.1743 \text{ K}$$

- Vapor pressures from Antoine equations:

$$P_1^{sat} = 91.71971 \text{ kPa}; \quad P_2^{sat} = 120.569 \text{ kPa}$$

- Vapor pressures of reference species (species 1):

$$P_1^{Sat} = P\left[(y_1\Phi_1/\gamma_1) + (y_2\Phi_2/\gamma_2)\left(P_1^{sat}/P_2^{Sat}\right)\right] = P\left[y_1 + y_2\left(P_1^{sat}/P_2^{Sat}\right)\right] = 110[0.37 + 0.63(91.71971/120.569)] = 93.4182 \text{ kPa}$$

- Improved value of temperature

$$T = \frac{B_1}{A_1 - logP_1^{Sat}} - C_1 = \frac{1580.08}{7.20277 - \log(93.4182)} - (-33.65) = 335.6335 \text{ K}$$

- Vapor pressures from Antoine equations:

$$P_1^{sat} = 93.418 \text{ kPa}; \quad P_2^{sat} = 122.438 \text{ kPa}$$

- Fugacity Coefficients of vapor phase using Equations (12.70) - (12.75)

The values of the relevant quantities are summarized in the following table. For cross-coefficients, Equations (12.63) – (12.67) are used.

Species	T_c, K	ω	T_r	B^o	B^1	Z_c	v_c	P_c(bar)	B_{ij}
1	512.6	0.564	0.654767	-0.74795	-0.87951	0.224	118	80.97	$B_{11} = -654.759$
2	506.6	0.331	0.662522	-0.73244	-0.83037	0.257	228	47.5	$B_{22} = -893.176$
Cross values	$T_{c12} = 509.5912$	$\omega_{12} = 0.4475$	0.658633	-0.74016	-0.85463	$Z_{c12} = 0.2405$	$v_{c12} = 167.01$	$P_{c12} = 61.01$	$B_{12} = -779.571$

v_c and B_{ij} are in units of cm³/mol; $\delta_{12} = 2B_{12} - B_{11} - B_{22} = -11.2068$ cm³/mol

$$ln\hat{\varphi}_1 = \frac{P}{RT}[B_{11} + y_2^2\delta_{12}] = -0.02599; \quad \hat{\varphi}_1 = 0.974349$$

$$ln\hat{\varphi}_2 = \frac{P}{RT}[B_{22} + y_1^2\delta_{12}] = -0.03527\ ; \ \ \hat{\varphi}_2 = 0.965345$$

Note that $P = 110$ kPa and $T = 335.6335$ K

For pure species, φ_i^{sat}'s are calculated using Equation (12.56). Thus,

$$\varphi_1^{sat} = exp\left(\frac{B_{11}P_1^{sat}}{RT}\right) = 0.978319\ ;\ \varphi_2^{sat} = exp\left(\frac{B_{22}P_2^{sat}}{RT}\right) = 0.961568$$

$$\Phi_1 = \left(\frac{\hat{\varphi}_1^V}{\varphi_1^{sat}}\right) = \frac{0.974349}{0.978319} = 0.99594\ ;\ \ \Phi_2 = \left(\frac{\hat{\varphi}_2^V}{\varphi_2^{sat}}\right) = \frac{0.965345}{0.961568} = 1.00393$$

- Calculation of liquid composition

$$x_1 = \frac{y_1\Phi_1 P}{\gamma_1 P_1^{sat}} = \frac{0.37 \times 0.99594 \times 110}{1 \times 93.418} = 0.43391$$

$$x_2 = \frac{y_2\Phi_2 P}{\gamma_2 P_2^{sat}} = \frac{0.63 \times 1.00393 \times 110}{1 \times 122.438} = 0.56823$$

$$\sum x_i = 1.00214$$

Normalize x values. Thus, $x_1 = \frac{0.43391}{1.00214} = 0.43298\ ;\ x_2 = \frac{0.56823}{1.00214} = 0.56702$

- Activity coefficients from the Wilson model:

$$\Omega_{12} = 0.579102\ ;\ \Omega_{21} = 0.534571\ ;\ \gamma_1 = 1.383136;\ \gamma_2 = 1.20392$$

- Vapor pressures of reference species (species 1):

$$P_1^{Sat} = P\left[(y_1\Phi_1/\gamma_1) + (y_2\Phi_2/\gamma_2)\left(P_1^{sat}/P_2^{Sat}\right)\right] = 110\left[\frac{0.37\times0.99594}{1.383136} + \left(\frac{0.63\times1.00395}{1.20392}\right)(93.418/122.438)\right] = 73.4 \text{ kPa}$$

- Improved value of temperature

$$T = \frac{B_1}{A_1 - logP_1^{Sat}} - C_1 = \frac{1580.08}{7.20277 - \log(73.4)} - (-33.65) = 329.7074 \text{ K}$$

- Vapor pressures from Antoine equations:

$$P_1^{sat} = 73.4 \text{ kPa}\ ;\ P_2^{sat} = 99.9997 \text{ kPa}$$

- Fugacity Coefficients of vapor phase using Equations (12.70) - (12.75)

The values of the relevant quantities are summarized in the following table. For cross-coefficients, Equations (12.63) – (12.67) are used.

Species	T_c, K	ω	T_r	B^o	B^1	Z_c	v_c	P_c(bar)	B_{ij}
1	512.6	0.564	0.643206	-0.77197	-0.95864	0.224	118	80.97	$B_{11} = -690.894$
2	506.6	0.331	0.650824	-0.75602	-0.90568	0.257	228	47.5	$B_{22} = -936.185$
Cross values	$T_{c12} =$ 509.5912	$\omega_{12} =$ 0.4475	0.647004	-0.76396	-0.93183	Z_{c12} =0.2405	v_{c12} =167.01	P_{c12} =61.01	$B_{12} = -820.087$

v_c and B_{ij} are in units of cm³/mol; $\delta_{12} = 2B_{12} - B_{11} - B_{22} = -13.0954$ cm³/mol

$$ln\hat{\varphi}_1 = \frac{P}{RT}[B_{11} + y_2^2\delta_{12}] = -0.027933\ ;\ \ \hat{\varphi}_1 = 0.972453$$

$$ln\hat{\varphi}_2 = \frac{P}{RT}[B_{22} + y_1^2\delta_{12}] = -0.03764\ ;\ \ \hat{\varphi}_2 = 0.96306$$

Note that $P = 110$ kPa and $T = 329.7074$ K

For pure species, φ_i^{sat}'s are calculated using Equation (12.56). Thus,

$$\varphi_1^{sat} = exp\left(\frac{B_{11}P_1^{sat}}{RT}\right) = 0.98167\ \ ;\ \varphi_2^{sat} = exp\left(\frac{B_{22}P_2^{sat}}{RT}\right) = 0.966424$$

$$\Phi_1 = \left(\frac{\hat{\varphi}_1^V}{\varphi_1^{sat}}\right) = \frac{0.972453}{0.98167} = 0.99061\ ;\ \ \Phi_2 = \left(\frac{\hat{\varphi}_2^V}{\varphi_2^{sat}}\right) = \frac{0.96306}{0.966424} = 0.99652$$

- Calculation of liquid composition

$$x_1 = \frac{y_1\Phi_1 P}{\gamma_1 P_1^{sat}} = \frac{0.37 \times 0.99061 \times 110}{1.383136 \times 73.4} = 0.3971$$

$$x_2 = \frac{y_2\Phi_2 P}{\gamma_2 P_2^{sat}} = \frac{0.63 \times 0.99652 \times 110}{1.20392 \times 99.9997} = 0.5736$$

$$\sum x_i = 0.9707$$

Normalize x values. Thus, $x_1 = \frac{0.3971}{0.9707} = 0.4091\ ;\ x_2 = \frac{0.5736}{0.9707} = 0.5909$

- Activity coefficients from the Wilson model:

$$\Omega_{12} = 0.566548\ ;\ \Omega_{21} = 0.53502\ ;\ \gamma_1 = 1.34922;\ \gamma_2 = 1.236965$$

- Vapor pressures of reference species (species 1):
-

$$P_1^{Sat} = P[(y_1\Phi_1/\gamma_1) + (y_2\Phi_2/\gamma_2)(P_1^{sat}/P_2^{Sat})] = 110\left[\frac{0.37\times 0.99061}{1.34922} + \left(\frac{0.63\times 0.99652}{1.236965}\right)(73.4/99.9997)\right] = 70.861 \text{ kPa}$$

- Improved value of temperature

$$T = \frac{B_1}{A_1 - logP_1^{Sat}} - C_1 = \frac{1580.08}{7.20277 - \log(70.861)} - (-33.65) = 328.862 \text{ K}$$

This value of temperature is very close to the previous value of 329.7074 K. Hence no more iterations are needed.

- Thus, the dew point temperature is 328.86 K and the liquid composition is as follows:

$$x_1 = \frac{y_1 \Phi_1 P}{\gamma_1 P_1^{sat}} = \frac{0.37 \times 0.99061 \times 110}{1.34922 \times 70.86} = 0.4217\,; \quad x_2 = \frac{y_2 \Phi_2 P}{\gamma_2 P_2^{sat}} = \frac{0.63 \times 0.99652 \times 110}{1.236965 \times 97.08} = 0.5751$$

$\sum x_i = 0.9968$. Normalized x values: $x_1 = \frac{0.4217}{0.9968} = 0.4231\,;\ x_2 = \frac{0.5751}{0.9968} = 0.5769$

Problem 7

A flash chamber operating at 89 °C and 101 kPa is separating a binary mixture of 1-propanol (species 1) and water (species 2). The feed composition is $z_1 = 0.30$ and $z_2 = 0.70$.

a) Would flashing occur at given temperature and pressure?
b) If flashing occurs, what is the fraction of feed that goes to the vapor distillate product and what are the compositions of the vapor and liquid products?

Due to low pressure, the vapor mixture can be treated as an ideal gas mixture. However, the liquid solution is non-ideal in nature.
The vapor pressures of the pure components are given by the following Antoine equations:

$$lnP_1^{sat} = 16.0692 - \frac{3448.66}{T - 69.06}$$
$$lnP_2^{sat} = 16.2620 - \frac{3799.89}{T - 46.8}$$

where P is in kPa and T is in K.
The activity coefficients of the components in liquid solution are given by the Wilson model (Equations (12.101) - (12.104)). The following data are given regarding the Wilson model:

$$v_1 = 75.14 \text{ cm}^3\text{/mol}\,;\ v_2 = 18.07 \text{ cm}^3\text{/mol}\,;\ a_{12} = 775.48 \text{ cal/mol}\,;\ a_{21} = 1351.90 \text{ cal/mol}$$

Solution

a) Would flashing occur at given temperature and pressure?

For flashing to occur, the flash P (given 101 kPa) must fall in between the dew point and bubble point pressures, that is, $P_{dew} < P < P_{bubl}$. The bubble point and dew point pressure calculations at 89 °C are given below.

BUBLP Calculation:

At 89 °C (362.15 K), the vapor pressures are calculated from the given Antoine equations:

$$P_1^{sat} = 73.89489 \text{ kPa}\ ;\ P_2^{sat} = 67.50798 \text{ kPa}$$

The bubble point pressure is:

$$P_{bubl} = \sum_i \frac{x_i \gamma_i P_i^{sat}}{\Phi_i} = \frac{x_1 \gamma_1 P_1^{sat}}{\Phi_1} + \frac{x_2 \gamma_2 P_2^{sat}}{\Phi_2} = x_1 \gamma_1 P_1^{sat} + x_2 \gamma_2 P_2^{sat}$$

As the vapor mixture is assumed to be an ideal gas mixture, the fugacity coefficients are 1.

Calculation of activity coefficients from the Wilson model (Equations (12.101) - (12.102)):

$$\Omega_{12} = \frac{v_2}{v_1} exp\left(-\frac{a_{12}}{RT}\right) = \frac{18.07}{75.14} exp\left(-\frac{775.48}{1.987 \times 362.15}\right) = 0.081858$$

$$\Omega_{21} = \frac{v_1}{v_2} exp\left(-\frac{a_{21}}{RT}\right) = \frac{75.14}{18.07} exp\left(-\frac{1351.9}{1.987 \times 362.15}\right) = 0.635335$$

$$\gamma_1 = 1.994082\ ;\ \gamma_2 = 1.298413$$

$$P_{bubl} = x_1 \gamma_1 P_1^{sat} + x_2 \gamma_2 P_2^{sat} = 0.3 \times 1.994082 \times 73.89489 + 0.7 \times 1.298413 \times 67.50798 = 105.563 \text{ kPa}$$

DEWP Calculation:

$$P = \frac{1}{\sum y_i \Phi_i / \gamma_i P_i^{sat}} = \frac{1}{\frac{y_1}{\gamma_1 P_1^{sat}} + \frac{y_2}{\gamma_2 P_2^{sat}}} = \frac{1}{\frac{0.3}{73.89489} + \frac{0.7}{67.50798}} = 69.305 \text{ kPa}$$

As the vapor mixture is assumed to be an ideal gas mixture, the fugacity coefficients are 1. In this initial calculation of dew point pressure, the activity coefficients are taken 1.

Liquid composition:

$$x_1 = \frac{y_1 P}{P_1^{sat}} = \frac{0.3 \times 69.305}{73.89489} = 0.2814\ ;\ x_2 = \frac{y_2 P}{P_2^{sat}} = \frac{0.7 \times 69.305}{67.50798} = 0.7186$$

Activity coefficients from Wilson model: $\gamma_1 = 2.100742\ ;\ \gamma_2 = 1.270988$

Improved dew point pressure:

$$P = \frac{1}{\sum y_i \Phi_i / \gamma_i P_i^{sat}} = \frac{1}{\frac{y_1}{\gamma_1 P_1^{sat}} + \frac{y_2}{\gamma_2 P_2^{sat}}} = \frac{1}{\frac{0.3}{2.100742 \times 73.89489} + \frac{0.7}{1.270988 \times 67.50798}} = 99.0992 \text{ kPa}$$

Liquid composition:

$$x_1 = \frac{y_1 P}{\gamma_1 P_1^{sat}} = \frac{0.3 \times 99.0992}{2.100742 \times 73.89489} = 0.1915\ ;\ x_2 = \frac{y_2 P}{\gamma_2 P_2^{sat}} = \frac{0.7 \times 99.0992}{1.270988 \times 67.50798} = 0.8085$$

Activity coefficients from Wilson model: $\gamma_1 = 2.888196\ ;\ \gamma_2 = 1.15299$

Improved dew point pressure:

$$P = \frac{1}{\sum y_i \Phi_i / \gamma_i P_i^{sat}} = \frac{1}{\frac{y_1}{\gamma_1 P_1^{sat}} + \frac{y_2}{\gamma_2 P_2^{sat}}} = \frac{1}{\frac{0.3}{2.888196 \times 73.89489} + \frac{0.7}{1.15299 \times 67.50798}} = 96.1638 \text{ kPa}$$

This value of dew point pressure is close to the previous value of 99.0992. However, more iterations are needed to improve the value.

Liquid composition:

$$x_1 = \frac{y_1 P}{\gamma_1 P_1^{sat}} = \frac{0.3 \times 96.1638}{2.888196 \times 73.89489} = 0.13517\ ;\ x_2 = \frac{y_2 P}{\gamma_2 P_2^{sat}} = \frac{0.7 \times 96.1638}{1.15299 \times 67.50798} = 0.86483$$

Activity coefficients from Wilson model: $\gamma_1 = 3.841376\ ;\ \gamma_2 = 1.091068$

Improved dew point pressure:

$$P = \frac{1}{\frac{y_1}{\gamma_1 P_1^{sat}} + \frac{y_2}{\gamma_2 P_2^{sat}}} = \frac{1}{\frac{0.3}{3.841376 \times 73.89489} + \frac{0.7}{1.091068 \times 67.50798}} = 94.6922 \text{ kPa}$$

This value is very close to the previous value of 96.1638. We will go for one more iteration.

Liquid composition:

$$x_1 = \frac{y_1 P}{\gamma_1 P_1^{sat}} = \frac{0.3 \times 94.6922}{3.841376 \times 73.89489} = 0.10008\ ;\ x_2 = \frac{y_2 P}{\gamma_2 P_2^{sat}} = \frac{0.7 \times 94.6922}{1.091068 \times 67.50798} = 0.89992$$

Activity coefficients from Wilson model: $\gamma_1 = 4.858554\ ;\ \gamma_2 = 1.057619$

Improved dew point pressure:

$$P = \frac{1}{\frac{y_1}{\gamma_1 P_1^{sat}} + \frac{y_2}{\gamma_2 P_2^{sat}}} = \frac{1}{\frac{0.3}{4.858554 \times 73.89489} + \frac{0.7}{1.057619 \times 67.50798}} = 93.9864 \text{ kPa}$$

This value is almost the same as the previous value of 94.6922. No more iteration is necessary. Thus, the dew point pressure is approximately 94 kPa.

Flashing occurs in the present case as the flash *P* (101 kPa) falls in between the dew point pressure (94 kPa) and bubble point pressure (105.563).

b) If flashing occurs, what is the fraction of feed that goes to the vapor distillate product and what are the compositions of the vapor and liquid products?

Refer to the iteration scheme shown in Figure 12.5.

- Initial estimates of activity coefficients and K_i's

The initial estimates of the activity coefficients are given by Equation (12.140).

$$\gamma_1 = \gamma_{1,dew} + \left(\frac{P - P_{dew}}{P_{bubl} - P_{dew}}\right)\left(\gamma_{1,bubl} - \gamma_{1,dew}\right)$$
$$= 4.858554 + \left(\frac{101 - 94}{105.563 - 94}\right)(1.994082 - 4.858554) = 3.1245$$

$$\gamma_2 = \gamma_{2,dew} + \left(\frac{P - P_{dew}}{P_{bubl} - P_{dew}}\right)\left(\gamma_{2,bubl} - \gamma_{2,dew}\right)$$
$$= 1.057619 + \left(\frac{101 - 94}{105.563 - 94}\right)(1.298413 - 1.057619) = 1.20339$$

$$K_1 = \frac{\gamma_1 P_1^{sat}}{\Phi_1 P} = \frac{3.1245 \times 73.89489}{1 \times 101} = 2.286$$

$$K_2 = \frac{\gamma_2 P_2^{sat}}{\Phi_2 P} = \frac{1.20339 \times 67.50798}{1 \times 101} = 0.80434$$

- Calculation of ψ and compositions

As the system is binary, we can solve the Rachford-Rice equation analytically and hence no iteration is required:

$$\sum \frac{(K_i - 1)z_i}{1 + \psi(K_i - 1)} = \frac{(K_1 - 1)z_1}{1 + \psi(K_1 - 1)} + \frac{(K_2 - 1)z_2}{1 + \psi(K_2 - 1)} = \frac{(2.286 - 1) \times 0.3}{1 + \psi(2.286 - 1)} + \frac{(0.80434 - 1) \times 0.7}{1 + \psi(0.80434 - 1)} = 0$$

The solution gives: $\psi = 0.98896$. The vapor and liquid compositions are:

$$y_1 = \frac{K_1 z_1}{1 + \psi(K_1 - 1)} = \frac{2.286 \times 0.30}{1 + 0.98896 \times 1.286} = 0.301875$$

$$y_2 = \frac{K_2 z_2}{1 + \psi(K_2 - 1)} = \frac{0.80434 \times 0.7}{1 + 0.98896 \times (-0.19566)} = 0.698125$$

$$x_1 = \frac{z_1}{1 + \psi(K_1 - 1)} = \frac{0.3}{1 + 0.98896 \times 1.286} = 0.132054$$

$$x_2 = \frac{z_2}{1 + \psi(K_2 - 1)} = \frac{0.7}{1 + 0.98896 \times (-0.19566)} = 0.867948$$

- Improved estimates of activity coefficients and K_i's

Activity coefficients are calculated from Wilson model

$\Omega_{12} = 0.081858$; $\Omega_{21} = 0.635335$; $x_1 = 0.132054$; $x_2 = 0.867948$; $\gamma_1 = 3.913868$; $\gamma_2 = 1.087927$

$$K_1 = \frac{\gamma_1 P_1^{sat}}{\Phi_1 P} = \frac{3.913868 \times 73.89489}{1 \times 101} = 2.8635\ ;\ K_2 = \frac{\gamma_2 P_2^{sat}}{\Phi_2 P} = \frac{1.087927 \times 67.50798}{1 \times 101} = 0.727166$$

- Calculation of ψ and compositions

$$\frac{(K_1 - 1)z_1}{1 + \psi(K_1 - 1)} + \frac{(K_2 - 1)z_2}{1 + \psi(K_2 - 1)} = \frac{(2.8635 - 1) \times 0.3}{1 + \psi(2.8635 - 1)} + \frac{(0.727166 - 1) \times 0.7}{1 + \psi(0.727166 - 1)} = 0$$

The solution gives: $\psi = 0.72393$. The vapor and liquid compositions are:

$$y_1 = \frac{K_1 z_1}{1 + \psi(K_1 - 1)} = \frac{2.8635 \times 0.30}{1 + 0.72393 \times 1.8635} = 0.3657$$

$$y_2 = \frac{K_2 z_2}{1 + \psi(K_2 - 1)} = \frac{0.727166 \times 0.7}{1 + 0.72393 \times (-0.272834)} = 0.6343$$

$$x_1 = \frac{z_1}{1+\psi(K_1-1)} = \frac{0.3}{1+0.72393\times 1.8635} = 0.12771$$

$$x_2 = \frac{z_2}{1+\psi(K_2-1)} = \frac{0.7}{1+0.72393\times(-0.272834)} = 0.87229$$

The ψ value of 0.72393 is significantly off from the previous ψ value of 0.98896. We need another iteration.

- Improved estimates of activity coefficients and $K_i's$

Activity coefficients are calculated from Wilson model

$\Omega_{12} = 0.081858$; $\Omega_{21} = 0.635335$; $x_1 = 0.12771$; $x_2 = 0.87229$; $\gamma_1 = 4.01975$; $\gamma_2 = 1.083601$

$$K_1 = \frac{\gamma_1 P_1^{sat}}{\Phi_1 P} = \frac{4.01975\times 73.89489}{1\times 101} = 2.94098\ ;\ K_2 = \frac{\gamma_2 P_2^{sat}}{\Phi_2 P} = \frac{1.083601\times 67.50798}{1\times 101} = 0.724274$$

- Calculation of ψ and compositions

$$\frac{(K_1-1)z_1}{1+\psi(K_1-1)} + \frac{(K_2-1)z_2}{1+\psi(K_2-1)} = \frac{(2.94098-1)\times 0.3}{1+\psi(2.94098-1)} + \frac{(0.724274-1)\times 0.7}{1+\psi(0.724274-1)} = 0$$

The solution gives: $\psi = 0.727394$. The vapor and liquid compositions are:

$$y_1 = \frac{K_1 z_1}{1+\psi(K_1-1)} = \frac{2.94098\ \times 0.30}{1+0.727394\times 1.94098} = 0.3658$$

$$y_2 = \frac{K_2 z_2}{1+\psi(K_2\quad 1)} = \frac{0.724274\times 0.7}{1+0.727394\times(-0.275726)} = 0.6342$$

$$x_1 = \frac{z_1}{1+\psi(K_1-1)} = \frac{0.3}{1+0.727394\times 1.94098} = 0.1244$$

$$x_2 = \frac{z_2}{1+\psi(K_2-1)} = \frac{0.7}{1+0.727394\times(-0.275726)} - 0.8756$$

The ψ value of 0.727394 is very close to the previous ψ value of 0.72393. Thus, no more iterations are needed, and these are the final values of ψ and vapor and liquid compositions.

Problem 8

Consider a binary system of 1-propanol (species 1) and water (species 2) at 70 °C. The vapor mixture is ideal whereas the liquid solution is non-ideal. The following form of the Raoult's law is applicable:

$$y_i P = x_i \gamma_i P_i^{sat}$$

Construct the P-x-y diagram. The vapor pressures of the pure components are given by the following Antoine equations:

$$lnP_1^{sat} = 16.0692 - \frac{3448.66}{T-69.06}$$

$$lnP_2^{sat} = 16.2620 - \frac{3799.89}{T-46.8}$$

where P is in kPa and T is in K. The activity coefficients of the components in liquid solution are given by the Wilson model (Equations (12.101) - (12.104)). The following data are given regarding the Wilson model:
$v_1 = 75.14$ cm^3/mol ; $v_2 = 18.07$ cm^3/mol ; $a_{12} = 775.48$ cal/mol ; $a_{21} = 1351.90$ cal/mol

Solution

The following procedure is used to construct the P-x-y diagram:

- Calculate P_1^{sat} and P_2^{sat} at T=70 °C using the given Antoine equations.
- Select a number of x_1 values covering the range 0 to1.
- For each liquid phase composition (x_1), estimate the activity coefficients γ_1 and γ_2 from the Wilson model
- Calculate bubble point pressure as: $P = \sum x_i \gamma_i P_i^{sat} = x_1\gamma_1 P_1^{sat} + x_2\gamma_2 P_2^{sat}$
- Calculate vapor phase composition at each x_1 values as: $y_1 = x_1\gamma_1 P_1^{sat}/P$
- Plot bubble point curve P vs x_1 and dew point curve P vs y_1

The following table gives the values of x_1, γ_1, γ_2, P, and y_1 at 70 °C. The vapor pressures of pure species are: $P_1^{sat} = 32.68716$ kPa and $P_2^{sat} = 31.17746$ kPa. Note that x_1, y_1 refer to mole fractions of 1-propanol in liquid and vapor phases, respectively.

x_1	γ_1	γ_2	P(kPa)	y_1
0	19.88299	1	31.17746	0
0.03	11.09854	1.008265	41.37549	0.26304
0.06	7.508266	1.026665	44.8137	0.328592
0.09	5.627792	1.050736	46.36703	0.357066
0.12	4.491465	1.078766	47.21478	0.373137
0.15	3.738355	1.110007	47.74554	0.383898
0.18	3.206124	1.144128	48.11406	0.392065
0.21	2.811899	1.181008	48.39019	0.398877
0.24	2.509292	1.220644	48.6082	0.404977
0.27	2.270448	1.26311	48.78573	0.410733
0.30	2.077692	1.308536	48.93194	0.416378
0.33	1.919302	1.357094	49.05127	0.42207
0.36	1.787211	1.408999	49.14536	0.42793
0.39	1.675695	1.464502	49.21402	0.434058
0.42	1.58059	1.523893	49.25575	0.440544
0.45	1.498795	1.587502	49.26797	0.447473
0.48	1.427959	1.655705	49.2472	0.454938
0.51	1.36627	1.728926	49.18906	0.463037
0.54	1.312311	1.807644	49.08824	0.471878
0.57	1.264959	1.892401	48.93843	0.481591
0.60	1.223319	1.98381	48.73216	0.492326
0.63	1.186667	2.082567	48.46071	0.504263
0.66	1.154416	2.189461	48.11384	0.517623
0.69	1.126087	2.30539	47.67954	0.53268
0.72	1.101288	2.431382	47.14375	0.549777
0.75	1.079699	2.568612	46.48991	0.569354
0.78	1.061058	2.71843	45.69853	0.591982
0.81	1.045153	2.882393	44.74657	0.618418
0.84	1.031815	3.0623	43.60673	0.649688
0.87	1.020912	3.260244	42.24653	0.687217
0.90	1.012346	3.47866	40.62724	0.733047

0.93	1.006048	3.720402	38.70241	0.790207
0.96	1.001979	3.988825	36.41622	0.8634
0.98	1.000496	4.184525	34.65856	0.924715
1.00	1	4.39518	32.68716	1

The P-x-y diagram and y-x equilibrium curve for 1-propanol/water system at 70 °C are shown below. The system forms an azeotrope at $x \approx 0.45$.

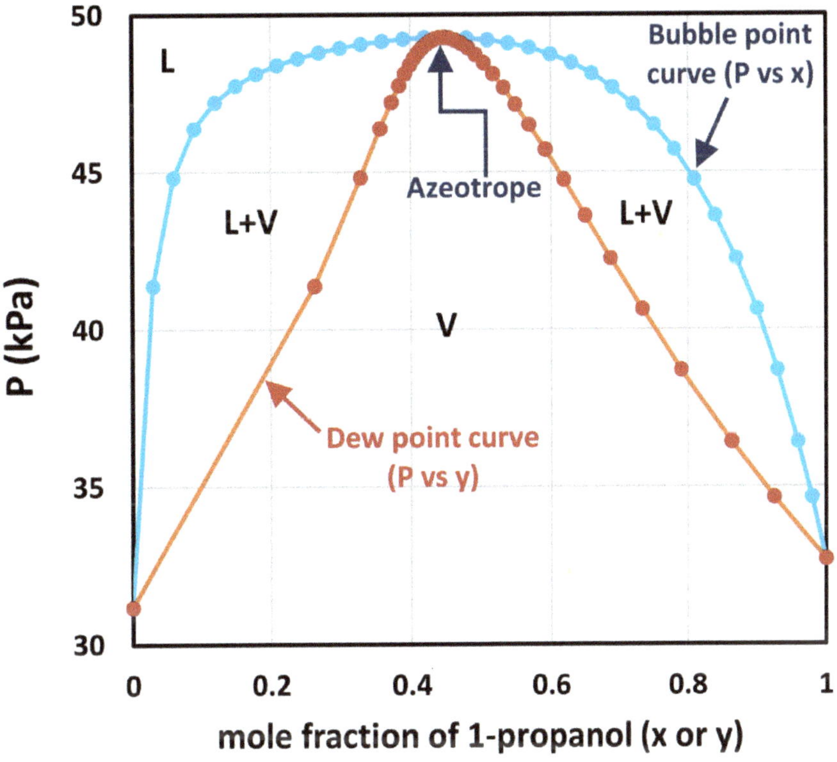

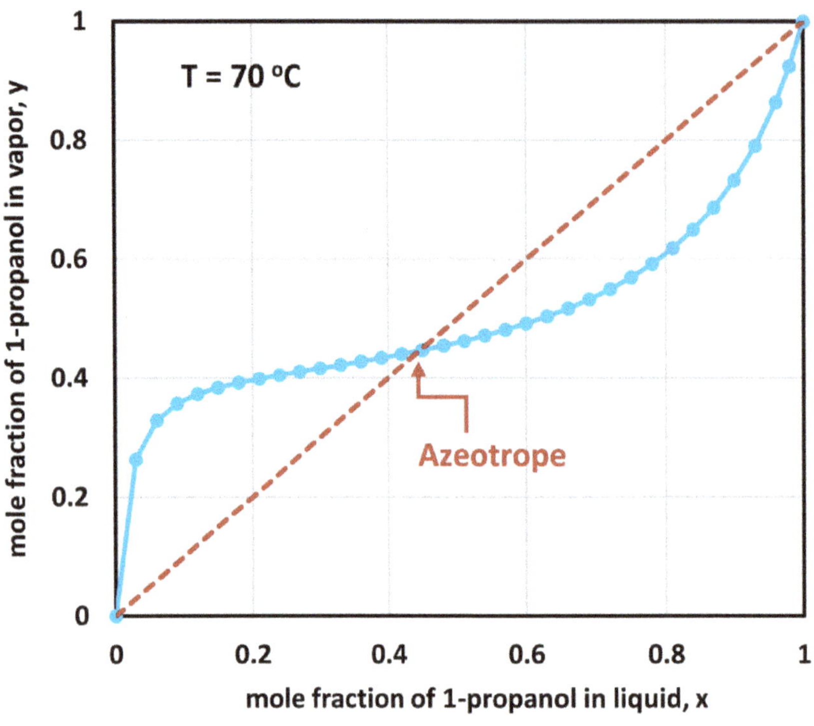

Problem 9

Consider a binary system of 1-propanol (species 1) and water (species 2) at 1 bar. The vapor mixture is ideal whereas the liquid solution is non-ideal. The following form of the Raoult's law is applicable:

$$y_i P = x_i \gamma_i P_i^{sat}$$

Construct the T-x-y diagram at 1 bar. The vapor pressures of the pure components are given by the Antoine equations:

$$lnP_1^{sat} = 16.0692 - \frac{3448.66}{T - 69.06}$$

$$lnP_2^{sat} = 16.2620 - \frac{3799.89}{T - 46.8}$$

where P is in kPa and T is in K. The activity coefficients of the components in liquid solution are given by the Wilson model (Equations (12.101) - (12.104)). The following data are given regarding the Wilson model:

$v_1 = 75.14$ cm^3/mol ; $v_2 = 18.07$ cm^3/mol ; $a_{12} = 775.48$ cal/mol ; $a_{21} = 1351.90$ cal/mol

Solution

The following procedure is used to construct the T-x-y diagram:

- Select a number of x_1 values covering the range 0 to1.
- For each liquid phase composition (x_1), calculate the bubble point temperature iteratively as described below:
  Calculate T_1^{sat} and T_1^{sat} at given total P using the Antoine equations. The initial guess of unknown temperature is $T = \sum x_i T_i^{sat}$. Calculate P_1^{sat}, P_2^{sat}, γ_1, and γ_2. The activity coefficients are estimated the Wilson model. Let species 2 be the reference species. Calculate the vapor pressure of reference species from $P_2^{Sat} = \frac{P}{x_1\gamma_1(P_1^{sat}/P_2^{Sat})+x_2\gamma_2}$. Calculate the improved temperature from the

Antoine equation for species 2 as: $T = \frac{B_2}{A_2 - lnP_2^{Sat}} - C_2$. Continue iteration until convergence is reached. This iteration scheme to estimate the temperature is similar to that shown in Figure 12.3 except that the vapor mixture is assumed ideal here.

- Once T is known, estimate the activity coefficients γ_1 and γ_2 from the Wilson model
- Calculate P_1^{sat} from the Antoine equation.
- Calculate the vapor phase composition at each x_1 values as: $y_1 = x_1\gamma_1 P_1^{sat}/P$
- Plot bubble point curve T vs x_1 and dew point curve T vs y_1

The following table gives the values of $x_1, T, \gamma_1, \gamma_2, P_1^{sat}$, and y_1 at $P = 1$ bar. Note that x_1, y_1 refer to mole fractions of 1-propanol in liquid and vapor phases, respectively.

x_1	T(°C)	γ_1	γ_2	P_1^{sat}(kPa)	y_1
0	372.7797	16.47322	1	111.5479	0
0.06	363.2691	6.924114	1.02549	77.27734	0.321046
0.09	362.3087	5.256278	1.048844	74.36694	0.351804
0.12	361.7909	4.227723	1.076128	72.83593	0.369516
0.15	361.4652	3.528364	1.106567	71.88639	0.380462
0.21	361.0671	2.682872	1.175674	70.73975	0.39855
0.27	360.8207	2.179674	1.255291	70.0377	0.41218
0.35	360.612	1.767874	1.378768	69.4476	0.429711
0.43	360.5128	1.509645	1.524898	69.16856	0.449006
0.5	360.5274	1.354811	1.674891	69.20957	0.468829
0.58	360.6976	1.228826	1.876959	69.68913	0.496688
0.65	361.0329	1.149353	2.086213	70.64196	0.527752
0.75	361.9684	1.071186	2.450185	73.35778	0.589349
0.85	363.7819	1.024345	2.907761	78.86962	0.686712
0.97	368.1626	1.000922	3.593361	93.61368	0.90889
1	369.8844	1	3.783294	100	1

The T-x-y diagram and y-x equilibrium curve for 1-propanol/water system at 1 bar are shown below. The system forms an azeotrope at $x \approx 0.45$.

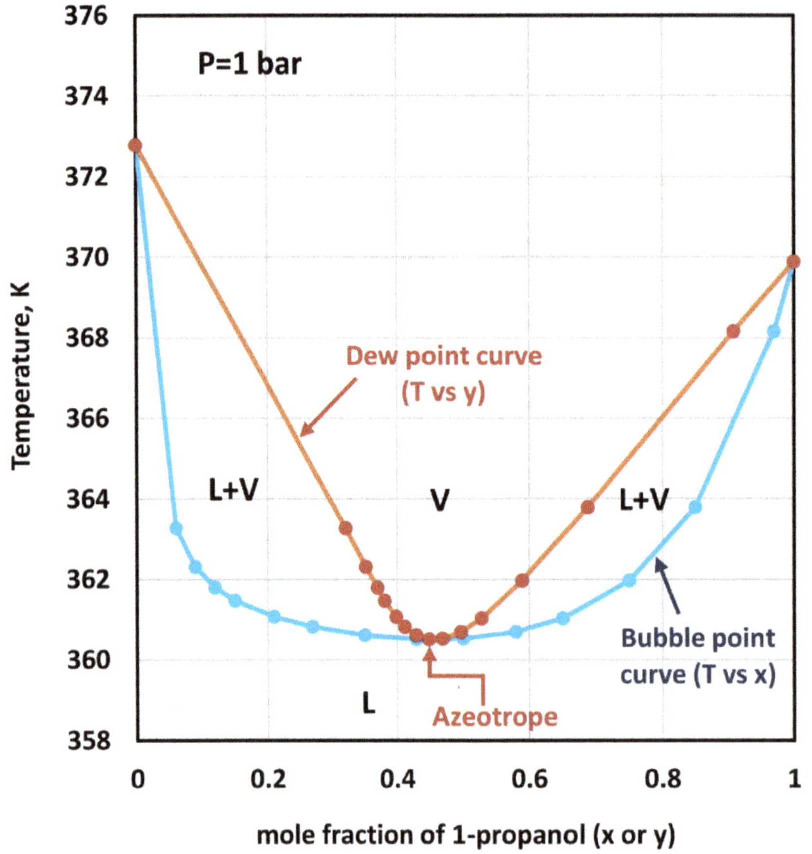
P=1 bar
Dew point curve
(T vs y)
L+V
V
L+V
Bubble point
curve (T vs x)
L
Azeotrope
Temperature, K
376
374
372
370
368
366
364
362
360
358
0
0.2
0.4
0.6
0.8
1
mole fraction of 1-propanol (x or y)

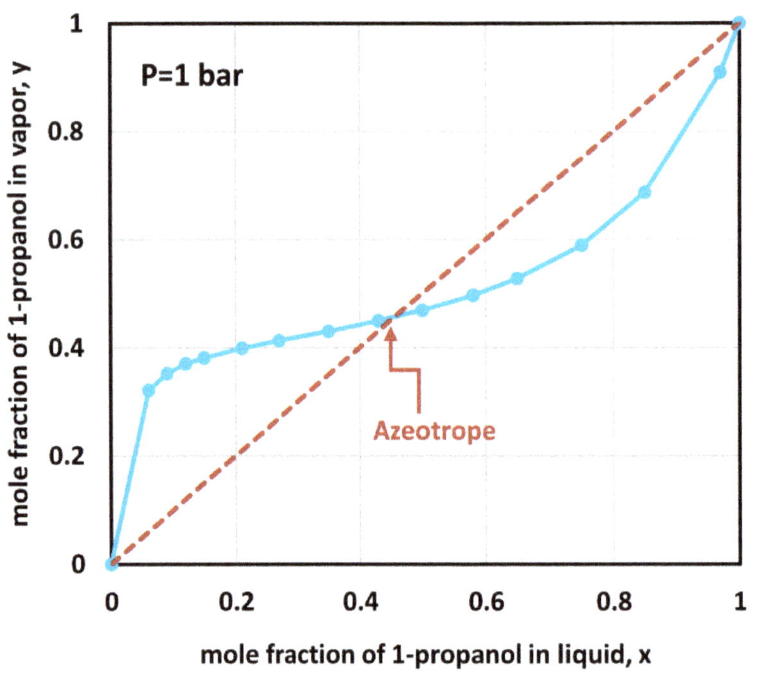
P=1 bar
Azeotrope
mole fraction of 1-propanol in vapor, y
1
0.8
0.6
0.4
0.2
0
0
0.2
0.4
0.6
0.8
1
mole fraction of 1-propanol in liquid, x

Problem 10

Consider a binary system of methanol (species 1) and benzene (species 2). The vapor pressures of the pure components are given by the following Antoine equations:

$$lnP_1^{sat} = 16.585 - \frac{3638.27}{T + 239.50}$$

$$lnP_2^{sat} = 13.8594 - \frac{2773.78}{T + 220.07}$$

where P is in kPa and T is in °C. The activity coefficients of the components in liquid solution are given by the Wilson model (Equations (12.101) - (12.104)). The following data are given regarding the Wilson model:

$v_1 = 40.73$ cm³/mol ; $v_2 = 89.41$ cm³/mol ; $a_{12} = 1734.42$ cal/mol ; $a_{21} = 183.04$ cal/mol

a) Determine if an azeotrope exists for this system at 58.3 °C.
b) If the answer to part (a) is yes, what is the composition of the azeotrope? Also determine the pressure at which this azeotrope is formed.

Solution

To solve this problem, we make use of the relative volatility, α_{12} , defined as:

$$\alpha_{12} = \frac{y_1/x_1}{y_2/x_2}$$

For an azeotrope, $\alpha_{12} = 1$ as the liquid and vapor compositions are the same, that is, $y_1 = x_1$ and $y_2 = x_2$.

It is well known that α_{12} is a continuous function of x_1. If within the limits of $x_1 = 0$ and $x_1 = 1$, the value of α_{12} changes from less than 1 to greater than 1, it is indicative of the existence of an azeotrope as at some intermediate composition, α_{12} will have a value of unity. Thus, in order to determine the existence of an azeotrope, we need to calculate: $(\alpha_{12})_{x_1=0}$ and $(\alpha_{12})_{x_1=1}$. If one value is less than 1 and the other value is greater than 1, an azeotrope does exist. From the following form of the Raoult's law:

$$y_i P = x_i \gamma_i P_i^{sat}$$

we can easily show that:

$$\alpha_{12} = \frac{\gamma_1 P_1^{sat}}{\gamma_2 P_2^{sat}}$$

In the limits of $x_1 = 0$ and $x_1 = 1$, we get:

$$(\alpha_{12})_{x_1=0} = \frac{\gamma_1 P_1^{sat}}{P_2^{sat}} \; ; \quad (\alpha_{12})_{x_1=1} = \frac{P_1^{sat}}{\gamma_2 P_2^{sat}}$$

To calculate the composition of the azeotrope, we make use of the fact that $\alpha_{12} = 1$ for azeotrope. Thus,

$$\left(\frac{\gamma_1}{\gamma_2}\right)_{az} = \frac{P_2^{sat}}{P_1^{sat}}$$

Since P_2^{sat}/P_1^{sat} is known, we can calculate the ratio of activity coefficients at azeotropic composition. As $(\gamma_1/\gamma_2)_{az}$ is related to the composition, we can determine the composition from the knowledge of

$(\gamma_1/\gamma_2)_{az}$. The azeotropic pressure can be calculated from the Raoult's law as: $(P)_{az} = (\gamma_1)_{az}P_1^{sat}$. Note that $y_1 = x_1$ for azeotropes.

At $T = 58.3^oC$, the Antoine equations give: $P_1^{sat} = 78.87315$ kPa and $P_2^{sat} = 49.15782$ kPa. The activity coefficients are calculated from the Wilson model (Equations (12.101) - (12.104)) at $T = 58.3 + 273.15 = 331.45$ K.

$$\Omega_{12} = \frac{v_2}{v_1} exp\left(-\frac{a_{12}}{RT}\right) = \frac{89.41}{40.73} exp\left(-\frac{1734.42}{1.987\times 331.45}\right) = 0.157668$$

$$\Omega_{21} = \frac{v_1}{v_2} exp\left(-\frac{a_{21}}{RT}\right) = \frac{40.73}{89.41} exp\left(-\frac{183.04}{1.987\times 331.45}\right) = 0.345006$$

The activity coefficients are summarized in the table below:

x_1	γ_1	γ_2	γ_1/γ_2
0	12.21003	1	12.21003
0.05	7.548833	1.011741	7.461228
0.1	5.33114	1.040273	5.124752
0.15	4.07706	1.080621	3.772885
0.20	3.285734	1.131025	2.905093
0.25	2.747479	1.191161	2.306557
0.30	2.361	1.261531	1.871536
0.35	2.072053	1.343259	1.542557
0.40	1.849249	1.438055	1.285938
0.45	1.673298	1.548278	1.080748
0.50	1.531772	1.677095	0.913348
0.55	1.416358	1.828734	0.774502
0.60	1.321336	2.008874	0.657749
0.65	1.242686	2.225246	0.558449
0.70	1.177561	2.488559	0.47319
0.75	1.12396	2.813975	0.399421
0.80	1.080525	3.223522	0.3352
0.85	1.04644	3.750182	0.279037
0.90	1.021392	4.445137	0.229777
0.95	1.00561	5.391246	0.186526
1.0	1	6.729666	0.148596

a) Determine if an azeotrope exists for this system at 58.3 °C

$$(\alpha_{12})_{x_1=0} = \frac{\gamma_1 P_1^{sat}}{P_2^{sat}} = \frac{12.21003 \times 78.87315}{49.15782} = 19.59$$

$$(\alpha_{12})_{x_1=1} = \frac{P_1^{sat}}{\gamma_2 P_2^{sat}} = \frac{78.87315}{6.729666 \times 49.15782} = 0.238$$

As $(\alpha_{12})_{x_1=0} > 1$ and $(\alpha_{12})_{x_1=1} < 1$, an azeotrope does exist at some intermediate composition where $\alpha_{12} = 1$.

b) If the answer to part (a) is yes, what is the composition of the azeotrope? Also determine the pressure at which this azeotrope is formed

The ratio of activity coefficients at azeotropic composition is given as:

$$\left(\frac{\gamma_1}{\gamma_2}\right)_{az} = \frac{P_2^{sat}}{P_1^{sat}} = \frac{49.15782}{78.87315} = 0.62325$$

From the table given above,

γ_1/γ_2	x_1
0.657749	0.60
0.558449	0.65

By interpolation, the azeotrope composition is:

$$x_1 = 0.65 + \frac{(0.60 - 0.65)}{(0.657749 - 0.558449)} \times (0.62325 - 0.558449) = 0.6174$$

The pressure at which this azeotrope is formed is:

$$(P)_{az} = (\gamma_1)_{az} P_1^{sat} = 1.292274 \times 78.87315 = 101.926 \text{ kPa}$$

where γ_1 is determined from the Wilson model at $x_1 = 0.6174$. The pressure is approximately 1 atm. Thus, the methanol/benzene system forms an azeotrope at 58.3 °C at a methanol mole fraction of 0.6174 under 1 atm pressure.

Problem 11

Consider a binary system of methanol (species 1) and water (species 2). The vapor pressures of the pure components are given by the following Antoine equations:

$$lnP_1^{sat} = 16.585 - \frac{3638.27}{T + 239.50}$$
$$lnP_2^{sat} = 16.2620 - \frac{3799.89}{T + 226.35}$$

where P is in kPa and T is in °C. The activity coefficients of the components in liquid solution are given by the Wilson model (Equations (12.101) - (12.104)). The following data are given regarding the Wilson model:

$v_1 = 40.73$ cm³/mol ; $v_2 = 18.07$ cm³/mol ; $a_{12} = 107.38$ cal/mol ; $a_{21} = 169.55$ cal/mol

Determine if an azeotrope exists for this system at $P = 1$ bar.

Solution

An azeotrope exists if the relative volatility α_{12} is equal to 1 at some liquid composition. The relative volatility α_{12} is given as:

$$\alpha_{12} = \frac{y_1/x_1}{y_2/x_2} = \frac{\gamma_1 P_1^{sat}}{\gamma_2 P_2^{sat}}$$

As α_{12} is a continuous function of liquid composition x_1, an azeotrope exists if one of the limiting relative volatilities $(\alpha_{12})_{x_1=0}$ and $(\alpha_{12})_{x_1=1}$ is > 1 and the other is < 1. The limiting relative volatilities can be evaluated as follows:

$(\alpha_{12})_{x_1=0} = \frac{(\gamma_1)_{x_1=0} P_1^{sat}}{P_2^{sat}}$ where $(\gamma_1)_{x_1=0}$ and P_1^{sat} are evaluated at T_2^{sat} and P_2^{sat} is equal to the given pressure P.

$(\alpha_{12})_{x_1=1} = \frac{P_1^{sat}}{(\gamma_2)_{x_1=1} P_2^{sat}}$ where $(\gamma_2)_{x_1=1}$ and P_2^{sat} are evaluated at T_1^{sat} and P_1^{sat} is equal to the given pressure P.

Note that T_1^{sat} and T_2^{sat} are boiling points of species 1 and 2 at the given pressure P.

From Antoine equations at $P = 1$ bar, $T_1^{sat} = 64.19964$ °C, $T_2^{sat} = 99.62971$ °C. The relevant vapor pressures are: P_1^{sat} at T_2^{sat} is 349.5844 kPa and P_2^{sat} at T_1^{sat} is 24.13634 kPa

From Wilson model, $(\gamma_1)_{x_1=0} = 2.142494$ at $T = T_2^{sat} = 99.62971$ °C and $(\gamma_2)_{x_1=1} = 1.665047$ at $T = T_1^{sat} = 64.19964$ °C.

Thus,

$$(\alpha_{12})_{x_1=0} = \frac{(\gamma_1)_{x_1=0} P_1^{sat}}{P_2^{sat}} = \frac{2.142494 \times 349.5844}{100} = 7.4898$$

$$(\alpha_{12})_{x_1=1} = \frac{P_1^{sat}}{(\gamma_2)_{x_1=1} P_2^{sat}} = \frac{100}{1.665047 \times 24.13634} = 2.488$$

As both limiting α_{12} values are greater than 1, there does not exist an azeotrope for the methanol/water system at $P = 1$ bar.

PRACTICE PROBLEMS

Problem 1

a) Estimate the fugacity of carbon dioxide gas at 400 K and 15 bar. Use the Pitzer correlation. The critical properties are: $T_c = 304.2$ K, $P_c = 73.83$ bar, $\omega = 0.224$.
b) Estimate the fugacity of carbon dioxide gas at 400 K and 100 bar. Use the Lee/Kesler correlation.

Problem 2

Estimate the fugacity of carbon dioxide gas at 400 K and 100 bar. Use the Peng-Robinson (PR) and Soave-Redlich-Kwong (SRK) equations of state and compare your results. The critical properties are: $T_c = 304.2$ K, $P_c = 73.83$ bar, $\omega = 0.224$.

Problem 3

Estimate the fugacity of liquid ethanol at 78.25 °C and 75 bar. At 78.25 °C, the vapor pressure of methanol is 1.013 bar. Repeat the calculation at 78.25 °C and 250 bar. Given that: $\omega = 0.645$, $T_c = 513.9$ K, $P_c = 61.48$ bar, $Z_c = 0.240$, $v_c = 167$ cm^3/mol.

Problem 4

Consider a gas mixture of two species 1 and 2 with the following composition $y_1 = 0.43$ and $y_2 = 0.57$. The mixture is at $T = 170$ °C and $P = 35$ bar. Estimate the fugacity coefficients of species 1 and 2 in the given gas mixture using Equations (12.70) to (12.75). The critical properties of the pure species 1 and 2 are given as:

species	T_c, K	P_c, bar	Z_C	ω	v_c, cm^3/mol
1	305.3	48.72	0.279	0.10	145.5
2	190.6	45.99	0.286	0.012	98.6

Problem 5

Consider a binary liquid solution of acetone (species 1) and water (species 2) at 45 °C. Plot γ_1 and γ_2 as functions of liquid composition x_1. Use Wilson model to calculate γ_1 and γ_2. Determine the limiting values of γ_1 and γ_2: $\gamma_1\ (x_1 = 0), \gamma_1\ (x_1 = 1.0), \gamma_2\ (x_1 = 0)$, and $\gamma_2\ (x_1 = 1.0)$.

Given that:

$v_1 = 74.05$ cm^3/mol ; $v_2 = 18.07$ cm^3/mol ; $a_{12} = 291.27$ cal/mol ; $a_{21} = 1448.01$ cal/mol

Problem 6

Consider a binary system of methanol (species 1) and water (species 2). The vapor pressures of the pure components are given by the following Antoine equations:

$$lnP_1^{sat} = 16.585 - \frac{3638.27}{T + 239.50}$$

$$lnP_2^{sat} = 16.2620 - \frac{3799.89}{T + 226.35}$$

where P is in kPa and T is in °C. The activity coefficients of the components in liquid solution are given by the Wilson model (Equations (12.101) - (12.104)). The following data are given regarding the Wilson model:

$v_1 = 40.73$ cm^3/mol ; $v_2 = 18.07$ cm^3/mol ; $a_{12} = 107.38$ cal/mol ; $a_{21} = 469.55$ cal/mol

For methanol (1)/water (2) system,

Chemical	ω	T_c, K	P_c, bar	Z_C	v_c, cm^3/mol
methanol (1)	0.564	512.6	80.97	0.224	118
water (2)	0.345	647.1	220.55	0.229	55.9

Do the following:

a) Given T=332 K, $x_1 = 0.50$, $x_2 = 0.50$, estimate the bubble point pressure and composition of bubble
b) Given T=332 K, $y_1 = 0.50$, $y_2 = 0.50$, estimate the dew point pressure and composition of liquid drop
c) Given P=125 kPa, $x_1 = 0.50$, $x_2 = 0.50$, estimate the bubble point temperature and composition of bubble
d) Given P=125 kPa, $y_1 = 0.50$, $y_2 = 0.50$, estimate the dew point temperature and composition of liquid drop

Do not assume the vapor mixture to be ideal.

Problem 7

A flash chamber operating at 80 °C and 101 kPa is separating a binary mixture of methanol (species 1) and water (species 2). The feed composition is $z_1 = 0.45$ and $z_2 = 0.55$.

a) Would flashing occur at given temperature and pressure?
b) If flashing occurs, what is the fraction of feed that goes to the vapor distillate product and what are the compositions of the vapor and liquid products?

Due to low pressure, the vapor mixture can be treated as an ideal gas mixture. However, the liquid solution is non-ideal in nature.

$$lnP_1^{sat} = 16.585 - \frac{3638.27}{T + 239.50}$$
$$lnP_2^{sat} = 16.2620 - \frac{3799.89}{T + 226.35}$$

where P is in kPa and T is in °C. The activity coefficients of the components in liquid solution are given by the Wilson model (Equations (12.101) - (12.104)). The following data are given regarding the Wilson model:

$v_1 = 40.73$ cm³/mol ; $v_2 = 18.07$ cm³/mol ; $a_{12} = 107.38$ cal/mol ; $a_{21} = 469.55$ cal/mol

Problem 8

Consider a binary system of methanol (species 1) and benzene (species 2) at 58.3 °C. The vapor mixture is ideal whereas the liquid solution is non-ideal. The following form of the Raoult's law is applicable:

$$y_i P = x_i \gamma_i P_i^{sat}$$

Construct the P-x-y diagram and y-x equilibrium curve. The vapor pressures of the pure components are given by the following Antoine equations:

$$lnP_1^{sat} = 16.585 - \frac{3638.27}{T + 239.50}$$
$$lnP_2^{sat} = 13.8594 - \frac{2773.78}{T + 220.07}$$

where P is in kPa and T is in °C. The activity coefficients of the components in liquid solution are given by the Wilson model (Equations (12.101) - (12.104)). The following data are given regarding the Wilson model:

$v_1 = 40.73$ cm³/mol ; $v_2 = 89.41$ cm³/mol ; $a_{12} = 1734.42$ cal/mol ; $a_{21} = 183.04$ cal/mol

Problem 9

Consider a binary system of methanol (species 1) and benzene (species 2) at 1 bar. The vapor mixture is ideal whereas the liquid solution is non-ideal. The following form of the Raoult's law is applicable:

$$y_i P = x_i \gamma_i P_i^{sat}$$

Construct the T-x-y diagram and y-x equilibrium curve at 1 bar. The vapor pressures of the pure components are given by the following Antoine equations:

$$lnP_1^{sat} = 16.585 - \frac{3638.27}{T + 239.50}$$
$$lnP_2^{sat} = 13.8594 - \frac{2773.78}{T + 220.07}$$

where P is in kPa and T is in °C. The activity coefficients of the components in liquid solution are given by the Wilson model (Equations (12.101) - (12.104)). The following data are given regarding the Wilson model:

$v_1 = 40.73$ cm³/mol ; $v_2 = 89.41$ cm³/mol ; $a_{12} = 1734.42$ cal/mol ; $a_{21} = 183.04$ cal/mol

Problem 10

Consider a binary system of ethanol (species 1) and toluene (species 2). The vapor pressures of the pure components are given by the following Antoine equations:

$$lnP_1^{sat} = 16.6758 - \frac{3674.49}{T + 226.45}$$
$$lnP_2^{sat} = 14.0098 - \frac{3103.01}{T + 219.79}$$

where P is in kPa and T is in °C. The activity coefficients of the components in liquid solution are given by the Wilson model (Equations (12.101) - (12.104)). The following data are given regarding the Wilson model:

$v_1 = 58.68$ cm³/mol ; $v_2 = 106.85$ cm³/mol ; $a_{12} = 1556.45$ cal/mol ; $a_{21} = 210.52$ cal/mol

a) Determine if an azeotrope exists for this system at 76.7 °C.
b) If the answer to part (a) is yes, what is the composition of the azeotrope? Also determine the pressure at which this azeotrope is formed.

Chapter 13

Chemical Reaction Equilibrium

13.1 LEARNING OUTCOMES

Upon completion of this chapter, students will be able to:

- Define important terms such as: mole of a reaction, extent of a reaction, standard heat of a reaction, standard heat of combustion, standard states, standard heat of formation of a compound, standard free energy change of a reaction, standard free energy of formation of a compound.
- Develop thermodynamic criterion for reaction equilibrium
- Analyze reaction equilibrium in a mixture of ideal gases. Express equilibrium constant in terms of mole fractions of species present in the gas mixture
- Analyze reaction equilibrium in an ideal liquid solution. Express equilibrium constant in terms of mole fractions of species present in the liquid solution
- Analyze reaction equilibrium in non-ideal systems such as non-ideal gas-phase reactions and non-ideal liquid-phase reactions. Define activities of species in multicomponent gas phase and liquid solution. Express equilibrium constant in terms of activities of species.
- Discuss the significance of standard Gibbs free energy of reaction
- Analyze reaction equilibrium in heterogeneous systems. Define activities of pure liquid and pure solid species. Develop expressions for equilibrium constant of heterogeneous systems
- Describe the significance of coupled reactions and membrane reactors
- Analyze multi-reaction equilibria
- Carry out thermodynamic analysis of fuel cells. Develop the expression for emf of a fuel cell
- Develop the expression for entropy production in chemical reactions
- Discuss the limitations of thermodynamics in dealing with chemical reaction systems
- Solve a variety of problems dealing with chemical reaction equilibrium

13.2 INTRODUCTION

Chemical reactions play a very important role in our lives. They are used:

- to transform raw materials into more valuable products.
- to generate heat energy as in combustion reactions.
- to produce a large number of chemicals that are utilized in the manufacturing of fertilizers, plastics, paints, detergents, pesticides, medicines, etc.

In general, the chemical reactions are *reversible* in nature meaning that the reactants react to generate the products and the products react simultaneously to generate the reactants. Thus, the reactants and products are never fully consumed. As an example, consider the following reversible reaction:

$$aA + bB \rightleftharpoons cC + dD \tag{13.1}$$

In the forward direction,

$$aA + bB \rightarrow cC + dD \tag{13.2}$$

In the reverse or backward direction,

$$cC + dD \rightarrow aA + bB \qquad (13.3)$$

These two reactions occur simultaneously. At the beginning of the reaction, only the forward reaction, Equation (13.2), occurs as there are no products (C and D) present. As soon as the products are formed, the reverse reaction, Equation (13.3), kicks in to produce the reactants (A and B). The rate of forward reaction gradually decreases while the rate of reverse reaction increases, as shown schematically in Figure 13.1. The rate of forward reaction is eventually balanced by the rate of backward reaction to reach a state of *equilibrium*. The concentrations of different chemical species also change with time as the reaction proceeds towards equilibrium. The concentration of the species A and B decrease while the concentrations of the species C and D increase. At equilibrium, the concentrations of all the species become constant.

Figure 13.2 shows the variation of the net rate of reaction in the forward direction with time. The net rate of reaction is the rate of reaction in the forward direction minus the rate of reaction in the reverse direction. The net rate of reaction decreases with time and becomes zero at equilibrium.

In this chapter, we will learn how to calculate the equilibrium conversion of reactions and how pressure, temperature, and initial composition affect the equilibrium conversion and composition.

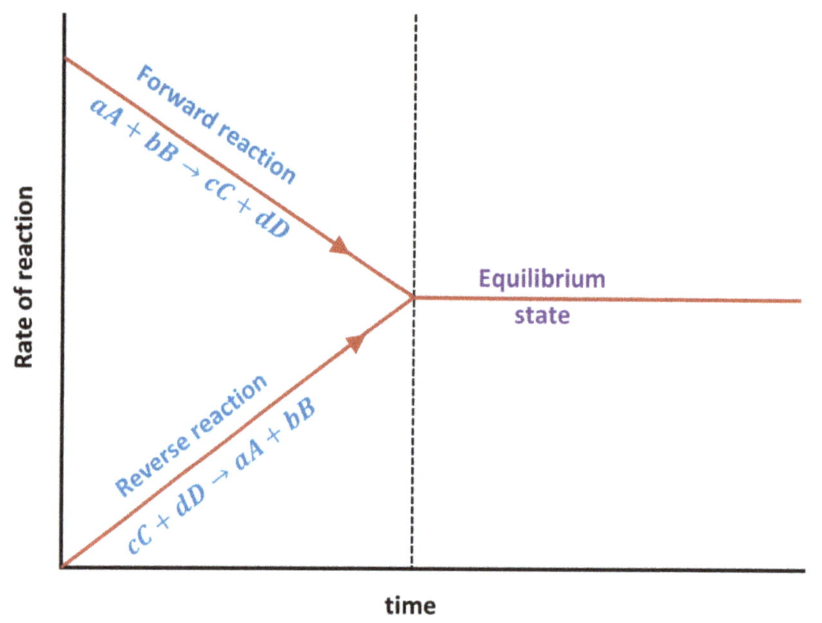

Figure 13.1 Schematic diagram showing the variation of rates of forward and backward reactions with time.

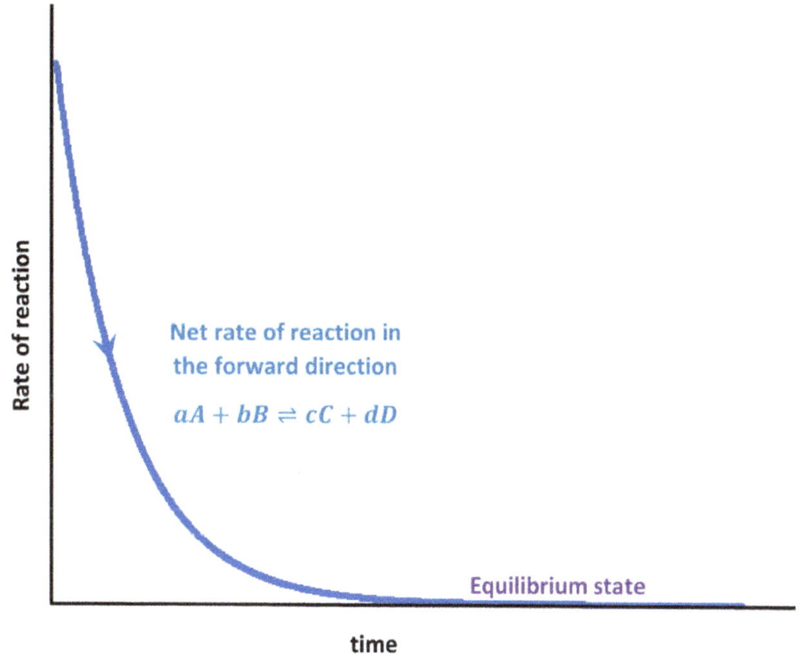

Figure 13.2 Variation of net rate of reaction in the forward direction with time.

13.3 SOME DEFINITIONS

We can represent a chemical reaction in the following condensed notation:

$$\sum_i v_i A_i = 0 \tag{13.4}$$

where v_i's are stoichiometric coefficients and A_i's are various chemical species involved in the reaction. The stoichiometric coefficient v_i is positive for products and negative for reactants. For the reaction given in Equation (13.1): $A_1 = A, v_1 = -a, A_2 = B, v_2 = -b, A_3 = C, v_3 = c, A_4 = D, v_4 = d$.

13.3.1 One Mole of a Reaction

Referring to reaction given in Equation (13.1), one mole of a reaction means "a" moles of species A react with "b" moles of species B to form "c" moles of species C and "d" moles of species D.
The number of moles of each reactant and product changes by stoichiometric (v_i) amount. We can also say that when one mole of a reaction occurs: $n_i - n_i^o = v_i$, where n_i^o is the initial number of moles of species i and n_i is the number of moles of species i after the reaction.

13.3.2 Extent of a Reaction

The extent of a reaction, denoted by ξ, is a quantity that measures the extent to which the reaction proceeds in terms of moles of reaction. Referring to reaction given in Equation (13.1), ξ moles of a reaction means "$a\xi$" moles of species A react with "$b\xi$" moles of species B to form "$c\xi$" moles of species C and "$d\xi$" moles of species D. The number of moles of each reactant and product changes by $v_i\xi$ amount. We can also define the *extent* or *degree of advancement* of a reaction as:

$$\xi = \frac{n_i - n_i^o}{v_i} \tag{13.5}$$

where n_i is the number of moles of species i at any time t and n_i^o is the initial number of moles of species i. Note that the extent of a reaction ξ is the same for each chemical species involved in a reaction.

13.3.3 Standard Heat of Reaction

The standard heat of a reaction at any temperature T, denoted as $\Delta H_{rxn}^o(T)$, is defined as the enthalpy change that results when the reactants, in their standard states at temperature T, react in stoichiometric amounts to form the products in their standard states at temperature T. It can be expressed as:

$$\Delta H_{rxn}^o(T) = \sum_i v_i\, h_i^o(T) \tag{13.6}$$

where $h_i^o(T)$ is the standard molar enthalpy of species i (reactant or product) at temperature T .

13.3.4 Standard Heat of Combustion

The standard heat of *combustion* of a compound (or an element) at any temperature T is defined as the enthalpy change that results when 1 mole of a compound, in its standard state, is burnt with oxygen (in its standard state) to form specified combustion products (in their standard states). For hydrocarbons (composed of C, H, O), the combustion products are CO_2 and H_2O.

Examples of combustion reactions are:

$$CH_4(g) + 2O_2(g) \rightarrow CO_2(g) + 2H_2O(g) \tag{13.7a}$$

$$C_3H_8(g) + 5O_2(g) \rightarrow 3CO_2(g) + 4H_2O(g) \tag{13.7b}$$

13.3.5 Standard States

- For *gases*, the standard state is a pure gas in the ideal-gas state at the given temperature T and pressure of 1 bar.
- For *liquids and solids*, the standard state is a pure liquid or pure solid at the given temperature T and pressure of 1 bar.

13.3.6 Standard Heat of Formation of a Compound

The standard heat of formation of a compound at any temperature T, denoted as $\Delta h_f^o(T)$, is defined as the enthalpy change that results when one mole of a compound, in its standard state, is formed from its elements in their standard states.

Examples of formation reactions are:

$$\frac{1}{2}N_2(g) + \frac{3}{2}H_2(g) \rightarrow NH_3(g) \tag{13.8a}$$

$$H_2(g) + \frac{1}{2}O_2(g) \rightarrow H_2O(l) \tag{13.8b}$$

A single compound is formed from its elements in a formation reaction. For the formation reactions given in Equations (13.8a) and (13.8b), the standard heats of formation of the compounds involved are as follows:

$$\Delta h_f^o(NH_3, g) = h^o(NH_3, g) - \frac{1}{2}h^o(N_2, g) - \frac{3}{2}h^o(H_2, g) \tag{13.9}$$

$$\Delta h_f^o(H_2O,l) = h^o(H_2O,l) - \frac{1}{2}h^o(O_2,g) - h^o(H_2,g) \tag{13.10}$$

The standard-state enthalpies of all elements in their stable states of aggregation are arbitrarily set equal to zero at 298.15 K. Consequently, the standard enthalpy of a compound is simply its standard heat of formation at 298.15 K, that is:

$$h_i^o(\text{compound}) = \Delta h_{fi}^o(\text{compound}) \tag{13.11}$$

Also note that the standard enthalpies of formation of all elements in their stable states of aggregation are zero at 298.15 K.

Interestingly, the standard heat of any reaction depends only on the standard heats of formation of compounds involved in the reaction:

$$\Delta H_{rxn}^o = \textstyle\sum_i v_i\, h_i^o = \sum_i v_i\, \Delta h_{fi}^o \text{ (compounds)} \tag{13.12}$$

This is because the sum of enthalpies of the elements composing the products is equal to the sum of enthalpies of the elements composing the reactants. Consequently, the enthalpies of the elements cancel out. The independence of ΔH_{rxn}^o of standard-state enthalpies of elements is valid at all temperatures. Thus, the standard heat of any reaction can be calculated from the standard heats of formation of compounds involved in the reaction.

13.3.7 Standard Gibbs Free Energy Change of Reaction

The standard Gibbs free energy change of a reaction at any temperature T, denoted as $\Delta G_{rxn}^o(T)$, is defined as the Gibbs free energy change that results when the reactants, in their standard states at temperature T, react in stoichiometric amounts to form the products in their standard states at temperature T. It can be expressed as:

$$\Delta G_{rxn}^o(T) = \textstyle\sum_i v_i\, g_i^o(T) \tag{13.13}$$

where $g_i^o(T)$ is the standard molar Gibbs free energy of species i (reactant or product) at temperature T .

13.3.8 Standard Gibbs Free Energy of Formation of a Compound

The standard Gibbs free energy of formation of a compound at any temperature T, denoted as $\Delta g_f^o(T)$, is defined as the Gibbs free energy change that results when one mole of a compound, in its standard state, is formed from its elements in their standard states.

The standard-state Gibbs free energies of all elements in their stable states of aggregation are arbitrarily set equal to zero at 298.15 K. Consequently, the standard Gibbs free energy of a compound is simply its standard Gibbs free energy of formation at 298.15 K, that is:

$$g_i^o(\text{compound}) = \Delta g_{fi}^o(\text{compound}) \tag{13.14}$$

Also note that the standard Gibbs free energies of formation of all elements in their stable states of aggregation are zero at 298.15 K.

Like the standard heat of reaction, the Gibbs free energy of any reaction depends only on the standard Gibbs free energies of formation of compounds involved in the reaction:

$$\Delta G^o_{rxn} = \sum_i v_i\, g^o_i = \sum_i v_i\, \Delta g^o_{fi} \text{ (compounds)} \quad (13.15)$$

Thus, the standard Gibbs free energy change of any reaction can be calculated from the standard Gibbs free energies of formation of compounds involved in the reaction.

13.4 THERMODYNAMIC CRITERION OF CHEMICAL REACTION EQUILIBRIUM

For a closed system at constant T and P, the condition of thermodynamic equilibrium is:

$$(dG)_{T,P} = 0 \quad (13.16)$$

Consider a chemical reaction taking place in a closed system at constant T and P. Let n_i be the number of moles of species i at any time t, n^o_i is the initial number of moles of species i, and ξ be the extent of reaction at any time t. Thus,

$$n_i = n^o_i + v_i\xi \quad (13.17)$$

For systems of variable composition, the differential changes in the Gibbs free energy at constant T and P is given as:

$$(dG)_{T,P} = \sum_i \mu_i\, dn_i \quad (13.18)$$

From Equations (13.17) and (13.18), we get:

$$(dG)_{T,P} = \sum_i \mu_i v_i\, d\xi \quad (13.19)$$

Upon using the condition of thermodynamic equilibrium, Equation (13.16), Equation (13.19) yields:

$$\sum_i \mu_i v_i = 0 \quad (13.20)$$

Equation (13.20) is the thermodynamic criterion for chemical reaction equilibrium.

13.5 CHEMICAL REACTION EQUILIBRIUM IN A MIXTURE OF IDEAL GASES

Consider a closed reaction system undergoing the following reaction at constant T and P. All the

$$\sum_i v_i\, A_i = 0 \quad (13.4)$$

species involved ($A_i's$) are ideal gases. Thus, the reaction system is an ideal gas mixture of various ideal gas species $A_i's$.

The chemical potential of an ideal gas species i in an ideal gas mixture is given as:

$$\mu^{ig}_{i,mix}(T,P) = \mu^{ig}_{i,pure}(T,P) + RT ln y_i \quad (13.21)$$

The effect of pressure on the chemical potential of a pure ideal gas is given as:

$$\mu^{ig}_{i,pure}(T,P) = \mu^{ig}_{i,pure}(T,P^o) + RT ln\left(\frac{P}{P^o}\right) \quad (13.22)$$

where P^o is the standard pressure (=1 bar). Combining Equations (13.21) and (13.22), we get:

$$\mu_{i,mix}^{ig}(T,P) = \mu_i^o + RTln\left(\frac{y_iP}{P^o}\right) = \mu_i^o + RTln\left(\frac{P_i}{P^o}\right) \tag{13.23}$$

where μ_i^o is the standard chemical potential of species i and P_i is the partial pressure of species i in the reaction mixture (ideal gas mixture).

We now apply the thermodynamic criterion for chemical reaction equilibrium, Equation (13.20), to the reaction system. Thus,

$$\sum_i \mu_i v_i = \sum_i v_i \left[\mu_i^o + RTln\left(\frac{P_i}{P^o}\right)_e\right] = 0 \tag{13.24}$$

$$\sum_i v_i\mu_i^o + RT\sum_i v_i \left[ln\left(\frac{P_i}{P^o}\right)_e\right] = 0 \tag{13.25}$$

$$\Delta G^o + RT\sum_i \left[ln\left(\frac{P_i}{P^o}\right)_e^{v_i}\right] = 0 \tag{13.26}$$

where subscript e indicates equilibrium value and $\Delta G^o = \sum_i v_i\mu_i^o$ is the standard Gibbs free energy change of the reaction, see Equation (13.13). Note that the subscript *rxn* is removed from ΔG_{rxn}^o and $g_i^o = \mu_i^o$. Equation (13.26) can also be written as:

$$\Delta G^o = -RTln\prod_i \left(\frac{P_i}{P^o}\right)_e^{v_i} \tag{13.27}$$

where $\prod$ is the product sign defined as:

$$\prod_i \left(\frac{P_i}{P^o}\right)_e^{v_i} = \left(\frac{P_1}{P^o}\right)_e^{v_1}\left(\frac{P_2}{P^o}\right)_e^{v_2}\left(\frac{P_3}{P^o}\right)_e^{v_3} \dots\dots\dots \tag{13.28}$$

The *equilibrium constant* for a chemical reaction equilibrium in a mixture of ideal gases is denoted as K_p and defined as:

$$K_p = \prod_i \left(\frac{P_i}{P^o}\right)_e^{v_i} \tag{13.29}$$

Thus, Equation (13.27) becomes:

$$\Delta G^o = -RTlnK_p \tag{13.30}$$

This is the key relation for a chemical reaction equilibrium in a mixture of ideal gases. As ΔG^o is a function of T only, the equilibrium constant K_p is a function of temperature only. The equilibrium constant could also be expressed in terms of the composition variables y_i's as follows:

$$K_p = \prod_i (y_i)_e^{v_i}\prod_i \left(\frac{P}{P^o}\right)_e^{v_i} = \left(\frac{P}{P^o}\right)_e^{\sum_i v_i}\prod_i (y_i)_e^{v_i} \tag{13.31}$$

As an example, consider the following reaction:

$$aA + bB \rightleftharpoons cC + dD \tag{13.1}$$

The equilibrium constant for this reaction is:

$$K_p = \prod_i \left(\frac{P_i}{P^o}\right)_e^{v_i} = \frac{\left(\frac{P_C}{P^o}\right)_e^c\left(\frac{P_D}{P^o}\right)_e^d}{\left(\frac{P_A}{P^o}\right)_e^a\left(\frac{P_B}{P^o}\right)_e^b} = \left(\frac{P}{P^o}\right)_e^{(c+d-a-b)}\frac{(y_C)_e^c(y_D)_e^d}{(y_A)_e^a(y_B)_e^b} \tag{13.32}$$

13.5.1 Effect of Temperature on Equilibrium Constant

The temperature-dependence of equilibrium constant K_p can be determined using Equation (13.30) as:

$$lnK_p = -\frac{\Delta G^o}{RT} \tag{13.33}$$

Taking the derivative of Equation (13.33) with respect of temperature,

$$\frac{d(lnK_p)}{dT} = -\frac{1}{R}\frac{d(\Delta G^o/T)}{dT} \tag{13.34}$$

The standard Gibbs free energy change of a reaction is:

$$\Delta G^o_{rxn}(T) = \sum_i v_i\, g^o_i(T) \tag{13.13}$$

From Equation (13.13), we get:

$$\frac{d(\Delta G^o/T)}{dT} = \sum_i v_i \frac{d(g^o_i/T)}{dT} \tag{13.35}$$

Note the subscript *rxn* has been removed from G^o_{rxn}. The derivative on the right hand side of Equation (13.35) can be expressed as:

$$\frac{d(g^o_i/T)}{dT} = -\frac{1}{T^2}\left(g^o_i - T\frac{dg^o_i}{dT}\right) \tag{13.36}$$

For pure substances, the derivative of g^o_i is:

$$dg^o_i = -s^o_i dT \tag{13.37}$$

where s^o_i is standard molar entropy of pure species i. From Equations (13.36) and (13.37), we get:

$$\frac{d(g^o_i/T)}{dT} = -\frac{1}{T^2}(g^o_i + Ts^o_i) \tag{13.38}$$

Using the definition of g^o_i, that is, $g^o_i = h^o_i - Ts^o_i$, Equation (13.38) can be re-written as:

$$\frac{d(g^o_i/T)}{dT} = -\frac{h^o_i}{T^2} \tag{13.39}$$

Upon substitution of this derivative into Equation (13.35), we obtain:

$$\frac{d(\Delta G^o/T)}{dT} = -\sum_i \frac{v_i h^o_i}{T^2} = -\frac{\Delta H^o}{T^2} \tag{13.40}$$

where ΔH^o is standard heat of reaction. From Equations (13.34) and (13.40),

$$\frac{d(lnK_p)}{dT} = \frac{\Delta H^o}{RT^2} \tag{13.41}$$

Upon integration, Equation (13.41) gives:

$$lnK_p = ln(K_p)_o - \frac{\Delta H^o}{R}\left(\frac{1}{T} - \frac{1}{T_o}\right) \tag{13.42}$$

Note that we have assumed that ΔH^o is independent of temperature. According to Equation (13.42), when the reaction is *exothermic* (ΔH^o is negative), the equilibrium constant decreases with the increase in temperature. For *endothermic* reaction (ΔH^o is positive), the equilibrium constant increases with the increase in temperature.

13.6 CHEMICAL REACTION EQUILIBRIUM IN AN IDEAL LIQUID SOLUTION

Consider a closed reaction system undergoing the following reaction at constant T and P. All the

$$\sum_i v_i A_i = 0 \tag{13.4}$$

species involved ($A_i's$) are liquid species. Thus, the reaction system is a liquid solution of various liquid species $A_i's$. We further assume that the reaction system is an *ideal liquid solution.*

Recall from Chapter 12 (see Equation (12.91)), the following relation that defines an *ideal liquid solution*:

$$\mu_{i,solu}^{id}(T,P) = \mu_{i,pure}^{L}(T,P) + RTln(x_i) \tag{13.43}$$

Neglecting the effect of pressure on the chemical potential of pure liquid,

$$\mu_{i,pure}^{L}(T,P) \approx \mu_i^o(T) \tag{13.44}$$

where μ_i^o is the standard chemical potential of species i (pure liquid at standard pressure of 1 bar and temperature T). Thus, Equation (13.43) becomes:

$$\mu_{i,solu}^{id}(T,P) = \mu_i^o(T) + RTln(x_i) \tag{13.45}$$

We now apply the thermodynamic criterion for chemical reaction equilibrium, Equation (13.20), to the reaction system. Thus,

$$\sum_i \mu_i v_i = \sum_i v_i \left[\mu_i^o + RTln(x_i)_e\right] = 0 \tag{13.46}$$

$$\sum_i v_i \mu_i^o + RT\sum_i v_i[ln(x_i)_e] = 0 \tag{13.47}$$

$$\Delta G^o + RT\sum_i\left[ln(x_i)_e^{v_i}\right] = 0 \tag{13.48}$$

where subscript e indicates equilibrium value and $\Delta G^o = \sum_i v_i \mu_i^o$ is the standard Gibbs free energy change of the reaction. Equation (13.48) can also be written as:

$$\Delta G^o = -RTln\prod_i (x_i)_e^{v_i} \tag{13.49}$$

The *equilibrium constant* for a chemical reaction equilibrium in a liquid solution can be denoted as K and defined as:

$$K = \prod_i (x_i)_e^{v_i} \tag{13.50}$$

Thus, Equation (13.48) becomes:

$$\Delta G^o = -RTlnK \tag{13.51}$$

This is the key relation for a chemical reaction equilibrium in a liquid solution. Assuming ideal liquid solution, the equilibrium constant K is given by Equation (13.50). Like K_p, K is a function of temperature only. The effect of temperature on K is given by Equation (13.42). Thus,

$$lnK = lnK_o - \frac{\Delta H^o}{R}\left(\frac{1}{T} - \frac{1}{T_o}\right) \tag{13.52}$$

13.7 CHEMICAL REACTION EQUILIBRIUM IN NON-IDEAL SYSTEMS

Consider a closed reaction system undergoing the reaction specified by Equation (13.4) at constant T and P. The reaction system is *non-ideal* in nature. It could be a non-ideal gaseous mixture of reactants and products or could be a non-ideal liquid solution of reactants and products in liquid forms.

Recall from Chapter 12, the defining equation for the fugacity of a species in a non-ideal system (gaseous mixture or liquid solution) is as follows:

$$d\mu_{i,mix} = RTdln\hat{f}_i \tag{13.53}$$

where $\hat{f}_i$ is the fugacity of species i in the mixture/solution at T and P and $\mu_{i,mix}$ is the chemical potential of species i in the mixture/solution at T and P. Upon integration of Equation (13.53) from species i in its standard state to species i in the mixture/solution, we get:

$$\mu_{i,mix} - \mu_i^o = RTln\frac{\hat{f}_i}{f_i^o} \tag{13.54}$$

where f_i^o is the standard fugacity of species i.

The activity of species i in the mixture/solution is defined as:

$$\hat{a}_i = \frac{\hat{f}_i}{f_i^o} \tag{13.55}$$

Thus, Equation (13.54) becomes

$$\mu_{i,mix} = \mu_i^o + RTln\hat{a}_i \tag{13.56}$$

We now apply the thermodynamic criterion for chemical reaction equilibrium, Equation (13.20), to the reaction system. Thus,

$$\textstyle\sum_i \mu_i v_i = \sum_i v_i\,[\mu_i^o + RTln(\hat{a}_i)_e] = 0 \tag{13.57}$$

$$\textstyle\sum_i v_i\mu_i^o + RT\sum_i v_i[ln(\hat{a}_i)_e] = 0 \tag{13.58}$$

$$\Delta G^o + RT\textstyle\sum_i\left[ln(\hat{a}_i)_e^{v_i}\right] = 0 \tag{13.59}$$

where subscript e indicates equilibrium value and $\Delta G^o = \sum_i v_i\mu_i^o$ is the standard Gibbs free energy change of the reaction. Equation (13.59) can also be written as:

$$\Delta G^o = -RTln\textstyle\prod_i(\hat{a}_i)_e^{v_i} \tag{13.60}$$

The *equilibrium constant* for a chemical reaction equilibrium in a non-ideal system can be denoted as K and defined as:

$$K = \prod_i (\hat{a}_i)_e^{v_i} \tag{13.61}$$

Thus, Equation (13.61) becomes:

$$\Delta G^o = -RTlnK \tag{13.51}$$

Once again, the equilibrium constant, defined in Equation (13.61), is a function of temperature only and the temperature dependence is given by Equation (13.52).

13.7.1 Gas Phase Reactions

The activity of species i in a gas mixture is as follows:

$$\hat{a}_i = \frac{\hat{f}_i}{f_i^o} = \frac{\hat{\varphi}_i y_i P}{P^o} \tag{13.62}$$

where $\hat{\varphi}_i$ is the fugacity coefficient of species i in a gas mixture. Note that the standard state of a gaseous species is pure gaseous species in ideal gas state at 1 bar and system temperature T. Thus, $f_i^o = P^o = 1$ bar.

The equilibrium constant for a chemical reaction equilibrium in a mixture of non-ideal gases can be expressed as:

$$K = \prod_i \left(\frac{\hat{\varphi}_i y_i P}{P^o}\right)_e^{v_i} = \prod_i (\hat{\varphi}_i y_i)_e^{v_i} \prod_i \left(\frac{P}{P^o}\right)_e^{v_i} = \left(\frac{P}{P^o}\right)_e^{\sum_i v_i} \prod_i (\hat{\varphi}_i y_i)_e^{v_i} \tag{13.63}$$

In the special case of ideal gas mixture,

$$\hat{\varphi}_i = 1 \tag{13.64}$$

Consequently, the equilibrium constant expressed in Equation (13.63) becomes equal to K_p given in Equation (13.31).

13.7.2 Liquid Phase Reactions

The activity of species i in a liquid solution is as follows:

$$\hat{a}_i = \frac{\hat{f}_i}{f_i^o} = \frac{\gamma_i x_i f_i^L}{f_i^o} \tag{13.65}$$

where γ_i is the activity coefficient of species i in a liquid solution and f_i^L is the fugacity of pure liquid species i at system T and P. The standard state fugacity of species i (fugacity of pure liquid species i at system T and pressure of 1 bar) is approximately equal to f_i^L as the effect of pressure on liquid fugacity is negligible:

$$f_i^o \approx f_i^L \tag{13.66}$$

Consequently,

$$\hat{a}_i = \gamma_i x_i \tag{13.67}$$

The equilibrium constant for a chemical reaction equilibrium in a non-ideal liquid solution can be expressed as:

$$K = \prod_i (\hat{a}_i)_e^{v_i} = \prod_i (\gamma_i x_i)_e^{v_i} \tag{13.68}$$

In the special case of ideal liquid solution,

$$\gamma_i = 1 \tag{13.69}$$

Consequently, the equilibrium constant expressed in Equation (13.68) becomes equal to K given in Equation (13.50).

13.8 GENERAL COMMENTS ON ΔG^o AND EQUILIBRIUM CONSTANT K

The equilibrium constant is related to the standard Gibbs free energy of reaction through the following relation:

$$lnK = -\Delta G^o/RT \tag{13.70}$$

Based on this relation, we can draw the following conclusions about the thermodynamics of the forward and reverse reactions:

- a large negative ΔG^o value gives a large value of equilibrium constant K ($K \gg 1$). In this case, the reaction in the forward direction is highly favored over the reaction in the reverse direction. Consequently, the products predominate at equilibrium and the concentration of products is much larger than the concentration of reactants at equilibrium. The reaction nearly goes to completion, that is, most of the reactants are used up to form the products.
- a large positive ΔG^o value gives a negligibly small value of equilibrium constant K ($K \ll 1$). In this case, the reaction in the reverse direction is highly favored over the reaction in the forward direction. Consequently, the reactants predominate at equilibrium and the concentration of reactants is much larger than the concentration of products at equilibrium. The reaction does not occur to any significant extent, that is, most of the reactants remain unchanged and there are hardly any products formed.
- a zero ΔG^o value gives $K = 1$. In this case, the forward and reverse reactions are equally favored. Consequently, the reactants and products have nearly the same concentrations at equilibrium.

As an example, consider the following gas phase reactions occurring at $T = 25^oC$ with different values of ΔG^o (large negative, large positive, and zero):

$$A \rightleftharpoons B\ (\Delta G^o = -25{,}000 \text{ J/mol}) \tag{13.71}$$

$$C \rightleftharpoons D\ (\Delta G^o = +20{,}000 \text{ J/mol}) \tag{13.72}$$

$$E \rightleftharpoons F\ (\Delta G^o = 0 \text{ J/mol}) \tag{13.73}$$

Let the number of moles of reactant be 1 initially. Let x be the equilibrium conversion of reaction (fraction of reactant reacted at equilibrium). For the given reactions, $x = K/(1+K)$. The following table shows the percent conversion for the three reactions:

Table 13.1 Equilibrium constant and equilibrium conversion of example reactions

Reaction	ΔG^o, J/mol	K	x (%)
$A \rightleftharpoons B$	−25,000	28644.52	99.99
$C \rightleftharpoons D$	+20,000	$2.72\text{x}10^{-4}$	0.027
$E \rightleftharpoons F$	0	1.0	50

As expected, the conversion is almost complete at equilibrium when we have a large negative value of ΔG^o. The conversion is negligible when ΔG^o has a large positive value. When $\Delta G^o = 0$, we have a 50% conversion at equilibrium.

For a reaction occurring in a closed system at constant T and P, the thermodynamic criterion for spontaneous changes is:

$$(dG)_{T,P} < 0 \tag{13.74}$$

At equilibrium, $(dG)_{T,P} = 0$, as indicated in Equation (13.16).

Based on the spontaneity criterion given in Equation (13.74), it would appear that the reaction should not occur at all when $\Delta G^o \geq 0$. However, Table 13.1 suggests that the reaction does occur when $\Delta G^o \geq 0$. In fact, the conversion is 50% at equilibrium when $\Delta G^o = 0$. Does this mean that the thermodynamic criterion for spontaneity, expressed in the form of Equation (13.74), is violated by some of the reactions given in Table13.1? The answer is "No". According to the thermodynamic criterion for spontaneity, $\Delta G < 0$ for spontaneous changes and ΔG is not the same as ΔG^o. The standard Gibbs free energy change of a reaction, ΔG^o, assumes that the reaction occurs between the species in their pure form (standard states). In reality, the reaction occurs between the species in a mixture. Thus, ΔG^o neglects the "mixing effect" of reactants and products on the Gibbs free energy change of the reaction.

The total Gibbs free energy of a closed reaction system at given T and P can be expressed as:

$$G = \sum_i G_{i,pure} + \Delta G_{mix} \tag{13.75}$$

where $G_{i,pure}$ is the Gibbs free energy of species i (reactant or product) in its pure, unmixed state at given T and P and ΔG_{mix} is Gibbs free energy of mixing of reactants and products at given T and P. Note that ΔG_{mix} is always negative ($\Delta G_{mix} < 0$).

The total Gibbs free energy G of the reaction system (Equation (13.75)) is a function of the extent of reaction ξ. As reaction proceeds, both terms on the right hand side of Equation (13.75) change. According to the criterion for spontaneity (see Equation (13.74)), the total Gibbs free energy of the reaction system must decrease for the reaction to occur. Thus, the reaction continues to proceed as long as the following condition is met:

$$\left(\frac{\partial G}{\partial \xi}\right)_{T,P} = \left(\frac{\partial \sum_i G_{i,pure}}{\partial \xi}\right)_{T,P} + \left(\frac{\partial \Delta G_{mix}}{\partial \xi}\right)_{T,P} < 0 \tag{13.76}$$

When equilibrium is reached, $(\partial G/\partial \xi)_{T,P} = 0$.

When the products in their pure (unmixed) state have a higher Gibbs free energy than the reactants in their pure state, $\sum_i G_{i,pure\ products} > \sum_i G_{i,pure\ reactants}$ and $\left(\partial \sum_i G_{i,pure}/\partial \xi\right)_{T,P} > 0$. In this case, $\Delta G^o > 0$ but the reaction will still proceed to some extent due to mixing effect, that is, $(\partial \Delta G_{mix}/\partial \xi)_{T,P} < 0$. The extent of reaction ξ is small at equilibrium and the equilibrium position is closer to the reactant side as shown in Figure 13.3.

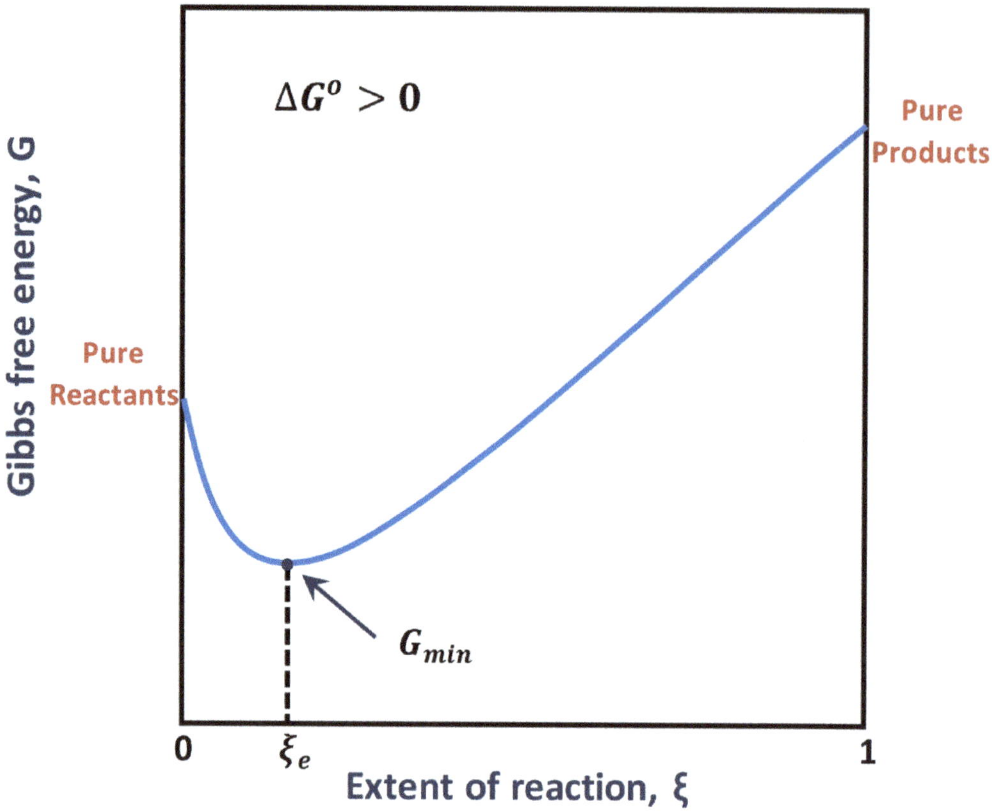

Figure 13.3 Variation of G of a reaction system with ξ when $\Delta G^o > 0$.

If the products and reactants in their pure (unmixed) states have the same Gibbs free energy, $\sum_i G_{i,pure\,products} = \sum_i G_{i,pure\,reactants}$ and $\left(\partial \sum_i G_{i,pure} / \partial \xi\right)_{T,P} = 0$. In this case, $\Delta G^o = 0$ but $(\partial \Delta G_{mix} / \partial \xi)_{T,P} < 0$ until some intermediate value of ξ (0.5 for binary mixture). Now the reaction is driven by the mixing effect and the extent of reaction ξ at equilibrium is close to some intermediate value (0.5 for binary mixture) as shown in Figure 13.4.

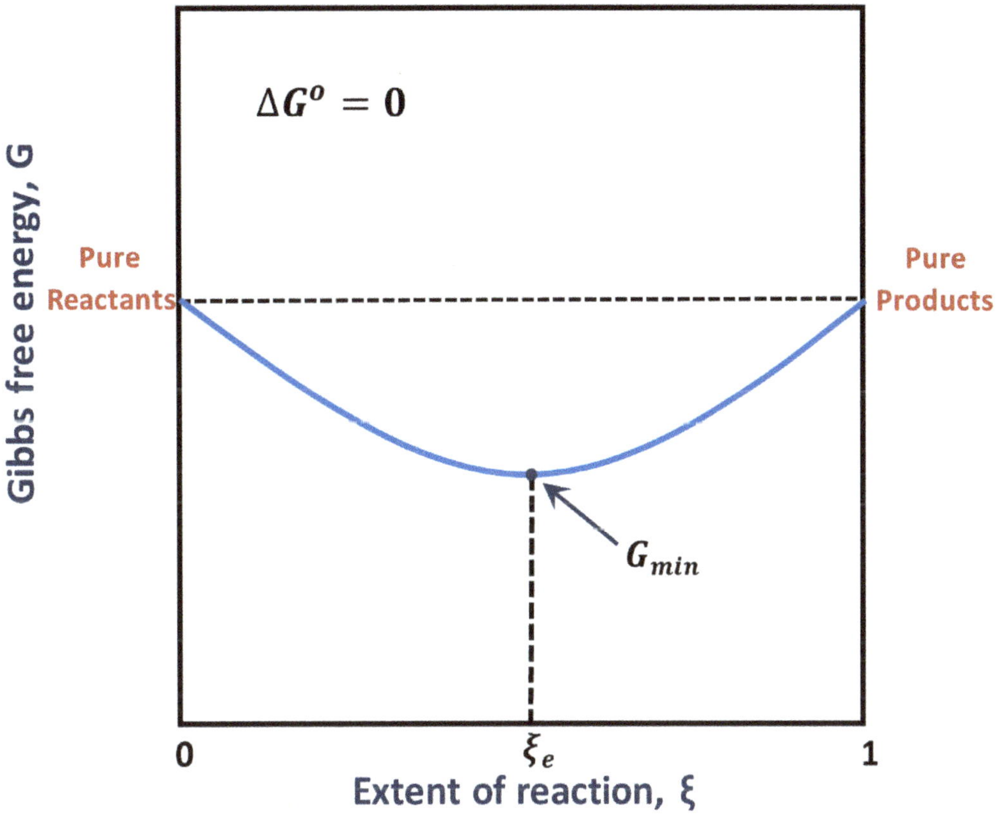

Figure 13.4 Variation of G of a reaction system with ξ when $\Delta G^o = 0$.

When the products in their pure (unmixed) state have a lower Gibbs free energy than the reactants in their pure state, $\sum_i G_{i,pure\ products} < \sum_i G_{i,pure\ reactants}$ and $\left(\partial \sum_i G_{i,pure} / \partial\xi\right)_{T,P} < 0$. In this case, $\Delta G^o < 0$ and both terms on the right hand side of Equation (13.76) are negative until some intermediate value of ξ (0.5 for binary mixture) is reached. While the mixing term becomes positive, $(\partial \Delta G_{mix}/\partial\xi)_{T,P} > 0$, at large values of ξ (ξ>0.5 for binary mixture), the other term $\left(\partial \sum_i G_{i,pure} / \partial\xi\right)_{T,P}$ continues to remain negative over the entire ξ range. Consequently, the extent of reaction ξ is large at equilibrium and the equilibrium position is closer to the product side as shown in Figure 13.5.

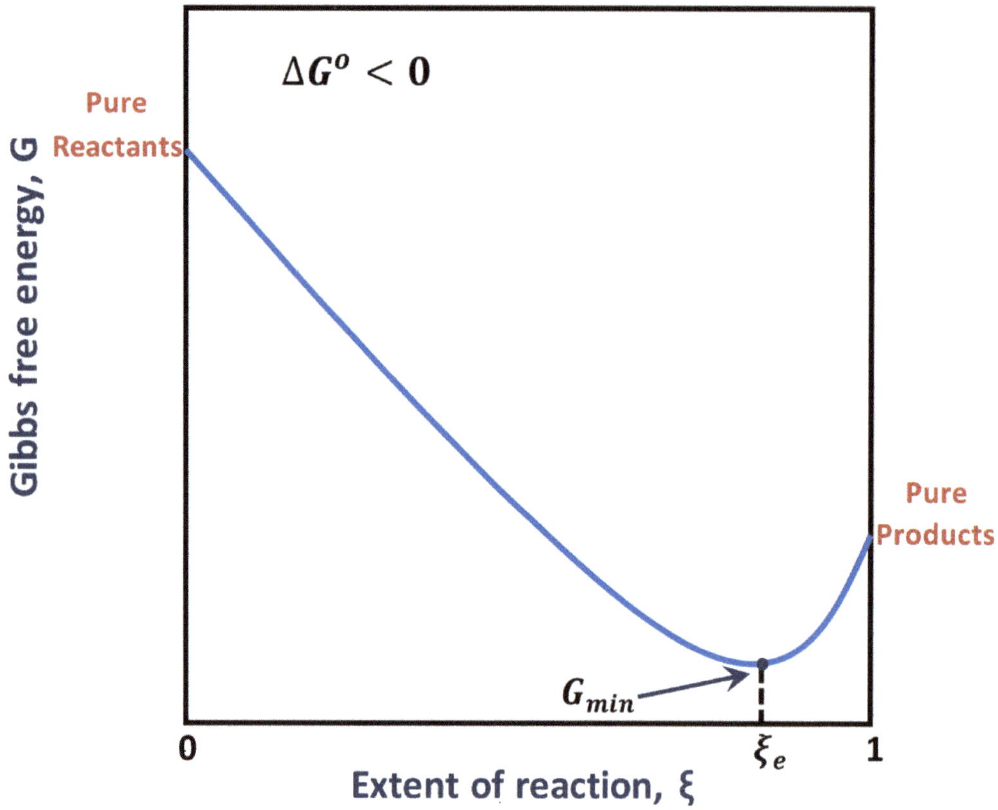

Figure 13.5 Variation of G of a reaction system with ξ when $\Delta G^o < 0$.

13.9 REACTIONS IN HETEROGENEOUS SYSTEMS

Thus far we have considered chemical reactions that occur in a single phase (gas or liquid). Such chemical reaction equilibrium systems are called *homogeneous equilibrium* systems. Now we will consider *heterogeneous* chemical reaction equilibrium systems where reactants, products, or both are in more than one phase. Some examples of heterogeneous chemical reactions are:

$$CO_2(g) + C(s) \rightleftharpoons 2CO(g) \tag{13.77}$$

$$Fe_3O_4(s) + 4H_2(g) \rightleftharpoons 3Fe(s) + 4H_2O(g) \tag{13.78}$$

$$CaCO_3(s) \rightleftharpoons CaO(s) + CO_2(g) \tag{13.79}$$

$$HgO(s) \rightleftharpoons Hg(l) + \frac{1}{2}O_2(g) \tag{13.80}$$

$$CH_4(g) \rightleftharpoons C(s) + 2H_2(g) \tag{13.81}$$

13.9.1 Activities of Pure Liquid and Pure Solid Species

The activities of reactant and products are needed in the calculation of the equilibrium constant for the reaction. The activities of pure liquid and pure solid species are given as:

$$a(l) = \frac{f^L(T,P)}{f^o} \quad (13.82)$$

$$a(s) = \frac{f^s(T,P)}{f^o} \quad (13.83)$$

where $f^L(T,P)$ is fugacity of pure liquid at given T and P, $f^s(T,P)$ is fugacity of pure solid at given T and P, and f^o is the fugacity of pure liquid or solid in its standard state. For pure liquids and solids, the effect of pressure on the fugacity can be neglected. Consequently,

$$f^L(T,P) \approx f^o(T,P^o) \quad (13.84)$$

$$f^s(T,P) \approx f^o(T,P^o) \quad (13.85)$$

where P^o is the standard pressure of 1 bar. From Equations (13.82) to (13.85), it follows that:

$$a(l) = 1 \quad (13.86)$$

$$a(s) = 1 \quad (13.87)$$

Thus, the activity of pure liquid or pure solid species is unity.

13.9.2 Equilibrium Constant for Heterogeneous Reaction Systems

Consider the following heterogeneous reaction:

$$A(s) + B(l) \rightleftharpoons C(s) + D(g) \quad (13.88)$$

The equilibrium constant for this reaction is (see Equation (13.61)):

$$K = \prod_i (\hat{a}_i)_e^{\nu_i} = \frac{(\hat{a}_C)_e(\hat{a}_D)_e}{(\hat{a}_A)_e(\hat{a}_B)_e} \quad (13.89)$$

We now assume that the gaseous, liquid, and solid species are mutually insoluble, that is, species A, B, C and D exist as pure species in the reaction system. Thus,

$$K = \frac{(a_C)_e(a_D)_e}{(a_A)_e(a_B)_e} = (a_D)_e \quad (13.90)$$

Note that for pure liquid and solid species (A, B, C), the activity is unity. The activity of pure gaseous species D is:

$$(a_D)_e = \frac{f_D}{f_D^o} = \frac{P}{P^o} \quad (13.91)$$

We have assumed that the gaseous species D is an ideal gas. Thus,

$$K = \frac{P}{P^o} \quad (13.92)$$

Consider another heterogeneous reaction:

$$A(g) \rightleftharpoons B(s) + C(g) \quad (13.93)$$

In this case, the equilibrium constant is:

$$K = \prod_i (\hat{a}_i)_e^{v_i} = \frac{(a_B)_e (\hat{a}_C)_e}{(\hat{a}_A)_e} = \frac{(\hat{a}_C)_e}{(\hat{a}_A)_e} \quad (13.94)$$

Note that the solid species B is assumed to be pure. Hence, $\hat{a}_B = a_B = 1$. The gaseous activities $\hat{a}_C$ and $\hat{a}_A$ can be expressed as:

$$\hat{a}_C = \frac{\hat{f}_C}{f_C^o} = \frac{y_C P}{P^o} \quad (13.95)$$

$$\hat{a}_A = \frac{\hat{f}_A}{f_A^o} = \frac{y_A P}{P^o} \quad (13.96)$$

Substituting these expressions of $\hat{a}_C$ and $\hat{a}_A$ into Equation (13.94), we get:

$$K = \frac{(\hat{a}_C)_e}{(\hat{a}_A)_e} = \frac{y_C}{y_A} \quad (13.97)$$

Finally, it should be noted that although the equilibrium constant involves only the gaseous species, the standard Gibbs free energy of reaction ΔG^o contains the standard Gibbs free energies of all the reactants and products, as indicated below for the reaction of Equation (13.93).

$$\Delta G^o = \sum_i v_i\, g_i^o = g_B^o(s) + g_C^o(g) - g_A^o(g) \quad (13.98)$$

13.10 COUPLED REACTIONS

Many reactions of practical interest have a large positive value of ΔG^o. As an example, consider the following reaction:

$$TiO_2(s) + 2Cl_2(g) \rightleftharpoons TiCl_4(l) + O_2(g) \qquad \Delta G^o(25^oC) = +153\ kJ/mol \quad (13.99)$$

This reaction produces titanium tetrachloride $TiCl_4$ from titanium dioxide TiO_2, a naturally occurring oxide of titanium. $TiCl_4$ is a dense colorless liquid that is used as an intermediate in the production of titanium metal and other titanium containing compounds. Since ΔG^o of this reaction has a large positive value of ΔG^o (+153 kJ/mol), only a very small amount of $TiCl_4$ is formed at equilibrium. The yield of $TiCl_4$ can be improved by increasing the temperature but the improvement is not enough to use this reaction for the production of $TiCl_4$.

One way to overcome the limitation imposed by a large positive value of ΔG^o of a reaction is to *couple* the given reaction, Equation (13.99), with another reaction with a large negative value of ΔG^o. The composite reaction should have a large negative value of ΔG^o so that it can proceed spontaneously with a good yield of the desired product ($TiCl_4$ in the present case).

In order to pull the Equation (13.99) reaction along, it is important that the coupling reaction uses one of the products of the Equation (13.99) reaction as its reactant. For example, the following combustion reaction that consumes oxygen (one of the products of the Equation (13.99) reaction) can be coupled with the Equation (13.99) reaction:

$$C(s) + O_2(g) \rightleftharpoons CO_2(g) \qquad \Delta G^o(25^oC) = -394.40\ kJ/mol \quad (13.100)$$

Note that this reaction has a larger ΔG^o in magnitude as compared with the ΔG^o of the reaction given by Equation (13.99). Table 13.2 shows the coupling of reaction given in Equation (13.99) with the reaction given in Equation (13.100). The composite reaction has a large negative value of ΔG^o. Hence it is highly spontaneous and gives a good yield of the desired product $TiCl_4$ at equilibrium.

Table 13.2 Coupling of reactions

Reaction	ΔG^o at 25°C, kJ/mol	Comments
$TiO_2(s) + 2Cl_2(g) \rightleftharpoons TiCl_4(l) + O_2(g)$	+153	Given reaction; desired product $TiCl_4$
$C(s) + O_2(g) \rightleftharpoons CO_2(g)$	−394.4	Coupling reaction
$C(s) + TiO_2(s) + 2Cl_2(g) \rightleftharpoons TiCl_4(l) + CO_2(g)$	−241.4	Composite reaction

In summary, two reactions can be coupled such that a nonspontaneous reaction ($\Delta G^o \gg 0$) is driven by a spontaneous reaction ($\Delta G^o \ll 0$). The product of a nonspontaneous reaction is a reactant of the spontaneous reaction. The overall composite reaction is spontaneous with a large negative value of ΔG^o.

13.11 MEMBRANE REACTORS

A membrane reactor is a device that combines a membrane separation process with a chemical reaction conversion process to achieve a much higher conversion of reactants into products than dictated by thermodynamic equilibrium.

Suppose we have the following gas-phase reaction with a low equilibrium conversion of say less than 5%:

$$A(g) \rightleftharpoons B(g) + C(g) \tag{13.101}$$

We can get a very high conversion of reactant into products if we remove one of the products (B or C) from the reaction mixture. The removal of B or C from the reaction mixture will cause more A to convert into products so as to achieve the equilibrium. The reaction continues to proceed in the forward direction (product side) as long as we continue to remove one of the products and the equilibrium is not established. In a membrane reactor, a membrane is placed around the reaction mixture to achieve the separation. The membrane is selected such that it is permeable to only one of the product species and impermeable to all other chemical species. Thus, the product species is removed from the reaction mixture by diffusion through the membrane.

As an example, consider the following dehydrogenation reaction for the production of butene (C_4H_8):

$$C_4H_{10}(g) \rightleftharpoons C_4H_8(g) + H_2(g) \tag{13.102}$$

In this reaction, one of the products, that is, H_2 is significantly smaller in size than the other molecules. The small molecules of H_2 can easily permeate through the membranes. Thus, this reaction which is thermodynamically unfavorable ($\Delta G^o = +87\ kJ/mol$ at 25°C) can be carried out in a membrane reactor with high conversion.

13.12 MULTIREACTION EQUILIBRIA

Thus far, we have generally considered only single reaction systems. In many reaction systems of practical importance, two or more independent reactions occur simultaneously. We are interested in calculating the composition and extents of different reactions at equilibrium.

13.12.1 Notation and Definitions

In a multiple reaction system involving two or more independent reactions, each reaction has a separate extent of reaction, denoted by ξ_j where the subscript j indicates the reaction number. If there are two independent reactions taking place simultaneously, j values are 1 and 2. The stoichiometric coefficients of species are indicated by double subscripts to specify their association with a species (ith species) and a reaction (jth reaction). Thus, $v_{i,j}$ specifies the stoichiometric coefficient of species i and reaction j.

Multiple reactions are expressed in condensed notation as:

$$\sum_i v_{i,j} A_{i,j} = 0 \quad (i = 1,2, \ldots\ldots N;\ j = 1,2, \ldots.r) \tag{13.103}$$

where N is the total number of species present and r is the total number of independent reactions. As an example, consider a multiple reaction system with three independent reactions and three species. According to the notation given in Equation (13.103), we can express the reactions as:

$$v_{1,1}A_{1,1} + v_{2,1}A_{2,1} + v_{3,1}A_{3,1} = 0 \tag{13.104}$$

$$v_{1,2}A_{1,2} + v_{2,2}A_{2,2} + v_{3,2}A_{3,2} = 0 \tag{13.105}$$

$$v_{1,3}A_{1,3} + v_{2,3}A_{2,3} + v_{3,3}A_{3,3} = 0 \tag{13.106}$$

The number of moles of species i at any time t, n_i, can be expressed as:

$$n_i = n_i^o + \sum_j v_{i,j}\xi_j \quad (j = 1,2, \ldots.r) \tag{13.107}$$

Note that species i may be consumed or produced in all the reactions (total of r). For a multiple reaction system with three independent reactions and three species, we have:

$$n_1 = n_1^o + (v_{1,1}\xi_1 + v_{1,2}\xi_2 + v_{1,3}\xi_3) \tag{13.108}$$

$$n_2 = n_2^o + (v_{2,1}\xi_1 + v_{2,2}\xi_2 + v_{2,3}\xi_3) \tag{13.109}$$

$$n_3 = n_3^o + (v_{3,1}\xi_1 + v_{3,2}\xi_2 + v_{3,3}\xi_3) \tag{13.110}$$

In this case, there are three extents of reactions (ξ_1, ξ_2, ξ_3) and three species (1,2,3) are involved in the reactions. Each species may be consumed or produced in all the reactions.

The total number of moles of all species at any time is:

$$n = \sum_i n_i \quad (i = 1,2, \ldots.N) \tag{13.111}$$

The total number of moles of all species initially is:

$$n^o = \sum_i n_i^o \quad (i = 1,2, \ldots.N) \tag{13.112}$$

From Equations (13.107), (13.111) and (13.112):

$$n = n^o + \sum_j \sum_i v_{i,j}\xi_j \quad (i = 1,2, \ldots\ldots N;\ j = 1,2, \ldots.r) \tag{13.113}$$

For a gas-phase reaction mixture, the mole fraction of any species i is:

$$y_i = \frac{n_i}{n} = \frac{n_i^o + \sum_j v_{i,j}\xi_j}{n^o + \sum_j \sum_i v_{i,j}\xi_j} \quad (i = 1,2, \ldots . N) \tag{13.114}$$

For a liquid-phase reaction mixture, the mole fraction of any species i is:

$$x_i = \frac{n_i}{n} = \frac{n_i^o + \sum_j v_{i,j}\xi_j}{n^o + \sum_j \sum_i v_{i,j}\xi_j} \quad (i = 1,2, \ldots . N) \tag{13.115}$$

13.12.2 Equilibrium Constant

For a multiple reaction system with r independent reactions, we have a separate equilibrium relation for each reaction. Thus, multi-reaction equilibria can be expressed as:

$$\sum_i v_{i,j}\mu_{i,j} = 0 \qquad (i = 1,2, \ldots . . N;\ j = 1,2, \ldots . r) \tag{13.116}$$

For a multiple reaction system with three independent reactions and three species, we have:

$$v_{1,1}\mu_{1,1} + v_{2,1}\mu_{2,1} + v_{3,1}\mu_{3,1} = 0 \tag{13.117}$$

$$v_{1,2}\mu_{1,2} + v_{2,2}\mu_{2,2} + v_{3,2}\mu_{3,2} = 0 \tag{13.118}$$

$$v_{1,3}\mu_{1,3} + v_{2,3}\mu_{2,3} + v_{3,3}\mu_{3,3} = 0 \tag{13.119}$$

For each independent chemical reaction, we have a separate equilibrium constant given as:

$$K_j = \prod_i (\hat{a}_i)_e^{v_{i,j}} \quad (i = 1,2, \ldots . . N;\ j = 1,2, \ldots . r) \tag{13.120}$$

Likewise, there is a separate standard Gibbs free of reaction for each independent reaction:

$$\Delta G_j^o = -RTlnK_j \quad (j = 1,2, \ldots . r) \tag{13.121}$$

For a multiple reaction system with three independent reactions and three species, we have:

$$K_1 = \prod_i (\hat{a}_i)_e^{v_{i,1}} = (\hat{a}_1)_e^{v_{1,1}} (\hat{a}_2)_e^{v_{2,1}} (\hat{a}_3)_e^{v_{3,1}} \tag{13.122}$$

$$K_2 = \prod_i (\hat{a}_i)_e^{v_{i,2}} = (\hat{a}_1)_e^{v_{1,2}} (\hat{a}_2)_e^{v_{2,2}} (\hat{a}_3)_e^{v_{3,2}} \tag{13.123}$$

$$K_1 = \prod_i (\hat{a}_i)_e^{v_{i,3}} = (\hat{a}_1)_e^{v_{1,3}} (\hat{a}_2)_e^{v_{2,3}} (\hat{a}_3)_e^{v_{3,3}} \tag{13.124}$$

$$\Delta G_1^o = -RTlnK_1 \tag{13.125}$$

$$\Delta G_2^o = -RTlnK_2 \tag{13.126}$$

$$\Delta G_3^o = -RTlnK_3 \tag{13.127}$$

For gas-phase reactions, Equation (13.120) can be written as:

$$K_j = \prod_i \left(\frac{\hat{\varphi}_i y_i P}{P^o}\right)_e^{v_{i,j}} \quad (i = 1,2, \ldots . . N;\ j = 1,2, \ldots . r) \tag{13.128}$$

If the gaseous reaction mixture is an ideal gas mixture

$$K_j = \prod_i (y_i)_e^{v_{i,j}} \left(\frac{P}{P^o}\right)_e^{v_{i,j}} \quad (i = 1,2, \ldots..N; \; j = 1,2, \ldots.r) \tag{13.129}$$

For liquid-phase reactions, Equation (13.120) can be written as:

$$K_j = \prod_i (x_i \gamma_i)_e^{v_{i,j}} \quad (i = 1,2, \ldots..N; \; j = 1,2, \ldots.r) \tag{13.130}$$

If the liquid solution is an ideal solution

$$K_j = \prod_i (x_i)_e^{v_{i,j}} \quad (i = 1,2, \ldots..N; \; j = 1,2, \ldots.r) \tag{13.131}$$

13.12.3 Multi-Reaction Equilibria Problems

Given ΔG_j^o for different reactions and initial composition, how to calculate the equilibrium composition? Calculate the equilibrium constants K_j from Equation (13.121). Express activities in terms of the mole fractions and mole fractions in terms of the extents of reactions (see Equations (13.114) and (13.115)) and substitute them into the K_j expressions (see Equations (13.128) to (13.131)). Solve for the extents of reactions and hence the equilibrium composition.

Calculations for multi-reaction equilibria are complicated, especially when non-ideal reaction systems are involved. More often than not, numerical techniques are required to solve multi-reaction equilibria problems.

13.13 FUEL CELLS

Fuel cells are devices which produce electric power by *direct conversion* of chemical energy of a fuel into electricity without any burning of fuel. Fuel cells are much more efficient than the power cycles which produce electric power from chemical energy of a fuel *indirectly* by first burning the fuel to generate the heat energy and then converting heat into electric power. Other advantages of fuel cells include: no pollution and quiet operation.

Figure 13.6 shows a schematic diagram of a PEM (proton exchange membrane) fuel cell. It consists of anode and cathode, separated by a solid proton-conducting polymer membrane (called polymer electrolyte membrane). The fuel cell is continuously fed with a fuel (assumed to be hydrogen gas) on the anode side and oxygen (air) on the cathode side. At the anode, the hydrogen gas is oxidized, that is, it gives up electrons:

$$H_2(g) \rightarrow 2H^+ + 2e^- \qquad \text{(half-cell reaction at anode)} \tag{13.132}$$

The electrons released at the anode pass through the external circuit (hence electricity is generated) and end up at the cathode. At the cathode, the electrons are taken up by the following reaction:

$$2H^+ + \frac{1}{2}O_2(g) + 2e^- \rightarrow H_2O(g) \qquad \text{(half-cell reaction at cathode)} \tag{13.133}$$

The overall reaction in the cell is a combination of the two half-cell reactions given in Equations (13.132) and (13.133). Thus,

$$H_2(g) + \frac{1}{2}O_2(g) \rightarrow H_2O(g) \qquad \text{(overall cell reaction)} \tag{13.134}$$

This is the combustion reaction of hydrogen that takes place electrochemically in the cell without any burning of hydrogen. The product of the reaction, that is, water vapor is continuously removed from the cell along with the unused oxygen.

It should be noted the polymer electrolyte membrane does not allow the passage of electrons through it. Only protons (H^+ ions) are able to diffuse through the membrane. The perfluorosulfonic acid polymer *Nafion* is used extensively as a material for proton exchange membrane in fuel cells. Also note that the electrodes (anode and cathode) are made from porous carbon impregnated with finely divided platinum as a catalyst.

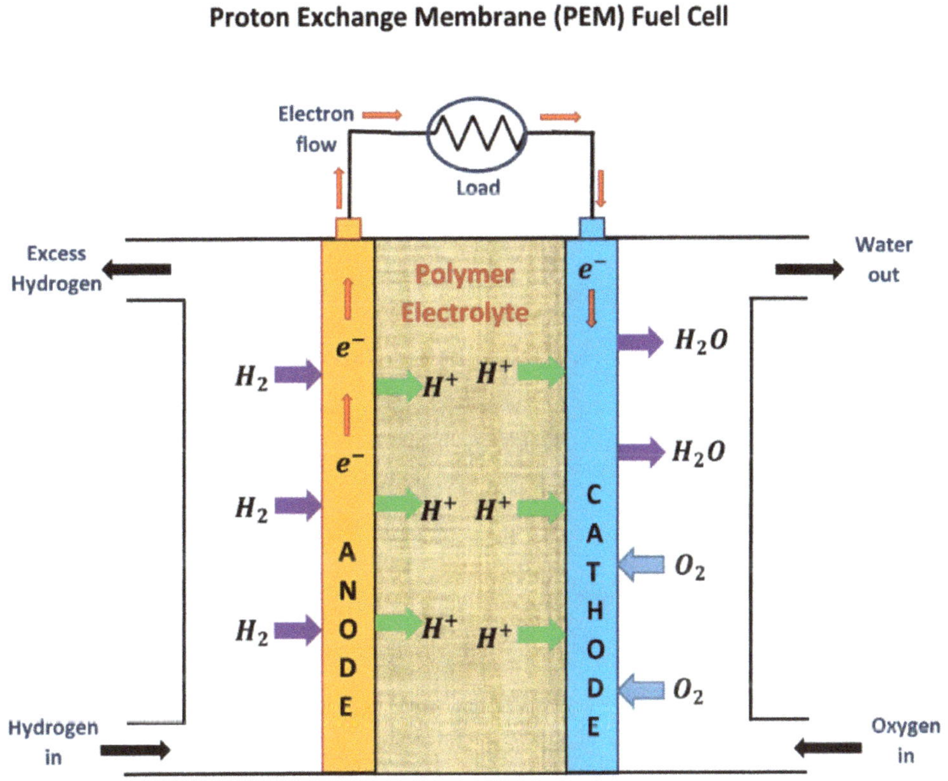

Figure 13.6 Schematic diagram of PEM fuel cell.

13.13.1 Thermodynamic Analysis of a Fuel Cell

Figure 13.7 shows the net effects in a fuel cell. Applying the first law of thermodynamics for open systems to the fuel cell:

$$\Sigma_{OUTLETS}\ (\dot{n}_j h_j) - \Sigma_{INLETS}\ (\dot{n}_j\, h_j) = \dot{Q} - \dot{W}_{elect} \tag{13.135}$$

where h_j is the molar enthalpy of fluid stream entering/leaving jth opening, $\dot{n}_j$ is the total molar flow rate of fluid stream entering/leaving jth opening, and $\dot{W}_{elect}$ is the rate of electrical work produced. We have neglected kinetic and potential energy changes. It should be noted that if a fluid stream is a mixture of different components, the molar enthalpy of the mixture at any jth opening is given as:

$$h_j = \Sigma_i\, y_i\, \bar{h}_i \tag{13.136}$$

where $\bar{h}_i$ is the partial molar enthalpy of species i in the mixture and y_i is the mole fraction of species i in the mixture. Equation (13.136) is the summability relation discussed in Chapter 11.

Application of the second law of thermodynamics for open systems to the fuel cell (Figure 13.7) gives:

$$\dot{S}_G = \Sigma_{OUTLETS}\ (\dot{n}_j s_j) - \Sigma_{INLETS}\ (\dot{n}_j\ s_j) - \frac{\dot{Q}}{T} \geq 0 \tag{13.137}$$

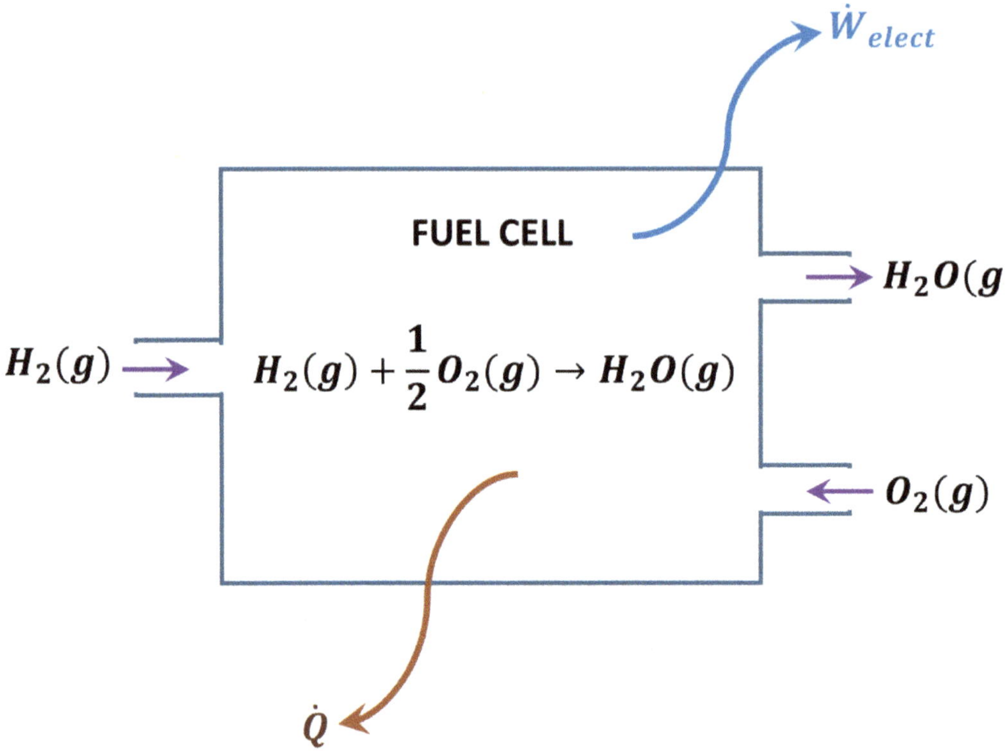

Figure 13.7 Net effects in a PEM fuel cell.

where $\dot{S}_G$ is the rate of entropy production in the universe. We have assumed that the fuel cell is an isothermal system with temperature T (the temperatures of fluid streams entering and leaving the cell and the temperature within the cell are all the same, that is, T). Equation (13.137) can be re-written as:

$$\Sigma_{OUTLETS}\ (\dot{n}_j s_j) - \Sigma_{INLETS}\ (\dot{n}_j\ s_j) - \frac{\dot{Q}}{T} - \dot{S}_G = 0 \tag{13.138}$$

Upon re-arrangement of Equation (13.138), $\dot{Q}$ can be expressed as:

$$\dot{Q} = T[\Sigma_{OUTLETS}\ (\dot{n}_j s_j) - \Sigma_{INLETS}\ (\dot{n}_j\ s_j)] - T\dot{S}_G \tag{13.139}$$

Substitution of Equation (13.139) into the first law, Equation (13.135), yields the following expression for $\dot{W}_{elect}$:

$$\dot{W}_{elect} = T\left[\sum_{OUTLETS} (\dot{n}_j s_j) - \sum_{INLETS} (\dot{n}_j\ s_j)\right] - T\dot{S}_G - [\Sigma_{OUTLETS}\ (\dot{n}_j h_j) - \Sigma_{INLETS}\ (\dot{n}_j\ h_j)] \tag{13.140}$$

Combining entropy and enthalpy terms, Equation (13.140) can be re-written as:

$$\dot{W}_{elect} = \left[\sum_{INLETS} \dot{n}_j (h_j - Ts_j) - \sum_{OUTLETS} \dot{n}_j (h_j - Ts_j)\right] - T\dot{S}_G \quad (13.141)$$

Since,

$$g_j = (h_j - Ts_j) \quad (13.142)$$

Equation (13.141) becomes:

$$\dot{W}_{elect} = \left[\sum_{INLETS} \dot{n}_j g_j - \sum_{OUTLETS} \dot{n}_j g_j\right] - T\dot{S}_G \quad (13.143)$$

We now assume *reversibility*, that is, all processes occurring in the cell are reversible. Thus $\dot{S}_G = 0$ and Equation (13.143) reduces to:

$$\dot{W}_{elect} = \left[\sum_{INLETS} \dot{n}_j g_j - \sum_{OUTLETS} \dot{n}_j g_j\right] \quad (13.144)$$

Application of Equation (13.144) to the fuel cell of Figure 13.7 gives:

$$\dot{W}_{elect} = \dot{n}_{H_2} g_{H_2} + \dot{n}_{O_2} g_{O_2} - \dot{n}_{H_2O} g_{H_2O} \quad (13.145)$$

We are assuming that the reactants (hydrogen and oxygen) and products (water vapor) are pure ideal gas streams and that the pressure of all the streams is the standard pressure $P^0 = 1$ bar. Furthermore, the reactants and products are in stoichiometric amounts. Thus,

$$\dot{n}_{O_2} = \frac{1}{2} \dot{n}_{H_2} \ ; \ \ \dot{n}_{H_2O} = \dot{n}_{H_2} \quad (13.146)$$

Consequently, Equation (13.145) simplifies to:

$$\dot{W}_{elect} = -\dot{n}_{H_2} \left[g^o(H_2O, g) - g^o(H_2, g) - \frac{1}{2} g^o(O_2, g)\right] \quad (13.147)$$

$$\dot{W}_{elect} = -\dot{n}_{H_2} [\Delta G^o] \quad (13.148)$$

Thus, the electrical work produced per mole of hydrogen is negative of the standard Gibbs free energy of the overall cell reaction given in Equation (13.134).

The amount of heat generated in the cell can be estimated from the first law, Equation (13.135), written as:

$$\dot{Q} = -\dot{n}_{H_2} [\Delta G^o] + \dot{n}_{H_2} \left[h^o(H_2O, g) - h^o(H_2, g) - \frac{1}{2} h^o(O_2, g)\right] \quad (13.149)$$

Equation (13.149) can be re-written as:

$$\dot{Q} = \dot{n}_{H_2} [-\Delta G^o + \Delta H^o] \quad (13.150)$$

where ΔH^o is the standard heat of the overall cell reaction given in Equation (13.134). The heat generated per unit mole of hydrogen is $[\Delta H^o - \Delta G^o]$.

The values of ΔG^o and ΔH^o for the cell reaction are -228.572 kJ/mol and -241.818 kJ/mol at 25°C, respectively. Assuming that the fuel cell is operated at 25°C and 1 bar, the electrical work produced, and the heat generated per mole of hydrogen are as follows:

$$W_{elect} = -\Delta G^o = 228.572\ kJ/mol \tag{13.151}$$

$$Q = [-\Delta G^o + \Delta H^o] = 228.572 - 241.818 = -13.246\ kJ/mol \tag{13.152}$$

Note that the electrical work output of the cell ($-\Delta G^o$) is approximately 95% of the heat that would be released (ΔH^o) if one mole of hydrogen is burnt. If this heat generated by burning one mole of hydrogen is used in a Carnot cycle to produce work, the efficiency would be much smaller. Thus fuel cells are inherently much more efficient as compared with heat engines due to direct conversion of chemical energy into work without any burning of fuel.

13.13.2 EMF of a Fuel Cell

The *emf* (electromotive force) of a cell, expressed in volts, is the electrical work done by the cell per coulomb of charge passing through it:

$$emf = \frac{W_{elect}}{q} \tag{13.153}$$

For 1 mol of hydrogen,

$$W_{elect} = -\Delta G^o \tag{13.154}$$

The amount of electric charge passed through the cell per mol of hydrogen is:

$$q = 2eN_A = 2F \tag{13.155}$$

where e is the electric charge on each electron ($e = 1.60218 \times 10^{-19}$ coulombs), N_A is the Avagadro number ($N_A = 6.02214 \times 10^{23}$ mol^{-1}), and F is the Faraday constant ($F = 96485$ coulombs). From Equations (13.153) to (13.155), it follows that:

$$emf = \frac{-\Delta G^o}{2F} \tag{13.156}$$

For hydrogen/oxygen fuel cell operating reversibly at 25^{0}C and 1 bar, Equation (13.156) gives:

$$emf = \frac{228.572 \times 10^3}{2 \times 96485} = 1.184 \text{ volts} \tag{13.157}$$

It should be noted that this is the maximum emf generated by the cell as we have assumed that the cell operation is reversible in nature. In actual operation, the emf developed by the cell is lower due to internal irreversibilities in the cell.

A single fuel cell generates just a small amount of emf, not enough in most practical applications. Therefore, the fuel cells are stacked together in series parallel combinations to increase the voltage and current outputs.

13.14 ENTROPY PRODUCTION IN CHEMICAL REACTIONS

In any spontaneous process, entropy is generated. Chemical reactions are no exception. Chemical reactions occurring spontaneously generate entropy.

Consider a chemical reaction taking place in a closed system at constant T and P (see Figure 13.8). The entropy balance for the system can be written as:

$$(Entropy\ in) - (Entropy\ out) + (Entropy\ generated) = (Entropy\ change) \tag{13.158}$$

$$\frac{\delta Q}{T} + \delta S_G = dS \tag{13.159}$$

where δS_G is the entropy generated.

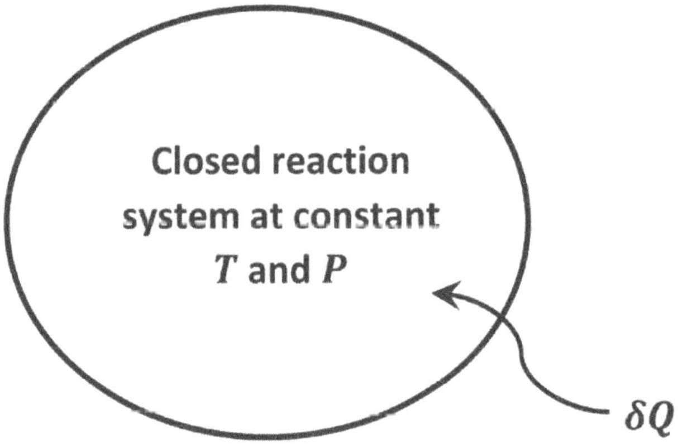

Figure 13.8 Closed reaction system at constant T and P.

Upon rearrangement, Equation (13.159) becomes:

$$\delta S_G = dS - \frac{\delta Q}{T} \tag{13.160}$$

For a constant pressure process,

$$\delta Q = dH \tag{13.161}$$

Upon substitution of Equation (13.161) into Equation (13.160), we obtain:

$$T\delta S_G = TdS - dH \tag{13.162}$$

As T is constant, Equation (13.162) can be re-written as:

$$T\delta S_G = -d(H - TS) = -dG \tag{13.163}$$

Thus, the rate of production entropy in the reaction system is:

$$\dot{S}_G = \frac{1}{T}\left(\frac{dG}{dt}\right) \tag{13.164}$$

For the reaction system,

$$(dG)_{T,P} = \left(\frac{\partial G}{\partial \xi}\right)_{T,P} d\xi = \sum_i v_i \mu_i d\xi \tag{13.165}$$

From Equations (13.164) and (13.165), it follows that:

$$\dot{S}_G = -\frac{1}{T}\left(\sum_i v_i \mu_i\right)\left(\frac{d\xi}{dt}\right) \tag{13.166}$$

Let A be the *affinity* of a chemical reaction defined as:

$$A = -\left(\frac{\partial G}{\partial \xi}\right)_{T,P} = -\sum_i v_i \, \mu_i \tag{13.167}$$

We can re-write Equation (13.166) in terms of A as:

$$\dot{S}_G = \left(\frac{A}{T}\right)\left(\frac{d\xi}{dt}\right) \tag{13.168}$$

According to the second law of thermodynamics,

$$\dot{S}_G = \left(\frac{A}{T}\right)\left(\frac{d\xi}{dt}\right) > 0 \tag{13.169}$$

Equation (13.169) is referred to as *de Donder's* inequality. According to the de Donder's inequality, the rate of entropy production in the reaction system is directly proportional to the thermodynamic force term (A/T) and the rate of reaction $(d\xi/dt)$. When both A and $d\xi/dt$ are positive, the reaction is spontaneous, and it occurs from left to right (forward direction). If both A and $d\xi/dt$ are negative, the reaction is still spontaneous but occurs from right to left (backward direction). If $A = 0$, the system is in thermodynamic equilibrium with $\dot{S}_G$=0. If $A \neq 0$ and $d\xi/dt = 0$, the system is in so-called *false equilibrium* meaning that the system is not in equilibrium as the thermodynamic force is not zero, but the rate of reaction is vanishingly small.

Equation (13.169) can be extended to multiple reactions occurring simultaneously. The rate of entropy production is now the sum of the entropy production due to all the reactions involved. For r reactions taking place at the same time, Equation (13.169) can be generalized as:

$$\dot{S}_G = \left(\frac{1}{T}\right)\sum_{j=1}^{j=r} A_j \left(\frac{d\xi_j}{dt}\right) > 0 \tag{13.170}$$

where A_j is the affinity of the jth reaction given as:

$$A_j = -\sum_i v_{i,j} \, \mu_{i,j} \quad (j = 1,2, \ldots .r) \tag{13.171}$$

In the equilibrium state, the affinities of all the reactions are zero, that is:

$$A_1 = A_2 = \cdots = A_r = 0 \tag{13.172}$$

13.15 LIMITATIONS OF THERMODYNAMICS

With respect to chemical reaction systems, thermodynamics allows us to predict: a) if a given reaction is feasible or not; b) the equilibrium extent of the reaction; and c) the equilibrium composition of the reaction mixture. However, thermodynamics is completely silent when it comes to the *speed* of the reaction. How long would it take for a reaction to attain equilibrium or undergo completion? This question is not answered by thermodynamics.

As an example, consider the following reaction at 25°C and 1 bar:

$$H_2(g) + \frac{1}{2}O_2(g) \rightarrow H_2O(g) \qquad \Delta G^o(25^oC) = -228.572\ kJ/mol \tag{13.173}$$

This reaction is thermodynamically feasible as it has a large negative value of ΔG^o. The equilibrium constant for this reaction is:

$$K = \exp\left(-\Delta G^o/RT\right) = 1.1125 \times 10^{40} \quad (13.174)$$

The equilibrium extent of this reaction is very close to unity. Thus, thermodynamically the reaction should go to completion (complete conversion of reactants to product).

If we place hydrogen and oxygen in stoichiometric amounts in a tank at pressure of 1 bar and temperature of 25°C and see what happens? *Nothing*, even if we wait for thousands of years. No measurable conversion occurs at all indicating that the rate of reaction is practically zero.

Although the reaction is thermodynamically highly favorable, it is kinetically useless. The kinetics of reactions is considered under a separate branch of physical chemistry known as "*Chemical Kinetics*".

In chemical kinetics, also known as reaction kinetics, we learn that in order to carry out the given water formation reaction, Equation (13.173), a *catalyst* is required. The job of the catalyst is to speed up the reaction without altering the thermodynamics of the reaction. For example, a platinum catalyst will make the reaction of water formation go to completion almost instantaneously.

NOTATION

A	affinity of a chemical reaction, see Equation (13.167)
A_j	affinity of jth chemical reaction, see Equation (13.171)
A_i	ith chemical species, see Equation (13.4)
$A_{i,j}$	ith chemical species of jth chemical reaction, see Equation (13.103)
$a(l)$	activity of pure liquid species
$a(s)$	activity of pure solid species
a_i	activity of pure species i
$\hat{a}_i$	activity of species i in mixture or solution
d	differential of exact quantity
e	electric charge on electron
e^-	electron
emf	electromotive force of a cell, expressed in volts
F	Faraday constant
f	fugacity
f^o	standard fugacity
f^L	fugacity of pure liquid
f^S	fugacity of pure solid
f_i^o	standard fugacity of species i
$\hat{f}_i$	fugacity of species i in mixture or solution
f_i^L	fugacity of pure liquid species i
G	Gibbs free energy
$G_{i,pure}$	Gibbs free energy of pure species i
g^o	standard molar Gibbs free energy
g_i^o	standard molar Gibbs free energy of species i
H	enthalpy
h_j	molar enthalpy of fluid stream at jth opening
h^o	standard molar enthalpy
h_i^o	standard molar enthalpy of species i
$\bar{h}_i$	partial molar enthalpy of species i in a mixture

K	equilibrium constant
K_j	equilibrium constant of jth reaction
K_p	equilibrium constant for a chemical reaction equilibrium in a mixture of ideal gases
N_A	Avagadro number
n	total number of moles
n^o	total number of moles initially
n_i	number of moles of species i
n_i^o	number of moles of species i initially
$\dot{n}_j$	total molar flow rate of fluid stream entering/leaving jth opening
P	pressure
P^o	standard pressure
P_i	partial pressure of species i
Q	heat
$\dot{Q}$	heat transfer rate
q	amount of electric charge
R	universal gas constant
S	entropy
S_G	entropy generated
$\dot{S}_G$	rate of entropy generation
s_j	molar entropy of fluid stream at jth opening
s_i^o	standard molar entropy of species i
T	temperature
t	time
v_i	stoichiometric coefficient of species i
$v_{i,j}$	stoichiometric coefficient of species i in jth reaction
W_{elect}	electrical work
$\dot{W}_{elect}$	electrical work rate
x_i	mole fraction of species i in liquid solution
y_i	mole fraction of species i in gas mixture

Greek Letters

γ_i	activity coefficient of species i
Δ	difference operator
ΔG^o	standard Gibbs free energy change of reaction
ΔG_j^o	standard Gibbs free energy change of jth reaction
ΔG_{rxn}^o	standard Gibbs free energy change of reaction
ΔG_{mix}	Gibbs free energy of mixing
Δg_{fi}^o	standard Gibbs free energy of formation of compound species i
ΔH^o	standard enthalpy change of reaction
ΔH_{rxn}^o	standard heat of reaction
Δh_{fi}^o	standard enthalpy of formation of compound species i
δ	differential of inexact quantity
ξ	extent of reaction
ξ_j	extent of reaction of jth reaction
μ_i	chemical potential of species i in a mixture or solution
$\mu_{i,j}$	chemical potential of species i of jth reaction
$\mu_{i,mix}$	chemical potential of species i in a mixture
μ_i^o	standard chemical potential
$\mu_{i,mix}^{ig}$	chemical potential of species i in an ideal gas mixture

$\mu_{i,pure}^{ig}$	chemical potential of pure ideal gas species i
$\mu_{i,solu}^{id}$	chemical potential of species i in an ideal solution
$\mu_{i,pure}^{L}$	chemical potential of pure liquid species i
$\hat{\varphi}_i$	fugacity coefficient of species i in a mixture or solution

TUTORIAL PROBLEMS WITH SOLUTION

Problem 1

Consider the following reaction dealing with dissociation of dinitrogen tetroxide to nitrogen dioxide:

$$N_2O_4(g) \rightleftharpoons 2NO_2(g)$$

The initial number of moles of species involved in the reaction are: $N_2O_4(g) = 1$ mole and $NO_2(g) = 0$ mole. The reaction takes place at constant T and P in a reaction vessel.

a) Calculate the equilibrium constant K_p at 25 °C. What is the equilibrium extent of the reaction and equilibrium composition at 25 °C and $P = 1$ bar?
b) Calculate the equilibrium constant K_p at 400 °C. What is the equilibrium extent of the reaction and equilibrium composition at 400 °C and $P = 1$ bar?
c) Does the extent of reaction increase or decrease with the increase in pressure at any given temperature? Plot the equilibrium extent of reaction ξ and $y_{N_2O_4}$ as functions of pressure at 50 °C.

Given that at 298.15 K:

Species	Δg_f^o, kJ/mol	Δh_f^o, kJ/mol
$N_2O_4(g)$	97.5	9.2
$NO_2(g)$	51.3	33.2

You can assume that the reaction mixture is an ideal gas mixture.

Solution

The standard Gibbs free energy and standard enthalpy of reaction at 298.15 K are:

$$\Delta G^o = \sum_i v_i \, \Delta g_{fi}^o \text{ (compounds)} = 2 \times 51.3 - 97.5 = 5.1 \text{ kJ/mol}$$

$$\Delta H^o = \sum_i v_i \, \Delta h_{fi}^o \text{ (compounds)} = 2 \times 33.2 - 9.2 = 57.2 \text{ kJ/mol}$$

a) Calculate the equilibrium constant K_p at 25 °C. What is the equilibrium extent of the reaction and equilibrium composition at 25 °C and $P = 1$ bar?

$\Delta G^o = -RTlnK_p$; $K_p = exp(-\Delta G^0/RT) = \exp(-5.1 \times 10^3/(8.314 \times 298.15)) = 0.12778$. Let the equilibrium extent of the reaction be ξ.

From the given data, we can express the mole fractions and partial pressures of each species at equilibrium in the table form as shown below:

	$N_2O_4(g)$	$NO_2(g)$	total
Initial number of moles, n_i^0	1	0	1
Moles at equilibrium, n_i	1-ξ	2ξ	1+ξ

Mole fractions at equilibrium, y_i	$\frac{1-\xi}{1+\xi}$	$\frac{2\xi}{1+\xi}$	1
Partial pressure at equilibrium, P_i	$\left(\frac{1-\xi}{1+\xi}\right)P$	$\left(\frac{2\xi}{1+\xi}\right)P$	P

$$K_p = \prod_i \left(\frac{P_i}{P^o}\right)_e^{v_i} = \frac{\left(P_{NO_2}/P^o\right)_e^2}{\left(P_{N_2O_4}/P^o\right)_e} = \left(\frac{2\xi}{1+\xi}\right)^2 \left(\frac{1+\xi}{1-\xi}\right)\left(\frac{P}{P^o}\right) = \left(\frac{4\xi^2}{1-\xi^2}\right)\left(\frac{P}{P^o}\right)$$

Upon re-arrangement:

$$\xi = \sqrt{\frac{K_p}{K_p + 4(P/P^o)}}$$

As $P = 1$ bar, standard pressure $P^o = 1$ bar, and $K_p = 0.12778$, the equilibrium extent of reaction is:

$$\xi = \sqrt{\frac{K_p}{K_p+4(P/P^o)}} = \sqrt{\frac{0.12778}{0.12778+4}} = 0.176$$

Equilibrium composition:

$$y_{N_2O_4} = \frac{1-\xi}{1+\xi} = \frac{1-0.176}{1+0.176} = 0.7007$$

$$y_{NO_2} = \frac{2\xi}{1+\xi} = \frac{2 \times 0.176}{1+0.176} = 0.2993$$

b) Calculate the equilibrium constant K_p at 400 °C. What is the equilibrium extent of the reaction and equilibrium composition at 400 °C and $P = 1$ bar?

Assuming ΔH^o to be independent of temperature, the effect of temperature on the equilibrium constant is given by Equation (13.42) as:

$$lnK_p = ln\left(K_p\right)_o - \frac{\Delta H^o}{R}\left(\frac{1}{T} - \frac{1}{T_o}\right)$$

Given $\left(K_p\right)_o = 0.12778$ at $T_o = 298.15$ K, and $\Delta H^o = 57.2$ kJ/mole, we get K_p at 400 °C as: $lnK_p = \ln(0.12778) - \frac{57.2\times10^3}{8.314}\left(\frac{1}{673.15} - \frac{1}{298.15}\right) = 10.7976$; $K_p = 48903.3$

The equilibrium extent of reaction is:

$$\xi = \sqrt{\frac{K_p}{K_p + 4(P/P^o)}} = \sqrt{\frac{48903.3}{48903.3 + 4}} = 0.99996$$

Equilibrium composition:

$$y_{N_2O_4} = \frac{1-\xi}{1+\xi} = \frac{1-0.99996}{1+0.99996} = 0.00002$$

$$y_{NO_2} = \frac{2\xi}{1+\xi} = \frac{2 \times 0.99996}{1 + 0.99996} = 0.99998$$

Thus, for all practical purposes, the reaction goes to completion.

c) Does the extent of reaction increase or decrease with the increase in pressure at any given temperature? Plot the equilibrium extent of reaction ξ and $y_{N_2O_4}$ as functions of pressure at 50 °C.

The equilibrium extent of reaction is:

$$\xi = \sqrt{\frac{K_p}{K_p + 4(P/P^o)}}$$

As K_p is a function of temperature only, ξ decreases with the increase in pressure P.

Given $(K_p)_o = 0.12778$ at $T_o = 298.15$ K, and $\Delta H^o = 57.2$ kJ/mole, we get K_p at 50 °C as: $lnK_p = \ln(0.12778) - \frac{57.2\times10^3}{8.314}\left(\frac{1}{323.15} - \frac{1}{298.15}\right) = -0.2722$; $K_p = 0.7617$

The equilibrium extent of reaction ξ and $y_{N_2O_4}$ as functions of pressure at 50 °C are plotted in the figure given below. Clearly, the equilibrium extent of reaction decreases from close to unity to zero when the pressure is increased from 0.01 to 250 bar. The equilibrium mole fraction of N_2O_4 increases from almost zero to unity over the same pressure range.

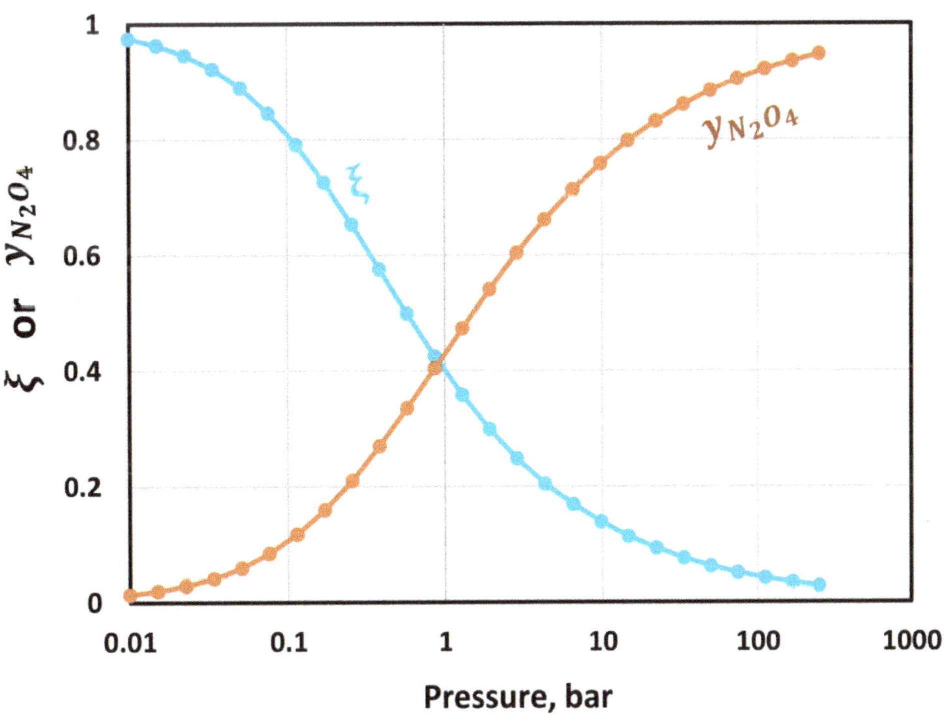

Problem 2

Consider the following reaction dealing with dissociation of phosphorous pentachloride to phosphorous trichloride and chlorine:

$$PCl_5(g) \rightleftharpoons PCl_3(g) + Cl_2(g)$$

The initial number of moles of species involved in the reaction are: $PCl_5(g) = 1$ mole, $PCl_3(g) = 0$ mole, and $Cl_2(g) = 0$ mole. The reaction takes place at constant T and P in a reaction vessel.

Plot the equilibrium extent of reaction ξ over the temperature range of 298.15 K to 800 K at pressures of 0.1, 1, and 10 bar. Also, plot the equilibrium mole fraction of different species over the temperature range of 298.15 K to 800 K at pressure of 0.1 bar.

Given that at 298.15 K:

Species	Δg_f^o, kJ/mol	Δh_f^o, kJ/mol
$PCl_5(g)$	-305	-375
$PCl_3(g)$	-268	-287
$Cl_2(g)$	0	0

You can assume that the reaction mixture is an ideal gas mixture.

Solution

The standard Gibbs free energy and standard enthalpy of reaction at 298.15 K are:

$$\Delta G^o = \sum_i v_i \Delta g_{fi}^o \text{ (compounds)}=(-268) - (-305) = 37 \text{ kJ/mol}$$

$$\Delta H^o = \sum_i v_i \Delta h_{fi}^o \text{ (compounds)}=(-287) - (-375) = 88 \text{ kJ/mol}$$

$\Delta G^o = -RTlnK_p$. The equilibrium constant at 298.15 K is:

$$K_p = exp(-\Delta G^o/RT) = \exp\left(-37 \times 10^3/(8.314 \times 298.15)\right) = 3.29 \times 10^{-7}.$$

Assuming ΔH^o to be independent of temperature, the equilibrium constant at any temperature T can be calculated from:

$$lnK_p = ln(K_p)_o - \frac{\Delta H^o}{R}\left(\frac{1}{T} - \frac{1}{T_o}\right) = \ln(3.29 \times 10^{-7}) - \frac{88 \times 10^3}{8.314}\left(\frac{1}{T} - \frac{1}{298.15}\right)$$

Let the equilibrium extent of the reaction be ξ.

From the given data, we can express the mole fractions and partial pressures of each species at equilibrium in the table form as shown below:

	$PCl_3(g)$	$Cl_2(g)$	$PCl_5(g)$	total
Initial number of moles, n_i^0	0	0	1	1
Moles at equilibrium, n_i	ξ	ξ	1- ξ	1+ξ
Mole fractions at equilibrium, y_i	$\frac{\xi}{1+\xi}$	$\frac{\xi}{1+\xi}$	$\frac{1-\xi}{1+\xi}$	1
Partial pressure at equilibrium, P_i	$\left(\frac{\xi}{1+\xi}\right)P$	$\left(\frac{\xi}{1+\xi}\right)P$	$\left(\frac{1-\xi}{1+\xi}\right)P$	P

$$K_p = \Pi_i \left(\frac{P_i}{P^o}\right)_e^{v_i} = \frac{\left(P_{PCl_3}/P^o\right)_e \left(P_{Cl_2}/P^o\right)_e}{\left(P_{PCl_5}/P^o\right)_e} = \left(\frac{\xi}{1+\xi}\right)^2 \left(\frac{1+\xi}{1-\xi}\right)\left(\frac{P}{P^o}\right) = \left(\frac{\xi^2}{1-\xi^2}\right)\left(\frac{P}{P^o}\right)$$

Upon re-arrangement:

$$\xi = \sqrt{\frac{K_p}{K_p + \left(\frac{P}{P^o}\right)}}$$

where the standard pressure P^o=1 bar. To plot equilibrium extent of reaction ξ over the temperature range of 298.15 K to 800 K, we first calculate K_p as a function temperature and then ξ as a function of temperature at a given pressure.

The equilibrium extent of reaction ξ and composition as functions of temperature are plotted in the figures given below. The equilibrium extent of reaction increases from close to zero to unity when the temperature is increased from 300 to 800 K. With the increase in pressure at any temperature, the extent of reaction decreases. The equilibrium mole fraction of PCl_5 decreases from unity to zero when temperature is raised from 300 to 600 K at a pressure of 0.1 bar. The mole fractions of PCl_3 and Cl_2 increase from 0 to 0.5 over the same temperature range at 0.1 bar.

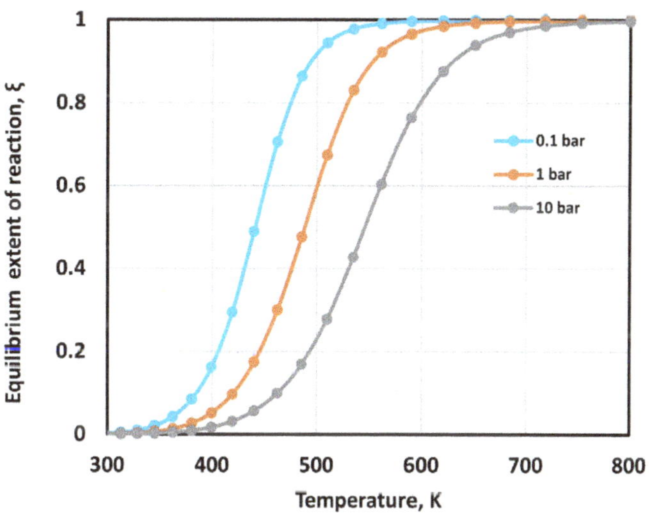

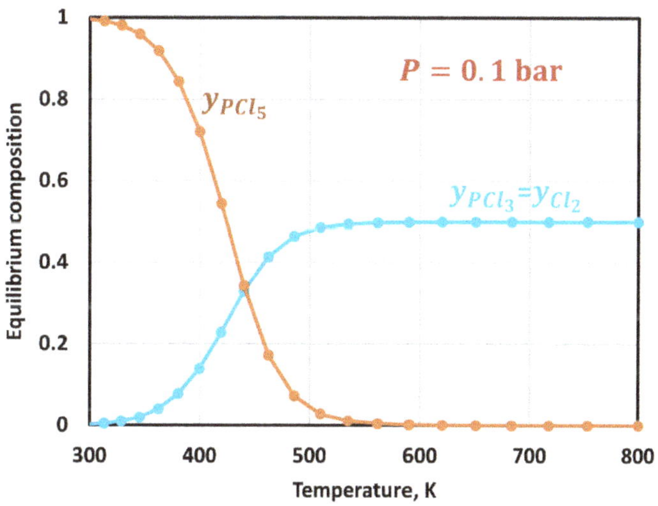

Problem 3

One mole of pure phosphorous pentachloride at a low temperature is mixed with 4 moles of an inert diluent and the mixture is heated to a temperature of 500 K at a pressure of 1 bar. The following dissociation reaction takes place at 500 K and 1 bar.

$$PCl_5(g) \rightleftharpoons PCl_3(g) + Cl_2(g)$$

Estimate the equilibrium extent of reaction at 500 K and 1 bar. What is the equilibrium composition? You can assume that the reaction mixture is an ideal gas mixture.

Solution

Assuming ΔH^o to be independent of temperature, the equilibrium constant at 500 K can be calculated as follows (see preceding problem solution):

$$lnK_p = ln(K_p)_o - \frac{\Delta H^o}{R}\left(\frac{1}{T} - \frac{1}{T_o}\right) = \ln(3.29 \times 10^{-7}) - \frac{88 \times 10^3}{8.314}\left(\frac{1}{T} - \frac{1}{298.15}\right)$$

At 500 K, this equation gives $K_p = 0.5513$.

From the given data, we can express the mole fractions and partial pressures of each species at equilibrium in the table form as shown below:

	$PCl_3(g)$	$Cl_2(g)$	$PCl_5(g)$	Inert diluent	total
Initial number of moles, n_i^o	0	0	1	4	5
Moles at equilibrium, n_i	ξ	ξ	1- ξ	4	5+ξ
Mole fractions at equilibrium, y_i	$\frac{\xi}{5+\xi}$	$\frac{\xi}{5+\xi}$	$\frac{1-\xi}{5+\xi}$	$\frac{4}{5+\xi}$	1
Partial pressure at equilibrium, P_i	$\left(\frac{\xi}{5+\xi}\right)P$	$\left(\frac{\xi}{5+\xi}\right)P$	$\left(\frac{1-\xi}{5+\xi}\right)P$	$\left(\frac{4}{5+\xi}\right)P$	P

$$K_p = \prod_i \left(\frac{P_i}{P^o}\right)_e^{\nu_i} = \frac{(P_{PCl_3}/P^o)_e (P_{Cl_2}/P^o)_e}{(P_{PCl_5}/P^o)_e} = \left(\frac{\xi}{5+\xi}\right)^2 \left(\frac{5+\xi}{1-\xi}\right)\left(\frac{P}{P^o}\right) = \left[\frac{\xi^2}{(5+\xi)(1-\xi)}\right]\left(\frac{P}{P^o}\right)$$

Since $P = P^o = 1$ bar,

$$K_p = \left[\frac{\xi^2}{(5+\xi)(1-\xi)}\right] = 0.5513$$

Upon re-arrangement, we get:

$$1.5513\xi^2 + 2.2052\xi - 2.7565 = 0$$

The solution of this quadratic equation gives $\xi = 0.79989$. Thus, the equilibrium extent of reaction is about 0.8.

The equilibrium composition of the reaction mixture is:

$y_{PCl_5} = \left(\frac{1-\xi}{5+\xi}\right) = 0.0345$; $y_{PCl_3} = \left(\frac{\xi}{5+\xi}\right) = 0.1379$; $y_{Cl_2} = \left(\frac{\xi}{5+\xi}\right) = 0.1379$; $y_{diluent} = \left(\frac{4}{5+\xi}\right) = 0.6897$

Comments: The equilibrium extent of reaction without the addition of inert diluent is $\xi = 0.596$ at 500 K and 1 bar. The addition of diluent shifts the equilibrium to the right leading to more dissociation of $PCl_5(g)$ as reflected in the increase in ξ to 0.8.

Problem 4

Consider the following liquid phase reaction:

$$A(l) + B(l) \rightleftharpoons C(l) + D(l)$$

The reaction is taking place at $T = 80$ °C and $P = 1$ bar. The initial number of moles of species involved in the reaction are: $A(l) = 1$ mole, $B(l) = 1$ mole, $C(l) = 0$ mole, and $D(l) = 0$ mole.
Assuming that the liquid solution is ideal, estimate the equilibrium extent of reaction ξ and the equilibrium composition.

Given that at 298.15 K:

Species	Δg_f^o, kJ/mol	Δh_f^o, kJ/mol
$A(l)$	-389.9	-484.5
$B(l)$	-174.8	-277.7
$C(l)$	-332.2	-480
$D(l)$	-237.1	-285.8

Solution

The standard Gibbs free energy and standard enthalpy of reaction at 298.15 K are:

$$\Delta G^o = \sum_i v_i \, \Delta g_{fi}^o(\text{compounds}) = (-332.2) + (-237.1) - (-389.9) - (-174.8) = -4.6 \text{ kJ/mol}$$

$$\Delta H^o = \sum_i v_i \, \Delta h_{fi}^o(\text{compounds}) = (-480) + (-285.8) - (-484.5) - (-277.7) = -3.6 \text{ kJ/mol}$$

$\Delta G^o = -RTlnK$. The equilibrium constant at 298.15 K is:

$$K = exp(-\Delta G^o/RT) = \exp\left(4.6 \times 10^3/(8.314 \times 298.15)\right) = 6.3963.$$

Assuming ΔH^o to be independent of temperature, the equilibrium constant at $T = 80 + 273.15 = 353.15$ K can be calculated from:

$$lnK = ln(K)_o - \frac{\Delta H^o}{R}\left(\frac{1}{T} - \frac{1}{T_o}\right) = \ln(6.3963) + \frac{3.6 \times 10^3}{8.314}\left(\frac{1}{353.15} - \frac{1}{298.15}\right) = 1.629536$$

Thus, K= 5.1015. Let the equilibrium extent of the reaction be ξ.

From the given data, we can express the mole fractions of each species at equilibrium in the table form as shown below:

	$A(l)$	$B(l)$	$C(l)$	$D(l)$	total
Initial number of moles, n_i^o	1	1	0	0	2
Moles at equilibrium, n_i	1-ξ	1-ξ	ξ	ξ	2
Mole fractions at equilibrium, x_i	$\frac{1-\xi}{2}$	$\frac{1-\xi}{2}$	$\frac{\xi}{2}$	$\frac{\xi}{2}$	1

$$K = \prod_i (x_i)_e^{v_i} = \left(\frac{\xi}{1-\xi}\right)^2 = 5.1015 \; ; \qquad \xi = 0.6931$$

Equilibrium composition:

$$x_C = x_D = \frac{\xi}{2} = 0.34655 \; ; \;\; x_A = x_B = \frac{1-\xi}{2} = 0.15345$$

Problem 5

Consider the following gas-phase reaction where ethylene is converted to ethanol using steam:

$$C_2H_4(g) + H_2O(g) \rightleftharpoons C_2H_5OH(g)$$

The reaction is taking place at a temperature of 210 °C and pressure of 28 bar.

a) Calculate the equilibrium constant at 210 °C.
b) Estimate the equilibrium extent of reaction and equilibrium composition if 4 moles of steam is used per mole of ethylene.

The reaction mixture is a non-ideal gas mixture. However, you can assume, as an approximation, that $\hat{\varphi}_i(T, P, composition) = \varphi_i(T, P)$, that is, the fugacity coefficient of a species in the reaction mixture is equal to the fugacity coefficient of a pure species at the same T and P.
Given that:

species	T_c, K	P_c, bar	ω	Δg_f^o, kJ/mol at 298.15 K	Δh_f^o, kJ/mol at 298.15 K
$C_2H_4(g)$	282.3	50.40	0.087	68.5	52.5
$H_2O(g)$	647.1	220.55	0.345	-228.6	-241.8
$C_2H_5OH(g)$	513.9	61.48	0.645	-168.5	-235.1

Solution

a) Calculate the equilibrium constant at 210 °C.

The standard Gibbs free energy and standard enthalpy of reaction at 298.15 K are:

$$\Delta G^o = \sum_i v_i \, \Delta g_{fi}^o \text{ (compounds)} = (-168.5) - (68.5) - (-228.6) = -8.4 \text{ kJ/mol}$$

$$\Delta H^o = \sum_i v_i \, \Delta h_{fi}^o \text{(compounds)} = (-235.1) - (52.5) - (-241.8) = -45.8 \text{ kJ/mol}$$

$\Delta G^o = -RTlnK$. The equilibrium constant at 298.15 K is:

$$K = exp(-\Delta G^o/RT) = \exp\,(8.4 \times 10^3/(8.314 \times 298.15)) = 29.6277.$$

Assuming ΔH^o to be independent of temperature, the equilibrium constant at $T = 210 + 273.15 = 483.15$ K can be calculated from:

$$lnK = ln(K)_o - \frac{\Delta H^o}{R}\left(\frac{1}{T} - \frac{1}{T_o}\right) = \ln(29.6277) + \frac{45.8 \times 10^3}{8.314}\left(\frac{1}{483.15} - \frac{1}{298.15}\right) = -3.686$$

Thus, K= 0.02507.

b) Estimate the equilibrium extent of reaction and equilibrium composition if 4 moles of steam is used per mole of ethylene.

Let the equilibrium extent of the reaction be ξ.

From the given data, we can express the mole fractions and partial pressures of each species at equilibrium in the table form as shown below:

	$C_2H_4(g)$	$H_2O(g)$	$C_2H_5OH(g)$	total
Initial number of moles, n_i^o	1	4	0	5
Moles at equilibrium, n_i	1-ξ	4-ξ	ξ	5-ξ
Mole fractions at equilibrium, y_i	$\frac{1-\xi}{5-\xi}$	$\frac{4-\xi}{5-\xi}$	$\frac{\xi}{5-\xi}$	1
Partial pressure at equilibrium, P_i	$\left(\frac{1-\xi}{5-\xi}\right)P$	$\left(\frac{4-\xi}{5-\xi}\right)P$	$\left(\frac{\xi}{5-\xi}\right)P$	P

$$K = \prod_i \left(\frac{\hat{\varphi}_i y_i P}{P^o}\right)_e^{v_i} = \prod_i (\hat{\varphi}_i y_i)_e^{v_i} \prod_i \left(\frac{P}{P^o}\right)_e^{v_i} = \left(\frac{P}{P^o}\right)_e^{\sum_i v_i} \prod_i (\hat{\varphi}_i y_i)_e^{v_i}$$

$$K = \left(\frac{P}{P^o}\right)^{-1} \left(\hat{\varphi}_{C_2H_5OH} y_{C_2H_5OH}\right)\left(\hat{\varphi}_{C_2H_4} y_{C_2H_4}\right)^{-1}\left(\hat{\varphi}_{H_2O} y_{H_2O}\right)^{-1} = \left(\frac{P}{P^o}\right)^{-1}\left(\frac{\hat{\varphi}_{C_2H_5OH}}{\hat{\varphi}_{C_2H_4}\hat{\varphi}_{H_2O}}\right)\left(\frac{y_{C_2H_5OH}}{y_{C_2H_4} y_{H_2O}}\right)$$
$$= \left(\frac{P}{P^o}\right)^{-1}\left(\frac{y_{C_2H_5OH}}{y_{C_2H_4} y_{H_2O}}\right)\left(\frac{\varphi_{C_2H_5OH}}{\varphi_{C_2H_4}\varphi_{H_2O}}\right)$$

Note that in the last term the hat has been removed from the fugacity coefficients as we are assuming that $\hat{\varphi}_i = \varphi_i$ where φ_i is the fugacity coefficient of a pure species.

Let us now calculate the fugacity coefficients of pure species at 210 °C and 28 bar. We use the Pitzer correlation for the fugacity coefficient, given as:

$$ln\varphi = \left(\frac{P_r}{T_r}\right)(B^o + \omega B^1)$$

where B^o and B^1 are as follows:

$$B^o = 0.083 - \frac{0.422}{T_r^{1.6}} \quad ; \quad B^1 = 0.139 - \frac{0.172}{T_r^{4.2}}$$

The calculations are summarized in the table below.

species	T_r	P_r	B^o	B^1	φ
$C_2H_4(g)$	1.711477	0.555556	-0.09562	0.120996	0.972758
$H_2O(g)$	0.746639	0.126955	-0.5905	-0.44776	0.881022
$C_2H_5OH(g)$	0.940163	0.455433	-0.38279	-0.08388	0.809258

Thus,

$$\left(\frac{y_{C_2H_5OH}}{y_{C_2H_4}y_{H_2O}}\right) = K\left(\frac{P}{P^o}\right)\left(\frac{\varphi_{C_2H_4}\varphi_{H_2O}}{\varphi_{C_2H_5OH}}\right) = 0.02507 \times 28 \times \left(\frac{0.972758 \times 0.881022}{0.809258}\right) = 0.7434$$

$$\left(\frac{\xi}{5-\xi}\right)\left(\frac{5-\xi}{1-\xi}\right)\left(\frac{5-\xi}{4-\xi}\right) = 0.7434 \; ; 1.7434\xi^2 - 8.717\xi + 2.9736 = 0 \; ; \xi = 0.3682$$

The equilibrium extent of reaction is 0.3682.

Equilibrium composition:

$$y_{C_2H_5OH} = \frac{\xi}{5-\xi} = \frac{0.3682}{5-0.3682} = 0.0795 \; ; y_{C_2H_4} = \frac{1-\xi}{5-\xi} = \frac{1-0.3682}{5-0.3682} = 0.1364 \; ; y_{H_2O} = \frac{4-\xi}{5-\xi} = \frac{4-0.3682}{5-0.3682} = 0.7841$$

Problem 6

Consider the production of carbon black in solid form from methane gas as:

$$CH_4(g) \rightleftharpoons C(s) + 2H_2(g)$$

The reaction is taking place at constant temperature and pressure. The feed consists of 1 mole of pure methane.

a) Plot ΔH^o_{rxn} as a function of temperature. Is it independent of temperature?
b) Plot the equilibrium constant as a function of temperature
c) Plot the equilibrium extent of the reaction as a function of temperature.
d) Plot the equilibrium mole fraction of hydrogen as a function of temperature.

Given that at 298.15 K:

Species	Δg^o_f, kJ/mol	Δh^o_f, kJ/mol
$CH_4(g)$	-50.5	-74.5
$C(s)$	0	0
$H_2(g)$	0	0

The gas mixture (CH_4+H_2) can be treated as an ideal gas mixture.

Solution

The standard Gibbs free energy and standard enthalpy of reaction at 298.15 K are:

$$\Delta G^o = \sum_i v_i \, \Delta g^o_{fi} \text{ (compounds)}=(-50.5) = 50.5 \text{ kJ/mol}$$

$$\Delta H^o = \sum_i v_i \, \Delta h^o_{fi} \text{ (compounds)}=-(-74.5) = 74.5 \text{ kJ/mol}$$

a) Plot ΔH^o_{rxn} as a function of temperature. Is it independent of temperature?

$$\Delta H^o_{rxn} = \sum v_i h^o_i$$

$$h^o_i(T) - h^o_i(T_o) = \int_{T_o}^{T} c^o_{p,i}\, dT$$

$$\frac{c^o_{p,i}}{R} = A_i + B_i T + C_i T^2 + D_i T^{-2}$$

From the above equations, it can be readily shown that:

$$\Delta H^o_{rxn}(T) = \Delta H^o_{rxn}(T_o) + R\left[(\Delta A)(T - T_o) + \left(\frac{\Delta B}{2}\right)(T^2 - T_o^2) + \left(\frac{\Delta C}{3}\right)(T^3 - T_o^3) - (\Delta D)(T^{-1} - T_o^{-1})\right]$$

where

$$\Delta A = \sum v_i A_i \; ; \Delta B = \sum v_i B_i \; ; \Delta C = \sum v_i C_i \; ; \Delta D = \sum v_i D_i$$

From the tables in Appendices G and H,

species	A	$B \times 10^3$	$C \times 10^6$	$D \times 10^{-5}$
$CH_4(g)$	1.702	9.081	-2.164	0
$H_2(g)$	3.249	0.422	0	0.083
$C(s)$	1.771	0.771	0	-0.867

$\Delta H^o_{rxn}(T_o) = 74500$ J/mol ; $\Delta A = 6.567$; $\Delta B = -7.466 \times 10^{-3}$; $\Delta C = 2.164 \times 10^{-6}$; $\Delta D = -0.781 \times 10^5$.

We can now generate $\Delta H^o_{rxn}(T)$ versus T data and plot it. The plot of $\Delta H^o_{rxn}(T)$ versus T is shown below. Clearly $\Delta H^o_{rxn}(T)$ is not independent of temperature. It increases with the increase in temperature and tends to level off at high temperatures.

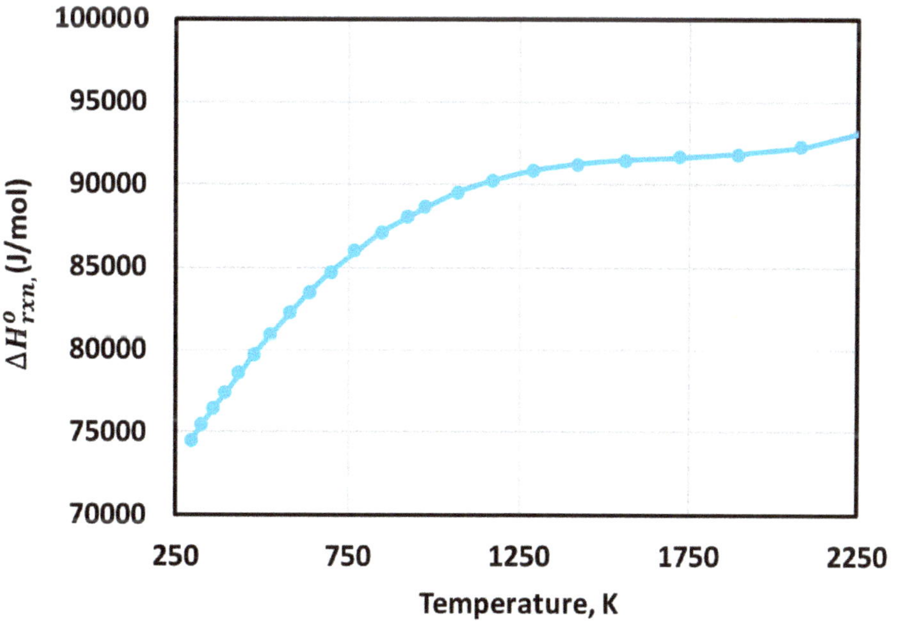

b) Plot the equilibrium constant as a function of temperature

The temperature dependence of equilibrium constant is governed by

$$\frac{d(lnK)}{dT} = \frac{\Delta H^o_{rxn}}{RT^2}$$

Upon integration, we get:

$$ln\left(\frac{K}{K_o}\right) = \int_{T_o}^{T}\left(\frac{\Delta H^o_{rxn}}{RT^2}\right)dT$$

The heat of reaction $\Delta H^o_{rxn}(T)$ is given as:

$$\Delta H^o_{rxn}(T) = \Delta H^o_{rxn}(T_o) + R\left[(\Delta A)(T - T_o) + \left(\frac{\Delta B}{2}\right)(T^2 - T_o^2) + \left(\frac{\Delta C}{3}\right)(T^3 - T_o^3) - (\Delta D)(T^{-1} - T_o^{-1})\right]$$

Thus,

$$ln\left(\frac{K}{K_o}\right) = \left[(\Delta A)ln\left(\frac{T}{T_o}\right) + \left(\frac{\Delta B}{2}\right)(T - T_o) + \left(\frac{\Delta C}{6}\right)(T^2 - T_o^2) + \left(\frac{\Delta D}{2}\right)(T^{-2} - T_o^{-2})\right] + \left[\frac{-\Delta H^o_{rxn}(T_o)}{R} + (\Delta A)T_o + \left(\frac{\Delta B}{2}\right)T_o^2 + \left(\frac{\Delta C}{3}\right)T_o^3 - \left(\frac{\Delta D}{T_o}\right)\right] \times (T^{-1} - T_o^{-1})$$

$$lnK_o = -\Delta G^o/RT = -50.5 \times 10^3/(8.314 \times 298.15) = -20.3726$$

We can now generate K versus T data from the above equations. The plot is shown below. The equilibrium constant increases rapidly with the increase in temperature at low temperatures. At high temperatures, the increase in equilibrium constant with the increase in temperature slows down.

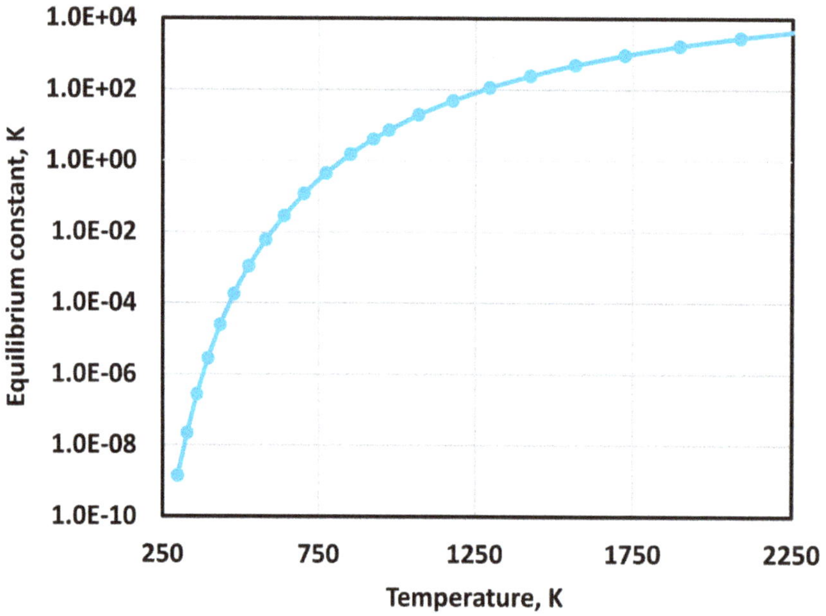

c) Plot the equilibrium extent of the reaction as a function of temperature.

This is a heterogeneous reaction. The equilibrium constant is:

$$K = \prod_i (\hat{a}_i)_e^{v_i} = \frac{(a_C)_e(\hat{a}_{H_2})_e^2}{(\hat{a}_{CH_4})_e} = \frac{(\hat{a}_{H_2})_e^2}{(\hat{a}_{CH_4})_e}$$

Note that the activity of pure solid species $C(s)$ is unity. The gaseous activities $\hat{a}_{H_2}$ and $\hat{a}_{CH_4}$ can be expressed as:

$$\hat{a}_{H_2} = \frac{\hat{f}_{H2}}{f_{H_2}^o} = \frac{y_{H_2}P}{P^o}$$

$$\hat{a}_{CH_4} = \frac{\hat{f}_{CH_4}}{f_{CH_4}^o} = \frac{y_{CH_4}P}{P^o}$$

Thus,

$$K = \frac{(\hat{a}_{H_2})_e^2}{(\hat{a}_{CH_4})_e} = \frac{(y_{H_2})^2}{y_{CH_4}}\left(\frac{P}{P^o}\right)$$

From the given data, we can express the equilibrium mole fractions of methane and hydrogen in the gas mixture in the table form as shown below:

	$CH_4(g)$	$H_2(g)$	total
Initial number of moles, n_i^o	1	0	1
Moles at equilibrium, n_i	1-ξ	2ξ	1+ξ
Mole fractions at equilibrium, y_i	$\frac{1-\xi}{1+\xi}$	$\frac{2\xi}{1+\xi}$	1

Thus,

$$K = \frac{(y_{H_2})^2}{y_{CH_4}}\left(\frac{P}{P^o}\right) = \left(\frac{2\xi}{1+\xi}\right)^2\left(\frac{1+\xi}{1-\xi}\right)\left(\frac{P}{P^o}\right) = \frac{4\xi^2}{1-\xi^2}\left(\frac{P}{P^o}\right)$$

$$\xi = \sqrt{\frac{K}{K + 4\left(\frac{P}{P^o}\right)}}$$

We can calculate the equilibrium extent of reaction as a function of temperature at any pressure using the above expression. Note that $P^o = 1$ bar. The plots of ξ versus T at different pressures are shown in the figure below. The extent of reaction increases rapidly with the increase in temperature at any given pressure. With the increase in pressure, ξ decreases at a given T.

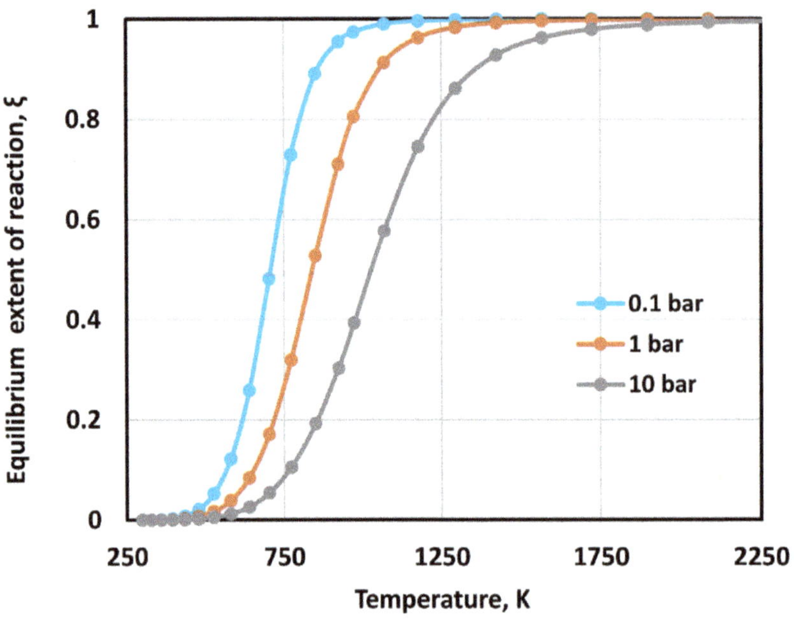

d) Plot the equilibrium mole fraction of hydrogen as a function of temperature.

The equilibrium mole fraction of hydrogen is given as:

$$y_{H_2} = \frac{2\xi}{1+\xi}$$

The plot of y_{H_2} as a function of temperature is shown below at a fixed pressure of 1 bar. As expected, y_{H_2} increases rapidly with the increase in temperature due to an increase in the extent of reaction.

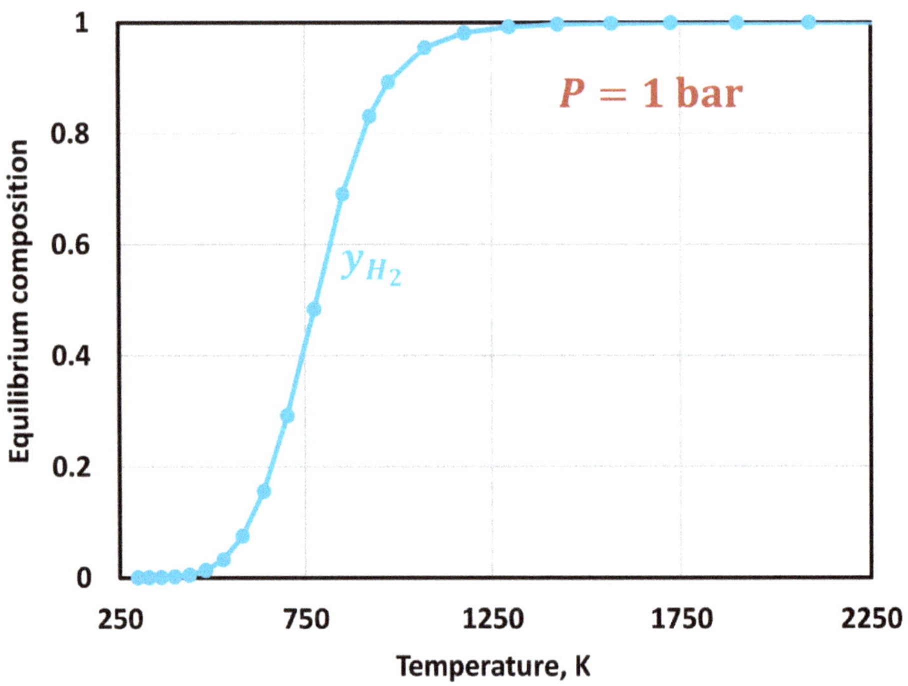

Problem 7

Consider the following set of gas-phase reactions occurring simultaneously:

$$A(g) + B(g) \rightleftharpoons C(g) + 4D(g) \qquad (rxn\ 1)$$
$$A(g) + 3B(g) \rightleftharpoons E(g) + 5D(g) \qquad (rxn\ 2)$$

The initial number of moles of species involved in the reaction are: $A(g) = 2$ moles, $B(g) = 4$ moles, $C(g) = 0$ mole, $D(g) = 0$ mole, and $E(g) = 0$ mole. The equilibrium extents of the reactions are: ξ_1 for reaction (1) and ξ_2 for reaction (2). Express the equilibrium composition of the reaction mixture in terms of the extents of reactions.

Solution

The solution is given in the following table.

species→	$A(g)$	$B(g)$	$C(g)$	$D(g)$	$E(g)$	total
n_i^o	2	4	0	0	0	$\sum n_i^o = 6$
$v_{i,j}\xi_j$ (j=1)	$-\xi_1$	$-\xi_1$	ξ_1	$4\xi_1$	0	$\sum v_{i,1}\xi_1 = 3\xi_1$
$v_{i,j}\xi_j$ (j=2)	$-\xi_2$	$-3\xi_2$	0	$5\xi_2$	ξ_2	$\sum v_{i,2}\xi_2 = 2\xi_2$
$\sum_j v_{i,j}\xi_j$	$-\xi_1 - \xi_2$	$-\xi_1 - 3\xi_2$	ξ_1	$4\xi_1 + 5\xi_2$	ξ_2	$\sum_j \sum_i v_{i,j}\xi_j = 3\xi_1 + 2\xi_2$
$n_i = n_i^o + \sum_j v_{i,j}\xi_j$	$2 - \xi_1 - \xi_2$	$4 - \xi_1 - 3\xi_2$	ξ_1	$4\xi_1 + 5\xi_2$	ξ_2	$n = \sum n_i^o + \sum_j \sum_i v_{i,j}\xi_j = 6 + 3\xi_1 + 2\xi_2$
$y_i = \frac{n_i}{n}$	$\frac{2 - \xi_1 - \xi_2}{6 + 3\xi_1 + 2\xi_2}$	$\frac{4 - \xi_1 - 3\xi_2}{6 + 3\xi_1 + 2\xi_2}$	$\frac{\xi_1}{6 + 3\xi_1 + 2\xi_2}$	$\frac{4\xi_1 + 5\xi_2}{6 + 3\xi_1 + 2\xi_2}$	$\frac{\xi_2}{6 + 3\xi_1 + 2\xi_2}$	$\sum y_i = 1$

Problem 8

Consider the following set of reactions occurring simultaneously:

$$A(s) + 2B(g) \rightleftharpoons C(g) + 2D(g) \qquad (rxn\ 1)$$
$$A(s) + B(g) \rightleftharpoons E(g) + D(g) \qquad (rxn\ 2)$$
$$A(s) + C(g) \rightleftharpoons 2E(g) \qquad (rxn\ 3)$$
$$A(s) + 2D(g) \rightleftharpoons F(g) \qquad (rxn\ 4)$$
$$E(g) + B(g) \rightleftharpoons C(g) + D(g) \qquad (rxn\ 5)$$

Are these reactions independent of each other?

Solution

All the reactions are not independent as:

$$(rxn\ 3) = 2(rxn\ 2) - (rxn\ 1)$$

$$2A(s) + 2B(g) \rightleftharpoons 2E(g) + 2D(g) \qquad (2 \times rxn\ 2)$$

$$A(s) + 2B(g) \rightleftharpoons C(g) + 2D(g) \qquad (rxn\ 1)$$

$$A(s) + C(g) \rightleftharpoons 2E(g) \quad (2 \times rxn\ 2 - rxn1 = rxn\ 3)$$

Likewise,

$$(rxn\ 5) = (rxn\ 1) - (rxn\ 2)$$

$$A(s) + 2B(g) \rightleftharpoons C(g) + 2D(g) \qquad (rxn\ 1)$$

$$A(s) + B(g) \rightleftharpoons E(g) + D(g) \quad (rxn\ 2)$$

$$E(g) + B(g) \rightleftharpoons C(g) + D(g) \qquad (rxn\ 1 - rxn\ 2 = rxn\ 5)$$

Thus, there are only three independent reactions:

$$A(s) + 2B(g) \rightleftharpoons C(g) + 2D(g) \qquad (rxn\ 1)$$

$$A(s) + B(g) \rightleftharpoons E(g) + D(g) \qquad (rxn\ 2)$$

$$A(s) + 2D(g) \rightleftharpoons F(g) \qquad (rxn\ 4)$$

Problem 9

Consider the following set of independent gas phase reactions occurring simultaneously at constant T and P :

$$A(g) \rightleftharpoons B(g) + C(g) \qquad (rxn\ 1)$$
$$A(g) \rightleftharpoons D(g) + E(g) \qquad (rxn\ 2)$$

The feed consists of 2 moles of pure $A(g)$.

a) Plot ΔH^o_{rxn} as a function of temperature for both reactions. Is ΔH^o_{rxn} independent of temperature?
b) Plot the equilibrium constant as a function of temperature for both reactions.
c) Calculate the equilibrium extents of the reactions ξ_1 and ξ_2 at 500 °C and 1 bar.
d) Calculate the equilibrium composition of the reaction mixture at 500 °C and 1 bar.

Given that at 298.15 K:

Species	Δg^o_f, kJ/mol	Δh^o_f, kJ/mol
$A(g)$	-16.6	-125.8
$B(g)$	68.5	52.5
$C(g)$	-31.9	-83.8
$D(g)$	62.2	19.7
$E(g)$	-50.5	-74.5

The standard heat capacity of species is given as:

$$\frac{c_{p,i}^{o}}{R} = a_i + b_i T + c_i T^2 + d_i T^{-2}$$

where the coefficients for different gaseous species are given as:

species	a	$b \times 10^3$	$c \times 10^6$	$d \times 10^{-5}$
$A(g)$	1.935	36.915	-11.402	0
$B(g)$	1.424	14.394	-4.392	0
$C(g)$	1.131	19.225	-5.561	0
$D(g)$	1.637	22.706	-6.915	0
$E(g)$	1.702	9.081	-2.164	0

The gas mixture can be treated as an ideal gas mixture.

Solution

The standard Gibbs free energy and standard enthalpy of reaction at 298.15 K are:

rxn 1:

$$\Delta G^o = \sum_i v_i \, \Delta g_{fi}^o \text{ (compounds)} = (68.5) + (-31.9) - (-16.6) = 53.2 \text{ kJ/mol}$$

$$\Delta H^o = \sum_i v_i \, \Delta h_{fi}^o \text{ (compounds)} = (52.5) + (-83.8) - (-125.8)) = 94.8 \text{ kJ/mol}$$

rxn 2:

$$\Delta G^o = \sum_i v_i \, \Delta g_{fi}^o \text{ (compounds)} = (62.2) + (-50.5) - (-16.6) = 28.3 \text{ kJ/mol}$$

$$\Delta H^o = \sum_i v_i \, \Delta h_{fi}^o \text{ (compounds)} = (19.7) + (-74.5) - (-125.8)) = 71 \text{ kJ/mol}$$

a) Plot ΔH_{rxn}^o as a function of temperature for both reactions. Is ΔH_{rxn}^o independent of temperature?

$$\Delta H_{rxn}^o(T) = \Delta H_{rxn}^o(T_o) + R\left[(\Delta a)(T - T_o) + \left(\frac{\Delta b}{2}\right)(T^2 - T_o^2) + \left(\frac{\Delta c}{3}\right)(T^3 - T_o^3) - (\Delta d)(T^{-1} - T_o^{-1})\right]$$

where

$T_o = 298.15$ K, $\Delta a = \sum v_i a_i$; $\Delta b = \sum v_i b_i$; $\Delta c = \sum v_i c_i$; $\Delta d = \sum v_i d_i$

For rxn 1:

$$\Delta H_{rxn}^o(T_o) = 94.8 \text{ kJ/mol}, \Delta a = 0.62, \Delta b = -3.296 \times 10^{-3}, \Delta c = 1.449 \times 10^{-6}, \Delta d = 0$$

For rxn 2:

$$\Delta H_{rxn}^o(T_o) = 71 \text{ kJ/mol}, \Delta a = 1.404, \Delta b = -5.128 \times 10^{-3}, \Delta c = 2.323 \times 10^{-6}, \Delta d = 0$$

The following figure shows the plots of $\Delta H_{rxn}^o(T)$ versus T. The $\Delta H_{rxn}^o(T)$ for reaction 1 is larger than the $\Delta H_{rxn}^o(T)$ for reaction 2. Also, the standard heat of reaction varies significantly with temperature.

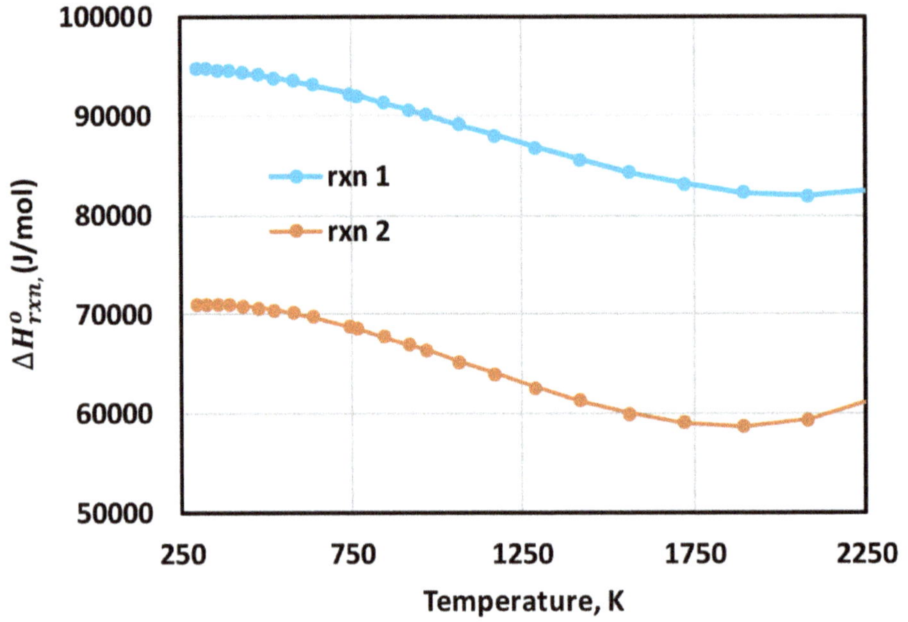

b) Plot the equilibrium constant as a function of temperature for both reactions.

The equilibrium constant is given as:

$$ln\left(\frac{K}{K_o}\right) = \left[(\Delta a)ln\left(\frac{T}{T_o}\right) + \left(\frac{\Delta b}{2}\right)(T - T_o) + \left(\frac{\Delta c}{6}\right)(T^2 - T_o^2) + \left(\frac{\Delta d}{2}\right)(T^{-2} - T_o^{-2})\right] + \left[\frac{-\Delta H^o_{rxn}(T_o)}{R} + (\Delta a)T_o + \left(\frac{\Delta b}{2}\right)T_o^2 + \left(\frac{\Delta c}{3}\right)T_o^3 - \left(\frac{\Delta d}{T_o}\right)\right] \times (T^{-1} - T_o^{-1})$$

$T_o = 298.15$ K, $\Delta a = \sum v_i a_i$; $\Delta b = \sum v_i b_i$; $\Delta c = \sum v_i c_i$; $\Delta d = \sum v_i d_i$

For rxn 1:

$$lnK_o = -\Delta G^o/RT = -53.2 \times 10^3/(8.314 \times 298.15) = -21.4618$$

$$\Delta a = 0.62, \Delta b = -3.296 \times 10^{-3}, \Delta c = 1.449 \times 10^{-6}, \Delta d = 0$$

For rxn 2:

$$lnK_o = -\Delta G^o/RT = -28.3 \times 10^3/(8.314 \times 298.15) = -11.4167$$

$$\Delta a = 1.404, \Delta b = -5.128 \times 10^{-3}, \Delta c = 2.323 \times 10^{-6}, \Delta d = 0$$

We can now generate K versus T data from the above equations. The plots are shown below. The equilibrium constant increases rapidly with the increase in temperature at low temperatures. At high temperatures, the increase in equilibrium constant with the increase in temperature slows down. The equilibrium constant of reaction 2 is higher than that of reaction 1 at the same temperature.

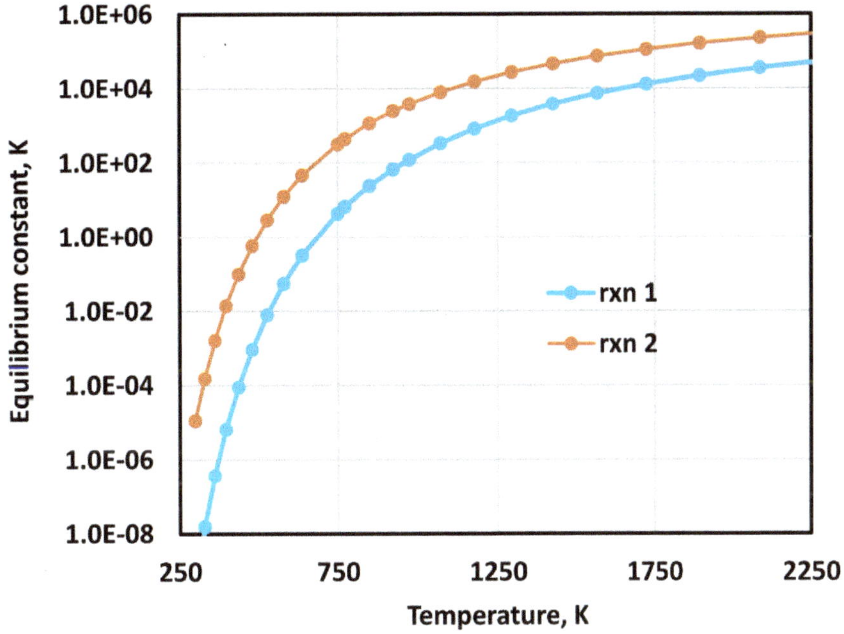

c) Calculate the equilibrium extents of the reactions ξ_1 and ξ_2 at 500 °C and 1 bar.

At 500 °C and 1 bar, $K_1 = 6.421$ and $K_1 = 429.337$.

$$K_1 = \prod_i \left(\frac{P_i}{P^o}\right)_e^{\nu_i} = \left(\frac{P_B}{P^0}\right)\left(\frac{P_C}{P^o}\right)\left(\frac{P_A}{P^o}\right)^{-1} = \left(\frac{y_B y_C}{y_A}\right)\left(\frac{P}{P^o}\right)$$

$$K_2 = \prod_i \left(\frac{P_i}{P^o}\right)_e^{\nu_i} = \left(\frac{P_D}{P^o}\right)\left(\frac{P_E}{P^o}\right)\left(\frac{P_A}{P^o}\right)^{-1} = \left(\frac{y_D y_E}{y_A}\right)\left(\frac{P}{P^o}\right)$$

The following table shows how to express the mole fractions y_i's in terms of the extents of reactions ξ_1 and ξ_2:

species→	$A(g)$	$B(g)$	$C(g)$	$D(g)$	$E(g)$	total
n_i^o	2	0	0	0	0	$\sum n_i^o = 2$
$\nu_{i,j}\xi_j$ (j=1)	$-\xi_1$	ξ_1	ξ_1	0	0	$\sum \nu_{i,1}\xi_1 = \xi_1$
$\nu_{i,j}\xi_j$ (j=2)	$-\xi_2$	0	0	ξ_2	ξ_2	$\sum \nu_{i,2}\xi_2 = \xi_2$
$\sum_j \nu_{i,j}\xi_j$	$-\xi_1 - \xi_2$	ξ_1	ξ_1	ξ_2	ξ_2	$\sum_j \sum_i \nu_{i,j}\xi_j = \xi_1 + \xi_2$
$n_i = n_i^o + \sum_j \nu_{i,j}\xi_j$	$2 - \xi_1 - \xi_2$	ξ_1	ξ_1	ξ_2	ξ_2	$n = \sum n_i^o + \sum_j \sum_i \nu_{i,j}\xi_j = 2 + \xi_1 + \xi_2$
$y_i = \frac{n_i}{n}$	$\frac{2 - \xi_1 - \xi_2}{2 + \xi_1 + \xi_2}$	$\frac{\xi_1}{2 + \xi_1 + \xi_2}$	$\frac{\xi_1}{2 + \xi_1 + \xi_2}$	$\frac{\xi_2}{2 + \xi_1 + \xi_2}$	$\frac{\xi_2}{2 + \xi_1 + \xi_2}$	$\sum y_i = 1$

Thus,

$$K_1\left(\frac{P}{P^o}\right)^{-1} = \left(\frac{y_B y_C}{y_A}\right) = \frac{\xi_1^2}{(2+\xi_1+\xi_2)(2-\xi_1-\xi_2)} \quad (1)$$

$$K_2\left(\frac{P}{P^o}\right)^{-1} = \left(\frac{y_D y_E}{y_A}\right) = \frac{\xi_2^2}{(2+\xi_1+\xi_2)(2-\xi_1-\xi_2)} \quad (2)$$

Upon dividing Equation (2) by Equation (1), we get:

$$\left(\frac{K_2}{K_1}\right) = \left(\frac{\xi_2}{\xi_1}\right)^2 = \left(\frac{429.337}{6.421}\right) = 66.8645$$

$$\xi_2 = 8.177\xi_1$$

Eliminating ξ_2 from Equation (2) gives:

$$K_2\left(\frac{P}{P^o}\right)^{-1} = 429.337 = \frac{66.8645\xi_1^2}{(2+\xi_1+8.177\xi_1)(2-\xi_1-8.177\xi_1)}$$

Note that $P = P^o = 1$ bar. Upon solving this equation, we get: $\xi_1 = 0.2177$. Thus, $\xi_2 = 8.177\xi_1 = 1.78$

d) Calculate the equilibrium composition of the reaction mixture at 500 °C and 1 bar.

$$\xi_1 = 0.2177, \ \xi_2 = 1.78.$$

$$y_A = \frac{2-\xi_1-\xi_2}{2+\xi_1+\xi_2} = \frac{2-0.2177-1.78}{2+0.2177+1.78} = 0.000575\ ;\ y_B = y_C = \frac{\xi_1}{2+\xi_1+\xi_2} = \frac{0.2177}{2+0.2177+1.78} = 0.054456\ ;\ y_D =$$

$$y_E = \frac{\xi_2}{2+\xi_1+\xi_2} = \frac{1.78}{2+0.2177+1.78} = 0.445256$$

PRACTICE PROBLEMS

Problem 1

A reaction system consisting of 3 moles of $NH_3(g)$ and 3 moles of $O_2(g)$ initially is undergoing the following reaction:

$$4NH_3(g) + 3O_2(g) \rightleftharpoons 4\,N_2(g) + 6H_2O(g)$$

Express the mole fractions of species as functions of the extent of reaction.

Problem 2

Consider the formation reaction of ammonia gas:

$$\frac{1}{2}N_2(g) + \frac{3}{2}\,H_2(g) \rightleftharpoons NH_3(g)$$

The standard heat of formation of $NH_3(g)$ at 25 °C is -46110 J/mol.

a) Estimate the standard heat of formation of $NH_3(g)$ at 600 °C.
b) Estimate the standard enthalpy of $NH_3(g)$ at 600 °C.
c) Is the standard enthalpy of $NH_3(g)$ at 600 °C equal to the standard heat of formation of $NH_3(g)$ at 600 °C ? Explain.

Given that:

$$\frac{c_{p,i}^{o}}{R} = A_i + B_iT + C_iT^2 + D_iT^{-2}$$

where T is in K and the coefficients are as follows:

species	A	$B \times 10^3$	$C \times 10^6$	$D \times 10^{-5}$
$NH_3(g)$	3.578	3.020	0	-0.186
$H_2(g)$	3.249	0.422	0	0.083
$N_2(g)$	3.280	0.593	0	0.04

Problem 3
Consider the following gas-phase reaction:

$$\frac{1}{2}H_2(g) + \frac{1}{2}I_2(g) \rightleftharpoons 2HI(g)$$

The reaction takes place at constant T and P in a reaction vessel. The initial number of moles of species involved in the reaction are: $H_2(g) = 0.5$ mol, $I_2(g) = 0.5$ mol, and $HI(g) = 0$ mol. Calculate:

a) The equilibrium constant K_p at 500 K.
b) The equilibrium extent of the reaction and equilibrium composition at 500 K and $P = 1$ bar?
c) The equilibrium extent of the reaction and equilibrium composition at 500 K and $P = 8$ bar?

Given that at 500 K

Species	h^o, kJ/mol	s^o, kJ/K.mol
$HI(g)$	32.41	0.22163
$H_2(g)$	5.88	0.14564
$I_2(g)$	69.75	0.27994

You can assume that the reaction mixture is an ideal gas mixture.

Problem 4
Consider the following reaction involving oxidation of nitric oxide to nitrogen dioxide:

$$NO(g) + \frac{1}{2}O_2(g) \rightleftharpoons NO_2(g)$$

One mole of nitric oxide is mixed with 20% excess oxygen. The reaction is carried out at 500 °C and 1 bar. Estimate the equilibrium extent of reaction at 500 °C and 1 bar. What is the equilibrium composition? You can assume that the reaction mixture is an ideal gas mixture and ΔH_{rxn}^{o} is independent of temperature.

Problem 5
Consider the following liquid-phase esterification reaction where acetic acid combines with methanol to give methyl acetate:

$$CH_3COOH(l) + CH_3OH(l) \rightleftharpoons CH_3COOCH_3(l) + H_2O(l)$$

The reaction is taking place at $T = 50$ °C, and $P = 1$ bar. The initial number of moles of species involved in the reaction are: $CH_3COOH(l) = 1$ mole, $CH_3OH(l) = 1$ mole, $CH_3COOCH_3(l) = 0$ mole, and $H_2O(l) = 0$ mole. Assuming that the liquid solution is ideal, estimate the equilibrium extent of reaction ξ and the equilibrium composition. You can also assume that ΔH_{rxn}^{o} is independent of temperature.

Given that at 298.15 K:

Species	Δg_f^o, kJ/mol	Δh_f^o, kJ/mol
$CH_3COOH(l)$	-389.9	-484.5
$CH_3OH(l)$	-166.3	-238.7
$CH_3COOCH_3(l)$	-328.4	-442.8
$H_2O(l)$	-237.1	-285.8

Problem 6
Consider the formation reaction of ammonia gas:

$$\frac{1}{2}N_2(g) + \frac{3}{2}H_2(g) \rightleftharpoons NH_3(g)$$

a) Plot ΔH_{rxn}^o of this reaction as a function of temperature over the temperature range of 300 to 1000 K. Is it independent of temperature?
b) Plot the equilibrium constant as a function of temperature over the temperature range of 300 to 1000 K.
c) The above reaction is carried out at 500 K and 500 bar. The initial number of moles of species involved in the reaction are: $N_2(g) = 0.5$ moles, $H_2(g) = 1.5$ moles, $NH_3(g) = 0$ moles. Estimate the equilibrium extent of reaction and equilibrium composition at 500 K and 500 bar. The reaction mixture is non-ideal at high pressure of 500 bar. However, you can assume, as an approximation, that $\hat{\varphi}_i(T, P, composition) = \varphi_i(T, P)$, that is, the fugacity coefficient of a species in the gas mixture is equal to the fugacity coefficient of a pure gas species at the same T and P.

Problem 7
Consider the following heterogeneous reaction:

$$FeO(s) + CO(g) \rightleftharpoons Fe(s) + CO_2(g)$$

a) Determine ΔH_{rxn}^o and ΔG_{rxn}^o at 25 °C
b) Estimate the equilibrium constant at 150 °C assuming ΔH_{rxn}^o is constant
c) Estimate the equilibrium extent of reaction and equilibrium composition of gas phase at 150 °C, if only one mole of $CO(g)$ is present initially with an excess of $FeO(s)$.

Given that at 298.15 K:

Species	Δg_f^o, kJ/mol	Δh_f^o, kJ/mol
$FeO(s)$	-251.46	-272
$CO(g)$	-137.2	-110.5
$Fe(s)$	0	0
$CO_2(g)$	-394.4	-393.5

Problem 8
As discussed in section 13.10, the following reaction has a large positive value of ΔG^o. Consequently, the equilibrium extent of reaction is small.

$$TiO_2(s) + 2Cl_2(g) \rightleftharpoons TiCl_4(l) + O_2(g)$$

$$\Delta G^0(25^oC) = +153\ kJ/mol\ ;\ \Delta H^0(25^oC) = 141\ kJ/mol$$

To increase the production of $TiCl_4(l)$, we can couple the above reaction with the following reaction having a large negative value of ΔG^o:

$$C(s) + O_2(g) \rightleftharpoons CO_2(g)$$

$$\Delta G^o(25^oC) = -394.40\ kJ/mol\ ;\quad \Delta H^0(25^oC) = -393.5\ kJ/mol$$

The coupled reaction is:

$$C(s) + TiO_2(s) + 2Cl_2(g) \rightleftharpoons TiCl_4(l) + CO_2(g)$$

$$\Delta G^o(25^oC) = -241.4\ kJ/mol\ ;\quad \Delta H^o(25^oC) = -252.5\ kJ/mol$$

Calculate the equilibrium extent of the coupled reaction and equilibrium gas composition at 350 K and 1 bar, if 1 mole of $TiO_2(s)$, 2 moles of $Cl_2(g)$, and 1 mole of $C(s)$ are present initially. You can assume that ΔH^0 is constant independent of temperature.

Problem 9

Consider the following set of independent gas phase reactions occurring simultaneously at 250 °C and 150 bar :

$$2A(g) + B(g) \rightleftharpoons C(g) \qquad (rxn\ 1)$$
$$A(g) + D(g) \rightleftharpoons B(g) + E(g) \qquad (rxn\ 2)$$

The feed to the reactor consists of 80 mol% $A(g)$, 10 mol% $B(g)$, and 10 mol% $D(g)$. Using 1 mole of feed as the basis for calculation,

a) Determine the equilibrium constants for both reactions at 250 °C.
b) Develop the expressions for the equilibrium extents of the reactions (ξ_1 and ξ_2) corresponding to reactions (1) and (2) at 250 °C and 150 bar.
c) Solve for the equilibrium extents of the reactions ξ_1 and ξ_2 at 250 °C and 150 bar.
d) Calculate the equilibrium composition of the reaction mixture at 250 °C and 150 bar.

Given that at 298.15 K:

Species	Δg_f^o, kJ/mol	Δh_f^o, kJ/mol
$A(g)$	0	0
$B(g)$	-137.2	-110.5
$C(g)$	-162.0	-200.7
$D(g)$	-394.4	-393.5
$E(g)$	-228.6	-241.8

The standard heat capacity of species is given as:

$$\frac{c_{p,i}^o}{R} = a_i + b_iT + c_iT^2 + d_iT^{-2}$$

where the coefficients for different gaseous species are given as:

species	a	$b \times 10^3$	$c \times 10^6$	$d \times 10^{-5}$
$A(g)$	3.249	0.422	0	0.083
$B(g)$	3.376	0.557	0	-0.031
$C(g)$	2.211	12.216	-3.450	0

$D(g)$	5.457	1.045	0	-1.157
$E(g)$	3.470	1.450	0	0.121

The gas mixture can be treated as an ideal gas mixture.

Problem 10

In a hydrogen/oxygen fuel cell, the overall reaction involved is the combustion of hydrogen gas as shown below:

$$H_2(g) + \frac{1}{2}\,O_2(g) \rightleftharpoons H_2O(g)$$

a) Plot ΔH^o_{rxn} of the reaction as a function of temperature over the temperature range of 250 to 500 K. Is it independent of temperature?
b) Plot ΔG^o_{rxn} of the reaction as a function of temperature over the temperature range of 250 to 500 K. Is it independent of temperature?
c) Plot the emf (electromotive force) of a reversible hydrogen/oxygen fuel cell as a function of temperature over the temperature range of 250 to 500 K. Comment on the effect of temperature on the performance of a fuel cell.

Note that:

$$\Delta H^o_{rxn}(T) = \Delta H^o_{rxn}(T_o) + R\left[(\Delta A)(T - T_o) + \left(\frac{\Delta B}{2}\right)(T^2 - T_o^2) + \left(\frac{\Delta C}{3}\right)(T^3 - T_o^3) - (\Delta D)(T^{-1} - T_o^{-1})\right]$$

$$\Delta S^o_{rxn}(T) = \Delta S^o_{rxn}(T_o) + R\left[(\Delta A)ln\left(\frac{T}{T_o}\right) + (\Delta B)(T - T_o) + \left(\frac{\Delta C}{2}\right)(T^2 - T_o^2) - \left(\frac{\Delta D}{2}\right)(T^{-2} - T_o^{-2})\right]$$

$$\Delta S^o_{rxn}(T_o) = [\Delta H^o_{rxn}(T_o) - \Delta G^o_{rxn}(T_o)]/T_o$$

where $\Delta A = \sum v_i A_i$; $\Delta B = \sum v_i B_i$; $\Delta C = \sum v_i C_i$; $\Delta D = \sum v_i D_i$ and coefficients A_i, B_i, C_i and D_i belong to the heat capacity expression:

$$\frac{c^o_{p,i}}{R} = A_i + B_i T + C_i T^2 + D_i T^{-2}$$

Once $\Delta H^o_{rxn}(T)$ and $\Delta S^o_{rxn}(T)$ are known at temperature T, $\Delta G^o_{rxn}(T)$ can be calculated from:

$$\Delta G^o_{rxn}(T) = \Delta H^o_{rxn}(T) - T\Delta S^o_{rxn}(T)$$

The values of ΔG^o_{rxn} and ΔH^o_{rxn} for the cell reaction are -228.572 kJ/mol and -241.818 kJ/mol at 25^{0}C, respectively.

Chapter 14

Work and Entropy Analysis of Systems and the Gouy-Stodola Theorem

14.1 LEARNING OUTCOMES

Upon completion of this chapter, students will be able to:

- Appreciate that work is lost in a process due to irreversibilities and that irreversibilities can be quantified in terms of entropy generation
- Develop Gouy-Stodola theorem for closed systems interacting thermally with single or multiple heat reservoirs
- Define thermodynamic efficiency of a process
- Develop Gouy-Stodola theorem for open (flow) systems interacting thermally with single or multiple heat reservoirs
- Apply Gouy-Stodola theorem to multi-unit flow processes and determine the contribution of each unit to the total entropy generated and the total work lost in the whole process.
- Solve problems related to work and entropy analysis of closed and open systems. Apply the Gouy-Stodola theorem to estimate the work lost in the process

14.2 INTRODUCTION

Irreversibilities in a process make the process inefficient from a thermodynamic perspective. If the process is producing work, the work produced is less than the reversible work when irreversibilities are present, for the same change in state of the system. If the process is consuming work, the work consumed is more than the reversible work when irreversibilities are present, for the same change in state of the system.

In order to improve the thermodynamic efficiency of the process, it is important that we are able to quantify the irreversibilities in the process. To that end, the Gouy-Stodola theorem plays an important role. It allows us to quantify irreversibilities in a process in terms of entropy generation. In this chapter, the work and entropy analysis of systems will be carried out and the basis for the Gouy-Stodola theorem will be discussed.

14.3 CLOSED SYSTEMS

14.3.1 Closed System with a Single Heat Reservoir

Consider a closed system interacting thermally with a single thermal energy reservoir (TER), namely the surroundings/environment at a uniform temperature of T_o. The system is receiving heat in the amount of Q_o and producing work in the amount of W, as shown in Figure 14.1.

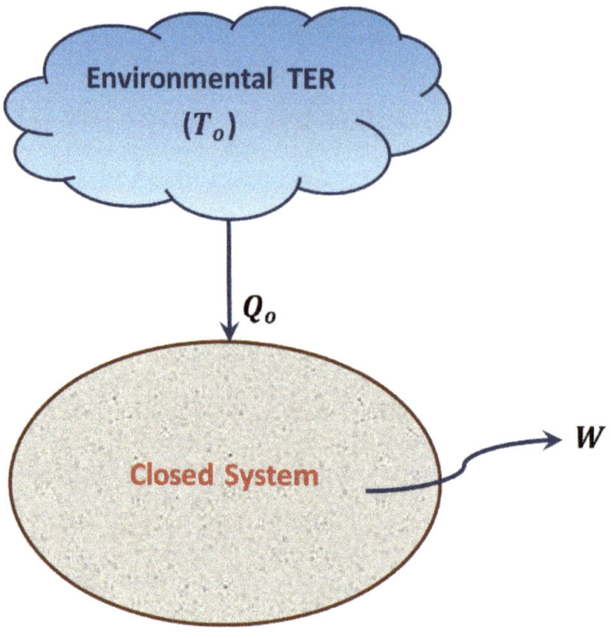

Figure 14.1 Closed system interacting with environment TER (thermal energy reservoir) at uniform temperature of T_o and producing work.

From the first law of thermodynamics for closed systems,

$$W = Q_o - \Delta U \tag{14.1}$$

The second law of thermodynamics for closed systems gives:

$$S_{G,total} = \Delta S - \frac{Q_o}{T_o} \geq 0 \tag{14.2}$$

For the process to be reversible, $S_{G,total} = 0$. Therefore, Equation (14.2) yields:

$$Q_{o,rev} = T_o \Delta S \tag{14.3}$$

Combination of Equations (14.1) and (14.3) gives:

$$W_{rev} = Q_{o,rev} - \Delta U \tag{14.4}$$

$$W_{rev} = T_o \Delta S - \Delta U \tag{14.5}$$

Equation (14.1) gives the actual work produced by the process. For the same change in the state of the process, Equation (14.5) gives the reversible work produced. W_{rev} is the maximum possible work produced for a given change in state. Note that Q_0 is the not the same in actual and reversible processes.

The work lost due to irreversibilities in the process is defined as:

$$W_{lost} = W_{rev} - W \tag{14.6}$$

Substituting the expressions for W from Equation (14.1) and W_{rev} from Equation (14.5) into Equation (14.6), we get:

$$W_{lost} = T_o \Delta S - Q_o = T_o \left(\Delta S - \frac{Q_o}{T_o} \right) \tag{14.7}$$

From Equation (14.2), the total entropy generated in the universe is the quantity inside the bracket on the right hand side of Equation (14.7). Thus, Equation (14.7) becomes:

$$W_{lost} = T_o S_{G,total} \tag{14.8}$$

This is the *Gouy-Stodola theorem* which states that the work lost due to irreversibilities in the process is directly proportional to the total entropy generated (internally and externally) due to the process.

14.3.2 Closed System with Three Heat Reservoirs

Now we consider a closed system interacting thermally with three thermal energy reservoirs (TERs), as shown in Figure 14.2.

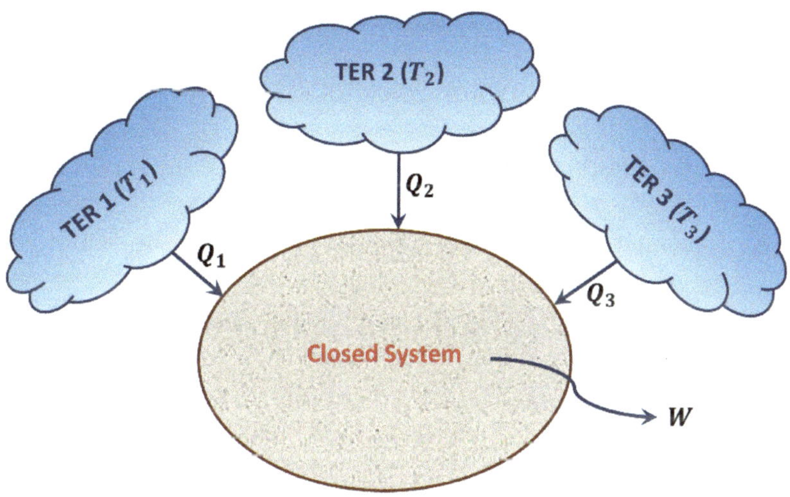

Figure 14.2 Closed system interacting with three TERs (thermal energy reservoirs) and producing work.

From the first law of thermodynamics for closed systems,

$$W = Q_1 + Q_2 + Q_3 - \Delta U \tag{14.9}$$

The second law of thermodynamics for closed systems gives:

$$S_{G,total} = \Delta S - \left(\frac{Q_1}{T_1} + \frac{Q_2}{T_2} + \frac{Q_3}{T_3} \right) \geq 0 \tag{14.10}$$

Comparison of Actual Process with Reversible Process

To compare the actual process with the reversible process, we will keep the change in states of the system to be the same. The amounts of heat exchanged between the system and the TERs will also be kept the same except for one TER. This TER is called the *reference TER*. The amount of heat exchanged between the system and the reference TER will depend on the degree of irreversibility in the process. The reference heat reservoir is also referred to as *floating reservoir* as the amount of heat exchanged varies with the degree of irreversibility and work output of the system. We can select any TER as the reference TER.

Reversible Work and Work Lost due to Irreversibilities

We take TER 1 (see Figure 14.2) as the floating reservoir. For the process to be reversible, $S_{G,total} = 0$. Therefore, Equation (14.10) yields:

$$\left(\frac{Q_{1,rev}}{T_1} + \frac{Q_2}{T_2} + \frac{Q_3}{T_3}\right) = \Delta S \tag{14.11}$$

Upon re-arrangement, Equation (14.11) gives:

$$Q_{1,rev} = T_1\left[\Delta S - \left(\frac{Q_2}{T_2} + \frac{Q_3}{T_3}\right)\right] \tag{14.12}$$

Upon substitution for $Q_{1,rev}$ from Equation (14.12) into Equation (14.9), the reversible work is obtained as:

$$W_{rev} = Q_2 + Q_3 + T_1\left[\Delta S - \left(\frac{Q_2}{T_2} + \frac{Q_3}{T_3}\right)\right] - \Delta U \tag{14.13}$$

Upon re-arrangement, Equation (14.13) can be re-written as:

$$W_{rev} = Q_2\left(1 - \frac{T_1}{T_2}\right) + Q_3\left(1 - \frac{T_1}{T_3}\right) + T_1\Delta S - \Delta U \tag{14.14}$$

Using Equation (14.6), the work lost due to irreversibilities in the process is:

$$W_{lost} = \left[Q_2\left(1 - \frac{T_1}{T_2}\right) + Q_3\left(1 - \frac{T_1}{T_3}\right) + T_1\Delta S - \Delta U\right] - [Q_1 + Q_2 + Q_3 - \Delta U] \tag{14.15}$$

Upon simplification, Equation (14.15) gives:

$$W_{lost} = T_1\left[\Delta S - \left(\frac{Q_1}{T_1} + \frac{Q_2}{T_2} + \frac{Q_3}{T_3}\right)\right] \tag{14.16}$$

The square bracketed term on the right hand side of Equation (14.16) is $S_{G,total}$ (see Equation (14.10)).

Thus,

$$W_{lost} = T_1 S_{G,total} \tag{14.17}$$

Once again, the work lost due to irreversibilities in the process is directly proportional to the total entropy generated (internally and externally) due to the process. This is the *Gouy-Stodola theorem.*
Note that when the system interacts with more than one TER, the W_{lost} is a relative quantity depending on the selection of the floating reservoir. For example, $W_{lost} = T_2 S_{G,total}$ if TER 2 is selected as the reference reservoir. However, the total entropy generation $S_{G,total}$ is independent of the choice of the floating reservoir. $S_{G,total}$ depends only on the internal and external irreversibilities associated with the process.

14.3.3 Closed System with *N* Heat Reservoirs

We can now further generalize the results for a closed system interacting thermally with *N* TERs and producing work (see Figure 14.3).

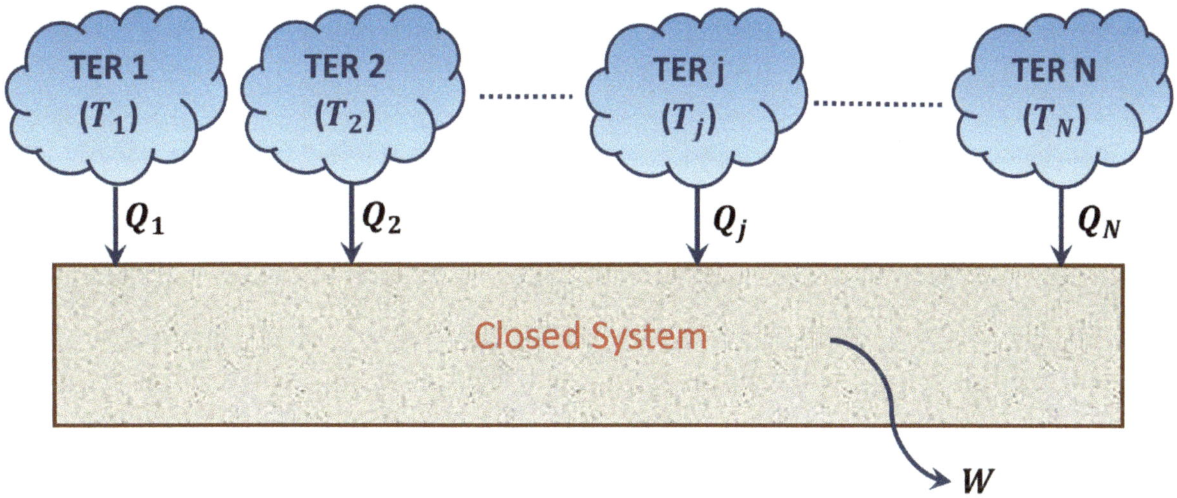

Figure 14.3 Closed system interacting with N TERs (thermal energy reservoirs) and producing work.

Using the first law of thermodynamics, the actual work can be expressed as:

$$W = \sum_{i=1}^{N} Q_i - \Delta U = Q_j + \sum_{i=1}^{j-1} Q_i + \sum_{i=j+1}^{N} Q_i - \Delta U \tag{14.18}$$

where jth TER is the reference/floating reservoir. Using the second law,

$$S_{G,total} = \Delta S - \sum_{i=1}^{N} \frac{Q_i}{T_i} \geq 0 \tag{14.19}$$

For a reversible process,

$$S_{G,total} = \Delta S - \frac{Q_{j,rev}}{T_j} - \sum_{i=1}^{j-1} \frac{Q_i}{T_i} - \sum_{i=j+1}^{N} \frac{Q_i}{T_i} = 0 \tag{14.20}$$

From Equation (14.20), we get:

$$\frac{Q_{j,rev}}{T_j} = \Delta S - \sum_{i=1}^{j-1} \frac{Q_i}{T_i} - \sum_{i=j+1}^{N} \frac{Q_i}{T_i} \tag{14.21}$$

Upon substitution of $Q_{j,rev}$ from Equation (14.21) into Equation (14.18), the following result is obtained:

$$W_{rev} = T_j \Delta S - \Delta U + \sum_{i=1}^{j-1} Q_i \left(1 - \frac{T_j}{T_i}\right) + \sum_{i=j+1}^{N} Q_i \left(1 - \frac{T_j}{T_i}\right) \tag{14.22}$$

where T_j is the absolute temperature of the reference reservoir chosen to be the jth TER. The heat interaction between the system and the jth TER is floating, that is, it varies with the degree of irreversibility in the process. All other heat interactions Q_i's are the same for the actual and reversible processes.

Using Equations (14.6), (14.18), and (14,22), the work lost due to irreversibilities in the process is:

$$W_{lost} = \left[T_j \Delta S - \Delta U + \sum_{i=1}^{j-1} Q_i \left(1 - \frac{T_j}{T_i}\right) + \sum_{i=j+1}^{N} Q_i \left(1 - \frac{T_j}{T_i}\right)\right] - \left[Q_j + \sum_{i=1}^{j-1} Q_i + \sum_{i=j+1}^{N} Q_i - \Delta U\right] \tag{14.23}$$

Upon simplification, Equation (14.23) gives:

$$W_{lost} = T_j \left[\Delta S - \sum_{i=1}^{N} \frac{Q_i}{T_i}\right] \tag{14.24}$$

The square bracketed term on the right hand side of Equation (14.24) is $S_{G,total}$ (see Equation (14.19)). Thus, Equation (14.24) becomes:

$$W_{lost} = T_j S_{G,total} \tag{14.25}$$

This is the *Gouy-Stodola theorem*. The work lost due to irreversibilities in the process is directly proportional to the total entropy generated (internally and externally) due to the process. The proportionality constant is the absolute temperature of the floating/reference reservoir (jth reservoir).

As noted earlier, W_{lost} is a relative quantity depending on the selection of the floating reservoir whereas the total entropy generation $S_{G,total}$ is independent of the choice of the floating reservoir. $S_{G,total}$ depends only on the irreversibilities related to the process. Based on Equation (14.25), we can have N different values of W_{lost} depending on the choice of the floating reservoir. Thus,

$$W_{lost,j} = T_j S_{G,total} \qquad (j = 1,2, \ldots \ldots . N) \tag{14.26}$$

We can also calculate the work lost based on any ith reservoir if the work lost based on any one reservoir (jth) is known, as shown below:

$$W_{lost,i} = T_i S_{G,total} \tag{14.27}$$

Dividing Equation (14.27) by Equation (14.26) gives:

$$W_{lost,i} = \left(\frac{T_i}{T_j}\right) W_{lost,j} \tag{14.28}$$

14.3.4 Thermodynamic Efficiency of a Process

The thermodynamic efficiency (ε) of a process can be defined as:

$$\varepsilon = \frac{W}{W_{rev}} \qquad \text{(work producing process)} \tag{14.29}$$

$$\varepsilon = \frac{W_{rev}}{W} \qquad \text{(work absorbing process)} \tag{14.30}$$

where W is the actual work produced or absorbed in a process and W_{rev} is the corresponding work for a reversible process. The change in states of the systems are kept the same in both the processes.
The thermodynamic efficiency defined in this way, that is, Equations (14.29) and (14.30), is also referred to as the *second law efficiency.*

The thermodynamic efficiency is always less than one. Note that in a work producing process, W_{rev} is the maximum work produced and hence $W_{rev} > W$. In a work absorbing process, W_{rev} is the minimum work required and hence $|W_{rev}| < |W|$.

We can also re-write the expressions for ε in terms of the work lost as follows:

$$\varepsilon = 1 - \frac{W_{lost}}{W_{rev}} \qquad \text{(work producing process)} \qquad (14.31)$$

$$\varepsilon = 1 + \frac{W_{lost}}{W} \qquad \text{(work absorbing process)} \qquad (14.32)$$

The work lost W_{lost}, defined in Equation (14.6), is always positive regardless of whether the process is producing work or absorbing work. For the work absorbing process, the thermodynamic efficiency is still less than one as W is negative.

14.4 OPEN SYSTEMS

14.4.1 Open System with a Single Heat Reservoir

Consider an open system interacting thermally with a single thermal energy reservoir (TER), namely the surroundings/environment at a uniform temperature of T_o. The system is receiving heat at a rate of $\dot{Q}_o$ and producing shaft work at a rate of $\dot{W}_{sh}$, as shown in Figure 14.4.

From the first law of thermodynamics for control volume:

$$\dot{m}\left[(h_2 - h_1) + \frac{1}{2}(V_2^2 - V_1^2) + g(Z_2 - Z_1)\right] = \dot{Q}_o - \dot{W}_{sh} \qquad (14.33)$$

$$\dot{m}\left[\Delta h + \frac{1}{2}\Delta V^2 + g\Delta Z\right] = \dot{Q}_o - \dot{W}_{sh} \qquad (14.34)$$

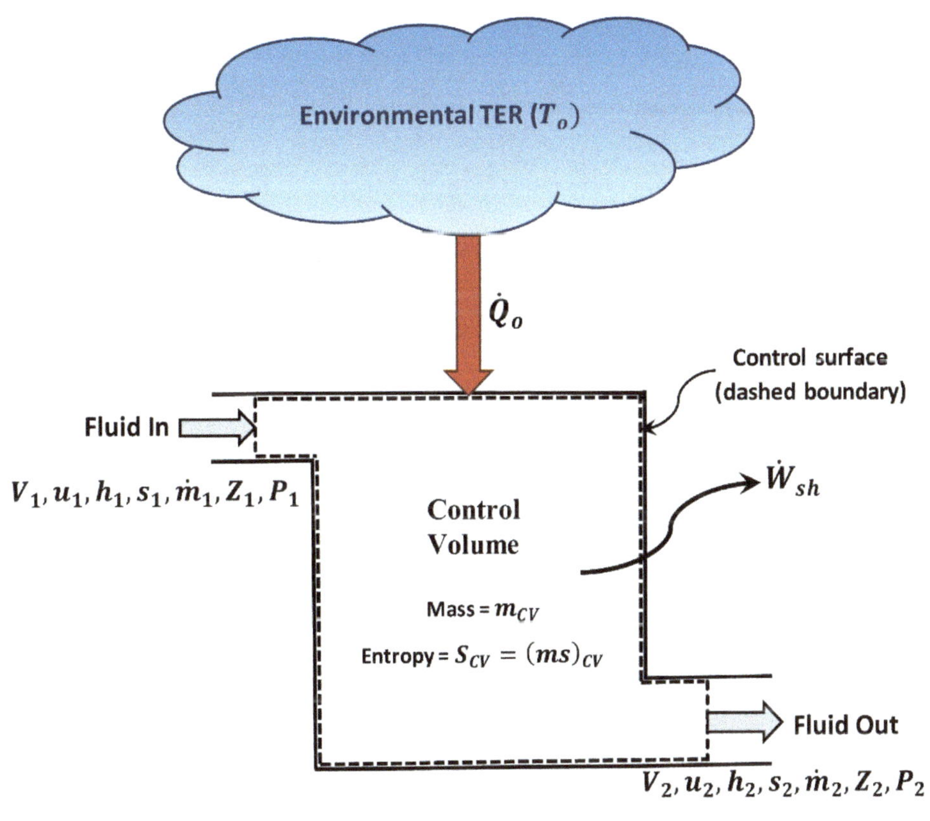

Figure 14.4 Control volume with one inlet and one outlet (P is pressure, V is velocity, u is specific internal energy, h is specific enthalpy, s is specific entropy, $\dot{m}$ is mass flow rate, Z is elevation, $\dot{Q}_o$ is rate of heat transfer, $\dot{W}_{sh}$ is rate of shaft work, m is mass, and S is total entropy).

Neglecting kinetic and potential energy changes, Equation (14.34) can be re-written as:

$$\dot{W}_{sh} = \dot{Q}_o - \dot{m}(\Delta h) \tag{14.35}$$

From the second law for open systems,

$$\dot{S}_{G,total} = \dot{m}(s_2 - s_1) - \frac{\dot{Q}_o}{T_o} \geq 0 \tag{14.36}$$

For a reversible process,

$$\dot{Q}_{o,rev} = T_o \dot{m}(s_2 - s_1) \tag{14.37}$$

Upon substitution of Equation (14.37) into Equation (14.35), we get:

$$\dot{W}_{sh,rev} = \dot{m}(T_o \Delta s - \Delta h) \tag{14.38}$$

The rate of shaft work lost due to irreversibilities (internal and external) associated with the process is:

$$\dot{W}_{sh,lost} = \dot{W}_{sh,rev} - \dot{W}_{sh} = \dot{m}(T_o \Delta s) - \dot{Q}_o \tag{14.39}$$

Upon re-arrangement, Equation (14.39) gives:

$$\dot{W}_{sh,lost} = T_o \left(\dot{m}\Delta s - \frac{\dot{Q}_o}{T_o} \right) \tag{14.40}$$

The bracketed term on the right hand side of Equation (14.40) is $\dot{S}_{G,total}$ (see Equation (14.36)). Thus Equation (14.40) can be re-written as:

$$\dot{W}_{sh,lost} = T_o \dot{S}_{G,total} \tag{14.41}$$

This is the *Gouy-Stodola theorem* for open (flow) systems. The rate of shaft work lost in a flow process is directly proportional to the total rate of entropy generation in the process. The proportionality constant is the absolute temperature of the heat reservoir with which the process is exchanging heat.

14.4.2 Open System with *N* Heat Reservoirs

We now consider an open system interacting thermally with N TERs and producing shaft work (see Figure 14.5). From the first law of thermodynamics for control volume:

$$\dot{W}_{sh} = \sum_{i=1}^{N} \dot{Q}_i - \dot{m}(\Delta h) \tag{14.42}$$

We have neglected kinetic and potential energy changes. Using the second law,

$$\dot{S}_{G,total} = \dot{m}\Delta s - \sum_{i=1}^{N} \frac{\dot{Q}_i}{T_i} \geq 0 \tag{14.43}$$

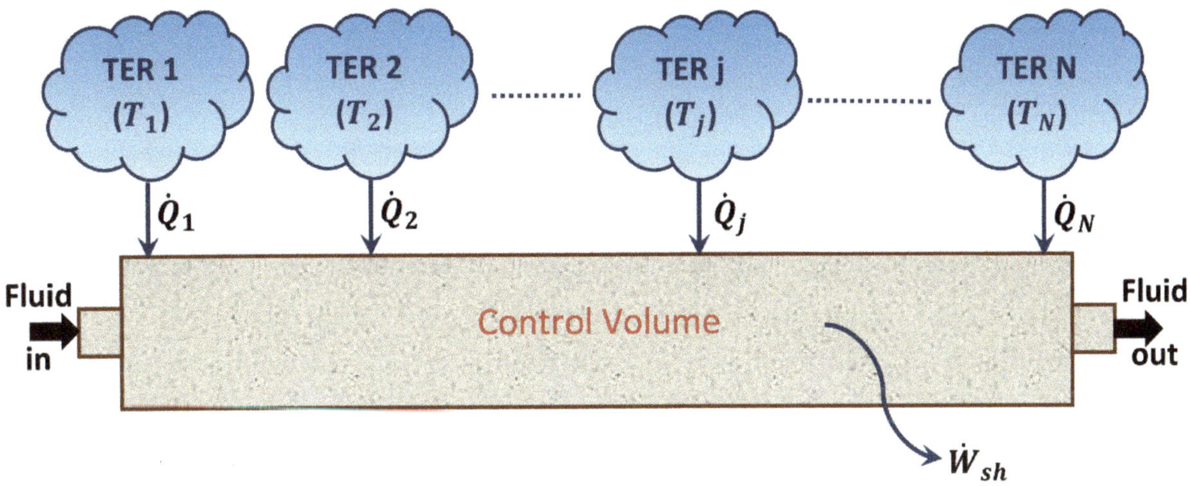

Figure 14.5 Open system interacting with N TERs (thermal energy reservoirs) and producing shaft work.

In order to compare the shaft work produced in the actual and reversible processes for the same change in state of the system, we need to select the floating TER. The heat exchanged between the floating reservoir and the process varies with the nature (reversible or irreversible) of the process. The heat exchanged between all other heat reservoirs and the process is the same, regardless of the nature of the process (reversible or irreversible). Let jth reservoir with absolute temperature of T_j be the floating reservoir.

We can re-write Equations (14.42) and (14.43) as:

$$\dot{W}_{sh} = \dot{Q}_j + \sum_{i=1}^{j-1} \dot{Q}_i + \sum_{i=j+1}^{N} \dot{Q}_i - \dot{m}(\Delta h) \tag{14.44}$$

$$\dot{S}_{G,total} = \dot{m}\Delta s - \frac{\dot{Q}_j}{T_j} - \sum_{i=1}^{j-1} \frac{\dot{Q}_i}{T_i} - \sum_{i=j+1}^{N} \frac{\dot{Q}_i}{T_i} \geq 0 \tag{14.45}$$

For a reversible process,

$$\dot{S}_{G,total} = \dot{m}\Delta s - \frac{\dot{Q}_{j,rev}}{T_j} - \sum_{i=1}^{j-1} \frac{\dot{Q}_i}{T_i} - \sum_{i=j+1}^{N} \frac{\dot{Q}_i}{T_i} = 0 \tag{14.46}$$

From Equation (14.46), we get:

$$\dot{Q}_{j,rev} = T_j \left[\dot{m}\Delta s - \sum_{i=1}^{j-1} \frac{\dot{Q}_i}{T_i} - \sum_{i=j+1}^{N} \frac{\dot{Q}_i}{T_i}\right] \tag{14.47}$$

Upon substitution of $\dot{Q}_{j,rev}$ from Equation (14.47) into Equation (14.44), the following result is obtained:

$$\dot{W}_{sh,rev} = T_j \dot{m}\Delta s - \dot{m}(\Delta h) + \sum_{i=1}^{j-1} \dot{Q}_i \left(1 - \frac{T_j}{T_i}\right) + \sum_{i=j+1}^{N} \dot{Q}_i \left(1 - \frac{T_j}{T_i}\right) \tag{14.48}$$

From Equations (14.44) and (14.48), it can be readily shown that the rate of shaft work lost due to irreversibilities in the process is as follows:

$$\dot{W}_{sh,lost} = \dot{W}_{sh,rev} - \dot{W}_{sh} = T_j \left[\dot{m}\Delta s - \sum_{i=1}^{N} \frac{\dot{Q}_i}{T_i}\right] \tag{14.49}$$

The square bracketed term on the right hand side of Equation (14.49) is $\dot{S}_{G,total}$ (see Equation (14.43)). Thus, Equation (14.49) becomes:

$$\dot{W}_{sh,lost} = T_j \dot{S}_{G,total} \tag{14.50}$$

Once again, this is the *Gouy-Stodola theorem*. The rate of shaft work lost due to irreversibilities in the process is directly proportional to the total rate of entropy generation (internally and externally) associated with the process. The proportionality constant is the absolute temperature of the floating/reference reservoir (jth reservoir).

14.5 APPLICATION OF THE GOUY-STODOLA THEOREM TO MULTI-UNIT FLOW PROCESSES

Industrial processes are usually multi-unit flow processes involving a number of units or components such as: pipes, pumps, compressors, valves and fittings, flowmeters, heat exchangers, reactors, separators, etc. In order to improve the efficiency of a multi-unit flow process from thermodynamic perspective, it is important to be able to quantify irreversibilities in different components of the process and identify the units which are most inefficient. To that end, the Gouy-Stodola theorem plays an important role.

As an example, consider the multi-unit flow process shown in Figure 14.6. There are four units connected serially. We assume that each unit is interacting thermally with the same thermal energy reservoir (TER) at a uniform temperature of T_o. Each unit is also producing (or absorbing) shaft work. Applying the Gouy-Stodola theorem to each unit and summing over all the units of the process yields:

$$\dot{W}_{sh,lost,i} = T_o \dot{S}_{G,total,i} \tag{14.51}$$

$$\sum_{i=1}^{4} \dot{W}_{sh,lost,i} = T_o \sum_{i=1}^{4} \dot{S}_{G,total,i} \tag{14.52}$$

Dividing Equation (14.51) with Equation (14.52), we get:

$$\frac{\dot{W}_{sh,lost,i}}{\sum_{i=1}^{4} \dot{W}_{sh,lost,i}} = \frac{\dot{S}_{G,total,i}}{\sum_{i=1}^{4} \dot{S}_{G,total,i}} \tag{14.53}$$

According to Equation (14.53), the fractional contribution of each unit to the shaft work lost in the whole process is equal to the ratio of total entropy generated in the unit to the total entropy generated in the whole process. When analyzing a process, it is important to calculate the contribution of each unit to the total entropy generated in the whole process. This way one can identify the units which need to be modified or redesigned to improve the overall thermodynamic efficiency of the process.

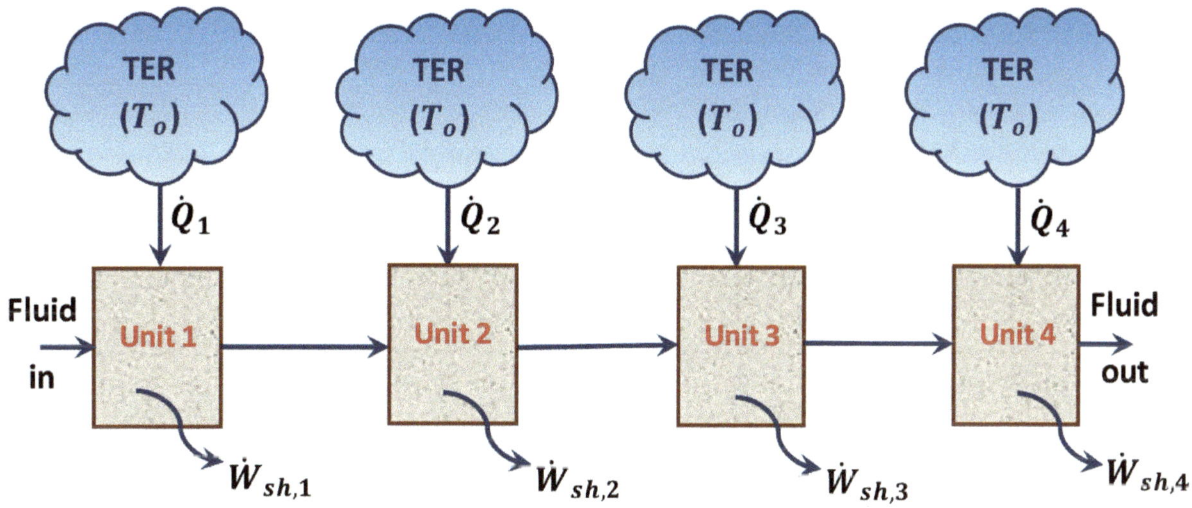

Figure 14.6 Multi-unit flow process where each unit is interacting with a TER (thermal energy reservoir) and producing shaft work.

NOTATION

h	specific enthalpy
$\dot{m}$	mass flow rate
N	total number of interacting thermal energy reservoirs
P	pressure
Q	heat
Q_o	heat transfer between environmental thermal energy reservoir and system
Q_i	heat transfer between ith thermal energy reservoir and system
Q_j	heat transfer between jth thermal energy reservoir and system
Q_{rev}	reversible heat transfer
$\dot{Q}$	rate of heat transfer
$\dot{Q}_o$	rate of heat transfer between environmental thermal energy reservoir and system
s	specific entropy
S	total entropy
$S_{G,total}$	total entropy generated
$\dot{S}_{G,total}$	rate of total entropy generation
$\dot{S}_{G,total,i}$	rate of total entropy generation in ith unit of a multi-unit process
T	temperature
T_o	temperature of environmental thermal energy reservoir
T_i	temperature of ith thermal energy reservoir
T_j	temperature of jth thermal energy reservoir
TER	thermal energy reservoir
U	total internal energy
u	specific internal energy
V	velocity
W	work
W_{lost}	work lost
$W_{lost,i}$	work lost based on ith thermal energy reservoir as floating reservoir

$W_{lost,j}$	work lost based on jth thermal energy reservoir as floating reservoir
W_{rev}	reversible work
$\dot{W}_{sh}$	rate of shaft work
$\dot{W}_{sh,lost}$	rate of shaft work lost
$\dot{W}_{sh,lost,i}$	rate of shaft work lost in ith unit
Z	elevation

Greek Letters

Δ	difference operator
ε	thermodynamic efficiency of a process

FURTHER READING

Pal, R., "Demystification of the Gouy-Stodola theorem of thermodynamics for closed systems", *Int. J. Mech. Eng. Edu.* 2017; 45: 142-153.

Pal, R., "On the Gouy-Stodola theorem of thermodynamics for open systems", *Int. J. Mech. Eng. Edu.* 2017; 45: 194-206.

TUTORIAL PROBLEMS WITH SOLUTION

Problem 1

A 1 kg metal block is heated to 800 °C in an oven. It is removed from the oven and allowed to cool to 27 °C by transferring heat to the surroundings air at 27 °C.

a) Estimate the total amount of entropy generated.
b) What is the lost work?
c) How much work can be extracted by a reversible heat engine connected between the metal block and surroundings air? The final temperatures of the metal block and surroundings air are the same, that is, 27 °C.

The specific heat capacity of metal block is 0.5 kJ/kg.K

Solution

a) total amount of entropy generated.

Entropy change of metal block:

$$\Delta S_{block} = mc ln\left(\frac{T_f}{T_i}\right) = 1 \times 0.5 \times ln\left(\frac{27+273.15}{800+273.15}\right) = -0.637 \text{ kJ/K}$$

Heat lost by metal block:

$$Q = mc(T_f - T_i) = 1 \times 0.5 \times (27 - 800) = -386.5 \text{ kJ}$$

Entropy change of surroundings (air):

$$\Delta S_{air} = \frac{|Q|}{T_o} = \frac{386.5}{273.15+27} = 1.2877 \text{ kJ/K}$$

Total entropy change:

$$\Delta S_{total} = -0.637 + 1.2877 = 0.6507 \text{ kJ/K}$$

Total entropy generated:

$$S_G = \Delta S_{total} = -0.637 + 1.2877 = 0.6507 \text{ kJ/K}$$

b) Lost work?

Using Gouy-Stodola theorem:

$$W_{lost} = T_o S_{G,total} = (27 + 273.15) \times 0.6507 = 195.3 \text{ kJ}$$

c) Work that can be extracted by a reversible heat engine connected between the metal block and surroundings air:

The work produced by a reversible (Carnot) heat engine on a differential basis is:

$$\delta W = \delta Q\left(1 - \frac{T_o}{T}\right)$$

where T_o is the surroundings temperature (27 °C) and T is the metal block temperature. The amount of heat input to the engine δQ is:

$$\delta Q = -mcdT$$

where dT is the change in metal block temperature. Thus,

$$\delta W = \delta Q\left(1 - \frac{T_o}{T}\right) = -mc\left(1 - \frac{T_o}{T}\right)dT$$

Upon integration from 800 °C to 27 °C, we get:

$$W = -\int_{800+273.15}^{27+273.15} mc\left(1 - \frac{T_o}{T}\right)dT = -1 \times 0.5\left[(300.15 - 1073.15) - (300.15) \times ln\left(\frac{300.15}{1073.15}\right)\right] = 195.3 \text{ kJ}$$

This is the same amount as lost work. Thus, all the lost work is recovered if the process is carried out reversibly.

Problem 2

100 moles of an ideal gas at 298 K and 50 bar expands to five times its initial volume as follows:

a) By a reversible isothermal process
b) By an irreversible isothermal process in which the expansion is against a restraining pressure of 5 bars

For each process, calculate the following: the final pressure, the work done by the gas, the heat transferred to the gas, the total entropy generated, the work lost, and the thermodynamic efficiency. You can assume that the surrounding atmosphere (environmental thermal energy reservoir) is at a temperature of 298 K.

Solution

a) Reversible isothermal process

$$P_2 = \frac{P_1V_1}{V_2} = \frac{50}{5} = 10 \text{ bar}$$

$$\Delta U = 0\,;\,Q = W = nRTln\left(\frac{V_f}{V_i}\right) = 100 \times 8.314 \times 298 \times ln(5) = 398.749831 \times 10^3 \text{ J}$$

$$\Delta S = nRln\left(\frac{V_f}{V_i}\right) = 100 \times 8.314 \times ln(5) = 1.338087 \times 10^3 \text{ J/K}$$

$$\Delta S_{surr} = -\frac{398.749831\times10^3}{298} = -1.338087 \times 10^3 \text{ J/K}$$

$$S_{G,total} = \Delta S + \Delta S_{surr} = 1.338087 \times 10^3 - 1.338087 \times 10^3 = 0$$

Using Gouy-Stodola theorem,

$$W_{lost} = T_o S_{G,total} = 0$$

Thus, no work is lost in a reversible process. The thermodynamic efficiency of the process is:

$$\varepsilon = 1 - \frac{W_{lost}}{W_{rev}} = 1 - 0 = 1$$

The reversible process is 100% efficient thermodynamically.

b) Irreversible isothermal process

$$P_2 = \frac{P_1V_1}{V_2} = \frac{50}{5} = 10 \text{ bar}$$

$$\Delta U = 0\,;\,Q = W = P_{ext}\Delta V = 4P_{ext}V_1 = 4P_{ext}\left(\frac{nRT}{P_1}\right) = 4 \times 5 \times \left(\frac{100\times8.314\times298}{50}\right) = 99.10288 \times 10^3 \text{ J}$$

The work produced in the irreversible process (99.10288 kJ) is much lower than that produced in the reversible process (398.749831 kJ) for the same change in state.

$$\Delta S = nRln\left(\frac{V_f}{V_i}\right) = 100 \times 8.314 \times ln(5) = 1.338087 \times 10^3 \text{ J/K}$$

$$\Delta S_{surr} = -\frac{99.10288\times10^3}{298} = -0.33256 \times 10^3 \text{ J/K}$$

$$S_{G,total} = \Delta S + \Delta S_{surr} = 1.338087 \times 10^3 - 0.33256 \times 10^3 = 1.05527 \times 10^3 \text{ J/K}$$

Using Gouy-Stodola theorem,

$$W_{lost} = T_o S_{G,total} = 298 \times 1.05527 \times 10^3 = 299.647 \times 10^3 \text{ J}$$

Note that:

$$W_{lost} = W_{rev} - W = 398.749831 \times 10^3 - 99.10288 \times 10^3 = 299.647 \times 10^3 \text{ J}$$

The thermodynamic efficiency of the process is:

$$\varepsilon = 1 - \frac{W_{lost}}{W_{rev}} = 1 - \frac{299.647 \times 10^3}{398.749831 \times 10^3} = 0.24853$$

The irreversible process is only 24.85% efficient thermodynamically.

Problem 3

Superheated steam enters the turbine at 35 bar and 450 °C. The exhaust from the turbine is saturated steam at 90 °C. There are negligible changes in kinetic and potential energies. The heat loss from the turbine is also negligible. The temperature of the surroundings is 298 K.
Calculate:

a) the actual work output from the turbine per kg of fluid flowing
b) the maximum work output from the turbine per kg of fluid flowing
c) the total amount of entropy production per kg of fluid flowing
d) the lost work per kg of fluid flowing
e) the thermodynamic efficiency of the turbine

Solution

We need specific enthalpies and entropies for the inlet and discharge streams. We obtain this information from the steam tables given in Appendices M and N.

Superheated steam at 35 bar and 450 °C:

$h = 3337.2$ kJ/kg ; $s = 7.0052$ kJ/kg.K

Saturated steam at 90 °C:

$h = 2660.1$ kJ/kg ; $s = 7.4799$ kJ/kg.K

a) the actual work output from the turbine per kg of fluid flowing

Apply first law for open systems:

$$\left[\Delta h + \frac{1}{2}\Delta V^2 + g\Delta Z\right] = Q - W_{sh}$$

Neglecting potential energy changes and re-arranging:

$$W_{sh} = -\Delta h = -(2660.1 - 3337.2) = 677.1 \text{ kJ/kg}$$

b) the maximum work output from the turbine per kg of fluid flowing

The maximum work is produced when the process is carried out reversibly

$$W_{sh,rev} = (T_o\Delta s - \Delta h) = 298 \times (7.4799 - 7.0052) - (2660.1 - 3337.2) = 818.5606 \text{ kJ/kg}$$

c) the total amount of entropy production per kg of fluid flowing

$$S_{G,total} = (s_2 - s_1) = (7.4799 - 7.0052) = 0.4747 \text{ kJ/kg.K}$$

d) the lost work per kg of fluid flowing

Using the Gouy-Stodola theorem,

$$W_{sh,lost} = T_o S_{G,total} = 298 \times 0.4747 = 141.4606 \text{ kJ/kg}$$

Note that:

$$W_{sh,lost} = W_{sh,rev} - W_{sh} = 818.5606 - 677.1 = 141.4606 \text{ kJ/kg}$$

e) the thermodynamic efficiency of the turbine

The thermodynamic efficiency of the process is:

$$\varepsilon = 1 - \frac{W_{sh,lost}}{W_{sh,rev}} = 1 - \frac{141.4606}{818.5606} = 0.827$$

Thus, the turbine is 82.7% efficient.

Problem 4

Consider a double pipe countercurrent heat exchanger. Cold water is to be heated from 20 °C to 80 °C using the geothermal water available at 160 °C. The mass flow rates of hot and cold fluids are 2 kg/s and 1.2 kg/s, respectively. The heat capacities of hot and cold fluids are 4310 J/kg.K and 4180 J/kg.K, respectively. Estimate the rate of lost work in the heat exchanger.
You can assume that the surrounding atmosphere (environmental thermal energy reservoir) is at a temperature of 298 K.

Solution

The first law for control volume (see Chapter 5) with multiple inlets and outlets gives:

$$\dot{m}_h c_{p,h}\left(T_{h,2} - T_{h,1}\right) + \dot{m}_c c_{p,c}\left(T_{c,2} - T_{c,1}\right) = 0$$

From this equation, we get:

$$2 \times 4310\left(T_{h,2} - 160\right) + 1.2 \times 4180(80 - 20) = 0 \ ; \ T_{h,2} = 125.086^o C$$

The total rate of entropy generation (assuming no heat loss to the surroundings) is:

$$\dot{S}_{G,total} = \dot{m}_h c_{p,h} ln\left(\frac{T_{h,2}}{T_{h,1}}\right) + \dot{m}_c c_{p,c} ln\left(\frac{T_{c,2}}{T_{c,1}}\right)$$

$$\dot{S}_{G,total} = 2 \times 4310 \times ln\left(\frac{125.086+273.15}{160+273.15}\right) + 1.2 \times 4180 \times ln\left(\frac{80+273.15}{20+273.15}\right) = 209.603 \text{ W/K}$$

Using the Gouy-Stodola theorem,

$$\dot{W}_{sh,lost} = T_o \dot{S}_{G,total} = 298 \times 209.603 = 62461.694 \text{ W} = 62.462 \text{ kW}$$

Problem 5

Saturated refrigerant vapor (HFC-134A) at -36°C is compressed to 5 bar and 30 °C. There are negligible changes in kinetic and potential energies. The heat loss from the compressor is also negligible. The temperature of the surroundings is 298 K.
Calculate the following:

a) the actual work required by the compressor per kg of fluid flowing
b) the minimum work required by the compressor per kg of fluid flowing
c) the total amount of entropy production per kg of fluid flowing
d) the lost work per kg of fluid flowing
e) the thermodynamic efficiency of the compressor

Solution

We need specific enthalpies and entropies for the inlet and discharge streams. We obtain this information from the tables given in Appendices K and L.

Saturated refrigerant vapor (HFC-134A) at -36^oC*:*

$h = 225.40$ kJ/kg ; $s = 0.9506$ kJ/kg.K

Superheated refrigerant vapor (HFC-134A) at 30^oC *and 5 bar:*

$h = 270.28$ kJ/kg ; $s = 0.9597$ kJ/kg.K

a) the actual work required by the compressor per kg of fluid flowing

Apply first law for open systems:

$$\left[\Delta h + \frac{1}{2}\Delta V^2 + g\Delta Z\right] = Q - W_{sh}$$

Neglecting potential energy changes and re-arranging:

$$W_{sh} = -\Delta h = -(270.28 - 225.40) = -44.88 \text{ kJ/kg}$$

b) the minimum work required by the compressor per kg of fluid flowing

The minimum work is required when the process is carried out reversibly

$$W_{sh,rev} = (T_o\Delta s - \Delta h) = 298 \times (0.9597 - 0.9506) - (270.28 - 225.40) = -42.168 \text{ kJ/kg}$$

c) the total amount of entropy production per kg of fluid flowing

$$S_{G,total} = (s_2 - s_1) = (0.9597 - 0.9506) = 0.0091 \text{ kJ/kg.K}$$

d) the lost work per kg of fluid flowing

Using the Gouy-Stodola theorem,

$$W_{sh,lost} = T_o S_{G,total} = 298 \times 0.0091 = 2.712 \text{ kJ/kg}$$

Note that:

$$W_{sh,lost} = W_{sh,rev} - W_{sh} = -42.168 - (-44.88) = 2.712 \text{ kJ/kg}$$

e) the thermodynamic efficiency of the compressor

$$\varepsilon = 1 + \frac{W_{sh,lost}}{W_{sh}} = 1 + \frac{2.712}{-44.88} = 0.9396$$

Thus, the compressor is 93.96% efficient.

Problem 6

Consider throttling of wet steam at 20 bar in a steady flow process to 3 bar. The temperature of the steam after throttling is 150 °C. The temperature of the surroundings is 298 K.
Calculate the following:

a) the quality of the wet steam at 20 bar
b) the total amount of entropy production per kg of fluid flowing
c) the lost work per kg of fluid flowing

Solution

We need specific enthalpies and entropies for the inlet and discharge streams. We obtain this information from the tables given in Appendices M and N.

Saturated steam at 20 bar:

$h^L = 908.80$ kJ/kg ; $h^v = 2799.5$ kJ/kg ;

$s^L = 2.4478$ kJ/kg.K ; $s^v = 6.3417$ kJ/kg.K

Superheated steam at $150^o C$ *and 3 bar:*

$h = 2761$ kJ/kg ; $s = 7.0778$ kJ/kg.K

a) the quality of the wet steam at 20 bar

As throttling is an isenthalpic process,

$$h_1 = h_2 \text{ where } h_1 = x^v h^v + (1 - x^v) h^L \text{ and } h_2 = 2761 \text{ kJ/kg}$$

Thus,

$$x^v(2799.5) + (1 - x^v)(908.8) = 2761 ; \; x^v = 0.97964$$

b) the total amount of entropy production per kg of fluid flowing

$$S_{G,total} = (s_2 - s_1)$$

$$s_2 = 7.0778 \text{ kJ/kg.K}$$

$$s_1 = x^v s^v + (1 - x^v) s^L = 0.97964 \times 6.3417 + (1 - 0.97964) \times 2.4478 = 6.26242 \text{ kJ/kg.K}$$

$$S_{G,total} = (s_2 - s_1) = 7.0778 - 6.26242 = 0.81538 \text{ kJ/kg.K}$$

c) the lost work per kg of fluid flowing

Using the Gouy-Stodola theorem,

$$W_{sh} = T_o S_{G,total} = 298 \times 0.81538 = 242.983 \text{ kJ/kg}$$

Problem 7

Consider a three-unit process shown in the figure below. Saturated vapor of refrigerant HFC-134A enters the compressor at 2 bar. The compressor exit stream is superheated vapor at 10 bar and 60 °C. The superheated vapor is condensed in a condenser to saturated liquid at 10 bar. The heat from the condenser is rejected to the surroundings at 298 K. The saturated liquid from the condenser is throttled to 2 bar. The discharge stream from the throttling valve is two-phase vapor-liquid mixture at 2 bar.

a) Carry out the thermodynamic analysis of the process in terms of entropy generation and lost work. Which unit is the most inefficient thermodynamically?
b) What is the thermodynamic efficiency of the overall process?

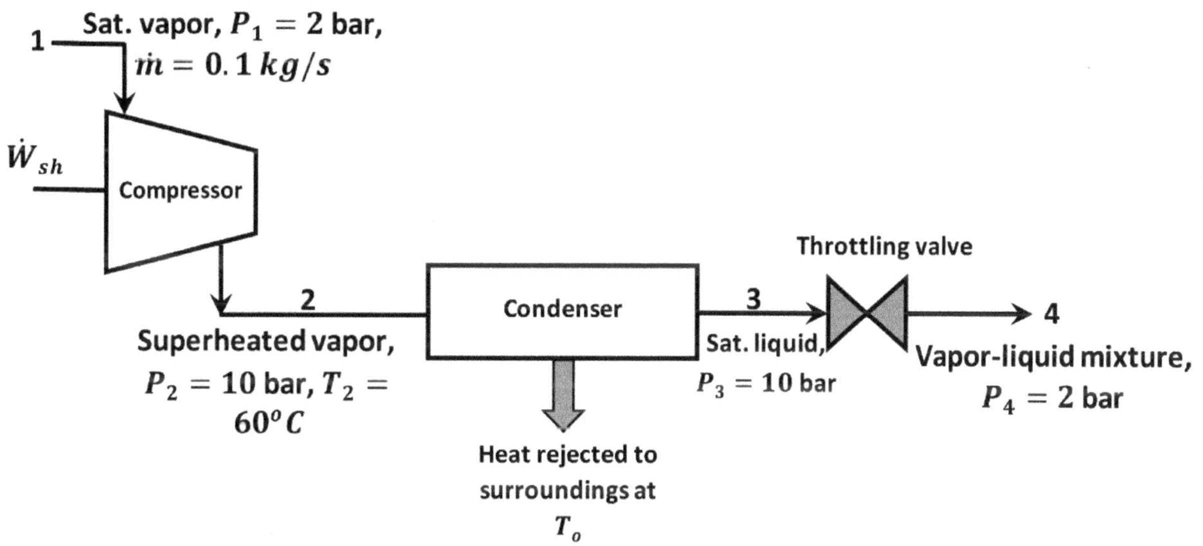

Solution

We need specific enthalpies and entropies of fluid stream at different locations. We obtain this information from the tables given in Appendices M and N.

State 1: Saturated vapor at 2 bar:

$h_1 - 241.30$ kJ/kg ; $s_1 = 0.9253$ kJ/kg.K

State 2: Superheated vapor at P=10 bar and T=60 °C:

$h_2 = 291.36$ kJ/kg ; $s_2 = 0.9768$ kJ/kg.K

State 3: Saturated liquid at 10 bar:

$h_3 = 105.29$ kJ/kg ; $s_3 = 0.3838$ kJ/kg.K

State 4: Vapor-liquid mixture at 2 bar:

$h_4^v = 241.30$ kJ/kg ; $h_4^L = 36.84$ kJ/kg

$s_4^v = 0.9253$ kJ/kg.K ; $s_4^L = 0.1481$ kJ/kg.K

As throttling process is isenthalpic, $h_4 = h_3 = 105.29$ kJ/kg

$$h_4 = x_4^v h_4^v + (1 - x_4^v)h_4^L = x_4^v \times 241.30 + (1 - x_4^v) \times 36.84 = 105.29\ ;\ x_4^v = 0.3348$$
$$s_4 = x_4^v s_4^v + (1 - x_4^v)s_4^L = 0.3348 \times 0.9253 + (1 - 0.3348) \times 0.1481 = 0.4083 \text{ J/kg.K}$$

a) Thermodynamic analysis of the process in terms of entropy generation and lost work

Reversible (ideal) work of the whole process:

$$\dot{W}_{sh,rev} = \dot{m}[T_o\Delta s - \Delta h] = 0.1 \times [298 \times (s_4 - s_1) - (h_4 - h_1)] = 0.1 \times [298 \times (0.4083 - 0.9253) - (105.29 - 241.30)] = -1.8056 \text{ kW}$$

Rate of entropy generation and lost work for each unit of the process:

- Compressor

$$\dot{W}_{sh,rev} = \dot{m}[T_o\Delta s - \Delta h] = 0.1 \times [298 \times (s_2 - s_1) - (h_2 - h_1)] = 0.1 \times [298 \times (0.9768 - 0.9253) - (291.36 - 241.30)] = -3.4713 \text{ kW}$$

$$\dot{S}_G = \dot{m}(s_2 - s_1) = 0.1 \times (0.9768 - 0.9253) = 5.15 \times 10^{-3} \text{ kW/K}$$

Using the Gouy-Stodola theorem,

$$\dot{W}_{sh,lost} = T_o\dot{S}_G = 298 \times 5.15 \times 10^{-3} = 1.5347 \text{ kW}$$

Note that the actual work required by the compressor is $\dot{W}_{sh,rev} - \dot{W}_{sh,lost} = -5.006$ kW

- Condenser

$$\dot{W}_{sh,rev} = \dot{m}[T_o\Delta s - \Delta h] = 0.1 \times [298 \times (s_3 - s_2) - (h_3 - h_2)] = 0.1 \times [298 \times (0.3838 - 0.9768) - (105.29 - 291.36)] = 0.9356 \text{ kW}$$

$$\dot{S}_G = \dot{m}(s_3 - s_2) - (\dot{Q}/T_o)\ ;\ \dot{Q} = \dot{m}\Delta h = 0.1 \times (105.29 - 291.36) = -18.607 \text{ kW}$$

$$\dot{S}_G = 0.1 \times (0.3838 - 0.9768) - (-18.607/298) = 3.1396 \times 10^{-3} \text{ kW/K}$$

Using the Gouy-Stodola theorem,

$$\dot{W}_{sh,lost} = T_o\dot{S}_G = 298 \times 3.1396 \times 10^{-3} = 0.9356 \text{ kW}$$

Note that in this case, all the potential work ($\dot{W}_{sh,rev}$) is lost as there is no actual work produced in the condenser

- Throttling valve

$$\dot{W}_{sh,rev} = \dot{m}[T_o\Delta s - \Delta h] = 0.1 \times [298 \times (s_4 - s_3) - (h_4 - h_3)] = 0.1 \times [298 \times (0.4083 - 0.3838) - (105.29 - 105.29)] = 0.7301 \text{ kW}$$

$$\dot{S}_G = \dot{m}(s_4 - s_3) = 0.1 \times (0.4083 - 0.3838) = 2.45 \times 10^{-3} \text{ kW/K}$$

Using the Gouy-Stodola theorem,

$$\dot{W}_{sh,lost} = T_o\dot{S}_G = 298 \times 2.45 \times 10^{-3} = 0.7301 \text{ kW}$$

Note that in this case too, all the potential work ($\dot{W}_{sh,rev}$) is lost as there is no actual work produced in the throttling process

Summary:

The calculations are summarized in the table given below.

Unit	$\dot{W}_{sh,rev}$, kW	$\dot{S}_G$, kW/K	$\frac{\dot{S}_G}{\sum \dot{S}_G}$	$\dot{W}_{sh,lost}$, kW	$\frac{\dot{W}_{sh,lost}}{\sum \dot{W}_{sh,lost}}$
Compressor	-3.4713	5.15 x 10^{-3}	0.47953	1.5347	0.47953
Condenser	0.9356	3.1396 x 10^{-3}	0.29234	0.9356	0.29234
Throttling valve	0.7301	2.45 x 10^{-3}	0.22813	0.7301	0.22813
Whole process	-1.8056	$\sum \dot{S}_G = 10.7396 \times 10^{-3}$	1.00000	$\sum \dot{W}_{sh,lost} = 3.2004$	1.00000

From the table, it is clear that the fraction of total work lost in the process is largest in the case of compressor. The fraction of total entropy generation in the process is also the largest in compressor. Thus, compressor is relatively the most inefficient unit in the given multi-unit process. As expected, $\dot{W}_{sh,lost}/\sum \dot{W}_{sh,lost} = \dot{S}_G/\sum \dot{S}_G$ for each unit.

b) What is the thermodynamic efficiency of the overall process?

The actual work of the process is:

$$\dot{W}_{sh} = \dot{W}_{sh,rev} - \sum \dot{W}_{sh,lost} = -1.8056 - 3.2004 = -5.006 \text{ kW}$$

The thermodynamic efficiency of the whole process is:

$$\varepsilon = 1 + \frac{\sum \dot{W}_{sh,lost}}{\dot{W}_{sh}} = 1 + \frac{3.2004}{-5.006} = 0.3607$$

Thus, the overall process has a thermodynamic efficiency of 36.07%.

PRATICE PROBLEMS

Problem 1

Consider 35 moles of ideal gas initially at 25°C and 1 bar within an insulated tank of fixed volume. The gas is stirred by an impeller until its pressure is reached to 1.5 bars. The temperature of the surroundings is 25°C.

Calculate:

a) The temperature of the gas in the final state
b) The work input to the gas through the impeller
c) The total amount of entropy production
d) The lost work

The specific heat capacities of the gas are: $c_p = 7R/2$ and $c_v = 5R/2$

Problem 2

An insulated vessel of total volume 2 m^3 consists of two equal volume compartments (each 1 m^3) separated by a valve. Initially, one of the compartments is filled with an ideal gas. The pressure and temperature of the gas in the compartment are 10 bar and 100 °C, respectively. The other compartment is empty (vacuum) free of any gas. The valve between the compartments is opened and the ideal gas is allowed to expand to fill the total volume (2 m^3) of the tank. The temperature of the surroundings is 25 °C.

Calculate:

a) The temperature and pressure of the gas in the final state
b) The total amount of entropy production
c) The lost work

Problem 3

Consider throttling of an ideal gas from 50 bar to 3 bar in a steady flow process. The molar flow rate of gas is 100 mol/s. The temperature of the surroundings is 25 °C.

Calculate:

a) The rate of total entropy production
b) The rate of lost work

Problem 4

Consider a double pipe co-current heat exchanger. Cold water is to be heated from 25 °C to 75 °C using the geothermal water available at 160 °C. The temperature of the hot fluid leaving the exchanger is 90 °C. The mass flow rate of the hot fluid stream is 2 kg/s. The heat capacities of hot and cold fluids are 4310 J/kg.K and 4180 J/kg.K, respectively.

Determine:

a) The mass flow rate of the cold fluid stream
b) The rate of total entropy production
c) The rate of lost work in the heat exchanger.

You can assume that the surrounding atmosphere (environmental thermal energy reservoir) is at a temperature of 298.15 K.

Problem 5

Consider a three-unit process shown in the figure below. Superheated vapor of a hydrocarbon enters the compressor at 1 bar and 300 K and exits at 60 bar and 300 K. The mass flow rate of the fluid is 10 kg/s. The superheated vapor leaving the compressor is cooled to 207 K by rejecting heat to the surroundings at 298.15 K. The superheated vapor stream leaving the cooler undergoes throttling. The discharge stream from the throttling valve is a two-phase vapor-liquid mixture ($x^v = 0.9514$) at 1 bar. The specific enthalpies and entropies of the fluid stream at different locations of the process are given in the table.

a) Carry out the thermodynamic analysis of the process in terms of entropy generation and lost work. Which unit is the most inefficient thermodynamically?

b) What is the thermodynamic efficiency of the overall process?

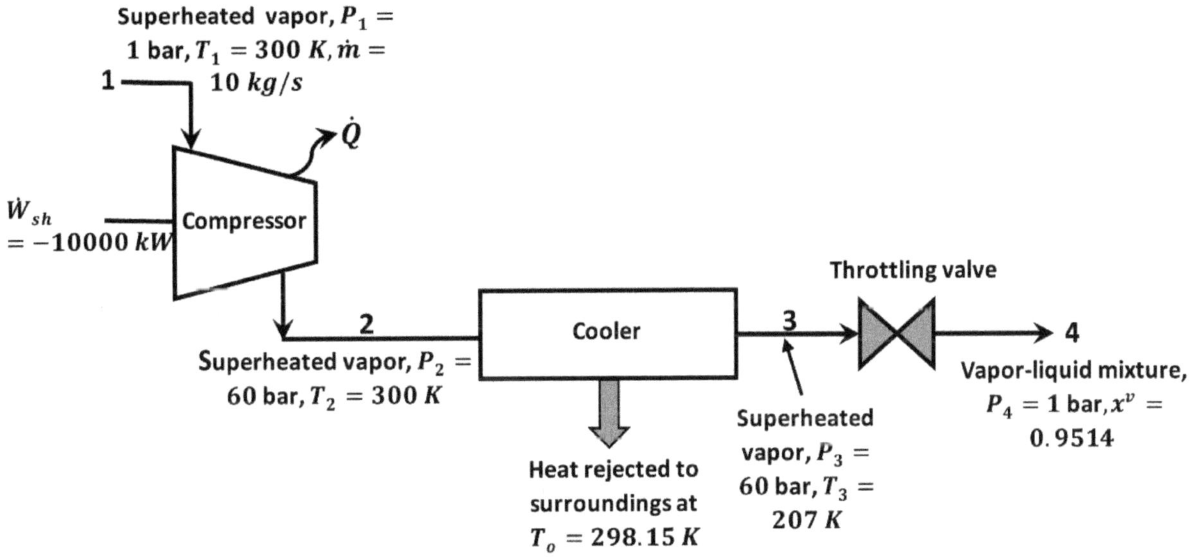

Location	State of fluid	$h\ (kJ/kg)$	$s\ (kJ/kg.K)$
1	Superheated vapor	1199.8	11.629
2	Superheated vapor	1140	8.959
3	Superheated vapor	772	7.798
4	Vapor-liquid mixture, $x^v = 0.9514$	$h^v = 796.9$, $h^L =$ 285.4	$s^v = 9.523$, $s^L =$ 4.962

Chapter 15

Introduction to Exergy and Exergy Analysis

15.1 LEARNING OUTCOMES

Upon completion of this chapter, students will be able to:

- Appreciate the practical significance of exergy analysis.
- Define exergy for closed systems. Develop the expression for exergy of closed systems
- Define exergy for open (flow) systems
- Explain different ways of exergy transfer
- Explain the principle of decrease of exergy
- Carry out exergy balances for closed and open systems
- Develop Gouy-Stodola theorem in terms of exergy destruction
- Carry out exergy analysis of multi-unit flow processes
- Solve problems related to exergy analysis of closed and open systems.

15.2 INTRODUCTION

In simple terms, exergy (also referred to as availability) is the work potential of a system, that is, the maximum possible work that can be extracted from the system. All real processes are irreversible in nature. Consequently, there always occurs a loss of potential power or work that could be extracted from the system. A loss of work potential implies destruction of exergy. Thus, there always occur destruction of exergy in real processes due to irreversibilities.

In the analysis and design of a process from a thermodynamics point of view, it is important for an engineer to quantify irreversiblilities in a process. The quantification of irreversibilities allows the design engineer to identify the portions of the process where improvement in the design is needed in order to minimize the irreversibilities. The irreversibilities in a process can be quantified using two different approaches: entropy generation analysis and exergy destruction analysis. The entropy generation analysis was covered in Chapter 14. Exergy and exergy analysis are introduced in this chapter. The exergy analysis of a process to quantify the irreversibilities is as powerful as the entropy analysis. In fact, it is advantageous over entropy analysis in several respects [1].

Exergy is receiving extensive coverage in scientific publications due to its vast applications in different scientific and engineering fields. Exergy analysis is becoming increasingly popular not only to improve the efficiency of processes and systems, but also to assess their environmental and economic impacts. The concept of exergy is also useful in ecological and sustainability studies as exergy could be interpreted as a measure of resource quality and quantity. In a recent publication [2] in *Science Europe* entitled "*In a resource-constrained World: Think Exergy, not Energy*", the authors state that "Exergy analysis can be applied not only to individual processes, but also to industries, and even to whole national economies. It provides a firm basis from which to judge the effect of policy measures taken towards energy, resource and climate efficiency. In the future, consumers could be informed about products and services in terms of their *exergy-destruction footprint* in much the same way as they are about their *carbon emissions*". Some of the key recommendations made by the authors are: a) the concepts related to exergy must be taught in universities; b) exergy destruction footprints should be introduced in industrial applications to give a useful basis for energy efficiency improvements; c) excess exergy destruction footprints should be taxed to promote improvements in industrial efficiency and to drive the development of more efficient technologies;

d) accounts of exergy destruction footprints should be created to assist the IPCC (Intergovernmental Panel on Climate Change) in developing measures to mitigate climate changes; and e) exergy should be integrated into policy, law, and everyday practice.

15.3 CONCEPTS AND DEFINITIONS

15.3.1 Useful Work

Useful work is the work produced by a system over and above that is required to push the surrounding atmosphere. As an example, consider an expansion of a gas contained in a piston/cylinder assembly shown in Figure 15.1. We assume that the piston, shown in black color in Figure 15.1, is mass-less and that there is no friction present between the piston and cylinder. The gas is expanded from some initial volume V_i to some final volume V_f. When the gas is expanded, the mass m is elevated by a vertical distance of L in the gravitational field. The total work produced by the system (gas) is:

$$W = (Work\ to\ lift\ the\ mass) + (Work\ to\ push\ the\ atmosphere) \tag{15.1}$$

$$W = mgL + P_{atm}A_{piston}L = mgL + P_{atm}(V_f - V_i) \tag{15.2}$$

where A is the cross-sectional area of the piston. In this case, the useful work is the work done by the gas to lift the mass, that is, mgL. The work done to push the atmosphere, that is, $P_{atm}(V_f - V_i)$, is useless work.

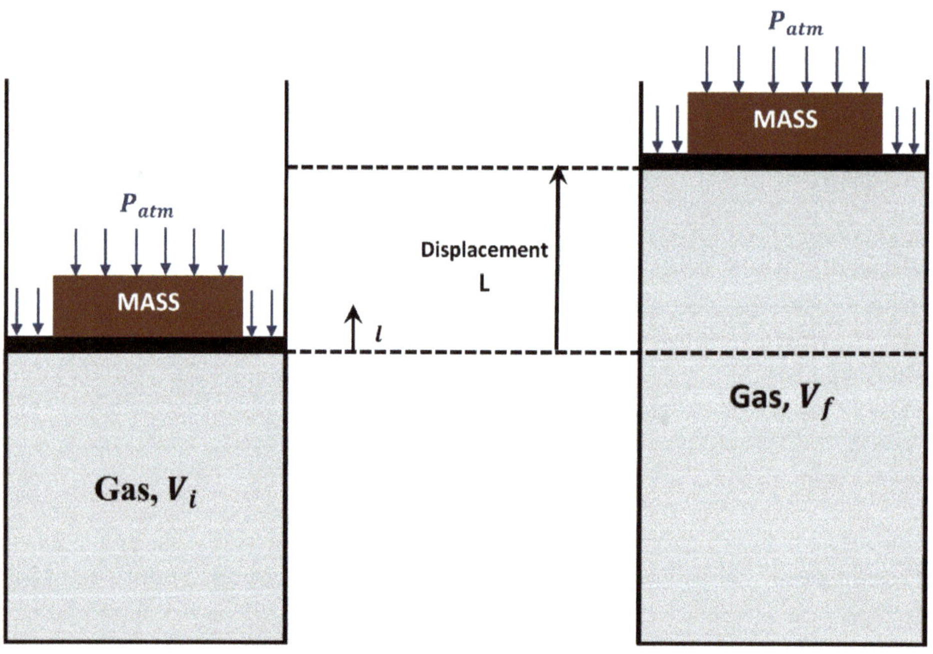

Figure 15.1 Expansion of gas against mass m and surrounding atmosphere.

15.3.2 Dead State

A system is said to be in dead state when it attains thermodynamic equilibrium with the environment (surroundings). In thermodynamic equilibrium, the conditions of mechanical, thermal, and chemical equilibrium between the system and the environment are satisfied. The existence of mechanical and thermal equilibrium between the system and the environment implies equality of pressure and temperature between the system and the environment. The condition of chemical equilibrium implies the chemical potential of each chemical species present in the system is the same in the system and the environment. If only thermo-mechanical effects are present, then the dead state is specified by the temperature T_o and pressure P_o of the environment (surroundings). Usually, $T_o = 25^0\text{C}$ and $P_o = 1\ atm$. Note that when the chemical composition of the system is constant, only thermo-mechanical effects are relevant in the thermodynamic analysis of the process.

An important point to note is that when the system and the environment are in different states, an opportunity exists to produce useful work upon interaction between the system and the environment. If the system is in the dead state, it has zero work potential. No work can be extracted from a system in the dead state.

15.3.3 Exergy

Exergy is defined as the maximum possible useful work that can be extracted from a system when the system undergoes a change in state from its current *live* state to a *dead* state. The *dead state*, as already defined, is one where the system attains thermodynamic equilibrium with the environment. It should be noted that the work produced is maximum only when the process is carried out *reversibly*. Hence exergy could also be defined as the reversible useful work that can be extracted from a system when the system undergoes a change in state from its current live state to a dead state.

Clearly, exergy is a measure of the *degree of departure* of the state of the system from its environment. When the system and the environment are not in thermodynamic equilibrium with each other, an opportunity exists to produce useful work upon interaction between the system and the environment. When the system and the environment are in thermodynamic equilibrium, the exergy of the system is zero and no useful work can be extracted from the system.

It is important to recognize that exergy is a *state property*. However, this state property is dependent on both the state of the system and the state of the environment. Thus exergy is a state property of the *combined system and environment*. If the environment of a given system is changed, its exergy will also change. For a given environment, the change in exergy of a system between two fixed states is the same regardless of how this change in state of the system is brought about. Also note that exergy is an extensive property.

15.3.4 Exergy of a Closed System

The *specific exergy* of a closed system can be expressed as follows:

$$\psi = (u - u_o) + P_o(v - v_o) - T_o(s - s_o) \tag{15.3}$$

where $u, v,$ and s are specific internal energy, specific volume, and specific entropy, respectively. The subscript "o" refers to the dead state where the system attains thermodynamic equilibrium with the environment. The environment pressure and temperature are P_o and T_o, respectively. Equation (15.3) assumes that the system is stationary (no bulk kinetic energy) and that its elevation relative to the environment is zero (no potential energy). It is further assumed that the chemical composition of the system is constant and hence only thermo-mechanical effects are relevant. As exergy is an extensive property, the total exergy of a closed system is:

$$\Psi = m\psi \tag{15.4}$$

where m is the mass of the system.

Proof of Equation (15.3)

Consider gas enclosed in a piston/cylinder assembly at P, V, and T as the system (see Figure 15.2). The environment is at P_o and T_o. The system is initially *not* in thermodynamic equilibrium ($T > T_o, P > P_o$). As the system is not in thermodynamic equilibrium, it has exergy and hence the potential to do work.

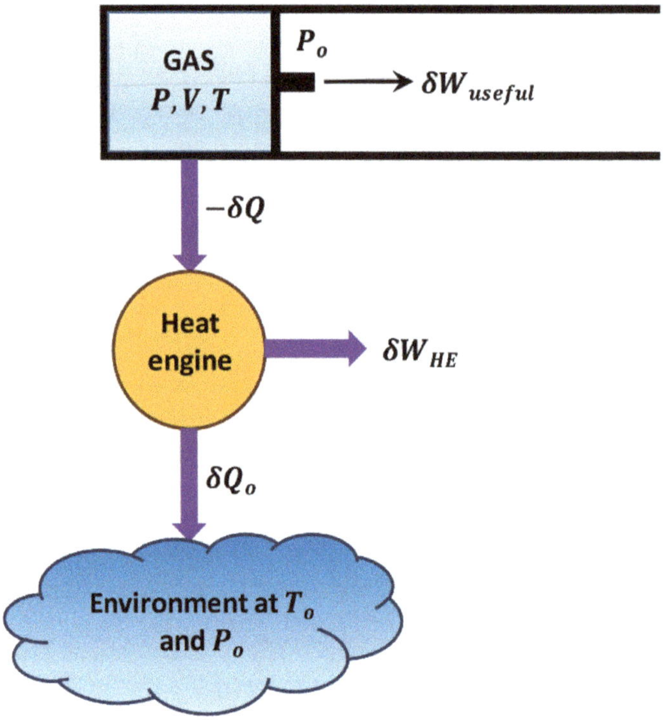

Figure 15.2 Set up to derive the expression for exergy of a closed system (gas contained in a piston/cylinder assembly).

Let the gas undergo a differential amount of expansion reversibly. During this differential expansion, the system produces δW amount of work and rejects heat in the amount of - δQ. As heat is rejected by the system, δQ is negative (- δQ is positive). The work produced is:

$$\delta W = PdV = (P - P_o)dV + P_o dV \tag{15.5}$$

$$\delta W_{useful} = (P - P_o)dV \tag{15.6}$$

The heat rejected by the system is utilized by a reversible (Carnot) heat engine to produce work δW_{HE} as:

$$\delta W_{HE} = (-\delta Q)\left(1 - \frac{T_o}{T}\right) \tag{15.7}$$

Thus, the total useful work produced in reversible expansion of gas is:

$$\delta W_{total\,useful} = (P - P_o)dV + (-\delta Q)\left(1 - \frac{T_o}{T}\right) \tag{15.8}$$

Application of the first law to the system (gas) gives:

$$dU = \delta Q - PdV \tag{15.9}$$

From Equations (15.8) and (15.9), we get:

$$\delta W_{total\ useful} = -dU - P_o dV + \frac{T_o}{T}\delta Q \tag{15.10}$$

As the expansion process is reversible,

$$\delta Q = TdS \tag{15.11}$$

From Equations (15.10) and (15.11), it follows that:

$$\delta W_{total\ useful} = -dU - P_o dV + T_o dS \tag{15.12}$$

Upon integration from system's initial state to its dead state:

$$W_{total\ useful} = (U - U_o) + P_o(V - V_o) - T_o\ (S - S_o) \tag{15.13}$$

By definition, Equation (15.13) gives the total exergy of the system:

$$\Psi = (U - U_o) + P_o(V - V_o) - T_o\ (S - S_o) \tag{15.14}$$

where $U, V,$ and S are total internal energy, total volume, and total entropy, respectively of the system. Equation (15.14) could also be expressed in terms of the specific exergy as:

$$\psi = (u - u_o) + P_o(v - v_o) - T_o\ (s - s_o) \tag{15.3}$$

15.3.5 Exergy of Open (Flow) System

The specific exergy of a fluid stream (flow system) is given as:

$$\psi = (h - h_o) - T_o(s - s_o) + \frac{1}{2}V^2 + gZ \tag{15.15}$$

where h is specific enthalpy, V is fluid velocity, g is acceleration due to gravity, and Z is the elevation of fluid stream.

Note that Equation (15.15) follows from the specific exergy expression of closed system, Equation (15.3), upon addition of kinetic and potential energies and the useful flow work. The useful flow work on a unit mass basis is given as:

$$w_{useful\ flow\ work} = Pv - P_o v = (P - P_o)v \tag{15.16}$$

15.3.6 Exergy Transfer

Exergy could be transferred to or from a system in three different ways: mass transfer, heat transfer, and work transfer.

Transfer of Exergy by Mass Transfer

Mass carries it with exergy, just like it carries with it energy, entropy, enthalpy, etc. Exergy is added to the system if the system receives mass. Exergy is removed from the system if mass is removed from the system. The exergy transferred due to transfer of mass in an open system can be expressed as:

$$\Psi_{mass} = m\psi = m\left[(h - h_o) - T_o(s - s_o) + \frac{1}{2}V^2 + gZ\right] \quad (15.17)$$

Transfer of Exergy by Heat Transfer

Heat can be converted into useful work. Thus, heat carries with it exergy. However, only a portion of heat can be converted into work according to the second law of thermodynamics. Using the Carnot heat engine formula, the portion of heat that can be converted into useful work is $\left(1 - \frac{T_o}{T}\right)$, where T is the temperature at which heat is available and T_o is the environment temperature. Thus, exergy associated with heat is as follows:

$$\Psi = Q\left(1 - \frac{T_o}{T}\right) \quad (15.18)$$

Transfer of Exergy by Work Transfer

Useful work is itself exergy. Thus, exergy transferred to or from a system is equal to the useful work transferred. For expansion and compression work, the exergy associated with work is:

$$\Psi_{work} = W_{useful} = W - W_{surr} \quad (15.19)$$

where

$$W_{surr} = P_o \Delta V \quad (15.20)$$

Note that V is the total volume of the system.

Other types of work such as shaft work or electric work are 100% useful work and, hence, they are equal to exergy:

$$\Psi_{work} = W \qquad \text{(non-expansion/non-compression work)} \quad (15.21)$$

15.4 DECREASE OF EXERGY PRINCIPLE

Consider an *isolated* system. The exergy balance over the system can be written as:

$$Input - Output + Generation - Destruction = Change \quad (15.22)$$

Exergy cannot be generated from nowhere. It can only be destroyed due to irreversibilities. For an isolated system, there is no inflow or outflow of exergy. Thus,

$$Destruction\ of\ Exergy = Decrease\ in\ Exergy\ of\ the\ System \quad (15.23)$$

$$\Psi_D = \Psi_1 - \Psi_2 \quad (15.24)$$

Since

$$\Psi = (U - U_o) + P_o(V - V_o) - T_o\,(S - S_o) \quad (15.25)$$

it follows that:

$$\Psi_D = \Psi_1 - \Psi_2 = (U_1 - U_2) + P_o(V_1 - V_2) - T_o\,(S_1 - S_2) \quad (15.26)$$

Note that U, V, S are total quantities, not specific values.

For an isolated system:

$$U_1 = U_2 \quad \text{and} \quad V_1 = V_2 \tag{15.27}$$

Thus:

$$\Psi_D = T_o\,(S_2 - S_1) = T_o \Delta S \tag{15.28}$$

where T_o is the absolute temperature of the environment or surroundings.

We now do entropy balance over an isolated system. As entropy cannot be destroyed, the balance equation, Equation (15.22), for entropy becomes:

$$Input - Output + Generation = Change \tag{15.29}$$

For an isolated system, there is no inflow or outflow of entropy. Thus, Equation (15.29) gives:

$$Entropy\ Generation = Increase\ in\ Entropy \tag{15.30}$$

$$S_G = \Delta S \tag{15.31}$$

From Equations (15.28) and (15.31):

$$\Psi_D = T_o S_G \tag{15.32}$$

Equation (15.32) tells us that the exergy destroyed in the system is directly proportional to the entropy generated in the system and the proportionality constant is the absolute temperature of the environment. According to the second law of thermodynamics, $S_G > 0$ for any real process. Therefore, for any spontaneous change taking place inside an isolated system:

$$\Psi_D = \Psi_1 - \Psi_2 = T_o S_G > 0 \tag{15.33}$$

This is the *principle of decrease of exergy*. Any natural process taking place inside an isolated system is always accompanied by a decrease in exergy of the system due to destruction of exergy. Note that all natural (real or spontaneous) processes are irreversible processes. In the limiting case of a reversible process, no exergy is destroyed ($\Psi_D = 0$) as no entropy is generated ($S_G = 0$) and therefore, the exergy of an isolated system remains constant.

15.5 EXERGY BALANCE FOR A CLOSED SYSTEM

Consider a closed system shown in Figure 15.3. The system is receiving heat from the environment TER at T_o and producing total work of W. Note that the TER is in direct contact with the system indicating that heat is received by the system at T_o. The temperature at the boundary of the system is the same as that of TER, that is, T_o.

Exergy balance for the system can be expressed as:

$$Exergy\ Input - Exergy\ Output - Exergy\ Destruction = Exergy\ Change \tag{15.34}$$

$$Q_o\left(1 - \frac{T_o}{T_o}\right) - W_{useful} - \Psi_D = \Psi_2 - \Psi_1 \tag{15.35}$$

Note that exergy associated with heat is zero as heat is supplied to the system at environment temperature of T_o. The useful work is work over and above the work needed to push the environment, that is,

$$W_{useful} = W - P_o(V_2 - V_1) \tag{15.36}$$

Upon simplification, Equation (15.35) gives:

$$W_{useful} = (\Psi_1 - \Psi_2) - \Psi_D \tag{15.37}$$

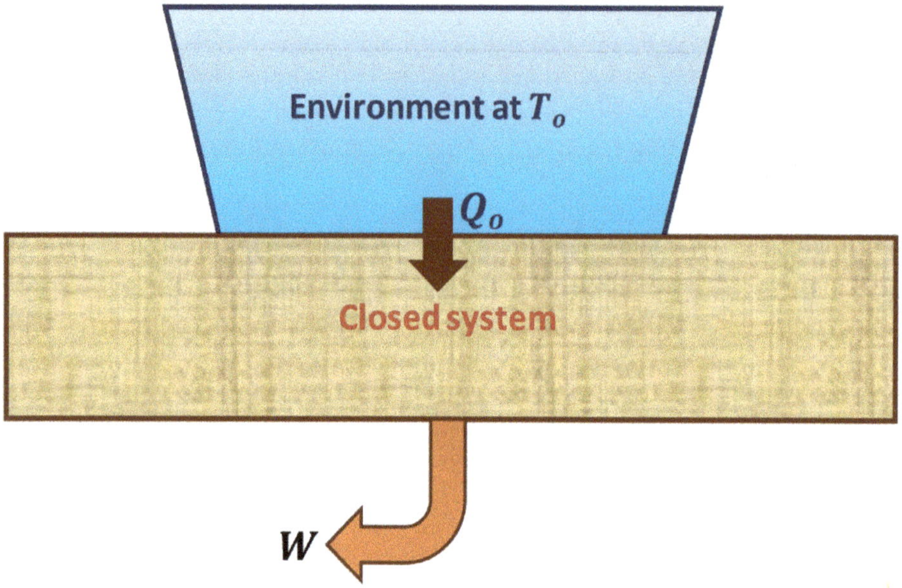

Figure 15.3 Closed system interacting thermally with environment TER (thermal energy reservoir) at T_o and producing work W.

For a reversible process, exergy destruction is zero and work produced is maximum. Thus,

$$W_{max,useful} = (\Psi_1 - \Psi_2) \tag{15.38}$$

The maximum useful work is simply the decrease in the exergy of the system. The work lost due to irreversibilities in the system can be expressed as:

$$W_{lost} = W_{rev} - W \tag{15.39}$$

$$W_{lost} = \Psi_D \tag{15.40}$$

This is an important result. It tells us that the work lost due to irreversibilities in a process is equal to the exergy destroyed in the process. Note that in the process considered here, there is no destruction of exergy outside the system as heat is supplied reversibly (the temperature at the system boundary is the same as the TER temperature). In general, exergy destruction occurs both internally and externally to the system and therefore,

$$W_{lost} = \Psi_{D,total} \tag{15.41}$$

where $\Psi_{D,total}$ is total exergy destroyed within and outside the system. Equation (15.41) represents the *Gouy-Stodola theorem* in terms of exergy destruction. The *Gouy-Stodola theorem* states that the work lost due to irreversibilities in the process is equal to total exergy destroyed (internally and externally) due to the process.

15.6 EXERGY BALANCE FOR OPEN SYSTEMS

Consider an open system shown in Figure 15.4. The system is receiving heat from the environment TER at T_o and producing shaft work of $\dot{W}_{sh}$. The TER is in direct contact with the system indicating that heat is received by the system at T_o. The temperature at the system boundary is the same as that of TER, that is, T_o. The fluid stream enters the control volume through the inlet port and leaves the control volume through the outlet port. The process is internally irreversible. The external irreversibility due to heat transfer is absent as heat is delivered at the boundary temperature of T_o.

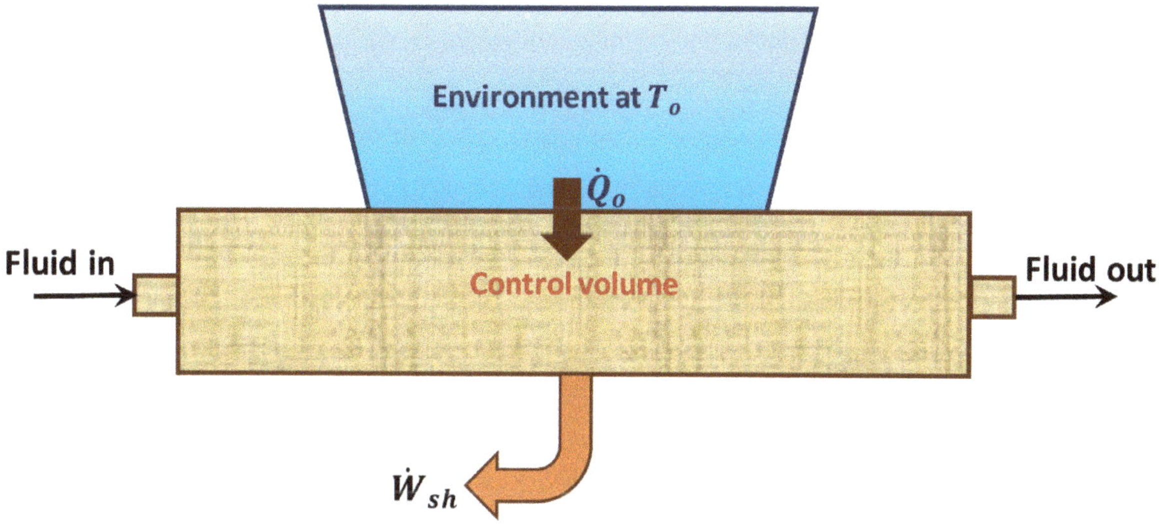

Figure 15.4 Open (flow) system interacting thermally with environment TER (thermal energy reservoir) at T_o and producing shaft work $\dot{W}_{sh}$.

Exergy balance over the control volume can be expressed as:

$$(Rate\ of\ Exergy\ Input) - (Rate\ of\ Exergy\ Output) - (Rate\ of\ Exergy\ Destruction) = (Rate\ of\ Exergy\ Accumulation) \quad (15.42)$$

The rate of exergy input to the control volume is:

$$(Rate\ of\ Exergy\ Input) = \dot{Q}_o\left(1 - \frac{T_o}{T_o}\right) + \dot{m}_1\psi_1 \quad (15.43)$$

The rate of exergy output from the control volume is:

$$(Rate\ of\ Exergy\ Output) = \dot{W}_{sh} + \dot{m}_2\psi_2 \quad (15.44)$$

The rate of exergy destruction is $\dot{\Psi}_{D,CV}$ and the rate of exergy accumulation is:

$$(Rate\ of\ Exergy\ Accumulation) = \frac{\partial \Psi_{CV}}{\partial t} \quad (15.45)$$

Upon substitution of various terms into Equation (15.42), we get:

$$\left[\dot{Q}_o\left(1 - \frac{T_o}{T_o}\right) + \dot{m}_1\psi_1\right] - \left[\dot{W}_{sh} + \dot{m}_2\psi_2\right] - \dot{\Psi}_{D,CV} = \frac{\partial \Psi_{CV}}{\partial t} \quad (15.46)$$

Assuming steady state and simplifying Equation (15.46) gives:

$$\dot{W}_{sh} = \dot{m}(\psi_1 - \psi_2) - \dot{\Psi}_{D,CV} \tag{15.47}$$

For a reversible process, the exergy destruction rate is zero and the rate of shaft work produced is maximum. Thus,

$$\dot{W}_{sh,max} = \dot{m}(\psi_1 - \psi_2) \tag{15.48}$$

The maximum rate of shaft work produced is simply the rate of decrease in the exergy of the fluid stream. The rate of shaft work lost due to irreversibilities is:

$$\dot{W}_{sh,lost} = \dot{W}_{sh,max} - \dot{W}_{sh} \tag{15.49}$$

$$\dot{W}_{sh,lost} = \dot{\Psi}_{D,CV} \tag{15.50}$$

This is an important result. It tells us that the loss of power (rate of shaft work) due to irreversibilities in a process is equal to the rate of exergy destruction in the process. Note that in the process considered here, there is no destruction of exergy outside the system as heat is supplied reversibly (the temperature at the system boundary is the same as the TER temperature). In general, exergy destruction occurs both internally and externally to the system and therefore,

$$\dot{W}_{sh,lost} = \dot{\Psi}_{D,total} \tag{15.51}$$

where $\dot{\Psi}_{D,total}$ is the rate of total exergy destruction within and outside the control volume. Equation (15.51) represents the *Gouy-Stodola theorem* for open systems in terms of exergy destruction. The *Gouy-Stodola theorem* states that the power (rate of shaft work) lost due to irreversibilities in the process is equal to the rate of total exergy destruction (internally and externally) due to the process.

15.7 ANALYSIS OF MULTI-UNIT FLOW PROCESSES IN TERMS OF EXERGY DESTRUCTION

We can apply the exergy based Gouy-Stodola theorem for open systems, Equation (15.51), to analyze multi-unit flow processes involving a number of different units or components. Consider the multi-unit flow process shown in Figure 15.5. There are four units connected serially. We assume that each unit is interacting thermally with the same thermal energy reservoir (TER) at a uniform temperature of T_o. Each unit is also producing (or absorbing) shaft work. Applying the Gouy-Stodola theorem in terms of exergy destruction to each unit and summing over all the units of the process yields:

$$\dot{W}_{sh,lost,i} = \dot{\Psi}_{D,total,i} \tag{15.52}$$

$$\sum_{i=1}^{4} \dot{W}_{sh,lost,i} = \sum_{i=1}^{4} \dot{\Psi}_{D,total,i} \tag{15.53}$$

Dividing Equation (15.52) with Equation (15.53), we get:

$$\frac{\dot{W}_{sh,lost,i}}{\sum_{i=1}^{4} \dot{W}_{sh,lost,i}} = \frac{\dot{\Psi}_{D,total,i}}{\sum_{i=1}^{4} \dot{\Psi}_{D,total,i}} \tag{15.54}$$

According to Equation (15.54), the fractional contribution of each unit to the total loss of power (rate of shaft work) for the whole process is equal to the ratio of rate of total exergy destruction in the unit to the rate of total exergy destruction for the whole process. When analyzing a process, it is important to calculate the contribution of each unit to the rate of total exergy destruction for the whole process. This way one can identify the units where exergy destruction is large and hence need to be modified or redesigned to improve the overall thermodynamic efficiency of the process.

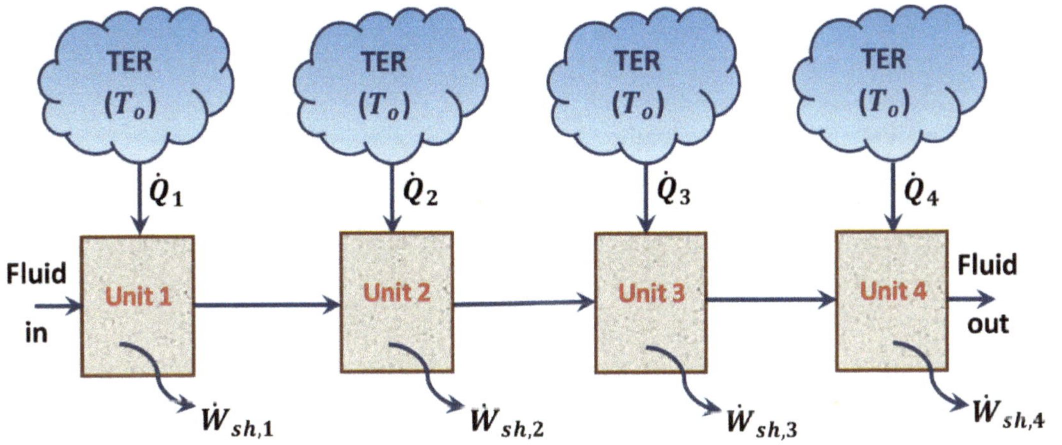

Figure 15.5 Multi-unit flow process where each unit is interacting with a TER (thermal energy reservoir) and producing shaft work.

NOTATION

A_{piston}	cross-sectional area of piston
g	acceleration due to gravity
h	specific enthalpy
h_o	specific enthalpy of system in dead state
L	displacement
m	mass
$\dot{m}$	mass flow rate
P	pressure
P_o	pressure of the system in dead state or pressure of environment
Q	heat
Q_o	heat transfer between system and environmental thermal energy reservoir
$\dot{Q}_o$	rate of heat transfer between system and environmental thermal energy reservoir
S	total entropy
S_G	entropy generated
S_o	total entropy in dead state
s	specific entropy
s_o	specific entropy of system in dead state
T	temperature
T_o	temperature of the system in dead state or temperature of environment
TER	thermal energy reservoir
t	time
U	total internal energy
U_o	total energy of system in dead state
u	specific internal energy
u_o	specific internal energy in dead state
V	total volume or velocity
V_o	total volume of system in dead state

v specific volume
v_o specific volume of system in dead state
W work
W_{HE} work produced by a heat engine
W_{lost} work lost
$W_{max,useful}$ maximum useful work
W_{rev} work in a reversible process
W_{surr} work done on the surroundings in pushing the surrounding atmosphere
W_{useful} useful work
$w_{useful\ flow\ work}$ useful flow work per unit mass of fluid
$\dot{W}_{sh}$ rate of shaft work
$\dot{W}_{sh,lost}$ rate of shaft work lost
$\dot{W}_{sh,lost,i}$ rate of shaft work lost in ith unit of a multi-unit process
$\dot{W}_{sh,max}$ maximum rate of shaft work
Z elevation

Greek Letters

Δ difference operator
δ differential of inexact quantity
ψ specific exergy
Ψ total exergy
Ψ_{mass} exergy associated with mass
Ψ_{work} exergy associated with work
Ψ_{CV} exergy in control volume
Ψ_D exergy destroyed
$\Psi_{D,CV}$ exergy destruction in control volume
$\Psi_{D,total}$ total exergy destruction internally and externally
$\dot{\Psi}_{D,CV}$ rate of exergy destruction in control volume
$\dot{\Psi}_{D,total}$ rate of total exergy destruction internally and externally
$\dot{\Psi}_{D,total,i}$ rate of total exergy destruction internally and externally in ith unit of a multi-unit process

REFERENCES

1. Pal, R., "Quantification of irreversibilities in practical cyclic processes using exergy analysis and Gouy-Stodola theorem", *Int. J. Mech. Eng. Edu.* 2018; 46: 118-137.
2. Brockway, P. E., Dewulf, P., Kjelstrup, S., Siebentritt, S., Valero, A., Whelan, C., "In a resource-constrained world: think exergy, not energy", Brussels: Science Europe, 2016.

TUTORIAL PROBLEMS WITH SOLUTION

Problem 1

A 5 kg metal block is heated to 800 °C in an oven. It is removed from the oven and allowed to cool to 25 °C by transferring heat to the surroundings air at 25 °C and 1 atm.

a) Determine the exergy of the metal block in the initial state.
b) What is the exergy of the metal block in the final state?
c) How much exergy is destroyed?

d) How much work can be extracted from the metal block if a reversible (Carnot) heat engine is connected between the metal block and surroundings at 25 °C?

The specific heat capacity (c) of metal block is 0.5 kJ/kg.K. The dead state is: $T_o = 25^oC$ and $P_o = 1$ atm.

Solution

a) By definition,

$$\Psi = m\psi = m[(u - u_o) + P_o(v - v_o) - T_o(s - s_o)]$$

Assuming the metal to be incompressible, $v = v_o$. Thus,

$$\Psi = m\left[c(T - T_o) - T_o c ln\left(\frac{T}{T_o}\right)\right] = 5 \times 0.5\left[(800 - 25) - 298.15 \times ln\left(\frac{800 + 273.15}{25 + 273.15}\right)\right] = 982.85 \text{ kJ}$$

b) As the metal block reaches equilibrium with the surroundings in the final state, the exergy of the metal block reduces to zero in the final (dead) state.

c) All the initial exergy of the metal block is wasted away (destroyed) as no useful work is extracted from the block. Thus, the amount of exergy destroyed is 982.85 kJ

d) Work that can be extracted from the metal block if a reversible (Carnot) heat engine is connected between the metal block and surroundings at 25 °C?

The work produced by a reversible (Carnot) heat engine on a differential basis is:

$$\delta W = \delta Q\left(1 - \frac{T_o}{T}\right)$$

where T_o is the surroundings temperature (25 °C) and T is the metal block temperature. The amount of heat input to the engine δQ is:

$$\delta Q = -mcdT$$

where dT is the change in metal block temperature. Thus,

$$\delta W = \delta Q\left(1 - \frac{T_o}{T}\right) = -mc\left(1 - \frac{T_o}{T}\right)dT$$

Upon integration from 800 °C to 25 °C, we get:

$$W = -\int_{800+273.15}^{25+273.15} mc\left(1 - \frac{T_o}{T}\right)dT = -5 \times 0.5\left[(298.15 - 1073.15) - (298.15) \times ln\left(\frac{298.15}{1073.15}\right)\right] = 982.85 \text{ kJ}$$

This is equal to the initial exergy of the metal block. Thus, all the exergy of the metal block is converted into useful work. No exergy is destroyed as the engine is reversible.

Problem 2

A 75 kg carbon-steel engine casting of specific heat capacity $c = 0.5$ kJ/kg°C is heated to 500 °C in an oven. It is removed from the oven and quenched by immersing in a closed tank containing 350 kg of oil of specific heat capacity $c = 2.5$ kJ/kg°C initially at 25 °C.

a) Calculate the final equilibrium temperature of casting and oil.
b) How much exergy is destroyed in this process?

You can assume that the materials (carbon-steel and oil) are incompressible. There is no heat loss to the surroundings. The dead state is: $T_o = 25^oC$ and $P_o = 1$ atm.

Solution

Refer to the figure given below.

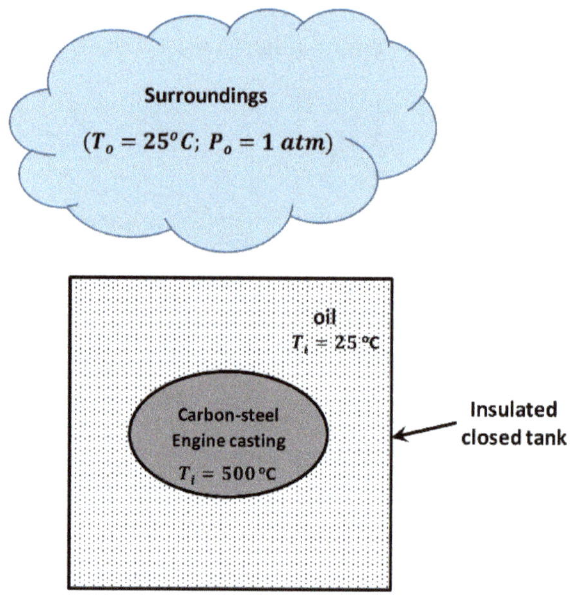

a) Calculate the final equilibrium temperature of casting and oil.

The engine casting and oil together form an isolated system. No heat or work is exchanged between the isolated system and the surroundings. From first law for closed systems,

$$\Delta U = \Delta U_{casting} + \Delta U_{oil} = 0$$

For incompressible materials, $\Delta U = mC\Delta T$. Thus,

$$75 \times 0.5(T_f - 773.15) + 350 \times 2.5(T_f - 298.15) = 0$$

The solution gives $T_f = 317.67$ K.

b) How much exergy is destroyed in this process?

From exergy balance:

Exergy In–Exergy Out–Exergy destroyed = Change in exergy

For an isolated system, Exergy In =0, Exergy Out = 0. Thus,

$$\Psi_D = -\Delta\Psi = -(\Delta\Psi_{casting} + \Delta\Psi_{oil})$$

By definition,

$$\Psi = (U - U_o) + P_o(V - V_o) - T_o(S - S_o)$$

Thus,

$$\Delta\Psi = (\Delta U) + P_o(\Delta V) - T_o(\Delta S)$$

Since the materials are incompressible, $\Delta V = 0$. Consequently,

$$\Delta\Psi = (\Delta U) - T_o(\Delta S)$$

For an isolated system, $Q = 0$, $W = 0$. From the first law for closed systems,

$$(\Delta U)_{isolated\ system} = 0$$

$$\Delta U_{casting} + \Delta U_{oil} = 0$$

$$\Delta\Psi_{casting} = \Delta U_{casting} - T_o\Delta S_{casting}$$

$$\Delta\Psi_{oil} = \Delta U_{oil} - T_o\Delta S_{oil}$$

$$\Psi_D = -(\Delta\Psi_{casting} + \Delta\Psi_{oil}) = -[(\Delta U_{casting} + \Delta U_{oil}) - T_o(\Delta S_{casting} + \Delta S_{oil})]$$
$$= T_o(\Delta S_{casting} + \Delta S_{oil})$$

$$\Delta S_{casting} = m_{casting}c_{casting}ln\left(\frac{T_f}{T_{i,casting}}\right) = 75 \times 0.5 \times ln\left(\frac{317.67}{773.15}\right) = -33.434 \text{ kJ/K}$$

$$\Delta S_{oil} = m_{oil}c_{oil}ln\left(\frac{T_f}{T_{i,oil}}\right) = 350 \times 2.5 \times ln\left(\frac{317.67}{298.15}\right) = 55.489 \text{ kJ/K}$$

Thus,

$$\Psi_D = T_o(\Delta S_{casting} + \Delta S_{oil}) = 298.15 \times (-33.434 + 55.489) = 6575.7 \text{ kJ}$$

The total amount of exergy destroyed in this irreversible process is 6575.7 kJ.

Problem 3

1000 moles of an ideal gas is contained in an insulated cylinder at a pressure of 15 bar and a temperature of 250 °C. Calculate the exergy of the gas. The heat capacities of the gas are: $c_p = 7R/2$ and $c_v = 5R/2$. The surroundings are at: $T_o = 25^oC$ and $P_o = 1$ atm.

Solution

The molar exergy of a gas is:

$$\psi = (u - u_o) + P_o(v - v_o) - T_o(s - s_o)$$

For ideal gas,

$$u - u_o = c_v(T - T_o) = 2.5 \times 8.314 \times (523.15 - 298.15) = 4676.625 \text{ J/mol}$$

$$s - s_o = c_p ln\left(\frac{T}{T_o}\right) - Rln\left(\frac{P}{P_o}\right) = 3.5 \times 8.314 \times ln\left(\frac{523.15}{298.15}\right) - 8.314 \times ln\left(\frac{15}{1.01325}\right) = -6.0437 \text{ J/mol.K}$$

$$P_o(v - v_o) = P_o\left(\frac{RT}{P} - \frac{RT_o}{P_o}\right) = 1.01325 \times 8.314 \times \left(\frac{523.15}{15} - \frac{298.15}{1.01325}\right) = -2185.01 \text{ J/mol}$$

Thus,

$$\psi = 4676.625 + (-2185.01) - 298.15 \times (-6.0437\,) = 4293.54 \text{ J/mol}$$

The total exergy is:

$$\Psi = n\psi = 1000 \times 4293.54 = 4293.54 \text{ kJ}$$

Problem 4

Consider the wall of an industrial furnace through which heat transfer is taking place at a rate of 4.25 kW. Measurements made during steady state operation reveal temperatures of 1400 and 1150 K at the inner and outer surfaces, respectively. The surroundings are at: $T_o = 25^oC$ and $P_o = 1$ atm. Calculate:

a) The exergy transfer rate accompanying heat at the inner and outer surfaces of the wall
b) The rate of exergy destruction due to irreversible heat transfer

Solution

Refer to the figure given below.

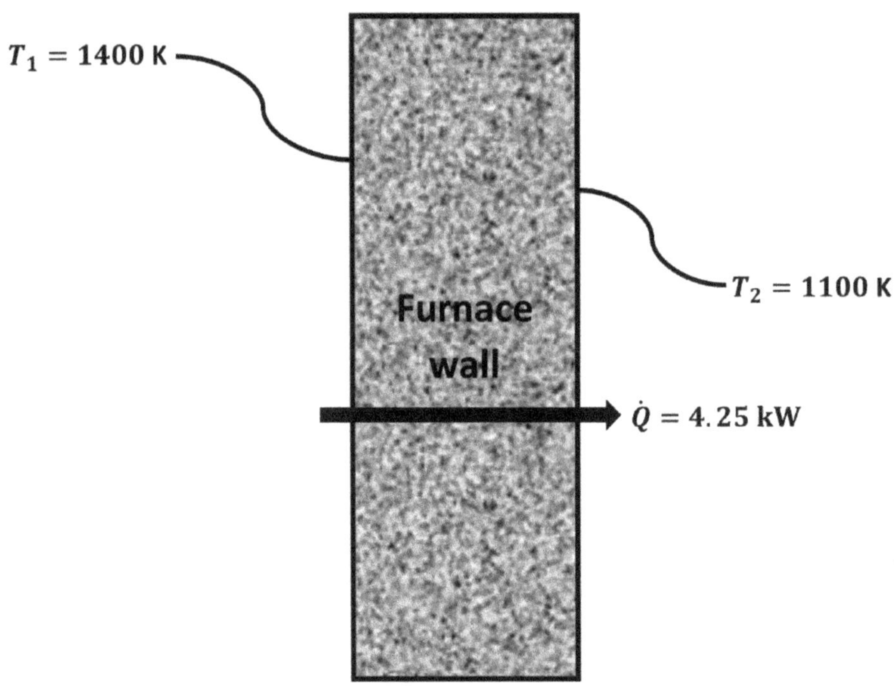

Exergy associated with heat is:

$$\Psi = Q\left(1 - \frac{T_o}{T}\right)$$

Exergy transfer rate associated with heat transfer rate $\dot{Q}$ is:

$$\dot{\Psi} = \dot{Q}\left(1 - \frac{T_o}{T}\right)$$

a) The exergy transfer rate accompanying heat at the inner and outer surfaces of the wall

$$\dot{\Psi}_1 = \dot{Q}_1\left(1 - \frac{T_o}{T_1}\right) = 4.25 \times \left(1 - \frac{298.15}{1400}\right) = 3.3449 \text{ kW}$$

$$\dot{\Psi}_2 = \dot{Q}_2\left(1 - \frac{T_o}{T_2}\right) = 4.25 \times \left(1 - \frac{298.15}{1100}\right) = 3.148 \text{ kW}$$

b) The rate of exergy destruction due to irreversible heat transfer

Exergy balance over the furnace wall gives:

$$Exergy\ Input - Exergy\ Output - Exergy\ Destruction = Exergy\ Accumulation$$

At steady state,

$$Exergy\ Destruction = Exergy\ Input - Exergy\ Output$$

$$\dot{\Psi}_D = \dot{\Psi}_1 - \dot{\Psi}_2 = 3.3449 - 3.148 = 0.1969 \text{ kW}$$

Thus, the rate of exergy destruction in the furnace wall due to irreversible heat transfer is 0.1969 kW

Problem 5

Superheated steam enters the turbine at 35 bar and 450 °C at a rate of 10 kg/s. The exhaust from the turbine is saturated steam at 90 °C. There are negligible changes in kinetic and potential energies. The heat loss from the turbine is also negligible. The surroundings are at: $T_o = 25^oC$ and $P_o = 1$ atm.
Calculate:

a) the rate at which exergy enters the turbine
b) the rate at which exergy leaves the turbine
c) the maximum possible power output of the turbine
d) the actual power output of the turbine
e) the rate of exergy destruction in the turbine

Solution

We need specific enthalpies and entropies for the inlet and discharge streams. We obtain this information from the steam tables given in Appendices M and N.

Superheated steam at 35 bar and 450 °C:

$h = 3337.2$ kJ/kg ; $s = 7.0052$ kJ/kg.K

Saturated steam at 90 °C.

$h = 2660.1$ kJ/kg ; $s = 7.4799$ kJ/kg.K

Dead state:

In the dead state ($T_o = 25^oC$ and $P_o = 1$ atm), the fluid is liquid water. Neglecting the effect of pressure on enthalpy and entropy of liquid water, we assume: $h_o(T_o, P_o) \cong h^{sat.liq}(T_o)$ and $s_0(T_o, P_o) \cong s^{sat.liq}(T_o)$. At 25^oC , the saturated steam tables give:

$h_o = h^L = 104.9$ kJ/kg ; $s_o = s^L = 0.3672$ kJ/kg

a) the rate at which exergy enters the turbine

$$\dot{\Psi} = \dot{m}\psi = \dot{m}\left[(h - h_o) - T_o(s - s_o) + \frac{1}{2}V^2 + gZ\right] = \dot{m}[(h - h_o) - T_o(s - s_o)]$$

$$\dot{\Psi}_1 = 10 \times [(3337.2 - 104.9) - 298.15 \times (7.0052 - 0.3672)] = 12531.803 \text{ kW}$$

b) the rate at which exergy leaves the turbine

$$\dot{\Psi}_2 = 10 \times [(2660.1 - 104.9) - 298.15 \times (7.4799 - 0.3672)] = 4345.485 \text{ kW}$$

c) the maximum possible power output of the turbine

$$\dot{W}_{sh,max} = \dot{W}_{sh,rev} = \dot{\Psi}_1 - \dot{\Psi}_2 = 12531.803 - 4345.485 = 8186.318 \text{ kW}$$

d) the actual power output of the turbine

$$\dot{W}_{sh} = -\dot{m}\Delta h = -10 \times (2660.1 - 3337.2) = 6771 \text{ kW}$$

e) rate of exergy destruction in the turbine

$$\dot{\Psi}_D = \dot{\Psi}_1 - \dot{\Psi}_2 - \dot{W}_{sh} = 8186.318 - 6771 = 1415.318 \text{ kW}$$

Problem 6

Consider a multi-unit process shown in the figure given below. Superheated steam at 125 bar and 550 °C enters the turbine. The mass flow rate of steam is 70 kg/s. The discharge stream from the turbine is wet steam at 0.08 bar and 41.5 °C with a quality of 0.9304. The wet steam passes through the condenser where it is completely condensed into saturated liquid water at 0.08 bar and 41.5 °C. The saturated liquid water from the condenser is pressurized to 125 bar using a pump. The temperature of the subcooled liquid water leaving the pump is 43.2 °C. The surroundings are at: $T_o = 25°C$ and $P_o = 1$ atm.
Calculate:

a) The specific exergy of fluid in states 1, 2, 3, and 4.
b) The rate of total exergy destruction in the turbine
c) The rate of total exergy destruction in the condenser
d) The rate of total exergy destruction in the pump
e) The rate of total exergy destruction in the whole process including all the units
f) The fraction of overall process exergy destruction that takes place in each unit
g) Which unit is the most inefficient thermodynamically?

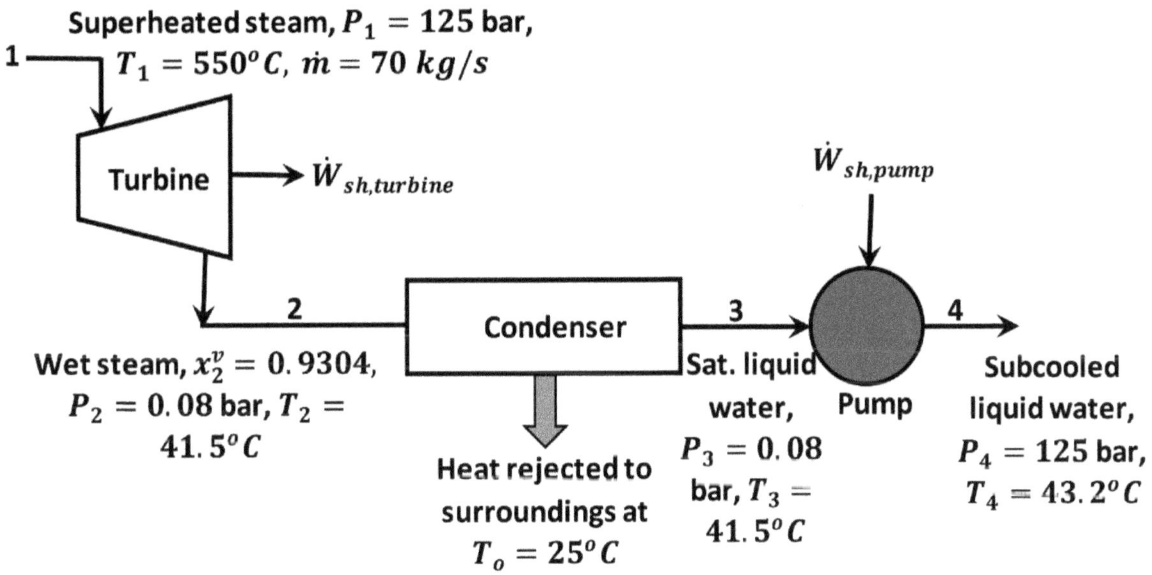

Solution

Refer to the figure above. We need specific enthalpies and entropies of fluid in different states including the dead state. We obtain this information from the steam tables and compressed liquid water tables given in Appendices M and N.

State 1- superheated steam at 125 bar and 550 °C:

$h_1 = 3475.2$ kJ/kg ; $s_1 = 6.6290$ kJ/kg.K

State 2-wet steam at 0.08 bar:

From saturated steam tables,

$h^L(kJ/kg)$	$h^v(kJ/kg)$	$s^L(kJ/kg.K)$	$s^v(kJ/kg.K)$
173.9	2577	0.5924	8.2295

$$h_2 = x_2^v h_2^v + (1 - x_2^v)h_2^L = 0.9304 \times 2577 + (1 - 0.9304) \times 173.9 = 2409.74 \text{ kJ/kg}$$

$$s_2 = x_2^v s_2^v + (1 - x_2^v)s_2^L = 0.9304 \times 8.2295 + (1 - 0.9304) \times 0.5924 = 7.698 \text{ kJ/kg}$$

State 3-saturated liquid water at 0.08 bar:

$h_3 = 173.9$ kJ/kg ; $s_3 = 0.5924$ kJ/kg.K

State 4-subcooled liquid water at 125 bar and 43.2 °C:

From compressed liquid water tables,

T, °C	$P = 100$ bar		$P = 150$ bar	
	h (kJ/kg)	s (kJ/kg.K)	h (kJ/kg)	s (kJ/kg.K)
40	176.38	0.5686	180.78	0.5666
60	259.49	0.8258	263.67	0.8232

We need double linear interpolation to estimate the properties at 125 bar and 43.2 °C. Using linear interpolation, first estimate properties at 125 bar. The values are given below.

T, °C	P = 125 bar	
	h (kJ/kg)	s (kJ/kg.K)
40	178.58	0.5676
60	261.58	0.8245

Now we estimate the properties at 125 bar and 43.2 °C using linear interpolation once again, as shown below.

$$h = 178.58 + \frac{(261.58-178.58)}{(60-40)}(43.2 - 40) = 191.86 \text{ kJ/kg}$$

$$s = 0.5676 + \frac{(0.8245-0.5676)}{(60-40)}(43.2 - 40) = 0.6087 \text{ kJ/kg.K}$$

Thus,

$h_4 = 191.86$ kJ/kg ; $s_4 = 0.6087$ kJ/kg.K

Dead state:

In the dead state ($T_o = 25^oC$ and $P_o = 1$ atm), the fluid is liquid water. Neglecting the effect of pressure on enthalpy and entropy of liquid water, we assume: $h_o(T_o, P_o) \cong h^{sat.liq}(T_o)$ and $s_o(T_o, P_o) \cong s^{sat.liq}(T_o)$. At 25^oC , the saturated steam tables give:

$h_o = h^L = 104.9$ kJ/kg ; $s_o = s^L = 0.3672$ kJ/kg

a) The specific exergy of fluid in states 1, 2, 3, and 4.

$$\psi = \left[(h - h_o) - T_0(s - s_o) + \frac{1}{2}V^2 + gZ\right] = [(h - h_o) - T_o(s - s_o)]$$

$$\psi_1 = [(h_1 - h_o) - T_o(s_1 - s_o)] = (3475.2 - 104.9) - 298.15 \times (6.6290 - 0.3672) = 1503.34 \text{ kJ/kg}$$

$$\psi_2 = [(h_2 - h_o) - T_o(s_2 - s_o)] = (2409.74 - 104.9) - 298.15 \times (7.698 - 0.3672) = 119.16 \text{ kJ/kg}$$

$$\psi_3 = [(h_3 - h_o) - T_o(s_3 - s_o)] = (173.9 - 104.9) - 298.15 \times (0.5924 - 0.3672) = 1.857 \text{ kJ/kg}$$

$$\psi_4 = [(h_4 - h_o) - T_o(s_4 - s_o)] = (191.86 - 104.9) - 298.15 \times (0.6087 - 0.3672) = 14.96 \text{ kJ/kg}$$

b) The rate of total exergy destruction in the turbine

Exergy balance on the turbine gives:

$$\dot{\Psi}_D = \dot{\Psi}_1 - \dot{\Psi}_2 - \dot{W}_{sh} = \dot{m}(\psi_1 - \psi_2) - \dot{W}_{sh}$$

From the first law for open systems,

$$\dot{W}_{sh} = -\dot{m}(\Delta h) = -70 \times (2409.74 - 3475.2) = 74582.2 \text{ kW}$$

Thus,

$$\dot{\Psi}_D = \dot{m}(\psi_1 - \psi_2) - \dot{W}_{sh} = 70 \times (1503.34 - 119.16) - 74582.2 = 22310.4 \text{ kW}$$

As there is no heat loss from the turbine, there is no external irreversibility. Thus the total rate of exergy destruction is the same as the rate of exergy destruction within the turbine.

$$\dot{\Psi}_{D,total} = \dot{\Psi}_D = 22310.4 \text{ kW}$$

c) The rate of total exergy destruction in the condenser

Exergy balance on the condenser gives:

$$\dot{\Psi}_D = \dot{\Psi}_2 - \left[\dot{\Psi}_3 + \left|\dot{Q}_{condener}\right|\left(1 - \frac{T_o}{T_{condenser}}\right)\right]$$

From the first law for open systems,

$$\dot{Q}_{condener} = \dot{m}(\Delta h) = \dot{m}(h_3 - h_2) = 70 \times (173.9 - 2409.74) = -156508.8 \text{ kW}$$

$$\left|\dot{Q}_{condener}\right| = 156508.8 \text{ kW}$$

Thus,

$$\dot{\Psi}_D = \dot{m}(\psi_2 - \psi_3) - \left[\left|\dot{Q}_{condener}\right|\left(1 - \frac{T_O}{T_{condenser}}\right)\right] = 70 \times (119.16 - 1.857\ ) - 156508.8\left(1 - \frac{298.15}{41.5+273.15}\right) = 4.01 \text{ kW}$$

This is the rate of exergy destruction within the condenser due to internal irreversibilities. Note that the condenser temperature is assumed to be that of saturated liquid water temperature at 0.08 bar, that is, 41.5 °C.

The heat leaving the condenser is transferred irreversibly to the surroundings at $T_o = 25^oC$. Thus, there occurs exergy destruction external to the condenser at a rate of

$$\dot{\Psi}_{D,external} = \left[\left|\dot{Q}_{condener}\right|\left(1 - \frac{T_O}{T_{condenser}}\right)\right] = 156508.8\left(1 - \frac{298.15}{41.5+273.15}\right) = 8207.2 \text{ kW}$$

Therefore, the rate of total exergy destruction in the condenser due to internal and external irreversibilities is:

$$\dot{\Psi}_{D,total} = 4.01 + 8207.2 = 8211.21 \text{ kW}$$

d) The rate of total exergy destruction in the pump

Exergy balance on the pump gives:

$$\dot{\Psi}_D = \dot{\Psi}_3 - \dot{\Psi}_4 - \dot{W}_{sh} = \dot{m}(\psi_3 - \psi_4) - \dot{W}_{sh}$$

From the first law for open systems,

$$\dot{W}_{sh} = -\dot{m}(\Delta h) = -70 \times (191.86 - 173.9) = -1257.2 \text{ kW}$$

Thus,

$$\dot{\Psi}_D = \dot{m}(\psi_3 - \psi_4) - \dot{W}_{sh} = 70 \times (1.857 - 14.96) - (-1257.2) = 339.99 \text{ kW}$$

As there is no heat loss from the pump, there is no external irreversibility. Thus, the total rate of exergy destruction is the same as the rate of exergy destruction within the pump.

$$\dot{\Psi}_{D,total} = \dot{\Psi}_D = 339.99 \text{ kW}$$

e) The rate of total exergy destruction in the whole process including all the units

$$\sum \dot{\Psi}_{D,total,i} = \dot{\Psi}_{D,total,turbine} + \dot{\Psi}_{D,total,condenser} + \dot{\Psi}_{D,total,pump}$$

$$\sum \dot{\Psi}_{D,total,i} = 22310.4 + 8211.21 + 339.99 = 30861.6 \text{ kW}$$

f) The fraction of overall process exergy destruction that takes place in each unit

Unit	$\dot{\Psi}_{D,total,i}$ (kW)	$\dot{\Psi}_{D,total,i} / \sum \dot{\Psi}_{D,total,i}$
turbine	22310.4	0.7229
condenser	8211.21	0.2661
pump	339.99	0.0110
process	30861.6	1.0000

g) Which unit is the most inefficient thermodynamically?

The largest fraction of the overall process exergy destruction is 0.7229 that occurs in the turbine. Hence turbine is the most inefficient unit in the given process. The fraction of the overall process exergy destruction in the pump is negligible.

PRATICE PROBLEMS

Problem 1

A 1.5 kg carbon-steel casting with an initial temperature of 615 °C is quenched in a 4 kg carbon-steel tank containing 50 kg of water. The carbon-steel tank and water are initially at 20 °C.

a) What is the final equilibrium temperature?
b) How much exergy is destroyed in this process?

You can assume that the materials (carbon-steel and water) are incompressible. The heat capacities of carbon-steel and water are 0.5 and 4.18 kJ/kg.K, respectively. There is no heat loss to the surroundings. The dead state is: $T_o = 25^oC$ and $P_o = 1$ atm.

Problem 2

One cubic meter of an ideal gas, initially at 42 °C and 1.5 bar, undergoes a process to a final state of 130 °C and 10 bar. Calculate:

a) The exergy of the gas in its initial state
b) The exergy of the gas in its final state
c) The change in exergy

The specific heat capacities of the gas are: $c_p = 7R/2$ and $c_v = 5R/2$. The dead state is: $T_o = 25^oC$ and $P_o = 1$ atm.

Problem 3

Two heat reservoirs are connected by a heat conducting rod as shown in the figure below. At steady state, the temperature distribution in the rod is linear:

$$T = ax + b$$

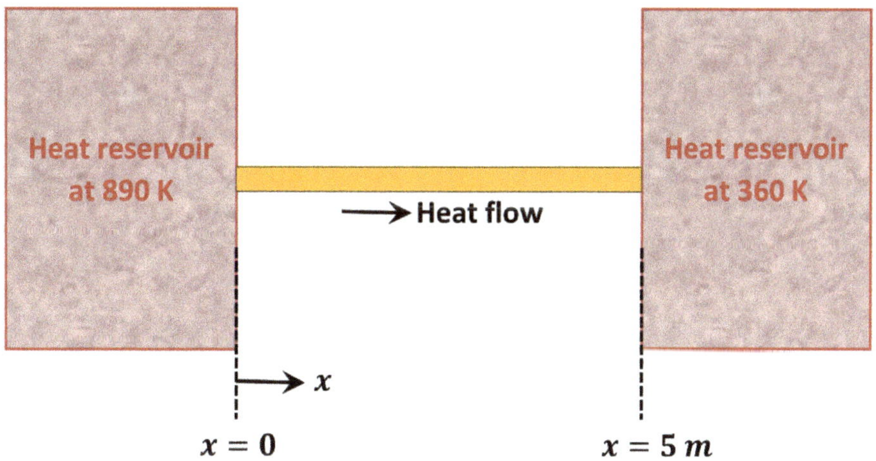

The heat transfer rate through the rod is 15 W. Calculate:

a) The rate of exergy transfer due to heat flow at $x = 0$.
b) The rate of exergy transfer due to heat flow at $x = 2.5$ m.
c) The rate of exergy transfer due to heat flow at $x = 5.0$ m.
d) The rate of exergy destruction in the rod due to irreversible heat transfer.

The dead state is: $T_o = 25^oC$ and $P_o = 1$ atm.

Problem 4

In a steady flow process shown below, 11.5 kg/s of superheated steam at 175 bar and 450 °C is continuously mixed with 1.65 kg/s of subcooled (compressed) liquid water at 100 bar and 20 °C. The product stream is saturated vapor (steam) at 100 bar. Determine:

a) The specific exergies of inlet and outlet fluid streams
b) The rate of total exergy destruction

The dead state is: $T_o = 25^oC$ and $P_o = 1$ atm.

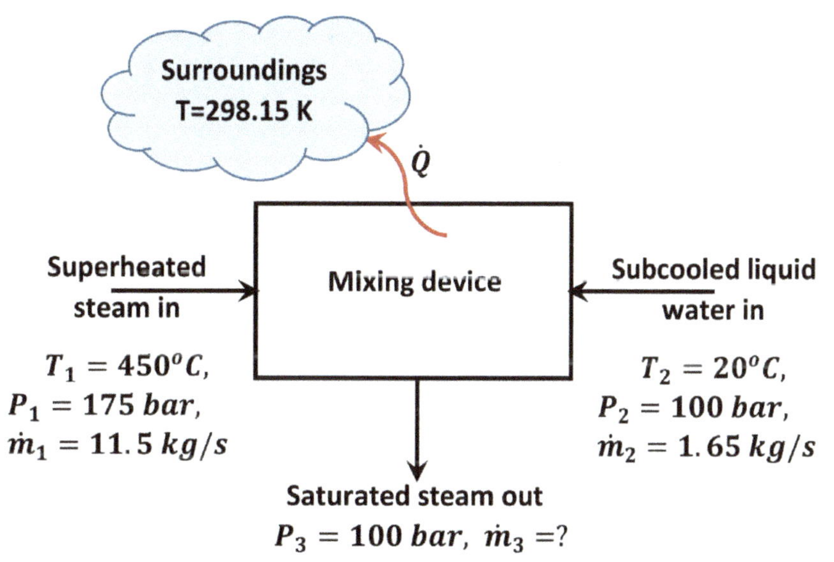

Chapter 16

Quantification of Irreversibilities in Practical Cyclic Processes Using Exergy Analysis and Gouy-Stodola Theorem

16.1 LEARNING OUTCOMES

Upon completion of this chapter, students will be able to:

- Carry out internal and external exergy balances on individual units of a multiple-unit cyclic process
- Determine total exergy destruction rates and exergetic efficiencies of individual units of a multiple-unit cyclic process
- Analyze a multiple-unit cyclic process showing the total exergy loss distribution among the individual units, and hence identifying those units which are contributing most to the overall process inefficiency
- Apply the Gouy-Stodola theorem to determine the distribution of the total entropy production among the individual units of a multiple-unit cyclic process
- Calculate the overall exergetic efficiency of a cyclic process.

16.2 INTRODUCTION

In Chapter 15, exergy was defined as the maximum possible useful work that can be extracted from a system when the system undergoes a change in state from its current *live* state to a *dead* state. The *dead state* is one where the system attains thermodynamic equilibrium with the environment. It was also noted that in order to extract the maximum possible useful work, the change in state must be carried out reversibly.

The exergy analysis is advantages over entropy analysis in that it provides the definition of efficiency of a process not supplied by entropy analysis [1]. For example, the exergetic efficiency of a process, defined in Equation (16.1) given below, is very useful in the quantification of irreversibilities in the process:

$$\varepsilon = \frac{Exergy\ output}{Exergy\ input} \tag{16.1}$$

Due to irreversibilities in a process, exergy is destroyed and as a consequence, exergy output is always less than the exergy input to the process. For a completely reversible process, the exergetic efficiency is unity. For example, $\varepsilon = 1$ for a Carnot cyclic process. The definition of exergetic efficiency, given in Equation (16.1), could also be interpreted as a ratio of exergy recovered from the process to exergy supplied to the process.

In this chapter, the quantification of irreversibilities in practical multi-unit cyclic processes is illustrated using exergy analysis. The theoretical background related to exergy and exergy analysis is first presented.

16.3 BACKGROUND

The exergy associated with a flow stream is given as:

$$\psi = (h - h_o) - T_o(s - s_o) + \frac{V^2}{2} + gZ \quad (16.2)$$

where ψ is the specific exergy, h and s are the specific enthalpy and entropy, respectively, V is the velocity, Z is the elevation, and T_o is the absolute temperature of the environment. The subscript "o" refers to the *dead state* where the system attains thermodynamic equilibrium with the environment. Neglecting kinetic energy and potential energy terms, the expression for ψ simplifies to:

$$\psi = (h - h_o) - T_o(s - s_o) \quad (16.3)$$

This expression of ψ is used extensively in exergy analysis as kinetic energy and potential energy changes are usually small in a majority of thermodynamic applications. It is important to recognize that exergy is a combined property of the system and the surroundings. The exergy of a system depends on its environment. Exergy of a system can be changed simply by changing the environment (surroundings) of the system. However, the environment is usually fixed in applications.

16.3.1 Exergy Balance (Work-Producing Process)

Consider a flow process shown schematically in Figure 16.1. The process is producing shaft work. The exergy balance on the control volume could be expressed as:

$$(Rate\ of\ Exergy\ In) - (Rate\ of\ Exergy\ Out) - (Rate\ of\ Exergy\ Destruction) = (Rate\ of\ Exergy\ Accumulation) \quad (16.4)$$

For the process shown in Figure 16.1:

$$(Rate\ of\ Exergy\ In) = \dot{Q}_o\left(1 - \frac{T_o}{T_b}\right) + \dot{m}_1\psi_1 \quad (16.5)$$

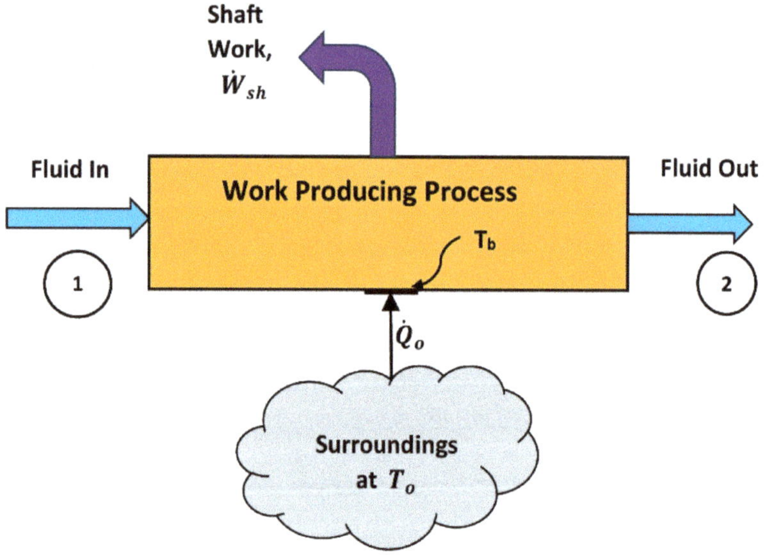

Figure 16.1 Flow process producing shaft work.

$$(Rate\ of\ Exergy\ Out) = \dot{W}_{sh} + \dot{m}_2\psi_2 \quad (16.6)$$

$$(Rate\ of\ Exergy\ Destruction) = \dot{\Psi}_{D,CV} \quad (16.7)$$

$$(Rate\ of\ Exergy\ Acccumulation) = \frac{\partial}{\partial t}(m\psi)_{CV} \tag{16.8}$$

Thus, the exergy balance on the control volume (CV) is as follows:

$$\dot{Q}_0\left(1 - \frac{T_o}{T_b}\right) + \dot{m}_1\psi_1 - \left(\dot{W}_{sh} + \dot{m}_2\psi_2\right) - \dot{\Psi}_{D,CV} = \frac{\partial}{\partial t}(m\psi)_{CV} \tag{16.9}$$

Assuming steady state, Equation (16.9) could be re-written as:

$$\dot{\Psi}_{D,CV} = \dot{Q}_o\left(1 - \frac{T_o}{T_b}\right) + \dot{m}(\psi_1 - \psi_2) - \dot{W}_{sh} \tag{16.10}$$

Note that at steady state, the mass flow rate is constant: $\dot{m}_1 = \dot{m}_2 = \dot{m}$.

Exergy balance on the surroundings can be carried out in a similar manner. It leads to the following result for the rate of exergy destruction in the surroundings:

$$\dot{\Psi}_{D,surr} = -\dot{Q}_o\left(1 - \frac{T_o}{T_b}\right) \tag{16.11}$$

Combining Equations (16.10) and (16.11) gives:

$$\dot{\Psi}_{D,total} = \dot{\Psi}_{D,CV} + \dot{\Psi}_{D,surr} = \dot{m}(\psi_1 - \psi_2) - \dot{W}_{sh} \tag{16.12}$$

where $\dot{\Psi}_{D,total}$ is the total rate of exergy destruction, within the control volume and outside the control volume (surroundings). For a *reversible process*, the total rate of exergy destruction is zero and the shaft work produced is *maximum*. Thus:

$$\dot{W}_{sh,max} = \dot{m}(\psi_1 - \psi_2) \tag{16.13}$$

For the actual process, the exergetic efficiency can be expressed as:

$$\varepsilon = \frac{\text{Exergy recovered}}{\text{Exergy supplied}} \tag{16.14}$$

The exergy is supplied to the process by the fluid stream and the exergy is recovered from the process as shaft work. Therefore, the exergetic efficiency is given as:

$$\varepsilon = \frac{\dot{W}_{sh}}{\dot{m}(\psi_1 - \psi_2)} \tag{16.15}$$

As noted earlier, the exergetic efficiency is unity when the process is internally and externally reversible.

16.3.2 Exergy Balance (Work Absorbing Process)

Consider a flow process shown schematically in Figure 16.2 where shaft work is absorbed. In a work absorbing process, the rates of exergy input to the control volume and exergy output from the control volume are as follows:

$$(Rate\ of\ Exergy\ In) = \dot{Q}_o\left(1 - \frac{T_o}{T_b}\right) + \dot{m}_1\psi_1 + \left|\dot{W}_{sh}\right| \tag{16.16}$$

$$(Rate\ of\ Exergy\ Out) = \dot{m}_2\psi_2 \tag{16.17}$$

Under steady state, the exergy balance for the control volume, Equation (16.4), becomes:

$$\dot{\Psi}_{D,CV} = \dot{Q}_o\left(1 - \frac{T_o}{T_b}\right) + \dot{m}(\psi_1 - \psi_2) + \left|\dot{W}_{sh}\right| \quad (16.18)$$

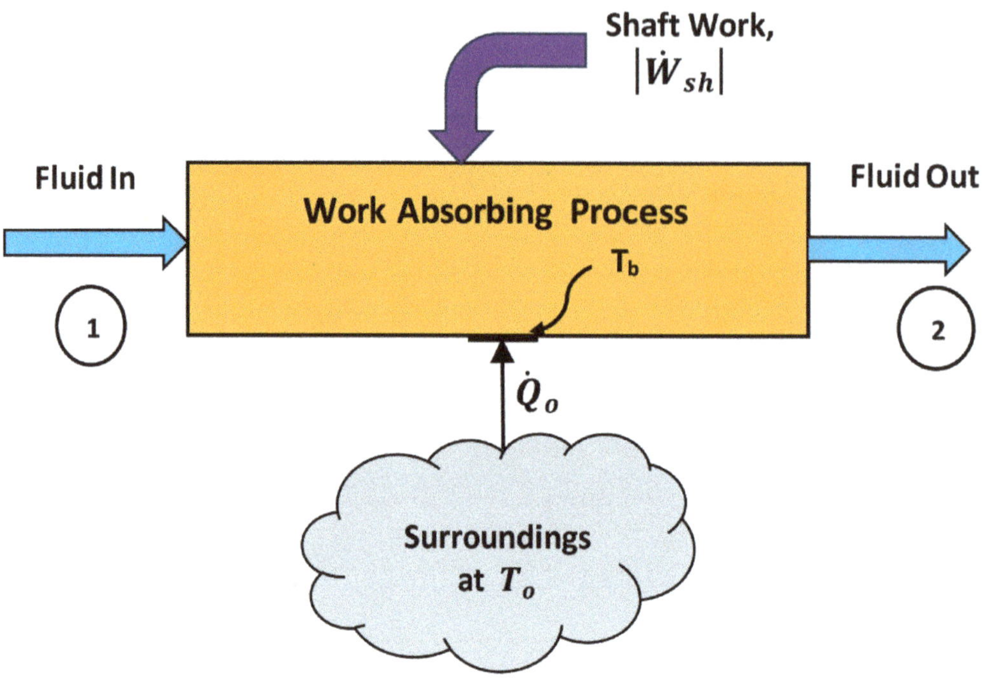

Figure 16.2 Flow process absorbing shaft work.

Exergy balance on the surroundings gives the same result as Equation (16.11). Thus,

$$\dot{\Psi}_{D,total} = \dot{\Psi}_{D,CV} + \dot{\Psi}_{D,surr} = \dot{m}(\psi_1 - \psi_2) + \left|\dot{W}_{sh}\right| \quad (16.19)$$

For a *reversible process*, the total rate of exergy destruction is zero and the shaft work absorbed is *minimum* given as:

$$\left|\dot{W}_{sh,min}\right| = \dot{m}(\psi_2 - \psi_1) \quad (16.20)$$

For a work-absorbing process, the exergy is supplied to the process by the shaft work and the exergy is recovered from the process by the fluid stream. The exergetic efficiency of a work-absorbing process is:

$$\varepsilon = \frac{\dot{m}(\psi_2 - \psi_1)}{|\dot{W}_{sh}|} \quad (16.21)$$

16.3.3 Relationship Between Exergy Destruction and Entropy Generation

It can be readily shown that:

$$\dot{\Psi}_{D,CV} = T_o\dot{S}_{G,CV} \quad (16.22)$$

$$\dot{\Psi}_{D,surr} = T_o\dot{S}_{G,surr} \quad (16.23)$$

$$\dot{\Psi}_{D,total} = \dot{\Psi}_{D,CV} + \dot{\Psi}_{D,surr} = T_o\dot{S}_{G,total} \quad (16.24)$$

Whenever and wherever exergy is destroyed, entropy is generated and vice versa. The rate of exergy destruction is directly proportional to the rate of entropy generation. The proportionality constant is the absolute temperature of the surroundings (environment). Thus, one can calculate the entropy generation rate once the exergy destruction rate is known.

We now present the proof of Equation (16.22). Consider the work-producing process of section 16.3.1. From exergy balance on the control volume of Figure 16.1 at steady state:

$$\dot{\Psi}_{D,CV} = \dot{Q}_o\left(1-\frac{T_o}{T_b}\right) + \dot{m}(\psi_1 - \psi_2) - \dot{W}_{sh} \quad (16.10)$$

Substituting ψ from Equation (16.3) into Equation (16.10), we get:

$$\dot{\Psi}_{D,CV} = \dot{Q}_o\left(1-\frac{T_o}{T_b}\right) - \dot{m}(\Delta h - T_o\Delta s) - \dot{W}_{sh} \quad (16.25)$$

From first law for open systems,

$$\dot{W}_{sh} = \dot{Q}_o - \dot{m}(\Delta h) \quad (16.26)$$

From Equations (16.25) and (16.26),

$$\dot{\Psi}_{D,CV} = \dot{Q}_o\left(1-\frac{T_o}{T_b}\right) - \dot{m}(\Delta h - T_o\Delta s) - \left[\dot{Q}_o - \dot{m}(\Delta h)\right] \quad (16.27)$$

Upon simplification, Equation (16.27) gives:

$$\dot{\Psi}_{D,CV} = T_o\left[\dot{m}\Delta s - \frac{\dot{Q}_o}{T_b}\right] = T_o\dot{S}_{G,CV} \quad (16.28)$$

This completes the proof of Equation (16.22).

16.3.4 Gouy-Stodola Theorem

The loss of power ($\dot{W}_{sh,lost}$) due to irreversibilities in the process can be determined through the Gouy-Stodola theorems expressed as:

$$\dot{W}_{sh,lost} = \dot{W}_{sh,rev} - \dot{W}_{sh} = \dot{\Psi}_{D,total} \quad (16.29)$$

$$\dot{W}_{sh,lost} = \dot{W}_{sh,rev} - \dot{W}_{sh} = T_o\dot{S}_{G,total} \quad (16.30)$$

where T_o is the absolute temperature of the environment/surroundings. The Gouy-Stodola theorems expressed in the form of Equations (16.29) and (16.30) are applicable to systems interacting thermally with the environment TER (thermal energy reservoir) at T_o. They are also applicable to systems interacting thermally with more than one TER with environment TER at T_o as the reference or floating reservoir.

16.4 EXERGY ANALYSIS OF A PRACTICAL STEAM POWER PLANT

Figure 16.3 shows a schematic flow diagram of a practical steam power plant [2]. The steam is generated in a furnace/boiler unit by burning methane completely to CO_2 and H_2O with 25% excess air. The flue gases leaving the furnace have a temperature of 460 K. The steam generated in the boiler leaves the boiler at a pressure of 8600 kPa and a temperature of 500 °C. The steam is fed to a turbine which produces shaft work. The exhaust from the turbine is condensed to a saturated liquid in a condenser at 10 kPa and the condensed liquid is pumped back to the boiler completing a full cycle. The surroundings are at a temperature of 25°C. The power rating of the plant is 100 MW. The states of steam and the corresponding properties at various locations of the steam cycle are given in Table 16.1. The location numbers given in Table 16.1 refer to Figure 16.3. For 1 mol of CH_4 burned, the enthalpy and entropy changes in the furnace are given as follows: $\Delta\tilde{h}_{1-2} = \tilde{h}_2 - \tilde{h}_1 = -737.64$ kJ/mol ; $\Delta\tilde{s}_{1-2} = \tilde{s}_2 - \tilde{s}_1 = 0.2082$ kJ/mol.K . Note that $\tilde{h}$, $\tilde{s}$ are molar quantities whereas h, s are specific quantities on a unit mass basis.

In what follows, a complete exergy analysis of the power plant is carried out. The exergy destruction rates are determined in various components as well as in the whole process. The exergetic efficiencies of the components and the whole cyclic process are also calculated. The Gouy-Stodola theorem is then used to determine the loss of power in various components and in the whole process.

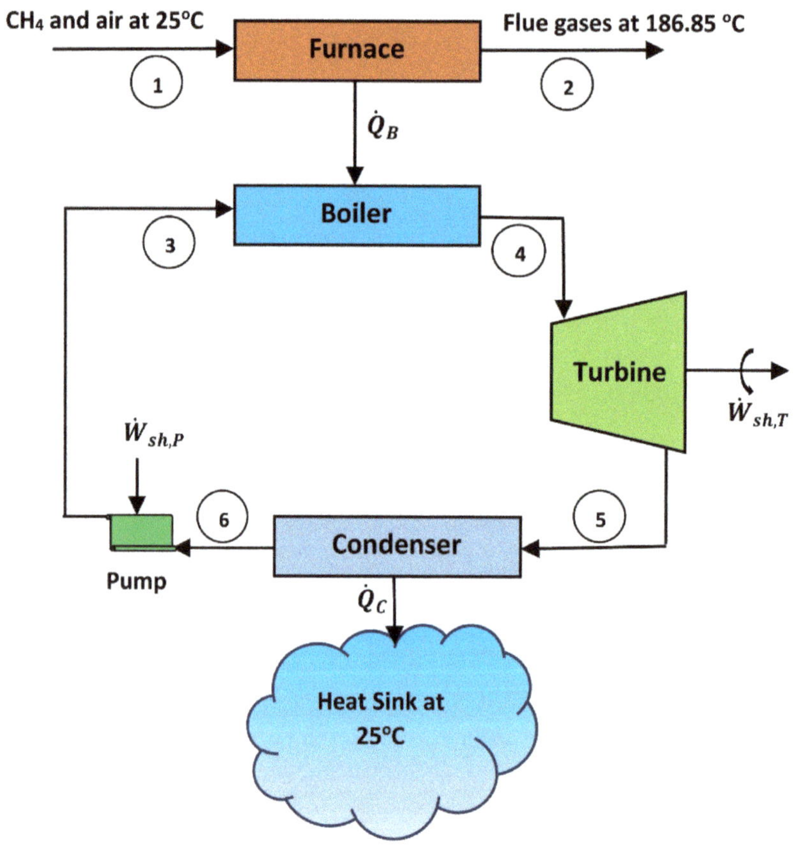

Figure 16.3 Practical steam power plant.

Table 16.1 States of fluid and the corresponding properties at various points in the steam cycle

Location point	State of fluid	Specific enthalpy (kJ/kg)	Specific entropy (kJ/kg.K)
3	Subcooled liquid water	203.4	0.6580
4	Superheated vapor	3391.6	6.6858
5	Wet vapor of quality 0.938	2436.0	7.6846
6	Saturated liquid	191.8	0.6493

Mass Flow Rate of Steam in the Loop

The mass flow rate of steam in the cycle can be obtained from the power rating (100 MW=100,000 kW) of the plant as:

$$\dot{m} = \dot{W}_{sh,net}/\left(W_{sh,turbine} - \left|W_{sh,pump}\right|\right) \tag{16.31}$$

Using the first law for open systems (neglecting kinetic and potential energy changes) and data given in Table 16.1:

$$W_{sh,turbine} = \dot{W}_{sh,turbine}/\dot{m} = -\Delta h_{4-5} = (h_4 - h_5) = 955.6 \text{ kJ/kg} \tag{16.32}$$

$$\left|W_{sh,pump}\right| = \left|\dot{W}_{sh,pump}\right|/\dot{m} = \Delta h_{6-3} = (h_3 - h_6) = 11.6 \text{ kJ/kg} \tag{16.33}$$

Thus, the mass flow rate in the loop is:

$$\dot{m} = \dot{W}_{sh,net}/\left(W_{sh,turbine} - \left|W_{sh,pump}\right|\right) = 100{,}000/(955.6 - 11.6) = 105.932 \text{ kg/s} \tag{16.34}$$

Molar Flow Rate of Methane in the Furnace

The molar flow rate of methane $\dot{n}_{CH_4}$ to be burned in the furnace can be obtained from energy balance around the combined furnace/boiler unit (see Figure 16.3) as:

$$\dot{n}_{CH_4}\left(\tilde{h}_1 - \tilde{h}_2\right) = \dot{m}_{steam}(h_4 - h_3) \tag{16.35}$$

$$\dot{n}_{CH_4} = \left[\dot{m}_{steam}(h_4 - h_3)/\left(\tilde{h}_1 - \tilde{h}_2\right)\right] = 105.932(3391.6 - 203.4)/737.64 = 457.85 \text{ mol/s} \tag{16.36}$$

Overall Process

Figure 16.4 shows the schematic of the overall process as a black box. The exergy balance on the system and the surroundings of this process gives:

$$\dot{\Psi}_{D,total} = \dot{n}_{CH_4}\left(\tilde{\psi}_1 - \tilde{\psi}_2\right) - \dot{W}_{sh,net} = \dot{n}_{CH_4}\left(T_o\Delta\tilde{s}_{1-2} - \Delta\tilde{h}_{1-2}\right) - \dot{W}_{sh,net} \tag{16.37}$$

The first term on the right hand side of Equation (16.37), that is, $\dot{n}_{CH_4}\left(\tilde{\psi}_1 - \tilde{\psi}_2\right)$, is the exergy supplied to the process and the second term, that is, $\dot{W}_{sh,net}$ is the exergy recovered from the process. Note that $\tilde{\psi}$ is molar exergy whereas ψ is exergy per unit mass. Using the given information, we obtain:

$$\dot{\Psi}_{D,total} = 457.85[298.15(0.2082) - (-737.64)] - 100{,}000 = 366149.4 - 100{,}000$$
$$= 266149.4 \text{ kW} \tag{16.38}$$

The exergetic efficiency of the overall process is:

$$\varepsilon = \frac{Exergy\ recovered}{Exergy\ supplied} = 1 - \frac{Exergy\ destroyed}{Exergy\ supplied} = 1 - \frac{266149.4}{366149.4} = 0.273 \tag{16.39}$$

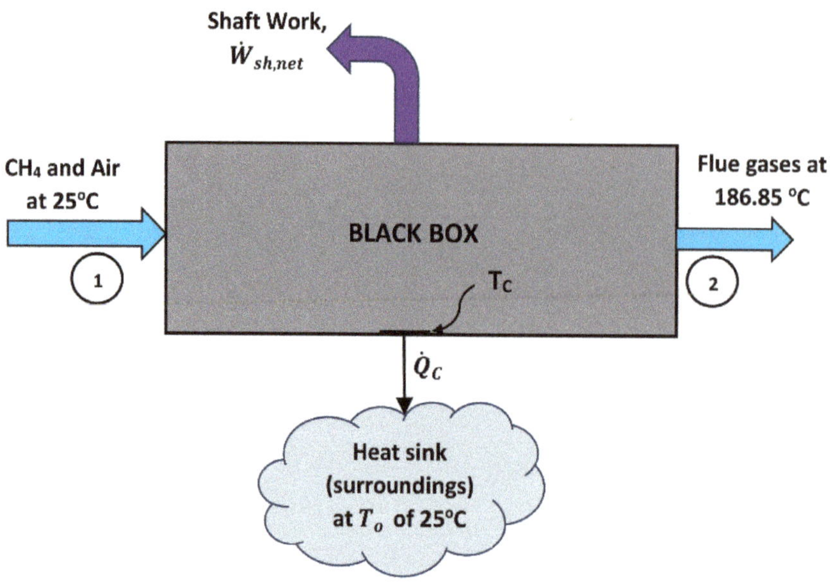

Figure 16.4 Steam power plant as a black box.

Note that the exergy supplied is $\dot{n}_{CH_4}(\tilde{\psi}_1 - \tilde{\psi}_2) = \dot{n}_{CH_4}(T_o\Delta\tilde{s}_{1-2} - \Delta\tilde{h}_{1-2}) = 366149.4\ \text{kW}$.

Exergy Destruction in Furnace/Boiler Combined Unit

Figure 16.5 shows the combined furnace/boiler unit over which the exergy balance is carried out. At steady state:

$$(Rate\ of\ Exergy\ Destruction) = (Rate\ of\ Exergy\ In) - (Rate\ of\ Exergy\ Out) \quad (16.40)$$

$$\dot{\Psi}_{D,furnace/boiler} = \dot{n}_{CH_4}(T_o\Delta\tilde{s}_{1-2} - \Delta\tilde{h}_{1-2}) - \dot{m}(\Delta h_{3-4} - T_o\Delta s_{3-4}) \quad (16.41)$$

Thus, the exergy destruction rate in the combined furnace/boiler unit is

$$\dot{\Psi}_{D,furnace/boiler} = 366149.4 - (105.932)[(3391.6 - 203.4) - 298.15(6.8658 - 0.6580)] = 218796.78\ \text{kW} \quad (16.42)$$

Note that this is the total exergy destruction in the combined furnace/boiler unit. There is no heat loss to the surroundings and hence no external irreversibilities and associated exergy destruction.

The exergetic efficiency of the furnace/boiler unit can be calculated as follows:

$$\varepsilon = \frac{\dot{m}(\psi_4-\psi_3)}{\dot{n}_{CH_4}(\tilde{\psi}_1-\tilde{\psi}_2)} = \frac{\dot{m}(\Delta h_{3-4}-T_o\Delta s_{3-4})}{\dot{n}_{CH_4}(T_o\Delta\tilde{s}_{1-2}-\Delta\tilde{h}_{1-2})} = \frac{(105.932)[(3391.6-203.4)-298.15(6.8658-0.6580)]}{366149.4} = \frac{147352.62}{366149.4} = 0.4024 \quad (16.43)$$

The numerator term in Equation (16.43) is the exergy recovered in the boiler and the denominator term is the exergy supplied by the furnace. Thus, the exergetic efficiency of the furnace/boiler unit is about 40%.

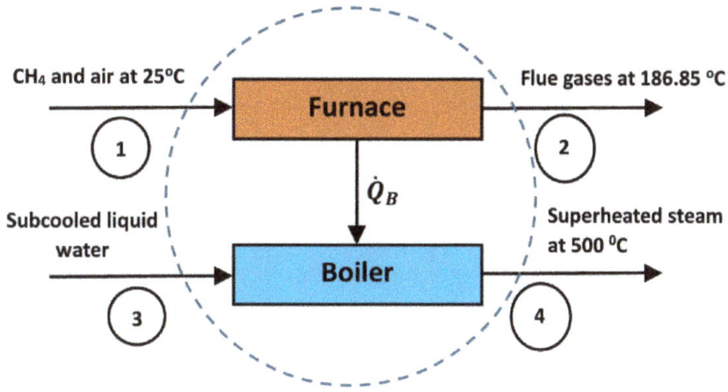

Figure 16.5 Furnace/boiler combined unit of the steam power plant.

Exergy Destruction in Turbine

The exergy balance on the turbine gives the following result for the exergy destruction rate:

$$\dot{\Psi}_{D,turbine} = \dot{m}(\psi_4 - \psi_5) - \dot{W}_{sh,turbine} = \dot{m}(T_o\Delta s_{4-5} - \Delta h_{4-5}) + \dot{m}\Delta h_{4-5} \tag{16.44}$$

$$= 105.932[298.15(7.6846 - 6.6858)] = 31545.73 \text{ kW}$$

This is the total exergy destruction in the turbine. There is no heat loss to the surroundings and hence no external irreversibilities and associated exergy destruction.

The exergetic efficiency of the turbine is:

$$\varepsilon = \frac{\dot{W}_{sh,turbine}}{\dot{m}(\psi_4-\psi_5)} = \frac{-\dot{m}\Delta h_{4-5}}{\dot{m}(T_o\Delta s_{4-5}-\Delta h_{4-5})} = \frac{(3391.6-2436)}{[298.15(7.6846-6.6858)-(2436-3391.6)]} = \frac{955.6}{1253.39} = 0.7624 \tag{16.45}$$

Exergy Destruction in Condenser

The exergy balance on the condenser gives the following result for the total exergy destruction rate (internally and externally):

$$\dot{\Psi}_{D,condenser} = \dot{m}(\psi_5 - \psi_6) = \dot{m}(T_o\Delta s_{5-6} - \Delta h_{5-6}) = 105.932[298.15(0.6493 - 7.6846) - (191.8 - 2436)] = 15532.31 \text{ kW} \tag{16.46}$$

As all the exergy supplied to the condenser (that is, 15532.31 kW) is wasted away to the environment, the exergetic efficiency of the condenser is zero.

Exergy Destruction in Pump

Assuming negligible heat loss from the pump, the exergy balance on the pump gives the following result for the total exergy destruction rate:

$$\dot{\Psi}_{D,pump} = \dot{m}(\psi_6 - \psi_3) + |\dot{W}_{sh,pump}| = \dot{m}(T_o\Delta s_{6-3} - \Delta h_{6-3}) + \dot{m}\Delta h_{6-3} = 105.932[298.15(0.6580 - 0.6493)] = 274.78 \text{ kW} \tag{16.47}$$

The exergetic efficiency of a pump is:

$$\varepsilon = \frac{\dot{m}(\psi_3-\psi_6)}{|\dot{W}_{sh,pump}|} = \frac{-\dot{m}(T_0\Delta s_{6-3}-\Delta h_{6-3})}{\dot{m}\Delta h_{6-3}} = \frac{(\Delta h_{6-3}-T_0\Delta s_{6-3})}{\Delta h_{6-3}} = \frac{(203.4-191.8)-298.15(0.6580-0.6493)}{(203.4-191.8)} = 0.776 \quad (16.48)$$

Summary of Results

Table 16.2 summarizes the results of the exergy analysis. The exergy destruction rates in different components are shown. For a given component, the percent of total exergy destruction rate that occurs in the component is also shown along with the exergetic efficiencies.

Table 16.2 Summary of the results of exergy analysis

Component	Exergy destruction rate, kW	Percent of total exergy destruction rate	Exergetic efficiency, percent
Furnace/boiler	218796.78	82.21	40.24
Turbine	31545.73	11.85	76.24
Condenser	15532.31	5.84	00.00
Pump	274.78	0.10	77.60
Whole Process	$\sum \dot{\Psi}_D = 266149.6$ kW **Exergy supplied =** $\dot{\Psi}_{sup} = 366149.4$ kW	**100**	**27.31**

The exergetic efficiency of the whole cyclic process is 27.31 % indicating that 72.69% of the exergy supplied to the process is destroyed in irreversibilities. Most of the exergy destruction occurs in the furnace/boiler unit. It is followed by the turbine and the smallest destruction of exergy occurs in the pump. Note that the thermal efficiency of the cycle (also called the first law efficiency), defined below, is 29.6%.

$$\eta = \frac{\dot{W}_{sh}}{\dot{Q}_{boiler}} = \frac{100000}{105.932\times(3391.6-203.4)} = 0.2961 \quad (16.49)$$

The first law efficiency of the power cycle (29.61%) is higher than the exergetic efficiency of the cycle (27.31%). This is unexpected as exergetic efficiency is usually higher than the first law efficiency. The reason for this discrepancy is the way we have defined the exergetic efficiency. We have taken exergy supplied to the cyclic process as that supplied by the combustion stream on the furnace side. In order to compare the two efficiencies, the exergy supplied to the cyclic process should be taken as that supplied by the boiler stream, that is,

$$\textit{Exergy supplied to the process} = \textit{Exergy supplied in the boiler} \quad (16.50)$$

The exergy supplied in the boiler is: $\dot{\Psi}_4 - \dot{\Psi}_3 = \dot{m}(\Delta h_{3-4} - T_0\Delta s_{3-4})$. Thus, the exergetic efficiency of the steam power plant is as follows:

$$\varepsilon = \frac{\dot{W}_{sh,net}}{\dot{m}(\Delta h_{3-4}-T_0\Delta s_{3-4})} = \frac{100000}{105.932\times[(3391.6-203.4)-298.15(6.6858-0.6580)]} = 0.6786 \quad (16.51)$$

As expected, the exergetic efficiency (67.86%) of the cycle is much higher than the first law efficiency (29.61%) of the cycle.

Application of the Gouy-Stodola Theorem

The loss of power, that is, the rate of work lost in the process, is identical to the rate of exergy destruction according to the exergy-form of the Gouy-Stodola theorem, that is, Equation (16.29). The entropy generation rates can be calculated for the various components of the process as well as for the whole process using the entropy-form of the Gouy-Stodola theorem, that is, Equation (16.30), once the loss of power is known. Table 16.3 shows the loss of power and entropy generation rates for various components and for the whole process calculated from the Gouy-Stodola theorem. Note that the surroundings temperature T_o = 298.15 K. As expected, most of the power loss and entropy generation occurs in the furnace/boiler unit.

Table 16.3 Loss of power and entropy generation rates based on the Gouy-Stodola theorem

Component	Loss of power, kW	Percent of total power loss	Entropy generation rate, kW/K	Percent of total entropy generation
Furnace/boiler	218796.78	82.21	733.85	82.21
Turbine	31545.73	11.85	105.80	11.85
Condenser	15532.31	5.84	52.10	5.84
Pump	274.78	0.10	0.92	0.10
Whole Process	$\sum \dot{W}_{sh,lost} = 266149.6$ kW	**100**	$\sum \dot{S}_G = 892.67$ kW/K	**100**

16.5 CONCLUDING REMARKS

The quantification of irreversibilities in a process is discussed in this chapter using exergy analysis and the Gouy-Stodola theorem. As an example, a practical cyclic steam power plant is considered. The rates of exergy destruction in the process components and in the process as a whole are calculated for the steam power plant. The exergetic efficiencies are also determined for the various components and for the whole process. The exergetic efficiency is found to be 27.31 % for the whole steam power plant considered in the chapter. Approximately 82% of the exergy supplied to the process is destroyed in irreversibilities in the furnace/boiler unit alone. The turbine destroys nearly 12 % of the supplied exergy. The smallest destruction of exergy occurs in the pump. The exergetic efficiency of the steam cycle is much higher (67.86%) than the first law efficiency of 29.61% if the exergy loss in the furnace is not considered. Using the Gouy-Stodola theorem, the power loss and entropy generation rates in different units and in the whole process are calculated.

NOTATION

g	acceleration due to gravity
h	specific enthalpy
h_o	specific enthalpy in dead state
$\tilde{h}$	molar enthalpy
m	mass
$\dot{m}$	mass flow rate
$\dot{n}$	molar flow rate
P	pressure
P_o	pressure in dead state
$\dot{Q}_o$	heat transfer rate between the process and the environmental Thermal energy reservoir

$\dot{Q}_{boiler}$	heat transfer rate in the boiler
$\dot{S}_G$	rate of entropy generation
$\dot{S}_{G,CV}$	rate of entropy generation inside control volume
$\dot{S}_{G,surr}$	rate of entropy generation in the surroundings
$\dot{S}_{G,total}$	rate of total entropy generation (internally and externally)
s	specific entropy
s_o	specific entropy in dead state
$\tilde{s}$	molar entropy
T_b	temperature at the system boundary
T_c	condenser temperature
T_o	temperature in dead state
t	time
V	velocity
TER	thermal energy reservoir
$\dot{W}_{sh}$	rate of shaft work
$\dot{W}_{sh,lost}$	rate of shaft work lost
$\dot{W}_{sh,max}$	maximum rate of shaft work
$\dot{W}_{sh,net}$	net rate of shaft work
$\dot{W}_{sh,pump}$	rate of shaft work for pump
$\dot{W}_{sh,rev}$	rate of shaft work in a reversible process
$\dot{W}_{sh,turbine}$	rate of shaft work for turbine
Z	elevation

Greek Letters

ε	exergetic efficiency of a process
η	thermal efficiency of a cyclic process
ψ	specific exergy
$\tilde{\psi}$	molar exergy
Ψ	total exergy
$\dot{\Psi}_{D,CV}$	rate of total exergy destruction inside the control volume
$\dot{\Psi}_{D,condenser}$	rate of total exergy destruction in a condenser
$\dot{\Psi}_{D,furnace/boiler}$	rate of total exergy destruction in combined furnace/boiler unit
$\dot{\Psi}_{D,pump}$	rate of total exergy destruction in a pump
$\dot{\Psi}_{D,surr}$	rate of total exergy destruction in the surroundings
$\dot{\Psi}_{D,total}$	rate of total exergy destruction internally and externally
$\dot{\Psi}_{D,turbine}$	rate of total exergy destruction in a turbine
$\dot{\Psi}_{sup}$	rate at which total exergy is supplied

REFERENCES

1. Pal, R., "Quantification of irreversibilities in practical cyclic processes using exergy analysis and Gouy-Stodola theorem", *Int. J. Mech. Eng. Edu.* 2018; 46: 118-137.
2. Smith, J.M., Van Ness, H.C., Abbott, M.M., *Introduction to Chemical Engineering Thermodynamics*, McGraw Hill, New York, 6th edition, 2001.

TUTORIAL PROBLEM WITH SOLUTION

Problem

Consider a vapor-compression refrigeration cyclic process shown in the figure below. The refrigerated cold region is maintained at 3.89°C and the cooling water for the condenser is available at 27.22 °C. The evaporation temperature is 5°C below the refrigerated region temperature and the condensation temperature is 5°C above the cooling water temperature. The refrigerant is ammonia and the refrigeration capacity is 3500 kW. The states of refrigerant and the corresponding properties at various locations of the refrigeration cycle are given in the table. The location numbers given in the table refer to the figure. For this cyclic process, determine the following:

a) the mass flow rate or circulation rate of the refrigerant
b) the total exergy destruction rate for the whole cyclic process
c) the exergetic efficiency of the whole cyclic process
d) the exergy destruction rates in individual units (evaporator, compressor, condenser, and expansion value)
e) the distribution of the total exergy destruction among the individual units of the cyclic process
f) the entropy production rates in the whole process and in the individual units using the Gouy-Stodola theorem.

The surroundings are at a temperature of 27.22°C.

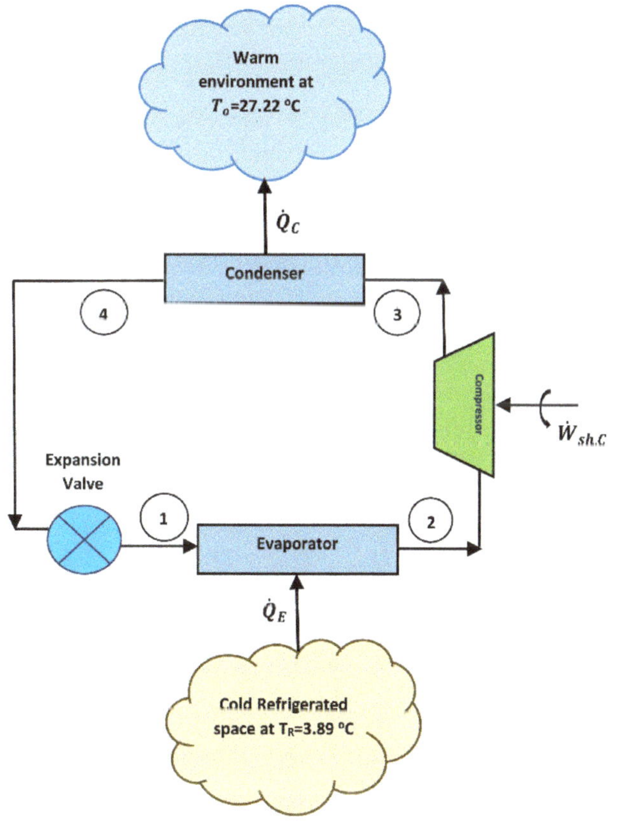

States of refrigerant and the corresponding properties at various points in the refrigeration cycle:

Location point	State of refrigerant	Temperature (°C)	Pressure (kPa)	Specific enthalpy (kJ/kg)	Specific entropy (kJ/kg.K)
1	Mixture of vapor and liquid, quality = 0.1244	-1.11	411.9	333.78	1.276
2	Saturated vapor	-1.11	411.9	1443.28	5.355
3	Superheated vapor	71	1245.2	1633.78	5.46
4	Saturated liquid	32.22	1245.2	333.78	1.238

Solution

a) the mass flow rate or circulation rate of the refrigerant

The mass flow rate of refrigerant in the cycle can be obtained from the refrigeration capacity (3500 kW) of the process as:

$$\dot{m} = \dot{Q}_E/(h_2 - h_1) = 3500/(1443.28 - 333.78) = 3.1546 \text{ kg/s}$$

b) the total exergy destruction rate for the whole cyclic process

To determine the total exergy loss for the whole cyclic process, the process can be represented by a black-box, as shown in the figure below.

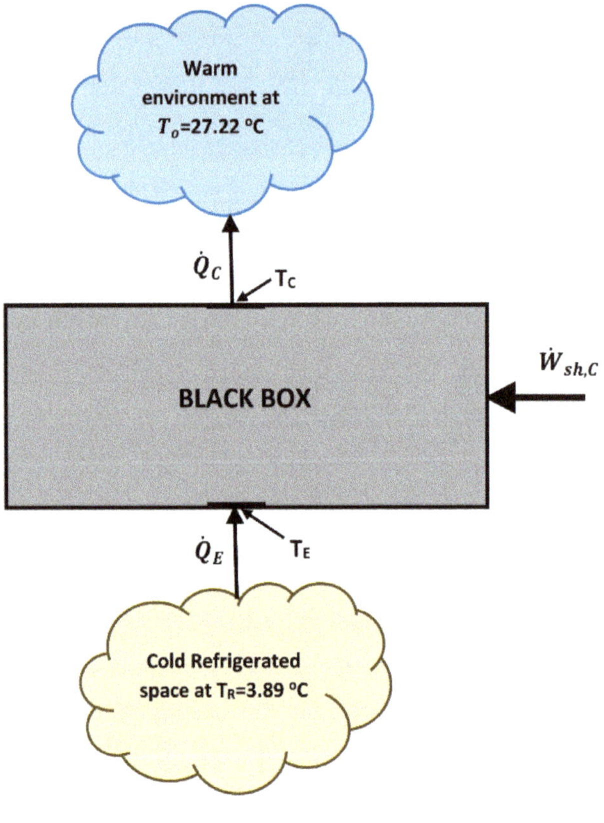

From exergy balance over the blackbox, the rate of exergy destruction within the blackbox (that is, the system) is given as:

$$\dot{\Psi}_{D,system} = \left|\dot{W}_{sh,C}\right| + \dot{Q}_E\left(1 - \frac{T_o}{T_E}\right) - \left|\dot{Q}_C\right|\left(1 - \frac{T_o}{T_C}\right)$$

The rate of exergy destruction in the surroundings (outside the blackbox) is given as:

$$\dot{\Psi}_{D,surr} = \left|\dot{Q}_C\right|\left(1 - \frac{T_o}{T_C}\right) - \dot{Q}_E\left(\frac{T_o}{T_R}\right)\left(1 - \frac{T_R}{T_E}\right)$$

where T_R is the absolute temperature of the refrigerated space.

The rate of total exergy destruction for the whole process is:

$$\dot{\Psi}_{D,total} = \dot{\Psi}_{D,system} + \dot{\Psi}_{D,surr} = \left|\dot{W}_{sh,C}\right| + \dot{Q}_E\left(1 - \frac{T_o}{T_R}\right)$$

$$\dot{\Psi}_{D,total} = \dot{m}(h_3 - h_2) + \dot{Q}_E\left(1 - \frac{T_o}{T_R}\right) = 3.1546 \times (1633.78 - 1443.28) + 3500 \times \left(1 - \frac{27.22+273.15}{3.89+273.15}\right) = 600.95 - 294.74 = 306.21\ \text{kW}$$

Thus, the total exergy destruction rate for the whole cyclic process due to internal and external irreversibilities is 306.21 kW. Note that the first term (600.95 kW) is simply the exergy supplied to the process through pump work and the second term (294.74) could be interpreted as the exergy recovered from the process by transferring heat from a low temperature T_R region to a high temperature T_o region.

c) the exergetic efficiency of the whole cyclic process

The exergetic efficiency of the whole cyclic process can be calculated as follows:

$$\varepsilon = \frac{Exergy\ recovered}{Exergy\ supplied} = \frac{-\dot{Q}_E\left(1 - \frac{T_o}{T_R}\right)}{\left|\dot{W}_{sh,C}\right|} = \frac{294.74}{600.95} = 0.4905$$

d) the exergy destruction rates in individual units (evaporator, compressor, condenser, and expansion value)

From exergy balances over the individual units, the total exergy destruction rates in the individual units (evaporator, compressor, condenser, and expansion value) are given as:

Evaporator:

$$\dot{\Psi}_{D,evaporator} = \dot{m}(\psi_1 - \psi_2) + \dot{Q}_E\left(1 - \frac{T_o}{T_R}\right) = \dot{m}[(h_1 - h_2) + T_o(s_2 - s_1)] + \dot{Q}_E\left(1 - \frac{T_o}{T_R}\right)$$

$$\dot{\Psi}_{D,evaporator} = 3.1546 \times [(333.78 - 1443.28) + 300.37 \times (5.355 - 1.276)] + 3500 \times \left(1 - \frac{300.37}{277.04}\right) = 70.275\ \text{kW}$$

Compressor:

$$\dot{\Psi}_{D,compressor} = \dot{m}(\psi_2 - \psi_3) + |\dot{W}_{sh,C}| = \dot{m}[(h_2 - h_3) + T_o(s_3 - s_2)] + |\dot{W}_{sh,C}|$$

$$\dot{\Psi}_{D,compressor} = 3.1546 \times [(1443.28 - 1633.78) + 300.37 \times (5.46 - 5.355)] + 600.95 = 99.49 \text{ kW}$$

Condenser:

$$\dot{\Psi}_{D,condenser} = \dot{m}(\psi_3 - \psi_4) = \dot{m}[(h_3 - h_4) + T_o(s_4 - s_3)]$$

$$\dot{\Psi}_{D,condenser} = 3.1546 \times [(1633.78 - 333.78) + 300.37 \times (1.238 - 5.46)] = 100.436 \text{ kW}$$

Expansion valve:

$$\dot{\Psi}_{D,expansion-valve} = \dot{m}(\psi_4 - \psi_1) = \dot{m}[(h_4 - h_1) + T_o(s_1 - s_4)]$$

$$\dot{\Psi}_{D,expansion-valve} = 3.1546 \times 300.37 \times [(1.276 - 1.238)] = 36 \text{ kW}$$

e) the distribution of the total exergy destruction among the individual units of the cyclic process

The distribution of the total exergy destruction among the individual units is evaluated as follows: $\dot{\Psi}_{D,i}/\sum \dot{\Psi}_{D,i}$. The values are shown in the table given below.

Unit	$\dot{\Psi}_{D,i}$ (kW)	$\dot{\Psi}_{D,i}/\sum \dot{\Psi}_{D,i}$	$\dot{S}_{G,i}$ (kW/K)
Evaporator	70.275	0.2295	0.234
Compressor	99.49	0.3249	0.331
Condenser	100.436	0.328	0.334
Expansion valve	36	0.1176	0.1198
Whole process	306.20	1.0	1.019

f) the entropy production rates in the whole process and in the individual units using the Gouy-Stodola theorem.

The entropy production rates are calculated using the Gouy-Stodola theorem as follows:

$$\dot{S}_{G,i} = \dot{\Psi}_{D,i}/T_o$$

$$\dot{S}_{G,whole\ process} = \sum \dot{\Psi}_{D,i}/T_o$$

where $T_o = 300.37$ K. The values are shown in the table given above.

Note that the compressor and condenser have the largest fractions of exergy destruction. The entropy production rates are also the highest in these devices.

PRACTICE PROBLEMS

Problem 1

Consider a vapor-compression refrigeration cyclic process shown in the figure below. The refrigerated cold region is maintained at a temperature 5°C above the evaporation temperature and the cooling water for the condenser is available at 25 °C. The evaporator operates at a pressure of 2 bar and the condenser operates at a pressure of 10 bar. The refrigerant is HFC-134A and the mass flow rate of refrigerant is 10 kg/s. The states of refrigerant and the corresponding properties at various locations of the refrigeration cycle are given in the table. The location numbers given in the table refer to the figure. For this cyclic process, determine the following:

a) the refrigeration rate in kW
b) the total exergy destruction rate for the whole cyclic process
c) the exergetic efficiency of the whole cyclic process
d) the exergy destruction rates in individual units (evaporator, compressor, condenser, and expansion value)
e) the distribution of the total exergy destruction among the individual units of the cyclic process
f) the entropy production rates in the whole process and in the individual units using the Gouy-Stodola theorem.

The surroundings are at a temperature of 25°C.

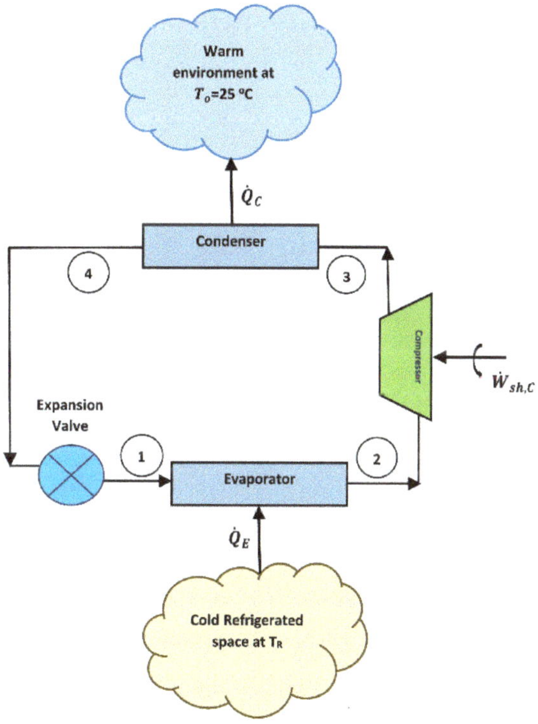

States of refrigerant and the corresponding properties at various points in the refrigeration cycle:

Location point	State of refrigerant	Temperature (°C)	Pressure (kPa)	Specific enthalpy (kJ/kg)	Specific entropy (kJ/kg.K)
1	Mixture of vapor and liquid, quality = 0.3348	-10.09	200	105.29	0.4083
2	Saturated vapor	-10.09	200	241.30	0.9253
3	Superheated vapor	56.05	1000	286.952	0.9634
4	Saturated liquid	39.39	1000	105.29	0.3838

Problem 2

Consider a steam power plant operating shown in the figure below. The boiler pressure is 12.5 MPa (125 bar) and the condenser pressure is 0.008 MPa (8 kPa). The temperature of the superheated steam leaving the boiler and entering the turbine is 550 °C. The plant is producing power at a rate of 150 MW. The states of fluid and the corresponding properties at various locations of the power cycle are given in the table. The location numbers given in the table refer to the figure.

The temperature of the heat-source reservoir is 600 °C. The surroundings are at: $T_o = 25^oC$ and $P_o =$ 1 atm.

Calculate:

a) The mass flow rate of fluid
b) The rates of total exergy destruction in boiler, turbine, condenser, and pump
c) The rate of total exergy destruction in the overall cyclic process
d) The distribution of the total exergy destruction among the individual units of the cyclic process
e) Exergetic efficiency of the cyclic process based on exergy supplied by the heat source.
f) Exergetic efficiency of the cyclic process based on exergy supplied in the boiler.
g) The thermal efficiency (first law efficiency) of the cyclic process
h) The entropy production rates in the individual units and the whole cyclic process using the Gouy-Stodola theorem

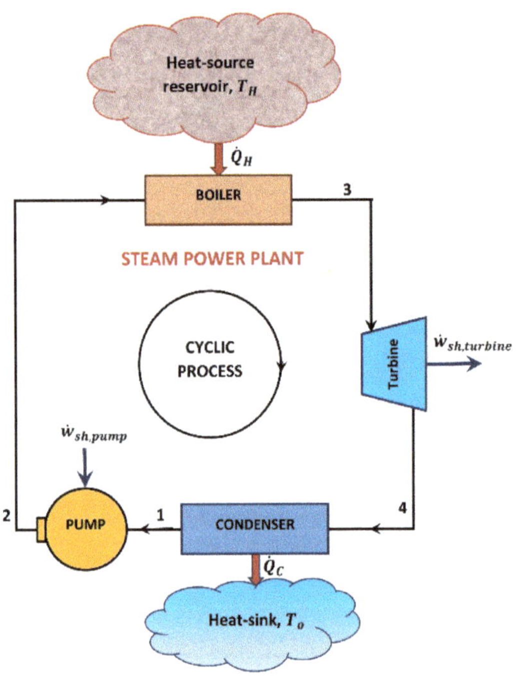

States of fluid and the corresponding properties at various points in the power cycle:

Location point	State of fluid	Temperature (°C)	Pressure (kPa)	Specific enthalpy (kJ/kg)	Specific entropy (kJ/kg.K)
1	Saturated liquid water	41.5	8	173.9	0.5924
2	Subcooled liquid water	43.2	12500	191.86	0.6087
3	Superheated vapor	550	12500	3475.2	6.6290
4	Wet steam, $x^v =$ 0.9304	41.5	8	2409.74	7.698

Chapter 17

Chemical Exergy and its Applications

17.1 LEARNING OUTCOMES

Upon completion of this chapter, students will be able to:

- Distinguish chemical exergy from thermo-mechanical exergy
- Define the two dead states and to distinguish the "restricted dead state" from the "complete dead state"
- Understand and appreciate the concept of "reference environment"
- Estimate the standard chemical molar exergies of substances not present in the reference environment
- Estimate the chemical exergies of ideal gas mixtures
- Estimate the chemical exergies of ideal liquid solutions
- Estimate the chemical exergies of non-ideal gas mixtures
- Estimate the chemical exergies of non-ideal liquid solutions
- Carry out chemical exergy analysis of processes

17.2 INTRODUCTION

Exergy (also referred to as availability) is the maximum amount of useful work obtainable when the system under consideration is brought from its initial state to the *dead* state. Alternatively, exergy could be defined as the minimum amount of useful work required to synthesize a quantity of matter from substances present in the environment and to deliver the matter to a specified state. The *dead state* is one where the system attains thermodynamic equilibrium with the environment. In thermodynamic equilibrium, the conditions of mechanical, thermal, and chemical equilibrium between the system and the environment are satisfied. The existence of mechanical and thermal equilibrium between the system and the environment implies equality of pressure and temperature between the system and the environment. The condition of chemical equilibrium implies that the chemical potentials of each chemical species is the same in the system and the environment. Clearly exergy is a measure of the *degree of departure* of the state of the system from its environment. When the system and the environment are in different states, an opportunity exists to produce useful work upon interaction between the system and the environment.

In exergy analysis, two types of dead states are often defined for the systems: *restricted dead state* and *complete dead state.* In the restricted dead state, only the conditions of mechanical and thermal equilibrium are satisfied whereas in the complete dead state, the condition of chemical equilibrium is also satisfied. The restricted dead state is relevant in exergy calculations dealing with constant composition systems.

The total exergy of a system can be divided into two parts, namely thermo-mechanical exergy (ψ_{TM}) and chemical exergy (ψ_C):

$$\psi = \psi_{TM} + \psi_C \tag{17.1}$$

The thermo-mechanical exergy ψ_{TM} of a system is defined as the maximum amount of useful work obtainable when the system under consideration is brought from its initial state to the *restricted dead state.* The chemical composition of the system remains the same in the initial state and in the *restricted dead state*. The thermo-mechanical exergy of a closed system at zero velocity and zero elevation relative to the environment is given by the expression:

$$\psi_{TM} = (u - u_o) + P_o(v - v_o) - T_o(s - s_o) \tag{17.2}$$

where $u, v, and\ s$ are the specific internal energy, specific volume, and specific entropy, respectively, of the system and T_o is the absolute temperature of the environment. The subscript "o" refers to the *restricted dead state* where the system attains thermo-mechanical equilibrium with the environment. For a fluid stream, the corresponding expression of thermo-mechanical exergy is given as:

$$\psi = (h - h_o) - T_o(s - s_o) + \frac{1}{2}V^2 + gZ \tag{17.3}$$

where h is specific enthalpy, V is fluid velocity, g is acceleration due to gravity, and Z is the elevation of fluid stream.

The chemical exergy ψ_C of a system is defined as the maximum amount of useful work that can be obtained when the system under consideration is brought from its *restricted dead state* to the *complete dead state* where the system is in complete thermodynamic equilibrium with the environment. Note that the temperature and pressure are the same in restricted and complete dead states.

Consider a turbine where the input stream consists of a pure gaseous species A at environment conditions of temperature T_o and pressure P_o. The gas is expanded isothermally and reversibly in a turbine to a discharge pressure of $y_{A,e}P_o$, corresponding to the partial pressure of species A in the environment, where $y_{A,e}$ is the mole fraction of species A in the environment. Thus, the discharge stream is in chemical equilibrium with the environment. The gas is discharged reversibly into the environment through a semi-permeable membrane (see Figure 17.1). For this reversible isothermal expansion of a gas in a turbine, the maximum shaft work produced is simply a decrease in the Gibbs free energy. Assuming gas to be an ideal gas, the maximum work obtainable or chemical exergy is thus given by:

$$\psi_C = -\Delta g = -RT_o \int_{P_o}^{y_{A,e}P_o} \frac{dP}{P} = -RT_o ln(y_{A,e}) \tag{17.4}$$

where g is the molar Gibbs free energy of gas and R is the universal gas constant.

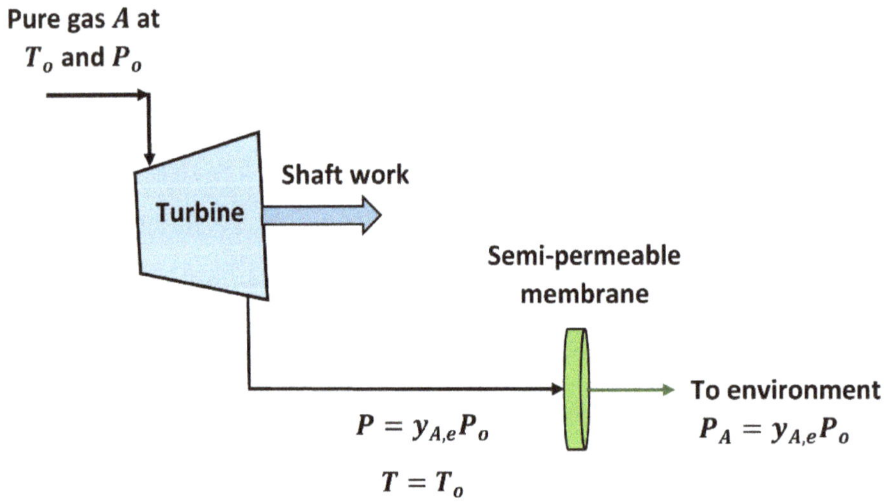

Figure 17.1 Method for evaluating the chemical exergy of a gas.

The focus of this chapter is the chemical exergy of matter. Understanding of chemical exergy and related concepts is important. In processes involving chemical reactions, and hence changes in chemical composition, chemical exergy plays a very important role in exergy analysis of the process. The key objectives of this chapter are as follows: 1) to briefly review the fundamental concepts of chemical exergy; 2) to develop the relations for chemical exergy of *ideal* gas mixtures and liquid solutions; 3) to develop the relations for chemical exergy of *non-ideal* gas mixtures and liquid solutions; and 4) to apply the concepts of chemical exergy in the analysis of practical processes.

17.3 MODEL FOR THE REFERENCE ENVIRONMENT

From the definition of exergy, it is clear that exergy of a system is relative to its environment. Thus, exergy is a combined property of the system and the environment. For the purpose of exergy calculations, the reference environment is modeled as a very large body in a state of complete thermodynamic equilibrium. There are no internal gradients of pressure, temperature, and chemical potentials. The exchanges of matter and energy with the systems do not alter its intensive properties such as temperature, pressure, and chemical potentials. Consequently, there is no possibility of producing work from interactions (physical or chemical) between different parts of the environment. Furthermore, all parts of the environment are considered to be at rest with respect to one another. Thus, the reference environment is a model atmosphere of infinite extent with uniform temperature, pressure, and composition. Figure 17.2 specifies the composition, pressure, and temperature of the reference environment used in this chapter. The reference composition is taken to be that of air with 60 % relative humidity.

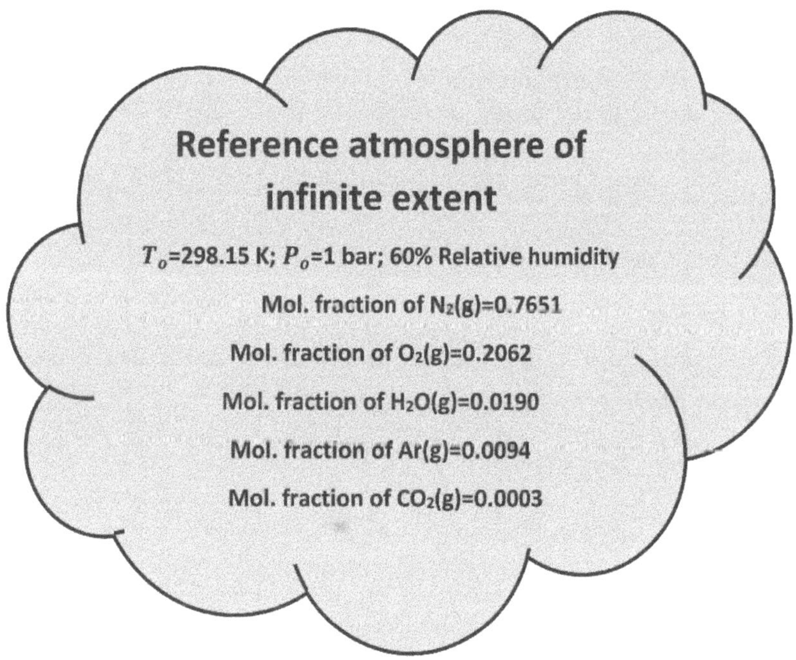

Figure 17.2 Reference environment for the purpose of exergy calculations.

17.4 STANDARD MOLAR CHEMICAL EXERGY

The standard molar chemical exergy of a substance is the chemical exergy of 1 mole of a substance in its standard state. The standard state of a gaseous species is a pure gas in the ideal gas state at 1 bar. The standard states for liquids and solids are the real pure liquid or solid at 1 bar. Note that the relevant temperature in chemical exergy calculations is the environment temperature T_0 fixed at 25°C. The standard molar chemical exergy of a gaseous species can be calculated using Equation (17.4). For example, the standard molar chemical exergy of nitrogen gas is: $-8.314 \times 298.15 \times ln(0.7651) = 0.664$ kJ/mol. Table 17.1 lists the standard molar chemical exergies of various species present in the reference environment, calculated using Equation (17.4).

Table 17.1 Standard molar chemical exergies of reference species

Species	**Mol. Fraction** $(y_{i,e})$	**Standard molar chemical exergy** $(\psi^o_{C,i})$**, kJ/mol.**
$N_2(g)$	0.7651	0.664
$O_2(g)$	0.2062	3.914
$H_2O(g)$	0.0190	9.824
$Ar(g)$	0.0094	11.57
$CO_2(g)$	0.0003	20.11

17.5 STANDARD CHEMICAL EXERGY OF A REACTION

The standard chemical exergy of a reaction ψ^o_{rxn} can be defined as:

$$\psi^o_{rxn} = \sum v_i \, \psi^o_{C,i} \tag{17.5}$$

where v_i is the stoichiometric coefficient and $\psi^o_{C,i}$ is the standard molar chemical exergy of chemical species i. The stoichiometric constant is positive for the products and negative for the reactants.

The standard molar chemical exergy of any species can be expressed in terms of the chemical potentials as:

$$\psi^o_{C,i} = \left(\mu^o_i - \mu_{i,e}\right) \tag{17.6}$$

where μ^o_i is the standard chemical potential of species i and $\mu_{i,e}$ is the chemical potential of species i in the reference environment. From Equations (17.5) and (17.6), it follows that:

$$\psi^o_{rxn} = \sum v_i \left(\mu^o_i - \mu_{i,e}\right) = \sum v_i \mu^o_i - \sum v_i \, \mu_{i,e} \tag{17.7}$$

By definition the standard Gibbs free energy of a reaction is:

$$\Delta G^o_{rxn} = \sum v_i \mu^o_i \tag{17.8}$$

As the reference environment is in a state of complete thermodynamic equilibrium:

$$\sum v_i \, \mu_{i,e} = 0 \tag{17.9}$$

Combining Equations (17.7) - (17.9), the following important result is obtained:

$$\psi^o_{rxn} = \Delta G^o_{rxn} \tag{17.10}$$

Thus, the standard chemical exergy of a reaction is equal to the standard Gibbs free energy of a reaction.

17.6 DETERMINATION OF STANDARD MOLAR CHEMICAL EXERGY OF SUBSTANCES NOT PRESENT IN THE REFERENCE ENVIRONMENT

If the substance is present in the reference environment, then the calculation of standard molar chemical exergy of a substance can be done readily. For example, Equation (17.4) can be used to calculate $\psi^o_{C,i}$ for any gaseous species. In general, $\psi^o_{C,i}$ is given as a difference in the chemical potentials of the species in the standard and reference environment states, as expressed by Equation (17.6). This raises the question as to what happens if the species of interest is not present in the reference environment. In this case, it so happens that one can still calculate the chemical exergy of the species provided that all the atoms present in the species are also present in the reference environment. However, we need to set up a chemical reaction to produce the species from the existing species in the reference environment. For example, consider liquid methanol as the species whose standard molar chemical exergy needs to be determined. The chemical reaction required to produce methanol from the existing species in the reference environment is as follows:

$$CO_2(g) + 2H_2O(g) \rightarrow CH_3OH(l) + (3/2)O_2(g) \tag{17.11}$$

The process for synthesizing methanol from the reference environment species is illustrated in Figure 17.3. Using Equations (17.5) and (17.10), one can determine the standard molar chemical exergy of liquid methanol as:

$$\psi^o_{C,methanol} = \Delta G^o_{rxn} - \textstyle\sum_{(other\ than\ methanol)} v_i\, \psi^o_{C,i} \tag{17.12}$$

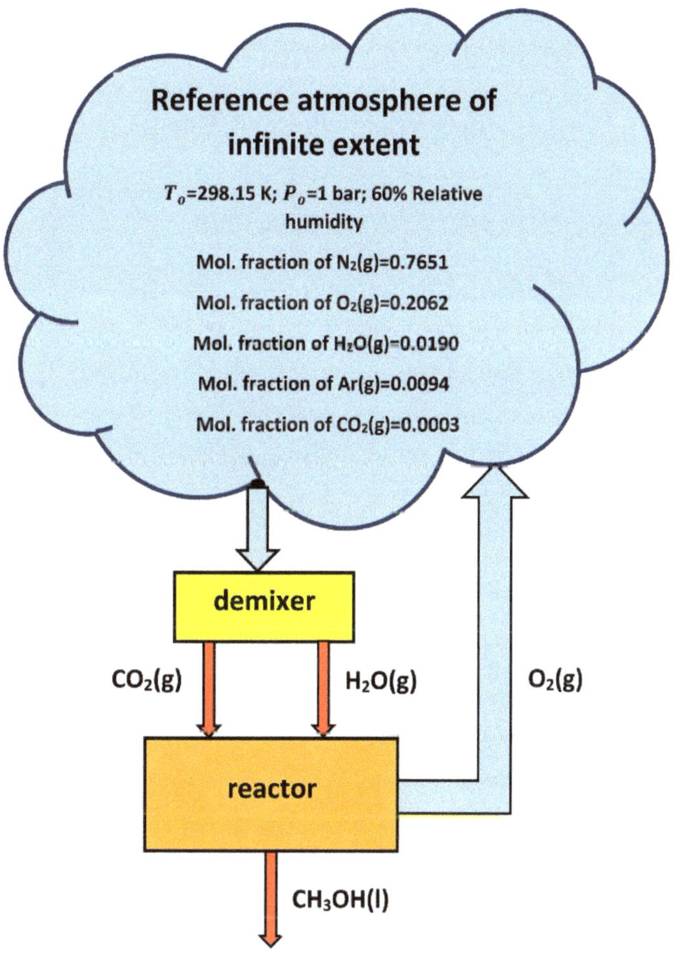

Figure 17.3 Process to synthesize liquid methanol from the reference environment species.

ΔG^o_{rxn} for the formation reaction of methanol (Equation (17.11)) can be determined from the standard Gibbs free energy of formation ΔG^o_f of various species as:

$$\Delta G^o_{rxn} = \Delta G^o_f(CH_3OH, l) + (3/2)\Delta G^o_f(O_2, g) - \Delta G^o_f(CO_2, g) - 2\Delta G^o_f(H_2O, g) \quad (17.13)$$

$$\Delta G^o_{rxn} = -166.4 + (3/2)(0) - (-394.36) - 2(-228.589) = 685.138 \text{ kJ/mol} \quad (17.14)$$

The standard molar chemical exergies of the species present in the reference atmosphere and involved in the reaction are as follows: $\psi^o_C(O_2, g) = 3.914$; $\psi^o_C(CO_2, g) = 20.11$; and $\psi^o_C(H_2O, g) = 9.824$ kJ/mol. Upon substitution of ΔG^o_{rxn} and $\psi^o_{C,i}$ values in Equation (17.12), the standard molar chemical exergy of liquid methanol is found to be 719.025 kJ/mol.

17.7 CHEMICAL EXERGY OF IDEAL GAS MIXTURES AND LIQUID SOLUTIONS

17.7.1 Ideal Gas Mixtures

The chemical potential of any species i in an ideal gas mixture at temperature T_o and pressure P_o, denoted as $\mu^{ig}_{i,mix}(T_o, P_o)$, is given by:

$$\mu^{ig}_{i,mix} = \mu^o_i + RTlny_i \quad (17.15)$$

where μ^o_i is the standard chemical potential of gas species i at T_o (standard state being pure ideal gas at pressure P_o). The chemical exergy of an ideal gas mixture is:

$$\Psi^{ig}_{C,mix} = \sum n_i(\mu^{ig}_{i,mix} - \mu_{i,e}) \quad (17.16)$$

where n_i is the number of moles of species i and $\mu_{i,e}$ is the chemical potential of species i in the reference environment. From Equations (17.15) and (17.16), it follows that:

$$\Psi^{ig}_{C,mix} = \sum n_i(\mu^o_i + RTlny_i) - \sum n_i(\mu^o_i + RTlny_{i,e}) = RT_o \sum n_i\, ln(y_i/y_{i,e}) \quad (17.17)$$

Thus, the molar chemical exergy of an ideal gas mixture is given as:

$$\psi^{ig}_{C,mix} = \Psi^{ig}_{C,mix}/N = RT_o \sum y_i\, ln(y_i/y_{i,e}) \quad (17.18)$$

where N is the total number of moles of the mixture including all the species. This result could be expressed in a more general form as:

$$\psi^{ig}_{C,mix} = \sum y_i \psi^0_{C,i} + RT_o \sum y_i\, lny_i \quad (17.19)$$

According to Equation (17.19), the molar chemical exergy of a mixture of ideal gases can be calculated from the knowledge of the standard molar chemical exergies of the species involved and the composition of the mixture.

17.7.2 Ideal Liquid Solutions

The chemical potential of any species i in an ideal liquid solution at temperature T_o and pressure P_o, denoted as $\mu^{id}_{i,solu}$, is given by:

$$\mu^{id}_{i,solu} = \mu^o_i + RT_o ln(x_i) \quad (17.20)$$

where μ_i^o is the standard chemical potential of liquid species i at T_o (standard state being pure liquid at pressure P_o). The chemical exergy of an ideal liquid solution is:

$$\Psi_C^{id} = \sum n_i(\mu_{i,solu}^{id} - \mu_{i,e}) \tag{17.21}$$

From Equations (20) and (21), it follows that:

$$\begin{aligned}\Psi_C^{id} &= \sum n_i[\mu_i^o + RT_o ln(x_i)] - \sum n_i\mu_{i,e} = \sum n_i(\mu_i^o - \mu_{i,e}) + RT_o \sum n_i\, ln(x_i) \\ &= \sum n_i\, \psi_{C,i}^o + RT_o \sum n_i\, ln(x_i)\end{aligned} \tag{17.22}$$

Thus, the molar chemical exergy of an ideal liquid solution is given as:

$$\psi_C^{id} = \Psi_C^{id}/N = \sum x_i\, \psi_{C,i}^o + RT_o \sum x_i\, ln(x_i) \tag{17.23}$$

17.8 CHEMICAL EXERGY OF NON-IDEAL GAS MIXTURES AND LIQUID SOLUTIONS

The chemical exergy of a non-ideal system (gas mixture or liquid solution) can be expressed as:

$$\Psi_C = \sum n_i(\mu_{i,solu} - \mu_{i,e}) \tag{17.24}$$

The change in the chemical potential of any species present in the mixture/solution is given in terms of the fugacity as:

$$d\mu_{i,solu} = RT_o dln(\hat{f}_i) \tag{17.25}$$

where $\hat{f}_i$ is the fugacity of species i in the mixture/solution. Upon integration, Equation (17.25) gives:

$$\mu_{i,solu} = \mu_i^o + RT_o ln\left(\frac{\hat{f}_i}{f_i^o}\right) \tag{17.26}$$

where f_i^o is the standard fugacity of species i. From Equations (17.24) and (17.26):

$$\Psi_C = \sum n_i(\mu_i^o - \mu_{i,e}) + RT_o \sum n_i\, ln\left(\frac{\hat{f}_i}{f_i^o}\right) \tag{17.27}$$

The chemical potential difference in the first summation term on the right hand side of equation (17.27) is the standard molar chemical exergy of species i and therefore, Equation (17.27) could be re-written as:

$$\Psi_C = \sum n_i \psi_{C,i}^o + RT_o \sum n_i\, ln\left(\frac{\hat{f}_i}{f_i^o}\right) \tag{17.28}$$

This is a very general expression for the total chemical exergy of a system, applicable to both non-ideal gas mixtures and non-ideal liquid solutions. The expression is, of course, applicable to ideal systems as well.

17.8.1 Non-Ideal Gas Mixtures

For non-ideal gas mixtures, the fugacity coefficient $\hat{\varphi}_i$ of any species is defined as:

$$\hat{\varphi}_i = \frac{\hat{f}_i}{y_i P} \tag{17.29}$$

Upon substitution of $\hat{\varphi}_i$ into Equation (17.28) and recognizing that $P = P_o$ and $f_i^o = P_o$, we obtain:

$$\Psi_C = \sum n_i \psi_{C,i}^o + RT_o \sum n_i \, ln(y_i \hat{\varphi}_i) \quad (17.30)$$

Thus, the molar chemical exergy of a non-ideal gas mixture is:

$$\psi_C = \sum y_i \psi_{C,i}^o + RT_o \sum y_i \, ln(y_i \hat{\varphi}_i) \quad (17.31)$$

Fugacity Coefficient of Pure Non-Ideal Gas

For a single-component (pure) non-ideal gas, Equation (17.31) reduces to:

$$\psi_C = \psi_C^o + RT_o ln(\varphi) \quad (17.32)$$

where φ is the fugacity coefficient of the pure gas. The first term on the right hand side of Equation (17.32), that is, the standard molar chemical exergy of gas, treats gas to be an ideal gas and the second term accounts for the non-ideal behavior. As the fugacity coefficient φ is a measure of non-ideality of gas, it is related to the compressibility factor Z defined as:

$$Z = Pv/RT \quad (17.33)$$

The fugacity coefficient is related to compressibility factor as:

$$ln\varphi = \int_0^P (Z-1)\frac{dP}{P} \quad (17.34)$$

For ideal gas, $Z = 1$ and $\varphi = 1$. In order to evaluate φ, we need an expression for the compressibility factor and evaluate the integral on the right-hand side of Equation (17.34). The two-term virial expansion of compressibility factor in P is:

$$Z = 1 + \frac{BP}{RT} \quad (17.35)$$

where B/RT is the second virial coefficient of Z-expansion. This is the simplest form of virial expansion with only 1st order correction in Z. As pressure involved in chemical exergy calculations is 1 bar, this simple truncated two-term virial expansion is good enough for the purpose of exergy calculation of gases. Upon substitution of the expression for Z in Equation (17.34) and integrating, we obtain the following expression for the fugacity coefficient of pure gas:

$$ln\varphi = \frac{BP}{RT} \quad (17.36)$$

The virial constant B can be expressed as follows:

$$B = \frac{RT_c}{P_c}(B^o + \omega B^1) \quad (17.37)$$

where T_c and P_c are critical temperature and pressure, respectively and ω is the acentric factor of the material. B^o and B^1 are functions of reduced temperature T_r $(= T/T_c)$ only and are given as:

$$B^o = 0.083 - \frac{0.422}{T_r^{1.6}} \quad (17.38)$$

$$B^1 = 0.139 - \frac{0.172}{T_r^{4.2}} \quad (17.39)$$

Thus, the fugacity coefficient of pure non-ideal gas, and hence its molar chemical exergy (see Equation (17.32)), can be calculated from Equation (17.36) in conjunction with Equations (17.38) - (17.39).

Fugacity Coefficients of Non-Ideal Gas Mixture

For non-ideal multicomponent gas mixtures, the fugacity coefficient expression given in the form of Equation (17.36) can be generalized to (see Chapter 12):

$$ln\hat{\varphi}_k = \frac{P}{RT}\left[B_{kk} + \frac{1}{2}\sum_i\sum_j y_i y_j\left(2\delta_{ik} - \delta_{ij}\right)\right] \tag{17.40}$$

where y's are mole fractions of the species, and δ_{ik} and δ_{ij} are defined below:

$$\delta_{ik} = 2B_{ik} - B_{ii} - B_{kk} \;\; ; \;\; \delta_{ij} = 2B_{ij} - B_{ii} - B_{jj} \tag{17.41}$$

Note that the dummy indices i and j in Equation (17.40) are to run over all species. Also,

$$\delta_{ii} = 0 \;\; ; \;\; \delta_{jj} = 0\, ; \;\; \delta_{ki} = \delta_{ik} \tag{17.42}$$

The coefficients B's are given as:

$$B_{ij} = \left(\frac{RT_{cij}}{P_{cij}}\right)\left(B^o + \omega_{ij}B^1\right) \tag{17.43}$$

where

$$\omega_{ij} = \frac{\omega_i + \omega_j}{2} \tag{17.44}$$

$$T_{cij} = \left(T_{ci}T_{cj}\right)^{1/2} \tag{17.45}$$

$$P_{cij} = \frac{Z_{cij}RT_{cij}}{v_{cij}} \tag{17.46}$$

$$Z_{cij} = \frac{Z_{ci} + Z_{cj}}{2} \tag{17.47}$$

$$v_{cij} = \left(\frac{v_{ci}^{1/3} + v_{cj}^{1/3}}{2}\right)^3 \tag{17.48}$$

Note that when $i = j$, the quantities B_{ii}, ω_{ij}, T_{cij}, P_{cij}, Z_{cij}, v_{cij} all become pure species i quantities.

For binary non-ideal gas mixtures, Equation (17.40) gives the following expressions for the fugacity coefficients:

$$ln\hat{\varphi}_1 = \frac{P}{RT}\left[B_{11} + y_2^2\delta_{12}\right] \tag{17.49}$$

$$ln\hat{\varphi}_2 = \frac{P}{RT}\left[B_{22} + y_1^2\delta_{12}\right] \tag{17.50}$$

where

$$\delta_{12} = 2B_{12} - B_{11} - B_{22} \tag{17.51}$$

Also note that:

$$B_{11} = B_1 = \left(\frac{RT_{c1}}{P_{c1}}\right)\left(B^o + \omega_1 B^1\right) \tag{17.52}$$

$$B_{22} = B_2 = \left(\frac{RT_{c2}}{P_{c2}}\right)(B^o + \omega_2 B^1) \tag{17.53}$$

$$B_{12} = \left(\frac{RT_{c12}}{P_{c12}}\right)(B^o + \omega_{12} B^1) \tag{17.54}$$

where B^o and B^1 are functions of reduced temperature, given in Equations (17.38) and (17.39), respectively.

Thus, the fugacity coefficients of different components of a non-ideal multi-component gas mixture, and hence the molar chemical exergy of the mixture (see Equation (17.42)), can be calculated from Equation (17.51) in conjunction with Equations (17.52) - (17.59). Note that the calculations require the critical properties (T_c, P_c, v_c and Z_c) and acentric factors (ω) of the pure components.

17.8.2 Non-Ideal Liquid Solutions

The fugacity of any species in an ideal liquid solution is given by the Lewis-Randall rule as:

$$\hat{f}_i^{id} = x_i f_i^L \tag{17.55}$$

where $\hat{f}_i^{id}$ is the fugacity of species i in an ideal liquid solution, x_i is the mole fraction of species i in the liquid solution, f_i^L is the fugacity of pure liquid species i at the solution temperature and pressure. The activity coefficient γ_i of species i in the liquid solution is defined as:

$$\gamma_i = \frac{\hat{f}_i}{x_i f_i^L} \tag{17.56}$$

where $\hat{f}_i$ is the fugacity of species i in the liquid solution. For an ideal liquid solution, the Lewis-Randall leads to $\gamma_i^{id} = 1$. From Equations (17.28) and (17.56), it follows that:

$$\Psi_C = \sum n_i \psi_{C,i}^o + RT_o \sum n_i \, ln\left(\frac{x_i f_i^L \gamma_i}{f_i^o}\right) \tag{17.57}$$

Since the pressure and temperature in chemical exergy calculations are the reference atmosphere pressure $P_o = 1$ bar and temperature $T_o = 298.15\ K$,

$$f_i^o = f_i^L \tag{17.58}$$

Thus, Equation (17.57) reduces to the following expression for molar chemical exergy of non-ideal liquid solutions:

$$\psi_C = \sum x_i \psi_{C,i}^o + RT_o \sum x_i \, ln(x_i \gamma_i) \tag{17.59}$$

According to Equation (17.59), we need the knowledge of activity coefficients in order to calculate the molar chemical exergy of non-ideal liquid solutions.

As discussed in Chapter 12, some of the commonly used empirical models for the activity coefficients of non-ideal binary liquid solutions are: Margules model, Van Laar model, and Wilson model. They are given as follows:

Margules Model

$$ln\gamma_1 = x_2^2[A_{12} + 2(A_{21} - A_{12})x_1] \tag{17.60}$$

$$ln\gamma_2 = x_1^2[A_{21} + 2(A_{12} - A_{21})x_2] \tag{17.61}$$

where A_{12} and A_{21} are constants for the binary system. However, they depend on the temperature of the system

Van Laar Model

$$ln\gamma_1 = A'_{12}\left[1 + \frac{A'_{12}x_1}{A'_{21}x_2}\right]^{-2} \tag{17.62}$$

$$ln\gamma_2 = A'_{21}\left[1 + \frac{A'_{21}x_2}{A'_{12}x_1}\right]^{-2} \tag{17.63}$$

where A'_{12} and A'_{21} are constants for the binary system. However, they depend on the temperature of the system

Wilson Model

$$ln\gamma_1 = -ln(x_1 + x_2\Omega_{12}) + x_2\left[\frac{\Omega_{12}}{x_1+x_2\Omega_{12}} - \frac{\Omega_{21}}{x_2+x_1\Omega_{21}}\right] \tag{17.64}$$

$$ln\gamma_2 = -ln(x_2 + x_1\Omega_{21}) - x_1\left[\frac{\Omega_{12}}{x_1+x_2\Omega_{12}} - \frac{\Omega_{21}}{x_2+x_1\Omega_{21}}\right] \tag{17.65}$$

where Ω_{12} and Ω_{21} are constants for the binary system. However, they depend on the temperature of the system. The temperature-dependence of Ω_{12} and Ω_{21} is further expressed as:

$$\Omega_{12} = \frac{v_2}{v_1} exp\left(-\frac{a_{12}}{RT}\right) \tag{17.66}$$

$$\Omega_{21} = \frac{v_1}{v_2} exp\left(-\frac{a_{21}}{RT}\right) \tag{17.67}$$

where a_{12} and a_{21} are constants independent of temperature and composition. They depend only on the chemical nature of the binary system. v_1 and v_2 are molar volumes of species 1 and 2, respectively. The Wilson model has the additional advantage that it can easily be generalized to multicomponent systems.

NOTATION

A_{12}	Margules constant, Equations (17.60) and (17.61)
A_{21}	Margules constant, Equations (17.60) and (17.61)
A'_{12}	Van Laar constant, Equations (17.62) and (17.63)
A'_{21}	Van Laar constant, Equations (17.62) and (17.63)
a_{12}	constant in Wilson model for binary systems, Equation (17.66)
a_{21}	constant in Wilson model for binary systems, Equation (17.67)
B	virial coefficient for pure gas
B_{ii}, B_{jj}, B_{kk}	virial coefficients of pure species i, j, k
B_{ij}, B_{ik}	virial cross cefficients
B^o	quantity given in Equation (17.38)
B^1	quantity given in Equation (17.39)
f_i^L	fugacity of pure liquid species i
f_i^o	standard fugacity of species i
$\hat{f}_i$	fugacity of species i in mixture or solution

$\hat{f}_i^{id}$	fugacity of species i in ideal solution
g	acceleration due to gravity or molar Gibbs free energy
h	molar or specific enthalpy
h_o	molar or specific enthalpy in dead state
n_i	moles of species i
N	total number of moles in a mixture
P	pressure
P_c	critical pressure
P_{cij}	critical pressure of mixture of species i and j
P_o	pressure in dead state
R	universal gas constant
s	molar or specific entropy
s_o	molar or specific entropy in dead state
T	temperature
T_c	critical temperature
T_{ci}	critical temperature of species i
T_{cj}	critical temperature of species j
T_{cij}	critical temperature of mixture of species i and j
T_o	temperature in dead state
T_r	reduced temperature
u	molar or specific internal energy
u_o	molar or specific internal energy in dead state
V	velocity
v	molar volume of pure substance
v_c	critical molar volume of pure substance
v_{ci}	critical molar volume of species i
v_{cj}	critical molar volume of species j
v_{cij}	critical molar volume of mixture of species i and j
v_i	stoichiometric coefficient
v_o	molar or specific volume in dead state
x_i	mole fraction of species i in liquid solution
y_i	mole fraction of species i in gas mixture
$y_{i,e}$	mole fraction of species i in reference environment
Z	compressibility factor or elevation
Z_c	critical compressibility factor
Z_{ci}	critical compressibility factor of species i
Z_{cj}	critical compressibility factor of species j
Z_{cij}	critical compressibility factor of mixture of species i and j

Greek Letters

γ_i	activity coefficient of species i
γ_i^{id}	activity coefficient of species i in ideal solution
Δ	difference operator
ΔG_f^o	standard Gibbs free energy of formation of compound
ΔG_{rxn}^o	standard Gibbs free energy of reaction
δ_{ii}	parameter defined in Equation (17.42)
δ_{ij}	parameter defined in Equation (17.41)
δ_{jj}	parameter defined in Equation (17.42)
δ_{ik}	parameter defined in Equation (17.41)
δ_{ki}	parameter defined in Equation (17.42)

$\mu_{i,e}$	chemical potential of species i in the reference environment
μ_i^o	standard chemical potential of species i
$\mu_{i,solu}^{id}$	chemical potential of species i in ideal liquid solution
$\mu_{i,mix}^{ig}$	chemical potential of species i in ideal gas mixture
φ	fugacity coefficient of pure substance
$\hat{\varphi}_i$	fugacity coefficient of species i in mixture or solution
$\hat{\varphi}_k$	fugacity coefficient of species k in mixture or solution
Ψ_C	total chemical exergy
Ψ_C^{id}	total chemical exergy of ideal solution
$\Psi_{C,mix}^{ig}$	total chemical exergy of ideal gas mixture
ψ	specific or molar total (chemical plus thermo-mechanical) exergy
ψ_C	specific or molar chemical exergy
ψ_{TM}	specific or molar thermo-mechanical exergy
ψ_C^{id}	specific or molar chemical exergy of ideal solution
$\psi_{C,i}^o$	standard molar chemical exergy of species i
ψ_{rxn}^o	standard chemical exergy of a reaction
$\psi_{C,mix}^{ig}$	specific or molar chemical exergy of ideal gas mixture
Ω_{12}	parameter in Wilson model, see Equation (17.66)
Ω_{21}	parameter in Wilson model, see Equation (17.67)
ω	acentric factor of pure substance
ω_i	acentric factor of pure species i
ω_j	acentric factor of pure species j
ω_{ij}	acentric factor of mixture of species i and j

FURTHER READING

Pal, R., "Chemical exergy of ideal and non-ideal gas mixtures and liquid solutions with applications", *Int. J. Mech. Eng. Edu.* 2019; 47: 44-72.

TUTORIAL PROBLEMS WITH SOLUTION

Problem 1

Determine the molar chemical exergy of a binary ideal-gas mixture of nitrogen and methane at 298.15 K and 1 bar. The mixture contains 40 mol. % nitrogen.

Solution

In order to solve this problem, we first need to determine the standard molar chemical exergy of methane by setting up the formation reaction of methane using the reference species shown in Figure 17.2. The formation reaction for methane from the reference species is:

$$CO_2(g) + 2H_2O(g) \rightarrow CH_4(g) + 2O_2(g)$$

From Equations (17.5) and (17.10):

$$\psi_{C,methane}^o = \Delta G_{rxn}^o - \sum_{(other\ than\ methane)} v_i\, \psi_{C,i}^o$$

ΔG^o_{rxn} for the formation reaction of methane can be determined from the standard Gibbs free energy of formation ΔG^o_f of various species as:

$$\Delta G^o_{rxn} = \Delta G^o_f(CH_4, g) + 2\Delta G^o_f(O_2, g) - \Delta G^o_f(CO_2, g) - 2\Delta G^o_f(H_2O, g)$$

$$\Delta G^o_{rxn} = -50.8 + 2(0) - (-394.36) - 2(-228.589) = 800.74 \text{ kJ/mol}$$

The standard molar chemical exergies of the species present in the reference atmosphere and involved in the reaction are as follows: $\psi^o_C(O_2, g) = 3.914$; $\psi^o_C(CO_2, g) = 20.11$; and $\psi^o_C(H_2O, g) = 9.824$ kJ/mol. Upon substitution of ΔG^o_{rxn} and $\psi^o_{C,i}$ values in the $\psi^o_{C,methane}$ expression above, the standard molar chemical exergy of methane gas is found to be 832.67 kJ/mol. The molar chemical exergy of the $CH_4(g) + N_2(g)$ mixture can now be calculated as Equation (17.19):

$$\psi^{ig}_{C,mix} = y_{CH_4}\psi^o_{C,CH_4} + y_{N_2}\psi^o_{C,N_2} + RT_o\left(y_{CH_4} ln y_{CH_4} + y_{N_2} ln y_{N_2}\right)$$

$$\psi^{ig}_{C,mix} = 0.60(832.67) + 0.40(0.664) + (8.314/1000)(298.15)[0.6 \ln(0.6) + 0.40 \ln(0.40)] = 498.20 \text{ kJ/mol.}$$

Problem 2

Determine the molar chemical exergy of a binary liquid solution of methanol and ethanol at 298.15 K and 1 bar. The mole fraction of methanol is 0.67. The solution can be assumed to be ideal.

Solution

As ethanol is not present in the reference atmosphere, we need to set up the formation reaction of ethanol using the reference species so as to determine the standard molar chemical exergy of ethanol. The formation reaction for ethanol from the reference species is:

$$2CO_2(g) + 3H_2O(g) \rightarrow CH_3CH_2OH(l) + 3O_2(g)$$

From Equations (17.5) and (17.10):

$$\psi^o_{C,ethanol} = \Delta G^o_{rxn} - \sum_{(other\ than\ ethanol)} v_i\, \psi^o_{C,i}$$

ΔG^o_{rxn} for the formation reaction of ethanol can be determined from the standard Gibbs free energy of formation ΔG^o_f of various species as:

$$\Delta G^o_{rxn} = \Delta G^o_f(CH_3CH_2OH, l) + 3\Delta G^o_f(O_2, g) - 2\Delta G^o_f(CO_2, g) - 3\Delta G^o_f(H_2O, g)$$

$$\Delta G^o_{rxn} = -174.9 + 3(0) - 2(-394.36) - 3(-228.589) = 1299.587 \text{ kJ/mol}$$

The standard molar chemical exergies of the species present in the reference atmosphere and involved in the reaction are as follows: $\psi^o_C(O_2, g) = 3.914$; $\psi^o_C(CO_2, g) = 20.11$; and $\psi^o_C(H_2O, g) = 9.824$ kJ/mol. Upon substitution of ΔG^o_{rxn} and $\psi^o_{C,i}$ values in the $\psi^o_{C,ethanol}$ expression above, the standard molar chemical exergy of ethanol is found to be 1416.48 kJ/mol. The molar chemical exergy of methanol-ethanol solution can now be calculated using Equation (17.23):

$$\psi_C^{id} = x_{methanol}\psi_{C,methanol}^o + x_{ethanol}\psi_{C,ethanol}^o + RT_o(x_{methanol}lnx_{methanol} + x_{ethanol}lnx_{ethanol}) = 0.67(719.025) + 0.33(1416.48) + (8.314/1000)(298.15)[0.67\ln(0.67) + 0.33\ln(0.33)] = 947.613 \text{ kJ/mol.}$$

Problem 3

Estimate and plot the activity coefficients γ_1 and γ_2 for binary solutions of methanol (1) and water (2) for different values of x_1. Also determine the molar chemical exergy of methanol (1)-water (2) solution with $x_1 = 0.10$ at 298.15 K and 1 bar. Use the Wilson model to estimate the activity coefficients. Given that: a_{12} =107.38 cal/mol., a_{21}=469.55 cal/mol., v_1=40.73 cm³/mol. and v_2=18.07 cm³/mol.

Solution

The values of Ω_{12} and Ω_{21} are calculated from Equations (17.66)-(17.67) as: $\Omega_{12} = 0.3701$ and $\Omega_{21} = 1.0203$. Using these values of Ω_{12} and Ω_{21}, the activity coefficients are calculated from Equations (17.64) – (17.65) and are plotted in the figure below. Clearly methanol-water solutions are non-ideal as the activity coefficients deviate from a value of unity.

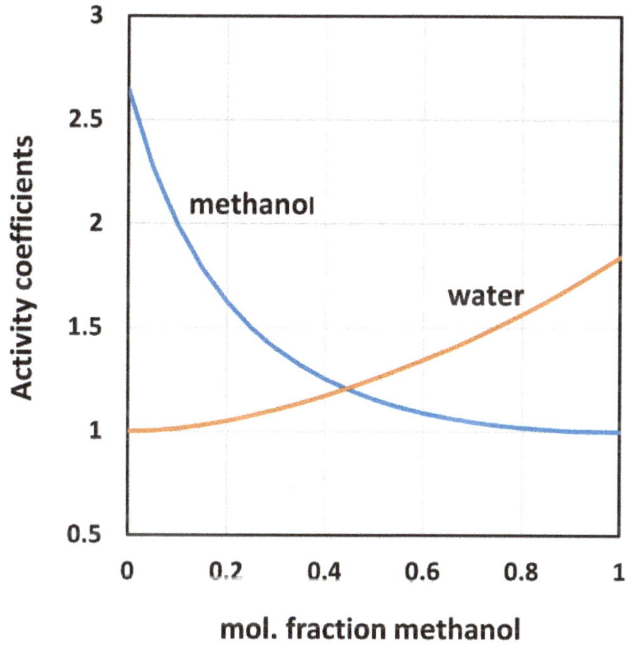

The molar chemical exergy of methanol-water solution can be estimated from Equation (17.59), that is,

$$\psi_C = x_1\psi_{C,1}^o + x_2\psi_{C,2}^o + RT_o[x_1 ln(x_1\gamma_1) + x_2 ln(x_2\gamma_2)]$$

The values of various quantities present in Equation (17.81) are as follows: $x_1 = 0.10$, $x_2 = 0.90$, $\gamma_1 = 1.9927$, $\gamma_1 = 1.0144$, $\psi_{C,1}^o = 719.025$ kJ/mol., $T_o = 298.15\ K$, $R = 8.314$ J/mol.K. The only unknown quantity on the right-hand side of the above equation is the standard molar exergy of liquid water $\psi_{C,2}^o$. To estimate $\psi_{C,2}^o$ (standard molar chemical exergy of liquid water), we need to set up the formation reaction for liquid water using the reference species as:

$$H_2O(g) \rightarrow H_2O(l)$$

For this reaction, $\Delta G_{rxn}^o = -8.589$ kJ/mol. The standard molar chemical exergy of liquid water can now be calculated as shown below:

$$\psi_C^o(H_2O,l) = \Delta G_{rxn}^o + \psi_C^o(H_2O,g) = -8.589 + 9.824 = 1.235 \text{ kJ/mol.}$$

Upon substitution of the values of various quantities on the right-hand side of ψ_C expression, the molar chemical exergy of methanol-water solution turns out to be 72.411 kJ/mol.

Problem 4

Consider a practical steam power plant where steam is generated in a furnace/boiler unit by burning methane completely to $CO_2(g)$ and $H_2O(g)$ with 25% excess air. Methane and air enter the furnace at 25°C and 1 bar and the flue gases leave the furnace at 186.85 °C and 1 bar. The molar flow rate of methane is 457.85 mol./s and the mass flow rate of steam in the steam cycle is 105.932 kg/s. The figure given below shows the furnace/boiler unit of the steam power plant. The reference environment is as shown in Figure 17.2.
Calculate the following:

a) exergy supplied by the furnace
b) exergy recovered by the boiler
c) exergy destroyed in the furnace/boiler unit
d) the exergetic efficiency of the furnace/boiler unit.

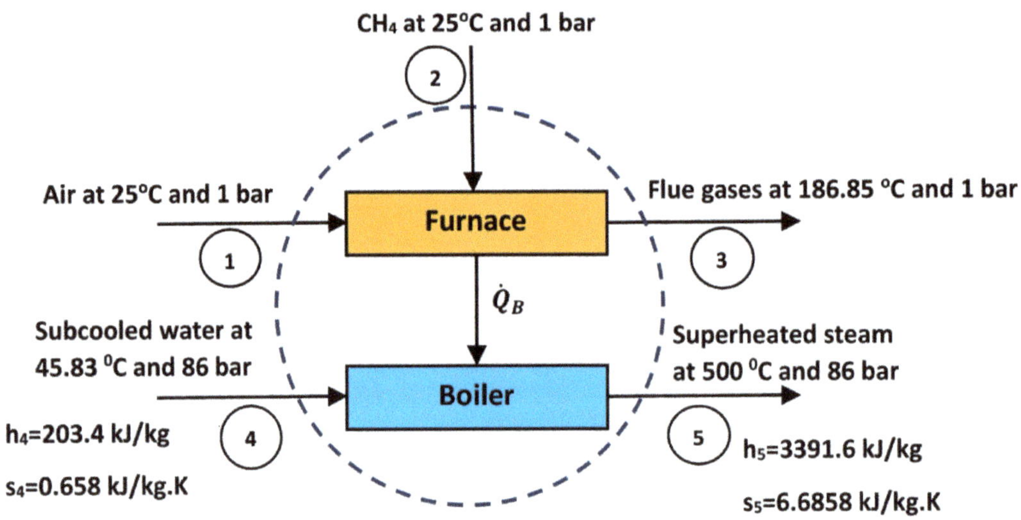

Solution

Compositions of the Furnace Streams:

The reaction occurring in the furnace is:

$$CH_4(g) + 2O_2(g) \rightarrow CO_2(g) + 2H_2O(g)$$

On the basis of 1 mol. of methane burned with 25% excess air, the compositions of the furnace streams can be calculated as follows:

Air entering the furnace: $O_2(g) = 2 \times (1.25) = 2.5$ mol.; $N_2(g) = 2.5 \times (79/21) = 9.405$ mol.; total air entering per mole of methane burnt = 11.905 mol. Thus, the composition of air is $y_{O_2} = 0.21$ and $y_{N_2} = 0.79$.

Flue gases leaving the furnace: $O_2(g) = 0.5$ mol.; $N_2(g) = 9.405$ mol.; $CO_2(g) = 1$ mol.; $H_2O(g) = 2$ mol.; total moles of flue gases per mole of methane burnt = 12.905 moles. Thus, the composition of flue gases is $y_{O_2} = 0.0387$; $y_{N_2} = 0.7288$; $y_{CO_2} = 0.0775$; $y_{H_2O} = 0.1550$.

Chemical Exergies of the Furnace Streams:

The chemical exergies of the furnace streams are determined on the basis of one mole of methane burnt in the furnace.

The chemical exergy of *methane* entering the furnace is calculated as follows:

$$\psi_{C,CH_4} = \psi^o_{C,CH_4} + RT_o ln\varphi_{CH_4} = 832.67 + (8.314/1000)(298.15)\ln(0.9983) = 832.66 \text{ kJ/mol.}$$

The fugacity coefficient of methane gas is calculated using Equation (17.36).

The chemical exergy of *air* is calculated as follows:

$$\psi_{C,Air} = y_{N_2}\psi^o_{C,N_2} + y_{O_2}\psi^o_{C,O_2} + RT_o\left[y_{N_2} ln\left(y_{N_2}\hat{\varphi}_{N_2}\right) + y_{O_2} ln\left(y_{O_2}\hat{\varphi}_{O_2}\right)\right]$$

where $y_{N_2} = 0.79$, $y_{O_2} = 0.21$, $\psi^o_{C,N_2} = 0.664$ kJ/mol., $\psi^o_{C,O_2} = 3.914$ kJ/mol., $\hat{\varphi}_{N_2} = 0.9998$, $\hat{\varphi}_{O_2} = 0.9994$, $T_o = 298.15\ K$, $R = 8.314$ J/mol.K. Thus, $\psi_{C,Air}$ =0.07179 kJ/mol.

The fugacity coefficients of a binary gas mixture were calculated using Equations (17.49) - (17.50). As the number of moles of air entering per mole of methane burnt in the furnace is 11.905, $\psi_{C,Air} = 11.905 \times (0.07179) = 0.85466$ kJ/mol.

The chemical exergy of the *flue gas* consisting of four components is calculated as follows:

$$\psi_{C,flue\ gas} = y_1\psi^o_{C,1} + y_2\psi^o_{C,2} + y_3\psi^o_{C,3} + y_4\psi^o_{C,4} + RT_o[y_1 ln(y_1\hat{\varphi}_1) + y_2 ln(y_2\hat{\varphi}_2) + y_3 ln(y_3\hat{\varphi}_3) + y_4 ln(y_4\hat{\varphi}_4)]$$

The fugacity coefficients of a quaternary gas mixture were calculated using Equation (17.40). They are nearly unity indicating that the flue gas is nearly ideal. Upon substitution of the quantities given in the table below into the above equation, the following value of the chemical exergy of flue gas (per mole of flue gas) is obtained: $\psi_{C,flue\ gas} = 1.625$ kJ/mol. As the number of moles of flue gas leaving the furnace per mole of methane burnt is 12.905, $\psi_{C,flue\ gas} = (12.905) \times 1.625 = 20.9706$ kJ/mol.

Species	Composition, y_i	Standard molar chemical exergy $\psi^o_{C,i}$, kJ/mol	Fugacity coefficient $\hat{\varphi}_i$
1= $O_2(g)$	0.0387	3.914	1.0002
2= $N_2(g)$	0.7288	0.664	1.0004
3= $CO_2(g)$	0.0775	20.11	0.9992
4= $H_2O(g)$	0.1550	9.824	0.9971

Thermo-Mechanical Exergy of the Flue Gases:

The thermo-mechanical exergy associated with the air and methane streams entering the furnace is zero. However, the flue gases leaving the furnace are at a temperature of 186.85 °C. Thus, they possess thermo-mechanical energy in addition to chemical exergy by virtue of having a temperature much higher than the reference environment temperature of 25 °C. The thermo-mechanical energy can be expressed as:

$$\psi_{TM,flue\ gas} = (h - h_o) - T_o(s - s_o)$$

As the flue gas is nearly ideal,

$$\psi_{TM,flue\ gas} = c_{pm}(T - T_o) - T_o c_{pm} ln\left(\frac{T}{T_o}\right) = 31.1 \times (460 - 298.15) - 298.15 \times (31.1) \times ln\left(\frac{460}{298.15}\right) = 1012.72 \text{ J/mol.}$$

where c_{pm}, the mean heat capacity of flue gas, is 31.1 J/mol.K. As the number of moles of flue gases is 12.905 per mole of methane burnt, the total thermo-mechanical exergy of flue gases is:
$\psi_{TM,flue\ gas} = 12.905 \times (1.0127) = 13.07$ kJ/mol.

Solution to the Problem:

a) The exergy supplied by the furnace per mole of methane burnt is as follows: $\psi_{supplied\ by furnace} = \psi_{C,methane} + \psi_{C,Air} - (\psi_{C,flue\ gas} + \psi_{TM,flue\ gas}) = 832.66 + 0.85466 - (20.9706 + 13.07) = 799.474$ kJ/mol.

This is the exergy supplied by the furnace per mole of methane burnt. The molar flow rate of methane is 457.85 mol./s. Thus, the total rate of exergy supplied in the furnace is:

$$\dot{\Psi}_{supplied\ by\ furnace} = 457.85 \times (799.474) = 366{,}039.2 \text{ kW.}$$

b) The exergy recovered in the boiler is equal to the increase in exergy of the boiler fluid stream (see figure given in the problem statement) given as:

$$\dot{\Psi}_{recovered\ in\ boiler} = \dot{m}[\psi_5 - \psi_4] = \dot{m}[(h_5 - h_4) - T_o(s_5 - s_4)] = (105.932) \times [(3391.6 - 203.4) - (298.15) \times (6.6858 - 0.658)] = 147{,}352.62 \text{ kW.}$$

c) The exergy destroyed in the furnace/boiler unit is as follows:

$$\dot{\Psi}_D = 366039.2 - 147352.62 = 218686.58 \text{ kW}$$

d) The exergetic efficiency of the furnace/boiler unit is the ratio of exergy recovered in the boiler to exergy supplied in the furnace:

$$\varepsilon = \frac{\dot{\Psi}_{recovered\ in\ boiler}}{\dot{\Psi}_{supplied\ by\ furnace}} = \frac{147352.62}{366039.2} = 0.4026$$

PRATICE PROBLEMS

Problem 1

Circle the correct answer

i) Exergy is also referred to as "Availability". True / False

ii) Exergy is a thermodynamic property of the system alone. True / False

iii) Exergy is the maximum possible useful work that can be extracted from a system till it reaches the state of thermodynamic equilibrium with its environment. True / False

iv) For a system to be in thermodynamic equilibrium with its environment, the only requirements are that the temperatures and pressures of the system and its environment are the same. True / False

v) The "restricted dead state" of a system is one where the system is in mechanical and thermal equilibrium with its environment. True / False

vi) A system is said to be in "complete dead state" when it is in thermodynamic equilibrium with its environment. True / False

vii) The chemical compositions of a system in its current state and in the restricted dead state are the same. True / False

viii) The thermo-mechanical exergy of a system is defined as the maximum possible useful work that can be extracted from the system till it reaches the restricted dead state. True/False

ix) The chemical exergy of a system is defined as the maximum possible useful work that can be extracted from the system when it goes from the "restricted" dead state to the "complete" dead state. True / False

x) As exergy is a property of the system relative to its reference environment, it is essential to define the "reference environment" in order to do any exergy calculations. True/False

xi) The fugacity of any species in an ideal liquid solution ($\hat{f}_i^{id}$) is given by the following Lewis-Randall rule: $\hat{f}_i^{id} = x_i f_i^L$ where x_i is the mol. fraction of species i in the solution and f_i^L is the fugacity of pure liquid species i. True/False

xii) The fugacity of any species i in an ideal gas mixture ($\hat{f}_i^{ig}$) is equal to the partial pressure of species i: $\hat{f}_i^{ig} = y_i P$ where y_i is the mol. fraction of species i in the gas mixture and P is the total pressure. True/False

xiii) The activity coefficient of any species in a non-ideal liquid solution is defined as: $\gamma_i = \hat{f}_i/\hat{f}_i^{id} = \hat{f}_i/x_i f_i^L$ where $\hat{f}_i$ is the fugacity of species i in the actual solution and $\hat{f}_i^{id}$ is the fugacity of species i in the solution if the solution was ideal. True/False

xiv) The fugacity coefficient of any species in a non-ideal gas mixture is defined as: $\hat{\varphi}_i = \hat{f}_i/\hat{f}_i^{ig} = \hat{f}_i/y_i P$ where $\hat{f}_i$ is the fugacity of species i in the actual gas mixture and $\hat{f}_i^{ig}$ is the fugacity of species i in the gas mixture if the gas mixture was ideal. True/False

xv) The liquid solutions are generally ideal in nature. True/False

Problem 2

Determine the standard molar chemical exergy of liquid octane $C_8H_{18}(l)$ at $T_o = 298.15$ K with respect to the reference environment specified in Figure 17.2.

Problem 3

(a) Calculate the molar chemical exergy of an equimolar *non-ideal* gas mixture of species 1 and 2 at 298.15 K and 1 bar. The virial coefficients are given as follows:
$B_{11} = -1500\ cm^3/mol$; $B_{22} = -2500\ cm^3/mol$; $B_{12} = -1900\ cm^3/mol$ and the parameter $\delta_{12} = 2B_{12} - B_{11} - B_{22} = 200\ cm^3/mol.$ The standard molar chemical exergies of species 1 and 2 are given as: $\psi_{C,1}^0 - 0.5\ kJ/mol$; $\psi_{C,1}^0 = 6.5\ kJ/mol.$

(b) What would be the molar chemical exergy of the gas mixture, if the mixture is *ideal*?

Problem 4

Determine the molar chemical exergy of binary *non-ideal* liquid solutions of 2-propanol (species 1) and water (species 2) at $T_o = 298.15$ K and 1 bar with the following compositions: (a) $x_1 = 0.05$; (b) $x_1 = 0.5$; and (c) $x_1 = 0.95$. Use Wilson model to estimate the activity coefficients. Given that: $a_{12} = 437.98$ cal/mol., $a_{21} = 1238$ cal/mol., $v_1 = 76.92$ cm^3/mol. and $v_2 = 18.07$ cm^3/mol. The reference environment is specified in Figure 17.2.

Chapter 18

Comparison of Entropic and Exergetic Methods of Quantification of Loss of Power due to Irreversibilities in Real Processes

18.1 LEARNING OUTCOMES

Upon completion of this chapter, students will be able to:

- Carry out entropy analysis of real processes involving chemical effects
- Quantify irreversibilities in real processes using entropy analysis
- Carry out exergy analysis of real processes involving chemical effects
- Quantify irreversibilities in real processes using exergy analysis
- Compare the entropic and exergetic methods of quantification of irreversiblilies in real processes.

18.2 INTRODUCTION

All real processes are irreversible in nature. As irreversibility in a process is related to the loss of potential power or work that could be extracted from the process, it is important to quantify and minimize irreversibilities in order to improve the efficiency of a process from thermodynamics perspective. The irreversibilities in a process can be quantified using two different approaches: entropy generation analysis and exergy destruction analysis. According to one form of the Gouy-Stodola theorem, the power loss in a process due to irreversibilities is directly proportional to rate of total entropy generation, that is:

$$\dot{I} = T_o \dot{S}_{G,total} \tag{18.1}$$

where $\dot{I}$ is the power loss due to irreversibilities in the process, T_o is the absolute temperature of the environment, and $\dot{S}_{G,total}$ is the rate of total entropy generation in the system and its surroundings (environment). According to another form of the Gouy-Stodola theorem, the power loss in a process due to irreversibilities is equal to the rate of total exergy destruction, that is:

$$\dot{I} = \dot{\Psi}_{D,total} \tag{18.2}$$

where $\dot{\Psi}_{D,total}$ is the rate of total (internal and external) exergy destruction in the process.

Thus, irreversibilities in a process can be quantified either by the rate of total entropy generation or by the rate of total exergy destruction. The entropic method is preferred when the process involves only thermo-mechanical effects whereas the exergetic method is preferred when chemical effects involving compositional changes due to chemical reactions, mixing of different chemical components, etc. are present. Although the calculation procedures involved in exergy analysis are quite different from those involved in entropy analysis, especially when chemical effects are present, the final results regarding the quantification of irreversibilities are expected to be the same, at least from a theoretical point of view.

This chapter illustrates quantification of irreversibilities in real processes using both entropic and exergetic methods. The results obtained from the two methods for the same processes are compared to determine if there are any discrepancies between the two approaches. The necessary theoretical background covering entropy analysis and exergy analysis is first presented for the benefit of the students.

18.3 BACKGROUND

18.3.1Entropy Analysis

Consider a flow process shown schematically in Figure 18.1. The process is producing shaft work and rejecting heat to the environment.

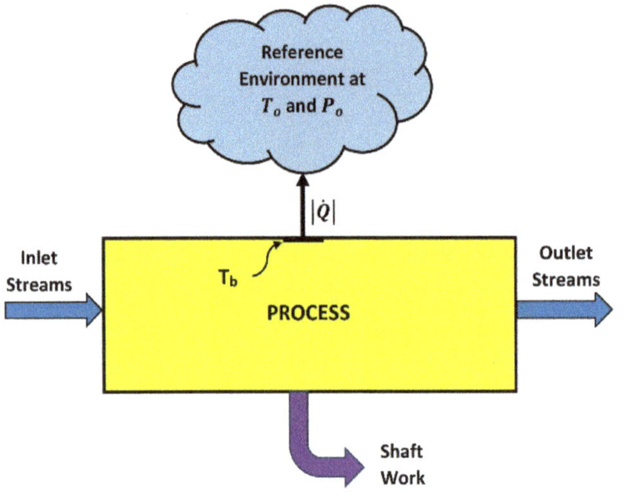

Figure 18.1 Schematic of a flow process producing shaft work and rejecting heat to the environment.

The entropy balance on the control volume could be expressed as:

$$(Rate\ of\ entropy\ in) - (Rate\ of\ entropy\ out) + (Rate\ of\ entropy\ generation) = (Rate\ of\ entropy\ accumulation) \tag{18.3}$$

For the process shown in Figure 18.1:

$$(Rate\ of\ entropy\ in) = \sum_{Inlet\ streams} \dot{m}_i\, s_i \tag{18.4}$$

$$(Rate\ of\ entropy\ out) = \sum_{Outlet\ streams} \dot{m}_i\, s_i + \frac{|\dot{Q}|}{T_b} \tag{18.5}$$

$$(Rate\ of\ entropy\ generation) = \dot{S}_{G,CV} \tag{18.6}$$

$$(Rate\ of\ entropy\ accumulation) = \frac{\partial}{\partial t}(ms)_{CV} \tag{18.7}$$

where $\dot{m}_i$ is the mass flow rate of flowing stream i, s_i is the specific entropy of flowing stream i, T_b is the absolute temperature at the boundary of the control volume, and $\dot{Q}$ is the rate of heat transfer ($\dot{Q}$ is considered positive if heat flows into the control volume). Thus, the entropy balance on the control volume (CV) is as follows:

$$\sum_{Inlet\ streams} \dot{m}_i\, s_i - \left(\sum_{Outlet\ streams} \dot{m}_i\, s_i + \frac{|\dot{Q}|}{T_b}\right) + \dot{S}_{G,CV} = \frac{\partial}{\partial t}(ms)_{CV} \tag{18.8}$$

Upon rearrangement, Equation (18.8) gives:

$$\dot{S}_{G,CV} = \frac{\partial}{\partial t}(ms)_{CV} + (\sum_{Outlet\ streams} \dot{m}_i\, s_i - \sum_{Inlet\ streams} \dot{m}_i\, s_i) + \frac{|\dot{Q}|}{T_b} \tag{18.9}$$

Assuming steady state, Equation (18.9) reduces to:

$$\dot{S}_{G,CV} = (\sum_{Outlet\ streams} \dot{m}_i\, s_i - \sum_{Inlet\ streams} \dot{m}_i\, s_i) + \frac{|\dot{Q}|}{T_b} \tag{18.10}$$

At steady state, the mass balance gives

$$\sum_{Outlet\ streams} \dot{m}_i = \sum_{Inlet\ streams} \dot{m}_i \tag{18.11}$$

Entropy balance on the surroundings can be carried out in a similar manner. It leads to the following result for the rate of entropy generation in the surroundings:

$$\dot{S}_{G,surr} = \frac{|\dot{Q}|}{T_o}\left(1 - \frac{T_o}{T_b}\right) \tag{18.12}$$

Combining Equations (18.10) and (18.12) gives:

$$\dot{S}_{G,total} = \dot{S}_{G,CV} + \dot{S}_{G,surr} = (\sum_{Outlet\ streams} \dot{m}_i\, s_i - \sum_{Inlet\ streams} \dot{m}_i\, s_i) + \frac{|\dot{Q}|}{T_o} \tag{18.13}$$

Using the entropic form of the Gouy-Stodola theorem, Equation (18.1), the power loss due to irreversibilities in a process is as follows:

$$\dot{I} = T_o\dot{S}_{G,total} = T_o(\sum_{Outlet\ streams} \dot{m}_i\, s_i - \sum_{Inlet\ streams} \dot{m}_i\, s_i) + |\dot{Q}| \tag{18.14}$$

From the first law of thermodynamics for open systems, neglecting kinetic and potential energy changes, it can be readily shown that:

$$|\dot{Q}| = (\sum_{Inlet\ streams} \dot{m}_i\, h_i - \sum_{Outlet\ streams} \dot{m}_i\, h_i) - \dot{W}_{sh} \tag{18.15}$$

where h_i is specific enthalpy of fluid stream i and $\dot{W}_{sh}$ is the rate of shaft work ($\dot{W}_{sh}$ is considered positive if work flows out of the control volume). Equations (18.14) - (18.15) give the following expression for the power loss due to irreversibilities in the process:

$$\dot{I} = T_o(\sum_{Outlet\ streams} \dot{m}_i\, s_i - \sum_{Inlet\ streams} \dot{m}_i\, s_i) - (\sum_{Outlet\ streams} \dot{m}_i\, h_i - \sum_{Inlet\ streams} \dot{m}_i\, h_i) - \dot{W}_{sh} \tag{18.16}$$

With one inlet and one outlet fluid streams and no shaft work, Equation (18.16) reduces to:

$$\dot{I} = \dot{m}(T_o\Delta s - \Delta h) \tag{18.17}$$

18.3.2 Exergy Analysis

Concept of Exergy

Exergy is the maximum amount of useful work that can be obtained when the system under consideration is taken from its current state to the *dead* state where the system attains complete thermodynamic equilibrium with the environment. In complete thermodynamic equilibrium, the conditions of mechanical, thermal, and chemical equilibrium between the system and the environment are satisfied. The total exergy of a system consists of two parts, namely thermo-mechanical exergy (ψ_{TM}) and chemical exergy (ψ_C):

$$\psi = \psi_{TM} + \psi_C \tag{18.18}$$

The thermo-mechanical exergy ψ_{TM} of a system is defined as the maximum amount of useful work that can be obtained when the system under consideration is taken from its current state to the *restricted dead state* while keeping the chemical composition of the system constant. In the restricted dead state, the system is in mechanical and thermal equilibrium with the environment. For a fluid stream, the thermo-mechanical exergy ψ_{TM} is given as:

$$\psi_{TM} = (h - h_o) - T_o(s - s_o) + \frac{1}{2}V^2 + gZ \tag{18.19}$$

where $h, s, V, g,$ and Z are the specific enthalpy, specific entropy, fluid velocity, acceleration due to gravity, and elevation, respectively. The quantities with "*o*" subscript refer to restricted dead state values where the temperature and pressure of the system are the same as that of the environment.

The chemical exergy ψ_C of a system is defined as the maximum amount of useful work obtainable when the system under consideration is taken from its *restricted dead state* to the *complete dead state* where the system is in complete thermodynamic equilibrium (mechanical, thermal, and chemical) with the environment.

Exergy is a combined property of the system and the environment. Therefore, in exergy calculations, it is necessary to specify the environment as exergy is relative to the assumed reference environment. For the purposes of this chapter, the reference environment is specified in Figure 18.2. It is a realistic model of the atmosphere as a thermal and pressure reservoir at 25 °C and 1 bar with the chemical composition as indicated in the figure.

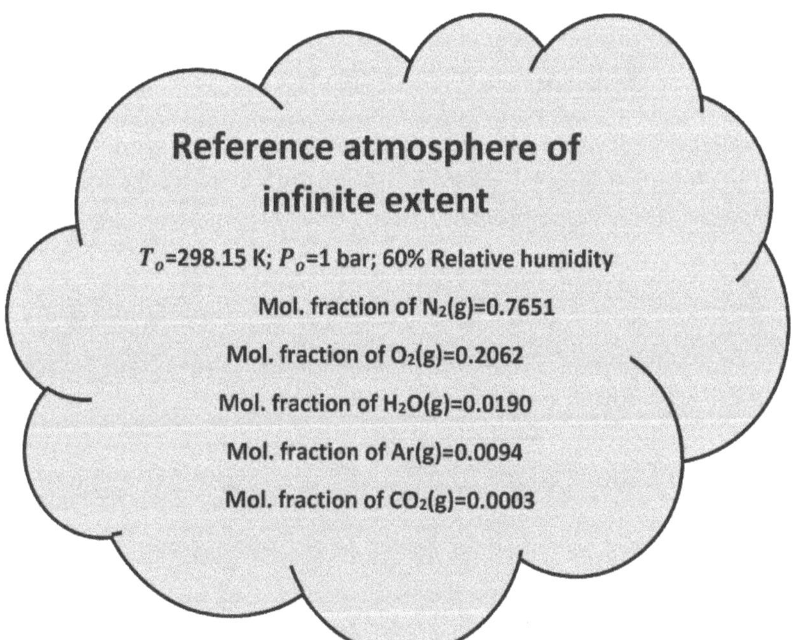

Figure 18.2 Reference environment for exergy calculations.

Exergy Balance

Consider the flow process of Figure 18.1. The exergy balance on the control volume could be expressed as:

$$(Rate\ of\ exergy\ in) - (Rate\ of\ exergy\ out) - (Rate\ of\ exergy\ destruction) = (Rate\ of\ exergy\ accumulation) \quad (18.20)$$

For the process shown in Figure 18.1:

$$(Rate\ of\ exergy\ in) = \sum_{Inlet\ streams} \dot{m}_i\, \psi_i \quad (18.21)$$

$$(Rate\ of\ exergy\ out) = \sum_{Outlet\ streams} \dot{m}_i\, \psi_i + |\dot{Q}|\left(1 - \frac{T_o}{T_b}\right) + \dot{W}_{sh} \quad (18.22)$$

$$(Rate\ of\ exergy\ destruction) = \dot{\Psi}_{D,CV} \quad (18.23)$$

$$(Rate\ of\ exergy\ accumulation) = \frac{\partial}{\partial t}(m\psi)_{CV} \quad (18.24)$$

Thus, the exergy balance on the control volume (CV) is as follows:

$$\sum_{Inlet\ streams} \dot{m}_i\, \psi_i - \left(\sum_{Outlet\ streams} \dot{m}_i\, \psi_i + |\dot{Q}|\left(1 - \frac{T_o}{T_b}\right) + \dot{W}_{sh}\right) - \dot{\Psi}_{D,CV} = \frac{\partial}{\partial t}(m\psi)_{CV} \quad (18.25)$$

Upon rearrangement, Equation (18.25) gives:

$$\dot{\Psi}_{D,CV} = \sum_{Inlet\ streams} \dot{m}_i\, \psi_i - \left(\sum_{Outlet\ streams} \dot{m}_i\, \psi_i + |\dot{Q}|\left(1 - \frac{T_o}{T_b}\right) + \dot{W}_{sh}\right) - \frac{\partial}{\partial t}(m\psi)_{CV} \quad (18.26)$$

Assuming steady state:

$$\dot{\Psi}_{D,CV} = \sum_{Inlet\ streams} \dot{m}_i\, \psi_i - \left(\sum_{Outlet\ streams} \dot{m}_i\, \psi_i + |\dot{Q}|\left(1 - \frac{T_o}{T_b}\right) + \dot{W}_{sh}\right) \quad (18.27)$$

Exergy balance on the surroundings gives:

$$\dot{\Psi}_{D,surr} = |\dot{Q}|\left(1 - \frac{T_o}{T_b}\right) \quad (18.28)$$

Combining Equations (18.27) and (18.28) gives:

$$\dot{\Psi}_{D,total} = \dot{\Psi}_{D,CV} + \dot{\Psi}_{D,surr} = \sum_{Inlet\ streams} \dot{m}_i\, \psi_i - \sum_{Outlet\ streams} \dot{m}_i\, \psi_i - \dot{W}_{sh} \quad (18.29)$$

Using the exergy form of the Gouy-Stodola theorem, Equation (18.2), the power loss due to irreversibilities in a process is as follows:

$$\dot{I} = \dot{\Psi}_{D,total} = \sum_{Inlet\ streams} \dot{m}_i\, \psi_i - \sum_{Outlet\ streams} \dot{m}_i\, \psi_i - \dot{W}_{sh} \quad (18.30)$$

In the special case where no shaft work is involved and there is one inlet and one outlet, Equation (18.30) gives the following expression for the power loss due to irreversibilities in a process:

$$\dot{I} = \dot{m}(-\Delta\psi) \tag{18.31}$$

In this special case the decrease in exergy of the fluid stream is entirely due to irreversibilities in the process.

Exergy of Materials

As noted earlier, the exergy of materials consists of two parts - thermo-mechanical exergy (ψ_{TM}) and chemical exergy (ψ_C). The total exergy of a material is the sum of these as indicated in Equation (18.18). Neglecting kinetic and potential energy effects, the thermo-mechanical exergy of a material at any given T and P is:

$$\psi_{TM}(P,T) = \Delta h - T_o\Delta s \tag{18.32}$$

where "Δ" refers to a change in state from "restricted dead state" to "current state". For an ideal gas, Equation (18.32) takes the form:

$$\psi_{TM}(P,T) = \bar{c}_{p,h}(T - T_o) - T_o\left[\bar{c}_{p,s}ln\left(\frac{T}{T_o}\right) - Rln\left(\frac{P}{P_o}\right)\right] \tag{18.33}$$

where $\bar{c}_{p,h}$ and $\bar{c}_{p,s}$ are the mean heat capacities of the material (ideal gas) over the given temperature range, and R is the universal gas constant. Note that the mean heat capacities for enthalpy calculation ($\bar{c}_{p,h}$) and entropy calculation ($\bar{c}_{p,s}$) are generally different, as they are defined differently. The temperature dependence of the heat capacity of materials is generally expressed as:

$$\frac{c_p}{R} = A + BT + CT^2 + DT^{-2} \tag{18.34}$$

where A, B, C, D are material constants and T is in K. The mean heat capacities for enthalpy and entropy calculations are defined and evaluated as follows:

$$\frac{\bar{c}_{p,h}}{R} = \frac{\int_{T_1}^{T_2}(c_p/R)dT}{T_2-T_1} = A + BT_{am} + \frac{C}{3}[4T_{am}^2 - T_1T_2] + \frac{D}{T_1T_2} \tag{18.35}$$

$$\frac{\bar{c}_{p,s}}{R} = \frac{\int_{T_1}^{T_2}(c_p/R)dT}{ln(T_2/T_1)} = A + BT_{lm} + CT_{am}T_{lm} + \frac{DT_{am}T_{lm}}{(T_1T_2)^2} \tag{18.36}$$

where

$$T_{am} = \frac{T_1+T_2}{2}; \qquad T_{lm} = \frac{T_2-T_1}{ln(T_2/T_1)} \tag{18.37}$$

Exergy calculations also require knowledge of the chemical exergies of the materials involved in the process. When chemical effects are present, it is the chemical exergy which is generally the main portion of the exergy. For pure single-component substances, chemical exergy is the same as standard chemical exergy where the standard state is defined as pure species (same phase, gas or liquid, as the actual material) at dead state conditions of temperature (T_o) and pressure (P_o). For pure ideal gas, the standard molar chemical exergy is given by the expression:

$$\psi_{C,i}^o = -RT_oln(y_{i,e}) \tag{18.38}$$

where $y_{i,e}$ is the mole fraction of species i in the reference environment. Table 18.1 gives the standard molar chemical exergies of various species present in the reference environment calculated from Equation (18.38).

Table 18.1 Standard molar chemical exergies of reference species

Species	**Mol. Fraction ($y_{i,e}$)**	**Standard molar chemical exergy ($\psi^o_{C,i}$), kJ/mol.**
$N_2(g)$	0.7651	0.664
$O_2(g)$	0.2062	3.914
$H_2O(g)$	0.0190	9.824
$Ar(g)$	0.0094	11.57
$CO_2(g)$	0.0003	20.11

If the chemical species in not present in the reference environment, its standard molar chemical exergy could still be determined provided that a formation reaction could be set up for the concerned species using the reference species present in the reference environment. This was illustrated in Chapter 17 with the help of several examples.

For mixtures of gases, the molar chemical exergy of a mixture can be calculated from the following relation assuming that the gas mixture behaves as an ideal gas mixture:

$$\psi^{ig}_{C,mix} = \Sigma y_i \psi^o_{C,i} + RT_o \Sigma y_i \, ln y_i \tag{18.39}$$

where y_i is the mole fraction of species i in gas mixture. Likewise, the molar chemical exergy of an ideal liquid solution is given as:

$$\psi^{id}_C = \Sigma x_i \, \psi^o_{C,i} + RT_o \Sigma x_i \, ln(x_i) \tag{18.40}$$

where x_i is the mole fraction of species i in liquid solution.

18.4 APPLICATION: De-Mixing of Gases

An equimolar mixture of methane and ethane is fed into a de-mixing device at a temperature of 150 °C and pressure of 5 bar, as shown in Figure 18.3. The product streams from the device are pure methane and ethane streams at 35 °C and 1 bar. The process is in steady state and the reference environment is as shown in Figure 18.2. Calculate the loss of power using both entropic and exergetic methods and compare the results.

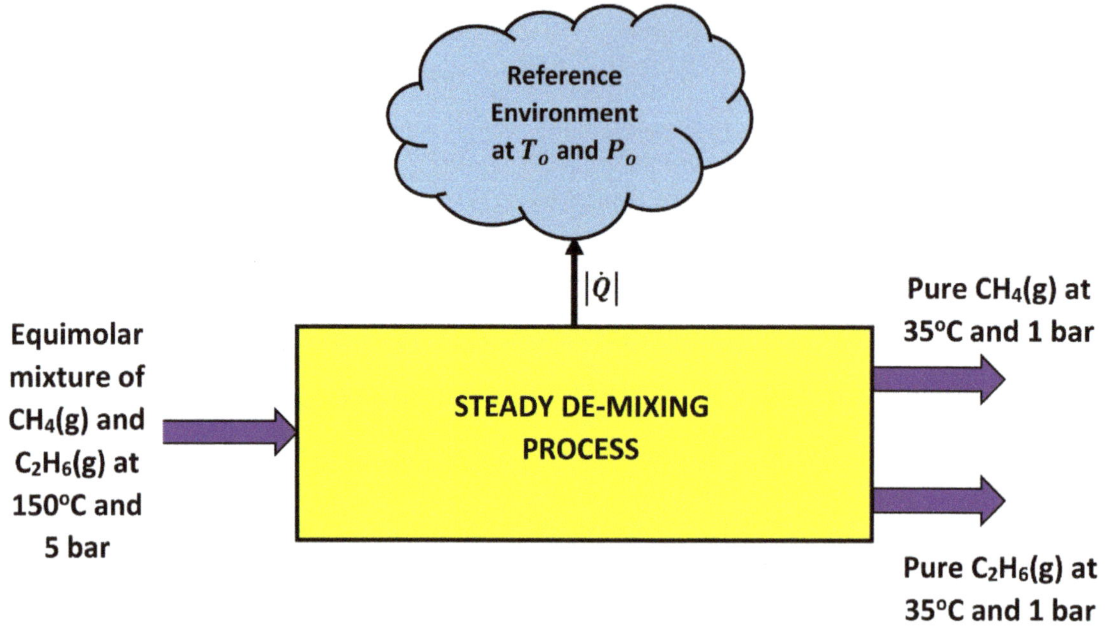

Figure 18.3 Steady state de-mixing of a binary mixture of gases.

Entropy Analysis

As a basis, we take 1 mole of equimolar mixture of methane and ethane entering the de-mixing process. Before we go into entropy analysis, it is important to recognize that work is required to de-mix a mixture of gases provided that P and T are kept constant. Figure 18.4 shows a scheme to estimate the minimum work required to de-mix a mixture of two gases (A and B) at the same P and T. The semi-permeable membrane allows the passage of only gas A and at equilibrium, the pressure on the pure A side of the membrane is P_A, the partial pressure of A in the mixture on the left side of the membrane. The pressure of pure gas A is then raised by a compressor from P_A to P while keeping the temperature constant. The minimum shaft work required by the compressor is related to Gibbs free energy change as $W_{sh,min} = -\Delta g$.

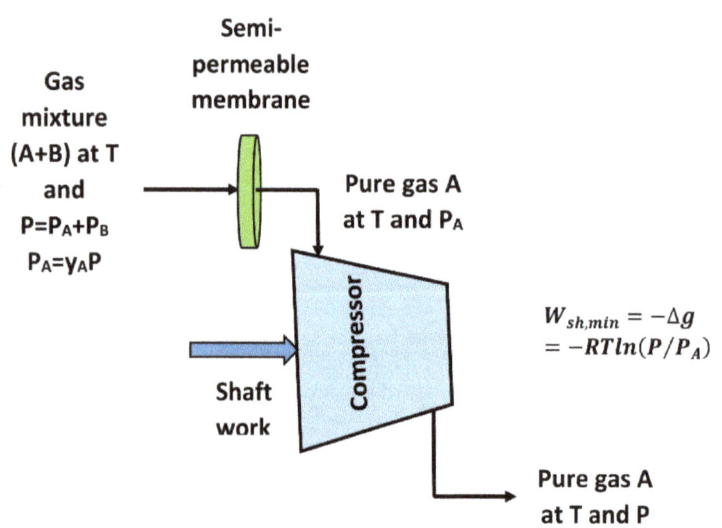

Figure 18.4 De-mixing of a gas mixture into pure components at the same P and T requires work.

In order to determine the loss of power in the de-mixing process shown in Figure 18.3, we need to calculate entropy and enthalpy changes (see Equation (18.17)). This requires a stepwise calculation scheme. The scheme to calculate entropy and enthalpy changes in the process is shown in Figure 18.5.

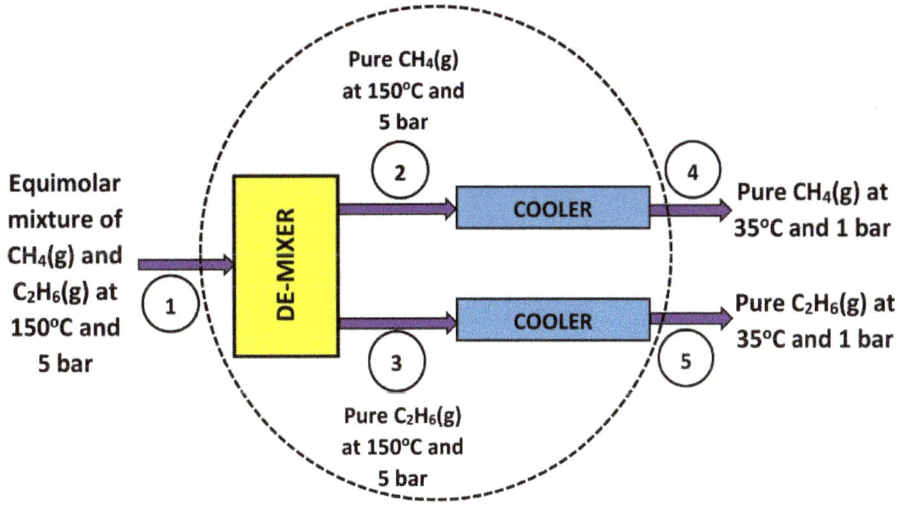

Figure 18.5 Stepwise calculation scheme to calculate entropy and enthalpy changes for the de-mixing process shown in Figure 18.3.

On the basis of 1 mole of gas mixture entering the process, the net entropy change of the flowing streams is as follows:

$$\Delta s = \Delta s_{De-mixing} + 0.5\Delta s_{2-4} + 0.5\Delta s_{3-5} \tag{18.41}$$

Each term on the right hand side of Equation (18.41) is calculated below:

$$\Delta s_{De-mixing} = R\sum y_i \ln y_i \tag{18.42}$$

$$\Delta s_{De-mixing} = 8.314[0.5 \times \ln(0.5) + 0.5 \times \ln(0.5)] = -5.7628 \text{ J/K} \tag{18.43}$$

$$\Delta s_{2-4} = \bar{c}_{p,s} ln\left(\frac{T_4}{T_2}\right) - Rln\left(\frac{P_4}{P_2}\right) \tag{18.44}$$

$$\Delta s_{3-5} = \bar{c}_{p,s} ln\left(\frac{T_5}{T_3}\right) - Rln\left(\frac{P_5}{P_3}\right) \tag{18.45}$$

where $\bar{c}_{p,s}$ is calculated from Equation (18.36) using the following coefficients for methane (Equation (18.46)) and ethane (Equation (18.47)):

methane: $$A = 1.702; B = 9.081 \times 10^{-3} K^{-1}; C = -2.164 \times 10^{-6} K^{-2}; D = 0 \tag{18.46}$$

ethane: $$A = 1.131; B = 19.225 \times 10^{-3} K^{-1}; C = -5.561 \times 10^{-6} K^{-2}; D = 0 \tag{18.47}$$

$$\Delta s_{2-4} = 39.1422 ln\left(\frac{35+273.15}{150+273.15}\right) - 8.314 ln\left(\frac{1}{5}\right) = 0.9673 \text{ J/K} \tag{18.48}$$

$$\Delta s_{3-5} = 61.2322 \times ln\left(\frac{35+273.15}{150+273.15}\right) - 8.314 \times ln\left(\frac{1}{5}\right) = -6.0383 \text{ J/K} \tag{18.49}$$

Thus, Equation (18.41) gives the following value of the net entropy change of the flowing streams:

$$\Delta s = -5.7628 + 0.5 \times (0.9673) + 0.5 \times (-6.0383) = -8.2984 \text{ J/K} \quad (18.50)$$

Using the same basis of 1 mole of gas mixture entering the process, the net enthalpy change of the process flow streams can be calculated as follows:

$$\Delta h = \Delta h_{De-mixing} + 0.5\Delta h_{2-4} + 0.5\Delta h_{3-5} \quad (18.51)$$

Each term on the right hand side of Equation (18.51) is calculated below:

$$\Delta h_{De-mixing} = 0 \quad (18.52)$$

$$\Delta h_{2-4} = \bar{c}_{p,h}(T_4 - T_2) = 39.3315 \times (35 - 150) = -4523.122 \text{ J} \quad (18.53)$$

$$\Delta h_{3-5} = \bar{c}_{p,h}(T_5 - T_3) = 61.6149 \times (35 - 150) = -7085.713 \text{ J} \quad (18.54)$$

where $\bar{c}_{p,h}$ is calculated from Equation (18.35) using the coefficients for methane (Equation (18.46)) and ethane (Equation (18.47)). Thus, Equation (18.51) gives the following value of the net enthalpy change of the process streams:

$$\Delta h = 0.5 \times (-4523.122) + 0.5 \times (-7085.713) = -5804.42 \text{ J} \quad (18.55)$$

From Equation (18.17), we find the loss of power due to irreversibilities in this de-mixing process, per mole of the entering gas mixture, to be:

$$I = 298.15 \times (-8.2984) - (-5804.42) = 3330.28 \text{ J} = 3.330 \text{ kJ} \quad (18.56)$$

Exergy Analysis

In order to carry out exergy analysis of a process, it is necessary to obtain the values of the standard molar chemical exergies of the substances involved. In the de-mixing process under consideration the species involved are methane and ethane gases. As these species are not part of the reference environment (see Figure 18.2), we need to set up the formation reactions for these species using the reference species. Thus:

$$CO_2(g) + 2H_2O(g) \rightarrow CH_4(g) + 2O_2(g) \quad (18.57)$$

$$2CO_2(g) + 3H_2O(g) \rightarrow C_2H_6(g) + (7/2)O_2(g) \quad (18.58)$$

The following relationships are then used to determine the standard molar chemical exergy of the two species:

$$\psi^o_{C,methane} = \Delta G^o_{rxn} - \sum_{(other\ than\ methane)} v_i \psi^o_{C,i} \quad (18.59)$$

$$\psi^o_{C,ethane} = \Delta G^o_{rxn} - \sum_{(other\ than\ ethane)} v_i \psi^o_{C,i} \quad (18.60)$$

Equation (18.59) refers to the formation reaction of methane gas (Equation (18.57)) and Equation (18.60) refers to the formation reaction of ethane gas (Equation (18.58)). The standard molar chemical exergies of all species, other than methane and ethane, are known (see Table 18.1). The standard Gibbs free energy of the formation reactions ΔG^o_{rxn} are given below for methane and ethane, respectively:

$$\Delta G^o_{rxn,methane} = \Delta G^o_f(CH_4, g) + 2\Delta G^o_f(O_2, g) - \Delta G^o_f(CO_2, g) - 2\Delta G^o_f(H_2O, g) \quad (18.61)$$

$$\Delta G^o_{rxn} = -50.8 + 2(0) - (-394.36) - 2(-228.589) = 800.738 \text{ kJ/mol} \tag{18.62}$$

$$\Delta G^o_{rxn,ethane} = \Delta G^o_f(C_2H_6, g) + (7/2)\Delta G^o_f(O_2, g) - 2\Delta G^o_f(CO_2, g) - 3\Delta G^o_f(H_2O, g) \tag{18.63}$$

$$\Delta G^o_{rxn} = -32.9 + (7/2)(0) - (-394.36) - 3(-228.589) = 1441.587 \text{ kJ/mol} \tag{18.64}$$

The standard molar chemical exergies of the species present in the reference atmosphere and involved in the formation reactions are as follows: $\psi^o_C(O_2, g) = 3.914$; $\psi^o_C(CO_2, g) = 20.11$; and $\psi^o_C(H_2O, g) = 9.824$ kJ/mol. Upon substitution of ΔG^o_{rxn} and $\psi^o_{C,i}$ values in Equations (18.59) – (18.60), the standard molar chemical exergy of methane gas is found to be 832.67 kJ/mol and that of ethane gas is 1497.58 kJ/mol.

We can now calculate the molar exergies of the inlet and outlet streams of the process. The inlet stream is an equimolar mixture of methane and ethane gases at 150 °C and 5 bar. This stream possesses both thermo-mechanical exergy (ψ_{TM}) and chemical exergy (ψ_C):

$$\psi = \psi_{TM} + \psi_C = \left[\langle\bar{c}_{p,h}\rangle(T_1 - T_o) - \langle\bar{c}_{p,s}\rangle T_o ln\left(\frac{T_1}{T_o}\right) - Rln\left(\frac{P_1}{P_o}\right)\right] + \psi_C \tag{18.65}$$

The square bracketed term on the right hand side of Equation (18.65) is the thermo-mechanical exergy of the inlet stream, $T_1 = 150^0C$ is the temperature and $P_1 = 5\ bar$ is the pressure of the stream, $T_o = 25^0C$ is the reference environment temperature and $P_o = 1\ bar$ is the reference environment pressure. The mean heat capacities for the inlet gas mixture stream can be calculated as follows:

$$\langle\bar{c}_{p,h}\rangle = \sum y_i\, \bar{c}_{p,h} \tag{18.66}$$

$$\langle\bar{c}_{p,s}\rangle = \sum y_i\, \bar{c}_{p,s} \tag{18.67}$$

where $\bar{c}_{p,h}$ is calculated from Equation (18.35) and $\bar{c}_{p,s}$ is calculated from Equation (18.36). For the inlet equimolar mixture of methane and ethane:

$$\psi_{TM} = \left[(49.9951)(150 - 25) - (49.651)(298.15)ln\left(\frac{150+273.15}{298.15}\right) - 8.314ln\left(\frac{5}{1}\right)\right] = 5055.76 \text{ J/mol} \tag{18.68}$$

The chemical exergy of the inlet gas mixture stream is calculated from Equation (18.39) as follows:

$$\psi_C = 0.5 \times 832.67 + 0.5 \times 1497.58 + (8.314 \times 298.15/1000)[0.5 \times \ln(0.5) + 0.5 \times \ln(0.5)] = 1163.4068 \text{ kJ/mol} \tag{18.69}$$

Thus, the total exergy of the inlet equimolar mixture of methane and ethane gases is:

$$\psi = \psi_{TM} + \psi_C = 5.0558 + 1163.4068 = 1168.4626 \text{ kJ/mol} \tag{18.70}$$

The molar exergies of the two outlet streams consisting of pure methane and pure ethane streams need to be calculated next.

For pure methane gas stream leaving the process at $T = 35^0C$ and $P = 1\ bar$:

$$\psi_{methane} = \psi_{TM} + \psi_C = \left[\bar{c}_{p,h}(T_1 - T_o) - \bar{c}_{p,s}T_o ln\left(\frac{T_1}{T_o}\right) - Rln\left(\frac{P}{P_o}\right)\right] + \psi^o_{C,methane} \tag{18.71}$$

$$\psi_{methane} = \left[35.3845 \times (35-25) - 35.3827 \times 298.15 \times ln\left(\frac{35+273.15}{298.15}\right)\right] + 832.67 \times 1000$$
$$= 5.822 + 832670 = 832675.8\ \text{J} = 832.6758\ \text{kJ/mol} \quad (18.72)$$

For pure ethane gas stream leaving the process at at $T = 35^0C$ and $P = 1\ bar$:

$$\psi_{ethane} = \psi_{TM} + \psi_C = \left[\bar{c}_{p,h}(T_1 - T_o) - \bar{c}_{p,s}T_o ln\left(\frac{T_1}{T_o}\right) - Rln\left(\frac{P}{P_o}\right)\right] + \psi^o_{C,ethane} \quad (18.73)$$

$$\psi_{ethane} = \left[53.6083 \times (35-25) - 53.6047 \times 298.15 \times ln\left(\frac{35+273.15}{298.15}\right)\right] + 1497.58 \times 1000$$
$$= 8.829 + 1497580 = 1497588.8\ \text{J} = 1497.589\ \text{kJ/mol} \quad (18.74)$$

Note that the thermo-mechanical exergies of the methane and ethane gas outlet streams are very small in comparison with their chemical exergies.

From Equation (18.31), we find the loss of power due to irreversibilities in this de-mixing process, per mole of the entering gas mixture, to be:

$$I = \psi_{inlet,mixture} - (0.5 \times \psi_{outlet,methane} + 0.5\psi_{outlet,ethane}) = 1168.4626 - (0.5 \times 832.6758 +$$
$$0.5 \times 1497.589) = 3.330\ \text{kJ} \quad (18.75)$$

This value is in agreement with the value obtained through entropy analysis.

18.5 CONCLUDING REMARKS

Entropic and exergetic methods of quantification of irreversibilities in real processes are reviewed. The loss of power due to irreversibilities is calculated and compared using the two methods for the same real process. The real process considered is steady de-mixing of a binary gas mixture into pure components. The loss of power due to irreversibilities obtained by the exergetic method is in agreement with the loss of power calculated by the entropic method although the calculation procedures involved are very different in the two methods. This confirms that the two methods of quantification of irreversibilities are equivalent. The exergetic method has the advantage that no stepwise calculation scheme is required to calculate the loss of power due to irreversibilities in the process. The only knowledge needed in the exergetic method is the values of the total exergy (thermo-mechanical and chemical) for various streams at the inlets and outlets of the process control volume. However, the entropic method requires the precise stepwise calculation scheme involved in going from inlet conditions to the outlet conditions of the process streams, in addition to entropy balances.

NOTATION

c_p	specific or molar constant-pressure heat capacity
$\bar{c}_{p,h}$	mean specific or molar constant-pressure heat capacity for enthalpy change calculation
$\bar{c}_{p,s}$	mean specific or molar constant-pressure heat capacity for entropy change calculation
$\langle\bar{c}_{p,h}\rangle$	mean specific or molar constant-pressure heat capacity of gas mixture for enthalpy change calculation
$\langle\bar{c}_{p,s}\rangle$	mean specific or molar constant-pressure heat capacity of gas mixture for entropy change calculation
g	acceleration due to gravity or molar Gibbs free energy
h	molar or specific enthalpy
h_i	specific enthalpy of fluid stream i

h_o	molar or specific enthalpy in dead state
I	power loss per mole
$\dot{I}$	power loss
m	mass
$\dot{m}$	mass flow rate
$\dot{m}_i$	mass flow rate of fluid stream i
P	pressure
P_A	partial pressure of species A
P_o	pressure in dead state
$\dot{Q}$	heat transfer rate
R	universal gas constant
$\dot{S}_G$	rate of entropy generation
$\dot{S}_{G,CV}$	rate of entropy generation inside the control volume
$\dot{S}_{G,surr}$	rate of entropy generation in the surroundings
$\dot{S}_{G,total}$	rate of total (internally and externally) entropy generation
s	molar or specific entropy
s_o	molar or specific entropy in dead state
s_i	specific entropy of fluid stream i
T	temperature
T_b	temperature at the boundary of the system
T_{am}	arithmetic mean temperature
T_{lm}	log-mean temperature
T_o	temperature in dead state
t	time
V	velocity
v_i	stoichiometric coefficient
$\dot{W}_{sh}$	rate of shaft work
$W_{sh,min}$	minimum shaft work
x_i	mole fraction of species i in liquid solution
y_i	mole fraction of species i in gas mixture
$y_{i,e}$	mole fraction of species i in reference environment
Z	elevation

Greek Letters

Δ	difference operator
ΔG_f^o	standard Gibbs free energy of formation of compound
ΔG_{rxn}^o	standard Gibbs free energy of reaction
$\Delta h_{De-mixing}$	enthalpy change of de-mixing
$\Delta s_{De-mixing}$	entropy change of de-mixing
$\dot{\Psi}_{D,CV}$	rate of total exergy destruction inside the control volume
$\dot{\Psi}_{D,surr}$	rate of total exergy destruction in the surroundings
$\dot{\Psi}_{D,total}$	rate of total exergy destruction internally and externally
ψ	specific or molar total (chemical plus thermo-mechanical) exergy
ψ_C	specific or molar chemical exergy
ψ_i	specific or molar chemical exergy of fluid stream i
ψ_{TM}	specific or molar thermo-mechanical exergy
$\psi_{C,i}^o$	standard specific or molar chemical exergy of species i
ψ_C^{id}	specific or molar chemical exergy of ideal solution
$\psi_{C,mix}^{ig}$	standard molar chemical exergy of species i

FURTHER READING

Pal, R., "Comparison of entropic and exergetic methods of quantification of loss of power due to irreversibilities in real processes using the Gouy-Stodola theorem", *Int. J. Mech. Eng. Edu.* 2019; 47: 73-96.

TUTORIAL PROBLEM WITH SOLUTION

Problem: Combustion of Carbon Monoxide

Consider a combustion chamber where combustion of carbon monoxide is being carried out (see figure given below). The reactants enter the chamber as pure carbon monoxide and as air containing the stoichiometric amount of oxygen. Both carbon monoxide and air enter at 25°C and 1 bar. The products of complete combustion of carbon monoxide leave the chamber at 25°C and 1 bar. The process is a steady state process and the reference environment is as shown in Figure 18.2. Calculate the loss of power using both entropic and exergetic methods and compare the results.

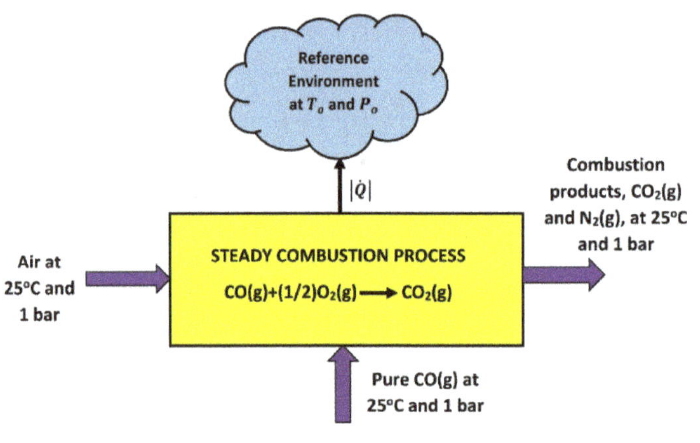

Steady state combustion of carbon monoxide

Solution

Entropy Analysis:

As a basis, we take 1 mole of pure carbon monoxide entering the combustion chamber. For 1 mole of pure CO(g) entering the combustion chamber, the stoichiometric amount of $O_2(g)$ is 0.5 mole. Thus the air should contain 0.5 mole of $O_2(g)$ and (79/21)(0.5)=1.881 moles of $N_2(g)$ with a total of 2.381 moles. In order to determine the loss of power in the combustion process shown in the figure above, we need to calculate entropy and enthalpy changes (see Equation (18.17)). This requires a stepwise calculation scheme. The scheme to calculate entropy and enthalpy changes of the flowing streams is shown in the figure below.

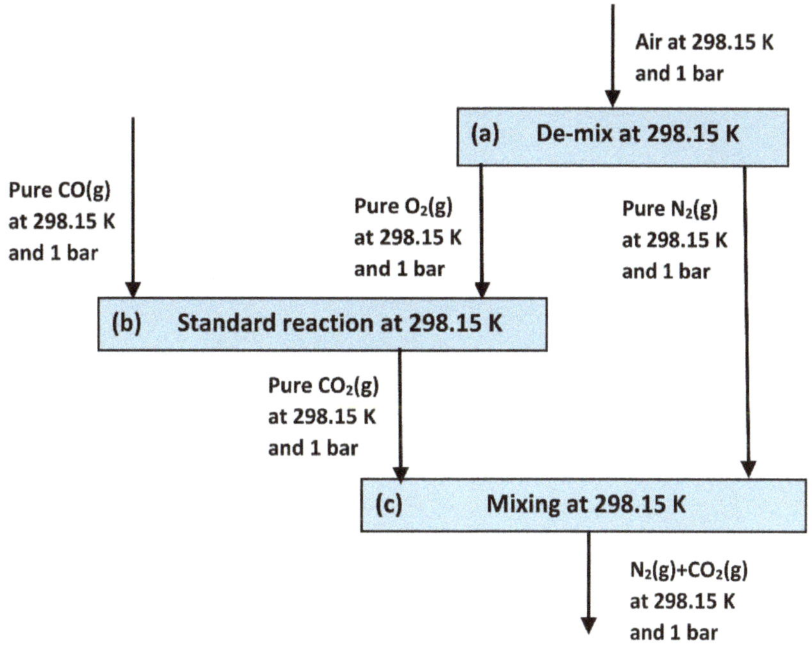

Stepwise calculation scheme to calculate entropy and enthalpy changes for the combustion process under consideration

The net entropy change of the process fluid streams is as follows:

$$\Delta s = (2.381)\Delta s_{De-mixing} + \Delta s^0_{rxn} + (2.881)\Delta s_{Mixing}$$

Note that that the combustion reaction occurs in stoichiometric amounts with 1 mole of carbon monoxide as the basis. In the last mixing step, 1 mole of $CO_2(g)$ is mixed with 1.881 moles of nitrogen gas resulting in a 2.881 moles of $CO_2(g)+N_2(g)$ gaseous mixture.

Each term on the right hand side of the above equation is calculated below:

$$\Delta s_{De-mixing} = R\sum y_i \, ln y_i$$

$$\Delta s_{De-mixing} = 8.314\left[\left(\frac{1.881}{2.381}\right) ln\left(\frac{1.881}{2.381}\right) + \left(\frac{0.5}{2.381}\right) ln\left(\frac{0.5}{2.381}\right)\right] = -4.273 \text{ J/K}$$

The standard entropy of reaction, Δs^o_{rxn} , can be determined from the standard entropies of various species involved in the reaction as:

$$\Delta s^o_{rxn} = s^o(CO_2, g) - s^o(CO, g) - (1/2)s^o(O_2, g) = 213.677 - 197.556 - 0.5 \times 205.037 = -86.4 \text{ J/K}$$

$$\Delta s_{Mixing} = -R\sum y_i \, ln y_i$$

$$\Delta s_{Mixing} = -8.314\left[\left(\frac{1}{2.881}\right) ln\left(\frac{1}{2.881}\right) + \left(\frac{1.881}{2.881}\right) ln\left(\frac{1.881}{2.881}\right)\right] = 5.368 \text{ J/K}$$

Thus, the net entropy change of the process streams is:

$$\Delta s = (2.381)(-4.273) + (-86.4) + (2.881)(5.368) = -81.11 \text{ J/K}$$

Using the same basis of 1 mole of pure carbon monoxide entering the combustion chamber, the net enthalpy change of the process streams can be calculated as follows:

$$\Delta h = (2.381)\Delta h_{De-mixing} + \Delta H^o_{rxn} + (2.881)\Delta h_{Mixing} = \Delta H^o_{rxn}$$

$$\Delta H^o_{rxn} = \Delta H^o_f(CO_2, g) - \Delta H^o_f(CO, g) - (1/2)\Delta H^o_f(O_2, g)$$

$$\Delta h = \Delta H^o_{rxn} = -393.51 - (-110.53) - (1/2)(0) = 282.98\ \text{kJ}$$

Note that the enthalpy of mixing (Δh_{Mixing}) and enthalpy of de-mixing ($\Delta h_{De-mixing}$) are zero for ideal gases. Therefore, only the standard enthalpy of reaction (ΔH^o_{rxn}), expressed in terms of the standard enthalpies of formation of compounds, is non zero. From Equation (18.17), we find the loss of power due to irreversibilities in this combustion process, per mole of the carbon monoxide entering, is:

$$I = 298.15 \times (-81.11) - (-282980) = 258797.05\ \text{J} = 258.80\ \text{kJ}$$

Exergy Analysis:

As noted earlier, in order to carry out an exergy analysis of a process, it is necessary to obtain the values of the standard molar chemical exergies of the substances involved. The only non-reference species (see Table 18.1 and Figure 18.2) involved in the combustion process under consideration is carbon monoxide. Thus, we need to set up a formation reaction for carbon monoxide using the reference species:

$$CO_2(g) \rightarrow CO(g) + (1/2)O_2(g)$$

The following relationship is then used to determine the standard molar chemical exergy of carbon monoxide:

$$\psi^o_{C,CO(g)} = \Delta G^o_{rxn} - \sum_{other\ than\ CO(g)} v_i\, \psi^0_{C,i}$$

The standard Gibbs free energy of the formation reaction ΔG^o_{rxn} is given below:

$$\Delta G^o_{rxn,CO(g)} = \Delta G^o_f(CO, g) + (1/2)\Delta G^o_f(O_2, g) - \Delta G^o_f(CO_2, g) = (-137.15) + (0.5 \times 0) - (-394.36) = 257.21\ \text{kJ/mol}$$

The standard molar chemical exergies of the species present in the reference atmosphere and involved in the formation reaction are as follows: $\psi^o_C(O_2, g) = 3.914$; $\psi^o_C(CO_2, g) = 20.11$ kJ/mol. Upon substitution of ΔG^o_{rxn} and $\psi^o_{C,i}$ values in the $\psi^o_{C,CO(g)}$ expression given above, the standard molar chemical exergy of carbon monoxide gas is found to be 275.363 kJ/mol.

We now need to calculate the molar exergies of all the inlet and outlet streams of the process. As all streams are at reference environment conditions of temperature and pressure, only chemical exergies are needed. Using 1 mole of pure carbon monoxide entering the combustion chamber as the basis, the air stream entering the process consists of 0.5 mole $O_2(g)$ and 1.881 moles $N_2(g)$. The combustion product stream is a mixture of 1 mole $CO_2(g)$ and 1.881 moles $N_2(g)$. The chemical exergy of air inlet stream is calculated using Equation (18.39) as follows:

$$\psi_{C,Air} = y_{O_2}\psi^o_{C,O_2} + y_{N_2}\psi^o_{C,N_2} + RT_o(\, y_{O_2}\ ln y_{O_2} + y_{N_2}\ ln y_{N_2})$$

$$\psi_{C,Air} = \left(\frac{0.5}{2.381}\right)(3.914) + \left(\frac{1.881}{2.381}\right)(0.664) + (8.314 \times 298.15/1000)\left[\left(\frac{0.5}{2.381}\right) ln \left(\frac{0.5}{2.381}\right) + \left(\frac{1.881}{2.381}\right) ln \left(\frac{1.881}{2.381}\right)\right] = 0.0725 \text{ kJ/mol}$$

The chemical exergy of the combustion outlet stream is calculated in a similar manner as follows:

$$\psi_{C,Outlet} = y_{CO_2}\psi^o_{C,CO_2} + y_{N_2}\psi^o_{C,N_2} + RT_o(\, y_{CO_2}\, lny_{CO_2} + y_{N_2}\, lny_{N_2})$$

$$\psi_{C,Air} = \left(\frac{1}{2.881}\right)(20.11) + \left(\frac{1.881}{2.881}\right)(0.664) + (8.314 \times 298.15/1000)\left[\left(\frac{1}{2.881}\right) ln \left(\frac{1}{2.881}\right) + \left(\frac{1.881}{2.881}\right) ln \left(\frac{1.881}{2.881}\right)\right] = 5.8133 \text{ kJ/mol}$$

From Equation (18.31), we find the loss of power due to irreversibilities in this combustion process, per mole of the entering carbon monoxide, to be:

$$I = (2.381)\psi_{C,Air} + (1)\psi_{C,CO(g)} - (2.881)\psi_{C,Outlet} = (2.381)(0.0725) + (1)(275.363) - (2.881)(5.8133) = 258.788 \text{ kJ}$$

This value is in agreement with the value obtained through entropy analysis.

PRACTICE PROBLEMS

Problem 1

Re-do the tutorial problem with pure oxygen (instead of air) in stoichiometric amount entering the combustion chamber.

Problem 2

Consider a combustion chamber where combustion of propane is being carried out. The propane is burned completely to $CO_2(g)$ and $H_2O(g)$ with 25% excess air. The propane and air enter the chamber as pure streams at 25°C and 1 bar. The products of complete combustion of propane leave the chamber at 200°C and 1 bar. The process is a steady state process and the reference environment is as shown in Figure 18.2. Calculate the loss of power using both entropic and exergetic methods and compare the results. You can base your calculations on per mole of propane entering the combustion chamber.

Problem 3

A three-component gaseous mixture of methane, ethane, and butane is fed into a de-mixing device at a temperature of 140 °C and pressure of 3 bar. The composition of the mixture in terms of mole fractions is: $y_{CH_4} = 0.25$, $y_{C_2H_6} = 0.35$, and $y_{C_4H_{10}} = 0.40$. The product streams from the device are pure methane, ethane, and butane streams at 30 °C and 1 bar. The process is in steady state and the reference environment is as shown in Figure 18.2. Calculate the loss of power using both entropic and exergetic methods and compare the results. You can base your calculations on per mole of gaseous mixture entering the de-mixing device.

Problem 4

Consider a reaction chamber where dissociation of dinitrogen tetroxide to nitrogen dioxide is being carried out at 25 °C and 1 bar:

$$N_2O_4(g) \rightleftharpoons 2NO_2(g)$$

The reactant (pure N_2O_4) enters the chamber at 25 °C and 1 bar. The products leave the chamber 25 °C and 1 bar. Assuming that the reaction reaches equilibrium, calculate the loss of power using both entropic and exergetic methods and compare the results. You can base your calculations on per mole of N_2O_4 entering the combustion chamber.

Given that at 298.15 K:

Species	Δg_f^o, kJ/mol	Δh_f^o, kJ/mol
$N_2O_4(g)$	97.5	9.2
$NO_2(g)$	51.3	33.2

You can assume that the reaction mixture is an ideal gas mixture.

Appendix A

Conversion Factors and Constants

Quantity	**Conversion**	
Mass	1 kg=2.2046 lb_m	1 lb_m=0.4536 kg
Length	1 m =3.2808 ft	1 ft = 0.3048 m
Force	1 N = 1 kg.m/s^2 1 N=0.22482 lb_f	1 lb_f=4.4482 N
Pressure	1 Pa=1.4504x10^{-4} lb_f/in^2 1 bar= 10^5 Pa=100 kPa	1 lb_f/in^2=6894.8 Pa
Volume	1 m^3=35.315 ft^3	1 ft^3=0.028317 m^3
Density	1 kg/m^3=0.06243 lb_m/ft^3	1 lb_m/ft^3=16.018 kg/m^3
Energy	1 kJ=0.9478 Btu 1 kJ=0.239 kcal	1 Btu=1.0551 kJ
Specific Energy	1 kJ/kg=0.42992 Btu/lb_m	1 Btu/lb_m = 2.326 kJ/kg
Power	1 W=1 J/s 1 W=3.413 Btu/h 1kW=1.341 hp	1 Btu/h=0.293 W 1 hp=550 ft.lb_f/s 1 hp= 2545 Btu/h 1 hp=0.7457 kW
Specific heat capacity	1 kJ/kg.K=0.238846 Btu/lb_m°R	1 Btu/lb_m°R=4.1868 kJ/kg.K 1 Btu/lb_m°R= 1 kcal/kg.K
Temperature	T (K)=T (°C)+273.15	T(°R)=T(°F)+459.67 T(°R)=1.8 T(K)
Constants		
Universal gas constant, R	R=8.314 J/mol.K=83.14 cm^3bar/mol.K=1.986 Btu/lbmol.°R=1545 ft.lb_f/lbmol.°R	
Standard acceleration due to gravity, g	g=9.80665 m/s^2 = 32.174 ft/s^2	
Standard atmospheric pressure	1 atm = 1.01325 bars=14.696 lb_f/in^2	

Appendix B

Review of Mathematical Background

Thermodynamics is a highly mathematical subject although the math involved is not difficult. In this appendix, the aspects of math relevant to thermodynamics are reviewed.

ORDINARY DERIVATIVE

Ordinary derivatives are relevant when we are dealing with functions of a single variable. Let $f(x)$ be some function of a single variable x. The ordinary derivative is defined as:

$$\frac{df}{dx} = \lim_{\Delta x \to 0} \left[\frac{f(x + \Delta x) - f(x)}{\Delta x} \right]$$

As an example,

$$f(x) = ax^2 + bx + c$$

The ordinary derivative of this $f(x)$ is:

$$\frac{df}{dx} = 2ax + b$$

PARTIAL DERIVATIVES

In thermodynamics, we often come across functions of two or more variables. Here partial derivatives become relevant. Let $f(x, y)$ be some function of two variables x and y. In this case, there are two *first order* partial derivatives and four *second order* partial derivatives.

The first order partial derivatives are:

a) first order partial derivative of f with respect to x , denoted as $\partial f/\partial x$. This partial derivative is determined by taking ordinary derivative of f with respect to x, treating variable y as a constant. This partial derivative could also be written as $(\partial f/\partial x)_y$ where the subscript specifies the variable that is held constant.

b) first order partial derivative of f with respect to y , denoted as $\partial f/\partial y$. This partial derivative is determined by taking ordinary derivative of f with respect to y, treating variable x as a constant. This partial derivative could also be written as $(\partial f/\partial y)_x$ where the subscript specifies the variable that is held constant.

As an example, consider the following function:

$$f(x, y) = ax^2 + by^2 + cxy$$

The first order partial derivatives of this function are:

$$(\partial f/\partial x)_y = 2ax + cy$$

$$(\partial f/\partial y)_x = 2by + cx.$$

The second order partial derivatives of function $f(x, y)$ include two *pure* partial derivatives and two *mixed* partial derivatives. The pure partial derivatives are:

$$\frac{\partial^2 f}{\partial x^2} = \left[\frac{\partial}{\partial x}\left(\frac{\partial f}{\partial x}\right)_y\right]_y$$

$$\frac{\partial^2 f}{\partial y^2} = \left[\frac{\partial}{\partial y}\left(\frac{\partial f}{\partial y}\right)_x\right]_x$$

The mixed partial derivatives are:

$$\frac{\partial^2 f}{\partial x \partial y} = \left[\frac{\partial}{\partial x}\left(\frac{\partial f}{\partial y}\right)_x\right]_y$$

$$\frac{\partial^2 f}{\partial y \partial x} = \left[\frac{\partial}{\partial y}\left(\frac{\partial f}{\partial x}\right)_y\right]_x$$

Note that the order of differentiation is different in the mixed derivatives. In $\partial^2 f/\partial x \partial y$, the function is first differentiated with respect to y (x held constant) and then with respect to x (y held constant). In $\partial^2 f/\partial y \partial x$, the function is first differentiated with respect to x (y held constant) and then with respect to y (x held constant).

As an example, consider the function $f(x, y) = ax^2 + by^2 + cxy$. The second order partial derivatives are as follows:

$$\partial^2 f/\partial x^2 = 2a$$

$$\partial^2 f/\partial y^2 = 2b$$

$$\partial^2 f/\partial x \partial y = \partial^2 f/\partial y \partial x = c$$

Note that the mixed second order partial derivatives are the same, that is, the order of differentiation does not matter.

Consider the ideal gas law:

$$P = RT/v$$

Thus, pressure is a function of two variable T and v: $P = P(T, v)$. The first order partial derivatives of P are:

$$(\partial P/\partial v)_T = -\,RT/v^2$$

$$(\partial P/\partial T)_v = R/v$$

The second order partial derivatives are:

$$\partial^2 P/\partial v^2 = 2RT/v^3$$

$$\partial^2 P/\partial T^2 = 0$$

$$\partial^2 P/\partial T \partial v = \partial^2 P/\partial v \partial T = -\,R/v^2$$

Once again, the order of differentiation does not matter in mixed second order partial derivatives.

Is it always true that for mixed second order partial derivatives, the order of differentiation does not matter?

As long as f is a well-behaved function of two variables and has continuous first and second order partial derivatives, that is, there are no discontinuities in either the function or its derivatives, the mixed second order partial derivatives are equal:

$$\partial^2 f/\partial x\partial y = \partial^2 f/\partial y\partial x$$

In thermodynamics, the state properties of pure substances are well-behaved functions of two variables and have continuous first and second order partial derivatives. Thus, the order of differentiation in mixed second order partial derivatives does not matter. For example, consider specific enthalpy which is a function of temperature and pressure:

$$h = h(T, P)$$

As the mixed second order partial derivatives are equal, it follows that:

$$\partial^2 h/\partial P\partial T = \partial^2 h/\partial T\partial P$$

TOTAL DERIVATIVE

The total derivative of $f(x, y)$ is given as:

$$df = \left(\frac{\partial f}{\partial x}\right)_y dx + \left(\frac{\partial f}{\partial y}\right)_x dy$$

Let the function be specific enthalpy $h(T, P)$. Thus,

$$dh = \left(\frac{\partial h}{\partial T}\right)_P dT + \left(\frac{\partial h}{\partial P}\right)_T dP$$

Now we consider a function of more than two independent variables. For example, the Gibbs free energy of a multicomponent system with variable composition can be expressed as:

$$G = G(T, P, n_1, n_2, n_3, \dots \dots .. n_N)$$

where $n_1, n_2, n_3, \dots \dots .. n_N$ are moles of different components and N is the total number of components. The total derivative of G is:

$$dG = \left(\frac{\partial G}{\partial T}\right)_{P,n_i} dT + \left(\frac{\partial G}{\partial P}\right)_{T,n_i} dP + \left(\frac{\partial G}{\partial n_1}\right)_{T,P,n_j} dn_1 + \left(\frac{\partial G}{\partial n_2}\right)_{T,P,n_j} dn_2 + \cdots \left(\frac{\partial G}{\partial n_N}\right)_{T,P,n_j} dn_N$$

where subscript n_i indicates moles of all species are held constant and subscript n_j indicates moles of all species, other than the one involved in the derivative, are held constant. It is important to note that in thermodynamics, the first order partial derivatives always specify the variables which are kept constant as a subscript.

EXACT AND INEXACT DIFFERENTIALS

In thermodynamics, we frequently come across differentials of the form:

$$df = M(x,y)dx + N(x,y)dy$$

We are interested in the integral of the differential quantity df over the limits (x_1, y_1) to (x_2, y_2), that is:

$$\int_{(x_1,y_1)}^{(x_2,y_2)} df = \int_{(x_1,y_1)}^{(x_2,y_2)} M(x,y)dx + \int_{(x_1,y_1)}^{(x_2,y_2)} N(x,y)dy$$

In order to evaluate the integrals on the right hand side, we need to know the *path of integration*, that is, how y varies with x. For example, the following figure shows two different paths of integration:

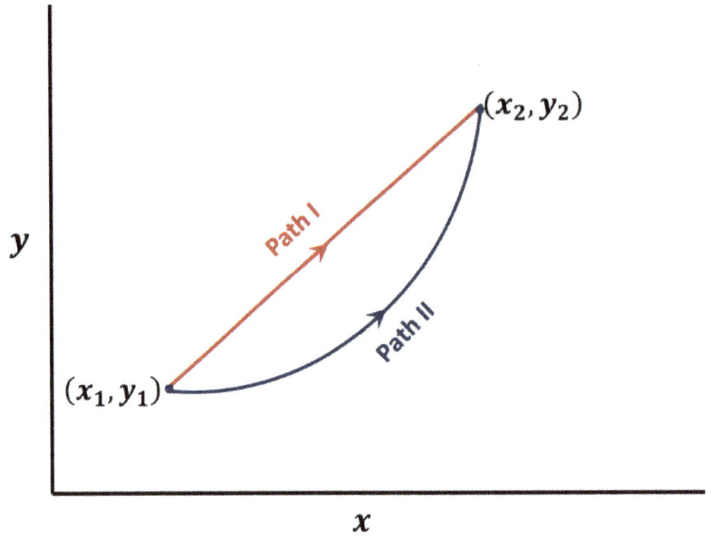

For a given path of integration, the variation of y as a function of x is known. Thus, one can eliminate one of the variables in the integrals on the right hand side of the above equation and carry out the integration in the usual manner.

Depending on the nature of the differential quantity df, we may obtain two different results:

- The integral $\int_{(x_1,y_1)}^{(x_2,y_2)} df$ is *path-independent*, that is, the value of the integral depends only on the initial and final integration limits and not on how we go from (x_1, y_1) to (x_2, y_2). The integration along all paths have the same value. In this case df is an *EXACT* differential. Thus, the integral of an exact differential is path-independent and depends only on the integration limits.
- The integral $\int_{(x_1,y_1)}^{(x_2,y_2)} df$ is *path-dependent*, that is, the value of the integral depends on how we go from (x_1, y_1) to (x_2, y_2). The integration along different paths give different values. In this case df is an *INEXACT* differential. Thus, the integral of an inexact differential is path-dependent, that is, the value of the integral depends on how we go from the initial limit to the final limit of integration.

As an example, consider the following differential:

$$df = 2xydx + x^2dy$$

What is the value of the integral $\int_{(0,0)}^{(1,1)} df$ along the following two different paths: a) $y = \sqrt{x}$ and b) $y = x$?

a) Solution with path of integration $y = \sqrt{x}$:

$$\int_{(0,0)}^{(1,1)} df = \int_{(0,0)}^{(1,1)} 2xydx + \int_{(0,0)}^{(1,1)} x^2dy$$

$$= \int_{(0)}^{(1)} 2x\sqrt{x}dx + \int_{(0)}^{(1)} y^4dy = 2\left(\frac{2}{5}\right) + \left(\frac{1}{5}\right) = 1$$

b) Solution with path of integration $y = x$:

$$\int_{(0,0)}^{(1,1)} df = \int_{(0,0)}^{(1,1)} 2xydx + \int_{(0,0)}^{(1,1)} x^2dy$$

$$= \int_{(0)}^{(1)} 2x^2dx + \int_{(0)}^{(1)} y^2dy = 2\left(\frac{1}{3}\right) + \left(\frac{1}{3}\right) = 1$$

The integral of given differential df is independent of the path of integration. Different paths of integration give the same values. Hence df is an *exact* differential.

Consider another example with the following differential:

$$df = y^2dx + x^2dy$$

What is the value of the integral $\int_{(0,0)}^{(1,1)} df$ along the following three different paths: a) $y = x$; b) $y = x^2$; and c) $y = x^3$?

a) Solution with path of integration $y = x$:

$$\int_{(0,0)}^{(1,1)} df = \int_{(0,0)}^{(1,1)} y^2dx + \int_{(0,0)}^{(1,1)} x^2dy$$

$$= \int_{(0)}^{(1)} x^2dx + \int_{(0)}^{(1)} y^2dy = \left(\frac{1}{3}\right) + \left(\frac{1}{3}\right) = \frac{2}{3}$$

b) Solution with path of integration $y = x^2$:

$$\int_{(0,0)}^{(1,1)} df = \int_{(0,0)}^{(1,1)} y^2 dx + \int_{(0,0)}^{(1,1)} x^2 dy$$

$$= \int_{(0)}^{(1)} x^4 dx + \int_{(0)}^{(1)} y dy = \left(\frac{1}{5}\right) + \left(\frac{1}{2}\right) = \frac{7}{10}$$

c) Solution with path of integration $y = x^3$:

$$\int_{(0,0)}^{(1,1)} df = \int_{(0,0)}^{(1,1)} y^2 dx + \int_{(0,0)}^{(1,1)} x^2 dy$$

$$= \int_{(0)}^{(1)} x^6 dx + \int_{(0)}^{(1)} y^{2/3} dy = \left(\frac{1}{7}\right) + \left(\frac{3}{5}\right) = \frac{26}{35}$$

The integral of given differential df is dependent of the path of integration. Different paths of integration give different values of the integral. Hence df is an *inexact* differential.

NOTATION FOR EXACT AND INEXACT DIFFERENTIALS

- The exact differential is denoted as df
- The inexact differential is denoted as δf

PROPERTIES OF EXACT DIFFERENTIALS

- $\oint df = 0$. The cyclic integral of an exact differential is always zero.
- If df is an exact differential, it implies the existence of function f. Consequently, $\int_{(x_1,y_1)}^{(x_2,y_2)} df = f(x_2, y_2) - f(x_1, y_1)$.

PROPERTIES OF INEXACT DIFFERENTIALS

- $\oint \delta f \neq 0$. The cyclic integral of an inexact differential is generally not zero.
- $\int_{(x_1,y_1)}^{(x_2,y_2)} \delta f \neq f(x_2, y_2) - f(x_1, y_1)$. In this case, the function f does not exist.
- $\int_{(x_1,y_1)}^{(x_2,y_2)} \delta f = F$, where F is the total quantity of f summed over small-small quantities of δf in going from initial limit to the final limit. The total quantity F depends on the path of integration.

IMPORTANCE OF EXACT AND INEXACT DIFFERENTIALS IN THERMODYNAMICS

In thermodynamics, all state properties such as internal energy, enthalpy, entropy, Gibbs free energy etc., possess *exact* differentials. Thus, the change in state property of the system in going from one state to another depends only on the initial and final states, regardless of the path or process followed by the system in bringing about the change. For example, the change in entropy of the system in going from one state to another can be expressed as:

$$\int_1^2 dS = S_2 - S_1 \quad \text{(regardless of the path/process)}$$

However, heat and work possess *inexact* differentials as they are path-dependent quantities. Consequently,

$$\int_1^2 \delta Q = Q \quad \text{(Q depends on the path/process)}$$

$$\int_1^2 \delta W = W \quad \text{(W depends on the path/process)}$$

CONDITION FOR EXACTNESS

Given the following differential df, what is the criterion for *exactness* of df?

$$df = M(x,y)dx + N(x,y)dy$$

Let us assume that df is an *exact* differential. Consequently there exists a well-behaved function f which is continuous and has continuous first order and second order partial derivatives, that is:

$$f = f(x,y)$$

The total differential of f can be written as:

$$df = \left(\frac{\partial f}{\partial x}\right)_y dx + \left(\frac{\partial f}{\partial y}\right)_x dy$$

Upon comparison of the above expressions of df, we can write:

$$M = \left(\frac{\partial f}{\partial x}\right)_y$$

$$N = \left(\frac{\partial f}{\partial y}\right)_x$$

Now we take the derivatives of M and N as follows:

$$\left(\frac{\partial M}{\partial y}\right)_x = \left[\frac{\partial}{\partial y}\left(\frac{\partial f}{\partial x}\right)_y\right]_x = \frac{\partial^2 f}{\partial y \partial x}$$

$$\left(\frac{\partial N}{\partial x}\right)_y = \left[\frac{\partial}{\partial x}\left(\frac{\partial f}{\partial y}\right)_x\right]_y = \frac{\partial^2 f}{\partial x \partial y}$$

For a well-behaved function f which is continuous and has continuous first order and second order partial derivatives, the order of differentiation in mixed second order partial derivatives does not matter, that is:

$$\frac{\partial^2 f}{\partial y \partial x} = \frac{\partial^2 f}{\partial x \partial y}$$

Consequently,

$$\left(\frac{\partial M}{\partial y}\right)_x = \left(\frac{\partial N}{\partial x}\right)_y$$

For this relation to hold, df must be an exact differential. Thus, this relation represents the criterion for exactness of $df = M(x,y)dx + N(x,y)dy$. This criterion for exactness of a differential is known as Euler's criterion for exactness.

Consider the following differential:

$$df = 2xydx + x^2dy$$

Thus,

$$M = 2xy$$

$$N = x^2$$

The partial derivatives are:

$$\left(\frac{\partial M}{\partial y}\right)_x = 2x$$

$$\left(\frac{\partial N}{\partial x}\right)_y = 2x$$

Thus, df is an exact differential as it meets the Euler criterion for exactness.

Consider another example,

$$df = y^2dx + x^2dy$$

Thus,

$$M = y^2$$

$$N = x^2$$

The partial derivatives are:

$$\left(\frac{\partial M}{\partial y}\right)_x = 2y$$

$$\left(\frac{\partial N}{\partial x}\right)_y = 2x$$

These derivatives are not the same, that is, the Euler criterion for exactness is not obeyed. Therefore, the given df is an inexact differential.

Appendix C

The Maxwell Relations of Thermodynamics

The thermodynamic properties can be divided into two groups as: *measurable properties* such as pressure, temperature and volume, and *non-measurable properties* such as internal energy, enthalpy, entropy, etc. which cannot be measured directly. The Maxwell relations allow us to calculate changes in non-measurable properties in terms of changes in measurable properties. The Maxwell relations are derived by applying the Euler criterion of exactness of differential to differentials of certain thermodynamic properties.

PRINCIPAL EXACT DIFFERENTIALS

For pure or fixed composition materials, the differential changes in internal energy and enthalpy can be expressed as:

$$du = Tds - Pdv$$

$$dh = Tds + vdP$$

The differential of Gibbs free energy ($g = h - Ts$) can be expressed as:

$$dg = -sdT + vdP$$

We now introduce the specific Helmholtz free energy (a), defined as:

$$a = u - Ts$$

The differential of a can be expressed as:

$$da = -sdT - Pdv$$

Since u, h, g and a are all state properties, their differentials are exact differentials. These differentials are also referred to as principal exact differentials.

MAXWELL RELATIONS

We now apply the Euler criterion of exactness to the principal differentials. Given the following differential df

$$df = M(x,y)dx + N(x,y)dy$$

T
he Euler criterion of exactness of df is:

$$\left(\frac{\partial M}{\partial y}\right)_x = \left(\frac{\partial N}{\partial x}\right)_y$$

Applying the Euler criterion of exactness of differential to du, dh, dg, and da gives:

$$\left(\frac{\partial T}{\partial v}\right)_s = -\left(\frac{\partial P}{\partial s}\right)_v$$

$$\left(\frac{\partial T}{\partial P}\right)_s = \left(\frac{\partial v}{\partial s}\right)_P$$

$$-\left(\frac{\partial s}{\partial P}\right)_T = \left(\frac{\partial v}{\partial T}\right)_P$$

$$\left(\frac{\partial s}{\partial v}\right)_T = \left(\frac{\partial P}{\partial T}\right)_v$$

These four partial derivative equations are known as the Maxwell relations. Note that the last two relations allow us to express changes in entropy with pressure and volume at constant temperature in terms of derivatives of measurable quantities.

APPLICATIONS OF THE MAXWELL RELATIONS

As noted earlier, the Maxwell relations allow us to calculate changes in non-measurable properties in terms of changes in measurable properties. Let us consider changes in internal energy, enthalpy, and entropy. None of these properties are directly measurable.

Internal Energy Changes

The internal energy change can be expressed as a function of T and v as:

$$du = \left(\frac{\partial u}{\partial T}\right)_v dT + \left(\frac{\partial u}{\partial v}\right)_T dv = c_v dT + \left(\frac{\partial u}{\partial v}\right)_T dv$$

The Tds relation gives:

$$du = Tds - Pdv$$

Expressing $s = s(T, v)$, we can re-write the Tds relation as:

$$du = T\left[\left(\frac{\partial s}{\partial T}\right)_v dT + \left(\frac{\partial s}{\partial v}\right)_T dv\right] - Pdv = T\left(\frac{\partial s}{\partial T}\right)_v dT + \left[T\left(\frac{\partial s}{\partial v}\right)_T - P\right] dv$$

Upon comparison of this du expression with the first du expression given above, we get:

$$c_v = T\left(\frac{\partial s}{\partial T}\right)_v \quad \text{and} \quad \left(\frac{\partial u}{\partial v}\right)_T = \left[T\left(\frac{\partial s}{\partial v}\right)_T - P\right]$$

Making use of the following Maxwell relation,

$$\left(\frac{\partial s}{\partial v}\right)_T = \left(\frac{\partial P}{\partial T}\right)_v$$

we can write:

$$\left(\frac{\partial u}{\partial v}\right)_T = \left[T\left(\frac{\partial P}{\partial T}\right)_v - P\right]$$

Consequently,

$$du = c_v dT + \left[T\left(\frac{\partial P}{\partial T}\right)_v - P\right] dv$$

All quantities on the right hand side of this equation are measurable quantities. Thus we can calculate the changes in internal energy in terms of measurable quantities.

Enthalpy Changes

The enthalpy change can be expressed as a function of T and P as:

$$dh = \left(\frac{\partial h}{\partial T}\right)_P dT + \left(\frac{\partial h}{\partial P}\right)_T dP = c_p dT + \left(\frac{\partial h}{\partial P}\right)_T dP$$

The Tds relation gives:

$$dh = Tds + vdP$$

Expressing $s = s(T, P)$, we can re-write the Tds relation as:

$$dh = T\left[\left(\frac{\partial s}{\partial T}\right)_P dT + \left(\frac{\partial s}{\partial P}\right)_T dP\right] + vdP = T\left(\frac{\partial s}{\partial T}\right)_P dT + \left[T\left(\frac{\partial s}{\partial P}\right)_T + v\right] dP$$

Upon comparison of this dh expression with the first dh expression given above, we get:

$$c_P = T\left(\frac{\partial s}{\partial T}\right)_P \quad \text{and} \quad \left(\frac{\partial h}{\partial P}\right)_T = \left[T\left(\frac{\partial s}{\partial P}\right)_T + v\right]$$

Making use of the following Maxwell relation,

$$\left(\frac{\partial s}{\partial P}\right)_T = -\left(\frac{\partial v}{\partial T}\right)_P$$

we can write:

$$\left(\frac{\partial h}{\partial P}\right)_T = \left[-T\left(\frac{\partial v}{\partial T}\right)_P + v\right]$$

Consequently,

$$dh = c_P dT + \left[v - T\left(\frac{\partial v}{\partial T}\right)_P\right] dP$$

All quantities on the right hand side of this equation are measurable quantities. Thus, we can calculate the changes in enthalpy in terms of measurable quantities.

Joule-Thomson Effect

When a gas passes through a throttling device (porous plug or partially-open valve), it undergoes a sharp decrease in pressure. In general, the temperature of the gas also changes but the enthalpy remains constant. This phenomenon of temperature change with the decrease in pressure at constant enthalpy is referred to as Joule-Thomson effect. The Joule-Thomson coefficient (μ_{JT}) is defined as:

$$\mu_{JT} = \left(\frac{\partial T}{\partial P}\right)_h$$

If $\mu_{JT} > 0$, the temperature decreases; If $\mu_{JT} < 0$, the temperature increases. There is no change in temperature when $\mu_{JT} = 0$.

As enthalpy is constant in the throttling process, we can write:

$$dh = c_P dT + \left[v - T\left(\frac{\partial v}{\partial T}\right)_P\right] dP = 0$$

Consequently,

$$\mu_{JT} = \left(\frac{\partial T}{\partial P}\right)_h = \frac{1}{c_p}\left[T\left(\frac{\partial v}{\partial T}\right)_P - v\right]$$

Note that $\mu_{JT} = 0$ for ideal gas, that is, no change in temperature occurs upon throttling of ideal gas.

Entropy Changes

The entropy change can be expressed as a function of T and P as:

$$ds = \left(\frac{\partial s}{\partial T}\right)_P dT + \left(\frac{\partial s}{\partial P}\right)_T dP = \frac{c_p}{T} dT + \left(\frac{\partial s}{\partial P}\right)_T dP$$

Making use of the following Maxwell relation,

$$\left(\frac{\partial s}{\partial P}\right)_T = -\left(\frac{\partial v}{\partial T}\right)_P$$

we can write:

$$ds = \frac{c_p}{T} dT - \left(\frac{\partial v}{\partial T}\right)_P dP$$

Alternatively, we can express entropy change as a function of T and v as:

$$ds = \left(\frac{\partial s}{\partial T}\right)_v dT + \left(\frac{\partial s}{\partial v}\right)_T dv = \frac{c_v}{T} dT + \left(\frac{\partial s}{\partial v}\right)_T dv$$

Making use of the following Maxwell relation,

$$\left(\frac{\partial s}{\partial v}\right)_T = \left(\frac{\partial P}{\partial T}\right)_v$$

we can write:

$$ds = \frac{c_v}{T} dT + \left(\frac{\partial P}{\partial T}\right)_v dv$$

Thus, entropy change can be calculated in terms of measurable quantities.

Relationship Between c_p and c_v

Subtracting the following relation

$$ds = \frac{c_v}{T} dT + \left(\frac{\partial P}{\partial T}\right)_v dv$$

from

$$ds = \frac{c_p}{T} dT - \left(\frac{\partial v}{\partial T}\right)_P dP$$

we obtain:

$$0 = (c_p - c_v)\frac{dT}{T} - \left(\frac{\partial v}{\partial T}\right)_P dP - \left(\frac{\partial P}{\partial T}\right)_v dv$$

Expressing P as $P(v, T)$, we can write:

$$dP = \left(\frac{\partial P}{\partial v}\right)_T dv + \left(\frac{\partial P}{\partial T}\right)_v dT$$

Consequently,

$$0 = (c_p - c_v)\frac{dT}{T} - \left(\frac{\partial v}{\partial T}\right)_P \left[\left(\frac{\partial P}{\partial v}\right)_T dv + \left(\frac{\partial P}{\partial T}\right)_v dT\right] - \left(\frac{\partial P}{\partial T}\right)_v dv$$

Upon re-arrangement, we get:

$$\left[(c_p - c_v) - T\left(\frac{\partial v}{\partial T}\right)_P \left(\frac{\partial P}{\partial T}\right)_v\right] dT - T\left[\left(\frac{\partial v}{\partial T}\right)_P \left(\frac{\partial P}{\partial v}\right)_T + \left(\frac{\partial P}{\partial T}\right)_v\right] dv = 0$$

Since T and v can be varied independently, it is necessary that each square-bracketed term in the above equation is zero. Thus,

$$(c_p - c_v) = T\left(\frac{\partial v}{\partial T}\right)_P \left(\frac{\partial P}{\partial T}\right)_v$$

$$\left(\frac{\partial v}{\partial T}\right)_P \left(\frac{\partial P}{\partial v}\right)_T = -\left(\frac{\partial P}{\partial T}\right)_v$$

Combining the above two relations, we get:

$$(c_p - c_v) = -T\left(\frac{\partial v}{\partial T}\right)_P^2 \left(\frac{\partial P}{\partial v}\right)_T$$

For ideal gas,

$$(c_p - c_v) = R$$

Appendix D

Derivations of Residual Property Relations

DERIVATION OF EQUATION (7.54)

Equation (7.54) is as follows:

$$\frac{h^R}{RT} = -T\int_0^P \left(\frac{\partial Z}{\partial T}\right)_P \frac{dP}{P} \qquad (7.54)$$

We begin the derivation with the following relation expressing h as a function of T and P:

$$dh = c_P dT + \left(\frac{\partial h}{\partial P}\right)_T dP$$

The Tds relation is

$$dh = Tds + vdP$$

From the Tds relation, it follows that:

$$\left(\frac{\partial h}{\partial P}\right)_T = T\left(\frac{\partial s}{\partial P}\right)_T + v$$

Using the following Maxwell relation:

$$\left(\frac{\partial s}{\partial P}\right)_T = -\left(\frac{\partial v}{\partial T}\right)_P$$

we get

$$\left(\frac{\partial h}{\partial P}\right)_T = v - T\left(\frac{\partial v}{\partial T}\right)_P$$

Consequently,

$$dh = c_P dT + \left[v - T\left(\frac{\partial v}{\partial T}\right)_P\right] dP$$

At constant T

$$dh = \left[v - T\left(\frac{\partial v}{\partial T}\right)_P\right] dP$$

Upon integration:

$$h(T,P) - h(T,0) = \int_0^P \left[v - T\left(\frac{\partial v}{\partial T}\right)_P\right] dP$$

As $P \to 0$, the fluid behaves as an ideal gas:

$$h(T,0) = h^{ig}(T,0)$$

Since the enthalpy of ideal gas is independent of pressure,

$$h(T,0) = h^{ig}(T,0) = h^{ig}(T,P)$$

Consequently,

$$h^R = h(T,P) - h^{ig}(T,P) = \int_0^P \left[v - T\left(\frac{\partial v}{\partial T}\right)_P\right] dP$$

Since

$$v = \frac{ZRT}{P}$$

$$\left(\frac{\partial v}{\partial T}\right)_P = \frac{R}{P}\left[Z + T\left(\frac{\partial Z}{\partial T}\right)_P\right]$$

$$\left[v - T\left(\frac{\partial v}{\partial T}\right)_P\right] = \frac{ZRT}{P} - \frac{RT}{P}\left[Z + T\left(\frac{\partial Z}{\partial T}\right)_P\right] = -\frac{RT^2}{P}\left(\frac{\partial Z}{\partial T}\right)_P$$

Therefore,

$$h^R = h(T,P) - h^{ig}(T,P) = -\int_0^P \frac{RT^2}{P}\left(\frac{\partial Z}{\partial T}\right)_P dP$$

$$\frac{h^R}{RT} = -T\int_0^P \left(\frac{\partial Z}{\partial T}\right)_P \frac{dP}{P} \tag{7.54}$$

This completes the derivation of Equation (7.54).

DERIVATION OF EQUATION (7.55)

Equation (7.55) is as follows:

$$\frac{s^R}{R} = -T\int_0^P \left(\frac{\partial Z}{\partial T}\right)_P \frac{dP}{P} - \int_0^P (Z-1)\frac{dP}{P} \tag{7.55}$$

We begin the derivation with the following relation expressing s as a function of T and P:

$$ds = \frac{c_P}{T}dT + \left(\frac{\partial s}{\partial P}\right)_T dP$$

Using the following Maxwell relation:

$$\left(\frac{\partial s}{\partial P}\right)_T = -\left(\frac{\partial v}{\partial T}\right)_P$$

we get

$$ds = \frac{c_P}{T}dT - \left(\frac{\partial v}{\partial T}\right)_P dP$$

At constant T

$$ds = -\left(\frac{\partial v}{\partial T}\right)_P dP$$

Upon integration:

$$s(T,P) - s(T,0) = -\int_0^P \left(\frac{\partial v}{\partial T}\right)_P dP$$

As $P \to 0$, the fluid behaves as an ideal gas:

$$s(T,0) = s^{ig}(T,0)$$

Thus,

$$s(T,P) - s^{ig}(T,0) = -\int_0^P \left(\frac{\partial v}{\partial T}\right)_P dP$$

For an ideal gas,

$$Pv = RT$$

$$\left(\frac{\partial v}{\partial T}\right)_P = \frac{R}{P}$$

Thus,

$$s^{ig}(T,P) - s^{ig}(T,0) = -\int_0^P \left(\frac{\partial v}{\partial T}\right)_P dP = -R\int_0^P \frac{dP}{P}$$

Consequently,

$$s^R = s(T,P) - s^{ig}(T,P) = -\int_0^P \left(\frac{\partial v}{\partial T}\right)_P dP + R\int_0^P \frac{dP}{P}$$

Making use of the relation,

$$Pv = ZRT$$

$$\left(\frac{\partial v}{\partial T}\right)_P = \frac{R}{P}\left[Z + T\left(\frac{\partial Z}{\partial T}\right)_P\right]$$

Hence, the residual entropy expression becomes:

$$s^R = -\int_0^P \frac{R}{P}\left[Z + T\left(\frac{\partial Z}{\partial T}\right)_P\right] dP + R\int_0^P \frac{dP}{P}$$

Upon simplification, we get:

$$\frac{s^R}{R} = -T\int_0^P \left(\frac{\partial Z}{\partial T}\right)_P \frac{dP}{P} - \int_0^P (Z-1)\frac{dP}{P} \qquad (7.55)$$

This completes the derivation of Equation (7.55).

Appendix E

Mollier Diagrams

Mollier diagrams refer to thermodynamic diagrams which use *enthalpy* as one of the co-ordinate axes. Commonly used Mollier diagrams are the enthalpy versus entropy (h vs s) diagram and pressure versus enthalpy (lnP vs h) diagram. Enthalpy versus entropy diagrams are useful in solving problems involving expansion and compression in turbines and compressors, respectively. Pressure (actually lnP) versus enthalpy diagrams are useful in solving refrigeration problems.

Enthalpy versus Entropy Diagram

The h vs s diagram is shown schematically in the figure below. Reversible adiabatic compression or expansion processes can be shown readily on this diagram. As a reversible adiabatic compression/expansion process is isentropic, it is represented as a vertical line. The figure below shows isentropic (reversible adiabatic) expansion in a turbine by a red line. Using the actual h vs s diagram, one can readily calculate the isentropic shaft work by reading enthalpy change in going from inlet pressure of a turbine to discharge pressure of turbine. Note that actual h vs s diagrams are available in the literature for a number of different materials.

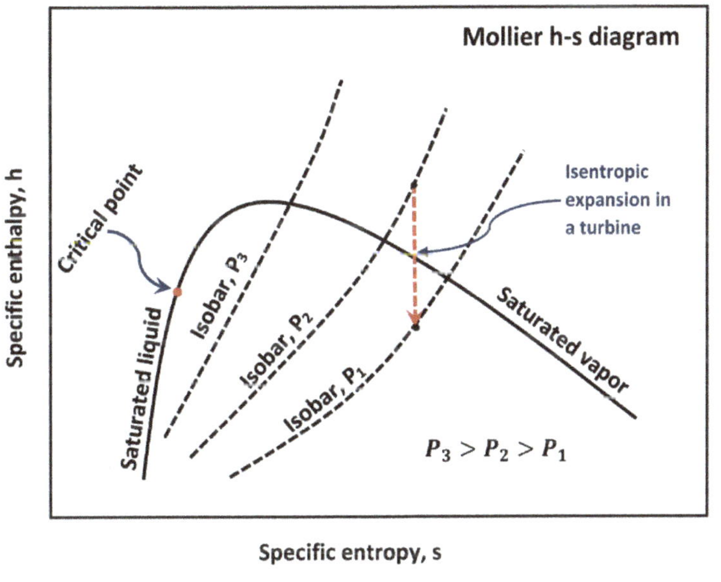

Pressure versus Enthalpy Diagram

The lnP vs h diagram is shown schematically in the figure below. The dome-shaped curve represents the two-phase vapor-liquid region. The isotherms are shown as blue dashed lines.

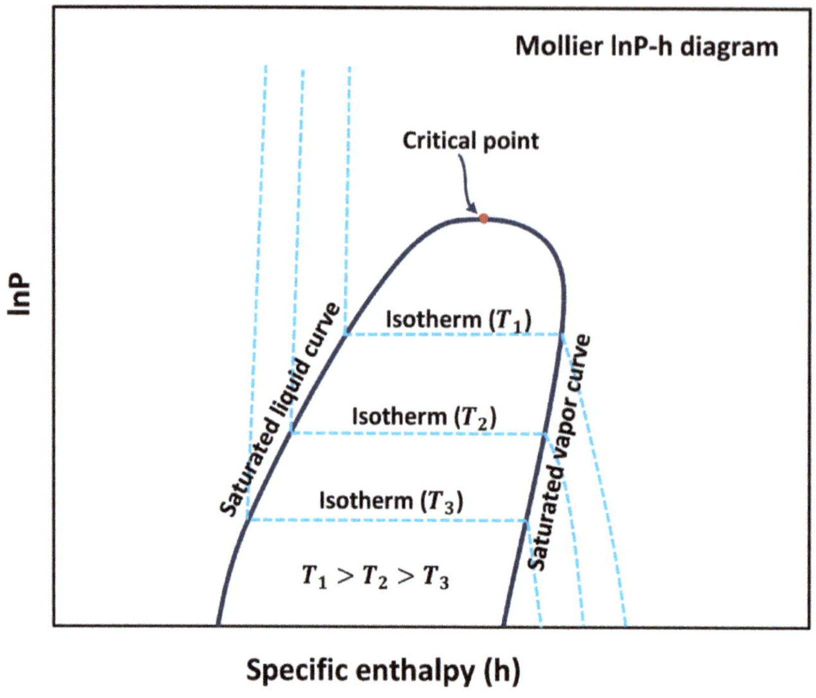

The following figure shows a refrigeration cycle using a throttling process (isenthalpic process) on a lnP versus h diagram.

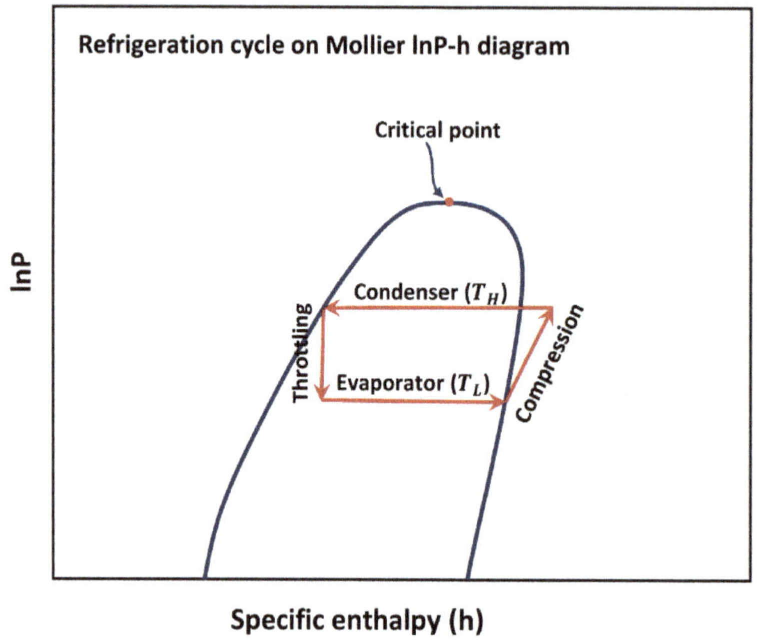

Appendix F

Physical Properties of Pure Substances

Substance	Molar mass	ω	$T_c(K)$	$P_c(bar)$	Z_c	$v_c(cm^3/mol)$	$T_n(K)$
Acetaldehyde	44.053	0.291	466.0	55.50	0.221	154	294
Acetic acid	60.053	0.467	592.0	57.86	0.211	179.7	391.1
Acetone	58.080	0.307	508.2	47.01	0.233	209	329.4
Acetonitrile	41.053	0.338	545.5	48.30	0.184	173	354.8
Acetylene	26.038	0.187	308.3	61.39	0.271	113	189.4
Air	28.97	0.035	132.2	37.45	0.289	84.8	
Ammonia	17.031	0.253	405.7	112.80	0.242	72.5	239.7
Argon	39.948	0.000	150.9	48.98	0.291	74.6	87.3
Benzene	78.114	0.210	562.2	48.98	0.271	259	353.1
Biphenyl	154.211	0.365	789.3	38.50	0.295	502	528.2
n-Butane	58.123	0.200	425.1	37.96	0.274	255	272.7
i-Butane	58.123	0.181	408.1	36.48	0.282	262.7	261.4
i-Butylene	56.108	0.194	417.9	40.00	0.275	238.9	266.3
1-Butanol	74.123	0.594	563.1	44.23	0.260	275	390.8
1-Butene	56.108	0.191	420.0	40.43	0.277	239.3	266.9
cis-2-Butene	56.108	0.205	435.6	42.43	0.273	233.8	276.9
trans-2-Butene	56.108	0.218	428.6	41.00	0.275	237.7	274.0
1,3-Butadiene	54.092	0.190	425.2	42.77	0.267	220.4	268.7
n-Butyric acid	88.106	0.681	615.7	40.64	0.232	291.7	436.4
Benzoic acid	122.123	0.603	751.0	44.70	0.246	344	522.4
Carbon monoxide	28.01	0.048	132.9	34.99	0.299	93.4	81.7
Carbon dioxide	44.01	0.224	304.2	73.83	0.274	94.0	
Carbon disulfide	76.143	0.111	552.0	79.0	0.275	160	319.4
Cyclopentane	70.134	0.196	511.8	45.02	0.273	258	322.4
Cyclohexane	84.161	0.210	553.6	40.73	0.273	308	353.9
Cyclohexene	82.145	0.212	560.4	43.50	0.272	291	356.1
Cumene	120.194	0.326	631.1	32.09	0.261	427	425.6
Carbon tetrachloride	153.822	0.193	556.4	45.60	0.272	276	349.8
Chloroform	119.377	0.222	536.4	54.72	0.293	239	334.3
Chlorobenzene	112.558	0.250	632.4	45.20	0.265	308	404.9
Chlorine	70.905	0.069	417.2	77.10	0.265	124	239.1
n-Decane	142.285	0.492	617.7	21.10	0.247	600	447.3
Diethyl ether	74.123	0.281	466.7	36.40	0.263	280	307.6
Dichloromethane	84.932	0.199	510.0	60.80	0.265	185	312.9
Ethane	30.070	0.100	305.3	48.72	0.279	145.5	184.6
Ethanol	46.069	0.645	513.9	61.48	0.240	167	351.4
Ethylene	28.054	0.087	282.3	50.40	0.281	131	169.4
Ethylbenzene	106.167	0.303	617.2	36.06	0.263	374	409.4
Ethyl acetate	88.106	0.366	523.3	38.80	0.255	286	350.2
Ethylene glycol	62.068	0.487	719.7	77.00	0.246	191.0	470.5
Ethylamine	45.084	0.285	456.2	56.20	0.307	207	289.7

Ethyl chloride	64.514	0.190	460.4	52.70	0.275	200	285.4
Formaldehyde	30.026	0.282	408.0	65.90	0.223	115	254.1
n-Hexane	86.177	0.301	507.6	30.25	0.266	371	341.9
n-Heptane	100.204	0.350	540.2	27.40	0.261	428	371.6
1-Hexene	84.161	0.280	504.0	31.40	0.265	354	336.3
1-Hexanol	102.177	0.579	611.4	35.10	0.263	381	430.6
Helium 4	4.003	-0.390	5.2	2.28	0.302	57.3	4.2
Hydrogen	2.016	-0.216	33.19	13.13	0.305	64.1	20.4
Hydrogen sulfide	34.082	0.094	373.5	89.63	0.284	98.5	212.8
Hydrogen chloride	36.461	0.132	324.7	83.10	0.249	81	188.2
Hydrogen cyanide	27.026	0.410	456.7	53.90	0.197	139	298.9
Krypton	83.8	0.000	209.4	55.02	0.288	91.2	119.8
Methane	16.043	0.012	190.6	45.99	0.286	98.6	111.4
Methylcyclopen-tane	84.161	0.230	532.8	37.85	0.272	319	345.0
Methylcyclohexane	98.188	0.235	572.2	34.71	0.269	368	374.1
Methyl acetate	74.079	0.331	506.6	47.50	0.257	228	330.1
Methyl ethyl ketone	72.107	0.323	535.5	41.50	0.249	267	352.8
Methyl t-butyl ether	88.15	0.266	497.1	34.30	0.273	329	328.4
Methyl chloride	50.488	0.153	416.3	66.80	0.276	143	249.1
Methanol	32.042	0.564	512.6	80.97	0.224	118	337.9
Methylamine	31.057	0.281	430.1	74.60	0.321	154	266.8
Nitrogen	28.014	0.038	126.2	34.00	0.289	89.2	77.3
Nitric oxide	30.006	0.583	180.2	64.80	0.251	58.0	121.4
Nitric acid	63.013	0.714	520.0	68.90	0.231	145	356.2
Nitrous oxide	44.013	0.141	309.6	72.45	0.274	97.4	184.7
Nitromethane	61.040	0.348	588.2	63.10	0.223	173	374.4
Naphthalene	128.174	0.302	748.4	40.51	0.269	413	
n-Nonane	128.258	0.444	594.6	22.90	0.252	544	424.0
Oxygen	31.999	0.022	154.6	50.43	0.288	73.4	90.2
n-Octane	114.231	0.400	568.7	24.90	0.256	486	398.8
i-octane	114.231	0.302	544.0	25.68	0.266	468	372.4
Propane	44.097	0.152	369.8	42.48	0.276	200	231.1
n-Pentane	72.150	0.252	469.7	33.70	0.270	313	309.2
Propylene	42.081	0.14	365.6	46.65	0.289	188.4	225.5
1-Propanol	60.096	0.622	536.8	51.75	0.254	219	370.4
2-Propanol	60.096	0.668	508.3	47.62	0.248	220	355.4
Phenol	94.113	0.444	694.3	61.30	0.243	229	455.0
Sulfur dioxide	64.065	0.245	430.8	78.84	0.269	122	263.1
Sulfur trioxide	80.064	0.424	490.9	82.10	0.255	127	317.9
Sulfuric acid	98.08		924.0	64.00	0.147	177	610.0
Styrene	104.152	0.297	636.0	38.40	0.256	352	418.3
Toluene	92.141	0.262	591.8	41.06	0.264	316	383.8
Tetrafluoroethane (Refrigerant 134A)	102.03	0.327	374.2	40.60	0.258	198	247.1
Water	18.015	0.345	647.1	220.55	0.229	55.9	373.2
Xenon	131.30	0.000	289.7	58.40	0.286	118	165.0
0-Xylene	106.167	0.310	630.3	37.34	0.263	369	417.6
m-Xylene	106.167	0.326	617.1	35.36	0.259	376	412.3
p-Xylene	106.167	0.322	616.2	35.11	0.260	379	411.5

Sources:
Daubert, T.E., Danner, R.P., Sibul, H.M., Stebbins, C.C., *Physical and thermodynamic properties of pure chemicals: data compilation*, Taylor and Francis, Bristol, PA, 1995.
Smith, J.M., Van Ness, H.C., Abbott, M.M., *Introduction to chemical engineering thermodynamics*, McGraw Hill, New York, 6th edition, 2001.

Appendix G

Heat Capacities of Gases in the Ideal Gas State ($P \rightarrow 0$)

Constants for the equation $c_p^{ig}/R = A + BT + CT^2 + DT^{-2}$, where T is in degrees Kelvin.

Substance	Chemical formula	A	$B \times 10^3$	$C \times 10^6$	$D \times 10^{-5}$	Temperature range, K
Acetaldehyde	C_2H_4O	1.693	17.978	-6.158	0	298-1000
Acetylene	C_2H_2	6.132	1.952	0	-1.299	298-1500
Air		3.355	0.575	0	-0.016	298-2000
Ammonia	NH_3	3.578	3.020	0	-0.186	298-1800
n-Butane	C_4H_{10}	1.935	36.915	-11.402	0	298-1500
i-Butane	C_4H_{10}	1.677	37.853	-11.945	0	298-1500
1-Butene	C_4H_8	1.967	31.630	-9.873	0	298-1500
Benzene	C_6H_6	-0.206	39.064	-13.301	0	298-1500
1,3-Butadiene	C_4H_6	2.734	26.786	-8.882	0	298-1500
Bromine	Br_2	4.493	0.056	0	-0.154	298-3000
Cyclohexane	C_6H_{12}	-3.876	63.249	-20.928		298-1500
Carbon monoxide	CO	3.376	0.557	0	-0.031	298-2500
Carbon dioxide	CO_2	5.457	1.045	0	-1.157	298-2000
Carbon disulfide	CS_2	6.311	0.805	0	-0.906	298-1800
Chlorine	Cl_2	4.442	0.089	0	-0.344	298-3000
Dinitrogen tetroxide	N_2O_4	11.660	2.257	0	-2.787	298-2000
Ethane	C_2H_6	1.131	19.225	-5.561	0	298-1500
Ethylene	C_2H_4	1.424	14.394	-4.392	0	298-1500
Ethanol	C_2H_6O	3.518	20.001	-6.002	0	298-1500
Ethylbenzene	C_8H_{10}	1.124	55.380	-18.476	0	298-1500
Ethylene oxide	C_2H_4O	-0.385	23.463	-9.296	0	298-1000
Formaldehyde	CH_2O	2.264	7.022	-1.877	0	298-1500
n-Heptane	C_7H_{16}	3.570	62.127	-19.486	0	298-1500
1-Hexene	C_6H_{12}	3.220	48.189	-15.157	0	298-1500
1-Heptene	C_7H_{14}	3.768	56.588	-17.847	0	298-1500
Hydrogen	H_2	3.249	0.422	0	0.083	298-3000
Hydrogen sulfide	H_2S	3.931	1.490	0	-0.232	298-2300
Hydrogen chloride	HCl	3.156	0.623	0	0.151	298-2000
Hydrogen cyanide	HCN	4.736	1.359	0	-0.725	298-2500
Methane	CH_4	1.702	9.081	-2.164	0	298-1500

Methanol	CH_4O	2.211	12.216	-3.450	0	298-1500
Nitrogen	N_2	3.280	0.593	0	0.040	298-2000
Nitrous oxide	N_2O	5.328	1.214	0	-0.928	298-2000
Nitric oxide	NO	3.387	0.629	0	0.014	298-2000
Nitrogen diox-ide	NO_2	4.982	1.195	0	-0.792	298-2000
n-Octane	C_8H_{18}	4.108	70.567	-22.208	0	298-1500
1-Octene	C_8H_{16}	4.324	64.960	-20.521	0	298-1500
Oxygen	O_2	3.639	0.506	0	-0.227	298-2000
Propane	C_3H_8	1.213	28.785	-8.824	0	298-1500
Propylene	C_3H_6	1.637	22.706	-6.915	0	298-1500
n-Pentane	C_5H_{12}	2.464	45.351	-14.111	0	298-1500
1-Pentene	C_5H_{10}	2.691	39.753	-12.447	0	298-1500
Styrene	C_8H_8	2.050	50.192	-16.662	0	298-1500
Sulfur dioxide	SO_2	5.699	0.801	0	-1.015	298-2000
Sulfur trioxide	SO_3	8.0694	1.056	0	-2.028	298-2000
Toluene	C_7H_8	0.290	47.052	-15.716	0	298-1500
Water	H_2O	3.470	1.450	0	0.121	298-2000

Sources:

Smith, J.M., Van Ness, H.C., Abbott, M.M., *Introduction to chemical engineering thermodynamics*, McGraw Hill, New York, 6th edition, 2001.

Spencer, H.M., *Ind. Eng. Chem.*, 1948; 40: 2152-2154.

Kelley, K.K., *U.S. Bur. Mines Bull. 584*, 1960.

Pankratz, L.B., *U.S. Bur. Mines Bull. 672*, 1982.

Appendix H

Heat Capacities of Liquids and Solids

Constants for the equation $c_p/R = A + BT + CT^2 + DT^{-2}$, where T is in degrees Kelvin.

Substance	Phase	A	$B \times 10^3$	$C \times 10^6$	$D \times 10^{-5}$	Temperature range, K
Ammonia	liquid	22.626	-100.75	192.71	0	273-373
Aniline	liquid	15.819	29.03	15.80	0	273-373
Benzene	liquid	-0.747	67.96	-37.78	0	273-373
1,3-Butadiene	liquid	22.711	-87.96	205.79	0	273-373
Carbon tetra-chloride	liquid	21.155	-48.28	101.14	0	273-373
Chlorobenzene	liquid	11.278	32.86	-31.90	0	273-373
Chloroform	liquid	19.215	-42.89	83.01	0	273-373
Cyclohexane	liquid	-9.048	141.38	-161.62	0	273-373
Ethanol	liquid	33.866	-172.60	349.17	0	273-373
Ethylene oxide	liquid	21.039	-86.41	172.28	0	273-373
Methanol	liquid	13.431	-51.28	131.13	0	273-373
n-Propanol	liquid	41.653	-210.32	427.20	0	273-373
Sulfur trioxide	liquid	-2.930	137.08	-84.73	0	273-373
Toluene	liquid	15.133	6.79	16.35	0	273-373
Water	liquid	8.712	1.25	-0.18	0	273-373
CaO	solid	6.104	0.443	0	-1.047	298-2000
$CaCO_3$	solid	12.572	2.637	0	-3.120	298-1200
$Ca(OH)_2$	solid	9.597	5.435	0	0	298-700
CaC_2	solid	8.254	1.429	0	-1.042	298-720
$CaCl_2$	solid	8.646	1.530	0	-0.302	298-1055
$C(graphite)$	solid	1.771	0.771	0	-0.867	298-2000
Cu	solid	2.677	0.815	0	0.035	298-1357
CuO	solid	5.780	0.973	0	-0.874	298-1400
$Fe(\alpha)$	solid	-0.111	6.111	0	1.150	298-1043
Fe_2O_3	solid	11.812	9.697	0	-1.976	298-960
Fe_3O_4	solid	9.594	27.112	0	0.409	298-850
FeS	solid	2.612	13.286	0	0	298-411
I_2	solid	6.481	1.502	0	0	298-386.8
$LiCl$	solid	5.257	2.476	0	-0.193	298-800
NH_4Cl	solid	5.939	16.105	0	0	298-458
Na	solid	1.988	4.688	0	0	298-371
$NaCl$	solid	5.526	1.963	0	0	298-1073
$NaOH$	solid	0.121	16.316	0	1.948	298-566
$NaHCO_3$	solid	5.128	18.148	0	0	298-400
$S(rhombic)$	solid	4.114	-1.728	0	-0.783	298-368.3
$SiO_2(quartz)$	solid	4.871	5.365	0	-1.001	298-847

Sources:
Smith, J.M., Van Ness, H.C., Abbott, M.M., *Introduction to chemical engineering thermodynamics*, McGraw Hill, New York, 6th edition, 2001.
Kelley, K.K., *U.S. Bur. Mines Bull. 584*, 1960.
Pankratz, L.B., *U.S. Bur. Mines Bull. 672*, 1982.
Miller, J.W., Schorr, G.R., Yaws, C.L., *Chem. Eng.*, 1976; 83:129

Appendix I

Standard Enthalpies and Gibbs Free Energies of Formation of Compounds at 298.15 K

Compound	Chemical formula	State	ΔH_f^o(kJ/mol)	ΔG_f^o(kJ/mol)
Methane	CH_4	(g)	-74.5	-50.5
Ethane	C_2H_6	(g)	-83.8	-31.9
Propane	C_3H_8	(g)	-104.7	-24.3
n-Butane	C_4H_{10}	(g)	-125.8	-16.6
n-Pentane	C_5H_{12}	(g)	-146.8	-8.7
n-Pentane	C_5H_{12}	(l)	-173.1	-9.2
n-Hexane	C_6H_{14}	(g)	-166.9	0.2
n-Heptane	C_7H_{16}	(g)	-187.8	8.3
n-Octane	C_8H_{18}	(g)	-208.8	16.3
n-Octane	C_8H_{18}	(l)	-255.1	-
Acetylene	C_2H_2	(g)	227.5	210.0
Ethylene	C_2H_4	(g)	52.5	68.5
Propylene	C_3H_6	(g)	19.7	62.2
1-Butene	C_4H_8	(g)	1.2	70.3
1-Pentene	C_5H_{10}	(g)	-21.3	78.4
1-Hexene	C_6H_{12}	(g)	-42.0	86.8
1-Heptene	C_7H_{14}	(g)	-62.8	-
1,3-Butadiene	C_4H_6	(g)	109.2	149.8
Benzene	C_6H_6	(g)	82.9	129.7
Benzene	C_6H_6	(l)	49.1	124.5
Ethylbenzene	C_8H_{10}	(g)	29.9	130.9
Naphthalene	$C_{10}H_8$	(s)	78.5	-
Styrene	C_8H_8	(g)	147.4	213.9
Toluene	C_7H_8	(g)	50.2	122.1
Toluene	C_7H_8	(l)	12.2	113.6
Cyclohexane	C_6H_{12}	(g)	-123.1	31.9
Cyclohexane	C_6H_{12}	(l)	-156.2	26.9
Cyclopropane	C_3H_6	(g)	53.3	104.5
Methylcyclohex-ane	C_7H_{14}	(g)	-154.8	27.5
Methylcyclohex-ane	C_7H_{14}	(l)	-190.2	20.6
Cyclohexene	C_6H_{10}	(l)	-	106.9
Acetaldehyde	C_2H_4O	(g)	-166.2	-128.9
Acetic acid	CH_3COOH	(l)	-484.5	-389.9
Acetic acid	CH_3COOH	(aq)	-486.1	-396.5
1,2-Ethanediol	$C_2H_6O_2$	(l)	-454.8	-323.1
Ethanol	C_6H_6O	(g)	-235.1	-168.5
Ethanol	C_6H_6O	(l)	-277.7	-174.8

Ethyl acetate	$CH_3COOC_2H_5$	(l)	-463.3	-318.4
Ethylene oxide	C_2H_4O	(g)	-52.6	-13.0
Formaldehyde	CH_2O	(g)	-108.6	-102.5
Formic acid	$HCOOH$	(l)	-424.7	-361.4
Methanol	CH_4O	(g)	-200.7	-162.0
Methanol	CH_4O	(l)	-238.7	-166.3
Phenol	C_6H_5OH	(s)	-165.0	-50.9
Ammonia	NH_3	(g)	-46.1	-16.5
Ammonia	NH_3	(aq)	-	-26.5
Calcium carbide	CaC_2	(s)	-59.8	-64.9
Calcium carbonate	$CaCO_3$	(s)	-1206.9	-1128.8
Calcium chloride	$CaCl_2$	(s)	-795.8	-748.1
Calcium chloride	$CaCl_2$	(aq)	-	-8101.9
Calcium chloride	$CaCl_2.6H_2O$	(s)	-2607.9	-
Calcium hydroxide	$Ca(OH)_2$	(s)	-986.1	-898.5
Calcium hydroxide	$Ca(OH)_2$	(aq)	-	-868.1
Calcium oxide	CaO	(s)	-635.1	-604
Carbon dioxide	CO_2	(g)	-393.5	-394.4
Carbon monoxide	CO	(g)	-110.5	-137.2
Hydrochloric acid	HCl	(g)	-92.3	-95.3
Hydrogen cyanide	HCN	(g)	135.1	124.7
Hydrogen sulfide	H_2S	(g)	-20.6	-33.6
Iron oxide	FeO	(s)	-272	
Iron oxide (hematite)	Fe_2O_3	(s)	-824.2	-742.2
Iron oxide (magnetite)	Fe_3O_4	(s)	-1118.4	-1015.4
Iron sulfide (pyrite)	FeS_2	(s)	-178.2	-166.9
Lithium chloride	$LiCl$	(s)	-408.6	-
Lithium chloride	$LiCl.H_2O$	(s)	-712.6	-
Lithium chloride	$LiCl.2H_2O$	(s)	-1012.7	-
Lithium chloride	$LiCl.3H_2O$	(s)	-1311.3	-
Nitric acid	HNO_3	(l)	-174.1	-80.7
Nitric acid	HNO_3	(aq)	-	-111.3
Nitric oxide	NO	(g)	90.3	86.6
Nitrogen dioxide	NO_2	(g)	33.2	51.3
Nitrous oxide	N_2O	(g)	82.1	104.2
Dinitrogen tetroxide	N_2O_4	(g)	9.2	97.5
Sodium carbonate	Na_2CO_3	(s)	-1130.7	-1044.4
Sodium carbonate	$Na_2CO_3.10H_2O$	(s)	-4081.3	
Sodium chloride	$NaCl$	(s)	-411.2	-384.1
Sodium chloride	$NaCl$	(aq)	-	-393.1
Sodium hydroxide	$NaOH$	(s)	-425.6	-379.5
Sodium hydroxide	$NaOH$	(aq)	-	-419.2
Sulfur dioxide	SO_2	(g)	-296.8	-300.2
Sulfur trioxide	SO_3	(g)	-395.7	-371.1
Sulfur trioxide	SO_3	(l)	-441	-

Sulfuric acid	H_2SO_4	(l)	-814	-690
Sulfuric acid	H_2SO_4	(aq)	-	-744.5
Water	H_2O	(g)	-241.8	-228.6
Water	H_2O	(l)	-285.8	-237.1

Notes: For solutes in aqueous solution, indicated by (aq), the standard state is a hypothetical ideal 1-molal solution in water at 1 bar. For gases, the standard state is the pure ideal gas at 1 bar. For liquids and solids, the standard state is pure substance in that state of aggregation at 1 bar. ΔH_f^o and ΔG_f^o are the changes in enthalpy and Gibbs free energy that occur when 1 mol of the listed compound is formed from its elements with each substance in its standard state. The table gives ΔH_f^o and ΔG_f^o at 298.15 K. Thus the temperature in the standard states is 298.15 K.

Sources:

TRC Thermodynamic Tables-Hydrocarbons, Thermodynamics Research Center, Texas A & M University System, College Station, Texas.

The NBS tables of chemical thermodynamic properties, *J. Physical and Chemical Reference Data* 1982; Vol. 11, Suppl. 2.

Smith, J.M., Van Ness, H.C., Abbott, M.M., *Introduction to chemical engineering thermodynamics*, McGraw Hill, New York, 6th edition, 2001.

Sandler, S.I., *Chemical and Engineering Thermodynamics*, John Wiley & Sons, New York, 3rd edition, 1999.

Appendix J

Lee/Kesler Tables

Table number	Description	Calculation of Property
J.1	Z^o values	$Z = Z^o + \omega Z^1$
J.2	Z^1 values	
J.3	$(h^R)^o/RT_c$ values	$h^R/RT_c = (h^R)^o/RT_c + \omega\,(h^R)^1/RT_c$
J.4	$(h^R)^1/RT_c$ values	
J.5	$(s^R)^o/R$ values	$s^R/R = (s^R)^o/R + \omega\,(s^R)^1/R$
J.6	$(s^R)^1/R$ values	
J.7	φ^o values	$ln\varphi = ln\varphi^o + \omega ln\varphi^1$ $\varphi = \varphi^o(\varphi^1)^\omega$
J.8	φ^1 values	

(*Sources:* Lee, I.B., Kesler, M.G., *AIChE J.* 1975; 21:510-527; Smith, J.M., Van Ness, H.C., Abbott, M.M., *Introduction to chemical engineering thermodynamics*, McGraw Hill, New York, 6th edition, 2001)

Table J.1: Z^o Values

Low $P_r (\leq 1.0)$

$P_r \rightarrow$ / $T_r \downarrow$	0.0100	0.0500	0.1000	0.2000	0.4000	0.6000	0.8000	1.0000
0.30	0.0029	0.0145	0.0290	0.0579	0.1158	0.1737	0.2315	0.2892
0.35	0.0026	0.0130	0.0261	0.0522	0.1043	0.1564	0.2084	0.2604
0.40	0.0024	0.0119	0.0239	0.0477	0.0953	0.1429	0.1904	0.2379
0.45	0.0022	0.0110	0.0221	0.0442	0.0882	0.1322	0.1762	0.2200
0.50	0.0021	0.0103	0.0207	0.0413	0.0825	0.1236	0.1647	0.2056
0.55	0.9804	0.0098	0.0195	0.0390	0.0778	0.1166	0.1553	0.1939
0.60	0.9849	0.0093	0.0186	0.0371	0.0741	0.1109	0.1470	0.1842
0.65	0.9881	0.9377	0.0178	0.0356	0.0710	0.1063	0.1415	0.1765
0.70	0.9904	0.9504	0.8958	0.0344	0.0687	0.1027	0.1366	0.1703
0.75	0.9922	0.9598	0.9165	0.0336	0.0670	0.1001	0.1330	0.1656
0.80	0.9935	0.9669	0.9319	0.8539	0.0661	0.0985	0.1307	0.1626
0.85	0.9946	0.9725	0.9436	0.8810	0.0661	0.0983	0.1301	0.1614
0.90	0.9954	0.9768	0.9528	0.9015	0.7800	0.1006	0.1321	0.1630
0.93	0.9959	0.9790	0.9573	0.9115	0.8059	0.6635	0.1359	0.1664
0.95	0.9961	0.9803	0.9600	0.9174	0.8206	0.6967	0.1410	0.1705
0.97	0.9963	0.9815	0.9625	0.9227	0.8338	0.7240	0.5580	0.1779
0.98	0.9965	0.9821	0.9637	0.9253	0.8398	0.7360	0.5887	0.1844
0.99	0.9966	0.9826	0.9648	0.9277	0.8455	0.7471	0.6138	0.1959
1.00	0.9967	0.9832	0.9659	0.9300	0.8509	0.7574	0.6355	0.2901
1.01	0.9968	0.9837	0.9669	0.9322	0.8561	0.7671	0.6542	0.4648
1.02	0.9969	0.9842	0.9679	0.9343	0.8610	0.7761	0.6710	0.5146
1.05	0.9971	0.9855	0.9707	0.9401	0.8743	0.8002	0.7130	0.6026
1.10	0.9975	0.9874	0.9747	0.9485	0.8930	0.8323	0.7649	0.6880
1.15	0.9978	0.9891	0.9780	0.9554	0.9081	0.8576	0.8032	0.7443
1.20	0.9981	0.9904	0.9808	0.9611	0.9205	0.8779	0.8330	0.7858
1.30	0.9985	0.9926	0.9852	0.9702	0.9396	0.9083	0.8764	0.8438
1.40	0.9988	0.9942	0.9884	0.9768	0.9534	0.9298	0.9062	0.8827
1.50	0.9991	0.9954	0.9909	0.9818	0.9636	0.9456	0.9278	0.9103
1.60	0.9993	0.9964	0.9928	0.9856	0.9714	0.9575	0.9439	0.9308
1.70	0.9994	0.9971	0.9943	0.9886	0.9775	0.9667	0.9563	0.9463
1.80	0.9995	0.9977	0.9955	0.9910	0.9823	0.9739	0.9659	0.9583
1.90	0.9996	0.9982	0.9964	0.9929	0.9861	0.9796	0.9735	0.9678
2.00	0.9997	0.9986	0.9972	0.9944	0.9892	0.9842	0.9796	0.9754
2.20	0.9998	0.9992	0.9983	0.9967	0.9937	0.9910	0.9886	0.9865
2.40	0.9999	0.9996	0.9991	0.9983	0.9969	0.9957	0.9948	0.9941
2.60	1.0000	0.9998	0.9997	0.9994	0.9991	0.9990	0.9990	0.9993
2.80	1.0000	1.0000	1.0001	1.0002	1.0007	1.0013	1.0021	1.0031
3.00	1.0000	1.0002	1.0004	1.0008	1.0018	1.0030	1.0043	1.0057
3.50	1.0001	1.0004	1.0008	1.0017	1.0035	1.0055	1.0075	1.0097
4.00	1.0001	1.0005	1.0010	1.0021	1.0043	1.0066	1.0090	1.0115

High $P_r (\geq 1.0)$

$P_r \rightarrow$ / $T_r \downarrow$	1.0000	1.2000	1.5000	2.0000	3.0000	5.0000	7.0000	10.000
0.30	0.2892	0.3479	0.4335	0.5775	0.8648	1.4366	2.0048	2.8507
0.35	0.2604	0.3123	0.3901	0.5195	0.7775	1.2902	1.7987	2.5539
0.40	0.2379	0.2853	0.3563	0.4744	0.7095	1.1758	1.6373	2.3211
0.45	0.2200	0.2638	0.3294	0.4384	0.6551	1.0841	1.5077	2.1338
0.50	0.2056	0.2465	0.3077	0.4092	0.6110	1.0094	1.4017	1.9801
0.55	0.1939	0.2323	0.2899	0.3853	0.5747	0.9475	1.3137	1.8520
0.60	0.1842	0.2207	0.2753	0.3657	0.5446	0.8959	1.2398	1.7440
0.65	0.1765	0.2113	0.2634	0.3495	0.5197	0.8526	1.1773	1.6519
0.70	0.1703	0.2038	0.2538	0.3364	0.4991	0.8161	1.1341	1.5729
0.75	0.1656	0.1981	0.2464	0.3260	0.4823	0.7854	1.0787	1.5047
0.80	0.1626	0.1942	0.2411	0.3182	0.4690	0.7598	1.0400	1.4456
0.85	0.1614	0.1924	0.2382	0.3132	0.4591	0.7388	1.0071	1.3943
0.90	0.1630	0.1935	0.2383	0.3114	0.4527	0.7220	0.9793	1.3496
0.93	0.1664	0.1963	0.2405	0.3122	0.4507	0.7138	0.9648	1.3257
0.95	0.1705	0.1998	0.2432	0.3138	0.4501	0.7092	0.9561	1.3108
0.97	0.1779	0.2055	0.2474	0.3164	0.4504	0.7052	0.9480	1.2968
0.98	0.1844	0.2097	0.2503	0.3182	0.4508	0.7035	0.9442	1.2901
0.99	0.1959	0.2154	0.2538	0.3204	0.4514	0.7018	0.9406	1.2835
1.00	0.2901	0.2237	0.2583	0.3229	0.4522	0.7004	0.9372	1.2772
1.01	0.4648	0.2370	0.2640	0.3260	0.4533	0.6991	0.9339	1.2710
1.02	0.5146	0.2629	0.2715	0.3297	0.4547	0.6980	0.9307	1.2650
1.05	0.6026	0.4437	0.3131	0.3452	0.4604	0.6956	0.9222	1.2481
1.10	0.6880	0.5984	0.4580	0.3953	0.4770	0.6950	0.9110	1.2232
1.15	0.7443	0.6803	0.5798	0.4760	0.5042	0.6987	0.9033	1.2021
1.20	0.7858	0.7363	0.6605	0.5605	0.5425	0.7069	0.8990	1.1844
1.30	0.8438	0.8111	0.7624	0.6908	0.6344	0.7358	0.8998	1.1580
1.40	0.8827	0.8595	0.8256	0.7753	0.7202	0.7761	0.9112	1.1419
1.50	0.9103	0.8933	0.8689	0.8328	0.7887	0.8200	0.9297	1.1339
1.60	0.9308	0.9180	0.9000	0.8738	0.8410	0.8617	0.9518	1.1320
1.70	0.9463	0.9367	0.9234	0.9043	0.8809	0.8984	0.9745	1.1343
1.80	0.9583	0.9511	0.9413	0.9275	0.9118	0.9297	0.9961	1.1391
1.90	0.9678	0.9624	0.9552	0.9456	0.9359	0.9557	1.0157	1.1452
2.00	0.9754	0.9715	0.9664	0.9599	0.9550	0.9772	1.0328	1.1516
2.20	0.9865	0.9847	0.9826	0.9806	0.9827	1.0094	1.0600	1.1635
2.40	0.9941	0.9936	0.9935	0.9945	1.0011	1.0313	1.0793	1.1728
2.60	0.9993	0.9998	1.0010	1.0040	1.0137	1.0463	1.0926	1.1792
2.80	1.0031	1.0042	1.0063	1.0106	1.0223	1.0565	1.1016	1.1830
3.00	1.0057	1.0074	1.0101	1.0153	1.0284	1.0635	1.1075	1.1848
3.50	1.0097	1.0120	1.0156	1.0221	1.0368	1.0723	1.1138	1.1834
4.00	1.0115	1.0140	1.0179	1.0249	1.0401	1.0747	1.1136	1.1773

Table J.2: Z^1 Values

Low $P_r (\leq 1.0)$

$P_r \rightarrow$ / $T_r \downarrow$	0.0100	0.0500	0.1000	0.2000	0.4000	0.6000	0.8000	1.0000
0.30	-0.0008	-0.0040	-0.0081	-0.0161	-0.0323	-0.0484	-0.0645	-0.0806
0.35	-0.0009	-0.0046	-0.0093	-0.0185	-0.0370	-0.0554	-0.0738	-0.0921
0.40	-0.0010	-0.0048	-0.0095	-0.0190	-0.0380	-0.0570	-0.0758	-0.0946
0.45	-0.0009	-0.0047	-0.0094	-0.0187	-0.0374	-0.0560	-0.0745	-0.0929
0.50	-0.0009	-0.0045	-0.0090	-0.0181	-0.0360	-0.0539	-0.0716	-0.0893
0.55	-0.0314	-0.0043	-0.0086	-0.0172	-0.0343	-0.0513	-0.0682	-0.0849
0.60	-0.0205	-0.0041	-0.0082	-0.0164	-0.0326	-0.0487	-0.0646	-0.0803
0.65	-0.0137	-0.0772	-0.0078	-0.0156	-0.0309	-0.0461	-0.0611	-0.0759
0.70	-0.0093	-0.0507	-0.1161	-0.0148	-0.0294	-0.0438	-0.0579	-0.0718
0.75	-0.0064	-0.0339	-0.0744	-0.0143	-0.0282	-0.0417	-0.0550	-0.0681
0.80	-0.0044	-0.0228	-0.0487	-0.1160	-0.0272	-0.0401	-0.0526	-0.0648
0.85	-0.0029	-0.0152	-0.0319	-0.0715	-0.0268	-0.0391	-0.0509	-0.0622
0.90	-0.0019	-0.0099	-0.0205	-0.0442	-0.1118	-0.0396	-0.0503	-0.0604
0.93	-0.0015	-0.0075	-0.0154	-0.0326	-0.0763	-0.1662	-0.0514	-0.0602
0.95	-0.0012	-0.0062	-0.0126	-0.0262	-0.0589	-0.1110	-0.0540	-0.0607
0.97	-0.0010	-0.0050	-0.0101	-0.0208	-0.0450	-0.0770	-0.1647	-0.0623
0.98	-0.0009	-0.0044	-0.0090	-0.0184	-0.0390	-0.0641	-0.1100	-0.0641
0.99	-0.0008	-0.0039	-0.0079	-0.0161	-0.0355	-0.0531	-0.0796	-0.0680
1.00	-0.0007	-0.0034	-0.0069	-0.0140	-0.0285	-0.0435	-0.0588	-0.0879
1.01	-0.0006	-0.0030	-0.0060	-0.0120	-0.0240	-0.0351	-0.0429	-0.0223
1.02	-0.0005	-0.0026	-0.0051	-0.0102	-0.0198	-0.0277	-0.0303	-0.0062
1.05	-0.0003	-0.0015	-0.0029	-0.0054	-0.0092	-0.0097	-0.0032	0.0220
1.10	0.0000	0.0000	0.0001	0.0007	0.0038	0.0106	0.0236	0.0476
1.15	0.0002	0.0011	0.0023	0.0052	0.0127	0.0237	0.0396	0.0625
1.20	0.0004	0.0019	0.0039	0.0084	0.0190	0.0326	0.0499	0.0719
1.30	0.0006	0.0030	0.0061	0.0125	0.0267	0.0429	0.0612	0.0819
1.40	0.0007	0.0036	0.0072	0.0147	0.0306	0.0477	0.0661	0.0857
1.50	0.0008	0.0039	0.0078	0.0158	0.0323	0.0497	0.0677	0.0864
1.60	0.0008	0.0040	0.0080	0.0162	0.0330	0.0501	0.0677	0.0855
1.70	0.0008	0.0040	0.0081	0.0163	0.0329	0.0497	0.0667	0.0838
1.80	0.0008	0.0040	0.0081	0.0162	0.0325	0.0488	0.0652	0.0814
1.90	0.0008	0.0040	0.0079	0.0159	0.0318	0.0477	0.0635	0.0792
2.00	0.0008	0.0039	0.0078	0.0155	0.0310	0.0464	0.0617	0.0767
2.20	0.0007	0.0037	0.0074	0.0147	0.0293	0.0437	0.0579	0.0719
2.40	0.0007	0.0035	0.0070	0.0139	0.0276	0.0411	0.0544	0.0675
2.60	0.0007	0.0033	0.0066	0.0131	0.0260	0.0387	0.0512	0.0634
2.80	0.0006	0.0031	0.0062	0.0124	0.0245	0.0365	0.0483	0.0598
3.00	0.0006	0.0029	0.0059	0.0117	0.0232	0.0345	0.0456	0.0565
3.50	0.0005	0.0026	0.0052	0.0103	0.0204	0.0303	0.0401	0.0497
4.00	0.0005	0.0023	0.0046	0.0091	0.0182	0.0270	0.0357	0.0443

High $P_r (\geq 1.0)$

$P_r \rightarrow$ / $T_r \downarrow$	1.0000	1.2000	1.5000	2.0000	3.0000	5.0000	7.0000	10.000
0.30	-0.0806	-0.0966	-0.1207	-0.1608	-0.2407	-0.3996	-0.5572	-0.7915
0.35	-0.0921	-0.1105	-0.1379	-0.1834	-0.2738	-0.4523	-0.6279	-0.8863
0.40	-0.0946	-0.1134	-0.1414	-0.1879	-0.2799	-0.4603	-0.6365	-0.8936
0.45	-0.0929	-0.1113	-0.1387	-0.1840	-0.2734	-0.4475	-0.6162	-0.8608
0.50	-0.0893	-0.1069	-0.1330	-0.1762	-0.2611	-0.4253	-0.5831	-0.8099
0.55	-0.0849	-0.1015	-0.1263	-0.1669	-0.2465	-0.3991	-0.5446	-0.7521
0.60	-0.0803	-0.0960	-0.1192	-0.1572	-0.2312	-0.3718	-0.5047	-0.6928
0.65	-0.0759	-0.0906	-0.1122	-0.1476	0.2160	-0.3447	-0.4653	-0.6346
0.70	-0.0718	-0.0855	-0.1057	-0.1385	-0.2013	-0.3184	-0.4270	-0.5785
0.75	-0.0681	-0.0808	-0.0996	-0.1298	-0.1872	-0.2929	-0.3901	-0.5250
0.80	-0.0648	-0.0767	-0.0940	-0.1217	-0.1736	-0.2682	-0.3545	-0.4740
0.85	-0.0622	-0.0731	-0.0888	-0.1138	-0.1602	-0.2439	-0.3201	-0.4254
0.90	-0.0604	-0.0701	-0.0840	-0.1059	-0.1463	-0.2195	-0.2862	-0.3788
0.93	-0.0602	-0.0687	-0.0810	-0.1007	-0.1374	-0.2045	-0.2661	-0.3516
0.95	-0.0607	-0.0678	-0.0788	-0.0967	-0.1310	-0.1943	-0.2526	-0.3339
0.97	-0.0623	-0.0669	-0.0759	-0.0921	-0.1240	-0.1837	-0.2391	-0.3163
0.98	-0.0641	-0.0661	-0.0740	-0.0893	-0.1202	-0.1783	-0.2322	-0.3075
0.99	-0.0680	-0.0646	-0.0715	-0.0861	-0.1162	-0.1728	-0.2254	-0.2989
1.00	-0.0879	-0.0609	-0.0678	-0.0824	-0.1118	-0.1672	-0.2185	-0.2902
1.01	-0.0223	-0.0473	-0.0621	-0.0778	-0.1072	-0.1615	-0.2116	-0.2816
1.02	-0.0062	-0.0227	-0.0524	-0.0722	-0.1021	-0.1556	-0.2047	-0.2731
1.05	0.0220	0.1059	0.0451	-0.0432	-0.0838	-0.1370	-0.1835	-0.2476
1.10	0.0476	0.0897	0.1630	0.0698	-0.0373	-0.1021	-0.1469	-0.2056
1.15	0.0625	0.0943	0.1548	0.1667	0.0332	-0.0611	-0.1084	-0.1642
1.20	0.0719	0.0991	0.1477	0.1990	0.1095	-0.0141	-0.0678	-0.1231
1.30	0.0819	0.1048	0.1420	0.1991	0.2079	0.0875	0.0176	-0.0423
1.40	0.0857	0.1063	0.1383	0.1894	0.2397	0.1737	0.1008	0.0350
1.50	0.0854	0.1055	0.1345	0.1806	0.2433	0.2309	0.1717	0.1058
1.60	0.0855	0.1035	0.1303	0.1729	0.2381	0.2631	0.2255	0.1673
1.70	0.0838	0.1008	0.1259	0.1658	0.2305	0.2788	0.2628	0.2179
1.80	0.0816	0.0978	0.1216	0.1593	0.2224	0.2846	0.2871	0.2576
1.90	0.0792	0.0947	0.1173	0.1532	0.2144	0.2848	0.3017	0.2876
2.00	0.0767	0.0916	0.1133	0.1476	0.2069	0.2819	0.3097	0.3096
2.20	0.0719	0.0857	0.1057	0.1374	0.1932	0.2720	0.3135	0.3355
2.40	0.0675	0.0803	0.0989	0.1285	0.1812	0.2602	0.3089	0.3459
2.60	0.0634	0.0754	0.0929	0.1207	0.1706	0.2484	0.3009	0.3475
2.80	0.0598	0.0711	0.0876	0.1138	0.1613	0.2372	0.2915	0.3443
3.00	0.0535	0.0672	0.0828	0.1076	0.1529	0.2268	0.2817	0.3385
3.50	0.0497	0.0591	0.0728	0.0949	0.1356	0.2042	0.2584	0.3194
4.00	0.0443	0.0527	0.0651	0.0849	0.1219	0.1857	0.2378	0.2994

Table J.3: $(h^R)^o/RT_c$ Values

Low $P_r (\leq 1.0)$

$P_r \rightarrow$ / $T_r \downarrow$	0.0100	0.0500	0.1000	0.2000	0.4000	0.6000	0.8000	1.0000
0.30	-6.045	-6.043	-6.040	-6.034	-6.022	-6.011	-5.999	-5.987
0.35	-5.906	-5.904	-5.901	-5.895	-5.882	-5.870	-5.858	-5.845
0.40	-5.763	-5.761	-5.757	-5.751	-5.738	-5.726	-5.713	-5.700
0.45	-5.615	-5.612	-5.609	-5.603	-5.590	-5.577	-5.564	-5.551
0.50	-5.465	-5.463	-5.459	-5.453	-5.440	-5.427	-5.414	-5.401
0.55	-0.032	-5.312	-5.309	-5.303	-5.290	-5.278	-5.265	-5.252
0.60	-0.027	-5.162	-5.159	-5.153	-5.141	-5.129	-5.116	-5.104
0.65	-0.023	-0.118	-5.008	-5.002	-4.991	-4.980	-4.968	-4.956
0.70	-0.02	-0.101	-0.213	-4.848	-4.838	-4.828	-4.818	-4.808
0.75	-0.017	-0.088	-0.183	-4.687	-4.679	-4.672	-4.664	-4.655
0.80	-0.015	-0.078	-0.160	-0.345	-4.507	-4.504	-4.499	-4.494
0.85	-0.014	-0.069	-0.141	-0.300	-4.309	-4.313	-4.316	-4.316
0.90	-0.012	-0.062	-0.126	-0.264	-0.596	-4.074	-4.094	-4.108
0.93	-0.011	-0.058	-0.118	-0.246	-0.545	-0.960	-3.920	-3.953
0.95	-0.011	-0.056	-0.113	-0.235	-0.516	-0.885	-3.763	-3.825
0.97	-0.011	-0.054	-0.109	-0.225	-0.490	-0.824	-1.356	-3.658
0.98	-0.010	-0.053	-0.107	-0.221	-0.478	-0.797	-1.273	-3.544
0.99	-0.010	-0.052	-0.105	-0.216	-0.466	-0.773	-1.206	-3.376
1.00	-0.010	-0.051	-0.103	-0.212	-0.455	-0.750	-1.151	-2.584
1.01	-0.010	-0.050	-0.101	-0.208	-0.445	-0.721	-1.102	-1.796
1.02	-0.010	-0.049	-0.099	-0.203	-0.434	-0.708	-1.060	-1.627
1.05	-0.009	-0.046	-0.094	-0.192	-0.407	-0.654	-0.955	-1.359
1.10	-0.008	-0.042	-0.086	-0.175	-0.367	-0.581	-0.827	-1.120
1.15	-0.008	-0.039	-0.079	-0.160	-0.334	-0.523	-0.732	-0.968
1.20	-0.007	-0.036	-0.073	-0.148	-0.305	-0.474	-0.657	-0.857
1.30	-0.006	-0.031	-0.063	-0.127	-0.259	-0.399	-0.545	-0.698
1.40	-0.005	-0.027	-0.055	-0.110	-0.224	-0.341	-0.463	-0.588
1.50	-0.005	-0.024	-0.048	-0.097	-0.196	-0.297	-0.400	-0.505
1.60	-0.004	-0.021	-0.043	-0.086	-0.173	-0.261	-0.350	-0.440
1.70	-0.004	-0.019	-0.038	-0.076	-0.153	-0.231	-0.309	-0.387
1.80	-0.003	-0.017	-0.034	-0.068	-0.137	-0.206	-0.275	-0.344
1.90	-0.003	-0.015	-0.031	-0.062	-0.123	-0.185	-0.246	-0.307
2.00	-0.003	-0.014	-0.028	-0.056	-0.111	-0.167	-0.222	-0.276
2.20	-0.002	-0.012	-0.023	-0.046	-0.092	-0.137	-0.182	-0.226
2.40	-0.002	-0.010	-0.019	-0.038	-0.076	-0.114	-0.150	-0.187
2.60	-0.002	-0.008	-0.016	-0.032	-0.064	-0.095	-0.125	-0.155
2.80	-0.001	-0.007	-0.014	-0.027	-0.054	-0.080	-0.105	-0.130
3.00	-0.001	-0.006	-0.011	-0.023	-0.045	-0.067	-0.088	-0.190
3.50	-0.001	-0.004	-0.007	-0.015	-0.029	-0.043	-0.056	-0.069
4.00	0.000	-0.002	-0.005	-0.009	-0.017	-0.026	-0.033	-0.041

High $P_r (\geq 1.0)$

$P_r \rightarrow$ $T_r \downarrow$	1.0000	1.2000	1.5000	2.0000	3.0000	5.0000	7.0000	10.000
0.30	-5.987	-5.975	-5.957	-5.927	-5.868	-5.748	-5.628	-5.446
0.35	-5.845	-5.833	-5.814	-5.783	-5.721	-5.595	-5.469	-5.278
0.40	-5.700	-5.687	-5.668	-5.636	-5.572	-5.442	-5.311	-5.113
0.45	-5.551	-5.538	-5.519	-5.486	-5.421	-5.288	-5.154	-5.950
0.50	-5.401	-5.388	-5.369	-5.336	-5.279	-5.135	-4.999	-4.791
0.55	-5.252	-5.239	-5.220	-5.187	-5.121	-4.986	-4.849	-4.638
0.60	-5.104	-5.091	-5.073	-5.041	-4.976	-4.842	-4.794	-4.492
0.65	-4.956	-4.949	-4.927	-4.896	-4.833	-4.702	-4.565	-4.353
0.70	-4.808	-4.797	-4.781	-4.752	-4.693	-4.566	-4.432	-4.221
0.75	-4.655	-4.646	-4.632	-4.607	-4.554	-4.434	-4.393	-4.095
0.80	-4.494	-4.488	-4.478	-4.459	-4.413	-4.303	-4.178	-3.974
0.85	-4.316	-4.316	-4.312	-4.302	-4.269	-4.173	-4.056	-3.857
0.90	-4.108	-4.118	-4.127	-4.132	-4.119	-4.043	-3.935	-3.744
0.93	-3.953	-3.976	-4.000	-4.020	-4.024	-3.963	-3.863	-3.678
0.95	-3.825	-3.865	-3.904	-3.940	-3.958	-3.910	-3.815	-3.634
0.97	-3.658	-3.732	-3.796	-3.853	-3.890	-3.856	-3.767	-3.591
0.98	-3.544	-3.652	-3.736	-3.806	-3.854	-3.829	-3.743	-3.569
0.99	-3.376	-3.558	-3.670	-3.758	-3.818	-3.801	-3.719	-3.548
1.00	-2.584	-3.441	-3.598	-3.706	-3.782	-3.774	-3.695	-3.526
1.01	-1.796	-3.283	-3.516	-3.652	-3.744	-3.746	-3.671	-3.505
1.02	-1.627	-3.039	-3.422	-3.595	-3.705	-3.718	-3.647	-3.484
1.05	-1.359	-2.034	-3.030	-3.398	-3.583	-3.632	-3.575	-3.420
1.10	-1.120	-1.487	-2.203	-2.965	-3.353	-3.484	-3.453	-3.315
1.15	-0.968	-1.239	-1.719	-2.479	-3.091	-3.329	-3.329	-3.211
1.20	-0.857	-1.076	-1.443	-2.079	-2.801	-3.166	-3.202	-3.107
1.30	-0.698	-0.860	-1.116	-1.560	-2.274	-2.825	-2.942	-2.899
1.40	-0.588	-0.716	-0.915	-1.253	-1.857	-2.486	-2.679	-2.692
1.50	-0.505	-0.611	-0.774	-1.046	-1.549	-2.175	-2.421	-2.486
1.60	-0.440	-0.531	-0.667	-0.894	-1.318	-1.904	-2.177	-2.285
1.70	-0.387	-0.446	-0.583	-0.777	-1.139	-1.672	-1.953	-2.091
1.80	-0.344	-0.413	-0.515	-0.683	-0.996	-1.476	-1.751	-1.908
1.90	-0.307	-0.368	-0.458	-0.606	-0.880	-1.309	-1.571	-1.736
2.00	-0.276	-0.330	-0.411	-0.541	-0.782	-1.167	-1.411	-1.577
2.20	-0.226	-0.269	-0.334	-0.437	-0.629	-0.937	-1.143	-1.295
2.40	-0.187	-0.222	-0.275	-0.359	-0.513	-0.761	-0.929	-1.058
2.60	-0.155	-0185	-0.228	-0.297	-0.422	-0.621	-0.756	-0.858
2.80	-0.130	-0.154	-0.190	-0.246	-0.348	-0.508	-0.614	-0.689
3.00	-0.109	-0.129	-0.159	-0.205	-0.288	-0.415	-0.495	-0.545
3.50	-0.069	-0.081	-0.099	-0.127	-0.174	-0.239	-0.270	-0.264
4.00	-0.041	-0.048	-0.058	-0.072	-0.095	-0.116	-0.110	-0.061

Table J.4: $(h^R)^1/RT_c$ Values

Low $P_r (\leq 1.0)$

$P_r \rightarrow$ / $T_r \downarrow$	0.0100	0.0500	0.1000	0.2000	0.4000	0.6000	0.8000	1.0000
0.30	-11.098	-11.096	-11.095	-11.091	-11.083	-11.076	-11.069	-11.062
0.35	-10.656	-10.655	-10.654	-10.653	-10.650	-10.646	-10.643	-10.640
0.40	-10.121	-10.121	-10.121	-10.120	-10.121	-10.121	-10.121	-10.121
0.45	-9.515	-9.515	-9.516	-9.517	-9.519	-9.521	-9.523	-9.525
0.50	-8.868	-8.869	-8.870	-8.872	-8.876	-8.880	-8.884	-8.888
0.55	-0.080	-8.211	-8.212	-8.215	-8.221	-8.226	-8.232	-8.238
0.60	-0.059	-7.508	-7.570	-7.573	-7.579	-7.585	-7.591	-7.596
0.65	-0.045	-0.247	-6.949	-6.952	-6.959	-6.966	-6.973	-6.980
0.70	-0.034	-0.185	-0.415	-6.360	-6.367	-6.373	-6.381	-6.388
0.75	-0.027	-0.142	-0.306	-5.796	-5.802	-5.809	-5.816	-5.824
0.80	-0.021	-0.110	-0.234	-0.542	-5.266	-5.271	-5.278	-5.285
0.85	-0.017	-0.087	-0.182	-0.401	-4.753	-4.754	-4.758	-4.763
0.90	-0.014	-0.070	-0.144	-0.308	-0.751	-4.254	-4.248	-4.249
0.93	-0.012	-0.061	-0.126	-0.265	-0.612	-1.236	-3.942	-3.934
0.95	-0.011	-0.056	-0.115	-0.241	-0.542	-0.994	-3.737	-3.712
0.97	-0.010	-0.052	-0.105	-0.219	-0.483	-0.837	-1.616	-3.470
0.98	-0.010	-0.050	-0.101	-0.209	-0.457	-0.776	-1.324	-3.332
0.99	-0.009	-0.048	-0.097	-0.200	-0.433	-0.722	-1.154	-3.164
1.00	-0.009	-0.046	-0.093	-0.191	-0.410	-0.675	-1.034	-2.471
1.01	-0.009	-0.044	-0.089	-0.183	-0.389	-0.632	-0.940	-1.375
1.02	-0.008	-0.042	-0.085	-0.175	-0.370	-0.594	-0.863	-1.180
1.05	-0.007	-0.037	-0.075	-0.153	-0.318	-0.498	-0.691	-0.877
1.10	-0.006	-0.030	-0.061	-0.123	-0.251	-0.381	-0.507	-0.617
1.15	-0.005	-0.025	-0.050	-0.099	-0.199	-0.296	-0.385	-0.459
1.20	-0.004	-0.020	-0.040	-0.080	-0.158	-0.232	-0.297	-0.349
1.30	-0.003	-0.013	-0.026	-0.052	-0.100	-0.142	-0.177	-0.203
1.40	-0.002	-0.008	-0.016	-0.032	-0.060	-0.083	-0.100	-0.111
1.50	-0.001	-0.005	-0.009	-0.018	-0.032	-0.042	-0.048	-0.049
1.60	-0.000	-0.002	-0.004	-0.007	-0.012	-0.013	-0.011	-0.005
1.70	-0.000	-0.000	-0.000	-0.000	0.003	0.009	0.017	0.027
1.80	0.000	0.001	0.003	0.006	0.015	0.025	0.037	0.051
1.90	0.001	0.003	0.005	0.011	0.023	0.037	0.053	0.070
2.00	0.001	0.003	0.007	0.015	0.030	0.047	0.065	0.085
2.20	0.001	0.005	0.010	0.020	0.040	0.062	0.083	0.106
2.40	0.001	0.006	0.012	0.023	0.047	0.071	0.095	0.120
2.60	0.001	0.006	0.013	0.026	0.052	0.078	0.104	0.130
2.80	0.001	0.007	0.014	0.028	0.055	0.082	0.110	0.137
3.00	0.001	0.007	0.014	0.029	0.058	0.086	0.114	0.142
3.50	0.002	0.008	0.016	0.031	0.062	0.092	0.122	0.152
4.00	0.002	0.008	0.016	0.032	0.064	0.096	0.127	0.158

High $P_r (\geq 1.0)$

$P_r \rightarrow$ / $T_r \downarrow$	1.0000	1.2000	1.5000	2.0000	3.0000	5.0000	7.0000	10.000
0.30	-11.062	-11.055	-11.044	-11.027	-10.992	-10.935	-10.872	-10.781
0.35	-10.640	-10.637	-10.632	-10.624	-10.609	-10.581	-10.554	-10.529
0.40	-10.121	-10.121	-10.121	-10.122	-10.123	-10.128	-10.135	-10.150
0.45	-9.525	-9.527	-9.531	-9.537	-9.549	-9.576	-9.611	-9.663
0.50	-8.888	-8.892	-8.899	-8.909	-8.932	-8.978	-9.030	-9.111
0.55	-8.238	-8.243	-8.252	-8.267	-8.298	-8.360	-8.425	-8.531
0.60	-7.596	-7.603	-7.614	-7.632	-7.669	-7.745	-7.824	-7.950
0.65	-6.980	-6.987	-6.997	-7.017	-7.059	-7.147	-7.239	-7.381
0.70	-6.388	-6.395	-6.407	-6.429	-6.475	-6.574	-6.677	-6.837
0.75	-5.824	-5.832	-5.845	-5.868	-5.918	-6.027	-6.142	-6.318
0.80	-5.285	-5.293	-5.306	-5.330	-5.385	-5.506	-5.632	-5.824
0.85	-4.763	-4.771	-4.784	-4.810	-4.872	-5.000	-5.149	-5.358
0.90	-4.249	-4.255	-4.268	-4.298	-4.371	-4.530	-4.688	-4.916
0.93	-3.934	-3.937	-3.951	-3.987	-4.073	-4.251	-4.422	-4.662
0.95	-3.712	-3.713	-3.730	-3.773	-3.873	-4.068	-4.248	-4.497
0.97	-3.470	-3.467	-3.492	-3.551	-3.670	-3.885	-4.077	-4.336
0.98	-3.332	-3.327	-3.363	-3.434	-3.568	-3.795	-3.992	-4.257
0.99	-3.164	-3.164	-3.223	-3.313	-3.464	-3.705	-3.909	-4.178
1.00	-2.471	-2.952	-3.065	-3.186	-3.358	-3.615	-3.825	-4.100
1.01	-1.375	-2.595	-2.880	-3.051	-3.251	-3.525	-3.742	-4.023
1.02	-1.180	-1.723	-2.650	-2.906	-3.142	-3.435	-3.661	-3.947
1.05	-0.877	-0.878	-1.496	-2.381	-2.800	-3.167	-3.418	-3.722
1.10	-0.617	-0.673	-0.617	-1.261	-2.167	-2.720	-3.023	-3.362
1.15	-0.459	-0.503	-0.487	-0.604	-1.497	-2.275	-2.641	-3.019
1.20	-0.349	-0.381	-0.381	-0.361	-0.934	-1.840	-2.273	-2.692
1.30	-0.203	-0.218	-0.218	-0.178	-0.300	-1.066	-1.592	-2.086
1.40	-0.111	-0.115	-0.128	-0.070	-0.044	-0.504	-1.012	-1.547
1.50	-0.049	-0.046	-0.032	0.008	0.078	-0.142	-0.556	-1.080
1.60	-0.005	0.004	0.023	0.065	0.151	0.082	-0.217	-0.689
1.70	0.027	0.040	0.063	0.109	0.202	0.223	0.028	-0.369
1.80	0.051	0.067	0.094	0.143	0.241	0.317	0.203	-0.112
1.90	0.070	0.088	0.117	0.169	0.271	0.381	0.330	0.092
2.00	0.085	0.105	0.136	0.190	0.295	0.428	0.424	0.255
2.20	0.106	0.128	0.163	0.221	0.331	0.493	0.551	0.489
2.40	0.120	0.144	0.181	0.242	0.356	0.535	0.631	0.645
2.60	0.130	0.156	0.194	0.257	0.376	0.567	0.687	0.754
2.80	0.137	0.164	0.204	0.269	0.391	0.591	0.729	0.836
3.00	0.142	0.170	0.211	0.278	0.403	0.611	0.763	0.899
3.50	0.152	0.181	0.224	0.294	0.425	0.650	0.827	1.015
4.00	0.158	0.188	0.233	0.306	0.442	0.680	0.874	1.097

Table J.5: $(s^R)^0/R$ Values

Low $P_r (\leq 1.0)$

$P_r \rightarrow$ / $T_r \downarrow$	0.0100	0.0500	0.1000	0.2000	0.4000	0.6000	0.8000	1.0000
0.30	-11.614	-10.008	-9.319	-8.635	-7.961	-7.574	-7.304	-7.099
0.35	-11.185	-9.579	-8.890	-8.205	-7.529	-7.140	-6.869	-6.663
0.40	-10.802	-9.196	-8.506	-7.821	-7.144	-6.755	-6.483	-6.275
0.45	-10.453	-8.847	-8.157	-7.472	-6.794	-6.404	-6.132	-5.924
0.50	-10.137	-8.531	-7.841	-7.156	-6.479	-6.089	-5.816	-5.608
0.55	-0.038	-8.245	-7.555	-6.870	-6.193	-5.803	-5.531	-5.324
0.60	-0.029	-7.983	-7.294	-6.610	-5.933	-5.544	-5.273	-5.066
0.65	-0.023	-0.122	-7.052	-6.368	-5.694	-5.306	-5.036	-4.830
0.70	-0.018	-0.096	-0.206	-6.140	-5.467	-5.082	-4.814	-4.610
0.75	-0.015	-0.078	-0.164	-5.917	-5.248	-4.866	-4.600	-4.399
0.80	-0.013	-0.064	-0.134	-0.294	-5.026	-4.694	-4.388	-4.191
0.85	-0.011	-0.054	-0.111	-0.239	-4.785	-4.418	-4.166	-3.976
0.90	-0.009	-0.046	-0.094	-0.199	-0.463	-4.145	-3.912	-3.738
0.93	-0.008	-0.042	-0.085	-0.179	-0.408	-0.750	-3.723	-3.569
0.95	-0.008	-0.039	-0.080	-0.168	-0.377	-0.671	-3.556	-3.433
0.97	-0.007	-0.037	-0.075	-0.157	-0.350	-0.607	-1.056	-3.259
0.98	-0.007	-0.036	-0.073	-0.153	-0.337	-0.580	-0.971	-3.142
0.99	-0.007	-0.035	-0.071	-0.148	-0.326	-0.555	-0.903	-2.972
1.00	-0.007	-0.034	-0.069	-0.144	-0.315	-0.532	-0.847	-2.178
1.01	-0.007	-0.033	-0.067	-0.139	-0.304	-0.510	-0.799	-1.391
1.02	-0.006	-0.032	-0.065	-0.135	-0.294	-0.491	-0.757	-1.225
1.05	-0.006	-0.030	-0.060	-0.124	-0.267	-0.439	-0.656	-0.965
1.10	-0.005	-0.026	-0.053	-0.108	-0.230	-0.371	-0.537	-0.742
1.15	-0.005	-0.023	-0.047	-0.096	-0.201	-0.319	-0.452	-0.607
1.20	-0.004	-0.021	-0.042	-0.085	-0.177	-0.277	-0.389	-0.512
1.30	-0.003	-0.017	-0.033	-0.068	-0.140	-0.217	-0.298	-0.385
1.40	-0.003	-0.014	-0.027	-0.056	-0.114	-0.174	-0.237	-0.303
1.50	-0.002	-0.011	-0.023	-0.046	-0.094	-0.143	-0.194	-0.246
1.60	-0.002	-0.010	-0.019	-0.039	-0.079	-0.120	-0.162	-0.204
1.70	-0.002	-0.008	-0.017	-0.033	-0.067	-0.102	-0.137	-0.172
1.80	-0.001	-0.007	-0.014	-0.029	-0.058	-0.088	-0.117	-0.147
1.90	-0.001	-0.006	-0.013	-0.025	-0.051	-0.076	-0.102	-0.127
2.00	-0.001	-0.006	-0.011	-0.022	-0.044	-0.067	-0.089	-0.111
2.20	-0.001	-0.004	-0.009	-0.018	-0.035	-0.053	-0.070	-0.087
2.40	-0.001	-0.004	-0.007	-0.014	-0.028	-0.042	-0.056	-0.070
2.60	-0.001	-0.003	-0.006	-0.012	-0.023	-0.035	-0.046	-0.058
2.80	0.000	-0.002	-0.005	-0.010	-0.020	-0.029	-0.039	-0.048
3.00	0.000	-0.002	-0.004	-0.008	-0.017	-0.025	-0.033	-0.041
3.50	0.000	-0.001	-0.003	-0.006	-0.012	-0.017	-0.023	-0.029
4.00	0.000	-0.001	-0.002	-0.004	-0.009	-0.013	-0.017	-0.021

High $P_r (\geq 1.0)$

$P_r \rightarrow$ / $T_r \downarrow$	1.0000	1.2000	1.5000	2.0000	3.0000	5.0000	7.0000	10.000
0.30	-7.099	-6.935	-6.740	-6.497	-6.180	-5.847	-5.683	-5.578
0.35	-6.663	-6.497	-6.299	-6.052	-5.728	-5.376	-5.194	-5.060
0.40	-6.275	-6.109	-5.909	-5.660	-5.330	-4.967	-4.772	-4.619
0.45	-5.924	-5.757	-5.557	-5.306	-4.974	-4.603	-4.401	-4.234
0.50	-5.608	-5.441	-5.240	-4.989	-4.656	-4.282	-4.074	-3.899
0.55	-5.324	-5.157	-4.956	-4.706	-4.373	-3.998	-3.788	-3.607
0.60	-5.066	-4.900	-4.700	-4.451	-4.120	-3.747	-3.537	-3.353
0.65	-4.830	-4.665	-4.467	-4.220	-3.892	-3.523	-3.315	-3.131
0.70	-4.610	-4.446	-4.250	-4.007	-3.684	-3.322	-3.117	-2.935
0.75	-4.399	-4.238	-4.045	-3.807	-3.491	-3.138	-2.939	-2.761
0.80	-4.191	-4.034	-3.846	-3.615	-3.310	-2.970	-2.777	-2.605
0.85	-3.976	-3.825	-3.646	-3.425	-3.135	-2.812	-2.629	-2.463
0.90	-3.738	-3.599	-3.434	-3.231	-2.964	-2.663	-2.491	-2.334
0.93	-3.569	-3.444	-3.295	-3.108	-2.860	-2.577	-2.412	-2.262
0.95	-3.433	-3.326	-3.193	-3.023	-2.790	-2.520	-2.362	-2.215
0.97	-3.259	-3.188	-3.081	-2.932	-2.719	-2.463	-2.312	-2.170
0.98	-3.142	-3.106	-3.019	-2.884	-2.682	-2.436	-2.287	-2.148
0.99	-2.972	-3.010	-2.953	-2.835	-2.646	-2.408	-2.263	-2.126
1.00	-2.178	-2.893	-2.879	-2.784	-2.609	-2.380	-2.239	-2.105
1.01	-1.391	-2.736	-2.798	-2.730	-2.571	-2.352	-2.215	-2.083
1.02	-1.225	-2.495	-2.706	-2.673	-2.533	-2.325	-2.191	-2.062
1.05	-0.965	-1.523	-2.328	-2.483	-2.415	-2.242	-2.121	-2.001
1.10	-0.742	-1.012	-1.557	-2.081	-2.202	-2.104	-2.007	-1.903
1.15	-0.607	-0.790	-1.126	-1.649	-1.968	-1.966	-1.897	-1.810
1.20	-0.512	-0.651	-0.890	-1.308	-1.727	-1.827	-1.789	-1.722
1.30	-0.385	-0.478	-0.628	-0.891	-1.299	-1.554	-1.581	-1.556
1.40	-0.303	-0.375	-0.478	-0.663	-0.990	-1.303	-1.386	-1.402
1.50	-0.246	-0.299	-0.381	-0.520	-0.777	-1.088	-1.208	-1.260
1.60	-0.204	-0.247	-0.312	-0.421	-0.628	-0.913	-1.050	-1.130
1.70	-0.172	-0.208	-0.261	-0.350	-0.519	-0.773	-0.915	-1.013
1.80	-0.147	-0.177	-0.222	-0.296	-0.438	-0.661	-0.799	-0.908
1.90	-0.127	-0.153	-0.191	-0.255	-0.375	-0.570	-0.702	-0.815
2.00	-0.111	-0.134	-0.167	-0.221	-0.625	-0.497	-0.620	-0.733
2.20	-0.087	-0.105	-0.130	-0.172	-0.251	-0.388	-0.492	-0.599
2.40	-0.070	-0.084	-0.104	-0.138	-0.201	-0.311	-0.399	-0.496
2.60	-0.058	-0.069	-0.086	-0.113	-0.164	-0.255	-0.329	-0.416
2.80	-0.048	-0.058	-0.072	-0.094	-0.137	-0.213	-0.277	-0.353
3.00	-0.041	-0.049	-0.061	-0.080	-0.116	-0.181	-0.236	-0.303
3.50	-0.029	-0.034	-0.042	-0.056	-0.081	-0.126	-0.166	-0.216
4.00	-0.021	-0.025	-0.031	-0.041	-0.059	-0.093	-0.123	-0.162

Table J.6: $(s^R)^1/R$ Values

Low $P_r (\leq 1.0)$

$P_r \rightarrow$ / $T_r \downarrow$	0.0100	0.0500	0.1000	0.2000	0.4000	0.6000	0.8000	1.0000
0.30	-16.782	-16.774	-16.764	-16.744	-16.705	-16.665	-16.626	-16.586
0.35	-15.413	-15.408	-15.401	-15.387	-15.359	-15.333	-15.305	-15.278
0.40	-13.990	-13.986	-13.981	-13.972	-13.953	-13.934	-13.915	-13.896
0.45	-12.564	-12.561	-12.558	-12.551	-12.537	-12.523	-12.509	-12.496
0.50	-11.202	-11.200	-11.197	-11.092	-11.082	-11.172	-11.162	-11.153
0.55	-0.115	-9.948	-9.946	-9.942	-9.935	-9.928	-9.921	-9.914
0.60	-0.078	-8.828	-8.826	-8.823	-8.817	-8.811	-8.806	-8.799
0.65	-0.055	-0.309	-7.832	-7.829	-7.824	-7.819	-7.815	-7.510
0.70	-0.040	-0.216	-0.491	-6.951	-6.945	-6.941	-6.937	-6.933
0.75	-0.029	-0.156	-0.340	6.173	-6.167	-6.162	-6.158	-6.155
0.80	-0.022	-0.116	-0.246	-0.578	-5.475	-5.468	-5.462	-5.458
0.85	-0.017	-0.088	-0.183	-0.400	-4.853	-4.841	-4.832	-4.826
0.90	-0.013	-0.068	-0.140	-0.301	-0.744	-4.269	-4.249	-4.238
0.93	-0.011	-0.058	-0.120	-0.254	-0.593	-1.219	-3.914	-3.894
0.95	-0.010	-0.053	-0.109	-0.228	-0.517	-0.961	-3.697	-3.658
0.97	-0.010	-0.048	-0.099	-0.206	-0.456	-0.797	-1.570	-3.406
0.98	-0.009	-0.046	-0.094	-0.196	-0.429	-0.734	-1.270	-3.264
0.99	-0.009	-0.044	-0.090	-0.186	-0.405	-0.680	-1.098	-3.093
1.00	-0.008	-0.042	-0.086	-0.177	-0.382	-0.632	-0.977	-2.399
1.01	-0.008	-0.040	-0.082	-0.169	-0.361	-0.590	-0.883	-1.306
1.02	-0.008	-0.039	-0.078	-0.161	-0.342	-0.552	-0.807	-1.113
1.05	-0.007	0.034	-0.069	-0.140	-0.292	-0.460	-0.642	-0.820
1.10	-0.005	-0.028	-0.055	-0.112	-0.229	-0.350	-0.470	-0.577
1.15	-0.005	-0.023	-0.045	-0.091	-0.183	-0.275	-0.361	-0.437
1.20	-0.004	-0.019	-0.037	-0.075	-0.149	-0.220	-0.286	-0.343
1.30	-0.003	-0.013	-0.026	-0.052	-0.102	-0.148	-0.190	-0.226
1.40	-0.002	-0.010	-0.019	-0.037	-0.072	-0.104	-0.133	-0.158
1.50	-0.001	-0.007	-0.014	-0.027	-0.053	-0.076	-0.097	-0.115
1.60	-0.001	-0.005	-0.011	-0.021	-0.040	-0.057	-0.073	-0.086
1.70	-0.001	-0.004	-0.008	-0.016	-0.031	-0.044	-0.056	-0.067
1.80	-0.001	-0.003	-0.006	-0.013	-0.024	-0.035	-0.044	-0.053
1.90	-0.001	-0.003	-0.005	-0.010	-0.019	-0.028	-0.036	-0.043
2.00	0.000	-0.002	-0.004	-0.008	-0.016	-0.023	-0.029	-0.035
2.20	0.000	-0.001	-0.003	-0.006	-0.011	-0.016	-0.021	-0.025
2.40	0.000	-0.001	-0.002	-0.004	-0.008	-0.012	-0.015	-0.019
2.60	0.000	-0.001	-0.002	-0.003	-0.006	-0.009	-0.012	-0.015
2.80	0.000	-0.001	-0.001	-0.003	-0.005	-0.008	-0.010	-0.012
3.00	0.000	-0.001	-0.001	-0.002	-0.004	-0.006	-0.008	-0.010
3.50	0.000	0.000	-0.001	-0.001	-0.003	-0.004	-0.006	-0.007
4.00	0.000	0.000	-0.001	-0.001	-0.002	-0.003	-0.005	-0.006

High $P_r (\geq 1.0)$

$P_r \rightarrow$ / $T_r \downarrow$	1.0000	1.2000	1.5000	2.0000	3.0000	5.0000	7.0000	10.000
0.30	-16.586	-16.547	-16.488	-16.390	-16.195	-15.837	-15.468	-14.925
0.35	-15.278	-15.251	-15.211	-15.144	-15.011	-14.751	-14.496	-14.153
0.40	-13.896	-13.877	-13.849	-13.803	-13.714	-13.541	-13.376	-13.144
0.45	-12.496	-12.482	-12.462	-12.430	-12.367	-12.248	-12.145	-11.999
0.50	-11.153	-11.143	-11.129	-11.107	-11.063	-10.985	-10.920	-10.836
0.55	-9.914	-9.907	-9.897	-9.882	-9.853	-9.806	-9.769	-9.732
0.60	-8.799	-8.794	-8.787	-8.777	-8.760	-8.736	-8.723	-8.720
0.65	-7.810	-7.807	-7.801	-7.794	-7.784	-7.779	-7.785	-7.811
0.70	-6.933	-6.930	-6.926	-6.922	-6.919	-6.929	-6.952	-7.002
0.75	-6.155	-6.152	-6.149	-6.147	-6.149	-6.174	-6.213	-6.285
0.80	-5.458	-5.455	-5.453	-5.452	-5.461	-5.501	-5.555	-5.648
0.85	-4.826	-4.822	-4.820	-4.822	-4.839	-4.898	-4.969	-5.082
0.90	-4.238	-4.232	-4.230	-4.236	-4.267	-4.351	-4.442	-4.578
0.93	-3.894	-3.885	-3.884	-3.896	-3.941	-4.046	-4.151	-4.300
0.95	-3.658	-3.647	-3.648	-3.669	-3.728	-3.851	-3.966	-4.125
0.97	-3.406	-3.391	-3.401	-3.437	-3.517	-3.661	-3.788	-3.957
0.98	-3.264	-3.247	-3.268	-3.318	-3.412	-3.569	-3.701	-3.875
0.99	-3.093	-3.082	-3.126	-3.195	-3.306	-3.477	-3.616	-3.796
1.00	-2.399	-2.868	-2.967	-3.067	-3.200	-3.387	-3.532	-3.717
1.01	-1.306	-2.513	-2.784	-2.933	-3.094	-3.297	-3.450	-3.640
1.02	-1.113	-1.655	-2.557	-2.790	-2.986	-3.209	-3.369	-3.565
1.05	-0.820	-0.831	-1.443	-2.283	-2.655	-2.949	-3.134	-3.348
1.10	-0.577	-0.640	-0.618	-1.241	-2.067	-2.534	-2.767	-3.013
1.15	-0.437	-0.489	-0.502	-0.654	-1.471	-2.138	-2.428	-2.708
1.20	-0.343	-0.385	-0.412	-0.447	-0.991	-1.767	-2.115	-2.430
1.30	-0.226	-0.254	-0.282	-0.300	-0.481	-1.147	-1.569	-1.944
1.40	-0.158	-0.178	-0.200	-0.220	-0.290	-0.730	-1.138	-1.544
1.50	-0.115	-0.130	-0.147	-0.166	-0.206	-0.479	-0.823	-1.222
1.60	-0.086	-0.098	-0.112	-0.129	-0.159	-0.334	-0.604	-0.969
1.70	-0.067	-0.076	-0.087	-0.102	-0.127	-0.248	-0.456	-0.775
1.80	-0.053	-0.060	-0.070	-0.083	-0.105	-0.195	-0.355	-0.628
1.90	-0.043	-0.049	-0.057	-0.069	-0.089	-0.160	-0.286	-0.518
2.00	-0.035	-0.040	-0.048	-0.058	-0.077	-0.136	-0.238	-0.434
2.20	-0.025	-0.029	-0.035	-0.043	-0.060	-0.105	-0.178	-0.322
2.40	-0.019	-0.022	-0.027	-0.034	-0.048	-0.086	-0.143	-0.254
2.60	-0.015	-0.018	-0.021	-0.028	-0.041	-0.074	-0.120	-0.210
2.80	-0.012	-0.014	-0.018	-0.023	-0.025	-0.065	-0.104	-0.180
3.00	-0.010	-0.012	-0.015	-0.020	-0.031	-0.058	-0.093	-0.158
3.50	-0.007	-0.009	-0.011	-0.015	-0.024	-0.046	-0.073	-0.122
4.00	-0.006	-0.007	-0.009	-0.012	-0.020	-0.038	-0.060	-0.100

Table J.7: φ^o Values

Low $P_r (\leq 1.0)$

$P_r \rightarrow$ / $T_r \downarrow$	0.0100	0.0500	0.1000	0.2000	0.4000	0.6000	0.8000	1.0000
0.30	0.0002	0.0000	0.0000	0.0000	0.0000	0.0000	0.0000	0.0000
0.35	0.0034	0.0007	0.0003	0.0002	0.0001	0.0001	0.0001	0.0000
0.40	0.0272	0.0055	0.0028	0.0014	0.0007	0.0005	0.0004	0.0003
0.45	0.1321	0.0266	0.0135	0.0069	0.0036	0.0025	0.0020	0.0016
0.50	0.4529	0.0912	0.0461	0.0235	0.0122	0.0085	0.0067	0.0055
0.55	0.9817	0.2432	0.1227	0.0625	0.0325	0.0225	0.0176	0.0146
0.60	0.9840	0.5383	0.2716	0.1384	0.0718	0.0497	0.0386	0.0321
0.65	0.9886	0.9419	0.5212	0.2655	0.1374	0.0948	0.0738	0.0611
0.70	0.9908	0.9528	0.9057	0.4560	0.2360	0.1626	0.1262	0.1045
0.75	0.9931	0.9616	0.9226	0.7178	0.3715	0.2559	0.1982	0.1641
0.80	0.9931	0.9683	0.9354	0.8730	0.5445	0.3750	0.2904	0.2404
0.85	0.9954	0.9727	0.9462	0.8933	0.7534	0.5188	0.4018	0.3319
0.90	0.9954	0.9772	0.9550	0.9099	0.8204	0.6823	0.5297	0.4375
0.93	0.9954	0.9795	0.9594	0.9183	0.8375	0.7551	0.6109	0.5058
0.95	0.9954	0.9817	0.9616	0.9226	0.8472	0.7709	0.6668	0.5521
0.97	0.9954	0.9817	0.9638	0.9268	0.8570	0.7852	0.7112	0.5984
0.98	0.9954	0.9817	0.9638	0.9290	0.8610	0.7925	0.7211	0.6223
0.99	0.9977	0.9840	0.9661	0.9311	0.8650	0.7980	0.7295	0.6442
1.00	0.9977	0.9840	0.9661	0.9333	0.8690	0.8035	0.7379	0.6668
1.01	0.9977	0.9840	0.9683	0.9354	0.8730	0.8110	0.7464	0.6792
1.02	0.9977	0.9840	0.9683	0.9376	0.8770	0.8166	0.7551	0.6902
1.05	0.9977	0.9863	0.9705	0.9441	0.8872	0.8318	0.7762	0.7194
1.10	0.9977	0.9886	0.9750	0.9506	0.9016	0.8531	0.8072	0.7586
1.15	0.9977	0.9886	0.9795	0.9572	0.9141	0.8730	0.8318	0.7907
1.20	0.9977	0.9908	0.9817	0.9616	0.9247	0.8892	0.8531	0.8166
1.30	0.9977	0.9931	0.9863	0.9705	0.9419	0.9141	0.8872	0.8590
1.40	0.9977	0.9931	0.9886	0.9772	0.9550	0.9333	0.9120	0.8892
1.50	1.0000	0.9954	0.9908	0.9817	0.9638	0.9462	0.9290	0.9141
1.60	1.0000	0.9954	0.9931	0.9863	0.9727	0.9572	0.9441	0.9311
1.70	1.0000	0.9977	0.9954	0.9886	0.9772	0.9661	0.9550	0.9462
1.80	1.0000	0.9977	0.9954	0.9908	0.9817	0.9727	0.9661	0.9572
1.90	1.0000	0.9977	0.9954	0.9931	0.9863	0.9795	0.9727	0.9661
2.00	1.0000	0.9977	0.9977	0.9954	0.9886	0.9840	0.9795	0.9727
2.20	1.0000	1.0000	0.9977	0.9977	0.9931	0.9908	0.9886	0.9840
2.40	1.0000	1.0000	1.0000	0.9977	0.9977	0.9954	0.9931	0.9931
2.60	1.0000	1.0000	1.0000	1.0000	1.0000	0.9977	0.9977	0.9977
2.80	1.0000	1.0000	1.0000	1.0000	1.0000	1.0000	1.0023	1.0023
3.00	1.0000	1.0000	1.0000	1.0000	1.0023	1.0023	1.0046	1.0046
3.50	1.0000	1.0000	1.0000	1.0023	1.0023	1.0046	1.0069	1.0093
4.00	1.0000	1.0000	1.0000	1.0023	1.0046	1.0069	1.0093	1.0116

High $P_r (\geq 1.0)$

$P_r \rightarrow$ / $T_r \downarrow$	1.0000	1.2000	1.5000	2.0000	3.0000	5.0000	7.0000	10.000
0.30	0.0000	0.0000	0.0000	0.0000	0.0000	0.0000	0.0000	0.0000
0.35	0.0000	0.0000	0.0000	0.0000	0.0000	0.0000	0.0000	0.0000
0.40	0.0003	0.0003	0.0003	0.0002	0.0002	0.0002	0.0002	0.0003
0.45	0.0016	0.0014	0.0012	0.0010	0.0008	0.0008	0.0009	0.0012
0.50	0.0055	0.0048	0.0041	0.0034	0.0028	0.0025	0.0027	0.0034
0.55	0.0146	0.0127	0.0107	0.0089	0.0072	0.0063	0.0066	0.0080
0.60	0.0321	0.0277	0.0234	0.0193	0.0154	0.0132	0.0135	0.0160
0.65	0.0611	0.0527	0.0445	0.0364	0.0289	0.0244	0.0245	0.0282
0.70	0.1045	0.0902	0.0759	0.0619	0.0488	0.0406	0.0402	0.0453
0.75	0.1641	0.1413	0.1188	0.0966	0.0757	0.0625	0.0610	0.0673
0.80	0.2404	0.2065	0.1738	0.1409	0.1102	0.0899	0.0867	0.0942
0.85	0.3319	0.2858	0.2399	0.1945	0.1517	0.1227	0.1175	0.1256
0.90	0.4375	0.3767	0.3162	0.2564	0.1995	0.1607	0.1524	0.1611
0.93	0.5058	0.4355	0.3656	0.2972	0.2307	0.1854	0.1754	0.1841
0.95	0.5521	0.4764	0.3999	0.3251	0.2523	0.2028	0.1910	0.2000
0.97	0.5984	0.5164	0.4345	0.3532	0.2748	0.2203	0.2075	0.2163
0.98	0.6223	0.5370	0.4529	0.3681	0.2864	0.2296	0.2158	0.2244
0.99	0.6442	0.5572	0.4699	0.3828	0.2978	0.2388	0.2244	0.2328
1.00	0.6668	0.5781	0.4875	0.3972	0.3097	0.2483	0.2328	0.2415
1.01	0.6792	0.5970	0.5047	0.4121	0.3214	0.2576	0.2415	0.2500
1.02	0.6902	0.6166	0.5224	0.4266	0.3334	0.2673	0.2506	0.2582
1.05	0.7194	0.6607	0.5728	0.4710	0.3690	0.2958	0.2773	0.2844
1.10	0.7586	0.7112	0.6412	0.5408	0.4285	0.3451	0.3228	0.3296
1.15	0.7907	0.7499	0.6918	0.6026	0.4875	0.3954	0.3690	0.3750
1.20	0.8166	0.7834	0.7328	0.6546	0.5420	0.4446	0.4150	0.4198
1.30	0.8590	0.8318	0.7943	0.7345	0.6383	0.5383	0.5058	0.5093
1.40	0.8892	0.8690	0.8395	0.7925	0.7145	0.6237	0.5902	0.5943
1.50	0.9141	0.8974	0.8730	0.8375	0.7745	0.6966	0.6668	0.6714
1.60	0.9311	0.9183	0.8995	0.8710	0.8222	0.7586	0.7328	0.7430
1.70	0.9462	0.9354	0.9204	0.8995	0.8610	0.8091	0.7907	0.8054
1.80	0.9572	0.9484	0.9376	0.9204	0.8913	0.8531	0.8414	0.8590
1.90	0.9661	0.9594	0.9506	0.9376	0.9162	0.8872	0.8831	0.9057
2.00	0.9727	0.9683	0.9616	0.9528	0.9354	0.9183	0.9183	0.9462
2.20	0.9840	0.9817	0.9795	0.9727	0.9661	0.9616	0.9727	1.0093
2.40	0.9931	0.9908	0.9908	0.9886	0.9863	0.9931	1.0116	1.0568
2.60	0.9977	0.9977	0.9977	0.9977	1.0023	1.0162	1.0399	1.0889
2.80	1.0023	1.0023	1.0046	1.0069	1.0116	1.0328	1.0593	1.1117
3.00	1.0046	1.0069	1.0069	1.0116	1.0209	1.0423	1.0740	1.1298
3.50	1.0093	1.0116	1.0139	1.0186	1.0304	1.0593	1.0914	1.1508
4.00	1.0116	1.0139	1.0162	1.0233	1.0375	1.0666	1.0990	1.1588

Table J.8: φ^1 Values

Low $P_r (\leq 1.0)$

$P_r \rightarrow$ / $T_r \downarrow$	0.0100	0.0500	0.1000	0.2000	0.4000	0.6000	0.8000	1.0000
0.30	0.0000	0.0000	0.0000	0.0000	0.0000	0.0000	0.0000	0.0000
0.35	0.0000	0.0000	0.0000	0.0000	0.0000	0.0000	0.0000	0.0000
0.40	0.0000	0.0000	0.0000	0.0000	0.0000	0.0000	0.0000	0.0000
0.45	0.0002	0.0002	0.0002	0.0002	0.0002	0.0002	0.0002	0.0002
0.50	0.0014	0.0014	0.0014	0.0014	0.0014	0.0014	0.0013	0.0013
0.55	0.9705	0.0069	0.0068	0.0068	0.0066	0.0065	0.0064	0.0063
0.60	0.9795	0.0227	0.0226	0.0223	0.0220	0.0216	0.0213	0.0210
0.65	0.9863	0.9311	0.0572	0.0568	0.0559	0.0551	0.0543	0.0535
0.70	0.9908	0.9528	0.9036	0.1182	0.1163	0.1147	0.1131	0.1116
0.75	0.9931	0.9683	0.9332	0.2112	0.2078	0.2050	0.2022	0.1994
0.80	0.9954	0.9772	0.9550	0.9057	0.3302	0.3257	0.3212	0.3168
0.85	0.9977	0.9863	0.9705	0.9375	0.4774	0.4708	0.4654	0.4500
0.90	0.9977	0.9908	0.9795	0.9594	0.9141	0.6323	0.6250	0.6165
0.93	0.9977	0.9931	0.9840	0.9705	0.9354	0.8953	0.7227	0.7144
0.95	0.9977	0.9931	0.9885	0.9750	0.9484	0.9183	0.7888	0.7797
0.97	1.0000	0.9954	0.9908	0.9795	0.9594	0.9354	0.9078	0.8413
0.98	1.0000	0.9954	0.9908	0.9817	0.9638	0.9440	0.9225	0.8729
0.99	1.0000	0.9954	0.9931	0.9840	0.9683	0.9528	0.9332	0.9036
1.00	1.0000	0.9977	0.9931	0.9863	0.9727	0.9594	0.9440	0.9311
1.01	1.0000	0.9977	0.9931	0.9885	0.9772	0.9638	0.9528	0.9462
1.02	1.0000	0.9977	0.9954	0.9908	0.9795	0.9705	0.9616	0.9572
1.05	1.0000	0.9977	0.9977	0.9954	0.9885	0.9863	0.9840	0.9840
1.10	1.0000	1.0000	1.0000	1.0000	1.0023	1.0046	1.0093	1.0163
1.15	1.0000	1.0000	1.0023	1.0046	1.0116	1.0186	1.0257	1.0375
1.20	1.0000	1.0023	1.0046	1.0069	1.0163	1.0280	1.0399	1.0544
1.30	1.0000	1.0023	1.0069	1.0116	1.0257	1.0399	1.0544	1.0716
1.40	1.0000	1.0046	1.0069	1.0139	1.0304	1.0471	1.0642	1.0815
1.50	1.0000	1.0046	1.0069	1.0163	1.0328	1.0496	1.0666	1.0865
1.60	1.0000	1.0046	1.0069	1.0163	1.0328	1.0496	1.0691	1.0865
1.70	1.0000	1.0046	1.0093	1.0163	1.0328	1.0496	1.0691	1.0865
1.80	1.0000	1.0046	1.0069	1.0163	1.0328	1.0496	1.0666	1.0840
1.90	1.0000	1.0046	1.0069	1.0163	1.0328	1.0496	1.0666	1.0815
2.00	1.0000	1.0046	1.0069	1.0163	1.0304	1.0471	1.0642	1.0815
2.20	1.0000	1.0046	1.0069	1.0139	1.0304	1.0447	1.0593	1.0765
2.40	1.0000	1.0046	1.0069	1.0139	1.0280	1.0423	1.0568	1.0716
2.60	1.0000	1.0023	1.0069	1.0139	1.0257	1.0399	1.0544	1.0666
2.80	1.0000	1.0023	1.0069	1.0116	1.0257	1.0375	1.0496	1.0642
3.00	1.0000	1.0023	1.0069	1.0116	1.0233	1.0352	1.0471	1.0593
3.50	1.0000	1.0023	1.0046	1.0023	1.0209	1.0304	1.0423	1.0520
4.00	1.0000	1.0023	1.0046	1.0093	1.0186	1.0280	1.0375	1.0471

High $P_r (\geq 1.0)$

$P_r \rightarrow$ / $T_r \downarrow$	1.0000	1.2000	1.5000	2.0000	3.0000	5.0000	7.0000	10.000
0.30	0.0000	0.0000	0.0000	0.0000	0.0000	0.0000	0.0000	0.0000
0.35	0.0000	0.0000	0.0000	0.0000	0.0000	0.0000	0.0000	0.0000
0.40	0.0000	0.0000	0.0000	0.0000	0.0000	0.0000	0.0000	0.0000
0.45	0.0002	0.0002	0.0002	0.0002	0.0001	0.0001	0.0001	0.0001
0.50	0.0013	0.0013	0.0013	0.0012	0.0011	0.0009	0.0008	0.0006
0.55	0.0063	0.0062	0.0061	0.0058	0.0053	0.0045	0.0039	0.0031
0.60	0.0210	0.0207	0.0202	0.0194	0.0179	0.0154	0.0133	0.0108
0.65	0.0536	0.0527	0.0516	0.0497	0.0461	0.0401	0.0350	0.0289
0.70	0.1117	0.1102	0.1079	0.1040	0.0970	0.0851	0.0752	0.0629
0.75	0.1995	0.1972	0.1932	0.1871	0.1754	0.1552	0.1387	0.1178
0.80	0.3170	0.3133	0.3076	0.2978	0.2812	0.2512	0.2265	0.1954
0.85	0.4592	0.4539	0.4457	0.4325	0.4093	0.3698	0.3365	0.2951
0.90	0.6166	0.6095	0.5998	0.5834	0.5546	0.5058	0.4645	0.4130
0.93	0.7145	0.7063	0.6950	0.6761	0.6457	0.5916	0.5470	0.4898
0.95	0.7798	0.7691	0.7568	0.7379	0.7063	0.6501	0.6026	0.5432
0.97	0.8414	0.8318	0.8185	0.7998	0.7656	0.7096	0.6607	0.5984
0.98	0.8730	0.8630	0.8492	0.8298	0.7962	0.7379	0.6887	0.6266
0.99	0.9036	0.8913	0.8790	0.8590	0.8241	0.7674	0.7178	0.6546
1.00	0.9311	0.9204	0.9078	0.8872	0.8531	0.7962	0.7464	0.6823
1.01	0.9462	0.9462	0.9333	0.9162	0.8831	0.8241	0.7745	0.7096
1.02	0.9572	0.9661	0.9594	0.9419	0.9099	0.8531	0.8035	0.7379
1.05	0.9840	0.9954	1.0186	1.0162	0.9886	0.9354	0.8872	0.8222
1.10	1.0162	1.0280	1.0593	1.0990	1.1015	1.0617	1.0186	0.9572
1.15	1.0375	1.0520	1.0814	1.1376	1.1858	1.1722	1.1403	1.0864
1.20	1.0544	1.0691	1.0990	1.1588	1.2388	1.2647	1.2474	1.2050
1.30	1.0715	1.0914	1.1194	1.1776	1.2853	1.3868	1.4125	1.4061
1.40	1.0814	1.0990	1.1298	1.1858	1.2942	1.4488	1.5171	1.5524
1.50	1.0864	1.1041	1.1350	1.1858	1.2942	1.4689	1.5740	1.6520
1.60	1.0864	1.1041	1.1350	1.1858	1.2883	1.4689	1.5996	1.7140
1.70	1.0864	1.1041	1.1324	1.1803	1.2794	1.4622	1.6033	1.7458
1.80	1.0839	1.1015	1.1298	1.1749	1.2706	1.4488	1.5959	1.7620
1.90	1.0814	1.0990	1.1272	1.1695	1.2618	1.4355	1.5849	1.7620
2.00	1.0814	1.0965	1.1220	1.1641	1.2503	1.4191	1.5704	1.7539
2.20	1.0765	1.0914	1.1143	1.1535	1.2331	1.3900	1.5346	1.7219
2.40	1.0715	1.0864	1.1066	1.1429	1.2190	1.3614	1.4997	1.6866
2.60	1.0666	1.0814	1.1015	1.1350	1.2023	1.3397	1.4689	1.6482
2.80	1.0641	1.0765	1.0940	1.1272	1.1912	1.3183	1.4388	1.6144
3.00	1.0593	1.0715	1.0889	1.1194	1.1803	1.3002	1.4158	1.5813
3.50	1.0520	1.0617	1.0789	1.0141	1.1561	1.2618	1.3614	1.5101
4.00	1.0471	1.0544	1.0691	1.0914	1.1403	1.2303	1.3213	1.4555

Appendix K

Thermodynamic Properties of Saturated Refrigerant HFC-134A

(Temperature table)

		Specific Volume (m^3/kg)		Specific Internal Energy (kJ/kg)		Specific Enthalpy (kJ/kg)		Specific Entropy (kJ/kg.K)	
$T(^oC)$	$P(bars)$	$v^L \times 10^3$	v^V	u^L	u^V	h^L	h^V	s^L	s^V
-40	0.5164	0.7055	0.3569	-0.04	204.45	0.00	222.88	0.0000	0.9560
-36	0.6332	0.7113	0.2947	4.68	206.73	4.73	225.40	0.0201	0.9506
-32	0.7704	0.7172	0.2451	9.47	209.01	9.52	227.90	0.0401	0.9456
-28	0.9305	0.7233	0.2052	14.31	211.29	14.37	230.38	0.0600	0.9411
-26	1.0199	0.7265	0.1882	16.75	212.43	16.82	231.62	0.0699	0.9390
-24	1.1160	0.7296	0.1728	19.21	213.57	19.29	232.85	0.0798	0.9370
-22	1.2192	0.7328	0.1590	21.68	214.70	21.77	234.08	0.0897	0.9351
-20	1.3299	0.7361	0.1464	24.17	215.84	24.26	235.31	0.0996	0.9332
-18	1.4483	0.7395	0.1350	26.67	216.97	26.77	236.53	0.1094	0.9315
-16	1.5748	0.7428	0.1247	29.18	218.10	29.30	237.74	0.1192	0.9298
-12	1.8540	0.7498	0.1068	34.25	220.36	34.39	240.15	0.1388	0.9267
-8	2.1704	0.7569	0.0919	39.38	222.60	39.54	242.54	0.1583	0.9239
-4	2.5274	0.7644	0.0794	44.56	224.84	44.75	244.90	0.1777	0.9213
0	2.9282	0.7721	0.0689	49.79	227.06	50.02	247.23	0.1970	0.9190
4	3.3765	0.7801	0.0600	55.08	229.27	55.35	249.53	0.2162	0.9169
8	3.8756	0.7884	0.0525	60.43	231.46	60.73	251.80	0.2354	0.9150
12	4.4294	0.7971	0.0460	65.83	233.63	66.18	254.03	0.2545	0.9132
16	5.0416	0.8062	0.0405	71.29	235.78	71.69	256.22	0.2735	0.9116
20	5.7160	0.8157	0.0358	76.80	237.91	77.26	258.36	0.2924	0.9102
24	6.4566	0.8257	0.0317	82.37	240.01	82.90	260.45	0.3113	0.9089
26	6.8530	0.8309	0.0298	85.18	241.05	85.75	261.48	0.3208	0.9082
28	7.2675	0.8362	0.0281	88.00	242.08	88.61	262.50	0.3302	0.9076
30	7.7006	0.8417	0.0265	90.84	243.10	91.49	263.50	0.3396	0.9070
32	8.1528	0.8473	0.0250	93.70	244.12	94.39	264.48	0.3490	0.9064
34	8.6247	0.8530	0.0236	96.58	245.12	97.31	265.45	0.3584	0.9058
36	9.1168	0.8590	0.0223	99.47	246.11	100.25	266.40	0.3678	0.9053
38	9.6298	0.8651	0.0210	102.38	247.09	103.21	267.33	0.3772	0.9047
40	10.164	0.8714	0.0199	105.30	248.06	106.19	268.24	0.3866	0.9041
42	10.720	0.8780	0.0188	108.25	249.02	109.19	269.14	0.3960	0.9035
44	11.299	0.8847	0.0177	111.22	249.96	112.22	270.01	0.4054	0.9030
48	12.526	0.8989	0.0159	117.22	251.79	118.35	271.68	0.4243	0.9017
52	13.851	0.9142	0.0142	123.31	253.55	124.58	273.24	0.4432	0.9004
56	15.278	0.9308	0.0127	129.51	255.23	130.93	274.68	0.4622	0.8990
60	16.813	0.9488	0.0114	135.82	256.81	137.42	275.99	0.4814	0.8973
70	21.162	1.0027	0.0086	152.22	260.15	154.34	278.43	0.5302	0.8918
80	26.324	1.0766	0.0064	169.88	262.14	172.71	279.12	0.5814	0.8827
90	32.435	1.1949	0.0046	189.82	261.34	193.69	276.32	0.6380	0.8655
100	39.742	1.5443	0.0027	218.60	248.49	224.74	259.13	0.7196	0.8117

(Pressure table)

		Specific Volume (m^3/kg)		Specific Internal Energy (kJ/kg)		Specific Enthalpy (kJ/kg)		Specific Entropy (kJ/kg.K)	
$P(bars)$	$T(^oC)$	$v^L \times 10^3$	v^V	u^L	u^V	h^L	h^V	s^L	s^V
0.6	**-37.07**	**0.7097**	**0.3100**	**3.41**	**206.12**	**3.46**	**224.72**	**0.0147**	**0.9520**
0.8	**-31.21**	**0.7184**	**0.2366**	**10.41**	**209.46**	**10.47**	**228.39**	**0.0440**	**0.9447**
1.0	**-26.43**	**0.7258**	**0.1917**	**16.22**	**212.18**	**16.29**	**231.35**	**0.0678**	**0.9395**
1.2	**-22.36**	**0.7323**	**0.1614**	**21.23**	**214.50**	**21.32**	**233.86**	**0.0879**	**0.9354**
1.4	**-18.80**	**0.7381**	**0.1395**	**25.66**	**216.52**	**25.77**	**236.04**	**0.1055**	**0.9322**
1.6	**-15.62**	**0.7435**	**0.1229**	**29.66**	**218.32**	**29.78**	**237.97**	**0.1211**	**0.9295**
1.8	**-12.73**	**0.7485**	**0.1098**	**33.31**	**219.94**	**33.45**	**239.71**	**0.1352**	**0.9273**
2.0	**-10.09**	**0.7532**	**0.0993**	**36.69**	**221.43**	**36.84**	**241.30**	**0.1481**	**0.9253**
2.4	**-5.37**	**0.7618**	**0.0834**	**42.77**	**224.07**	**42.95**	**244.09**	**0.1710**	**0.9222**
2.8	**-1.23**	**0.7697**	**0.0719**	**48.18**	**226.38**	**48.39**	**246.52**	**0.1911**	**0.9197**
3.2	**2.48**	**0.7770**	**0.0632**	**53.06**	**228.43**	**53.31**	**248.66**	**0.2089**	**0.9177**
3.6	**5.84**	**0.7839**	**0.0564**	**57.54**	**230.28**	**57.82**	**250.58**	**0.2251**	**0.9160**
4.0	**8.93**	**0.7904**	**0.0509**	**61.69**	**231.97**	**62.00**	**252.32**	**0.2399**	**0.9145**
5.0	**15.74**	**0.8056**	**0.0409**	**70.93**	**235.64**	**71.33**	**256.07**	**0.2723**	**0.9117**
6.0	**21.58**	**0.8196**	**0.0341**	**78.99**	**238.74**	**79.48**	**259.19**	**0.2999**	**0.9097**
7.0	**26.72**	**0.8328**	**0.0292**	**86.19**	**241.42**	**86.78**	**261.85**	**0.3242**	**0.9080**
8.0	**31.33**	**0.8454**	**0.0255**	**92.75**	**243.78**	**93.42**	**264.15**	**0.3459**	**0.9066**
9.0	**35.53**	**0.8576**	**0.0226**	**98.79**	**245.88**	**99.56**	**266.18**	**0.3656**	**0.9054**
10.0	**39.39**	**0.8695**	**0.0202**	**104.42**	**247.77**	**105.29**	**267.97**	**0.3838**	**0.9043**
12.0	**46.32**	**0.8928**	**0.0166**	**114.69**	**251.03**	**115.76**	**270.99**	**0.4164**	**0.9023**
14.0	**52.43**	**0.9159**	**0.0140**	**123.98**	**253.74**	**125.26**	**273.40**	**0.4453**	**0.9003**
16.0	**57.92**	**0.9392**	**0.0121**	**132.52**	**256.00**	**134.02**	**275.33**	**0.4714**	**0.8982**
18.0	**62.91**	**0.9631**	**0.0105**	**140.49**	**257.88**	**142.22**	**276.83**	**0.4954**	**0.8959**
20.0	**67.49**	**0.9878**	**0.0093**	**148.02**	**259.41**	**149.99**	**277.94**	**0.5178**	**0.8934**
25.0	**77.59**	**1.0562**	**0.0069**	**165.48**	**261.84**	**168.12**	**279.17**	**0.5687**	**0.8854**
30.0	**86.22**	**1.1416**	**0.0053**	**181.88**	**262.16**	**185.30**	**278.01**	**0.6156**	**0.8735**

Sources:

Wilson, D.P., Basu, R.S., "Thermodynamic properties of a new stratospherically safe working fluid-Refrigerant 134A", *ASHRAE Trans.* 1988; 94(2): 2095-2118.

Potter, M.C., Somerton, C.W., *Thermodynamics for Engineers,* McGraw Hill, New York, 2nd edition, 2006.

Appendix L

Thermodynamic Properties of Superheated Vapor of Refrigerant HFC-134A

$T\ (^oC)$	v	u	h	s	v	u	h	s	v	u	h	s
	$P = 0.06$ MPa $(T^{sat} = -37.07^oC)$				$P = 0.10$ MPa $(T^{sat} = -26.43^oC)$				$P = 0.14$ MPa $(T^{sat} = -18.80^oC)$			
Sat.	0.31003	206.12	224.72	0.9520	0.19170	212.18	231.35	0.9395	0.13945	216.52	236.04	0.9322
-20	0.33536	217.86	237.98	1.0062	0.19770	216.77	236.54	0.9602				
-10	0.34992	224.97	245.96	1.0371	0.20686	224.01	244.70	0.9918	0.14549	223.03	243.40	0.9606
0	0.36433	232.24	254.10	1.0675	0.21587	231.41	252.99	1.0227	0.15219	230.55	251.86	0.9922
10	0.37861	239.69	262.41	1.0973	0.22473	238.96	261.43	1.0531	0.15875	238.21	260.43	1.0230
20	0.39279	247.32	270.89	1.1267	0.23349	246.67	270.02	1.0829	0.16520	246.01	269.13	1.0532
30	0.40688	255.12	279.53	1.1557	0.24216	254.54	278.76	1.1122	0.17155	253.96	277.97	1.0828
40	0.42091	263.10	288.35	1.1844	0.25076	262.58	287.66	1.1411	0.17783	262.06	286.96	1.1120
50	0.43487	271.25	297.34	1.2126	0.25930	270.79	296.72	1.1696	0.18404	270.32	296.09	1.1407
60	0.44879	279.58	306.51	1.2405	0.26779	279.16	305.94	1.1977	0.19020	278.74	305.37	1.1690
70	0.46266	288.08	315.84	1.2681	0.27623	287.70	315.32	1.2254	0.19633	287.32	314.80	1.1969
80	0.47650	296.75	325.34	1.2954	0.28464	296.40	324.87	1.2528	0.20241	296.06	324.39	1.2244
90	0.49031	305.58	335.00	1.3224	0.29302	305.27	334.57	1.2799	0.20846	304.95	334.14	1.2516
100									0.21449	314.01	344.04	1.2785

(v in m^3/kg , u and h in kJ/kg, s in $kJ/kg.K$)

$T\ (^oC)$	v	u	h	s	v	u	h	s	v	u	h	s
	$P = 0.18$ MPa ($T^{sat} = -12.73^oC$)				$P = 0.20$ MPa ($T^{sat} = -10.09^oC$)				$P = 0.24$ MPa ($T^{sat} = -5.37^oC$)			
Sat.	0.10983	219.94	239.71	0.9273	0.09933	221.43	241.30	0.9253	0.08343	224.07	244.09	0.9222
-10	0.11135	222.02	242.06	0.9362	0.09938	221.50	241.38	0.9256				
0	0.11678	229.67	250.69	0.9684	0.10438	229.23	250.10	0.9582	0.08574	228.31	248.89	0.9399
10	0.12207	237.44	259.41	0.9998	0.10922	237.05	258.89	0.9898	0.08993	236.26	257.84	0.9721
20	0.12723	245.33	268.23	1.0304	0.11394	244.99	267.78	1.0206	0.09399	244.30	266.85	1.0034
30	0.13230	253.36	277.17	1.0604	0.11856	253.06	276.77	1.0508	0.09794	252.45	275.95	1.0339
40	0.13730	261.53	286.24	1.0898	0.12311	261.26	285.88	1.0804	0.10181	260.72	285.16	1.0637
50	0.14222	269.85	295.45	1.1187	0.12758	269.61	295.12	1.1094	0.10562	269.12	294.47	1.0930
60	0.14710	278.31	304.79	1.1472	0.13201	278.10	304.50	1.1380	0.10937	277.67	303.91	1.1218
70	0.15193	286.93	314.28	1.1753	0.13639	286.74	314.02	1.1661	0.11307	286.35	313.49	1.1501
80	0.15672	295.71	323.92	1.2030	0.14073	295.53	323.68	1.1939	0.11674	295.18	323.19	1.1780
90	0.16148	304.63	333.70	1.2303	0.14504	304.47	333.48	1.2212	0.12037	304.15	333.04	1.2055
100	0.16622	313.72	343.63	1.2573	0.14932	313.57	343.43	1.2483	0.12398	313.27	343.03	1.2326

(v in m^3/kg , u and h in kJ/kg, s in $kJ/kg.K$)

$T\ (^oC)$	v	u	h	s	v	u	h	s	v	u	h	s
	$P = 0.28$ MPa ($T^{sat} = -1.23^oC$)				$P = 0.32$ MPa ($T^{sat} = 2.48^oC$)				$P = 0.40$ MPa ($T^{sat} = 8.93^oC$)			
Sat.	0.07193	226.38	246.52	0.9197	0.06322	228.43	248.66	0.9177	0.05089	231.97	252.32	0.9145
0	0.07240	227.37	247.64	0.9238								
10	0.07613	235.44	256.76	0.9566	0.06576	234.61	255.65	0.9427	0.05119	232.87	253.35	0.9182
20	0.07972	243.59	265.91	0.9883	0.06901	242.87	264.95	0.9749	0.05397	241.37	262.96	0.9515
30	0.08320	251.83	275.12	1.0192	0.07214	251.19	274.28	1.0062	0.05662	249.89	272.54	0.9837
40	0.08660	260.17	284.42	1.0494	0.07518	259.61	283.67	1.0367	0.05917	258.47	282.14	1.0148
50	0.08992	268.64	293.81	1.0789	0.07815	268.14	293.15	1.0665	0.06164	267.13	291.79	1.0452
60	0.09319	277.23	303.32	1.1079	0.08106	276.79	302.72	1.0957	0.06405	275.89	301.51	1.0748
70	0.09641	285.96	312.95	1.1364	0.08392	285.56	312.41	1.1243	0.06641	284.75	311.32	1.1038
80	0.09960	294.82	322.71	1.1644	0.08674	294.46	322.22	1.1525	0.06873	293.73	321.23	1.1322
90	0.10275	303.83	332.60	1.1920	0.08953	303.50	332.15	1.1802	0.07102	302.84	331.25	1.1602
100	0.10587	312.98	342.62	1.2193	0.09229	312.68	342.21	1.2076	0.07327	312.07	341.38	1.1878
110	0.10897	322.27	352.78	1.2461	0.09503	322.00	352.40	1.2345	0.07550	321.44	351.64	1.2149
120	0.11205	331.71	363.08	1.2727	0.09774	331.45	362.73	1.2611	0.07771	330.94	362.03	1.2417
130									0.07991	340.58	371.54	1.2681
140									0.08208	350.35	383.18	1.2941

(v in m^3/kg, u and h in kJ/kg, s in $kJ/kg.K$)

$T\ (^oC)$	v	u	h	s	v	u	h	s	v	u	h	s
	$P = 0.50$ MPa ($T^{sat} = 15.74^oC$)				$P = 0.60$ MPa ($T^{sat} = 21.58^oC$)				$P = 0.70$ MPa ($T^{sat} = 26.72^oC$)			
Sat.	0.04086	235.64	256.07	0.9117	0.03408	238.74	259.19	0.9097	0.02918	241.42	261.85	0.9080
20	0.04188	239.40	260.34	0.9264								
30	0.04416	248.20	270.28	0.9597	0.03581	246.41	267.89	0.9388	0.02979	244.51	265.37	0.9197
40	0.04633	256.99	280.16	0.9918	0.03774	255.45	278.09	0.9719	0.03157	253.83	275.93	0.9539
50	0.04842	265.83	290.04	1.0229	0.03958	264.48	288.23	1.0037	0.03324	263.08	286.35	0.9867
60	0.05043	274.73	299.95	1.0531	0.04134	273.54	298.35	1.0346	0.03482	272.31	296.69	1.0182
70	0.05240	283.72	309.92	1.0825	0.04304	282.66	308.48	1.0645	0.03634	281.57	307.01	1.0487
80	0.05432	292.80	319.96	1.1114	0.04469	291.86	318.67	1.0938	0.03781	290.88	317.35	1.0784
90	0.05620	302.00	330.10	1.1397	0.04631	301.14	328.93	1.1225	0.03924	300.27	327.74	1.1074
100	0.05805	311.31	340.33	1.1675	0.04790	310.53	339.27	1.1505	0.04064	309.74	338.19	1.1358
110	0.05988	320.74	350.68	1.1949	0.04946	320.03	349.70	1.1781	0.04201	319.31	348.71	1.1637
120	0.06168	330.30	361.14	1.2218	0.05099	329.64	360.24	1.2053	0.04335	328.98	359.33	1.1910
130	0.06347	339.98	371.72	1.2484	0.05251	339.38	370.88	1.2320	0.04468	338.76	370.04	1.2179
140	0.06524	349.79	382.42	1.2746	0.05402	349.23	381.64	1.2584	0.04599	348.66	380.86	1.2444
150					0.05550	359.21	392.52	1.2844	0.04729	358.68	391.79	1.2706
160					0.05698	369.32	403.51	1.3100	0.04857	368.82	402.82	1.2963

(v in m^3/kg , u and h in kJ/kg, s in $kJ/kg.K$)

$T\ (^oC)$	v	u	h	s	v	u	h	s	v	u	h	s
	$P = 0.80$ MPa ($T^{sat} = 31.33^oC$)				$P = 0.90$ MPa ($T^{sat} = 35.53^oC$)				$P = 1.00$ MPa ($T^{sat} = 39.39^oC$)			
Sat.	**0.02547**	**243.78**	**264.15**	**0.9066**	**0.02255**	**245.88**	**266.18**	**0.9054**	**0.02020**	**247.77**	**267.97**	**0.9043**
40	**0.02691**	**252.13**	**273.66**	**0.9374**	**0.02325**	**250.32**	**271.25**	**0.9217**	**0.02029**	**248.39**	**268.68**	**0.9066**
50	**0.02846**	**261.62**	**284.39**	**0.9711**	**0.02472**	**260.09**	**282.34**	**0.9566**	**0.02171**	**258.48**	**280.19**	**0.9428**
60	**0.02992**	**271.04**	**294.98**	**1.0034**	**0.02609**	**269.72**	**293.21**	**0.9897**	**0.02301**	**268.35**	**291.36**	**0.9768**
70	**0.03131**	**280.45**	**305.50**	**1.0345**	**0.02738**	**279.30**	**303.94**	**1.0214**	**0.02423**	**278.11**	**302.34**	**1.0093**
80	**0.03264**	**289.89**	**316.00**	**1.0647**	**0.02861**	**288.87**	**314.62**	**1.0521**	**0.02538**	**287.82**	**313.20**	**1.0405**
90	**0.03393**	**299.37**	**326.52**	**1.0940**	**0.02980**	**298.46**	**325.28**	**1.0819**	**0.02649**	**297.53**	**324.01**	**1.0707**
100	**0.03519**	**308.93**	**337.08**	**1.1227**	**0.03095**	**308.11**	**335.96**	**1.1109**	**0.02755**	**307.27**	**334.82**	**1.1000**
110	**0.03642**	**318.57**	**347.71**	**1.1508**	**0.03207**	**317.82**	**346.68**	**1.1392**	**0.02858**	**317.06**	**345.65**	**1.1286**
120	**0.03762**	**328.31**	**358.40**	**1.1784**	**0.03316**	**327.62**	**357.47**	**1.1670**	**0.02959**	**326.93**	**356.52**	**1.1567**
130	**0.03881**	**338.14**	**369.19**	**1.2055**	**0.03423**	**337.52**	**368.33**	**1.1943**	**0.03058**	**336.88**	**367.46**	**1.1841**
140	**0.03997**	**348.09**	**380.07**	**1.2321**	**0.03529**	**347.51**	**379.27**	**1.2211**	**0.03154**	**346.92**	**378.46**	**1.2111**
150	**0.04113**	**358.15**	**391.05**	**1.2584**	**0.03633**	**357.61**	**390.31**	**1.2475**	**0.03250**	**357.06**	**389.56**	**1.2376**
160	**0.04227**	**368.32**	**402.14**	**1.2843**	**0.03736**	**367.82**	**401.44**	**1.2735**	**0.03344**	**367.31**	**400.74**	**1.2638**
170	**0.04340**	**378.61**	**413.33**	**1.3098**	**0.03838**	**378.14**	**412.68**	**1.2992**	**0.03436**	**377.66**	**412.02**	**1.2895**
180	**0.04452**	**389.02**	**424.63**	**1.3351**	**0.03939**	**388.57**	**424.02**	**1.3245**	**0.03528**	**388.12**	**423.40**	**1.3149**

(v in m^3/kg , u and h in kJ/kg, s in $kJ/kg.K$)

$T\ (^oC)$	v	u	h	s	v	u	h	s	v	u	h	s
	$P = 1.20$ MPa ($T^{sat} = 46.32^oC$)				$P = 1.40$ MPa ($T^{sat} = 52.43^oC$)				$P = 1.60$ MPa ($T^{sat} = 57.92^oC$)			
Sat.	0.01663	251.03	270.99	0.9023	0.01405	253.74	273.40	0.9003	0.01208	256.00	275.33	0.8982
50	0.01712	254.98	275.52	0.9164								
60	0.01835	265.42	287.44	0.9527	0.01495	262.17	283.10	0.9297	0.01233	258.48	278.20	0.9069
70	0.01947	275.59	298.96	0.9868	0.01603	272.87	295.31	0.9658	0.01340	269.89	291.33	0.9457
80	0.02051	285.62	310.24	1.0192	0.01701	283.29	307.10	0.9997	0.01435	280.78	303.74	0.9813
90	0.02150	295.59	321.39	1.0503	0.01792	293.55	318.63	1.0319	0.01521	291.39	315.72	1.0148
100	0.02244	305.54	332.47	1.0804	0.01878	303.73	330.02	1.0628	0.01601	301.84	327.46	1.0467
110	0.02335	315.50	343.52	1.1096	0.01960	313.88	341.32	1.0927	0.01677	312.20	339.04	1.0773
120	0.02423	325.51	354.58	1.1381	0.02039	324.05	352.59	1.1218	0.01750	322.53	350.53	1.1069
130	0.02508	335.58	365.68	1.1660	0.02115	334.25	363.86	1.1501	0.01820	332.87	361.99	1.1357
140	0.02592	345.73	376.83	1.1933	0.02189	344.50	375.15	1.1777	0.01887	343.24	373.44	1.1638
150	0.02674	355.95	388.04	1.2201	0.02262	354.82	386.49	1.2048	0.01953	353.66	384.91	1.1912
160	0.02754	366.27	399.33	1.2465	0.02333	365.22	397.89	1.2315	0.02017	364.15	396.43	1.2181
170	0.02834	376.69	410.70	1.2724	0.02403	375.71	409.36	1.2576	0.02080	374.71	407.99	1.2445
180	0.02912	387.21	422.16	1.2980	0.02472	386.29	420.90	1.2834	0.02142	385.35	419.62	1.2704
190					0.02541	396.96	432.53	1.3088	0.02203	396.08	431.33	1.2960
200					0.02608	407.73	444.24	1.3338	0.02263	406.90	443.11	1.3212

(v in m^3/kg, u and h in kJ/kg, s in $kJ/kg.K$)

Sources:

Wilson, D.P., Basu, R.S., "Thermodynamic properties of a new stratospherically safe working fluid-Refrigerant 134A", *ASHRAE Trans.* 1988; 94(2): 2095-2118.

Potter, M.C., Somerton, C.W., *Thermodynamics for Engineers,* McGraw Hill, New York, 2nd edition, 2006.

Appendix M

Thermodynamic Properties of Saturated Water and Steam

(Temperature table)

		Specific Volume (m^3/kg)		Specific Internal Energy (kJ/kg)		Specific Enthalpy (kJ/kg)		Specific Entropy (kJ/kg.K)	
$T(^oC)$	$P(MPa)$	v^L	v^V	u^L	u^V	h^L	h^V	s^L	s^V
0.01	0.0006113	0.001000	206.1	0.0	2375.3	0.0	2501.3	0.0000	9.1571
2	0.0007056	0.001000	179.9	8.4	2378.1	8.4	2505.0	0.0305	9.1043
5	0.0008721	0.001000	147.1	21.0	2382.2	21.0	2510.5	0.0761	9.0266
10	0.001228	0.001000	106.4	42.0	2389.2	42.0	2519.7	0.1510	8.9016
15	0.001705	0.001001	77.93	63.0	2396.0	63.0	2528.9	0.2244	8.7822
20	0.002338	0.001002	57.79	83.9	2402.9	83.9	2538.1	0.2965	8.6680
25	0.003169	0.001003	43.36	104.9	2409.8	104.9	2547.2	0.3672	8.5588
30	0.004246	0.001004	32.90	125.8	2416.6	125.8	2556.2	0.4367	8.4541
35	0.005628	0.001006	25.22	146.7	2423.4	146.7	2565.3	0.5051	8.3539
40	0.007383	0.001008	19.52	167.5	2430.1	167.5	2574.3	0.5723	8.2578
45	0.009593	0.001010	15.26	188.4	2436.8	188.4	2583.2	0.6385	8.1656
50	0.01235	0.001012	12.03	209.3	2443.5	209.3	2592.1	0.7036	8.0771
55	0.01576	0.001015	9.569	230.2	2450.1	230.2	2600.9	0.7678	7.9921
60	0.01994	0.001017	7.671	251.1	2456.6	251.1	2609.6	0.8310	7.9104
65	0.02503	0.001020	6.197	272.0	2463.1	272.0	2618.2	0.8934	7.8318
70	0.03119	0.001023	5.042	292.9	2469.5	293.0	2626.8	0.9549	7.7561
75	0.03858	0.001026	4.131	313.9	2475.9	313.9	2635.3	1.0155	7.6833
80	0.04739	0.001029	3.407	334.8	2482.2	334.9	2643.7	1.0754	7.6130
85	0.05783	0.001032	2.828	355.8	2488.4	355.9	2651.9	1.1344	7.5453
90	0.07013	0.001036	2.361	376.8	2494.5	376.9	2660.1	1.1927	7.4799
95	0.08455	0.001040	1.982	397.9	2500.6	397.9	2668.1	1.2503	7.4167
100	0.1013	0.001044	1.673	418.9	2506.5	419.0	2676.0	1.3071	7.3557
110	0.1433	0.001052	1.210	461.1	2518.1	461.3	2691.5	1.4188	7.2395
120	0.1985	0.001060	0.8919	503.5	2529.2	503.7	2706.3	1.5280	7.1304
130	0.2701	0.001070	0.6685	546.0	2539.9	546.3	2720.5	1.6348	7.0277
140	0.3613	0.001080	0.5089	588.7	2550.0	589.1	2733.9	1.7395	6.9307
150	0.4758	0.001090	0.3928	631.7	2559.5	632.2	2746.4	1.8422	6.8387
160	0.6178	0.001102	0.3071	674.9	2568.4	675.5	2758.1	1.9431	6.7510
170	0.7916	0.001114	0.2428	718.3	2576.5	719.2	2768.7	2.0423	6.6672
180	1.002	0.001127	0.1941	762.1	2583.7	763.2	2778.2	2.1400	6.5866
190	1.254	0.001141	0.1565	806.2	2590.0	807.5	2786.4	2.2363	6.5087
200	1.554	0.001156	0.1274	850.6	2595.3	852.4	2793.2	2.3313	6.4331
210	1.906	0.001173	0.1044	895.5	2599.4	897.7	2798.5	2.4253	6.3593
220	2.318	0.001190	0.08620	940.9	2602.4	943.6	2802.1	2.5183	6.2869
230	2.795	0.001209	0.07159	986.7	2603.9	990.1	2804.0	2.6105	6.2155
240	3.344	0.001229	0.05977	1033.2	2604.0	1037.3	2803.8	2.7021	6.1446
250	3.973	0.001251	0.05013	1080.4	2602.4	1085.3	2801.5	2.7933	6.0738
260	4.688	0.001276	0.04221	1128.4	2599.0	1134.4	2796.9	2.8844	6.0028
270	5.498	0.001302	0.03565	1177.3	2593.7	1184.5	2789.7	2.9757	5.9310

280	**6.411**	**0.001332**	**0.03017**	**1227.4**	**2586.1**	**1236.0**	**2779.6**	**3.0674**	**5.8579**
290	**7.436**	**0.001366**	**0.02557**	**1278.9**	**2576.0**	**1289.0**	**2766.2**	**3.1600**	**5.7830**
300	**8.580**	**0.001404**	**0.02168**	**1332.0**	**2563.0**	**1344.0**	**2749.0**	**3.2540**	**5.7053**
310	**9.856**	**0.001447**	**0.01835**	**1387.0**	**2546.4**	**1401.3**	**2727.3**	**3.3500**	**5.6239**
320	**11.27**	**0.001499**	**0.01549**	**1444.6**	**2525.5**	**1461.4**	**2700.1**	**3.4487**	**5.5370**
330	**12.84**	**0.001561**	**0.01300**	**1505.2**	**2499.0**	**1525.3**	**2665.9**	**3.5514**	**5.4425**
340	**14.59**	**0.001638**	**0.01080**	**1570.3**	**2464.6**	**1594.2**	**2622.1**	**3.6601**	**5.3366**
350	**16.51**	**0.001740**	**0.008815**	**1641.8**	**2418.5**	**1670.6**	**2564.0**	**3.7784**	**5.2122**
360	**18.65**	**0.001892**	**0.006947**	**1725.2**	**2351.6**	**1760.5**	**2481.2**	**3.9154**	**5.0536**
370	**21.03**	**0.002213**	**0.004931**	**1844.0**	**2229.0**	**1890.5**	**2332.7**	**4.1114**	**4.7990**
374.136	**22.088**	**0.003155**	**0.003155**	**2029.6**	**2029.6**	**2099.3**	**2099.3**	**4.4305**	**4.4305**

(Pressure Table)

		Specific Volume (m³/kg)		Specific Internal Energy (kJ/kg)		Specific Enthalpy (kJ/kg)		Specific Entropy (kJ/kg.K)	
$P(MPa)$	$T(^oC)$	v^L	v^V	u^L	u^V	h^L	h^V	s^L	s^V
0.0006113	0.01	0.001000	206.1	0.0	2375.3	0.0	2501.3	0.0000	9.1571
0.0008	3.8	0.001000	159.7	15.8	2380.5	15.8	2508.3	0.0575	9.0582
0.001	7.0	0.001000	129.2	29.3	2385.0	29.3	2514.2	0.1059	8.9765
0.0012	9.7	0.001000	108.7	40.6	2388.7	40.6	2519.1	0.1460	8.9099
0.0014	12.0	0.001001	93.92	50.3	2391.9	50.3	2523.4	0.1802	8.8538
0.0016	14.0	0.001001	82.76	58.9	2394.7	58.9	2527.1	0.2101	8.8053
0.0018	15.8	0.001001	74.03	66.5	2397.2	66.5	2530.5	0.2367	8.7626
0.002	17.5	0.001001	67.00	73.5	2399.5	73.5	2533.5	0.2606	8.7245
0.003	24.1	0.001003	45.67	101.0	2408.5	101.0	2545.5	0.3544	8.5784
0.004	29.0	0.001004	34.80	121.4	2415.2	121.4	2554.4	0.4225	8.4754
0.006	36.2	0.001006	23.74	151.5	2424.9	151.5	2567.4	0.5208	8.3312
0.008	41.5	0.001008	18.10	173.9	2432.1	173.9	2577.0	0.5924	8.2295
0.01	45.8	0.001010	14.67	191.8	2437.9	191.8	2584.6	0.6491	8.1510
0.012	49.4	0.001012	12.36	206.9	2442.7	206.9	2591.0	0.6961	8.0871
0.014	52.6	0.001013	10.69	220.0	2446.9	220.0	2596.6	0.7365	8.0333
0.016	55.3	0.001015	9.433	231.5	2450.5	231.5	2601.4	0.7719	7.9868
0.018	57.8	0.001016	8.445	241.9	2453.8	241.9	2605.8	0.8034	7.9459
0.02	60.1	0.001017	7.649	251.4	2456.7	251.4	2609.7	0.8319	7.9093
0.03	69.1	0.001022	5.229	289.2	2468.4	289.2	2625.3	0.9439	7.7695
0.04	75.9	0.001026	3.993	317.5	2477.0	317.6	2636.7	1.0260	7.6709
0.06	85.9	0.001033	2.732	359.8	2489.6	359.8	2653.5	1.1455	7.5328
0.08	93.5	0.001039	2.087	391.6	2498.8	391.6	2665.7	1.2331	7.4354
0.10	99.6	0.001043	1.694	417.3	2506.1	417.4	2675.5	1.3029	7.3602
0.12	104.8	0.001047	1.428	439.2	2512.1	439.3	2683.5	1.3611	7.2980
0.14	109.3	0.001051	1.237	458.2	2517.3	458.4	2690.4	1.4112	7.2472
0.16	113.3	0.001054	1.091	475.2	2521.8	475.3	2696.5	1.4553	7.2025
0.18	116.9	0.001058	0.9775	490.5	2525.9	490.7	2701.8	1.4948	7.1631
0.20	120.2	0.001061	0.8857	504.5	2529.5	504.7	2706.6	1.5305	7.1280
0.30	133.5	0.001073	0.6058	561.1	2543.6	561.5	2725.3	1.6722	6.9927
0.40	143.6	0.001084	0.4625	604.3	2553.6	604.7	2738.5	1.7770	6.8967
0.60	158.9	0.001101	0.3157	669.9	2567.4	670.6	2756.8	1.9316	6.7609
0.80	170.4	0.001115	0.2404	720.2	2576.8	721.1	2769.1	2.0466	6.6636
1.00	179.9	0.001127	0.1944	761.7	2583.6	762.8	2778.1	2.1391	6.5873
1.20	188.0	0.001139	0.1633	797.3	2588.8	798.6	2784.8	2.2170	6.5242
1.40	195.1	0.001149	0.1408	828.7	2592.8	830.3	2790.0	2.2847	6.4701
1.60	201.4	0.001159	0.1238	856.9	2596.0	858.8	2794.0	2.3446	6.4226
1.80	207.2	0.001168	0.1104	882.7	2598.4	884.8	2797.1	2.3986	6.3802
2.0	212.4	0.001177	0.09963	906.4	2600.3	908.8	2799.5	2.4478	6.3417
3.0	233.9	0.001216	0.06668	1004.8	2604.1	1008.4	2804.1	2.6462	6.1878
4.0	250.4	0.001252	0.04978	1082.3	2602.3	1087.3	2801.4	2.7970	6.0709
6.0	275.6	0.001319	0.03244	1205.4	2589.7	1213.3	2784.3	3.0273	5.8900
8.0	295.1	0.001384	0.02352	1305.6	2569.8	1316.6	2758.0	3.2075	5.7440
10	311.1	0.001452	0.01803	1393.0	2544.4	1407.6	2724.7	3.3603	5.6149
12	324.8	0.001527	0.01426	1472.9	2513.7	1491.3	2684.9	3.4970	5.4933

14	**336.8**	**0.001611**	**0.01149**	**1548.6**	**2476.8**	**1571.1**	**2637.6**	**3.6240**	**5.3726**
16	**347.4**	**0.001711**	**0.009307**	**1622.7**	**2431.8**	**1650.0**	**2580.7**	**3.7468**	**5.2464**
18	**357.1**	**0.001840**	**0.007491**	**1698.9**	**2374.4**	**1732.0**	**2509.2**	**3.8722**	**5.1054**
20	**365.8**	**0.002036**	**0.005836**	**1785.6**	**2293.2**	**1826.3**	**2410.0**	**4.0146**	**4.9281**
22.088	**374.136**	**0.003155**	**0.003155**	**2029.6**	**2029.6**	**2099.3**	**2099.3**	**4.4305**	**4.4305**

Sources:

Keenan, J.H., Keyes, F.G., Hill, P.G., Moore, J.G., *Steam Tables*, John Wiley & Sons, New York, 1969.
Van Wylen, G.J., Sonntag, R.E., *Fundamentals of Classical Thermodynamics*, John Wiley & Sons, New York, 1973.
Potter, M.C., Somerton, C.W., *Thermodynamics for Engineers,* McGraw Hill, New York, 2nd edition, 2006.

Appendix N

Thermodynamic Properties of Superheated Steam

$T\ (^oC)$	v	u	h	s	v	u	h	s	v	u	h	s
	$P = 0.010$ MPa ($T^{sat} = 45.81^oC$)				$P = 0.050$ MPa ($T^{sat} = 81.33^oC$)				$P = 0.10$ MPa ($T^{sat} = 99.63^oC$)			
Sat.	14.674	2437.9	2584.7	8.1502	3.240	2483.9	2645.9	7.5939	1.6940	2506.1	2675.5	7.3594
50	14.869	2443.9	2592.6	8.1749								
100	17.196	2515.5	2687.5	8.4479	3.418	2511.6	2682.5	7.6947	1.6958	2506.7	2676.2	7.3614
150	19.512	2587.9	2783.0	8.6882	3.889	2585.6	2780.1	7.9401	1.9364	2582.8	2776.4	7.6134
200	21.825	2661.3	2879.5	8.9038	4.356	2659.9	2877.7	8.1580	2.172	2658.1	2875.3	7.8343
250	24.136	2736.0	2977.3	9.1002	4.820	2735.0	2976.0	8.3556	2.406	2733.7	2974.3	8.0333
300	26.445	2812.1	3076.5	9.2813	5.284	2811.3	3075.5	8.5373	2.639	2810.4	3074.3	8.2158
400	31.063	2968.9	3279.6	9.6077	6.209	2968.5	3278.0	8.8642	3.103	2967.9	3278.2	8.5435
500	35.679	3132.3	3489.1	9.8978	7.134	3132.0	3488.7	9.1546	3.565	3131.6	3488.1	8.8342
600	40.295	3302.5	3705.4	10.1608	8.057	3302.2	3705.1	9.4178	4.028	3301.9	3704.7	9.0976
700	44.911	3479.6	3928.7	10.4028	8.981	3479.4	3928.5	9.6599	4.490	3479.2	3928.2	9.3398
800	49.526	3663.8	4159.0	10.6281	9.904	3663.6	4158.9	9.8852	4.952	3663.5	4158.6	9.5652
900	54.141	3855.0	4396.4	10.8396	10.828	3854.9	4396.3	10.0967	5.414	3854.8	4396.1	9.7767
1000	58.757	4053.0	4640.6	11.0393	11.751	4052.9	4640.5	10.2964	5.875	4052.8	4640.3	9.9764
1100	63.372	4257.5	4891.2	11.2287	12.674	4257.4	4891.1	10.4859	6.337	4257.3	4891.0	10.1659
1200	67.987	4467.9	5147.8	11.4091	13.597	4467.8	5147.7	10.6662	6.799	4467.7	5147.6	10.3463
1300	72.602	4683.7	5409.7	11.5811	14.521	4683.6	5409.6	10.8382	7.260	4683.5	5409.5	10.5183

(v in m^3/kg, u and h in kJ/kg, s in $kJ/kg.K$)

$T\ (^oC)$	v	u	h	s	v	u	h	s	v	u	h	s
	$P = 0.20$ MPa ($T^{sat} = 120.23^oC$)				$P = 0.30$ MPa ($T^{sat} = 133.55^oC$)				$P = 0.40$ MPa ($T^{sat} = 143.63^oC$)			
Sat.	0.8857	2529.5	2706.7	7.1272	0.6058	2543.6	2725.3	6.9919	0.4625	2553.6	2738.6	6.8959
150	0.9596	2576.9	2768.8	7.2795	0.6339	2570.8	2761.0	7.0778	0.4708	2564.5	2752.8	6.9299
200	1.0803	2654.4	2870.5	7.5066	0.7163	2650.7	2865.6	7.3115	0.5342	2646.8	2860.5	7.1706
250	1.1988	2731.2	2971.0	7.7086	0.7964	2728.7	2967.6	7.5166	0.5951	2726.1	2964.2	7.3789
300	1.3162	2808.6	3071.8	7.8926	0.8753	2806.7	3069.3	7.7022	0.6548	2804.8	3066.8	7.5662
400	1.5493	2966.7	3276.6	8.2218	1.0315	2965.6	3275.6	8.0330	0.7726	2964.4	3273.4	7.8985
500	1.7814	3130.8	3487.1	8.5133	1.1867	3130.0	3486.0	8.3251	0.8893	3129.2	3484.9	8.1913
600	2.013	3301.4	3704.0	8.7770	1.3414	3300.8	3703.2	8.5892	1.0055	3300.2	3702.4	8.4558
700	2.244	3478.8	3927.6	9.0194	1.4957	3478.4	3927.1	8.8319	1.1215	3477.9	3926.5	8.6987
800	2.475	3663.1	4158.2	9.2449	1.6499	3662.9	4157.8	9.0576	1.2372	3662.4	4157.3	8.9244
900	2.706	3854.5	4395.8	9.4566	1.8041	3854.2	4395.4	9.2692	1.3529	3853.9	4395.1	9.1362
1000	2.937	4052.5	4640.0	9.6563	1.9581	4052.3	4639.7	9.4690	1.4685	4052.0	4639.4	9.3360
1100	3.168	4257.0	4890.7	9.8458	2.1121	4256.5	4890.4	9.6585	1.5840	4256.5	4890.2	9.5256
1200	3.399	4467.5	5147.3	10.0262	2.2661	4467.2	5147.1	9.8389	1.6996	4467.0	5146.8	9.7060
1300	3.630	4683.2	5409.3	10.1982	2.4201	4683.0	5409.0	10.011	1.8151	4682.8	5408.8	9.8780

(v in m^3/kg, u and h in kJ/kg, s in $kJ/kg.K$)

$T\ (^oC)$	v	u	h	s	v	u	h	s	v	u	h	s
	$P = 0.50$ MPa ($T^{sat} = 151.86^oC$)				$P = 0.60$ MPa ($T^{sat} = 158.85^oC$)				$P = 0.80$ MPa ($T^{sat} = 170.43^oC$)			
Sat.	0.3749	2561.2	2748.7	6.8213	0.3157	2567.4	2756.8	6.7600	0.2404	2576.8	2769.1	6.6628
200	0.4249	2642.9	2855.4	7.0592	0.3520	2638.9	2850.1	6.9665	0.2608	2630.6	2839.3	6.8158
250	0.4744	2723.5	2960.7	7.2709	0.3938	2720.9	2957.2	7.1816	0.2931	2715.5	2950.0	7.0384
300	0.5226	2802.9	3064.2	7.4599	0.4344	2801.0	3061.6	7.3724	0.3241	2797.2	3056.5	7.2328
350	0.5701	2882.6	3167.7	7.6329	0.4742	2881.2	3165.7	7.5464	0.3544	2878.2	3161.7	7.4089
400	0.6173	2963.2	3271.9	7.7938	0.5137	2962.1	3270.3	7.7079	0.3843	2959.7	3267.1	7.5716
500	0.7109	3128.4	3483.9	8.0873	0.5920	3127.6	3482.8	8.0021	0.4433	3126.0	3480.6	7.8673
600	0.8041	3299.6	3701.7	8.3522	0.6697	3299.1	3700.9	8.2674	0.5018	3297.9	3699.4	8.1333
700	0.8969	3477.5	3925.9	8.5952	0.7472	3477.0	3925.3	8.5107	0.5601	3476.2	3924.2	8.3770
800	0.9896	3662.1	4156.9	8.8211	0.8245	3661.8	4156.5	8.7367	0.6181	3661.1	4155.6	8.6033
900	1.0822	3853.6	4394.7	9.0329	0.9017	3853.4	4394.4	8.9486	0.6761	3852.8	4393.7	8.8153
1000	1.1747	4051.8	4639.1	9.2328	0.9788	4051.5	4638.8	9.1485	0.7340	4051.0	4638.2	9.0153
1100	1.2672	4256.3	4889.9	9.4224	1.0559	4256.1	4889.6	9.3381	0.7919	4255.6	4889.1	9.2050
1200	1.3596	4466.8	5146.6	9.6029	1.1330	4466.5	5146.3	9.5185	0.8497	4466.1	5145.9	9.3855
1300	1.4521	4682.5	5408.6	9.7749	1.2101	4682.3	5408.3	9.6906	0.9076	4681.8	5407.9	9.5575

(v in m^3/kg, u and h in kJ/kg, s in $kJ/kg.K$)

$T\ (^oC)$	v	u	h	s	v	u	h	s	v	u	h	s
	$P = 1.00$ MPa ($T^{sat} = 179.91^oC$)				$P = 1.20$ MPa ($T^{sat} = 187.99^oC$)				$P = 1.40$ MPa ($T^{sat} = 195.07^oC$)			
Sat.	0.1944	2583.6	2778.1	6.5865	0.1633	2588.8	2784.8	6.5233	0.14084	2592.8	2790.0	6.4693
200	0.2060	2621.9	2827.9	6.6940	0.1693	2612.8	2815.9	6.5898	0.14302	2603.1	2803.3	6.4975
250	0.2327	2709.9	2942.6	6.9247	0.1923	2704.2	2935.0	6.8294	0.16350	2698.3	2927.2	6.7467
300	0.2579	2793.2	3051.2	7.1229	0.2138	2789.2	3045.8	7.0317	0.18228	2785.2	3040.4	6.9534
350	0.2825	2875.2	3157.7	7.3011	0.2345	2872.2	3153.6	7.2121	0.2003	2869.2	3149.5	7.1360
400	0.3066	2957.3	3263.9	7.4651	0.2548	2954.9	3260.7	7.3774	0.2178	2952.5	3257.5	7.3026
500	0.3541	3124.4	3478.5	7.7622	0.2946	3122.8	3476.3	7.6759	0.2521	3321.1	3474.1	7.6027
600	0.4011	3296.8	3697.9	8.0290	0.3339	3295.6	3696.3	7.9435	0.2860	3294.4	3694.8	7.8710
700	0.4478	3475.3	3923.1	8.2731	0.3729	3474.4	3922.0	8.1881	0.3195	3473.6	3920.8	8.1160
800	0.4943	3660.4	4154.7	8.4996	0.4118	3659.7	4153.8	8.4148	0.3528	3659.0	4153.0	8.8431
900	0.5407	3852.2	4392.9	8.7118	0.4505	3851.6	4392.2	8.6272	0.3861	3851.1	4391.5	8.5556
1000	0.5871	4050.5	4637.6	8.9119	0.4892	4050.0	4637.0	8.8274	0.4192	4049.5	4636.4	8.7559
1100	0.6335	4255.1	4888.6	9.1017	0.5278	4254.6	4888.0	9.0172	0.4524	4254.1	4887.5	8.9457
1200	0.6798	4465.6	5145.4	9.2822	0.5665	4465.1	5144.9	9.1977	0.4855	4464.7	5144.4	9.1262
1300	0.7261	4681.3	5407.4	9.4543	0.6051	4680.9	5407.0	9.3698	0.5186	4680.4	5406.5	9.2984

(v in m^3/kg, u and h in kJ/kg, s in $kJ/kg.K$)

$T\ (^oC)$	v	u	h	s	v	u	h	s	v	u	h	s
	$P = 1.60$ MPa ($T^{sat} = 201.41^oC$)				$P = 1.80$ MPa ($T^{sat} = 207.15^oC$)				$P = 2.00$ MPa ($T^{sat} = 212.42^oC$)			
Sat.	0.12380	2596.0	2794.0	6.4218	0.11042	2598.4	2797.1	6.3794	0.09963	2600.3	2799.5	6.3409
200	0.13287	2644.7	2857.3	6.5518	0.13673	2636.6	2846.7	6.4808	0.10377	2628.3	2835.8	6.4147
250	0.14184	2692.3	2919.2	6.6732	0.12497	2686.0	2911.0	6.6066	0.11144	2679.6	2902.5	6.5453
300	0.15862	2781.1	3034.8	6.8844	0.14021	2776.9	3029.2	6.8226	0.12547	2772.6	3023.5	6.7664
350	0.17456	2866.1	3145.4	7.0694	0.15457	2863.0	3141.2	7.01	0.13857	2859.8	3137.0	6.9563
400	0.19005	2950.1	3254.2	7.2374	0.16847	2947.7	3250.9	7.1794	0.15120	2945.2	3247.6	7.1271
500	0.2203	3119.5	3472.0	7.5390	0.19550	3117.9	3469.8	7.4825	0.17568	3116.2	3467.6	7.4317
600	0.2500	3293.3	3693.2	7.8080	0.2220	3292.1	3691.7	7.7523	0.19960	3290.9	3690.1	7.7024
700	0.2794	3472.7	3919.7	8.0535	0.2482	3471.8	3918.5	7.9983	0.2232	3470.9	3917.4	7.9487
800	0.3086	3658.3	4152.1	8.2808	0.2742	3657.6	4151.2	8.2258	0.2467	3657.0	4150.3	8.1765
900	0.3377	3850.5	4390.8	8.4935	0.3001	3849.9	4390.1	8.4386	0.2700	3849.3	4389.4	8.3895
1000	0.5668	4049.0	4635.8	8.6938	0.3260	4048.5	4635.2	8.6391	0.2933	4048.0	4634.6	8.5901
1100	0.3958	4253.7	4887.0	8.8837	0.3518	4253.2	4886.4	8.8290	0.3166	4252.7	4885.9	8.7800
1200	0.4248	4464.2	5143.9	9.0643	0.3776	4463.7	5143.4	9.0096	0.3398	4463.3	5142.9	8.9607
1300	0.4538	4679.9	5406.0	9.2364	0.4034	4679.5	5405.6	9.1818	0.3631	4679.0	5405.1	9.1329

(v in m^3/kg, u and h in kJ/kg, s in $kJ/kg.K$)

$T\ (^oC)$	v	u	h	s	v	u	h	s	v	u	h	s
	$P = 2.50$ MPa ($T^{sat} = 223.99^oC$)				$P = 3.00$ MPa ($T^{sat} = 233.90^oC$)				$P = 3.50$ MPa ($T^{sat} = 242.60^oC$)			
Sat.	0.07998	2603.1	2803.1	6.2575	0.06668	2604.1	2804.2	6.1869	0.05707	2603.7	2803.4	6.1253
225	0.08027	2605.6	2806.3	6.2639								
250	0.08700	2662.6	2880.1	6.4085	0.07058	2644.0	2855.8	6.2872	0.05872	2623.7	2829.2	6.1749
300	0.09890	2761.6	3008.8	6.6438	0.08114	2750.1	2993.5	6.5390	0.06842	2738.0	2977.5	6.4461
350	0.10976	2851.9	3126.3	6.8403	0.09053	2843.7	3115.3	6.7428	0.07678	2835.3	3104.0	6.6579
400	0.12010	2939.1	3239.3	7.0148	0.09936	2932.8	3230.9	6.9212	0.08453	2926.4	3222.3	6.8405
450	0.13014	3025.5	3350.8	7.1746	0.10787	3020.4	3344.0	7.0834	0.09196	3015.3	3337.2	7.0052
500	0.13998	3112.1	3462.1	7.3234	0.11619	3108.0	3456.5	7.2338	0.09918	3103.0	3450.9	7.1572
600	0.15930	3288.0	3686.3	7.5960	0.13243	3285.0	3682.3	7.5085	0.11324	3282.1	3678.4	7.4339
700	0.17832	3468.7	3914.5	7.8435	0.14838	3466.5	3911.7	7.7571	0.12699	3464.3	3908.8	7.6837
800	0.19716	3655.3	4148.2	8.0720	0.16414	3653.5	4145.9	7.9862	0.14056	3651.8	4143.7	7.9134
900	0.21590	3847.9	4387.6	8.2853	0.17980	3846.5	4385.9	8.1999	0.15402	3845.0	4384.1	8.1276
1000	0.2346	4046.7	4633.1	8.4861	0.19541	4045.4	4631.6	8.4009	0.16743	4044.1	4630.1	8.3288
1100	0.2532	4251.5	4884.6	8.6762	0.21098	4250.3	4883.3	8.5912	0.18080	4249.2	4881.9	8.5192
1200	0.2718	4462.1	5141.7	8.8569	0.22652	4460.9	5140.5	8.7720	0.19415	4459.8	5139.3	8.7000
1300	0.2905	4677.8	5404.0	9.0291	0.24206	4676.6	5402.8	8.9442	0.20749	4675.5	5401.7	8.8723

(v in m^3/kg, u and h in kJ/kg, s in $kJ/kg.K$)

$T\ (^oC)$	v	u	h	s	v	u	h	s	v	u	h	s
	$P = 4.0$ MPa ($T^{sat} = 250.40^oC$)				$P = 4.5$ MPa ($T^{sat} = 257.49^oC$)				$P = 5.0$ MPa ($T^{sat} = 263.99^oC$)			
Sat.	0.04978	2602.3	2801.4	6.0701	0.04406	2600.1	2798.3	6.0198	0.03944	2597.1	2794.3	5.9734
275	0.05457	2667.9	2886.2	6.2285	0.04730	2650.3	2863.2	6.1401	0.04141	2631.3	2838.3	6.0544
300	0.05884	2725.3	2960.7	6.3615	0.05135	2712.0	2943.1	6.2828	0.04532	2698.0	2924.5	6.2084
350	0.06645	2826.7	3092.5	6.5821	0.05840	2817.8	3080.6	6.5131	0.05194	2808.7	3068.4	6.4493
400	0.07341	2919.9	3213.6	6.7690	0.06475	2913.3	3204.7	6.7047	0.05781	2906.6	3195.7	6.6459
450	0.08002	3010.2	3330.3	6.9363	0.07074	3005.0	3323.3	6.8746	0.06330	2999.7	3316.2	6.8186
500	0.08643	3099.5	3445.3	7.0901	0.07651	3095.3	3439.6	7.0301	0.06857	3091.0	3433.8	6.9759
600	0.09885	3279.1	3674.4	7.3688	0.08765	3276.0	3670.5	7.3110	0.07869	3273.0	3666.5	7.2589
700	0.11095	3462.1	3905.9	7.6198	0.09847	3459.9	3903.0	7.5631	0.08849	3457.6	3900.1	7.5122
800	0.12287	3650.0	4141.5	7.8502	0.10911	3648.3	4139.3	7.7942	0.09811	3646.6	4137.1	7.7440
900	0.13469	3843.6	4382.3	8.0647	0.11965	3842.2	4380.6	8.0091	0.10762	3840.7	4378.8	7.9593
1000	0.14645	4042.9	4628.7	8.2662	0.13013	4041.6	4627.2	8.2108	0.11707	4040.4	4625.7	8.1612
1100	0.15817	4248.0	4880.6	8.4567	0.14056	4246.8	4879.3	8.4015	0.12648	4245.6	4878.0	8.3520
1200	0.16987	4458.6	5138.1	8.6376	0.15098	4457.5	5136.9	8.5825	0.13587	4456.3	5135.7	8.5331
1300	0.18156	4674.3	5400.5	8.8100	0.16139	4673.1	5399.4	8.7549	0.14526	4672.0	5398.2	8.7055

(v in m^3/kg , u and h in kJ/kg, s in $kJ/kg.K$)

$T\ (^oC)$	v	u	h	s	v	u	h	s	v	u	h	s
	$P = 6.0$ MPa ($T^{sat} = 275.64^oC$)				$P = 7.0$ MPa ($T^{sat} = 285.88^oC$)				$P = 8.0$ MPa ($T^{sat} = 295.06^oC$)			
Sat.	0.03244	2589.7	2784.3	5.8892	0.02737	2580.5	2772.1	5.8133	0.02352	2569.8	2758.0	5.7432
300	0.03616	2667.2	2884.2	6.0674	0.02947	2632.2	2838.4	5.9305	0.02426	2590.9	2785.0	5.7906
350	0.04223	2789.6	3043.0	6.3335	0.03524	2769.4	3016.0	6.2283	0.02995	2747.7	2987.3	6.1301
400	0.04739	2892.9	3177.2	6.5408	0.03993	2878.6	3158.1	6.4478	0.03432	2863.8	3138.3	6.3634
450	0.05214	2988.9	3301.8	6.7193	0.04416	2978.0	3287.1	6.6327	0.03817	2966.7	3272.0	6.5551
500	0.05665	3082.2	3422.2	6.8803	0.04814	3073.4	3410.3	6.7975	0.04175	3064.3	3398.3	6.7240
550	0.06101	3174.6	3540.6	7.0288	0.05195	3167.2	3530.9	6.9486	0.04516	3159.8	3521.0	6.8778
600	0.06525	3266.9	3658.4	7.1677	0.05565	3260.7	3650.3	7.0894	0.04845	3254.4	3642.0	7.0206
700	0.07352	3453.1	3894.2	7.4234	0.06283	3448.5	3888.3	7.3476	0.05481	3443.9	3882.4	7.2812
800	0.08160	3643.1	4132.7	7.6566	0.06981	3639.5	4128.2	7.5822	0.06097	3636.0	4123.8	7.5173
900	0.08958	3837.8	4375.3	7.8727	0.07669	3835.0	4371.8	7.7991	0.06702	3832.1	4368.3	7.7351
1000	0.09749	4037.8	4622.7	8.0751	0.08350	4035.3	4619.8	8.0020	0.07301	4032.8	4616.9	7.9384
1100	0.10536	4243.3	4875.4	8.2661	0.09027	4240.9	4872.8	8.1933	0.07896	4238.6	4870.3	8.1300
1200	0.11321	4454.0	5133.3	8.4474	0.09703	4451.7	5130.9	8.3747	0.08489	4449.5	5128.5	8.3115
1300	0.12106	4669.6	5396.0	8.6199	0.10377	4667.3	5393.7	8.5473	0.09080	4665.0	5391.5	8.4842

(v in m^3/kg , u and h in kJ/kg, s in $kJ/kg.K$)

$T\ (^oC)$	v	u	h	s	v	u	h	s	v	u	h	s
	$P = 9.0$ MPa ($T^{sat} = 303.40^oC$)				$P = 10.0$ MPa ($T^{sat} = 311.06^oC$)				$P = 12.5$ MPa ($T^{sat} = 327.89^oC$)			
Sat.	**0.02048**	**2557.8**	**2742.1**	**5.6772**	**0.018026**	**2544.4**	**2724.7**	**5.6141**	**0.013495**	**2505.1**	**2673.8**	**5.4624**
325	**0.02327**	**2646.6**	**2856.0**	**5.8712**	**0.019861**	**2610.4**	**2809.1**	**5.7568**				
350	**0.02580**	**2724.4**	**2956.6**	**6.0361**	**0.02242**	**2699.4**	**2923.4**	**5.9443**	**0.016126**	**2624.6**	**2826.2**	**5.7118**
400	**0.02993**	**2848.4**	**3117.8**	**6.2854**	**0.02641**	**2832.4**	**3096.5**	**6.2120**	**0.02000**	**2789.3**	**3039.3**	**6.0417**
450	**0.03350**	**2955.2**	**3256.6**	**6.4844**	**0.02975**	**2943.4**	**3240.9**	**6.4190**	**0.02299**	**2912.5**	**3199.8**	**6.2719**
500	**0.03677**	**3055.2**	**3336.1**	**6.6576**	**0.03279**	**3045.8**	**3373.7**	**6.5966**	**0.02560**	**3021.7**	**3341.8**	**6.4618**
550	**0.03987**	**3152.2**	**3511.0**	**6.8142**	**0.03564**	**3144.6**	**3500.9**	**6.7561**	**0.02801**	**3125.0**	**3475.2**	**6.6290**
600	**0.04285**	**3248.1**	**3633.7**	**6.9589**	**0.03837**	**3241.7**	**3625.3**	**6.9029**	**0.03029**	**3225.4**	**3604.0**	**6.7810**
650	**0.04574**	**3343.6**	**3755.3**	**7.0943**	**0.04101**	**3338.2**	**3748.2**	**7.0398**	**0.03248**	**3324.4**	**3730.4**	**6.9218**
700	**0.04857**	**3439.3**	**3876.5**	**7.221**	**0.04358**	**3434.7**	**3870.5**	**7.1687**	**0.03460**	**3422.9**	**3855.3**	**7.0536**
800	**0.05409**	**3632.5**	**4119.3**	**7.4596**	**0.04859**	**3628.9**	**4114.8**	**7.4077**	**0.03869**	**3620.0**	**4103.6**	**7.2965**
900	**0.05950**	**3829.2**	**4364.3**	**7.6783**	**0.05349**	**3826.3**	**4361.2**	**7.6272**	**0.04267**	**3819.1**	**4352.5**	**7.5182**
1000	**0.06485**	**4030.3**	**4614.0**	**7.8821**	**0.05832**	**4027.8**	**4611.0**	**7.8315**	**0.04658**	**4021.6**	**4603.8**	**7.7237**
1100	**0.07016**	**4236.3**	**4867.7**	**8.0740**	**0.06312**	**4234.0**	**4865.1**	**8.0237**	**0.05045**	**4228.2**	**4858.8**	**7.9165**
1200	**0.07544**	**4447.2**	**5126.2**	**8.2556**	**0.06789**	**4444.9**	**5123.8**	**8.2055**	**0.05430**	**4439.3**	**5118.0**	**8.0987**
1300	**0.08072**	**4662.7**	**5389.2**	**8.4284**	**0.07265**	**4460.5**	**5387.0**	**8.3783**	**0.05813**	**4654.8**	**5381.4**	**8.2717**

(v in m^3/kg , u and h in kJ/kg, s in $kJ/kg.K$)

$T\ (^oC)$	v	u	h	s	v	u	h	s	v	u	h	s
	$P = 15.0$ MPa ($T^{sat} = 342.24^oC$)				$P = 17.5$ MPa ($T^{sat} = 354.75^oC$)				$P = 20.0$ MPa ($T^{sat} = 365.81^oC$)			
Sat.	0.010337	2455.5	2610.5	5.3098	0.007920	2390.2	2528.8	5.1419	0.005834	2293.0	2409.7	4.9269
350	0.011470	2520.4	2692.4	5.4421								
400	0.015649	2740.7	2975.5	5.8811	0.012447	2685.0	2902.9	5.7213	0.009942	2619.3	2818.1	5.5540
450	0.018445	2879.5	3156.2	6.1404	0.015174	2844.2	3109.7	6.0184	0.012695	2806.2	3060.1	5.9017
500	0.02080	2996.6	3308.6	6.3443	0.017358	2970.3	3274.1	6.2383	0.014768	2942.9	3238.2	6.1401
550	0.02293	3104.7	3448.6	6.5199	0.019288	3083.9	3421.4	6.4230	0.016555	3062.4	3393.5	6.3348
600	0.02491	3208.6	3582.3	6.6776	0.02106	3191.5	3560.2	6.5866	0.018178	3174.0	3537.6	6.5048
650	0.02680	3310.3	3712.3	6.8224	0.02274	3296.0	3693.9	6.7357	0.019693	3281.4	3675.3	6.6582
700	0.02861	3410.9	3840.1	6.9572	0.02434	3398.7	3824.6	6.8736	0.02113	3386.4	3809.0	6.7993
800	0.03210	3610.9	4092.4	7.2040	0.02738	3601.8	4081.1	7.1244	0.02385	3592.7	4069.7	7.0544
900	0.03546	3811.9	4343.8	7.4279	0.03031	3804.7	4335.1	7.3507	0.02645	3797.5	4326.4	7.2830
1000	0.03875	4015.4	4596.6	7.6348	0.03316	4009.3	4589.5	7.5589	0.02897	4003.1	4582.5	7.4925
1100	0.04200	4222.6	4852.6	7.8283	0.03597	4216.9	4846.4	7.7531	0.03145	4211.3	4840.2	7.6874
1200	0.04523	4433.8	5112.3	8.0108	0.03876	4428.3	5106.6	7.9360	0.03391	4422.8	5101.0	7.8707
1300	0.04845	4649.1	5376.0	8.1840	0.04154	4643.5	5370.5	8.1093	0.03636	4638.0	5365.1	8.0442

(v in m^3/kg, u and h in kJ/kg, s in $kJ/kg.K$)

$T\ (^oC)$	v	u	h	s	v	u	h	s	v	u	h	s
	$P = 25.0$ MPa				$P = 30.0$ MPa				$P = 35.0$ MPa			
375	0.001973	1798.7	1848.0	4.0320	0.001789	1737.8	1791.5	3.9305	0.001700	1702.9	1762.4	3.8722
400	0.006004	2430.1	2580.2	5.1418	0.002790	2067.4	2151.1	4.4728	0.002100	1914.1	1987.6	4.2126
425	0.007881	2609.2	2806.3	5.4723	0.005303	2455.1	2614.2	5.1504	0.003428	2253.4	2373.4	4.7747
450	0.009162	2720.7	2949.7	5.6744	0.006735	2619.3	2821.4	5.4424	0.004961	2498.7	2672.4	5.1962
500	0.011123	2884.3	3162.4	5.9592	0.008678	2820.7	3081.1	5.7905	0.006927	2571.9	2994.4	5.6282
550	0.012724	3017.5	3335.6	6.1765	0.010168	2970.3	3275.4	6.0342	0.008345	2921.0	3213.0	5.9026
600	0.014137	3137.9	3491.4	6.3602	0.001446	3100.5	3443.9	6.2331	0.009527	3062.0	3395.5	6.1179
650	0.015433	3251.6	3637.4	6.5229	0.012596	3221.0	3598.9	6.4058	0.010575	3189.8	3559.9	6.3010
700	0.016646	3361.3	3777.5	6.6707	0.013661	3335.8	3745.6	6.5606	0.011533	3309.8	3713.5	6.4631
800	0.018912	3574.3	4047.1	6.9345	0.015623	3555.5	4024.2	6.8332	0.013278	3536.7	4001.5	6.7450
900	0.021045	3783.0	4309.1	7.1680	0.017448	3768.5	4291.9	7.0718	0.014883	3754.0	4274.9	6.9886
1000	0.02310	3990.9	4568.5	7.3802	0.019196	3978.8	4554.7	7.2867	0.016410	3966.7	4541.1	7.2064
1100	0.02512	4200.2	4828.2	7.5705	0.020903	4189.2	4816.3	7.4845	0.017895	4178.3	4804.6	7.4057
1200	0.02711	4412.0	5089.9	7.7605	0.022589	4401.3	5079.0	7.6692	0.019360	4390.7	5068.3	7.5910
1300	0.02910	4626.9	5354.4	7.9342	0.024266	4616.0	5344.0	7.8432	0.020815	4605.1	5333.6	7.7653

(v in m^3/kg , u and h in kJ/kg, s in $kJ/kg.K$)

$T\ (^oC)$	v	u	h	s	v	u	h	s	v	u	h	s
	$P = 40.0$ MPa				$P = 50.0$ MPa				$P = 60.0$ MPa			
375	0.001641	1677.1	1742.8	3.8290	0.001559	1638.6	1716.6	3.7639	0.001503	1609.4	1699.5	3.7141
400	0.001908	1854.6	1930.9	4.1135	0.001731	1788.1	1874.6	4.0031	0.001634	1745.4	1843.4	3.9318
425	0.002532	2096.9	2198.1	4.5029	0.002007	1959.7	2060.0	4.2734	0.001817	1892.7	2001.7	4.1626
450	0.003693	2365.1	2512.8	4.9459	0.002486	2159.6	2284.0	4.5884	0.002085	2053.9	2179.0	4.4121
500	0.005622	2678.4	2903.3	5.4700	0.003892	2525.5	2720.1	5.1726	0.002956	2390.6	2567.9	4.9321
550	0.006984	2869.7	3149.1	5.7785	0.005118	2763.6	3019.5	5.5485	0.003956	2658.8	2896.2	5.3441
600	0.008094	3022.6	3346.4	6.0114	0.006112	2942.0	3247.6	5.8178	0.004834	2861.1	3151.2	5.6452
650	0.009063	3158.0	3520.6	6.2054	0.006966	3093.5	3441.8	6.0342	0.005595	3028.8	3364.5	5.8829
700	0.009941	3283.6	3681.2	6.3750	0.007727	3230.5	3616.8	6.2189	0.006272	3177.2	3553.5	6.0824
800	0.011523	3517.8	3978.7	6.6662	0.009076	3479.8	3933.6	6.5290	0.007459	3441.5	3889.1	6.4109
900	0.012962	3739.4	4257.9	6.9150	0.010283	3710.3	4224.4	6.7882	0.008508	3681.0	4191.5	6.6805
1000	0.014324	3954.6	4527.6	7.1356	0.011411	3930.5	4501.1	7.0146	0.009480	3906.4	4475.2	6.9127
1100	0.015642	4167.4	4793.1	7.3364	0.012496	4145.7	4770.5	7.2184	0.010409	4124.1	4748.6	7.1195
1200	0.016940	4380.1	5057.7	7.5224	0.013561	4359.1	5037.2	7.4058	0.011317	4338.2	5017.2	7.3083
1300	0.018229	4594.3	5323.5	7.6969	0.014616	4572.8	5303.6	7.5808	0.012215	4551.4	5284.3	7.4837

(v in m^3/kg , u and h in kJ/kg, s in $kJ/kg.K$)

Sources:

Keenan, J.H., Keyes, F.G., Hill, P.G., Moore, J.G., *Steam Tables*, John Wiley & Sons, New York, 1969.

Van Wylen, G.J., Sonntag, R.E., *Fundamentals of Classical Thermodynamics*, John Wiley & Sons, New York, 1973.

Potter, M.C., Somerton, C.W., *Thermodynamics for Engineers,* McGraw Hill, New York, 2nd edition, 2006.

Sandler, S.I., *Chemical and Engineering Thermodynamics*, John Wiley & Sons, New York, 3rd edition, 1999.

Appendix O

Thermodynamic Properties of Compressed Liquid Water

$T\ (^oC)$	v	u	h	s	v	u	h	s	v	u	h	s
	$P = 5.0$ MPa ($T^{sat} = 263.99^oC$)				$P = 10.0$ MPa ($T^{sat} = 311.06^oC$)				$P = 15.0$ MPa ($T^{sat} = 342.24^oC$)			
Sat.	0.0012859	1147.8	1154.2	2.9202	0.0014524	1393.0	1407.6	3.3596	0.0016581	1585.6	1610.5	3.6848
0	0.0009977	0.04	5.04	0.0001	0.0009952	0.09	10.04	0.0002	0.0009928	0.15	15.05	0.0004
20	0.0009995	83.65	88.65	0.2956	0.0009972	83.36	93.33	0.2945	0.0009950	83.06	97.99	0.2934
40	0.0010056	166.95	171.97	0.5705	0.0010034	166.35	176.38	0.5686	0.0010013	165.76	180.78	0.5666
60	0.0010149	250.23	255.30	0.8285	0.0010127	249.36	259.49	0.8258	0.0010105	248.51	263.67	0.8232
80	0.0010268	333.72	338.85	1.0720	0.0010245	332.59	342.83	1.0688	0.0010222	331.48	346.81	1.0656
100	0.0010410	417.52	422.72	1.3030	0.0010385	416.12	426.50	1.2992	0.0010361	414.74	430.28	1.2955
120	0.0010576	501.80	507.09	1.5233	0.0010549	500.08	510.64	1.5189	0.0010522	498.40	514.19	1.5145
140	0.0010768	586.76	592.15	1.7343	0.0010737	584.68	595.42	1.7292	0.0010707	582.66	598.72	1.7242
160	0.0010988	672.62	678.12	1.9375	0.0010953	670.13	681.08	1.9317	0.0010918	667.71	684.09	1.9260
180	0.0011240	759.63	765.25	2.1341	0.0011199	756.65	767.84	2.1275	0.0011159	753.76	770.50	2.1210
200	0.0011530	848.1	853.9	2.3255	0.0011480	844.5	856.0	2.3178	0.0011433	841.0	858.2	2.3104
220	0.0011866	938.4	944.4	2.5128	0.0011805	934.1	945.9	2.5039	0.0011748	929.9	947.5	2.4953
240	0.0012264	1031.4	1037.5	2.6979	0.0012187	1026.0	1038.1	2.6872	0.0012114	1020.8	1039.0	2.6771
260	0.0012749	1127.9	1134.3	2.8830	0.0012645	1121.1	1133.7	2.8699	0.0012550	1114.6	1133.4	2.8576
280					0.0013216	1220.9	1234.1	3.0548	0.0013084	1212.5	1232.1	3.0393
300					0.0013972	1328.4	1342.3	3.2469	0.0013770	1316.6	1337.3	3.2260
320									0.0014724	1431.1	1453.2	3.4247
340									0.0016311	1567.5	1591.9	3.6546

v in m^3/kg, u and h in kJ/kg, s in $kJ/kg.K$

$T\ (^oC)$	v	u	h	s	v	u	h	s	v	u	h	s
	$P = 20.0$ MPa ($T^{sat} = 365.81^oC$)				$P = 30.0$ MPa				$P = 50.0$ MPa			
Sat.	0.002036	1785.0	1826.3	4.0139								
0	0.0009904	0.19	20.01	0.0004	0.0009856	0.25	29.82	0.0001	0.0009766	0.20	49.03	0.0014
20	0.0009928	82.77	102.62	0.2923	0.0009886	82.17	111.84	0.2899	0.0009804	81.00	130.02	0.2848
40	0.0009992	165.17	185.16	0.5646	0.0009951	164.04	193.89	0.5607	0.0009872	161.86	211.21	0.5527
60	0.0010084	247.68	267.85	0.8206	0.0010042	246.06	276.19	0.8154	0.0009962	242.98	292.79	0.8502
80	0.0010199	330.40	350.80	1.0624	0.0010156	328.30	358.77	1.0561	0.0010073	324.34	374.70	1.0440
100	0.0010337	413.39	434.06	1.2917	0.0010290	410.78	441.66	1.2844	0.0010201	405.88	456.89	1.2703
120	0.0010496	496.76	517.76	1.5102	0.0019445	493.59	524.93	1.5018	0.0010348	487.65	539.39	1.4857
140	0.0010678	580.69	602.04	1.7193	0.0010621	576.88	608.75	1.7098	0.0010515	569.77	622.35	1.6915
160	0.0010885	665.35	687.12	1.9204	0.0010821	660.82	693.28	1.9096	0.0010703	652.41	705.92	1.8891
180	0.0011120	750.95	773.20	2.1147	0.0011047	745.59	778.73	2.1024	0.0010912	735.69	790.25	2.0794
200	0.0011388	837.7	860.5	2.3031	0.0011302	831.4	865.3	2.2893	0.0011146	819.7	875.5	2.2634
220	0.0011693	925.9	949.3	2.4870	0.0011590	918.3	953.1	2.4711	0.0011408	904.7	961.7	2.4419
240	0.0012046	1016.0	1040.0	2.6674	0.0011920	1006.9	1042.6	2.6490	0.0011072	990.7	1049.2	2.6158
260	0.0012462	1108.6	1133.5	2.8459	0.0012303	1097.4	1134.3	2.8243	0.0012034	1078.1	1138.2	2.7860
280	0.0012965	1204.7	1230.6	3.0248	0.0012755	1190.7	1229.0	2.9986	0.0012415	1167.2	1229.3	2.9537
300	0.0013596	1306.1	1333.3	3.2071	0.0013304	1287.9	1327.8	3.1741	0.0012860	1258.7	1323.0	3.1200
320	0.0014437	1415.7	1444.6	3.3979	0.0013997	1390.7	1432.7	3.3539	0.0013388	1353.3	1420.2	3.2868
340	0.0015684	1539.7	1571.0	3.6075	0.0014920	1501.7	1546.5	3.5426	0.0014032	1452.0	1522.1	3.4557
360	0.0018226	1702.8	1739.3	3.8772	0.0016265	1626.6	1675.4	3.7494	0.0014838	1556.0	1630.2	3.6291
380					0.0018691	1781.4	1837.5	4.0012	0.0015884	1667.2	1746.6	3.8101

v in m^3/kg , u and h in kJ/kg, s in $kJ/kg.K$

Sources:

Keenan, J.H., Keyes, F.G., Hill, P.G., Moore, J.G., *Steam Tables*, John Wiley & Sons, New York, 1969.

Van Wylen, G.J., Sonntag, R.E., *Fundamentals of Classical Thermodynamics*, John Wiley & Sons, New York, 1973.

Potter, M.C., Somerton, C.W., *Thermodynamics for Engineers,* McGraw Hill, New York, 2nd edition, 2006.

Sandler, S.I., *Chemical and Engineering Thermodynamics*, John Wiley & Sons, New York, 3rd edition, 1999.

CPSIA information can be obtained
at www.ICGtesting.com
Printed in the USA
LVHW071655220222
711737LV00019B/500

9 781922 617255